Advanced X-ray Techniques in Research and Industry

Advanced X-ray Techniques in Research and Industry

Editor
Ashok Kumar Singh

Capital Publishing Company
NEW DELHI KOLKATA BANGALORE

Capital Publishing Company
7/28, Mahaveer Street, Ansari Road,
Daryaganj, New Delhi 110 002

Advanced X-ray Techniques in Research and Industry

Editor: Ashok Kumar Singh

ISBN: 81-85589-41-0

Typeset by Innovative Processors, New Delhi
Printed in India at Chaman Enterprises, New Delhi

Foreword

Advanced X-ray Techniques form a major segment of characterization tools for the development of materials covering all branches of science—metallurgy, materials science, physics, chemistry, mineralogy, biological and medical sciences. The application of X-ray analysis has become important in new fields of research and industry.

This volume is a collection of papers presented at a seminar on "Advanced X-ray Techniques in Research and Industry (XTRI-2003)". This book covers various aspects of X-ray techniques such as crystallography, macro and microtexture, texture mapping, absorption tomography, small angle X-ray scattering, X-ray line profile analysis and structural characterization by Rietveld refinement technique. Apart from this, it also covers the application of XRD techniques in coatings, thin films and forensic examination. The organiser of the conference and editor, Dr. A.K. Singh, has done a marvellous job in putting together a collection of articles which I believe will be of enormous value to researchers worldwide working in the field of materials characterization, as a source of basic understanding of new ideas, information and references.

Dr. Dipankar Banerjee
Chief Controller, Research & Development (AMS)
Defence Research and Development Organization
New Delhi, India

Preface

The importance of X-ray diffraction to the development of modern science can be appreciated best by considering its widespread influence on metallurgy, materials science, physics, chemistry, mineralogy, biological and medical sciences. It is amazing that an analytical technique developed in the early part of the 20th century is still so important and advancing so rapidly. Modern crystallography is no longer just structure determination per se nor is it just fundamental knowledge. Crystal structure analysis is conducted keeping its application in mind. It provides answers to specific problems, leads to new products and opens new perspectives. The application of X-ray techniques leads to ample opportunities and challenges.

With the advent of new processing techniques, which lead to wide range of materials from macro to micro to nano sizes for a variety of applications from structural to bio-medical applications, it is necessary to create an awareness amongst the scientists regarding the scope and potential of advanced X-ray techniques in research and industry. It was with this view in mind that a seminar on Advanced X-ray Techniques in Research and Industry (XTRI-03) was organized by Defence Metallurgical Research Laboratory, Hyderabad. The seminar was supported by DRDO, CSIR, NAL, M/s PANalytical-New Delhi, M/s Icon Analytical-Mumbai, M/s Gannon Dunkerley-Mumbai and M/s Techsciences Services-Chennai. The present volume includes all the papers presented in the seminar after proper reviewing. It is hoped that this publication will be received by the users of X-ray techniques throughout the world with interest and will be a source of inspiration and encouragement for X-ray users in India.

This conference was organized on the occasion of 60th birth year of Professor Shrikant Lele, Rector, Banaras Hindu University, Varanasi, India. Although the theme of the conference forms only early stage of his research activities, this is only a small tribute to his outstanding contribution to the education and research of metallurgy and materials science in India. Most of the members of organizing committee are his associates, collaborators and former students who have made the occasion and the conference a great success. I was fortunate enough to have an opportunity to work with him as a Ph.D. student and our interaction continues till today. What strikes me most about him are his sterling qualities of head and heart, patience and reticence and above all a rare combination of modesty and wisdom.

I would like to take this opportunity to thank the members of the organizing committee and authors of the papers for their cooperation. Special thanks are due to Dr. D. Banerjee, Chief Controller, R&D (AMS), DRDO and Dr. A.M. Srirama Murty, Director, DMRL, for their encouragement and help. The support given by Dr. G. Malakondaiah, Dr. M. Srinivas, Dr. T.K. Nandy, Dr. P. Ghosal, and Shri V.V. Rama Rao is also gratefully appreciated. My sincere thanks are due to my colleagues in XRD laboratory, Ms. S. Banumathy, Shri V. Ramakrishna, Shri M.Yadgiri and Md. Ghouse for their assistance during the conference and preparation of the book. Finally, I would like to thank Shri P. Nagendra for his help in preparing the manuscript of the book.

September, 2005 **A.K. Singh**
Hyderabad, India

Contents

Crystal Geometry and Symmetry

Shrikant Lele

Rector, Banaras Hindu University
Varanasi – 221 005, India

INTRODUCTION

Crystallography is not a new subject and the main results have been known for a long time. In fact, in the preface of his book on 'Mathematical Crystallography and the Theory of Groups of Movements', Hinton (1903) observed that "*The object of this book is to collect for the use of English readers those results of the mathematical theory of crystallography which are not proved in the modern textbooks on that subject in the English language. No claim to originality is made; the theorems proved are in most cases to be found in the works of Shoenflies, Liebisch, &c. ...*"

A comprehensive presentation of the subject is available in the International Tables for Crystallography, Volume A : Space-group Symmetry (Hahn, 1989) as also in a large number of textbooks (e.g., Buerger, 1971). We shall attempt to outline the salient features of the subject in this brief account.

SYMMETRY OPERATIONS

Spaces and Motions

Crystals exist in three dimensional space which is designated as point space. In addition to the point space E^3 in which crystals exist, we also consider a vector space $\mathbf{V}^3$ associated with the point space. Distances and angles in point space can thus be calculated using the metric of $\mathbf{V}^3$.

In terms of an origin O and a vector basis $\mathbf{a}_1$, $\mathbf{a}_2$, $\mathbf{a}_3$, any point X of E^3 is assigned a unique vector $\overrightarrow{OX}$ of $\mathbf{V}^3$ which can be uniquely expressed as

$$X \leftrightarrow \overrightarrow{OX} = \mathrm{x} = x_1\mathbf{a}_1 + x_2\mathbf{a}_2 + x_3\mathbf{a}_3 = \begin{bmatrix}\mathbf{a}_1 & \mathbf{a}_2 & \mathbf{a}_3\end{bmatrix}\begin{bmatrix}x_1\\x_2\\x_3\end{bmatrix}$$

The real numbers x_i are the coefficients of the vector $\mathbf{x}$ of $\mathbf{V}^3$ as well as the coordinates of the point X of E^3.

The scalar product of two vectors is defined by

$$\mathbf{x} \cdot \mathbf{y} = \mathbf{x}^T \mathbf{G}\, \mathbf{y} = \begin{bmatrix} x_1 & x_2 & x_3 \end{bmatrix} \begin{bmatrix} g_{11} & g_{12} & g_{13} \\ g_{21} & g_{22} & g_{23} \\ g_{31} & g_{32} & g_{33} \end{bmatrix} \begin{bmatrix} y_1 \\ y_2 \\ y_3 \end{bmatrix}$$

and the length of a vector by

$$|\mathbf{x}| = +\sqrt{\mathbf{x}^T \mathbf{G}\, \mathbf{x}}$$

The metric tensor $\mathbf{G} = [g_{ij}]$ is a symmetric tensor with

$$g_{ij} = |\mathbf{a}_i|\ |\mathbf{a}_j| \cos \alpha_{ij}$$

where α_{ij} is the angle between the basis vectors $\mathbf{a}_i$ and $\mathbf{a}_j$.

An operation which uniquely maps each point X of E^3 to a point $\tilde{X}$ such that all the distances are left invariant is called an isometry or a (rigid) motion M : $X \rightarrow \tilde{X}$ of E^3. There also exists a unique inverse motion M^{-1} : $\tilde{X} \rightarrow X$. Any motion M : $X \rightarrow \tilde{X}$ may be described by

$$\begin{bmatrix} \tilde{x}_1 \\ \tilde{x}_2 \\ \tilde{x}_3 \end{bmatrix} = \begin{bmatrix} W_{11} & W_{12} & W_{13} \\ W_{21} & W_{22} & W_{23} \\ W_{31} & W_{32} & W_{33} \end{bmatrix} \begin{bmatrix} x_1 \\ x_2 \\ x_3 \end{bmatrix} + \begin{bmatrix} w_1 \\ w_2 \\ w_3 \end{bmatrix}$$

or $$\tilde{\mathbf{x}} = \mathbf{W}\mathbf{x} + \mathbf{w} = (\mathbf{W}, \mathbf{w})\ \mathbf{x}$$

Here (**W**, **w**) is the Frobenius or Seitz symbol. The linear or matrix part of the motion, **W**, represents a rotation, while the vector or column part corresponds to a translation. For a given motion, the matrix **W** depends only on the choice of the basis vectors whereas the vector **w**, in general, depends on the choice of the basis vectors as well as the origin. The inverse motion of the motion (**W**, **w**) is described by $(\mathbf{W}, \mathbf{w})^{-1} = (\mathbf{W}^{-1}, -\mathbf{W}^{-1}\mathbf{w})$. Successive application of two motions $(\mathbf{W}_1, \mathbf{w}_1)$ and $(\mathbf{W}_2, \mathbf{w}_2)$ results in another motion $(\mathbf{W}_3, \mathbf{w}_3)$ with $(\mathbf{W}_3, \mathbf{w}_3) = (\mathbf{W}_2\mathbf{W}_1, \mathbf{W}_2\mathbf{w}_1 + \mathbf{w}_2)$. For a rigid motion, we have

$$|\mathbf{y} - \mathbf{x}|^2 = |\tilde{\mathbf{y}} - \tilde{\mathbf{x}}|^2$$

$$\Rightarrow \quad (\mathbf{y} - \mathbf{x})^T\ \mathbf{G}\ (\mathbf{y} - \mathbf{x}) = (\tilde{\mathbf{y}} - \tilde{\mathbf{x}})^T\ \mathbf{G}\ (\tilde{\mathbf{y}} - \tilde{\mathbf{x}})$$

$$= (\mathbf{y} - \mathbf{x})^T \mathbf{W}^T \mathbf{G}\, \mathbf{W} (\mathbf{y} - \mathbf{x})$$

$$\Rightarrow \quad \mathbf{G} = \mathbf{W}^T \mathbf{G}\, \mathbf{W}$$

Thus the determinants of the matrices on both sides of the equation must be equal and as a consequence

$$\det(\mathbf{W}) = \pm 1 \quad \Rightarrow \quad \mathbf{W} \text{ is unimodular.}$$

A point X is called a fixed point of the mapping M if it is invariant under the mapping, i.e., $\tilde{X} = X$. The fixed points of a motion lead to the concept of symmetry elements. The symmetry element of a symmetry operation is the set of its fixed points along with a geometric description of the motion. For symmetry operations without fixed points, the fixed points of the corresponding rotations or reflections are taken to define the symmetry element. Thus, the following types of motions can be distinguished depending on the values of **W** and **w** as mentioned below.

1. **W** = **I**, **w** = **0** : Identity (**I** is the 3 × 3 unit matrix and **0** is the null vector irrespective of the coordinate system).
2. **W** ≠ **I**, **w** = **0** : Fixed point motions: In these motions, at least one point does not move.
   (a) det (**W**) = +1 : Proper rotations.
   An n-fold rotation axis represents a counter-clockwise rotation by an angle $\alpha = 2\pi/n$ around the axis when viewed from the positive end of the axis. All the points lying on the axis are invariant and constitute the symmetry element while the rotation is the symmetry operation. This operation generates a sequence of motif units arranged along the circumference of a circle at intervals equal to the arc α, the orientations of successive motif units differing by the angle α as illustrated in Fig. 1. All the motifs are congruent and rotational operations are therefore called operations of the first sort. The symbol for an n-fold rotation axis is n. A more complete notation includes a superscript + or – which specifies the sense of rotation (not needed for $n = 2$). This is followed by the location of the rotation axis. For example, 6^+ 0, 0, z represents a counter clockwise rotation of $2\pi/6$ about the line 0, 0, z when viewed from the point 001 towards the point 000. The common designations for two-, three-, four- and six-fold axes are diad, triad, tetrad and hexad, respectively. As noted above, the values of the elements of **W** for a chosen rotation also depend on the coordinate system used. As an example, the values of **W** for a hexagonal rotational axis are given below in two coordinate systems. In the Cartesian coordinate system with the rotation axis along the z-axis, **W** is given by

$$W = \begin{bmatrix} \cos\alpha & -\sin\alpha & 0 \\ \sin\alpha & \cos\alpha & 0 \\ 0 & 0 & 1 \end{bmatrix} = \begin{bmatrix} \tfrac{1}{2} & -\tfrac{\sqrt{3}}{2} & 0 \\ \tfrac{\sqrt{3}}{2} & \tfrac{1}{2} & 0 \\ 0 & 0 & 1 \end{bmatrix}$$

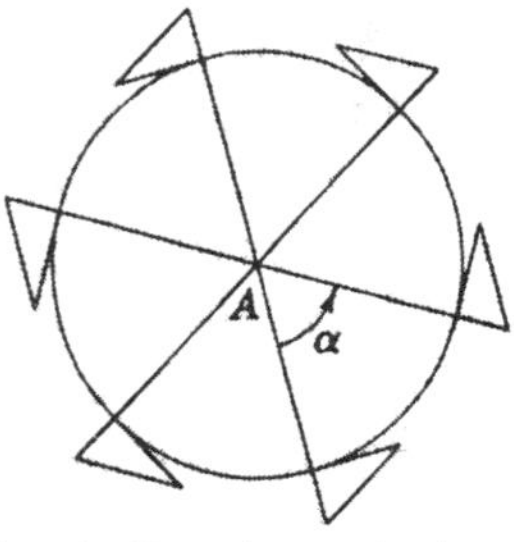

Fig. 1: Rotation axis A_α (after Buerger, 1971).

In the crystal coordinate system, two of the axes are at an angle of $2\pi/3$ while the third is perpendicular to both. With the rotation axis along the third, we have

$$\mathbf{W} = \begin{bmatrix} 1 & -1 & 0 \\ 1 & 0 & 0 \\ 0 & 0 & 1 \end{bmatrix}$$

(b) det (**W**) = –1: Improper rotations: Any operation which causes an object to come into coincidence with its enantiomorph is called an operation of the second sort.

(i) **W** = –**I**: This symmetry operation is designated as centre of symmetry or inversion centre and has the symbol $\bar{1}$ followed by the location of the symmetry centre. The symmetry operation consists of inversion through a point which is the symmetry element. This operation generates an enantiomorph of a motif unit as illustrated in Fig. 2. For an inversion centre at the origin, the symbol is $\bar{1}$ 0, 0, 0 for which we have

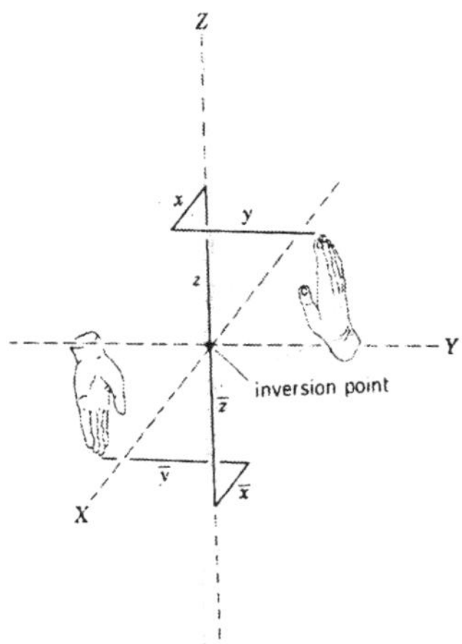

Fig. 2: Enantiomorphic objects related by $\bar{1}$ (after Buerger, 1971).

$$\mathbf{W} = \begin{bmatrix} -1 & 0 & 0 \\ 0 & -1 & 0 \\ 0 & 0 & -1 \end{bmatrix}$$

(ii) **W** ≠ –**I**, $\mathbf{W}^2$ = **I**: The symmetry element is designated as a reflection or mirror plane and has the symbol *m* followed by the location of the mirror plane. For example, *m x*, *y*, 0 denotes a mirror parallel to the xy-plane and passing through the origin. The symmetry operation is reflection through a plane. This operation generates an enantiomorphic object as illustrated in Fig. 3. For a mirror plane passing through the origin and perpendicular to the z-axis in Cartesian coordinates, we have

$$\mathbf{W} = \begin{bmatrix} 1 & 0 & 0 \\ 0 & 1 & 0 \\ 0 & 0 & -1 \end{bmatrix}$$

(iii) **W** ≠ –**I**, $\mathbf{W}^2$ ≠ **I**: These symmetry elements are designated as rotoinversions and have the symbol $\bar{n}$. In the complete symbol, this is followed by the superscript + or –, the location part of the axis and the location of the inversion point after a semicolon. The symmetry operation is a composite operation consisting of counter-clockwise rotation of $2\pi/n$ around an axis followed by inversion through a point on the axis. The angle and sense of rotation in the symbol refer to the pure rotation part. Note that $\bar{1}$ and $\bar{2}$ (= m) are special cases which by convention have distinct designations/symbols. In the crystal coordinate system described above for a hexad, **W** for a $\bar{6}^{+}0, 0, z; 0, 0, 0$ axis is given by

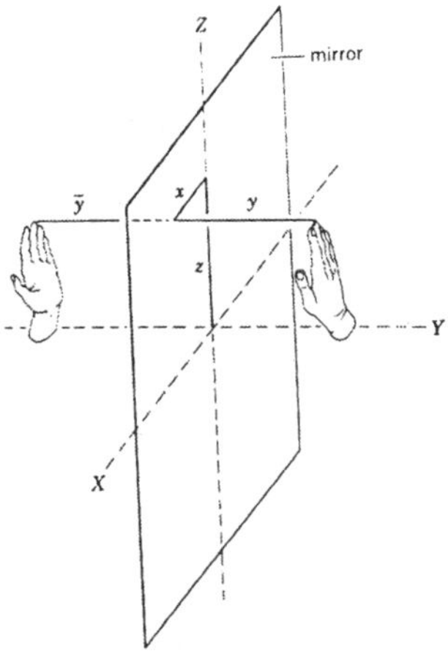

Fig. 3: Enantiomorphic objects related by m (after Buerger, 1971).

$$\mathbf{W} = \begin{bmatrix} -1 & 1 & 0 \\ -1 & 0 & 0 \\ 0 & 0 & -1 \end{bmatrix}$$

3. $\mathbf{W} = \mathbf{I}$, $\mathbf{w} \neq \mathbf{0}$: The vector $\mathbf{t} = \mathbf{w} = w_1\mathbf{a}_1 + w_2\mathbf{a}_2 + w_3\mathbf{a}_3$ is called a translation vector. An operation of pure translation and its multiples generate a sequence of parallel motif units at intervals equal to the translation **t** as shown in Fig. 4. For example, the symbol $\mathbf{t}(\frac{1}{2},\frac{1}{2},0)$ represents a translation by the vector $\frac{1}{2}\mathbf{a}_1 + \frac{1}{2}\mathbf{a}_2$.
4. $\mathbf{W} \neq \mathbf{I}$, $\mathbf{w} \neq 0$: Fixed point free motions which are not translations.
   (a) det (**W**) = +1: Screw rotations involve right handed screw rotations of $2\pi/n$ around an axis with screw vector (pitch) $\tau = (p/n)\mathbf{t}$, where **t** is the shortest translation along the direction of the screw axis. These can be considered as composite operations of a proper rotation and a translation. The pattern generated by this screw motion, designated $A_{\alpha,\tau}$, is a sequence of motifs evenly distributed along a helical path as shown in Fig. 5. The symbol is n_p. A more complete symbol is similar to that for a rotation except for the inclusion of the screw part in parenthesis. Thus, $6^{+}(0, 0, 1/6)\ 0, 0, z$ indicates a rotation of $2\pi/6$ around the line $00z$ combined with a translation of $(1/6)\mathbf{c}$, where **c** is the translation vector along this axis. As an example, the values of **W** and **w** for the hexagonal screw axis 6_1 in terms of crystal coordinates are given below

$$\mathbf{W} = \begin{bmatrix} 1 & -1 & 0 \\ 1 & 0 & 0 \\ 0 & 0 & 1 \end{bmatrix}, \quad \mathbf{w} = \begin{bmatrix} 0 \\ 0 \\ \frac{1}{6} \end{bmatrix}$$

Fig. 4: Translation **t** (after Buerger, 1971).

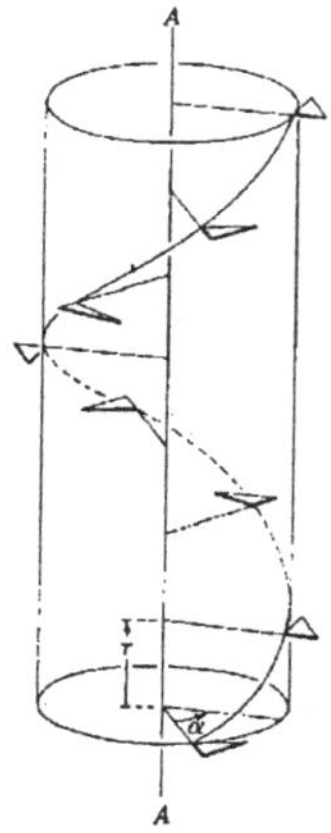

Fig. 5: Screw axis $A_{\alpha,\tau}$ (after Buerger, 1971).

(b) det (**W**) = −1: Glide reflections involve composite operations of a mirror and a translation through a vector τ. This operation is an operation of the second sort and generates patterns of the type illustrated in Fig. 6. Three types of glide planes are possible. Axial glide involves glide through a vector of **a**/2, **b**/2 or **c**/2 parallel to the glide plane. The corresponding symbols are *a*, *b* and *c* glide. Diagonal glide has the symbol *n* and involves glide through a vector (**a** + **b**)/2, (**b** + **c**)/2, (**c** + **a**)/2 or (±**a**±**b**±**c**)/2 parallel to the glide plane. Diamond glide occurs only in some crystal structures and has the symbol *d*. The complete symbol includes the translation part of the glide in parenthesis (not needed for axial glide) and the location of the glide plane.

Fig. 6: Glide reflection m_τ (after Buerger, 1971).

For example, for *a x*, 0, *z* glide, the glide plane is parallel to the *xz*-plane passing through the origin and the glide vector is equal to **a**/2. The values of **W** and **w** are then given by

$$\mathbf{W} = \begin{bmatrix} 1 & 0 & 0 \\ 0 & -1 & 0 \\ 0 & 0 & 1 \end{bmatrix}, \quad \mathbf{w} = \begin{bmatrix} \tfrac{1}{2} \\ 0 \\ 0 \end{bmatrix}$$

Symmetry Operations and Symmetry Groups

A symmetry operation of a given object in point space E^3 is a motion of E^3 which maps this object (set of points, crystal pattern, etc.) onto itself. The set of all symmetry operations of an object satisfy the group axioms and thus form a group **G**, the symmetry group of the object or its symmetry.

CRYSTAL PATTERNS, VECTOR LATTICES AND POINT LATTICES

A crystal or a crystal pattern can be idealized as an infinite three-dimensional periodic arrangement of atoms in point space. Three dimensional periodicity implies the existence of three translations among the symmetry operations of the crystal with translation vectors $\mathbf{t}_1$, $\mathbf{t}_2$, $\mathbf{t}_3$ which are linearly independent. Successive application of two translations of a crystal pattern results in another translation. Thus all vectors $\mathbf{t}\ (u_1\ u_2\ u_3) = u_1\ \mathbf{t}_1 + u_2\ \mathbf{t}_2 + u_3\ \mathbf{t}_3$ (u_1, u_2, u_3 being arbitrary integers) are translation vectors of the pattern.

The periodicity of crystal patterns is represented by their lattices. Similar to the distinction between vector space and point space, a distinction can be made between two kinds of lattices, namely vector and point lattices. The infinite set of all translation vectors of a crystal pattern is called the lattice of translation vectors or the vector lattice **L** of this crystal pattern. On the other hand, the point lattice is defined as follows. Given an arbitrary point X_0 in point space and a vector lattice **L** consisting of vectors $\mathbf{t}_j$, the set of all points X_j with $\overrightarrow{X_0X_j} = \mathbf{t}_j$ is called the point lattice belonging to X_0 and **L**. Thus, the point lattice can be visualized as the set of end points of all vectors of **L** with the initial point of each vector being an arbitrary point X_0 of point space. Since each point X could be chosen as X_0, an infinite set of interpenetrating point lattices belongs to each vector lattice. The origin of the coordinate system is often chosen as the point X_0.

Although any three linearly independent vectors may serve as a basis of the vector space $\mathbf{V}^3$, special bases called crystallographic bases, which offer simplicity in application are preferred. A basis of three vectors $\mathbf{a}_1$, $\mathbf{a}_2$, $\mathbf{a}_3$ of $\mathbf{V}^3$ is called a crystallographic basis of the vector lattice **L** if every integral linear combination $\mathbf{t} = u_1\mathbf{a}_1 + u_2\mathbf{a}_2 + u_3\mathbf{a}_3$ is a lattice vector of **L**. Further, a crystallographic basis is called a primitive (crystallographic) basis if the converse is also true, i.e., if every lattice vector **t** of **L** can be expressed as an integral linear combination of the basis vectors. It follows that the coefficients for each lattice vector must be rational numbers (including integers) when referred to a crystallographic basis. On the other hand, for a primitive basis, the coefficients must be integers. Primitive bases exist for every vector lattice **L**. However, non-primitive bases are used conventionally for the description of some lattices, since they reflect the symmetry of the lattice in a more transparent manner.

There are an infinite number of primitive bases for each vector lattice **L**. A procedure called basis reduction can be used to select one unique primitive basis, thus facilitating unambiguous description and comparison of vector lattices. The term primitive is used not only for bases of lattices but also for lattices themselves if their conventional bases are primitive.

An important feature of a lattice is its unit cell defined as follows. If $\mathbf{a}_1$, $\mathbf{a}_2$, $\mathbf{a}_3$ is a crystallographic basis of a vector lattice **L**, the set of all vectors $x_1\mathbf{a}_1 + x_2\mathbf{a}_2 + x_3\mathbf{a}_3$ with $0 \le x_i \le 1$ constitutes the unit cell of the vector lattice. Analogously, in point space, a unit cell of the crystal pattern is the set of all points X with coordinates x_i such that the equation $0 \le x_i - x_{0i} \le 1$ (i = 1, 2, 3) holds for a chosen crystallographic coordinate system (O, $\mathbf{a}_1$, $\mathbf{a}_2$, $\mathbf{a}_3$) and a point X_0 with coordinates x_{0i}.

A crystal structure may be described in two alternative ways. In the first, we start with a point X_i representing the centre of an atom. An infinite number of translationally equivalent points X_j ($\overrightarrow{X_iX_j}$ are lattice vectors) are associated with the point X_i. A point lattice of translationally equivalent points can thus be obtained from each of the points X_i (i = 1 to N) within the unit cell of a point lattice. We note that only a finite number N of atoms can lie within the unit cell of a crystal since the volumes of the unit cell as well as those of atoms are finite. The crystal structure is thus described by a finite number of interpenetrating infinite point lattices. Alternatively, the crystal structure consists of an infinite number of "building blocks", the building block being the unit cell.

A three dimensional lattice is fully described by the shape and size of a unit cell, i.e., by the lengths a, b and c of the basis vectors **a**, **b** and **c** (equivalently $\mathbf{a}_1$, $\mathbf{a}_2$ and $\mathbf{a}_3$) and by the three interaxial angles α, β and γ (between **b** and **c**, **c** and **a** and **a** and **b**, respectively). These are known as the lattice parameters of both the vector lattice and the associated point lattice of the crystal structure.

CRYSTALLOGRAPHIC CALCULATIONS

Crystallographic calculations using matrix methods are facilitated by the so-called structure matrix. The crystallographic basis vectors **a**, **b**, **c** can be represented in the orthonormal basis $\mathbf{e}_i$ (i = 1 to 3) as a linear combination of these basis vectors.

$$\mathbf{a} = \mathbf{e}_1\, s_{11} + \mathbf{e}_2\, s_{21} - \mathbf{e}_3\, s_{31}$$
$$\mathbf{b} = \mathbf{e}_1\, s_{12} + \mathbf{e}_2\, s_{22} + \mathbf{e}_3\, s_{32}$$
$$\mathbf{c} = \mathbf{e}_1\, s_{13} + \mathbf{e}_2\, s_{23} + \mathbf{e}_3\, s_{33}$$

The structure matrix **S** is defined as

$$\mathbf{S} = \begin{bmatrix} s_{11} & s_{12} & s_{13} \\ s_{21} & s_{22} & s_{23} \\ s_{31} & s_{32} & s_{33} \end{bmatrix}$$

As observed earlier, the lattice is completely described by six lattice parameters. The structure matrix contains nine elements. The three additional constants are independent and represent the degrees of freedom needed to choose the orientation of the unit cell within the orthonormal coordinate system. If we orient the crystal so that **c** is parallel to the *z*-axis and **a** lies in the *xz*-plane, then the orientation of **b** is fixed. For the chosen orientation, it is easy to show that

$$s_{13} = s_{23} = s_{21} = 0$$

$$s_{33} = c$$

$$s_{11} = a \sin \beta, \; s_{31} = a \cos \beta$$

$$s_{32} = b \cos \alpha$$

For determining the remaining two elements of **S**, we find the scalar products **a.b** and **b.b** in two ways and equate them. Thus, we have

$$ab \cos\gamma = a \sin\beta \;\; s_{12} + a \cos\beta \;\; b \cos\alpha$$

and $$s_{12} = \frac{b(\cos\gamma - \cos\alpha \;\; \cos\beta)}{\sin\beta}$$

Finally, from

$$b^2 = (s_{12})^2 + (s_{22})^2 + b^2 \cos^2\alpha$$

we obtain after simplification

$$s_{22} = \frac{b}{\sin\beta} D$$

where $D = \det(\mathbf{S}) = (1 + 2 \cos \alpha \cos \beta \cos \gamma - \cos^2 \alpha - \cos^2 \beta - \cos^2 \gamma)^{1/2}$.

The metric tensor introduced earlier can be related to the structure matrix by taking scalar products of the basis vectors

$$\mathbf{G} = \mathbf{S}^T \, \mathbf{I} \, \mathbf{S} = \mathbf{S}^T \, \mathbf{S}$$

As mentioned earlier, the metric tensor is represented by a symmetric matrix and contains only six independent elements. The metric tensor gives information concerning lengths and interaxial angles only.

THE RECIPROCAL LATTICE

For applications related to diffraction from crystals, it is useful to define the so-called reciprocal lattice. The basis vectors $\mathbf{a}_i^*$ (i = 1 to 3) for the reciprocal lattice are defined through the orthogonality relations

$$\mathbf{a}_i \cdot \mathbf{a}_j^* = \delta_{ij}$$

where δ_{ij} are elements of a 3 × 3 unit matrix. The basis vectors of the reciprocal lattice $\mathbf{a}_i^*$ can be expressed in the orthonormal basis by using the structure matrix $\mathbf{S}^*$.

$$\mathbf{a}^* = \mathbf{S}^{*T}\ \mathbf{e}$$

Taking transpose of this equation, forming scalar products and using the orthogonality relations, we have

$$\mathbf{S}^* = (\mathbf{S}^T)^{-1} = (\mathbf{S}^{-1})^T$$

The reciprocal basis vectors can also be expressed in an alternative manner. Since the vector $\mathbf{a}_i^*$ is normal to $\mathbf{a}_j$ and $\mathbf{a}_k$, it follows that it is parallel to $\mathbf{a}_j \times \mathbf{a}_k$. Since $\mathbf{a}_i \cdot \mathbf{a}_i^* = 1$, we have

$$\mathbf{a}_i^* = \frac{\mathbf{a}_j \times \mathbf{a}_k}{\mathbf{a}_i\ \cdot\ \mathbf{a}_j \times \mathbf{a}_k}$$

It may be noted that the denominator is the volume of the crystal unit cell. The metric tensor for the reciprocal basis is given by

$$\mathbf{G}^* = \mathbf{S}^{*T}\ \mathbf{S}^* = \mathbf{S}^{-1}\ (\mathbf{S}^T)^{-1} = (\mathbf{S}^T\ \mathbf{S})^{-1} = \mathbf{G}^{-1}$$

CRYSTALLOGRAPHIC SYMMETRY OPERATIONS

A motion is called a crystallographic symmetry operation if a crystal pattern exists for which it is a symmetry operation. Owing to the fact that every translation $(\mathbf{I}, \mathbf{t})$ of the translation group $\mathbf{T}^3$ is a crystallographic symmetry operation, there are some restrictions on the possible crystallographic symmetry operations. This can be shown as follows. Take the conjugate of $(\mathbf{I}, \mathbf{t})$ with respect to $(\mathbf{W}, \mathbf{w}) \in \mathbf{G}$

$$(\mathbf{W}, \mathbf{w})\ (\mathbf{I}, \mathbf{t})\ (\mathbf{W}, \mathbf{w})^{-1} = (\mathbf{W}, \mathbf{W}\ \mathbf{t} + \mathbf{w})\ (\mathbf{W}^{-1}, -\mathbf{W}^{-1}\ \mathbf{w})$$
$$= (\mathbf{I}, -\mathbf{W}\ \mathbf{W}^{-1}\ \mathbf{w} + \mathbf{W}\ \mathbf{t} + \mathbf{w}) = (\mathbf{I}, \mathbf{W}\ \mathbf{t})$$

We require that $(\mathbf{I}, \mathbf{W}\ \mathbf{t}) \in \mathbf{T}^3$. Thus $\mathbf{W}$ maps the lattice onto itself and leaves the origin fixed. Thus, $\mathbf{W}$ carries the crystallographic basis $\mathbf{a}_1$, $\mathbf{a}_2$, $\mathbf{a}_3$ into the lattice vectors $\mathbf{t}_1$, $\mathbf{t}_2$, $\mathbf{t}_3$. Since every lattice vector has integer components with reference to this basis, the representation matrix has integer coefficients. As the elements of $\mathbf{W}$ are integers, its trace is also an integer.

$$\text{tr}\ (\mathbf{W}) = W_{11} + W_{22} + W_{33} = \text{integer}$$

Since the trace is invariant (independent of coordinate system), using Cartesian basis, we have

$$\text{tr}\ (\mathbf{W}) = \pm\ (1 + 2\cos\alpha)$$

Combining these results, we have

$$\cos\alpha = m/2$$

where m is an integer. A graphical representation of this argument is shown in Fig. 7. Since the cosine of an angle must lie in the range -1 to $+1$, the fold of an axis is restricted to the values given in Table 1.

Every rotation $\mathbf{W} \in \mathbf{G}$ maps a lattice vector $\mathbf{t}$ onto a new lattice vector $\mathbf{t}'$ of the same length $|\mathbf{t}|$. Only a finite number of vectors of length $|\mathbf{t}|$ exist

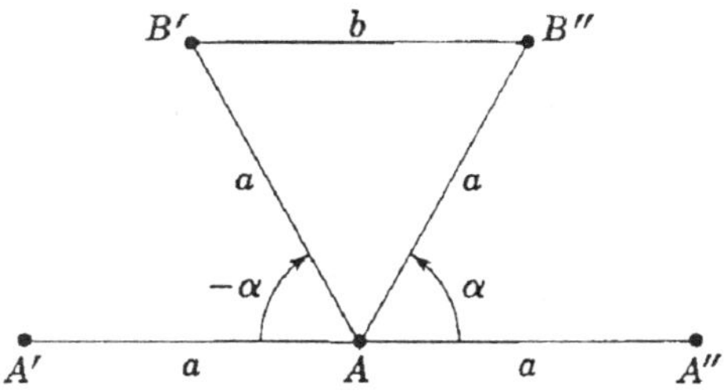

Fig. 7 Restriction on rotation angle owing to translational periodicity (after Buerger, 1971).

within a ball of radius $|\mathbf{t}|$. Thus, the original set of basis vectors must be identically restored after a finite number of repetitions of **W**. The smallest such number n is the order of rotation **W** and

$$\mathbf{W}^n = \mathbf{I} \quad \text{with} \quad n = 1, 2, 3, 4 \text{ or } 6.$$

Similarly

$$(\mathbf{W}, \mathbf{w})^n = (\mathbf{I}, (\mathbf{W}^{n-1} + \ldots + \mathbf{W} + \mathbf{I})\, \mathbf{w}) = (\mathbf{I}, \mathbf{t})$$

Thus

$$(\mathbf{W}^{n-1} + \ldots + \mathbf{W} + \mathbf{I})\, \mathbf{w} = \mathbf{t}$$

This imposes restrictions on the shift vector **w** of a screw/glide.

$$\mathbf{w} = (1/n)\mathbf{t}$$

This implies that an n-fold application of the motion (**W**, **w**) results in a translation **t** of the crystal pattern. The fractional translation $(1/n)\mathbf{t}$ is called the intrinsic translation part (screw or glide part) of the symmetry operation. As mentioned earlier, the translation part of a motion depends on the choice of the origin. However, the intrinsic translation part of a motion is unique. The intrinsic translation vector $\mathbf{w}_g = (1/n)\mathbf{t}$ is the shortest translation vector of the motion for any choice of origin. Since $\mathbf{w}_g$ is restricted to the values $(1/n)\mathbf{t}$, the possible screw axes are 2_1, 3_1, 3_2, 4_1, 4_2, 4_3, 6_1, 6_2, 6_3, 6_4 and 6_5 with screw vectors $(p/n)\mathbf{t}$ for a n_p screw axis where **t** is the shortest translation vector parallel to the axis. The glide vector is equal to $(1/2)\mathbf{t}$ where **t** is one of the three shortest translation vectors parallel to the plane. If $\mathbf{t} = \mathbf{0}$, the symmetry operation has at least one fixed point and is a rotation, inversion, reflection or rotoinversion.

Table 1: Restrictions on α and n due to translations

m	$\cos \alpha = m/2$	α	$n = 2\pi/\alpha$
2	1	0, 2π	1
1	½	π/3	6
0	0	π/2	4
−1	−½	2π/3	3
−2	−1	π	2

DERIVATION OF SYMBOLS FOR SYMMETRY OPERATIONS

As is obvious from the foregoing discussion, the symmetry operation can be alternatively given by the coordinate triplet $\tilde{x}\ \tilde{y}\ \tilde{z}$ since the symmetry operation maps the point *xyz* onto the point $\tilde{x}\ \tilde{y}\ \tilde{z}$. In the following we shall describe a procedure for deriving the symbol for the symmetry operation from coordinate triplets. Two cases need to be distinguished.

1. The matrix **W** is the unit matrix. It follows that the symmetry operation is a translation with translation vector **w**. As an example, consider an operation which maps *x y z* onto ½+*x* ½+*y z*. It follows that **W** is the unit matrix and **w** = ½**a** − ½**b**. The symbol for this symmetry operation is **t**(½, ½, 0) and stands for C centering.
2. The matrix **W** is not equal to the unit matrix. The rotation part of the symmetry operation can be identified by calculating the trace and determinant of **W** from Table 2. There are three subcases which need to be considered.

Table 2: Determination of the rotation part of a symmetry operation

tr*(W)* → *det(W)*↓	*−3*	*−2*	*−1*	*0*	*1*	*2*	*3*
1	–	–	2	3	4	6	1
−1	$\bar{1}$	$\bar{6}$	$\bar{4}$	$\bar{3}$	m		

(a) **W** corresponds to a rotoinversion. The coordinates of the inversion point can be obtained from the equation $\mathbf{x} = (\mathbf{I} - \mathbf{W})^{-1}\ \mathbf{w}$.
For a rotoinversion other than $\bar{1}$, the location of the axis follows from the equation $(\mathbf{W}, \mathbf{w})^2\ \mathbf{x} = (\mathbf{W}^2, (\mathbf{W} + \mathbf{I})\ \mathbf{w})\ \mathbf{x} = \mathbf{x}$. The sense of rotation may be found either by geometrical inspection of a pair of points related by the rotoinversion axis or by an algebraic procedure (Hahn, 1989).

(b) **W** corresponds to an *n*-fold rotation. (**W**, **w**) is thus either a rotation or a screw rotation. The two alternatives can be distinguished by calculating **t** from

$$\mathbf{t} = (\mathbf{W}^{n-1} + \dots + \mathbf{W} + \mathbf{I})\ \mathbf{w}.$$

For **t** = **0**, the operation (**W**, **w**) corresponds to a pure rotation. For **t** ≠ **0**, (**W**, **w**) corresponds to a screw rotation with screw part $\mathbf{w}_g = (1/n)\mathbf{t}$. The location of the screw axis is found from $\mathbf{x} = (\mathbf{I} - \mathbf{W})^{-1}\ \mathbf{w}_l$, where $\mathbf{w}_l = \mathbf{w} - \mathbf{w}_g$. The sense of the rotation can be found as in the earlier case.

(c) **W** corresponds to a reflection. (**W**, **w**) is thus either a reflection or a glide. To distinguish between these alternatives, **t** is calculated from $\mathbf{t} = (\mathbf{W} + \mathbf{I})\ \mathbf{w}$. For **t** = **0**, (**W**, **w**) corresponds to a pure reflection. The location of the plane can be obtained from $\mathbf{x} = (\mathbf{I} - \mathbf{W})^{-1}\,\mathbf{w}$. For $\mathbf{t} \neq \mathbf{0}$, $(\mathbf{W}, \mathbf{w})$ corresponds to a glide reflection with glide part $\mathbf{w}_g = \frac{1}{2}\mathbf{t}$. The location of the glide plane follows from

$$\mathbf{x} = (\mathbf{I} - \mathbf{W})^{-1}\,\mathbf{w}_l, \text{ where } \mathbf{w}_l = \mathbf{w} - \mathbf{w}_g = \mathbf{w} - \tfrac{1}{2}\mathbf{t}.$$

DERIVATION OF COORDINATE TRIPLETS FROM SYMBOLS OF SYMMETRY OPERATIONS

The coordinates of the image point $\tilde{x}\ \tilde{y}\ \tilde{z}$ and $(\mathbf{W}, \mathbf{w})$ can be found from the symbol of the symmetry operation by following procedures converse to those illustrated in the preceding section.

(a) For all symmetry operations of space groups, the rotation parts $\mathbf{W}$ referring to conventional coordinate systems are given in Tables 11.2 and 11.3 of Hahn (1989) as matrices of point group operations.
(b) The location part $\mathbf{w}_l$ of $\mathbf{w}$ may be found from $\mathbf{w}_l = (\mathbf{I} - \mathbf{W})\,\mathbf{x}_0$ where $\mathbf{x}_0 = \{x_0, y_0, z_0\}$ and x_0, y_0, z_0 are the coordinates of the inversion point of a rotoinversion or an arbitrary fixed point of any other symmetry operation. The intrinsic translation part $\mathbf{w}_g$ of $\mathbf{w}$ is given explicitly in the symbol of the symmetry operation. Thus, the translation part $\mathbf{w}$ can be found from $\mathbf{w} = \mathbf{w}_g + \mathbf{w}_l$
(c) The coordinate triplet can now be found from $\tilde{\mathbf{x}} = (\mathbf{W}, \mathbf{w})\,\mathbf{x} = \mathbf{W}\,\mathbf{x} + \mathbf{w}$.

COMBINATIONS OF ROTATIONS

Intersecting rotation axes can lead to homogeneous patterns if they satisfy certain angular relations. Consider the result of combining a 4-fold rotation axis A perpendicular to the plane of the paper (along z) with a 2-fold rotation axis B (along x) perpendicular to A and lying in the plane of the paper as shown in Fig. 8. If a triangle motif (black on one side and white on the other) is placed in the neighbourhood of these axes, eight congruent triangles arranged as in the figure (z and $\bar{z}$ indicate the distances of the triangles above and below the plane of the paper) are required to satisfy the symmetry. Each of these triangles is related to the rest of the pattern in an identical manner, so the pattern is homogeneous. Further, the pattern of triangles shows that a new 2-fold rotation axis C arises as a result of the combination. The matrices representing the 4-fold and 2-fold axes respectively are

$$\begin{bmatrix} 0 & -1 & 0 \\ 1 & 0 & 0 \\ 0 & 0 & 1 \end{bmatrix}, \begin{bmatrix} 1 & 0 & 0 \\ 0 & -1 & 0 \\ 0 & 0 & -1 \end{bmatrix}$$

The two products of these matrices ($\mathbf{W}_1\,\mathbf{W}_2$ and $\mathbf{W}_2\,\mathbf{W}_1$) are

$$\begin{bmatrix} 0 & 1 & 0 \\ 1 & 0 & 0 \\ 0 & 0 & -1 \end{bmatrix}, \begin{bmatrix} 0 & -1 & 0 \\ -1 & 0 & 0 \\ 0 & 0 & -1 \end{bmatrix}$$

These correspond to 2-fold axes along C [110] and D [$1\bar{1}0$] directions, respectively in Fig. 8. In general, when two operations are combined, a third arises as a consequence of the combination. The type and location of the new operations which arise as a consequence of combining rotational operations can be found by using Euler's construction.

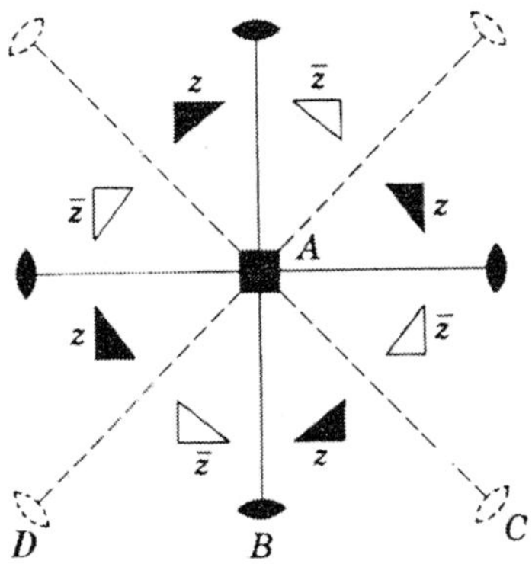

Fig. 8: The result of combining a 4-fold rotation axis A with a 2-fold rotation axis B is a new 2-fold rotation axis C (after Buerger, 1971).

Euler's Construction

Consider an axis A with a throw of α and another axis B with a throw of β. The rotational axes are represented by the points A and B where the axes cut the surface of a sphere centred at the intersection of the axes as shown in Fig. 9. The segment AB is part of a great circle. Then the operation of A takes the segment AM (on the surface of the sphere) making an angle of α/2 with AB to AM'. Similarly, the segment BN at an angle of β/2 from AB is taken to BN' by a rotation around B.

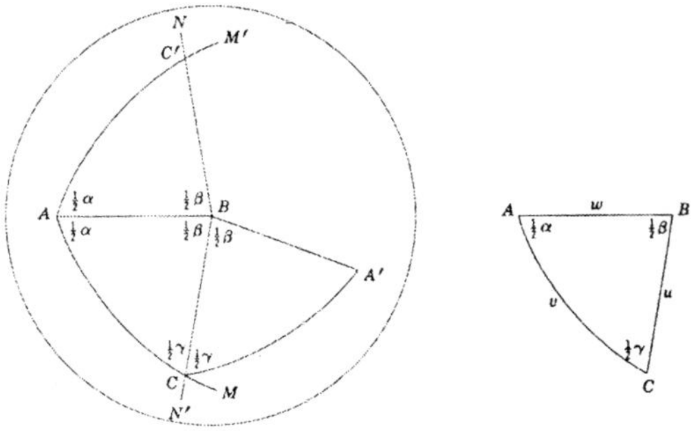

Fig. 9: Euler's construction for finding the result of combining two rotations (after Buerger, 1971).

$$AM \xrightarrow{A_\alpha} AM', \qquad BN \xrightarrow{B_\beta} BN'$$

The intersection point C of AM and BN' is carried to the intersection point C' of AM' and BN by the rotation A_α and then C' is returned to C by the rotation B_β. The net result of the two operations in succession is therefore a motion which leaves the axis passing through the point C and the centre of the sphere fixed. This corresponds to a rotation about C. The angle of rotation about C can be found by constructing a mirror image of the spherical triangle ABC across BC at $A'BC$. Applying the two rotations to the point A takes it to A'. Thus the rotation γ about C must bring A to A' as depicted below.

$$A \xrightarrow{A_\alpha} A \xrightarrow{B_\beta} A' \;\Rightarrow\; A \xrightarrow{C_{-\gamma}} A'$$

The result can be written as

$$A_\alpha \times B_\beta = C_{-\gamma}$$

The negative sign indicates that the sense of rotation around C is opposite to those around A and B. The magnitudes of α, β and γ and the angles between each pair of the three axes, namely,

$$u = B \wedge C, \quad v = C \wedge A, \quad w = A \wedge B$$

constitute the elements of a spherical triangle and are accordingly related. The following relation is useful for our purpose.

$$\cos\,(A \wedge B) = \cos w = \frac{\cos\frac{1}{2}\gamma + \cos\frac{1}{2}\alpha\,\cos\frac{1}{2}\beta}{\sin\frac{1}{2}\alpha\,\sin\frac{1}{2}\beta}$$

The right hand side of the above relation can be evaluated for all combinations of crystallographically permissible values of α, β and γ. If the right hand side is less than –1 or greater than +1, the chosen combination of α, β and γ is not permitted. Otherwise, the relation can be solved for w. The other angles can then be found in a similar manner.

A systematic application of this procedure shows that the only possible combinations of proper rotations are as follows

222, 32(2), 422, 622, 23(3), 432,

where each digit represents a particular rotation axis. In the case of 322 and 233, the standard notation is 32 and 23, respectively. The symbol for the third axis is omitted in these cases, because the location of the axis to which it refers coincides with one of the several axes which are equivalent. These are shown in Fig. 10(i). Since each of the axes of a combination multiplies the other axes, the resulting consistent sets of axes are those shown in Fig. 10(ii).

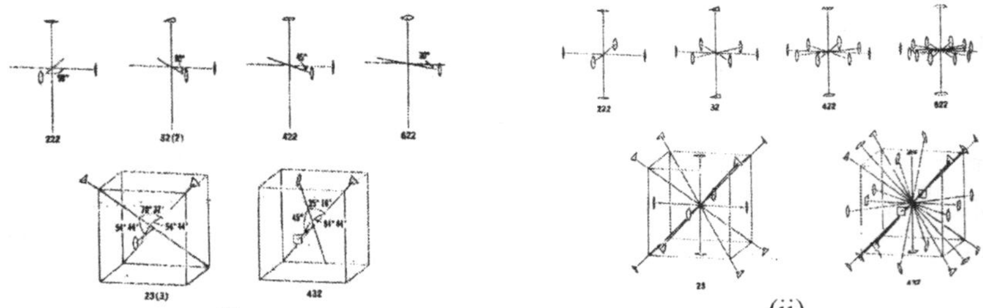

(i) (ii)

Fig. 10: The arrangements of axes in the six possible combinations of crystallographic axes (after Buerger, 1971).

Combinations Involving Improper Rotations

Euler's construction is equally valid for improper rotations. In Fig. 11, suppose A_α is a proper rotation and $\overline{B}_\beta$ is an improper rotation. Then A_α produces a congruent object ($R \rightarrow R$) and $\overline{B}_\beta$ produces an enantiomorphic one ($R \rightarrow L$). Consequently, the rotation about C must produce an enantiomorphic object, so that this rotation $\overline{C}_\gamma$ is an improper one. It is clear from this argument that there must be two or no improper rotation axes in a triplet. The angular relations are the same as in the case of corresponding proper rotation axes. However, a doublet combination of a proper (P) and an improper (I) rotation axis is possible when two axes n and $\overline{n}$ coincide. One can also have combinations of such doublets, namely, P/I P/I P/I.

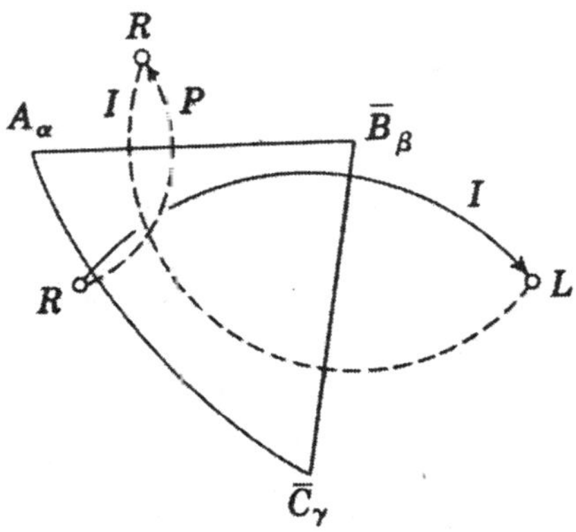

Fig. 11: Euler's construction for combining proper and improper rotations (after Buerger, 1971).

THE POINT GROUPS

As mentioned earlier, crystallographic symmetry operations form a group. The combinations of all fixed point symmetry operations are called point groups. ,The simplest point groups are the groups of operations of the individual n-fold symmetry axes. These are the point groups 1, 2, 3, 4 and

6. The only other point groups involving proper rotations can be found by using Euler's construction and are 222, 32, 422, 622, 23 and 432. Thus, the number of crystallographic groups involving only proper rotations is 11.

The point groups containing improper rotations can only be one of the six general types: I ($\bar{n}$), P/I ($n/\bar{n}$), P I I, I P I, I I P and P/I P/I P/I. The 21 point groups of this type are $\bar{1}$, m, $\bar{3}$, $\bar{4}$, $\bar{6}$, $2/m$, $4/m$, $6/m$, $2mm$, $3m$, $4mm$, $6mm$, $2/m\ \bar{3}$, $4/m\,\bar{3}\ 2/m$, $\bar{4}\ 2m$, $\bar{6}\,2m$, $\bar{4}\ 3m$, $2/m\ 2/m\ 2/m$, $\bar{3}\,2/m$, $4/m\ 2/m\ 2/m$ and $6/m\ 2/m\ 2/m$. The symbols written as fractions indicate that a proper rotation axis (whose symbol is in the numerator) is perpendicular to a mirror (represented by m in the denominator). Further, the symbols are transformed by substitution of m for $\bar{2}$ and also by making use of the facts that (i) $n/\bar{n}$ always contains an inversion centre and (ii) when n is even, $n/\bar{n}$ has m normal to the axis.

LATTICE TYPES AND COORDINATE SYSTEMS

Crystallographic bases can be chosen in an infinite number of ways. Conventionally, a basis is preferred if it leads to indices of planes and directions which reflect their symmetry. This feature can often be attained by taking as many of the three basis vectors as possible along the directions of symmetry. It has been shown earlier that translational symmetry restricts the types of possible rotational symmetries. Conversely, we will see that, the symmetry of a pattern specializes the lattice and limits it to a few types. These specialized lattices lead to specialized coordinate systems which are conventionally used in describing the geometry of crystal patterns.

Let us see in detail the restrictions imposed on the lattice by the point group 4. At first consider a plane lattice and place a 4-fold axis at each lattice point (filled squares in Fig. 12). The pattern is seen to have additional symmetry elements (open digons and squares). These are located either half

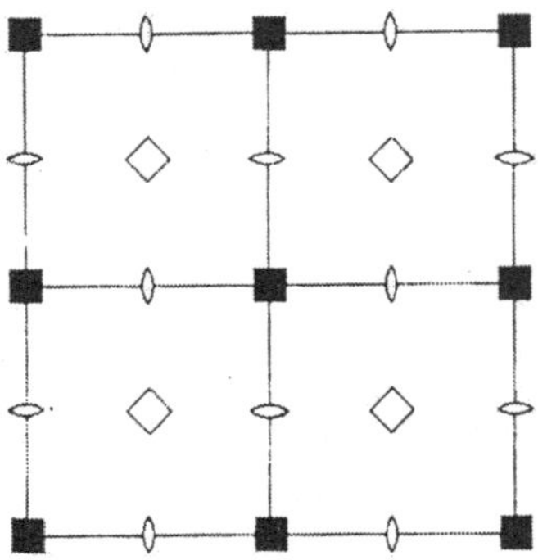

4

Fig. 12: Symmetries resulting from requiring a lattice to have a 4-fold axis (filled squares) at a lattice point. New symmetry elements result from combination of the rotation and translation, shown as open digons and squares (after Buerger, 1971).

way along the cell edges or in the centre of the cell. Clearly, the basis vectors are equal in magnitude and are mutually perpendicular.

Next we consider the restrictions imposed on a three dimensional or space lattice by the point group 4. We have already seen that the corresponding plane lattice has a 4-fold axis at each lattice point as well as at the centre of the square unit cell. To construct the space lattice, we can place another identical plane lattice at the next level so that the 4-fold axes of the two coincide. This can be done in two ways as shown in Fig. 13. In both cases, the separation between the two levels of the space lattice is $\mathbf{t}_3$. In the diagram at the left in Fig. 13, $\mathbf{t}_1$, $\mathbf{t}_2$ and $\mathbf{t}_3$ are mutually orthogonal and correspond to a cell having the shape of a square prism. On the other hand, in the diagram at the right in Fig. 13, $\mathbf{t}_1$ and $\mathbf{t}_2$ are orthogonal but $\mathbf{t}_3$ is oblique to these. The three translations lead to a cell which does not have the 4-fold symmetry of the lattice. This can be achieved by noting that a displacement of 2 $\mathbf{t}_3$ brings a lattice point of the lower level to the 4-fold axis again. Thus, a new cell with orthogonal axes $\mathbf{T}_1$, $\mathbf{T}_2$ and $\mathbf{T}_3$ can be chosen as follows.

$$\mathbf{T}_1 = \mathbf{t}_1$$
$$\mathbf{T}_2 = \mathbf{t}_2$$
$$\mathbf{T}_3 = \mathbf{t}_1 + \mathbf{t}_2 + 2\,\mathbf{t}_3$$

The shape of this cell is a square prism and has the symmetry of the lattice. However, this cell has a lattice point at its body centre, in addition to a lattice point at each vertex. Such a cell is called a body centred cell and denoted by the symbol I. This cell has a multiplicity (that is the number of lattice points per cell) equal to 2. There are no other distinct space lattices with a 4-fold axis. Thus, a 4-fold axis requires a space lattice whose cell has the shape of a square prism and is either primitive (P) or body-centred (In-centred, I).

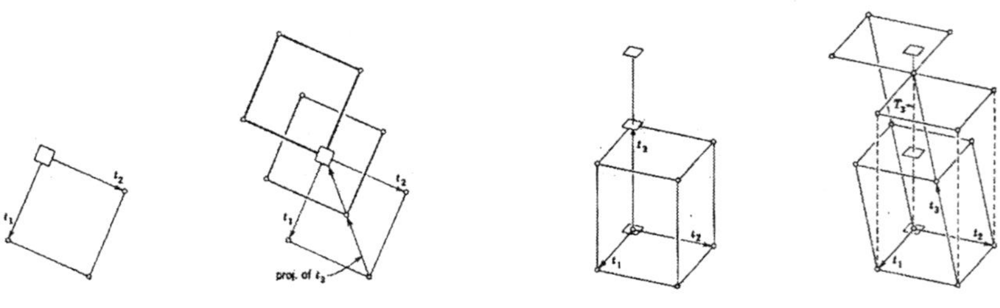

Fig. 13: Derivation of the space lattice types consistent with a 4-fold axis (after Buerger, 1971).

Derivation of Space Lattices Consistent with Other Symmetries

Space lattices consistent with other symmetries can be derived by a procedure similar to that used for a 4-fold axis. The procedure is outlined in Table 3 and uses the following arguments.

1. Space lattices are centrosymmetrical; thus, a point group without an inversion centre requires the same space lattice as one obtained by adding an inversion centre to the original point group. It is therefore sufficient to find space lattices consistent with the 11 centrosymmetrical point groups given in column 2 of Table 3.

Table 3: Outline of derivation of symmetry-specialized lattice types (after Buerger, 1971)

Point group	*Point group with inversion centre*	*Point group having the dimensional symmetry of lattice*	*Space lattice type*	*Minimum symmetry of space lattice*
$1, \bar{1}$	$\bar{1}$	$\bar{1}$	P	1
$2, m, \frac{2}{m}$	$\frac{2}{m}$	$\frac{2}{m}$	P B (A, I)	2
222, $2mm$, $\frac{2}{m}\frac{2}{m}\frac{2}{m}$	$\frac{2}{m}\frac{2}{m}\frac{2}{m}$	$\frac{2}{m}\frac{2}{m}\frac{2}{m}$	P I C(A, B) F	222
$4, \bar{4}, \frac{4}{m}$ 422, $4mm$, $\bar{4}2m$, $\frac{4}{m}\frac{2}{m}\frac{2}{m}$	$\frac{4}{m}$ $\frac{4}{m}\frac{2}{m}\frac{2}{m}$	$\frac{4}{m}\frac{2}{m}\frac{2}{m}$	P(C) I (F)	4
$3, \bar{3}$ 32, $3m$, $\bar{3}\frac{2}{m}$	$\bar{3}$ $\bar{3}\frac{2}{m}$	$\bar{3}\frac{2}{m}$	R P*	3
$6, \bar{6}, \frac{6}{m}$ 622, $6mm$ $\bar{6}m2$ $\frac{6}{m}\frac{2}{m}\frac{2}{m}$	$\frac{6}{m}$ $\frac{6}{m}\frac{2}{m}\frac{2}{m}$	$\frac{6}{m}\frac{2}{m}\frac{2}{m}$	P*	3
$23, \frac{2}{3}\bar{3}$ $432, \bar{4}3m$ $\frac{4}{m}\bar{3}\frac{2}{m}$	$\frac{2}{m}\bar{3}$ $\frac{4}{m}\bar{3}\frac{2}{m}$	$\frac{4}{m}\bar{3}\frac{2}{m}$	P I F	23

Note: The two Ps with * in the superscript represent the same lattice.

2. If the fold n of an axis is greater than 2, then the dimensional symmetry of the lattice already has mirrors parallel to the rotation axis. Thus, the lattices consistent with n are identical to those consistent with n and mirrors. Thus all symmetries except the seven listed in column 3 of Table 3 need not be considered separately.

For any particular symmetry, each plane lattice type consistent with that symmetry can be found. These are then placed with respect to the symmetry of the principal symmetry axis in alternative permissible ways. This results in the 14 distinct space lattices called Bravais lattices which are listed in column 4 of Table 3. The types given in parenthesis are equivalent and represent the same lattice type in a different orientation. Similarly, the two Ps with * in superscript represent the same lattice type. All the cells are either primitive or centred in one of only distinct ways as displayed in Table 4.

Table 4: General designations of space lattice types (after Buerger 1971)

Symbol	*Name*	*Locations of lattice points*	*Multiplicity of cell*
P	Primitive	000	1
I	Body centred	000 ½½½	2
A	End centred (A-centred)	000 0½½	2
B	End centred (B-centred)	000 ½0½	2
C	End centred (C-centred)	000 ½½0	2
F	Face centred	000 0½½ ½0½ ½½0	4
R	Rhombohedral	000 ⅔⅓⅓ ⅓⅔⅔	3

Coordinate Systems/Crystallographic Basis Vectors

The edges of the unit cells described above constitute a convenient crystallographic basis for each of the lattices. If the multiplicities of these cells are ignored, the shapes of these cells can be categorized into six general types, so that only six general types of crystallographic basis vectors are required to describe the geometries of all patterns. These are known as crystal systems. Table 5 shows the distribution of the 32 point groups among these six crystal systems. Restrictions on the lattice parameters owing to symmetry are listed in the last column of Table 5.

CHARACTERISTICS OF SPACE GROUPS

The symmetry of a pattern which involves translational repetition is called a space group. As is clear from the earlier discussion, the symmetry operations which must be considered for patterns involving translations are the following.

Table 5: The appropriate coordinate systems for the various point groups (after Buerger, 1971)

Point groups	*Space lattice types permitted*	*Symmetry of space lattice*	*Crystal system*	*Axes*
1, $\bar{1}$	P	$\bar{1}$	triclinic	a ≠ b ≠ c α ≠ β ≠ γ
2, m, $\frac{2}{m}$	P B (A, I)	$\frac{2}{m}$	monoclinic	a ≠ b ≠ c. γ ≠ α ≡ β = 90°
222, 2 mm, $\frac{2}{m}\frac{2}{m}\frac{2}{m}$	P I C(A, B), F	$\frac{2}{m}\frac{2}{m}\frac{2}{m}$	orthorhombic	a ≠ b ≠ c α ≡ β ≡ γ = 90°
4, $\bar{4}$, $\frac{4}{m}$; 422, 4mm, $\bar{4}$ 2m, $\frac{4}{m}\frac{2}{m}\frac{2}{m}$	P, I	$\frac{4}{m}\frac{2}{m}\frac{2}{m}$	tetragonal	a ≡ b ≠ c α ≡ β ≡ γ = 90°
3, $\bar{3}$; 32, 3m, $\bar{3}\frac{2}{m}$ 6, $\bar{6}$, $\frac{6}{m}$ 622, 6mm, $\bar{6}$ 2m, $\frac{6}{m}\frac{2}{m}\frac{2}{m}$	R P P	$\bar{3}\frac{2}{m}$ $\frac{6}{m}\frac{2}{m}\frac{2}{m}$ $\frac{6}{m}\frac{2}{m}\frac{2}{m}$	hexagonal	a ≡ b ≠ c α ≡ β = 90° γ = 120°
23, $\frac{2}{3}\bar{3}$; 432, $\bar{4}$ 3m, $\frac{4}{m}\bar{3}\frac{2}{m}$	PIF	$\frac{4}{m}\bar{3}\frac{2}{m}$	Isometric (cubic)	a ≡ b ≡ c α ≡ β ≡ γ = 90°

(i) The permissible crystallographic screw axes n_p.

(ii) The permissible crystallographic rotation axes *n*, which are special cases of screw axes having p = 0.

(iii) The glide planes *a*, *b*, *c*, *d* and *n*.

(iv) The reflection plane *m*, which is a special case of the glide planes having no translation.

(v) The permissible crystallographic rotoinversion axes, $\bar{2}$, $\bar{3}$, $\bar{4}$, and $\bar{6}$. However $\bar{2}$, is identical to reflection.

(vi) The centre of inversion ($\bar{1}$ or i) which is a special case of the rotoinversion axes having no rotation.

Two screw axes are said to be isogonal if they have the same rotational angle α. For example, the rotation axis 3 and the screw axes 3_1 and 3_2 are isogonal. This characteristic leads to the result that the angle between two different screw axes is identical to the angle between the two corresponding isogonal rotation axes, because the two situations differ only by translations which have no effect on angular relationships. Thus, the orientations of all axes in a space group are identical to those of the isogonal axes of some point group. As a consequence of this, if the translations of a space group are disregarded, all that remain are the symmetry elements of a particular point group, each of whose symmetry elements is isogonal with a set of parallel symmetry elements of the space group. Conversely, to every point group, there exists a set of isogonal space groups, since every space group is a member of a set of space groups which are isogonal with some point group. This relation makes it possible to discover all the space groups by systematically exploring all the point groups.

COMBINATIONS OF SYMMETRY ELEMENTS AND TRANSLATIONS

It follows from the isogonal relations that if (for example) a point group has an n-fold rotation axis, each of the related space groups has one isogonal screw axis n_p parallel to it. However, the space group also includes translational symmetry consistent with the n-fold axis. Since the symmetry operations form a group, the combinations of each operation of the n_p screw with each of its translations must also occur. We shall illustrate this feature with a few examples.

Combination of a 2-fold Rotation or 2_1 Screw Axis and a Translation

Three pairs of diagrams are shown in Fig. 14 which presents the possible combinations. In each figure, the upper diagram is a view along the direction of the axis Q, while the lower diagram is a view perpendicular to the plane formed by the axis Q and the translation T. The subscript in the symbol of the translation indicates whether it is perpendicular to the axis ($T_\perp$) or has components which are parallel as well as perpendicular to the axis ($T_{\perp+\parallel}$).

In Fig. 14(i), the symmetry element 2 at Q, designated 2(Q), is repeated by the translation $T_\perp$ as the symmetry element 2(Q′). It is obvious from the figure that the points related by 2(Q) and those related by 2(Q′) are also interrelated by a symmetry element 2 located at S at a translation of $(1/2)T_\perp$ from Q. This result can be written as

$$2(Q) \cdot T_\perp = 2(S)$$

In a similar way, Fig. 14(ii) shows that

$$2_1(Q) \cdot T_\perp = 2_1(S)$$

If the translation also has a component equal to half the translation vector parallel to the axis, there is a different outcome. In Fig. 14(iii), the translation $T_{\perp+\parallel}$ repeats 2(Q) as 2(Q′) such that all points related by 2(Q′) are shifted half a translation interval along the axis Q′. Thus the points related by 2(Q) and those related by 2(Q′) are also interrelated by a screw 2_1 at S. Thus,

$$2(Q)\ .\ T_{\perp+\parallel} = 2_1(S)$$

If the same argument is applied to the screw 2_1 at S, it is seen that this is repeated by the translation as a 2_1 screw axis at S′, but the points around S and those around S′ are interrelated by a 2-fold rotation at Q′. Thus,

$$2_1(S)\ .\ T_{\perp+\parallel} = 2_1(Q')$$

The triplet of symmetry operations in the last two cases is the same.

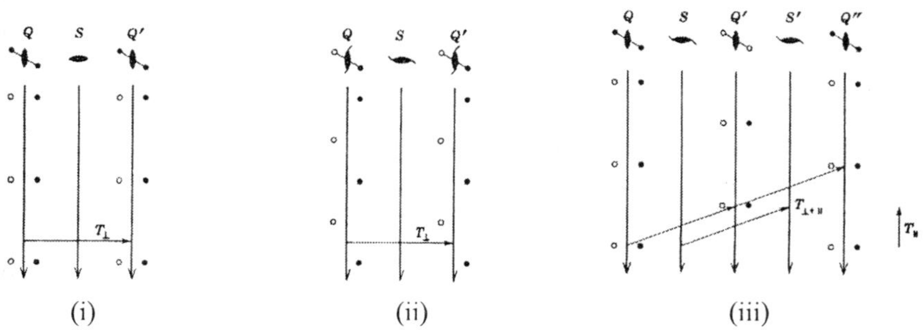

Fig. 14: Results of combining various kinds of screw axes with translations (after Buerger, 1971).

Combination of a Glide Plane and a Translation

The possible combinations of a glide and a translation are shown in Fig. 15, which are views normal to a plane determined by the lattice translation T and (only in the case of glide planes) the intrinsic translation τ_1 of the glide. The results differ depending on whether the translation T has a component parallel to the glide plane or not. If it has, there are two distinct cases corresponding to $T_\parallel$ being parallel to the intrinsic translation τ_1 of the glide or oblique to τ_1 with another component τ_2. The four resulting combinations are shown in Fig. 15 as listed below.

m(Q)	m(S)	$T_\perp$	Fig. 15(i)
$m_\tau(Q)$	$m_\tau(S)$	$T_\perp$	Fig. 15(ii)
$m_{\tau1}(Q)$	m(S)	$T_{\perp,\ \tau1}$	Fig. 15(iii)
$m_{\tau1}(Q)$	$m_{\tau2}(S)$	$T_{\perp,\ \tau1,\ \tau2}$	Fig. 15(iv)

Combination of an Inversion Centre and a Translation

The combination of an inversion centre i at Q with a translation T is shown in Fig. 16. The translation duplicates i at Q′. A little consideration shows that the points related by i(Q) and those related by i(Q′) are interrelated by

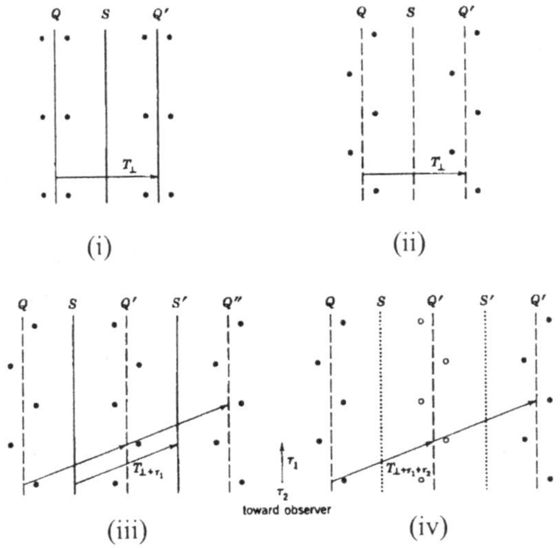

Fig. 15: Results of combining various kinds of glide planes with translations (after Buerger, 1971).

another inversion centre at S, halfway between Q and Q′. The resulting combination is thus,

i(Q) i(S) T

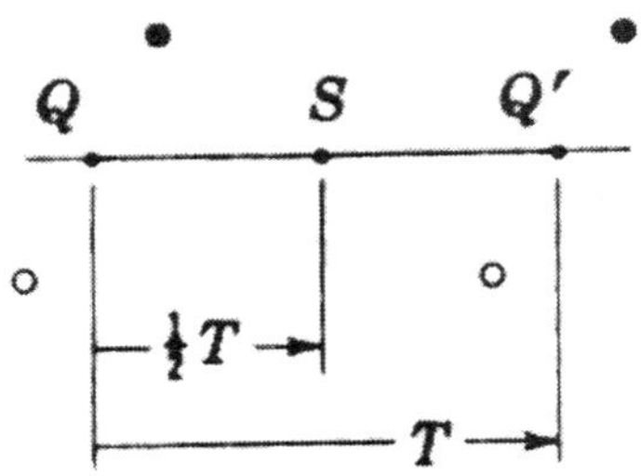

Fig. 16: The result of combining an inversion centre with a translation (after Buerger, 1971).

Combinations of Symmetry Elements with Each Other

If two symmetry elements occur in a point group, corresponding symmetry elements with the same angular relations occur in each isogonal space group. The only difference between the space group and point group geometries is the presence of translations in the former. These translations include displacements of the symmetry elements from one another. This is illustrated by considering two combinations in the following.

Combination of a Screw and a Glide

It is known that the combination of a 2-fold rotation and reflection plane perpendicular to it implies an inversion centre at the point of their intersection.

From the isogonal relation, it follows that the combination of a 2-fold screw and a perpendicular glide plane implies an inversion centre. The location of the inversion centre can be found as follows.

In Fig. 17, a 2-fold screw axis AA is shown normal to a glide plane P. The origin of coordinates is assumed to be at the intersection of the axis and the plane. The y-axis is taken along the glide direction and the z-axis is along the screw. A point with coordinates $x\ y\ z$ is also shown in the figure. The screw requires the point at $x\ y\ z$ to be related to another point whose coordinates have x and y reversed and z reduced by half a lattice translation in the direction z due to the intrinsic translation of the screw. The coordinates of this point are therefore $\bar{x}\ \bar{y}\ z$-½. The glide operation is performed in two successive steps. The intrinsic translation of the glide changes the y-component of this second point by ½, while reflection reverses its z-coordinate. The resulting coordinates of the third point are therefore, $\bar{x}\ \bar{y}$ + ½ $\bar{z}$ + ½. Figure 17 shows that the first and third points require an inversion centre located away from the axis and plane by components equal to half of τ_1 (the intrinsic translation of the screw) and half of τ_2 (the intrinsic translation of the glide). The combination is displayed in Fig. 18. The projection of this result, expressed in conventional symbols, is shown at the top left in Fig. 19. The specific cases for $\tau_1 = 0$, $\tau_2 = ½$; $\tau_1 = ½$, $\tau_2 = 0$ and $\tau_1 = \tau_2 = 0$ are shown in the other three diagrams of Fig. 19.

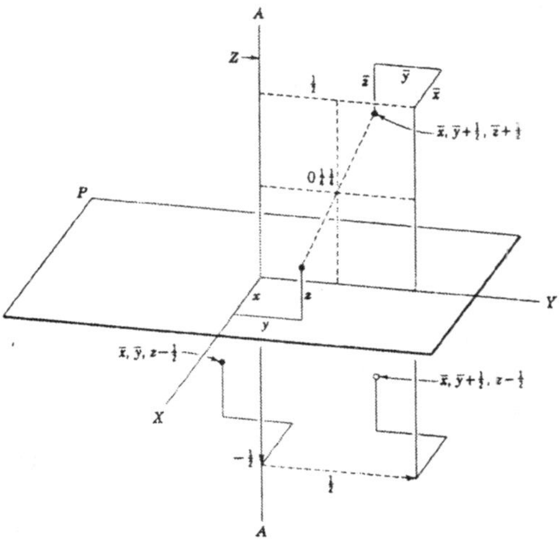

Fig. 17: The result of combining a 2_1 screw at AA with a perpendicular glide plane (after Buerger, 1971).

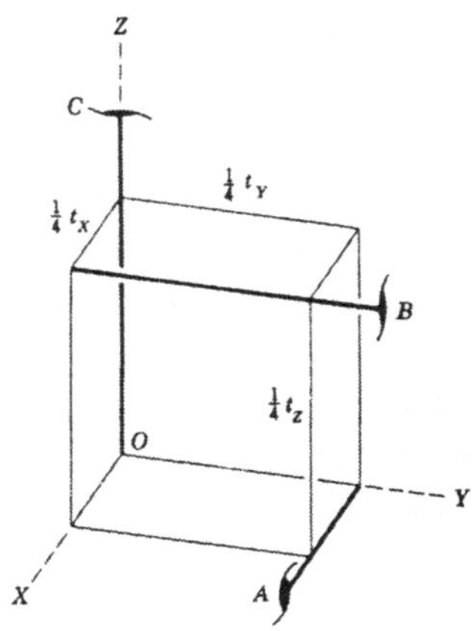

Fig. 18: Abbreviation of the result obtained from Fig. 23 (after Buerger, 1971).

Combination of Non-intersecting Screws

Consider two non-intersecting orthogonal 2_1 screw axes A and B. The coordinate axes are orthogonal with the x-axis being parallel to A, the y-axis to B and the origin as shown in Fig. 20. The edges of the block shown in

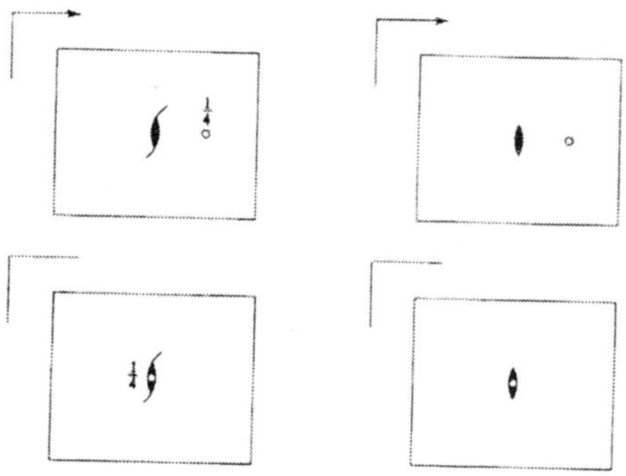

Fig. 19: Results of combining a 2-fold symmetry axis and a perpendicular symmetry plane for particular values of their translation components (after Buerger, 1971).

the figure have the dimensions of a quarter of a lattice translation along the screws A and B. It will turn out that this combination involves a 2_1 screw C and the height of the block is a quarter of a lattice translation along C.

A view of the combination of axes along C is shown in Fig 21. A point with coordinates $x\ y\ z$ is displaced by an operation of A to $\frac{1}{2} + x\ \frac{1}{2} - y\ \bar{z}$. This new point is displaced by an operation of B to a point with coordinates $\frac{1}{2} - (\frac{1}{2} + x) - \frac{1}{2} + (\frac{1}{2} - y)\ \frac{1}{2} - \bar{z} = \bar{x}\ \bar{y}\ \frac{1}{2} + z$. The coordinates of the first and last points are related by a 2_1 screw along C perpendicular to A as well as B. The geometry of Fig. 20 when projected along C is reproduced in the top left diagram of Fig. 22 as in a conventional space group diagram. The other possible cases corresponding to $\tau_1 = \tau_2 = 0$ and $\tau_1 = \tau_2 = \frac{1}{2}$ are shown in the other three diagrams of Fig. 22, depending on whether the two axes intersect or not.

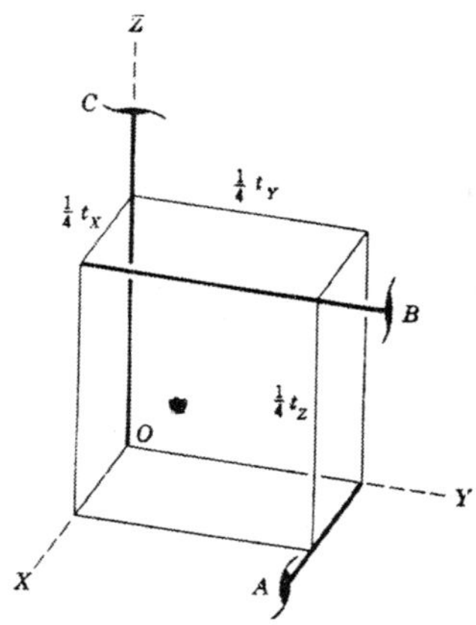

Fig. 20: General appearance of two orthogonal non-intersecting screws (after Buerger, 1971).

DERIVATION OF SPACE GROUPS

The space groups can be derived by considering one point group at a time, substituting systematically all possible space group symmetry elements for the corresponding operations of the isogonal point group and combining these with the translations of each of the lattice types possible for that point group. Each space group symmetry element is repeated at the end of every translation vector of the lattice. Further, as shown above, each translation

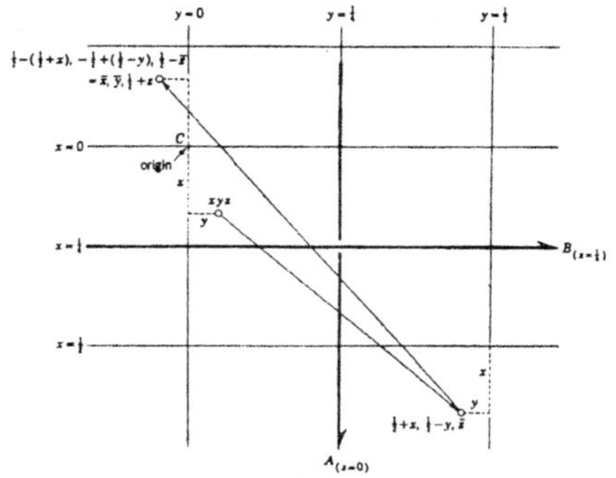

Fig. 21: The result of combining two orthogonal non-intersecting screws A and B (after Buerger, 1971).

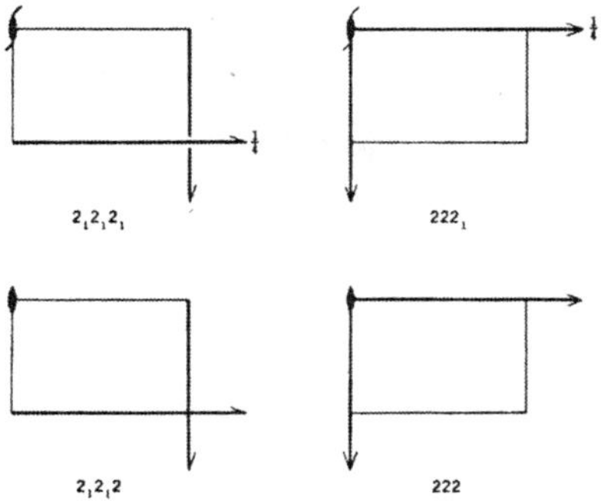

Fig. 22: The result of combining two horizontal screws for particular values of their translation components. The fraction indicates the height of axis B above axis A as a fraction of the lattice translation along C (after Buerger, 1971).

combines with the original symmetry element to yield another space group symmetry element. For forming such combinations, it is sufficient to consider only those lattice translations which radiate from one particular lattice point and terminate within the eight cells to which that lattice point is common.

The space group symbol consists of a capital letter indicating the lattice type and then a sequence of symbols which indicate the symmetry operations of the space group. We note that a point group is designated by a series of three simple or compound symbols which represent the combination of axes in the point group. If proper and improper rotations occur parallel to the same axis, the symbol for the pair is in the form of a fraction whose numerator is the symbol for the proper rotation and whose denominator is $\bar{2} \equiv m$. The latter part of the space group symbol is similar except for possible substitution of point group symmetry operations by corresponding isogonal space group symmetry operations.

We shall derive space groups isogonal with the point groups 2, *m* and 222 (only for C-centred lattices). All the other space groups numbering a total of 230 types can be derived in a similar manner.

Space Groups Isogonal with Point Group 2

This point group can have lattice types P and B as shown in Table 3. There is a single symmetry operation 2, whose isogonal operations are 2 and 2_1. Thus, the space groups isogonal with point group 2 are P2, P2_1, B2 and B2_1. However, the combinations B2 and B2_1 correspond to the combinations illustrated in Fig. 14(iii). These combinations are identical requiring alternating 2 and 2_1 axes along the translation. Therefore, the apparently different space groups B2 and B2_1 are the same with the conventional designation B2. The set of symmetry elements corresponding to these space groups is shown in Fig. 23. The isogonal point group 2 is shown at the top.

Each space group is shown in two ways. The location of symmetry elements in one unit cell is shown in the left. An atom in a general position (that is, not lying on a symmetry element) accompanied by equivalent atoms required by symmetry are shown on the right. The sign + (–) indicates that the z-coordinate of the atom is $+z$ ($-z$). Similarly, ½ ± indicates that the z-coordinate is $\frac{1}{2} \pm z$.

Space Groups Isogonal with Point Group *m*

The lattice types possible for point group *m* are P and B. The symmetry elements isogonal with *m* are m and a glide plane, say b. The possible space groups are thus, Pm, Pb, Bm and Bb as shown in Fig. 24. The diagrams show the unit cell projected along c and also an angle-like symbol near the top left corner. Each of these is the symbol for a pair of symmetry planes of some kind parallel to the plane of the paper located ½c apart. If no fraction appears near the symbol for the symmetry plane, the heights are 0 and ½. The fraction ¼ implies the two heights to be ¼ and ¼ + ½ = ¾. An arrow indicates that the symmetry plane is a glide plane with a translation component in the direction of the arrow. In the combinations involving B lattices, alternative symmetry planes are reflection and glide planes as required by Figs. 15(iii) and (iv).

Space Groups Isogonal with Point Group 222

The point group 222 can have lattice types P, C, I or F. We shall restrict consideration of space groups based on C type lattices. The combinations of

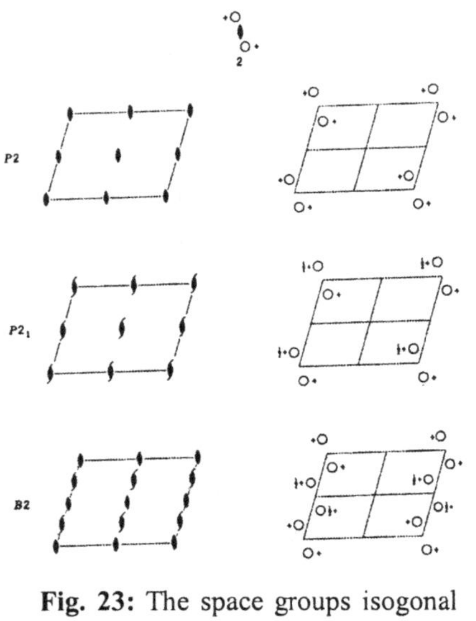

Fig. 23: The space groups isogonal with the point group 2 (after Buerger, 1971).

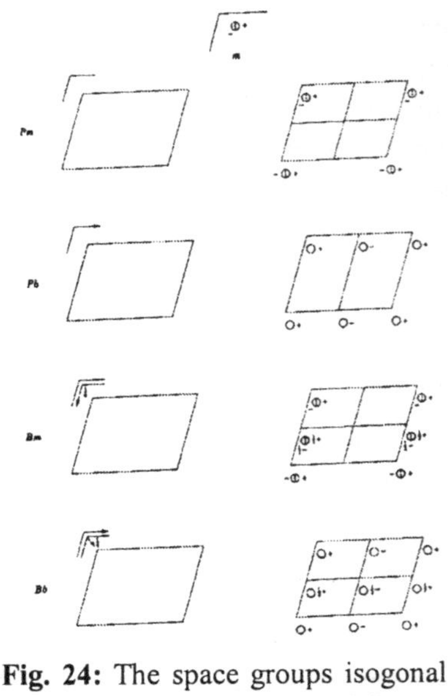

Fig. 24: The space groups isogonal with the point group *m* (after Buerger, 1971).

symmetry elements isogonal with 222 are 222, 222_1, 2_12_12 and $2_12_12_1$. Thus, the space groups C222, $C222_1$, ($C2_12_12$) and ($C2_12_12_1$) appear to be possible. However, the latter two combinations shown in parenthesis are not distinct because the C-centring translations require alternate 2 and 2_1 axes along these translations. The symbols without parenthesis are the conventional designations for the distinct space groups which are illustrated in Fig. 25.

Choice of Cell for Monoclinic Space Groups

Complications arise due to different settings only for the cases of monoclinic and orthorhombic space groups. These are discussed in the following. Two of the three monoclinic point groups (*m* and 2/*m*) have reflection planes. In the isogonal space groups, the symmetry operations corresponding to these reflection planes are either reflection or glide planes. The translation component τ of the glide plane can occur only in the direction of a lattice translation, but there is no restriction as to which lattice translation this could be. In Fig. 24, the translation component of the glide plane is that of the cell edge b for a primitive cell. However, this is simply an illustration of one of several possible cases. These are discussed below.

The standard choice of cell for any crystal is the one based upon the three shortest non-coplanar translations. This is called the reduced cell. By adopting the first setting for a monoclinic crystal, the cell axes a and b are chosen in the symmetry plane while c is chosen along the unique axis normal to this plane. Thus, the choice of c is fixed while a and b are translations in the symmetry plane. For monoclinic cells, the choice of a reduced cell is a two dimensional problem. The cell edges a and b are simply the two shortest translations in the symmetry plane. These can be found by noting that a plane cell is reduced only if each of the two cell edges is shorter than each of the two cell diagonals. The reduced cell designated 1 and two other choices of cells designated 2 and 3 with corresponding basis vectors $\mathbf{a}_i$, $\mathbf{b}_i$ (i = 1 to 3) are shown in Fig.26(i).

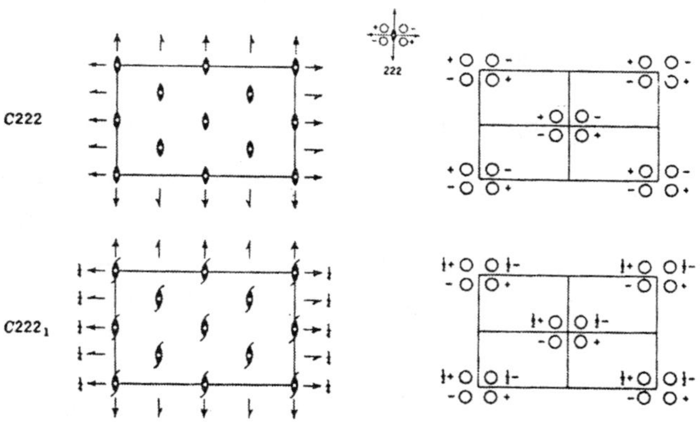

Fig. 25: The space groups isogonal with the point group 222 (after Buerger, 1971).

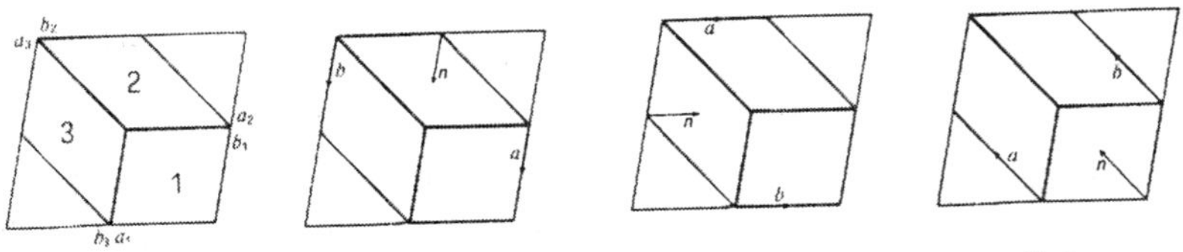

(i) (ii) (iii) (iv)

Fig. 26: (i) The reduced cell labeled 1 and two unreduced cells 2 and 3 as seen parallel to the glide plane for a monoclinic lattice, (ii), (iii) and (iv) glides *a*, *b* and *n* for the reduced cell and the corresponding designations of the same glides for the unreduced cells.

This choice of cell is governed by convention and has nothing to do with the glide direction which can be any translation within the plane. Of all the possible translations in the plane, only three are distinct for our purpose. These are the translations radiating from one lattice point of a cell to three other lattice points of the cell, that is, the translations a, b and [110]. These correspond to glide symbols *a*, *b* and *n*, respectively.

From this discussion, it might appear that the point group *m* has three isogonal space groups with glide planes and primitive cells, namely Pa, Pb and Pn. But by different choices of the cell and interchange of labeling for directions a and b, any of those three space group designations can be transformed into one another. An *a* glide for the cell choice 1 corresponds to *n* and *b* glides, respectively for cell choices 2 and 3 as illustrated in Fig. 26 (ii). Similar changes occur for *b* and *n* glides in cell 1. These are shown in Fig. 26 (iii) and (iv), respectively. Thus, Pa, Pb and Pn are not distinct symmetries since the labels are determined by the coordinate system, so the different symbols simply show how the glide direction is described in the coordinate system of the reduced cell. It may be noted that the conventional space group symbol for this space group is Pc, which is based on a setting with unique b axis and a coordinate system in which the c axis is chosen parallel to the glide direction and which may be different from a choice based on the reduced cell.

Alternative Settings for Orthorhombic Space Groups

As mentioned above, the symmetry of a crystal is independent of the choice of a coordinate system, which in turn depends on the choice of a cell and the labeling of its edges. The latter choice is made on the basis of convention. The setting of the symmetry axes (along *x*, *y* or *z*) of an orthorhombic crystal with respect to the cell edges (a, b or c) can be arranged in six ways as follows.

$Cmcm$	D_{2h}^{17}	mmm	Orthorhombic
No. 63	$C\,2/m\,2/c\,2_1/m$		Patterson symmetry $Cmmm$

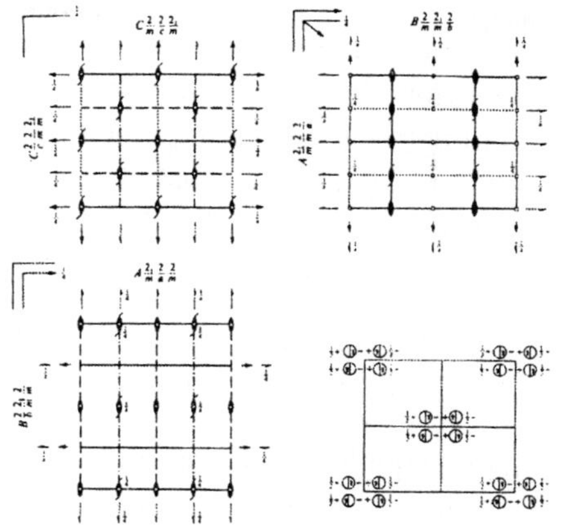

Origin at centre ($2/m$) at $2/m\,c\,2_1$

Asymmetric unit $0 \le x \le \frac{1}{2};\ 0 \le y \le \frac{1}{2};\ 0 \le z \le \frac{1}{4}$

Symmetry operations

For (0,0,0)+ set

(1) 1	(2) $2(0,0,\frac{1}{2})$ 0,0,z	(3) 2 0,y,$\frac{1}{4}$	(4) 2 x,0,0
(5) $\bar{1}$ 0,0,0	(6) m $x,y,\frac{1}{4}$	(7) c x,0,z	(8) m 0,y,z

For $(\frac{1}{2},\frac{1}{2},0)$+ set

(1) $t(\frac{1}{2},\frac{1}{2},0)$	(2) $2(0,0,\frac{1}{2})$ $\frac{1}{4},\frac{1}{4},z$	(3) $2(0,\frac{1}{2},0)$ $\frac{1}{4},y,\frac{1}{4}$	(4) $2(\frac{1}{2},0,0)$ $x,\frac{1}{4},0$
(5) $\bar{1}$ $\frac{1}{4},\frac{1}{4},0$	(6) $n(\frac{1}{2},\frac{1}{2},0)$ $x,y,\frac{1}{4}$	(7) $n(\frac{1}{2},0,\frac{1}{2})$ $x,\frac{1}{4},z$	(8) b $\frac{1}{4},y,z$

Generators selected (1); $t(1,0,0)$; $t(0,1,0)$; $t(0,0,1)$; $t(\frac{1}{2},\frac{1}{2},0)$; (2); (3); (5)

Positions

Multiplicity, Wyckoff letter, Site symmetry			Coordinates (0,0,0)+ $(\frac{1}{2},\frac{1}{2},0)$+				Reflection conditions
							General:
16	h	1	(1) x,y,z (5) $\bar{x},\bar{y},\bar{z}$	(2) $\bar{x},\bar{y},z+\frac{1}{2}$ (6) $x,y,\bar{z}+\frac{1}{2}$	(3) $\bar{x},y,\bar{z}+\frac{1}{2}$ (7) $x,\bar{y},z+\frac{1}{2}$	(4) $x,\bar{y},\bar{z}$ (8) $\bar{x},y,z$	hkl: $h+k=2n$; $0kl$: $k=2n$; $h0l$: $h,l=2n$; $hk0$: $h+k=2n$; $h00$: $h=2n$; $0k0$: $k=2n$; $00l$: $l=2n$
							Special: as above, plus
8	g	..m	$x,y,\frac{1}{4}$	$\bar{x},\bar{y},\frac{3}{4}$	$\bar{x},y,\frac{1}{4}$	$x,\bar{y},\frac{3}{4}$	no extra conditions
8	f	m..	$0,y,z$	$0,\bar{y},z+\frac{1}{2}$	$0,y,\bar{z}+\frac{1}{2}$	$0,\bar{y},\bar{z}$	no extra conditions
8	e	2..	$x,0,0$	$\bar{x},0,\frac{1}{2}$	$\bar{x},0,0$	$x,0,\frac{1}{2}$	hkl: $l=2n$
8	d	$\bar{1}$	$\frac{1}{4},\frac{1}{4},0$	$\frac{3}{4},\frac{3}{4},\frac{1}{2}$	$\frac{3}{4},\frac{1}{4},\frac{1}{2}$	$\frac{1}{4},\frac{3}{4},0$	hkl: $k,l=2n$
4	c	$m2m$	$0,y,\frac{1}{4}$	$0,\bar{y},\frac{3}{4}$			no extra conditions
4	b	$2/m$..	$0,\frac{1}{2},0$	$0,\frac{1}{2},\frac{1}{2}$			hkl: $l=2n$
4	a	$2/m$..	$0,0,0$	$0,0,\frac{1}{2}$			hkl: $l=2n$

Symmetry of special projections

Along [001] $c2mm$	Along [100] $p2gm$	Along [010] $p2mm$
$\mathbf{a}'=\mathbf{a}$ $\mathbf{b}'=\mathbf{b}$	$\mathbf{a}'=\frac{1}{2}\mathbf{b}$ $\mathbf{b}'=\mathbf{c}$	$\mathbf{a}'=\frac{1}{2}\mathbf{c}$ $\mathbf{b}'=\frac{1}{2}\mathbf{a}$
Origin at 0,0,z	Origin at x,0,0	Origin at 0,y,0

Maximal non-isomorphic subgroups

I	[2]$C222_1$	(1; 2; 3; 4)+
	[2]$C112_1/m$ ($P2_1/m$)	(1; 2; 5; 6)+
	[2]$C12/c1$ ($C2/c$)	(1; 3; 5; 7)+
	[2]$C2/m11$ ($C2/m$)	(1; 4; 5; 8)+
	[2]$Cmc2_1$	(1; 2; 7; 8)+
	[2]$Cm2m$ ($Amm2$)	(1; 3; 6; 8)+
	[2]$C2cm$ ($Ama2$)	(1; 4; 6; 7)+
IIa	[2]$Pmcm$ ($Pmma$)	1; 2; 3; 4; 5; 6; 7; 8
	[2]$Pbnn$ ($Pnna$)	1; 2; 3; 4; (5; 6; 7; 8)+$(\frac{1}{2},\frac{1}{2},0)$
	[2]$Pbnm$ ($Pnma$)	1; 2; 5; 6; (3; 4; 7; 8)+$(\frac{1}{2},\frac{1}{2},0)$
	[2]$Pbcn$	1; 3; 5; 7; (2; 4; 6; 8)+$(\frac{1}{2},\frac{1}{2},0)$
	[2]$Pmnn$ ($Pnnm$)	1; 4; 5; 8; (2; 3; 6; 7)+$(\frac{1}{2},\frac{1}{2},0)$
	[2]$Pmcn$ ($Pnma$)	1; 2; 7; 8; (3; 4; 5; 6)+$(\frac{1}{2},\frac{1}{2},0)$
	[2]$Pmnm$ ($Pmmn$)	1; 3; 6; 8; (2; 4; 5; 7)+$(\frac{1}{2},\frac{1}{2},0)$
	[2]$Pbcm$	1; 4; 6; 7; (2; 3; 5; 8)+$(\frac{1}{2},\frac{1}{2},0)$
IIb	none	

Maximal isomorphic subgroups of lowest index

IIc [3]$Cmcm$ ($\mathbf{a}'=3\mathbf{a}$); [3]$Cmcm$ ($\mathbf{b}'=3\mathbf{b}$); [3]$Cmcm$ ($\mathbf{c}'=3\mathbf{c}$)

Minimal non-isomorphic supergroups

I [3]$P6_3/mcm$; [3]$P6_3/mmc$

II [2]$Fmmm$; [2]$Pmcm$ ($2\mathbf{a}'=\mathbf{a}$, $2\mathbf{b}'=\mathbf{b}$)($Pmma$); [2]$Cmmm$ ($2\mathbf{c}'=\mathbf{c}$)

Fig. 27: Space group Cmcm (after Hahn, 1989).

a b c	a b c	a b c	a b c	a b c	a b c
x y z	*z x y*	*y z x*	*x z y*	*y x z*	*z y x*

In the general case, the symmetry axes along x, y and z are different. If the symbols for these symmetry axes do not depend on their relations to a, b and c, the problem of deriving the space group symbol for various settings is trivial. For example, the space group Pmn2_1 has $m \perp x$, $n \perp y$ and $2_1 \parallel z$. If these correspondences are substituted in the list given above, the resulting space group symbols are as follows.

a b c	a b c	a b c	a b c	a b c	a b c
Pmn2_1	P2_1mn	Pn2_1m	Pm2_1n	Pnm2_1	P2_1nm

The situation is more complicated for space groups such as Cmcm where some symbols change with orientation. This can be represented diagrammatically as Cm $\rightarrow$ m (and one of its reversed permutations C $\leftarrow$ mm) where the arrow implies that the symbol in that position takes the letter of the following (or preceding) axial location. Further, the centering can also change depending on the setting in an obvious way. The correspondence for this space group accordingly is as follows.

a b c	a b c	a b c	a b c	a b c	a b c
Cmcm	Amma	Bbmm	Bmmb	Ccmm	Amam

These are illustrated in Fig. 27. Each figure can be viewed in two orientations. The symbol for each setting of the symmetry elements is given above the diagram when viewed in that direction. Figure 27 also gives detailed information regarding the symmetry operations as well as other characteristics of the space group. Similar information is provided for all the 230 types of space groups in Hahn (1989).

REFERENCES

Buerger, M. (1971). Introduction to Crystal Geometry (New York: McGraw-Hill).

Hahn, T., Ed. (1989). International Tables for Crystallography, Volume A: Space-Group Symmetry, Second Revised Edition, Reprinted with corrections (Dordecht: Kluwer).

Hinton, H. (1903). Mathematical Crystallography and the Theory of Groups of Movements (Oxford: Clarendon Press).

Local Texture Measurements with High-Energy Synchrotron Radiation

W. Skrotzki, B. Klöden, N. Scheerbaum, R. Tamm, C.-G. Oertel, U. Garbe[1], E. Rybacki[2] and H.-G. Brokmeier[3]

Institute of Structural Physics, Division of Metal Physics
Dresden University of Technology, D-01062 Dresden, Germany
[1]HASYLAB at DESY, Notkestr. 85, D-22603 Hamburg, Germany
[2]Geo Research Centre Potsdam, Telegrafenberg, D-14473 Potsdam, Germany
[3]GKSS Research Centre, Max-Planck-Straße, D-21494 Geesthacht, Germany

INTRODUCTION

To measure texture gradients as they exist in samples deformed in torsion and by equal channel angular pressing (ECAP), methods with high spatial resolution have to be applied. The method nowadays widely used is electron backscatter diffraction (EBSD) allowing orientation measurements of single grains. Although spatial resolution has been steadily improved in the past, limits are given for very small grain sizes and/or high strains and poor grain statistics. In this paper a method is presented which is suitable to go beyond these limitations. High-energy synchrotron radiation, which is available at certain accelerators all over the world, offers the possibility of measuring textures with a high spatial resolution and good grain statistics. This is due to the high energy, i.e. large penetration depth, and low angular divergence allowing a strong collimation of the beam. Thus, there is a possibility of measuring textures strain-resolved in torsion deformed cylinders and in ECAP deformed bars.

The first example given is on the analysis of texture development in NiAl. This intermetallic compound has special properties, like high melting point, low density, good corrosion resistance and moderate creep strength, making this material interesting for high-temperature applications. However, a widespread acceptance as structural material was hindered so far by the inadequate low-temperature toughness and ductility. Attempts to increase the ductility of NiAl by alloying have not been successful (Noebe et al., 1996). Another way to approach this problem is to change the grain structure

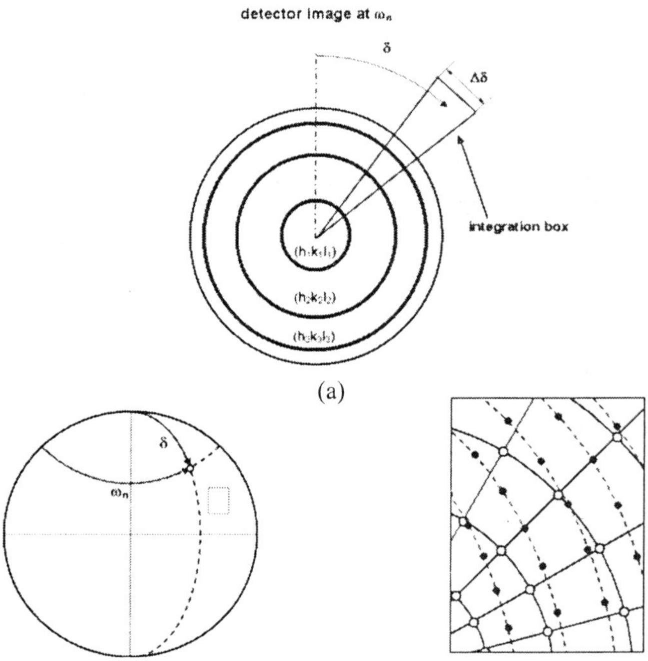

(b) (c)

Fig. 2: Data analysis of texture measurements with synchrotron radiation. (a) Read out of the Debye-Scherrer rings by moving an integration box around the centre of rings in steps $\delta = 5°$. (b) The intensity values of each box $\Delta\delta$ are converted into the corresponding pole figures ($h_ik_il_i$) (simplified sketch for small Bragg angles). The arrows indicate the direction of increasing values of δ and ω_n (ω_n = angle after rotation of *n* steps of 3° around pin axis). (c) Interpolation on a regular grid: Magnification of the rectangular small area indicated in (d). The original and regular grid are shown by solid and open symbols, respectively.

Torsion of NiAl

The starting material was high-temperature extruded polycrystalline NiAl with two different initial textures. The initial texture (Figs. 6b and c) of the torsion samples has been measured by neutron diffraction (Tamm et al., 1998). It represents the global texture of the whole cylinder. NiAl extruded through a rectangular die had an initial preferred orientation of <100> parallel to the torsion axis while that extruded through a round die had a <111> fibre parallel to the torsion axis. The equiaxed grain structure had an average grain size of about 100 µm (Fig. 5a). The diameter and height of the cylindrical samples were 10 mm. These cylinders were deformed in torsion at 1000 K in a Paterson-type rock deformation machine (Paterson and Olgaard, 2000) under 400MPa argon confining pressure.

The specimens have been deformed between alumina pistons, the whole assembly being enclosed in a thin steel jacket. The applied torsion leads to simple shear in the tangential direction in a plane normal to the torsion axis (Fig. 3).

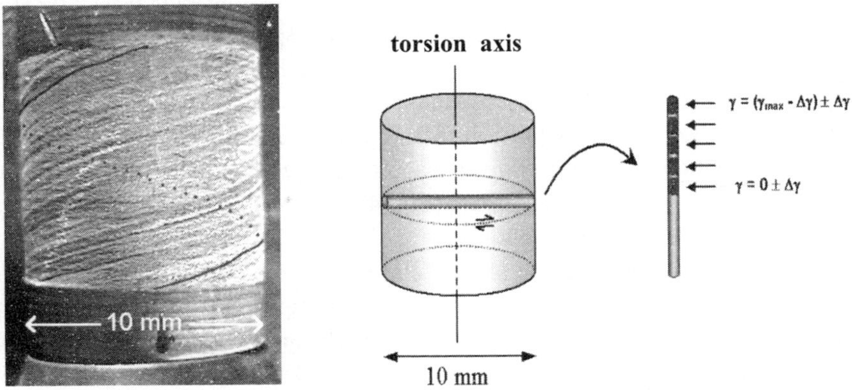

(a) (b)

Fig. 3: (a) Photograph of NiAl after torsion. The rolling traces on the steel jacket were initially parallel to the cylinder axis and serve as passive strain markers (indicated by the solid-dashed line) to illustrate the simple shear deformation in any tangential plane. (b) A sketch of the position of the pin taken for texture analysis is shown with different positions for texture measurements indicated (dark areas).

The shear strain and the shear strain rate in the samples increase linearly from zero at the torsion axis to a maximum at the sample edge. According to the sample dimensions and including sample shortening during torsion (Swift effect (Swift, 1947)) three revolutions yield $\gamma_{max} \approx 15$. To investigate the local textures between the torsion axis and the edge, small pins with a diameter of 1 mm were prepared in the radial direction for each of the deformed samples (Fig. 3b). The texture was measured at different positions along the pin axis. A range of $\Delta\gamma$ ($\pm$ 0.93 for the <111> and $\pm$0.56 for the <100> sample) is covered due to the extension of the synchrotron beam used in z-direction. To calculate the ODF the sample coordinate system was chosen with X, Y, Z parallel to the shear direction, shear plane normal and transverse direction, respectively.

ECAP of Copper

4N copper used was received in the form of extruded bars. Prior to ECAP, the as-received material was annealed for 1 h at 350°C in order to obtain a strain-free microstructure with equiaxed grains of about 70 μm grain size (Fig. 9a). The specimens with 10 mm × 10 mm square cross section and 100 mm length were machined from these annealed bars. ECAP was carried out at a crosshead speed of 1 mms^{-1} at room temperature using a Zwick 200 kN screw driven machine and a die set with rectangular intersection of the extrusion channels without any rounding of the corner region. The specimens were extruded in series; as they left the die they showed a more and more sheared shape (Figs. 4a and b). The number of ECAP passes was limited to three. Between individual passes the specimen was introduced in the same sense (route A, that is no rotation around the specimen longitudinal axis between intermediate passes). To reduce friction at the channel walls during ECAP lubrication was made by MoS_2.

The pins taken for synchrotron texture measurements were cut along the normal direction. Textures were measured in five steps from the top to the bottom surface of the extruded bars (Fig. 4b). The sample coordinate system to calculate the ODF was chosen with X, Y, Z parallel to the extrusion direction, normal and transverse directions, respectively.

The initial texture (Fig. 10b) of the ECAP samples has been measured by neutron diffraction on a cube of 10 mm × 10 mm × 10 mm. It represents the global texture of the as-received and annealed bars.

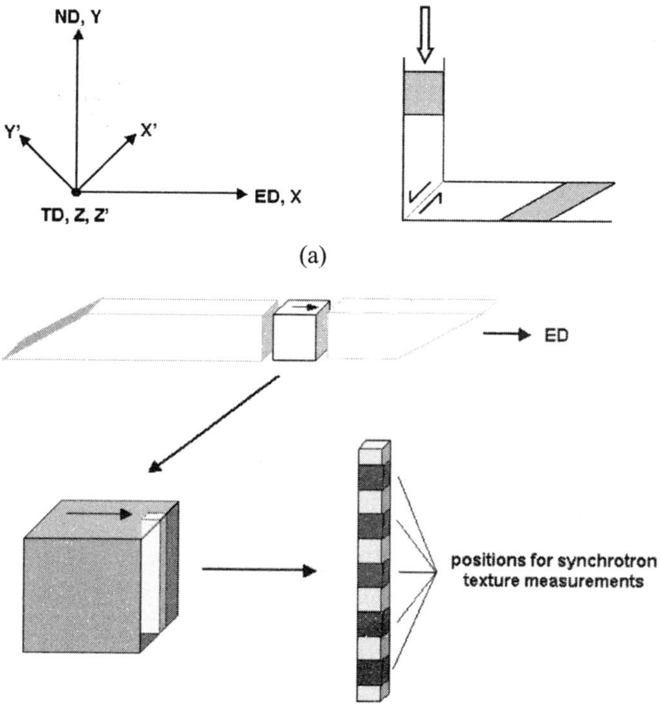

Fig. 4: (a) Idealized simple shear model of ECAP showing the shape change of a material element with shear by $\gamma = 2$. (b) Pin for local texture measurements by synchrotron radiation and its position within the ECAP bar. (ND = normal direction, TD = transverse direction, ED = extrusion direction, X,Y,Z = ECAP reference system; X',Y',Z' = simple shear reference system).

RESULTS

Torsion of NiAl

The grain structure of the initial and torsion deformed state is shown in Fig. 5. During torsion the shape of the grains changes from being equiaxed to elongated with the grain long axis aligned at a certain angle with respect to the shear direction. Due to continuous dynamic recrystallization, which is characterized by grain fragmentation and weak grain growth, a steady state grain size in the order of 10 μm develops (Skrotzki et al., 2003).

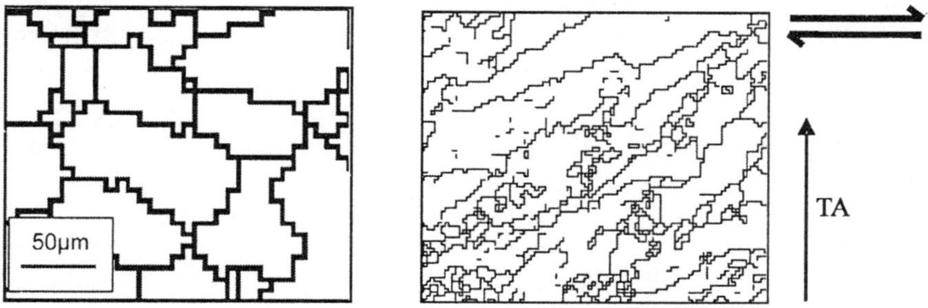

(a) (b)

Fig. 5: (a) Initial and (b) sheared grain structure of NiAl (shear strain γ = 2.8, TA = torsion axis). Due to simple shear the grains get ellipsoidal with the grain long axis preferentially inclined towards the shear direction. Pictures have been obtained by EBSD, only large angle grain boundaries (misorientation angles > 15°) are shown.

The texture after torsion was measured as a function of shear strain γ, i.e. in radial direction of the torsion sample. It is represented in Fig. 6 by the $\varphi_2 = 0°$ ODF section, which for NiAl contains all major shear components. During torsion the <111> sample develops a pronounced {110}<100> shear texture (notation: {shear plane}<shear direction>). For the <100> sample, there is no similar texture development. At low strains, the {100}<100> component develops, whereas for higher strains {110}<100> forms as a second component, too, and becomes dominant. The intensity of both components increases with strain, the increase being strongest for the {110}<100> component in the <111> sample (Fig. 7). With increasing strain this component is aligned parallel to the shear system, while for the {100}<100> component the shift in negative φ_1 direction is distinctly stronger and seems to stay constant at –10° (Fig. 8).

ECAP of Copper

The grain structure in the initial and ECAP deformed state is shown in Fig. 9. Initially, the grains are equiaxed and have an average size of about 70 μm. After ECAP the microstructure is strongly elongated with the long axis inclined towards the extrusion direction. The inclination angle decreases

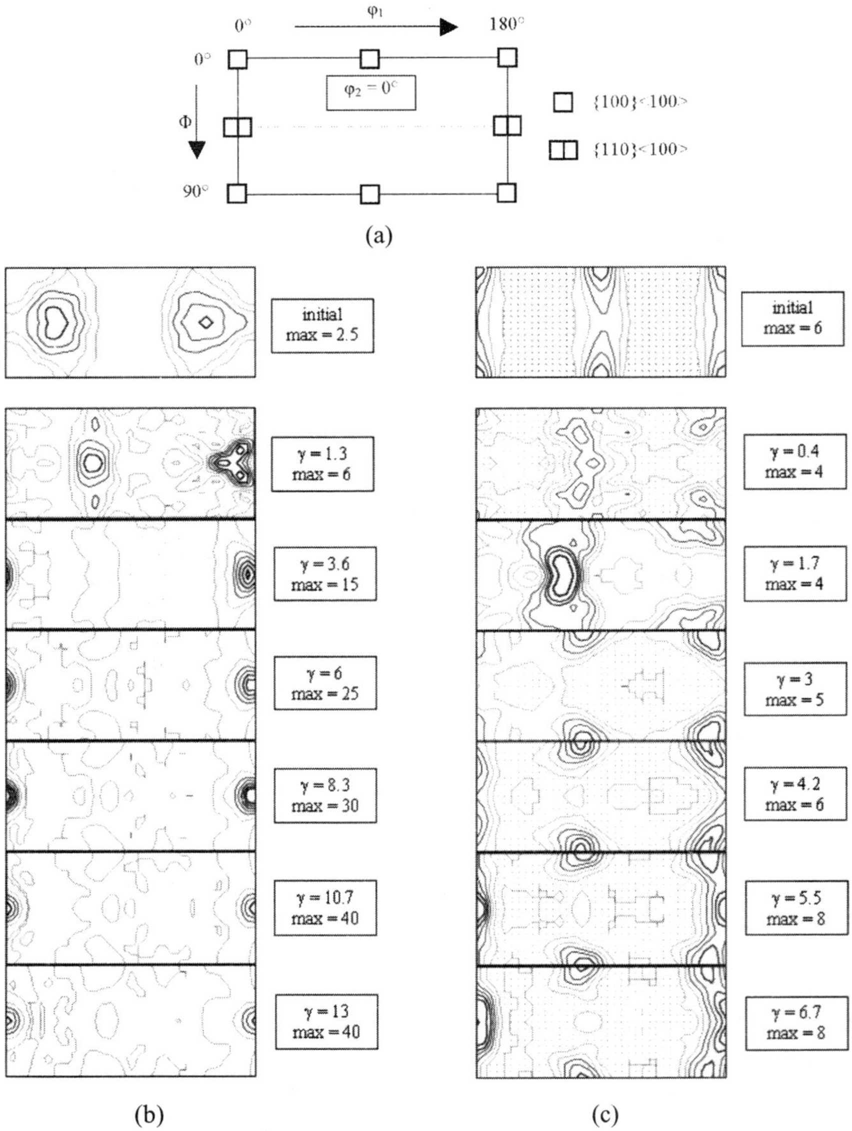

Fig. 6: (a) Ideal components of simple shear textures of NiAl and their positions within the $\varphi_2 = 0°$ ODF section. Experimental $\varphi_2 = 0°$ sections for samples with (b) <111> and (c) <100> initial preferred orientation, respectively, subject to the shear strain γ. The γ-values refer to the middle of the measured sample volume. Initial textures have been measured by neutron diffraction. Maximum values are given in multiples of a random orientation, mrd.

with the number of passes. Within each pass it also slightly decreases from the top to the bottom of the bar.

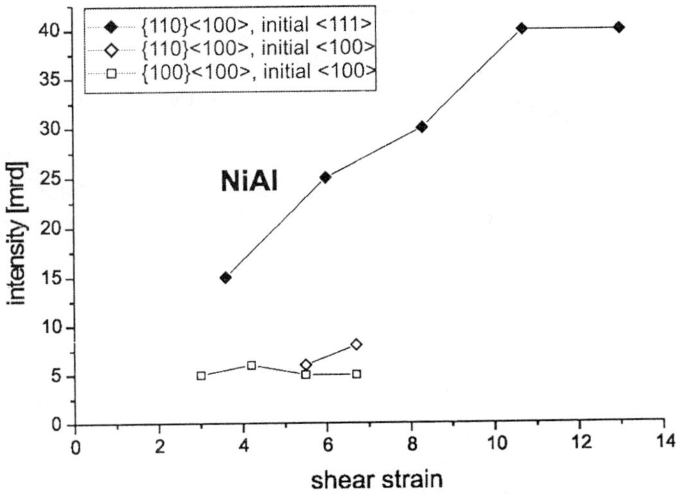

Fig. 7: Intensity of different components subject to the shear strain γ.

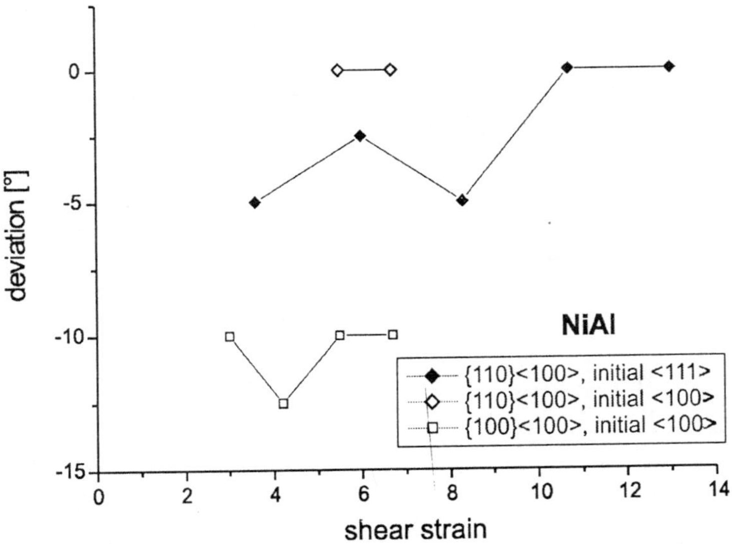

Fig. 8: Deviation in φ_1 of texture components from ideal orientations according to the ODF sections in Fig. 6. Negative values indicate an anticlockwise rotation.

The texture of the ECAP deformed Cu samples measured after 1, 2 and 3 passes, as a function of position from the top to the bottom of the extruded bars, is shown in Figs. 10 to 12. It is represented by $\varphi_2 = 0°$ and $\varphi_2 = 45°$ ODF sections, which for face-centered cubic (fcc) metals contain all major ECAP texture components. The preferred orientations after ECAP textures are compared with the ideal components of simple shear in –x′ direction on the x′-z′ plane (Table 1), i.e. with negative simple shear along the intersection

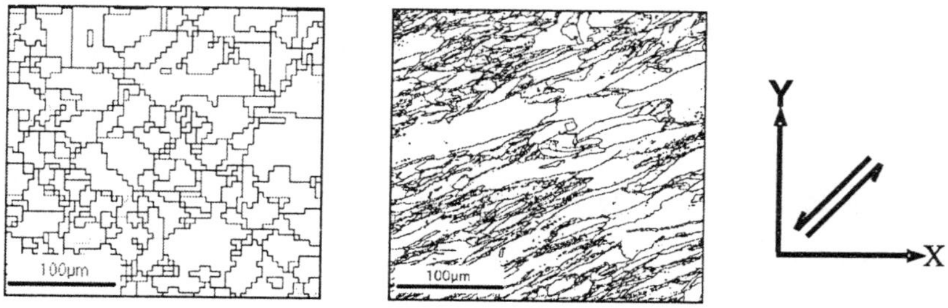

(a) (b)

Fig. 9: (a) Initial and (b) sheared grain structure of Cu after one pass.

plane of the bent channel, as suggested by Segal (1995) (Fig. 4b). The transformation to the ECAP reference system X, Y, Z has been given by Toth et al. (2004) and is presented in Fig. 10a and Table 2. However, it should be emphasized that the shape change does not say anything about the strain path producing it. Thus, intuitively in ECAP positive simple shear in x direction in the x-z plane may be assumed, too. In reality, none of these ideal strain paths seem to be applicable.

Table 1: Ideal components of fcc simple shear textures (Toth et al., 2004)

Component Designation	*Miller indices (shear plane)(shear direction)*
A	$(11\bar{1})[1\bar{1}0]$
$\bar{A}$	$(\bar{1}\bar{1}1)[\bar{1}10]$
B	$(11\bar{2})[1\bar{1}0]$
$\bar{B}$	$(\bar{1}\bar{1}2)[\bar{1}10]$
C	$(100)(0\bar{1}1)$
A_1^*	$(11\bar{1})(\bar{2}\bar{1}1)$
A_2^*	(111) [211]
{111} fibre	{111} <uvw>
<110> fibre	{hkl}<110>

To find out the actual strain path, texture measurements are useful. The sequence of the ODF sections from the top to the bottom of the bars shows that there is a texture gradient with respect to intensity and position of the components (Figs. 10 to 12). The strongest change is in the lowest section. Moreover, intensity as well as position changes from pass to pass (Figs. 13 and 14). In the first pass components A_{IE}^* and C_E are strongest, while in the second and third pass, it is A_{IE}^* and B_E/B_E. Similar to the torsion experiments, the deviation from the ideal components is different for different texture components.

Table 2: Ideal orientations of fcc ECAP textures (Toth et al., 2004)

	Miller indices			*Euler angles [°]*		
	ED	*ND*	*TD*			
A_{2E}	$[25\ \bar{3}\ 3]$	$[6\ 25\ \overline{25}]$	[011]	9.74	45	0
	$[\bar{3}\ 3\ 25]$	$[25\ \overline{25}\ 6]$	[110]	99.74	90	45
A_{1E}	$[\bar{6}\ \overline{25}\ 25]$	$[25\ 3\ \bar{3}]$	[011]	80.37	45	0
	$[\overline{25}\ 25\ 6]$	$[3\ \bar{3}\ 25]$	[110]	170.37	90	45
C_E	$[\overline{100}\ \overline{71}\ 71]$	$[100\ \overline{71}\ 71]$	[011]	135	45	0
	$[71\ \overline{71}\ 100]$	$[71\ \overline{71}\ \overline{100}]$	[110]	45	90	45
A_E	$[\bar{2}\ 20\ 9]$	$[20\ \bar{2}\ \bar{9}]$	[112]	45	35.26	45
$\bar{A}_E$	$[\bar{2}\ 20\ \bar{9}]$	$[\overline{20}\ 2\ 9]$	[112]	225	35.26	45
B_E	$[27\ \overline{100}\ 73]$	$[100\ \overline{27}\ \overline{73}]$	[111]	45	54.74	45
	$[\overline{100}\ 73\ 27]$	$[\overline{27}\ \overline{73}\ 100]$	[111]	165	54.74	45
$\bar{B}_E$	$[\overline{73}\ \overline{27}\ 100]$	$[73\ \overline{100}\ 27]$	[111]	105	54.74	45
	$[\overline{27}\ 100\ \overline{73}]$	$[\overline{100}\ 27\ 73]$	[111]	225	54.74	45

DISCUSSION

It has been demonstrated that local texture measurements by synchrotron radiation can be done on samples containing texture gradients. Because of the large penetration depth of high-energy synchrotron radiation disc-like volumes can be radiated which are large enough in two dimensions to give a statistically founded and therefore representative picture of the local texture with respect to intensity and position of the texture components.

For NiAl deformed in torsion at high temperatures it is shown that the main texture component developed in simple shear is {110}<100>. This is a likely type of texture component, since it represents the alignment of the primary slip system of NiAl onto the shear plane and shear direction. The tendency of formation of this texture is seen for both initial orientations. As the shear stress-shear strain curves reach a steady state with increasing strain, a steady state texture is expected, too, and this is {110}<100>. For the <100> sample the texture development also reflects the stability of {100}<100>.

The textures of the ECAP deformed copper show that ECAP is a rather complicated deformation process leading to a complex texture development. Due to friction there exists a texture gradient from the top to the bottom of the bar. Compared to the side and top channel walls, friction is much higher at the bottom wall because of the normal force acting on it during passage

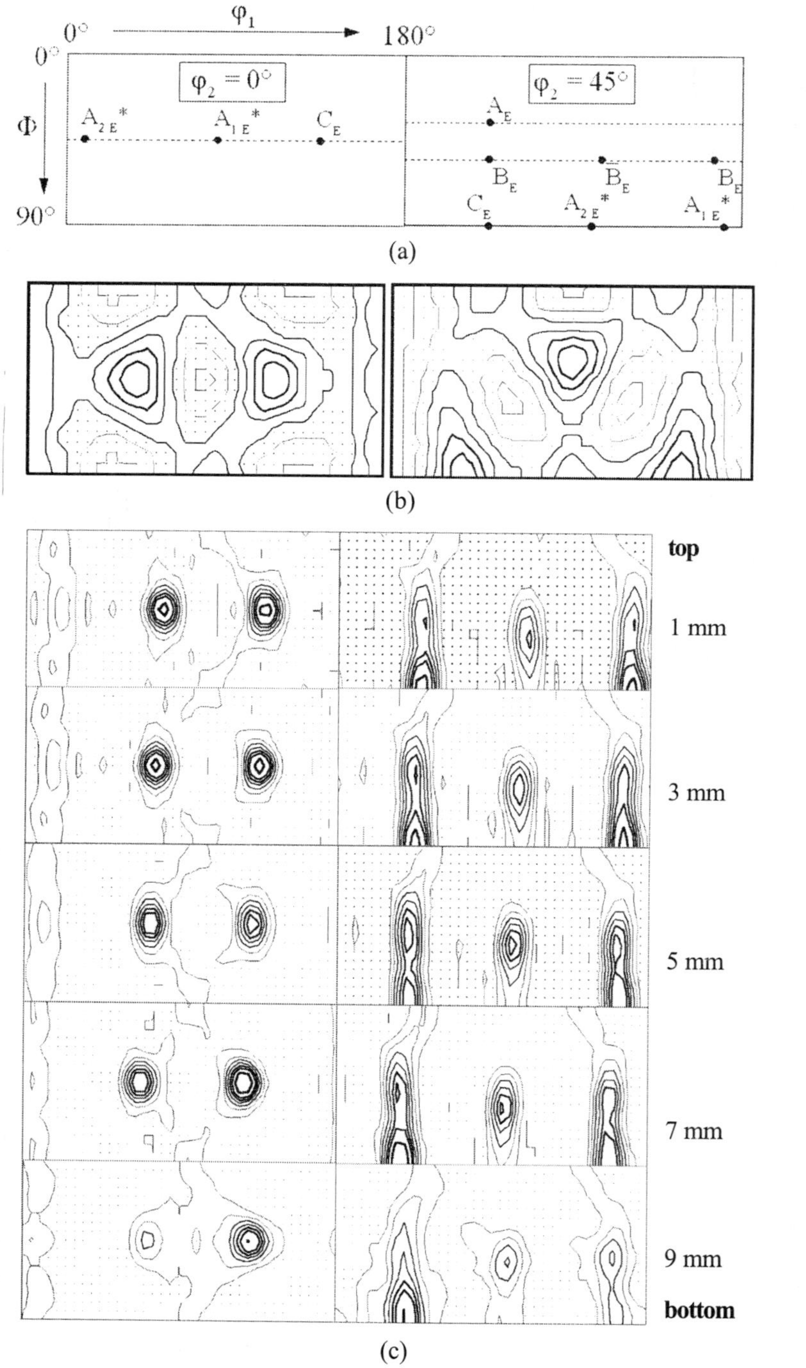

Fig. 10: (a) Ideal components of simple shear textures in fcc metals and their positions within the φ_2 = 0° and 45° ODF sections. Experimental ODF-sections of copper: (b) initial texture and (c) after one pass subject to the distance from top of the pin.

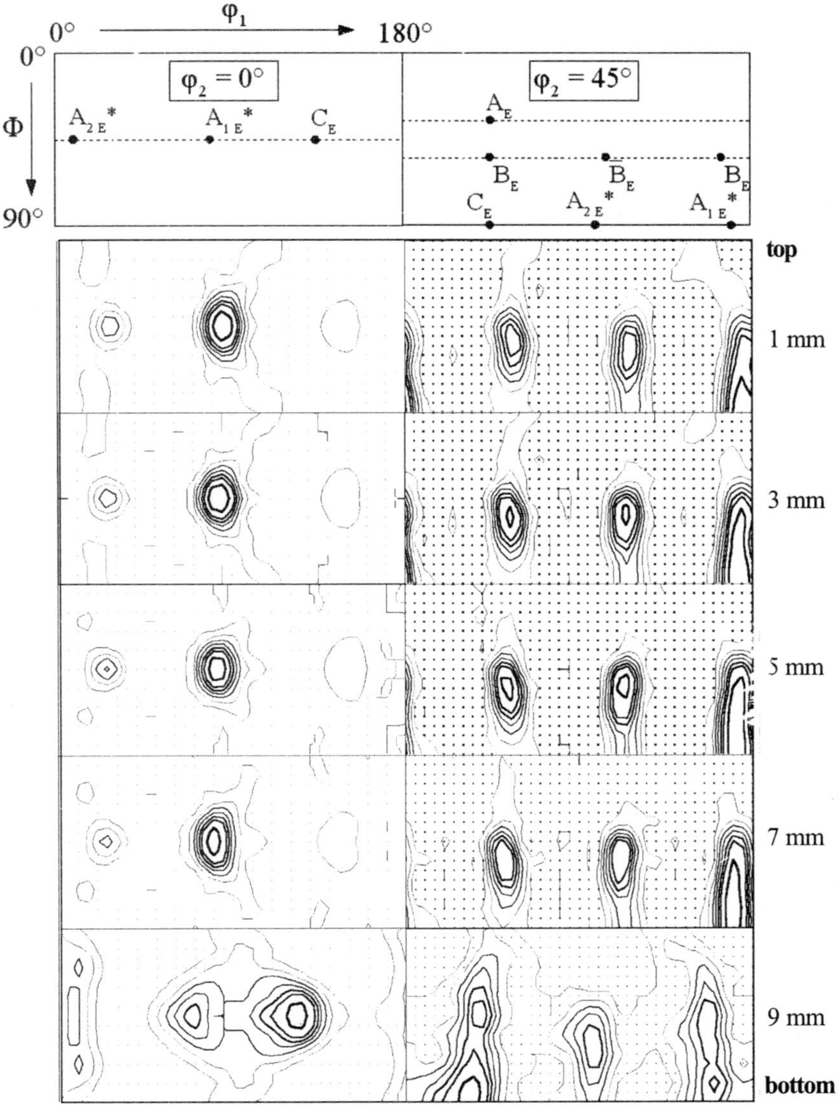

Fig. 11: As Fig. 10 after two passes.

of the sample through the intersecting plane. This effect has been demonstrated by own experiments with coloured plasticene pressed through a transparent channel. It has also been shown by visioplasticity on Al (Bowen, 2000). Friction not only depends on the lubricant used but also on the strength of the material determining the force necessary to extrude the sample. The latter parameter changes from pass to pass because of microstructure refinement leading to a pronounced hardening effect. The hardness increase is largest during the first pass and seems to saturate during further passes (Toth, 2004). The changes in texture from pass to pass may be partly due

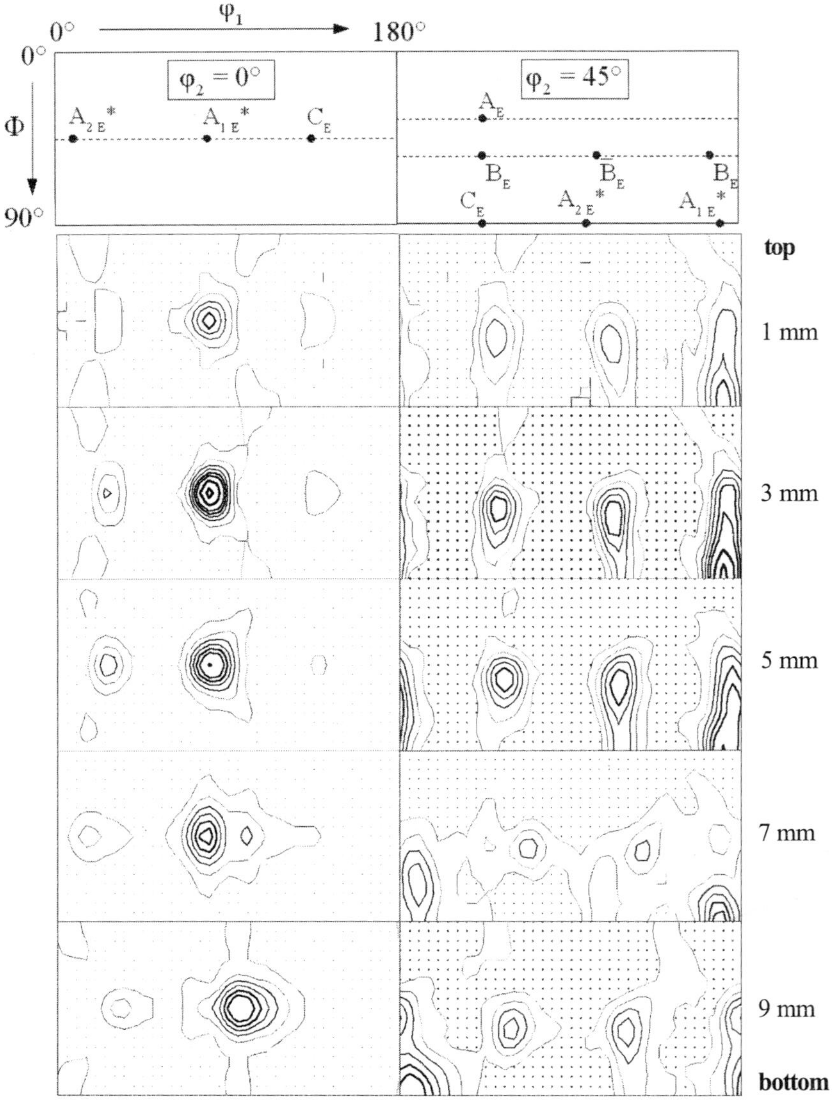

Fig. 12: As Fig. 10 after three passes.

to this effect. Another reason may be the change in starting texture and microstructure from pass to pass. The conditions are becoming even more complicated by changing the ECAP route.

Recently texture development during ECAP has been simulated by Toth et al. (2004) and compared with results obtained from X-ray texture measurements made on the x-z midplane of the same ECAP deformed copper samples. The best agreement between experiment and modelling is given by Toth's flow line model using the Viscoplastic Self Consistent Polycrystal Plasticity model (Toth, 2004). The only fitting parameter used is the exponent

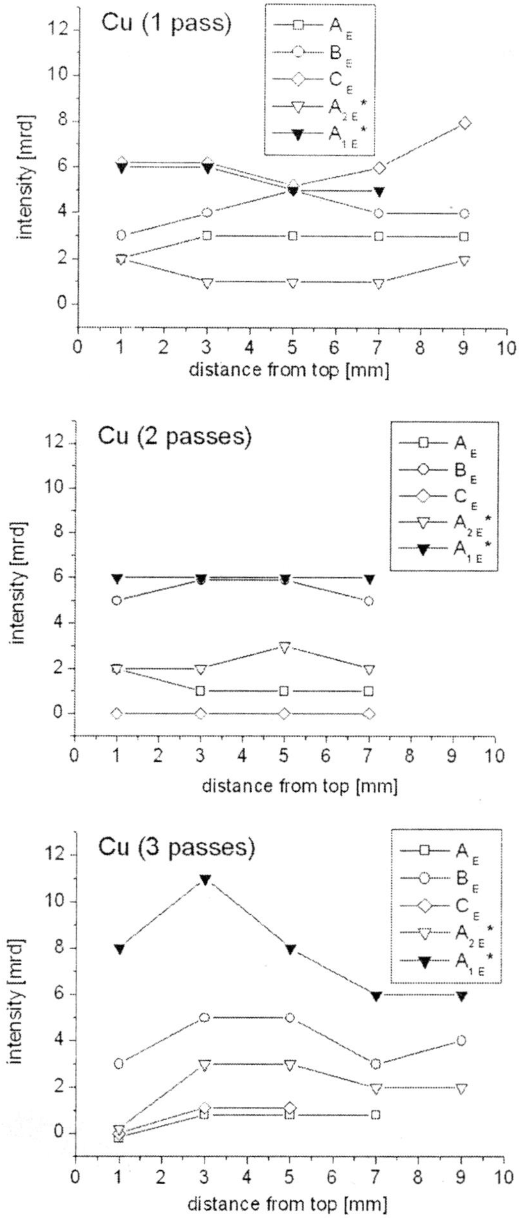

Fig. 13: Intensity of the components of Figs. 10 to 12 subject to position in the bar from the top to the bottom.

of the flow line function which determines the angular strain rate changes during passage through the channel intersection. Future simulations will show if changing this parameter and therefore accounting for the flow field due to friction may explain the texture gradient observed here.

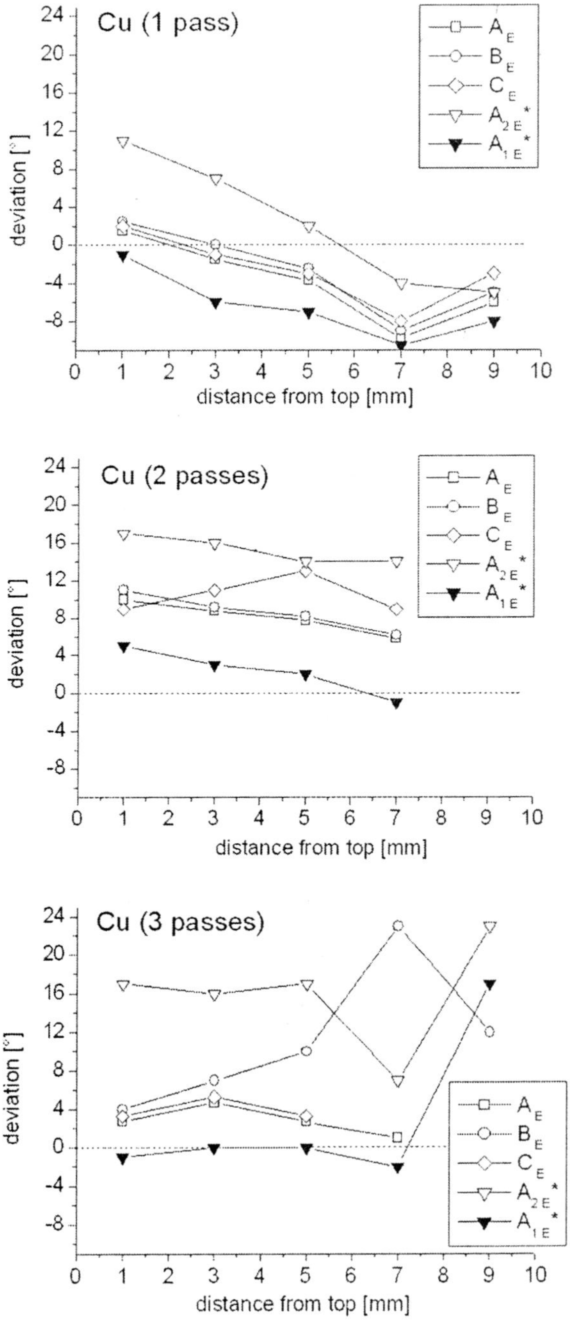

Fig. 14: Deviation in φ_1 of texture components from ideal positions as a function of position in the ECAP copper bars. Negative values indicate a shift towards decreasing values of φ_1 in the corresponding ODF section.

Acknowledgements: Thanks are due to the Deutsche Forschungsgemeinschaft for funding (Sk 21/19-1) and DESY-HASYLAB for beamtime and financial support of travel and accommodation. The technical assistance by Mr. T. Reiter (TU Dresden) is gratefully acknowledged. The ECAP tests have been done by Dr. S. Suwas (University of Metz). Discussions with Prof. L. Toth and Dr. S. Suwas have been very helpful.

REFERENCES

Bouchard, R., Hupfeld, D., Lippmann, T., Neuefeind, J., Neumann, H.B., Poulsen, H.F., Rütt, U., Schmidt, T., Schneider, J.R., Süssenbach J. and Von Zimmermann, M. (1998). *J. Synchrotron Rad.*, **5**: 90.

Bowen, J.R. (2000). PhD thesis, University of Manchester.

Bunge, H.J. and Klein, H. (1996). *Z. Metallkd.*, **87**: 465.

Bunge, H.J. (1993). Texture Analysis in Materials Science, Mathematical Methods, Culliver Verlag Göttingen.

Cottrell, A.H. (1958). *Trans. AIME*, **212**: 192.

Dahms M. and Eschner, T. (1996). Quantitative Texturanalyse durch iterative Reihenzerlegung von Beugungs-Polfiguren (software manual).

Noebe, R.D., Bowman, R.R. and Nathal, M.V. (1996). *In:* Physical Metallurgy and Processing of Intermetallic Compounds, Eds. Stoloff, N.S. and Sikka, V.K., Chapman and Hall, New York, 212.

Paterson, M.S. and Olgaard, D.L. (2000). *J. Struct. Geol.*, **22**: 1341.

Segal, V.M. (1995). *Mater. Sci. Eng.*, **A197**: 157.

Skrotzki, W., Klöden, B., Tamm, R., Oertel, C.-G., Garbe, U. and Rybacki, E. (2003). *Textures Microstruct.*, **35**: 163.

Swift, H.W. (1947). Engineering, **163**: 253.

Tamm, R., Lemke, M., Oertel, C.-G. and Skrotzki, W. (1998). *Mater. Sci. Forum*, **273-275**: 411.

Toth, L. (2004). Comp. Mater. Sci., in press.

Toth, L.S., Arruffat Massion, R., Germain, L., Baik, S.C. and Suwas, S. (2004). *Acta Mater.*, **52**: 1885.

Valiev, R.Z., Islamgaliev, R.K. and Alexandrov, I.V. (2000). *Progress Mater. Sci.*, **45**: 103.

Wcislak, L., Klein, H., Bunge, H.J., Garbe, U., Tschentscher, T. and Schneider, J.R. (2002). *J. Appl. Cryst.*, **35**: 82.

Texture Mapping by Scanning X-ray Diffraction and Related Methods

R.A. Schwarzer

Institut für Physik und Physikalische Technologien der TU, Leibnizstr. 4
D-38678 Clausthal-Zellerfeld, Germany

INTRODUCTION

Almost all X-ray analysis in the laboratory is *integrating* over macroscopic scales and is *non-imaging*, i.e. the acquired diffraction patterns as well as the fluorescence X-ray spectra, and hence the information on crystallography and element composition are an average over the whole illuminated wide sample area. Material properties of solids depend on the shape and arrangement of the crystallites constituting the microstructure and on the concentration of elements, but in particular on the spatial distribution of phases, on the distribution of grain orientations ("crystal texture"), on the character of grain respectively phase boundaries, on lattice defects and residual stress. Inhomogeneities in the sub-millimetre range of morphology, texture, residual stress and element composition may result in similar local variations of materials properties. They are quite common in natural and in man-made materials. Texture has to be accessed in quantitative microstructural analysis since it bears information about the history of materials processing and use, of rock-forming and sedimentation.

Material's homogeneity is often decisive to the reliability and performance of an industrial product. Therefore a microstructural analysis is required beyond spatial resolution of conventional X-ray diffraction (XRD) or X-ray fluorescence analysis (XFA). In many applications the knowledge of the spatial distribution of certain *crystal lattice directions* (i.e. pole densities rather than the full *crystal lattice orientation* of each grain), of residual stress gradients and estimates of the element distribution are already sufficient. If furthermore only a moderate resolution in the specimen surface is acceptable, electron microscopy with its specific limitations can be avoided. Although electron diffraction, in particular backscatter Kikuchi diffraction in the SEM (known as "ACOM" = Automated Crystal Orientation Mapping and "Automated EBSD") has made significant contributions to the characterization of microstructure during recent years (Schwarzer, 1997), it cannot fully replace XRD and XFA techniques. The dispersion, $\Delta\vartheta/\vartheta$, in XRD by far exceeds dispersion in electron diffraction, because spectral linewidth, $\Delta\lambda/\lambda$, is extremely small and the characteristic X-ray wavelengths, λ, are in the range of the lattice spacings, thus resulting in large Bragg angles, ϑ. A high accuracy in the determination of lattice constants and grain orientation is the consequence. Continuous background is almost negligible in XFA which enables a high peak-to-background signal and excellent sensitivity. X-ray techniques are comparatively gentle since the sample need not stay in a vacuum atmosphere nor is it exposed to a beam of energetic electrons. Hence radiation damage is of no concern. The requirements of sample preparation are moderate. Insulating, magnetic or wet materials are well suited for the investigation with X-rays.

THE PRINCIPLES OF ENERGY DISPERSIVE X-RAY DIFFRACTION

Energy dispersive (ED) X-ray diffraction is based on Bragg diffraction of a collimated "white" primary X-ray beam. If a grain fulfills the Bragg condition

$$E_{\mathrm{hkl}} = \frac{n \cdot h \cdot c}{(2 \cdot d_{\mathrm{hkl}} \cdot \sin \vartheta_{\mathrm{hkl}})} \tag{1}$$

for a photon energy E_{hkl} of the primary beam at an angle of incidence, ϑ_1, due to an appropriate orientation of lattice planes {hkl} with spacings d_{hkl}, a partial ray of this beam is reflected at an angle ϑ_2. The $\vartheta_{\mathrm{hkl}} = (\vartheta_1 + \vartheta_2)/2$ is known as Bragg angle (Fig. 1). If the illuminated specimen volume is composed of some grains with almost the same orientation, each crystallite may reflect its own partial beam E_{hkl} into the take-off angle, ϑ_2. If the "white" spectrum of primaries is sufficiently wide, it may contain several energies $E_{\mathrm{hkl}}{}^{1,2,\ldots}$ which may fulfill the Bragg condition at the same time and at the same angles of incidence and reflection, but for different sets of lattice planes $\{\mathrm{hkl}\}^{1,2,\ldots}$ belonging to other grains with different orientations. In this case, all the partial beams from the irradiated volume of the polycrystal simultaneously strike a solid state (Si(Li)) detector which is positioned at a fixed take-off angle, ϑ_2. The beams merge in the ED spectrum to form several peaks which are centered at $E_{\mathrm{hkl}}{}^{1,2,\ldots}$ in the same way as the hkl diffraction peaks are in a conventional ϑ-2ϑ scan with monochromatic radiation. They are separated in the spectrum and identified by their specific energies.

In a first approximation, the intensity of an hkl diffraction peak is a measure of the volumes of all grains which have fulfilled the Bragg condition for the corresponding [hkl] lattice direction, irrespective of their arrangement

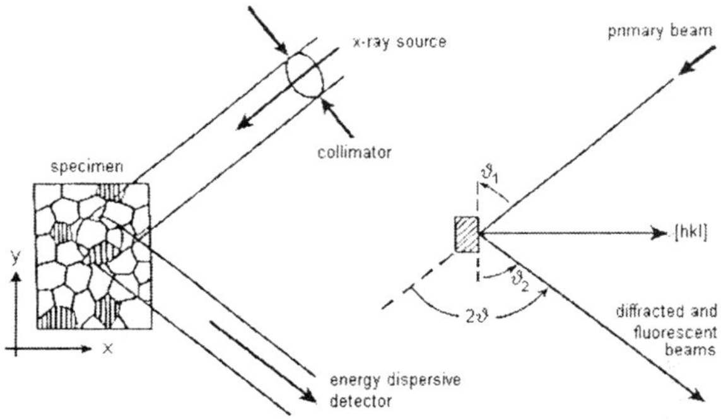

Fig. 1: Schematics of the microbeam X-ray scanning instrument.

and shapes in the irradiated sample volume as well as their rotation about [hkl]. However, the measured intensities are proportional not only to the volume fractions of the related grains, but also to the intensities falling in the corresponding spectral windows about $E_{hkl}^{1,2,\cdots}$ of primary X-ray radiation. They are further affected by the (spectral) absorption of the X-rays on their way from the source to the detector and by possible non-linearities of the detector system. Absorption is in particular significant within the sample. So corrections have to be applied to the measured pole intensities to relate them quantitatively to the corresponding volume fractions in quantitative texture analysis.

Pole figure measurement using a one-dimensional detector, such as a proportional counter or an ED detector, is performed by tilting the sample in small incremental steps about an axis in the diffraction plane through the angle α from 0° to about 70° and rotating the specimen about its surface normal through β from 0° to 360° relative to a coordinate system fixed to the sample. For this purpose, a special computer-controlled texture goniometer, named Euler cradle, is used. The corrected intensity values from an hkl peak are named hkl pole densities. They are stored on the hkl reference sphere of texture corresponding to the particular polar angular settings (α, β) of the specimen. An hkl pole figure is the stereographic projection of the hkl reference sphere. Since the angle of specimen tilt is limited due to excessive absorption of the beam traversing the specimen at high tilt angles, only incomplete pole figures are obtained in both modes of diffraction, in transmission as well as in backscattering. The advantage of ED XRD is in the acquisition of *several* pole intensities $E_{hkl}^{1,2,\cdots}$, respectively hkl pole figures *at the same time* and under the same reference direction (α, β), whereas conventional pole figure measurement with monochromatic radiation and a proportional counter as the detector allows only one hkl reflection to be acquired at a time.

Lattice strain is analyzed by evaluating the shifts and widths of the diffraction peaks in the ED spectrum (Fischer and Schwarzer, 1998). The angular shift, $\Delta\vartheta$, of a diffraction peak, ϑ_{hkl}, as compared to the peak position ϑ_0 from a relaxed specimen with lattice spacing d_0, follows from Bragg's equation:

$$\Delta\vartheta = \vartheta_{hkl} - \vartheta_0 = -\tan\vartheta_0 \frac{(d_{hkl} - d_0)}{d_0}, \tag{2}$$

or transformed to the scale of photon energy, E, which is more adequate for energy dispersive diffraction:

$$\Delta E = E_{hkl} - E_0 = -E_0 \frac{(d_{hkl} - d_0)}{d_0} \tag{2a}$$

The energy shift is extracted on-line from the diffraction pattern after background correction by line profile analysis which is performed by fitting Gauss peaks to the measured peaks. The width of a diffraction peak reflects the variations of lattice spacing, Δd, respectively residual strain and strain gradients across grains or microstructural imperfections in the grains (residual stress of the second and third kind) in the diffracting volume. For a (semi-) quantitative estimate it is common practice to take, under the same experimental conditions, the differences of the full widths at half maximum (FWHM) of the line profiles and centre line positions from the strained and from an annealed reference specimen, respectively, as a measure of residual strain. The primary beam, however, has to supply a continuous energy spectrum with constant intensity at least over a width ΔE according to Eq. (2a), since the width of the measured diffraction peak in the ED spectrum is the convolution integral of the width of the diffracted reflection ΔE due to lattice strain with the energy profile of the primary beam, the angular profile of the primary beam, the acceptance aperture of the detector and the energy resolution of the detector. The experimental parameters have to be well adjusted by using continuous background radiation for exciting the reflections, a sufficient collimation of the primary and diffracted beams and a high energy resolution of the detector. A well collimated parallel primary beam and a narrow (Soller slit) collimator on the diffracted beam side in front of the detector can drastically improve $\Delta d/d$ resolution, but at the expense of intensity and necessary dwell time per pixel.

In the present version of the X-ray scanning instrument, the sensitivity and the $\Delta d/d$ resolution of residual strain measurement are finally limited by the spectral resolution of the solid-state spectrometer system. The standard Si(Li) detector used here has a nominal energy resolution of 140 eV at 8.4 keV (MnKαline). Furthermore, a tradeoff has to be made between intensity and peak width by using wide primary beam and detector apertures. They are adjusted by using appropriate pinhole diaphragms in the primary beam and an appropriate distance of the Si(Li) detector from the specimen. The apertures, $\Delta\vartheta$, of the primary and secondary beams are not crucial, unless lattice strain mapping is of concern. Since texture in most materials is hardly ever sharper than a few degree, a width of diffraction peaks in this range, according to $\Delta\vartheta$ in Eq. (2), is quite acceptable. At 2ϑ=40° (corresponding to E_{hkl}≈10 to 13 keV for low index planes of metals) apertures of 0.35° produce a FWHM between 400eV and 500eV which is distributed over about 100 spectral channels. Peak position is not affected by the aperture size in a symmetric goniometer setup ($\vartheta_1 = \vartheta_2$). By a Gauss fit, the peak centre can so be located to $\Delta E \leq 0.5$ eV. Although the diffraction peaks in the ED spectrum are considerably broadened by the acquisition system, residual strains of the first and second kind are well resolved down to $\Delta d/d \approx 4 \cdot 10^{-4}$.

A noticeable improvement in lattice strain analysis is expected from a higher energy resolution of the detector system and a reduction of the energy interval per spectral channel. But a practical limitation still rests on the need of large apertures for gaining sufficient peak intensity. A position sensitive detector (PSD) with a standard Bragg-Brentano setup and monochromatic radiation is still superior over an ED detector by one order of magnitude or more as far as a high dispersion $\Delta d/d$ is of concern for residual strain analysis. A PSD setup enables a reliable peak-width analysis and so it will allow the mapping of residual strain distributions even of the third kind. The replacement of the ED system, however, by a PSD excludes simultaneous acquisition of several hkl peaks at the same time. Furthermore, a PSD does not enable the acquisition of X-ray fluorescence lines nor XFA nor element mapping, and imposes stringent restrictions on the flatness as well as the adjustment of the sample position since the focusing condition for the Bragg-Brentano geometry has to be matched.

X-ray fluorescence lines in the ED secondary spectrum are discriminated from diffraction peaks by their characteristic energy. If the Bragg angle is moved from ϑ_{hkl} to ϑ_{hkl}^*, the centres of the diffraction peaks on the ED spectrum shift from energy channels E_{hkl} to E_{hkl}^*, according to Eq. (1), whereas characteristic X-ray lines, which are scattered from the primary radiation and in particular fluorescence lines from the specimen, stay on fixed positions on the energy scale. In addition, diffraction peaks are typically much broader than characteristic lines which are not affected by the apertures of neither the primary nor the diffracted beams, nor by lattice strain. Diffraction peaks can so easily be discriminated from characteristic lines. A specific fluorescence line can be enhanced by using an X-ray tube with an anode containing an element about two positions higher in the periodic table than is the element to be detected in the sample. Likewise, a weak diffraction peak can be considerably enhanced for texture mapping as well as for pole figure measurement – but not for strain mapping – by adjusting the Bragg angle such that the energy of the diffraction peak coincides with the energy of a characteristic line in the primary beam spectrum. With these characteristics in mind, diffraction peaks, lattice strain, and X-ray fluorescence lines for element analysis can be acquired and evaluated simultaneously with an energy dispersive system. Windows (Regions of Interest, ROI) of appropriate widths are placed on the centres of selected hkl diffraction peaks and characteristic lines in the same way as is common practice in EDS element analysis to integrate the count rates and map element distributions.

Figure 2a shows the energy dispersive X-ray spectrum from a copper tube at 35 kV accelerating voltage, and Fig. 2b the ED diffraction spectrum from a cross rolled AlMn1 specimen using the same primary beam radiation. In both spectra the characteristic copper Kα and Kβ lines are present. Characteristic fluorescence lines from aluminum are not contained in the spectrum of Fig. 2b although aluminum is weakly excited by the primary beam, since this soft X-ray radiation of 1.5 keV is absorbed by air on the path from the sample to the detector. The small content of 1% manganese, however, is strongly excited by copper radiation to emit MnKα and MnKβ fluorescence lines of 5.9 keV respectively, 6.5 keV which are seen to form two sharp peaks on the low energy side of the spectrum. At the high energy side, starting from the characteristic copper K lines, a set of relatively broad diffraction peaks can be identified. The low-index peaks are separated in the spectrum, and no peak overlap occurs for 200, 220, 400, 331 and 422 reflections, whereas the 222 and 311 peaks are coincident with each other. The individual peak intensities differ significantly from tabulated powder diffraction intensities. The differences are an indication of preferred crystal orientations in the rolled aluminum sheet. The 111 peak at a position close to MnKβ, for instance, is missing at all due to the specific texture.

Spatial resolution of ED XRD and fluorescence analysis is limited by the size of the X-ray spot on the tilted specimen surface. Synchrotron radiation is an ideal source for obtaining an extremely bright and small X-ray probe (Wang et al., 1997), but the limited availability of synchrotron facilities still demands for using conventional X-ray generators as a more feasible alternative in the laboratory. Since no conventional lenses are available for white X-rays, a beam collimator has been used for the present XSI which consists of two small circular diaphragms. The main drawback of a diaphragm collimator is the loss of intensity. An increase in spot brightness is expected from glass capillary collimators (Maslen, 1958; Larsson and Engström, 1992) which would enable a higher spatial resolution or a shorter acquisition time. Capillary optics, however, have a filtering effect on the primary beam energy which might be adverse to ED diffraction.

The oblique incidence of the primary beam results in an elliptical foreshortening of the spot on the sample surface. Spatial resolution decreases with angular tilt, γ, of the specimen with respect to the incoming beam. A circular primary X-ray spot is elongated by projection on the tilted specimen surface to an ellipse of aspect ratio $1/\cos\gamma$, whereas at shallow incidence the depth of penetration beneath the surface is reduced and surface sensitivity is improved. A practical limit of spatial resolution is presently at 0.1 mm for

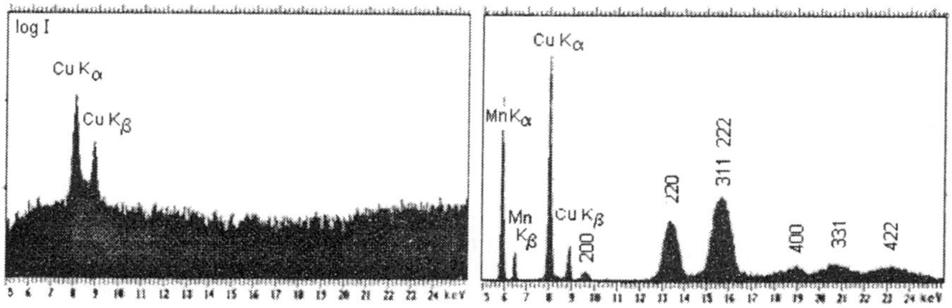

(a) Primary beam spectrum (Cu, 35 kV) (logarithmic intensity scale)

(b) Cross-rolled aluminum sheet (linear intensity scale)

Fig. 2: Energy dispersive X-ray spectra.

pole-figure measurement and lattice strain mapping, and 0.05 mm for element and texture mapping, due to the low intensity of the collimated primary beam.

Absorption in air must be considered for quantum energies $E_x < 3$ keV. Hence, a helium atmosphere or an evacuated specimen chamber will be indispensable if materials with large lattice constants have to be studied by ED XRD as well as light elements have to be detected by XFA. Very high quantum energies, on the other hand, are not adequate, since with hard X-rays Bragg´s angles decrease unduly up to grazing incidence. Then the spot size is extremely foreshortened and spatial resolution is impaired. A high excitation of fluorescence lines is obtained with an overvoltage of the primary beam of about twice the energy of the characteristic lines of the elements to be detected. A good compromise for most samples in reflection mode is obtained with an angle of incidence of $\vartheta_1 \sim 45°$ and quantum energies of the primary beam between 8 keV and 15 keV.

THE X-RAY SCANNING INSTRUMENT BASED ON A CONVENTIONAL X-RAY SOURCE AND ED SPECTROMETRY

The X-ray Scanning Instrument (XSI) (Schwarzer, 1989; Schwarzer, 1993; Fischer and Schwarzer, 1998) combines three techniques based on ED diffraction that are spatially resolved pole figure measurement as well as, on a grain-specific level, texture mapping and lattice strain mapping. The fourth technique is element mapping by X-ray micro-fluorescence analysis. Although applications have initially been focused on metals and their response to plastic deformation, a wide field regarding deformation of rocks and minerals in general can be tackled as well.

To obtain images of the morphology, the spatial distributions of elements, the type of crystal lattice (i.e. phases), grain orientations and residual lattice strain from the same localized specimen area requires a mapping technique whereby an X-ray beam of small spot size is generated, and the specimen is mechanically scanned under the stationary beam spot. In our design, shown in Fig. 3, the XSI consists of a conventional X-ray source (Philips generator PW 1830/25) equipped with a fine-focus tube. Most applications have been carried out with a tungsten anode in order to obtain a high intensity of continuous radiation. The "white" primary radiation is not filtered. Therefore almost no practical restrictions have been imposed on lattice spacings for diffraction, and a wide range of elements can be excited to fluorescence emission. A circular pinhole collimator is used to produce a small primary beam spot. On the two-circle goniometer base (Philips PW 1835), an open Euler cradle (Philips X'Pert MRD goniometer) with a stepping-motor driven x-y-z sample stage is mounted which allows the sample to be translated in 1 μm wide steps over an area of 100 mm by 100 mm. The diffracted secondary spectrum is acquired with an energy dispersive solid state"spectrometer system (Spectrace 6100 and Tracor Noran Si(Li) detector) which contains diffraction and X-ray fluorescence peaks for mapping. As an alternative, particularly for pole figure measurement and lattice strain mapping, a conventional proportional counter, respectively a position sensitive detector can be used, with a Kβ filter either in the primary beam or in the diffracted beam right in front of the detector. The proportional counter and the position sensitive detector have to be set to the appropriate ϑ–2ϑ positions for each measured hkl pole figure separately.

The instrument is under the full control of a personal computer. The specimen is translated step by step with the x-y stage across a user defined grid pattern, e.g. on scans along a line or on regular raster grids of rectangular or polygon shape up to 100 mm by 100 mm wide, in order to match the scanned field to the area of interest. Usually constant step widths in x and y direction are chosen. The current sample position under the beam spot is monitored with a video camera, and so the corner points of the sampled region are defined conveniently.

Pole-figure measurement has to precede texture mapping on the same specimen area in order to find out the significant texture components. Thinned angular grids, equal area grids or relevant sections on the pole sphere can be selected to reduce acquisition time. Pole-figure measurement can be combined with a controlled spatial x-y translation of the specimen and integration of the pole intensity for this specimen area.

Large specimen tilts require the correction of peak intensities for background. They may also result in inconsistent pole-figure data if texture is not uniform in depth beneath the shallow surface, since XRD becomes more surface sensitive with decreasing angle of incidence. Any specimen reference direction (α, β) can be chosen for mapping provided that pole density (i.e. diffracted intensity) is high enough – which of course depends on local texture. It is worth noting that texture as well as residual strain is probed exclusively for those grains only, which satisfy the Bragg condition for the selected specimen-fixed reference direction. There is no intensity at all acquired from grains with other orientations.

Fig. 3: The experimental setup of the X-ray scanning instrument.

A comprehensive PC software package has been developed (Fischer and Schwarzer, 1998). One part contains the hardware control of the goniometer, the Eulerian cradle, the x-y-z sample stage and the proportional counter as well as the EDS system. The other part has been developed for the graphical representation of measured pole figures and for mapping the microstructure by attributing pseudo-colours to the pole densities ("texture mapping"), the shift and width of the diffraction peaks ("lattice strain mapping"), respectively the intensities of selected fluorescence lines ("element mapping").

Modern X-ray texture goniometers are often equipped with a computer-controlled x-y specimen stage on the Eulerian cradle. They enable, in principle, texture mapping, if appropriate software is available for controlling the mechanical x-y sample scan, the acquisition, interpretation and representation of the diffraction data. Spatially resolved texture analysis has been demonstrated on a commercial X-ray texture goniometer by calculating off-line orientation densities, texture symmetry, texture index and the Taylor factor in several raster points (Klein and Bunge, 1999). X-ray microbeam scanning instruments for imaging XFA have demonstrated their high potential since many years (Horowith and Howell, 1972; Jones and Gordon, 1989; Rindby et al., 1989; Carpenter 1989), and commercial systems have been developed (Werry and Cross, 1989; Rammlmair et al., 2001). The combination of these techniques in one instrument, however, is a unique feature of the XSI.

EXAMPLES OF APPLICATION

Texture Mapping

Compression induces a change of texture in the material, which depends on the degree and rate of plastic deformation. By impressing the jaws of a mechanical testing machine into the coarse grain surface of a recrystallized aluminum sheet metal, a rhombic relief pattern has been formed. The sample was then mechanically ground and polished to a flat mirror such that the pattern was no longer visible. The compression process, however, has changed local texture. In the texture maps (Fig. 4) high pole densities can be seen in regions where the corrugating teeth of the jaws caused a high local deformation, whereas deformation in the grooves of the jaws was low enough to preserve the original texture of the metal sheet (Schwarzer, 1993). The texture maps of Fig. 4 were acquired at symmetrical diffraction settings ($\vartheta_1 = \vartheta_2$) simultaneously from the same sample area. Hence the spatial distributions of pole densities of the lattice planes (200), (220), (311) and (222), respectively (331) are shown. The sheet texture did not contribute to these selected diffraction peaks. If a certain texture component of the metal sheet is of interest, the specimen has to be tilted away from the symmetrical geometry to an appropriate pole-figure point into which the sheet is diffracting. The corresponding angles of tilt can be read from X-ray pole figures that have to be acquired in advance.

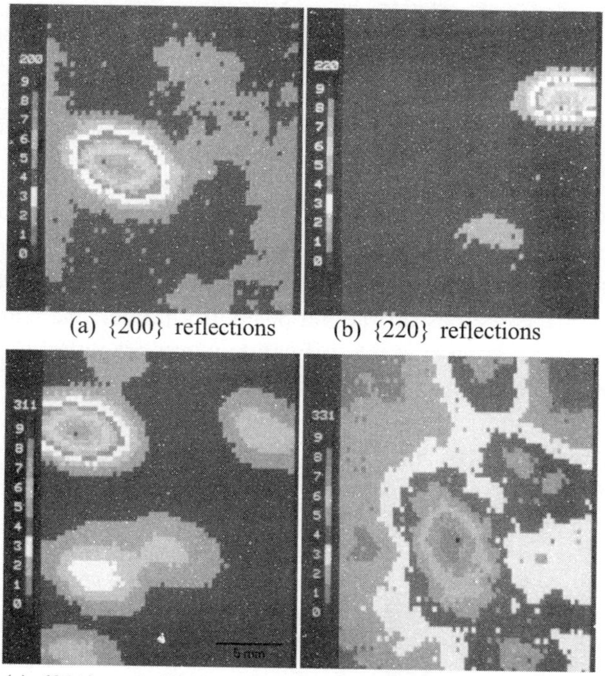

(a) {200} reflections (b) {220} reflections

(c) {311} and {222} reflections (d) {331} reflections.

Fig. 4: Texture maps of a compression pattern in aluminum revealing local plastic deformation.

An important industrial application of zinc is hot galvanizing of steel sheet metal. In the zinc bath a layer of Fe-Zn alloy is formed on the surface of the work piece, which is covered by liquid zinc. When this coating solidifies, a characteristic pattern known as "zinc flower" is visible which is composed of zinc grains of varying orientations. The crystal texture of the zinc coating has an effect on mechanical properties as well as corrosion resistance of the galvanized material. Hence local texture is of some practical importance. In texture maps of the surface the spatial variation of pole densities can be recognized (Fig. 5) (Schwarzer, 1993). The zinc flower contains clusters of grains which are stretched out in the rolling direction of the sheet substrate. Grain clusters with (10-10), (10-11) and (0002) planes parallel with the specimen surface are clearly visible in the texture maps.

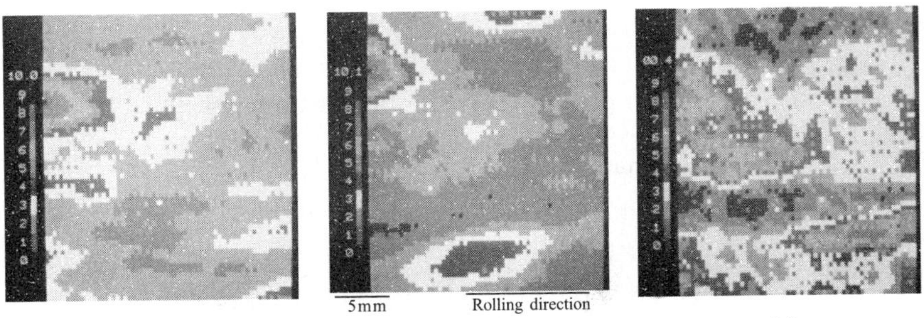

(a) (b) (c)

Fig. 5: Texture maps of a hot-galvanized steel sheet. (a) (10-10), (b) (10-11) and (c) (0004) reflections from planes which are parallel with the specimen surface.

Figure. 6 shows the ED X-ray spectrum from an image point of Fig. 5. It is composed of characteristic K lines from copper (i.e. scattered primary radiation) and from zinc (fluorescenced by the hard fraction of continuous primary radiation) as well as diffraction peaks from zinc. The characteristic lines form narrow peaks, whilst the diffraction peaks are rather broad again, due to the wide beam aperture (see Eq. (2a)) and can thus easily be discriminated. Two windows are marked by hatching which were used for the collection of the (10-10) and (10-11) texture maps. The (10-10) reflection

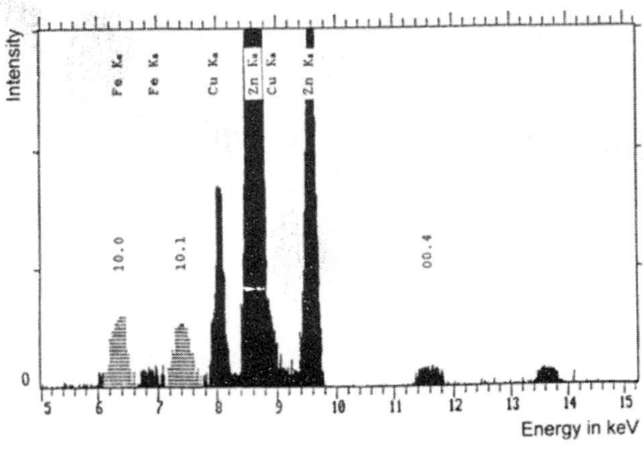

Fig. 6: Energy dispersive X-ray spectrum of the zinc coating of Fig. 5.

is superimposed on the FeKα peak. This characteristic fluorescence line from iron, however, is only induced very weakly beneath the zinc coating, a fact which is corroborated by the absence of the FeKβ? peak in the spectrum. Hence the FeK fluorescence background does not interfere with texture measurement in this example.

The process of metal forming is mainly dependent on the macroscopic parameters to which the workpiece is subjected such as distribution of strain, the rate and speed of plastic deformation, temperature, friction between the workpiece and the forming tool, but also on the microstructure, texture and (anisotropic) properties of the material. Finite element methods are now capable to model rather complicated forming processes, and spike forming (Fig. 7a) is a critical benchmark test in numerical modeling (Schönauer et al., 1993).

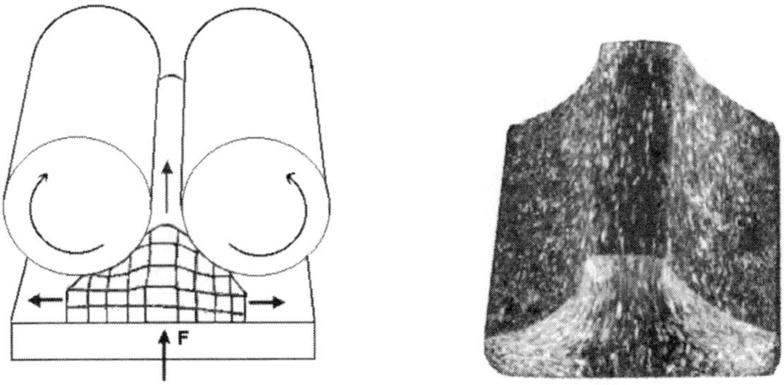

(a) Schematics of the experimental setup.

(b) Aluminum sample after friction test.

Fig. 7: The friction test by back-rolling.

Initially, the test specimen had a strong and uniform cube texture across the aluminum bar. After friction test rolling (Fig. 7b), pronounced texture inhomogeneities in the cross section plane are observed (Fischer, 1998). The cube texture is still maintained in the centre plane which is parallel with the rolls (Fig. 8a). Due to the friction with the rolls and the force *F*, the material is subjected to heavy plastic deformation on both sides of the centre plane. The grains are elongated and rotated to gradually accommodate the shape of the rolls. Crystal texture follows this change in microstructure from the centre to the surface of contact with the rolls (Figs. 8b and c), and a clear inhomogeneity of local texture develops (Figs. 9a and 9b). The texture maps have been acquired in the symmetric goniometer setup ($\vartheta_1 = \vartheta_2$) (central pole-figure point P in Fig. 8a). The local distributions of grains with (200) (Fig. 9a), respectively (111) (Fig. 9b) lattice planes parallel with the cross section plane are displayed.

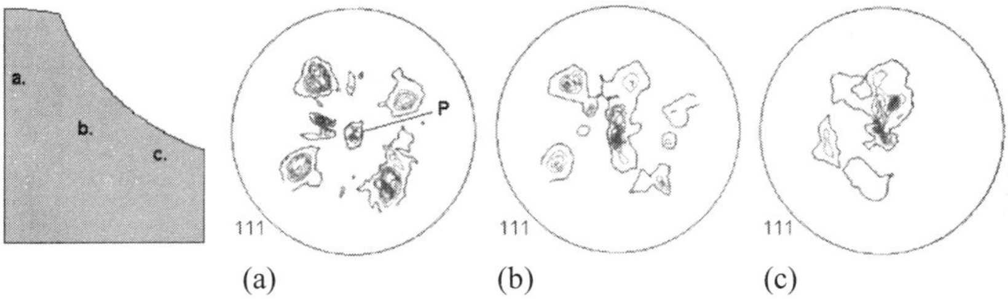

Fig. 8: The texture of an aluminum sample after friction test. Experimental 111 pole figures : (a) at the mid plane of the cross section, (b) at quarter depth and (c) close to the roll.

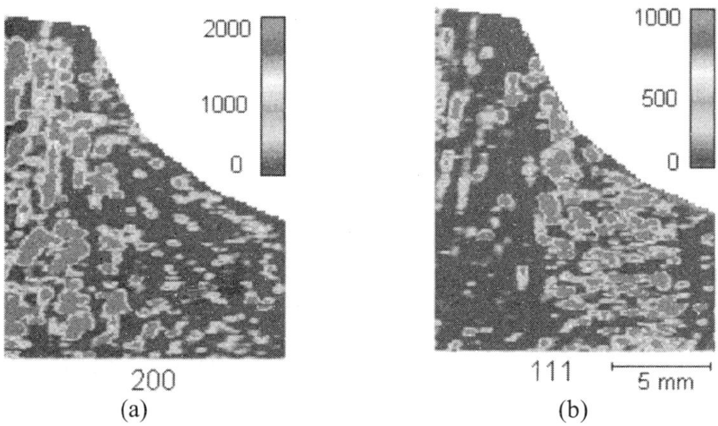

Fig. 9: Texture maps of the aluminum sample after friction test show heterogeneous texture: (a) Texture map for (hkl) = (200) (b) Texture map for (hkl) = (111).

Pole Figure Measurement and Lattice Strain Mapping

The cross section along the mid plane of an aluminum rivet was carefully ground and polished to a flat. A final electrochemical polishing was applied in order to remove any mechanical damage introduced during mechanical preparation. Before mapping, several pole figures have been acquired in steps along lines across the head and the bolt of the rivet to recognize relevant texture components. A fibre texture has been found in the bolt region (Fig. 10) (Fischer, 1998). The experimental pole-figure data can be used for ODF analysis and for the calculation of average tensorial material properties.

The (220) diffraction peak P (ϑ_1=ϑ_2=20.1°, E_{220}=12.62 keV; α=35° and β=83°) has been used for evaluating both, the spatial distributions of pole density (texture map) and residual lattice strain (residual strain map), at the same time. Because of the low diffracted intensity, dwell time had to be extended to 10 sec/pixel to obtain a sufficient count statistics. 2,232 image points at a step width of 120 μm have been acquired. A clear heterogeneity

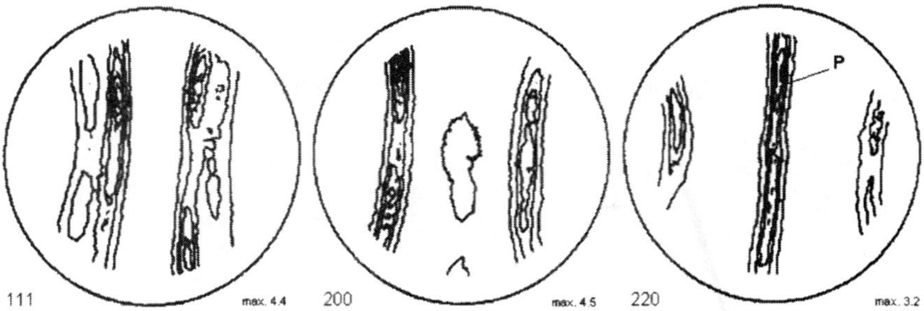

Fig. 10: Pole figures acquired across the rivet bolt. The polar specimen direction P (α=35°, β=83°) has been used for texture and residual strain mapping (cf. Fig. 11).

is revealed in the texture (Fig. 11a) as well as in the strain maps (Fig. 11b). Two almost parallel stripes of high pole density run along the axis of the bolt with a slight shift to the right hand side. They form a tube of about 1/3 of the bolt diameter. The head of the rivet shows a much lower density of this pole. It is worth noting that a butterfly-shaped minimum is observed which might result from punching the head during production.

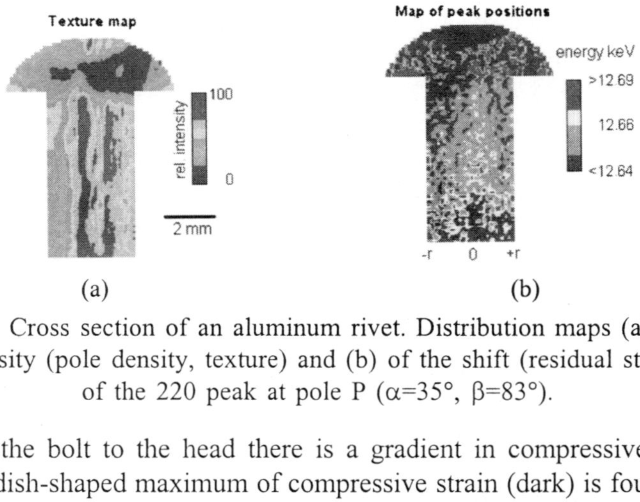

(a) (b)

Fig. 11: Cross section of an aluminum rivet. Distribution maps (a) of the intensity (pole density, texture) and (b) of the shift (residual strain) of the 220 peak at pole P (α=35°, β=83°).

Along the bolt to the head there is a gradient in compressive residual strain. A dish-shaped maximum of compressive strain (dark) is found in the head which is related to the butterfly-shaped minimum in the 220 texture map. There is an increased compressive residual strain in the interior of the "texture tube" (dark) and a tensile strain (bright) to the bolt surface.

Imaging Micro-XFA

A test specimen for imaging micro X-ray fluorescence analysis and for checking spatial resolution has been made by sputter-depositing gold through a TEM support grid on a glass substrate to form regular patches 420 μm by 420 μm wide. The annular primary beam apertures had a diameter of 30 μm. Using the AuLα fluorescence lines for imaging, the free bars (< 60 μm wide gaps) between the patches are clearly resolved in two perpendicular directions in the XFA map (Fig. 12) (Schwarzer and Wehrhahn, 1997). A spatial resolution of <50 μm is thus demonstrated.

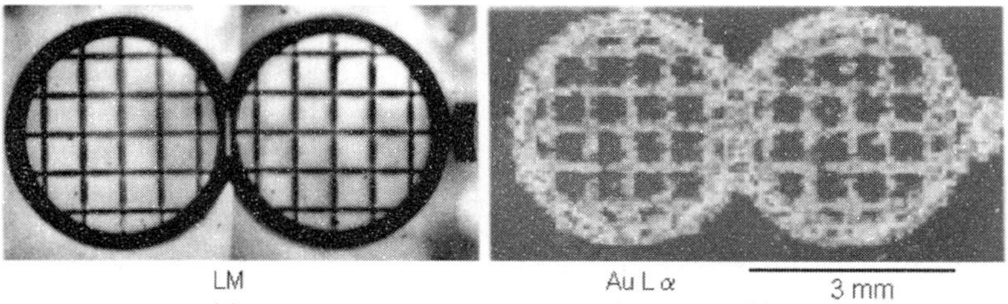

(a) (b)

Fig. 12 Gold pattern from sputtering through a TEM support grid. (a) Image in the light microscope; (b) X-ray micro-fluorescence map with the AuLα line.

CONCLUSIONS

Scanning X-ray mapping based on energy dispersive spectrometry is a powerful technique for a comprehensive materials characterization. It combines, on a grain specific scale, simultaneous texture mapping, residual strain mapping and imaging X-ray micro-fluorescence analysis by translating the sample step by step on a user defined grid under the stationary beam spot. Pole figures are measured from selected small areas. All reflections are measured simultaneously at the same angular setting. For ED diffraction it is not necessary to adjust the sample on a focusing circle. Measurement is not affected by the degree of plastic deformation. Due to the large depth of focus with a collimated primary beam, the sample surface need not be prepared to a plane and sample preparation is not difficult. Often a thorough cleaning of the surface is adequate. X-ray mapping is a very gentle method, since the specimen is only exposed to X-rays and stays in the atmospheric environment. Bulks can be studied irrespective of their conductivity.

Residual lattice strain as low as $\Delta a/a = 4 \cdot 10^{-4}$ is analyzed by evaluating the shifts and profiles of diffraction peaks in the spectrum using a standard ED detector. Spatial resolution of pole-figure measurement and residual lattice strain mapping is presently limited to 0.1 mm, whilst resolution of micro X-ray fluorescence element maps and of texture maps is limited to about 0.05 mm, as a consequence of the low intensity of a collimated primary beam.

REFERENCES

Carpenter, D.A. (1989). *X-ray Spectrometry*, **18:** 253-257.

Fischer, A. (1998). Ortsaufgelöste Polfigurmessung, Texturtopographie und Mikro-Röntgenfluoreszenzanalyse. PhD Thesis, TU Clausthal (Germany).

Fischer, A.H. and Schwarzer, R.A. (1998). *Materials Science Forum*, **273-275:** 673-677.

Fischer, A.H. and Schwarzer, R.A. (1998). *Materials Science Forum*, **273-275:** 255-262.

Horowith, P. and Howell, J.A. (1972). *Science*, **178:** 608-611.

Jones, K.W. and Gordon, B.M. (1989). *Anal. Chem.* **61:** 341A-356A.

Klein, H. and Bunge, H.J. (1999). *Z. Metallkunde*, **90:** 103-110.

Larsson, S. and Engström, P. (1992). *Advances in X-ray Analysis*, **35:** 1019-1025.

Maslen, E.N. (1958). *Sci. Instruments*, **35:** 110-111.

Rammlmair, D., Tacke, K.D. and Jung, H. (2001). *In:* Securing the Future. Proceedings of the International Conference on Mining and the Environment, Skellefteå, **2:** 683-692.

Rindby, A., Engström, P., Larsson, S. and Stocklass, B. (1989). *X-ray Spectrometry*, **18:** 109-112.

Schönauer, M., Rodiæ, T. and Owen, D.R.J. (1993). *J. Physique* IV, **3:** (colloque 7) 1199-1209.

Schwarzer, R.A. (1989). Proc. 12 Intern. Congr. on X-ray Optics and Microanalysis (12 ICXOM), Cracow, **1:** 205-208.

Schwarzer, R.A. (1993). *Steel Research*, **64:** 570-574.

Schwarzer, R.A. and Wehrhahn, M. (1997). *Textures and Microstructures*, **29:** 65-76.

Schwarzer, R.A. (1997). *Micron*, **28:** 249-265.

Wang, P.-C., Cargill III, G.S., Noyan, I.C., Liniger, E.G., Hu, C.-K. and Lee, K.Y. (1997). *Mat. Res. Soc. Symp. Proc.*, **473:** 273-278.

Werry, D. and Cross, B. (1989). *Spectroscopy International*, **1:** 58-59.

Deformation Bands in Texture Formation in Cold and Warm Rolling Interstitial-Free Steel

B.J. Duggan, M.Z. Quadir and Y.Y. Tse

Mechanical Engineering Department
The University of Hong Kong, Pokfulam, Hong Kong

INTRODUCTION

The performance of steel to make car body pressing depends on a number of factors, which are determined by the physics and material responses to strains. Among key parameters are the work hardening rate, for this affects the strain at which diffuse necking occurs (Wilson, 1966) and the way in which strain distributes in the straining process. The first parameter correlates with the strain hardening exponents, *n* in the Ludwig equation,

$$\sigma = K\varepsilon^{n} \tag{1}$$

where σ and ε are the current values in a tensile test and K is a constant. n should be as large as possible. The second parameter is measured by the Lankford parameter, R, given by,

$$R = \frac{\varepsilon_{w}}{\varepsilon^{t}} \tag{2}$$

where w and t refer to the width and thickness in a tensile test at about 15% elongation. R has been found to correlate with the Limiting Drawing Ratio (LDR) in a Swift Cupping Test and both R and LDR depend on the texture of the material. For steel R is maximum when the preponderance of the polycrystalline sheet has <111> parallel to the sheet normal direction, i.e. <111>ND. R has been found to vary in the plane of a sheet, and this gives rise to earing, an undesirable property since it indicates different ease of flow in certain directions causing variations in sheet thickness after forming. Hence, ΔR should be minimized in the plane of the sheet.

Interstitial free (IF) steel can be economically processed in a continuous annealing plant to give high n and R values, and ΔR is acceptable.

Commercially the material is cold rolled to 80-85% and heated at a heating rate of ~hundreds of degrees per minute up to ~800°C for a few minutes to give the desirable properties. In ground breaking research, Barnett and Jonas (1997) varied the temperature of hot rolling, which is conventionally performed and finished in the austenite range, and found that warm rolling in the ferrite range produced technologically interesting results. The warm rolled material has similar textures and microstructure at 700°C to those found in cold rolling. Annealing of this warm rolled IF steel produced recrystallization textures very similar to those formed after cold rolling; in other words warm rolling and annealing produced a formable product. Barnett and Jonas (1997) considered that the formation of the desirable annealing textures came from fine shear bands.

Conventionally, textures are represented by either ordinary or inverse pole figures. 3D information can be derived from them but this is experimentally cumbersome. Vigin (1960) tackled the 3D problem in the following manner. Mathematically, g represents an orientation and $f(g)$ is the Orientation Distribution Function (ODF) which represents the frequency of certain orientations as a function of three orientation parameters (Bitter, 1937). In the simplest way $f(g)$ can be defined as,

$$\frac{\Delta V(g)}{V} = f(g)\, dg \tag{3}$$

where $\Delta V(g)$ is the volume of all crystallites having orientations within a range g to $g + \Delta g$ orientation. V is the volume of the whole sample. ODFs are calculated from experimentally measured pole figures, each of which is an integral over the orientation distribution function since an individual pole density value comprises all crystals, which are rotated about the common direction of the pole. Several mathematical methods have been developed recently in order to obtain an orientation distribution function from pole density values. The way in which ODFs are derived from pole figures has been largely resolved by the work of Bunge (1969) and Roe (1965). In recent years the Bunge version has become the most widely used to represent rolling textures where the values of three continuous variables φ_1, Φ, φ_2 varies from 0° to 90° for a cubic system. To deal with a finite number of functions a frequently used value method to sample the continuous function is to use a "step" approach, the size of the step is typically 5-6°.

It is well established that both warm and cold rolling of a randomly oriented sheet of BCC metal, reorients the crystals into two fibres in the orientation space "sets". The first has <110> parallel to the rolling direction, RD, and is known as the α fibre, the second has <111> parallel to the sheet plane normal, and is known as the γ fibre. These can be most easily seen in the φ_2 = 45° section of the Bunge version of Euler space (Fig. 1b) (Bunge, 1969). Thus cold and warm rolled IF steel shows both the α and γ fibres, and annealing destroys most of the α fibres. Recrystallization also reorients metal in the γ fibres, from being centered on {111}<110> to orientations including {111}<112> and {554}<225>.

This paper presents the results of the two investigations of the annealing of both warm and cold rolled IF steel. It concentrates on the recrystallization phenomenon observed rather than the competing theories of Orientation Nucleation (ON) and Oriented Growth (OG) as explanation of what was seen. In the course of the work Deformation Bands (DB), in addition to Shear Bands (SB) were observed and their roles in recrystallization were critically evaluated. In discussing the results some theory of deformation is introduced.

EXPERIMENTAL PROCEDURE

Two initial hot bands were used and both had random textures throughout their thicknesses. The material of composition shown in Table 1 had 35 μm average grain size and was homogeneously cold rolled to 85% by 10% reduction per pass. The grain size of warm rolling material was 70 μm and had the chemical composition shown in Table 2. This was rolled at 700 °C to 75% reduction in a single pass using the maximum load capacity of the machine. The annealing was done in an air circulation pre-heated furnace at 710 °C. The mid plane section away from the edge of the material was selected for measuring global texture by X-ray radiation using a Co tube and the ODFs were generated from measured 110, 200 and 211 pole figures using the Bate software. Microstructural examinations were done under SEM, which is integrated with an EBSP detector for measuring the local orientation of a selected area. TEM was used to see the further detailed structure and Convergent Beam Electron Diffraction (CBED) pattern was used for orientation measurement.

Table 1: Composition of the Hot Band (weight %)

C	*Si*	*Mn*	*P*	*S*	*Al*	*Ti*	*N*
0.0019	0.010	0.13	0.014	0.014	0.041	0.046	0.0026

Table 2: Composition of the Hot Band (weight %)

C	*Si*	*Mn*	*P*	*S*	*Al*	*Ti*	*O*	*N*	*Cu*
0.0043	<0.01	0.1	0.006	0.0065	0.035	0.057	0.0031	0.002	0.04

RESULTS

Figure 1a is the $\varphi_2 = 45°$ ODF section of the Bunge (1969) version of Euler space, which contains the important rolling and annealing textures of BCC metals. The texture of cold and warm rolled material is shown in Figs. 1b and c and both have characteristic α and γ fibre textures. In both, the

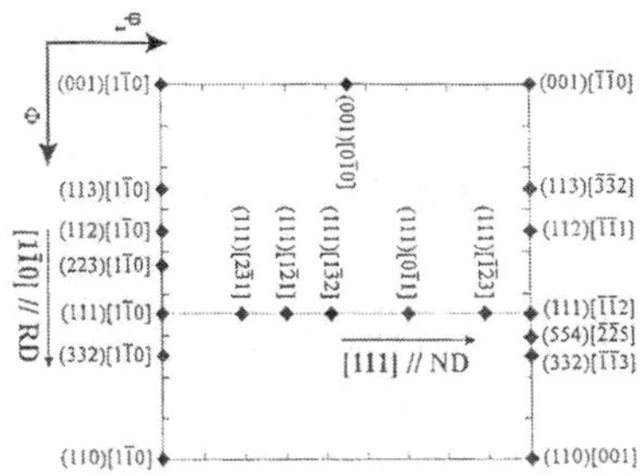

(a) φ_2 = 45° ODF sections.

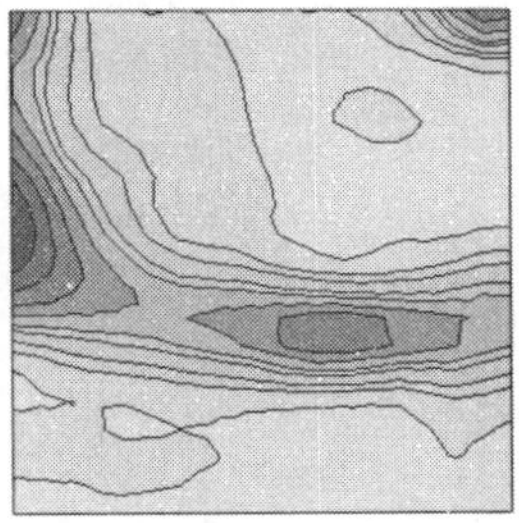

(b) 85% Cold Rolled
Max intensity 13.77R

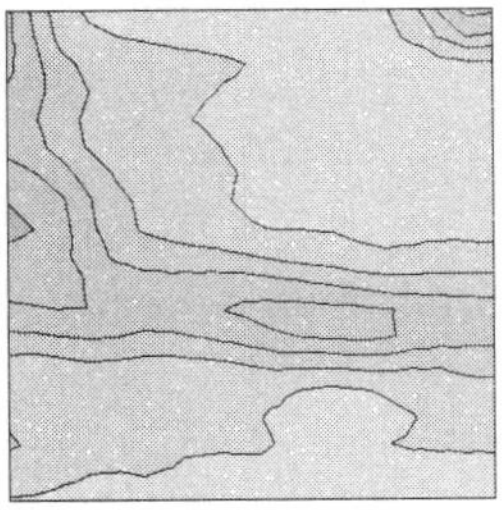

(c) 75% Warm Rolled
Max intensity 7R

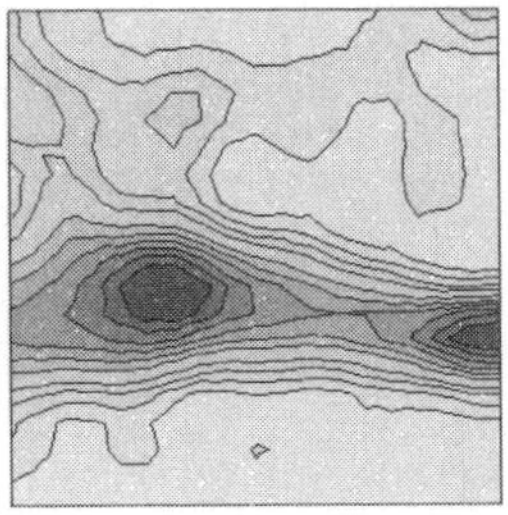

(d) 85% CR+ Ann.
Max intensity 12.06R

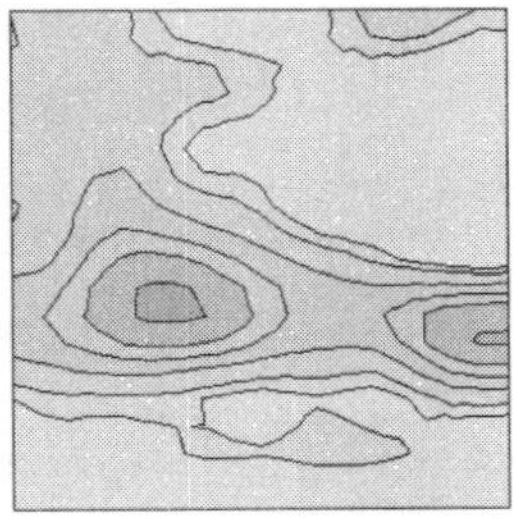

(e) 75% WR + Ann.
Max intensity 11R

Fig. 1: φ_2 = 45° ODF sections of cold/warm rolled and annealed IF steel.

recrystallization annealing eliminates the α fibre leaving a single γ fibre having the peak intensity at {554}<225>, a component about 8° away from the {111}<112> γ fibre component (Figs. 1d and e).

Cold Rolling Microstructure

The microstructural study was started with the investigation on longitudinal section (LS). Figure 2 is taken from a large montage and is a representative area containing all the features of 85% cold rolled material. During rolling all the grains became elongated from equiaxed but there are significantly different internal substructures. The structure is determined by the orientations of the grains which was measured by EBSP. The grains, which have a gradual change of contrast over larger distances, belong to the α fibre. In contrast the grains of γ fibre have sharp changes of contrast within very short distances, which give them a finer substructural appearance. The γ grains are full of microbands inclined ~30° or less with the rolling direction and are rarely cut by in-grain shear bands. Figure 2 shows such a rare occurrence.

In the rolling plane investigation the differences between α and γ grains are still obvious and a group of grains are shown in Fig. 3. The whole image covering four grains, A-D, was scanned point by point by 1 μm step size to measure the orientation. The 200 pole figures show how orientation

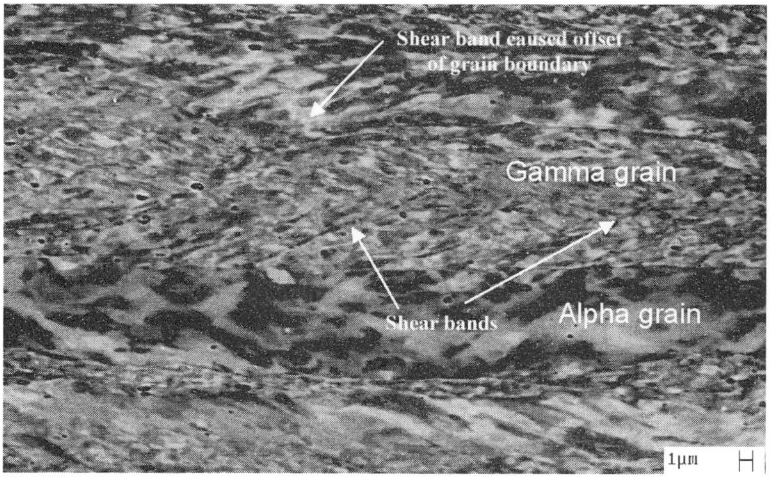

Fig. 2: SEM CC image of longitudinal section of 85% cold rolled IF steel.

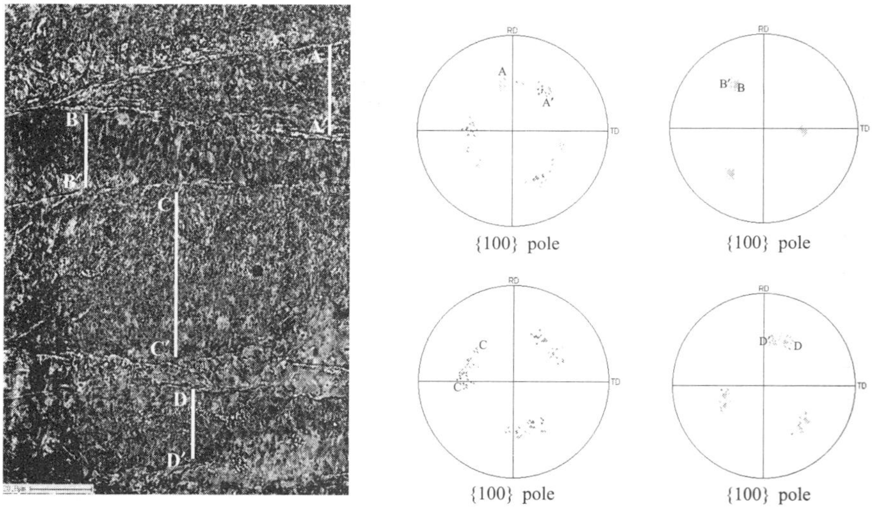

(a) Channeling contrast, RD → (b) 200 pole figures of the marked

Fig. 3: SEM CC image of RP section and corresponding 200 pole figures of the lines printed on the image showing DBs of 85% cold rolled IF steel.

varies along the lines in each grain. The orientation of grain B is uniform and centered at {225}<110> and belongs to the α fibre. Grains A, C and D belong to the γ fibre and have a large rotation from one grain edge to the other.

The partial recrystallization of this material shows the first recrystallized grains appear in the γ fibre grains and grow within the grain envelope, having the orientation of the γ spread. The α deformed grains do not recrystallize but are consumed at the last stage by invasion of recrystallized grains formed in the γ fibre. A detailed examination on the LS at the early recrystallization stage failed to reveal the origins of the nucleation. However, the investigation was successful when the rolling plane section was observed, as shown in Fig. 4. The orientation mapping of the channelling contrast (CC) image is shown in Fig. 4b, where it is revealed that the grain has been fragmented into several segments. The orientations of the corresponding segments are shown in the 200 pole figure in Fig. 4c. The principal boundary between the segments is along the rolling direction. In addition, there are other boundaries, which further fragment the right segment. Clearly recrystallized nuclei have formed on these boundaries, which are clearly the result of deformation banding (DB). This is one of many examples found in the cold rolling experiment. No shear band nucleation was found.

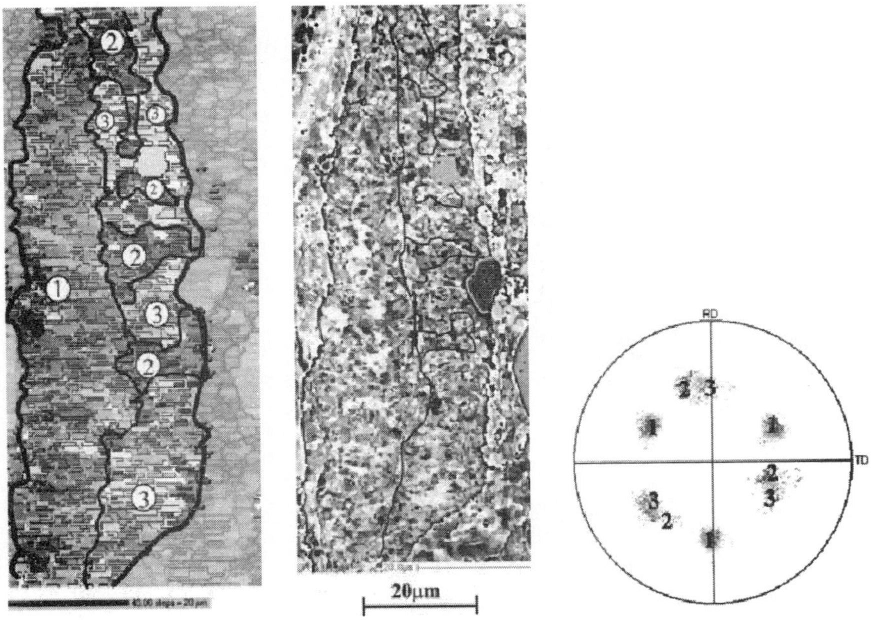

(a) CC image, RD↑. (b) OIM (c) 200 pole figure.

Fig. 4: SEM CC image with corresponding OIM and 200 pole figure showing the formation of recrystallized grains in RP section.

Warm Rolling Microstructure

The same procedures for investigation of the cold rolled material were adopted for the single pass 75% warm rolling material. The general microstructure observed in the LS is almost identical to that found in the cold rolling case, except that shear bands were more common. A SEM CC image is shown in Fig. 5a and in this shear bands are very fine, of less than a micron in thickness, and almost always terminated at the grain boundaries. The offsets created by the shear bands at the grain boundaries are visible in longitudinal section, LS as has been reported by many others. Since shear banding is three-dimensional, a detailed investigation of the offsets created by shear bands at grain boundaries involved inspection of the rolling plane. Offsets were also found in RP sections roughly parallel to TD. The OIM of the CC image, shown in Fig. 5a, is shown in Fig. 5b and in here any material measured to be more than 10° away from its neighbour, a line was plotted. These clearly outline the shear bands in the OIM. It was found during the data processing that all these parallel lines revealing shear bands in the OIM disappeared at a 12° misorientation setting, which implies the maximum magnitude of these boundaries to be around 10°. The 200 pole figure in Fig. 5c shows that the orientation of this grain is centered at ~{111}<112> and the orientation relationship between shear band and matrix can be seen in Fig. 5d, which represents the orientations along the line scan across shear band material 'S' and matrix 'M' in which they formed. In this particular figure the rotation axis is not along any of the sample reference axes, RD, TD or ND. Moreover further investigation failed to find any fixed crystallographic rotation axes between shear bands and matrix orientations.

(a) Channeling contrast image, RD→.

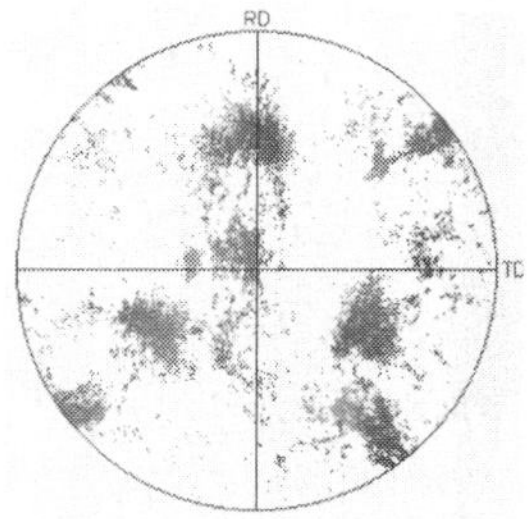

(c) 200 Pole figure, whole area

(b) OIM, misorientation boundary>10°.

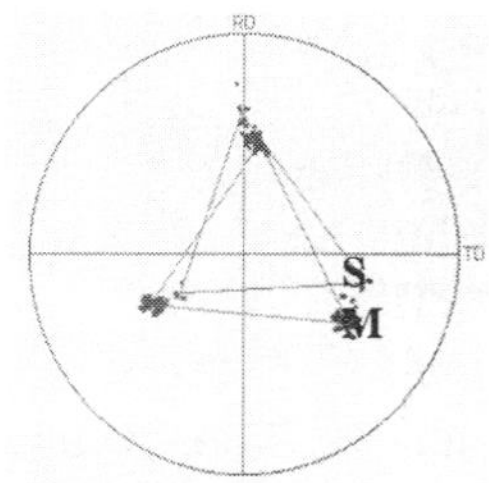

(d) 200 Pole figure, across the line SMS.

Fig. 5: SEM CC image with OIM and 200 Pole figures of LS of 75% WR IF steel, rolled at 700°C in single pass.

A detailed TEM investigation was done of the shear bands, an example of which is shown in Fig. 6a, which shows shear bands passing through the microbands. The orientations of the points marked in the image were measured using Kikuchi patterns derived from Convergent Beam Electron Diffraction (CBED). The pattern taken at 1 is identical to 2 and 3. In the same way 4 is identical to 5 and 6, and 7 is identical to 8 and 9. This means the material in the either side of the shear bands are homogeneous in

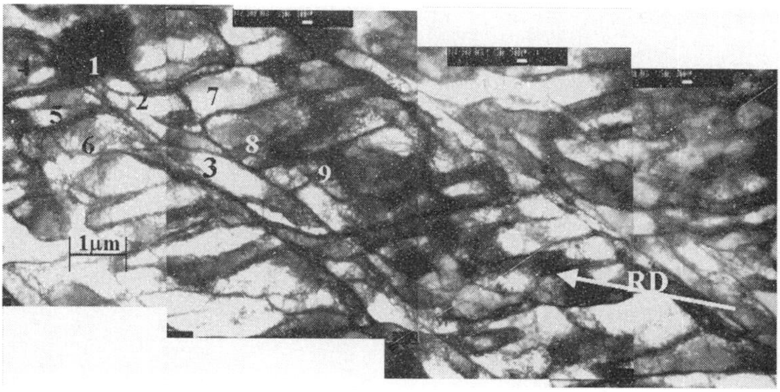

(a) TEM image

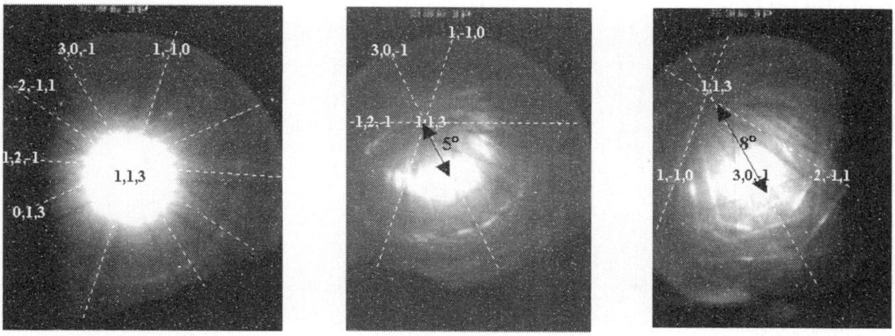

(b) Kikuchi pattern of 1≈2 and 3 (c) Kikuchi pattern of 4 ≈5 and 6. (d) Kikuchi pattern of 7 ≈ 8 and 9.

Fig. 6: TEM image of longitudinal section of 75% warm rolled IF steel.

orientation although they are separated by microband boundaries. The shear band material is misoriented by ~5° from the left side and by ~8° from the right side material.

As was done in cold rolling, the next part of the investigation concentrated on the rolling plane section. This section also shows identical α and γ microstructural contrasts to those found in longitudinal sections. Figure 7b is the OIM of the CC image of Fig. 7a and covers three grains. Grain X has α microstructural contrast with little misorientation across the grain and is centered at {511}<110>. The OIM of grain Y shows many small angle boundaries, which rotated the grain from {111}<112> to {111}<110> across the line scan Y1-Y2. Grain Z is divided by a sharp misorientation boundary between the segments Z1 and Z2 having the orientations ~{111}<123> and {111}<112>. Although comprising identical CC microstructures, grain Y has many small angle boundaries which are missing in the OIM of grain Z.

In the partially recrystallized sample it was found that the progress of recrystallization was identical to that observed after cold rolling i.e. the γ grains recrystallize preferentially and consume the α grains at the later stages of recrystallization. In this regard the role of shear bands appeared to

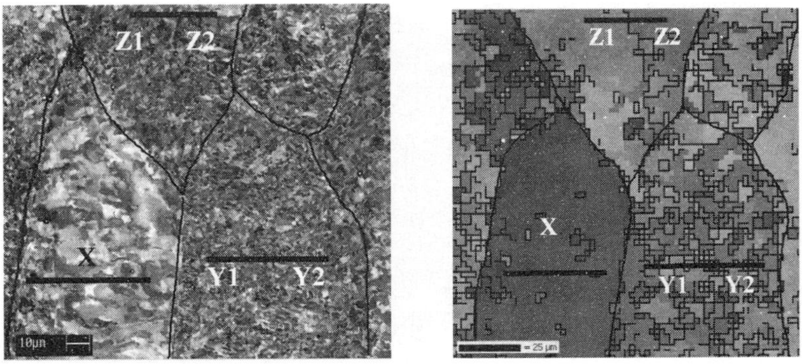

(a) Channelling contrast image, (RD↑) (b) OIM, step size 2 μm

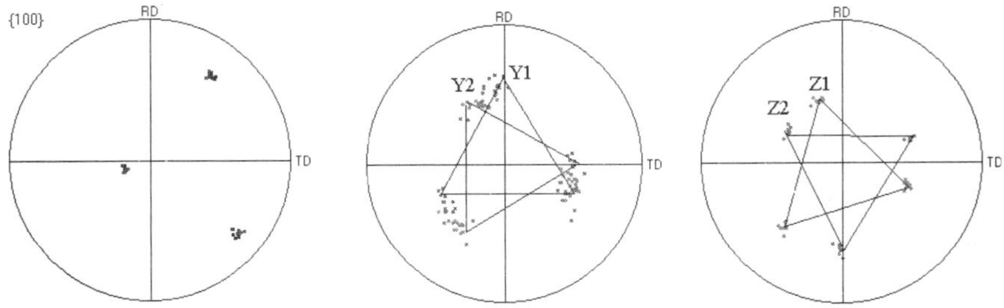

(c) Grain X (d) Grain Y (e) Grain Z

Fig. 7: Channelling contrast image and corresponding 200 pole figures of rolling plane section of 75% warm rolled IF steel.

be insignificant, which was proved by an intensive investigation of the shear bands. They appeared to be inactive as far as nucleation is concerned. Thereafter the investigation turned to focus on the rolling plane section. One example is shown in Fig. 8. The newly formed recrystallized grains labeled as A, B and C are located inside the middle of the deformed grain. The large montage, of which Fig. 8 is a detail, shows the neighbouring grains having the α microstructure to be without any sign of nucleation. The OIM in Fig. 8b shows the neighbouring grains are of α orientations and the grain where recrystallization occurred has its material rotated from {111}<110> to {111}<123> along the width of the grain. The nuclei are positioned at the centre of the lattice curvature with an orientation of {111}<123>.

DISCUSSION

The 85% cold rolling and 75% warm rolling textures are essentially the same comprising the typical α and γ orientation fibres. This observation is quite consistent with the early work by Schläfer and Bunge (1974) for low carbon steel and later confirmed by Hutchinson (1994) for IF steel. There are two variables involved in the cold and warm rolling process included in this investigation, the temperature and the strain rate. The results show that very large differences of these two concurrently operating variables do not

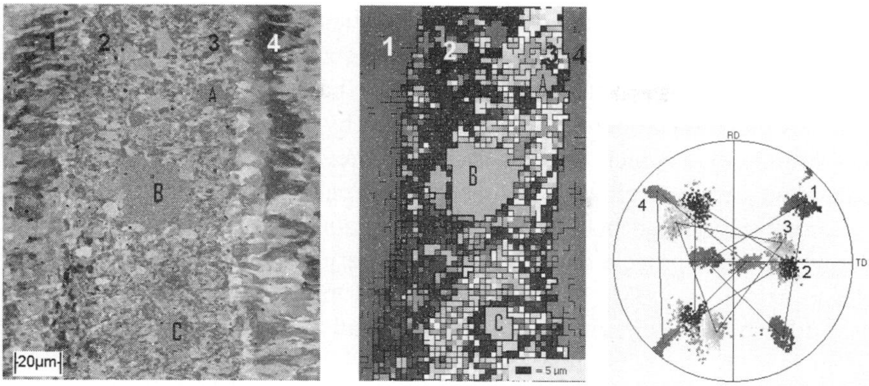

(a) CC image, RD↑. (b) OIM (c) 200 pole figure.

Fig. 8: SEM CC image with corresponding OIM and 200 pole figure of 75% warm rolled IF steel showing the formation of recrystallized grains in RP section.

affect the rolling textures. However, the individual effect of these parameters has not been considered in this experiment. Annealing of these materials also produces the same textures, essentially a γ fibre spread, which might imply that similar mechanisms are involved in the recrystallization processes. The same recrystallization results have been reported from numerous researchers in cold rolled (Hutchinson, 1994; Tse et al., 2000) and for warm rolled (Barnett and Jonas, 1997) of IF steels.

The observation of the partially recrystallized samples showed the early recrystallization process to be both very selective to occur inside the γ deformed grains. This is a confirmation of the repeated finding that only γ grains participate in the early recrystallization process and determine the final annealing texture by their growth to consume the rest of the deformed material. This was found and reported in a review by Hutchinson (1994) who, by using a large number of OIM measurements of recrystallized grains, was able to separate the developing recrystallization texture from the declining deformation texture, which they followed by macroscopic X-ray determined ODFs. Clearly, the α fibre was the last component to disappear, at a point where the γ fibre recrystallization component had already formed. Significantly the warm rolling process, which is a relatively new field of study, behaves identically to cold rolled material. Therefore, the central issue of this paper is to study the locations of recrystallized nuclei inside γ grains. Figures 4 and 8 are of the very early stages of recrystallization and show that recrystallized nuclei form at the deformation induced boundary where the grain has split into differently oriented segments. The boundary is either sharp or diffuse as shown in this investigation in Figs. 3 and 7.

From an extensive investigation of cold rolling Duggan et al. (2002) developed model shown in Fig. 9, which is based on two fundamentally necessary elements of nucleation, the formation of a high angle boundary, for this gives mobility and the driving force coming from stored energy. In this model, nucleation at pre-existing grain boundaries is assumed to be unlikely since such boundaries are decorated by Ti and Nb carbide particles (DeArdo, 1997), which create a dragging force against the grain boundary migration i.e. the well-known Zener drag (Zener and Smith, 1948). If the deformation induced boundary inside the grain is sharp, Fig. 9a, then the recrystallization process can follow the boundary migration process following the principle of SIBM (Bailey and Hirsch, 1962). On the other hand, if it is a diffuse boundary, showing relatively small cumulative lattice rotations, as in Fig. 9b, the nucleation can take place by the in-situ recrystallization mechanism proposed by Dillmaore et al. (1972). In either case, the mechanism is powered by the difference of stored energy, which is a function of the deformation microstructure.

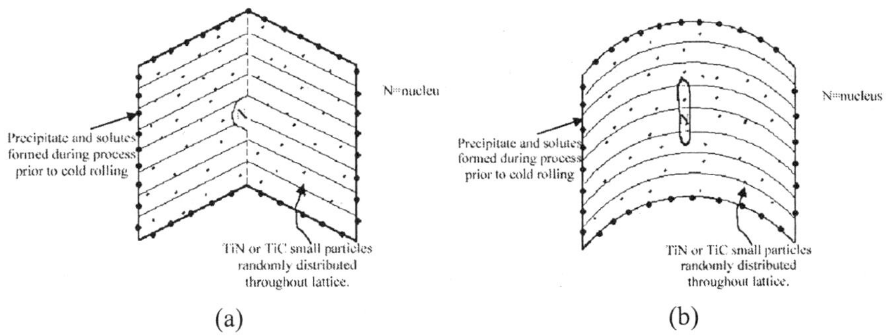

Fig. 9: Schematic diagram showing the nucleation occurs (a) at sharp interface by Bailey and Hirsch (1962) mechanism or (b) at the diffused interface following the mechanism of Dillamore et al. (1972).

In the recent work by Rajmohan et al. (1997), they measured the stored energy for γ components among others. After rolling, {111}<112> components have higher stored energy than {111}<110>, but this order is reversed during recovery annealing i.e. {111}<110> has higher stored energy than {111}<112>. This work repeats the earlier investigations of Takechi et al. (1968) who found essentially the same result, but from X-ray scattering rather than from neutron scattering. Therefore, logically this has a direct effect on the nucleation process, which should occur in lower stored energy volumes of material. It is to be stressed here that most probably the high intensity of {111}<112> component in the annealing texture might originate from the early nucleation of this component at the deformation induced boundary interface between the fragmented segments. Hence the formation of γ recrystallized grains in γ deformed metal arises from orientation splitting inside γ grains, to give mobile or potentially mobile interfaces and differential recovery in γ.

In this investigation fine in-grain shear bands found frequently in the warm rolled material are rare in the cold rolled material. The probable reason is the difference in the grain sizes, as noted in early work which showed that shear banding is more likely to occur in larger grains (Lee and Duggan, 1990). The material used for cold rolling has a 35 μm grain size, which are ~10% of the volume of 70 μm diameter grains. Whatever the reason, the detailed investigation of shear bands shows that they are extremely fine, are formed in grains having two sets of microbands and have remarkable similarities with the microbands. For example, they are of similar width, parallel to the second set of microbands and confined to the grain containing them. This indicates that these particular types of shear bands occurring in warm rolling could be considered as a special kind of microband which forms in well developed microbanded grains when the grains cannot continue deforming in the same way.

The pileup group of dislocations at the first set of microband boundaries release in the form of shear bands, resembling the dislocation release at grain boundaries in sample scale shear bands (Lee and Duggan, 1990). The details are not known. It is clearly shown in Figs. 5 and 6, however, that the material inside the shear bands is misoriented by 5-10° with the material surrounding them. This magnitude is known to be insufficient to provide a mobile interface in the recrystallization process. It is widely quoted that 15-20° is necessary for a boundary to be mobile although there is no sharp demarcation between high and intermediate angle boundaries. However, it is well known that recrystallization is a competitive process and low angle boundaries are significantly less mobile than high angle boundaries (Doherty, 1978). Therefore, although shear bands have been found in the warm rolled material, their role was not significant in the nucleation process. Rather the grain fragmentation evident in the rolling plane section is the essential requirement for nucleation. In Fig. 7, the γ grains Y and Z have been divided into segments by sharp and continuous rotations, respectively. It is clear that the same recrystallization mechanism is likely to operate in warm rolled material as was found in cold rolling. This contradicts the theory of Barnett and Jonas (1997).

The grains showing orientation splitting, or deformation bands (DB) found in the present investigation are of special kind; the interface between the segments is parallel to the RD and ND and therefore is only visible in the rolling plane section. The origin of such a special kind of splitting was explained by Liu and Duggan (2001) in the texture simulation of BCC metals using DB theory. DB theory is principally based on the Taylor (1938) framework and Fig. 10 contains the relevant information. An additional assumption in DB theory is that the grains do not adopt the same deformation as the whole sample, rather the grains split into segments, by allowing the e_{32} shear (Fig. 10a). In order to maintain the average shape of the whole grain, as required by the Taylor axiom, mutual cancelling of the negative and positive shear components between the segments occurs. A schematic diagram of this process is shown in the Fig. 10c. This process is supported with an energy criterion after considering the additional energy required to create an interface between the segments.

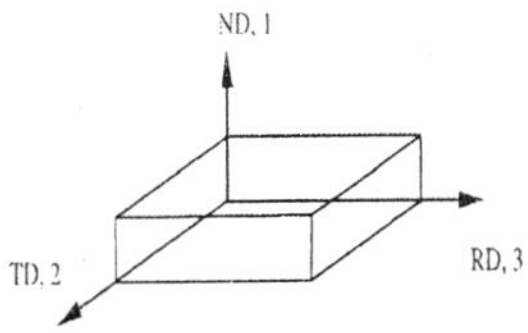

(a) Rolling geometry.

$$E = \begin{bmatrix} -d\varepsilon & 0 & 0 \\ e_{21} & 0 & 0 \\ e_{31} & e_{32} & d\varepsilon \end{bmatrix}$$

(b) Displacement gradient tensor.

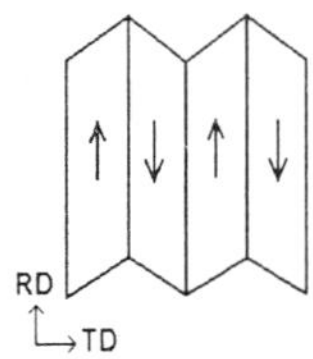

(c) e_{32} relaxed DB model

Fig. 10: Rolling geometry and displacement gradient tensor. (Liu and Duggan, 2001)
FC: $e_{21} = e_{31} = e_{32} = 0$; $RC_{lath} = e_{31} \neq 0$, $e_{21} = e_{31} = 0$; $RC_{pancake} = e_{31}$ and $e_{21} \neq 0$, $e_{32} = 0$; DB $= e_{31}$, e_{21} and $e_{32} \neq 0$.

The simulation results were compared with the RC lath model (Van Houtte, 1982), which is also based on the Taylor framework and known to be very good in predicting the BCC rolling texture. It was assumed that deformation banding will occur if a grain requires less work to deform, otherwise it will deform following the RC mode. The calculation was carried for 1000 initial orientations and it was found that the orientations subject to DB were close to γ, and gave γ components. The overall results are very encouraging. As examples three starting orientations are shown in Fig. 11, one close to α, the others in the γ circle. In line with the experimental results α orientations are without deformation banding and stable, but the γ grains are fragmented in this mode. {111}<112> is also a stable component, but orientations a little away from the ideal position produces DB elements. In Fig. 11b, if they are ND rotated from {111}<112> then they rotate to the more stable {111}<110> component and if they are TD rotated {111}<112> they rotate around TD and split into two components by rotation around ND, giving components on the γ circle in the 200 pole figure. This is something also found in the deformed state in Figs. 3 and 4 for cold rolling and in Figs. 7 and 8 for warm rolling. However, this model does not provide any criterion as to whether DB boundaries are diffuse or sharp. In the experimental observations both kinds of boundary were found and the clear presence of recrystallized nuclei at these interfaces is convincing evidence that they are the origin of γ nucleation.

The deformation banding theory, although crude in that it is solely energy based and ignored grain/grain interactions has given rise to a new research direction. This is based on the simulations shown in Figs. 11b and c. Clearly if a γ texture is rolled, the theory predicts that it will produce DB γ, which favours nucleation in γ. In practical terms this means processing steel so that the material prior to final cold rolling must have a well developed γ texture.

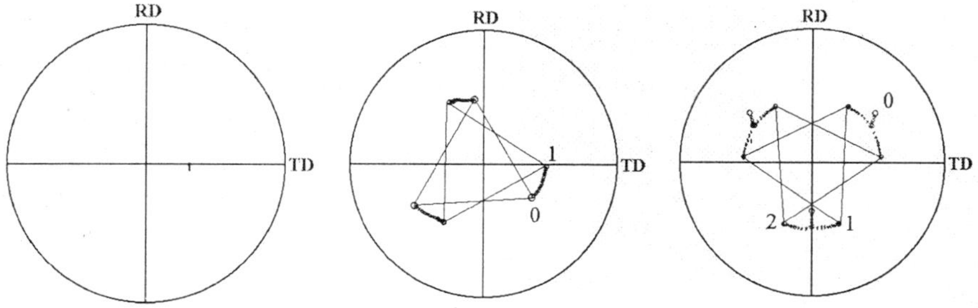

(a) {112}<110>

(b) ND rotated {111}<112>

(c) TD rotated {111}<112>

Fig. 11: 200 pole figures showing the rotations from specific initial orientations "0" to a single "1" or to a pair of orientations "1" and "2" calculated by deformation banding theory with e_{32} shear relaxation. (Duggan et al., 2002)

Now Fig. 1c shows the results of lowering the hot rolling temperature so that only ferrite is concerned, i.e. warm rolling. Annealing of this warm rolled material produces the texture shown in Fig. 1e, which is a well developed γ recrystallization texture. To test the theory, this warm rolled material was processed to give the annealed texture shown in Fig. 12a, cold rolled 80%, Fig. 12b and then annealed to give the texture shown in Fig. 12c. This is an excellent texture for deep drawing, for it has a rather flat γ fibre, and because of this D*R* should be small compared with peak type textures found in conventionally processed IF steel. The intensity maximum at ~16R is also approximately 70% higher than normal and, therefore, *R* should also be significantly better than conventionally processed material. The success of this experiment does not necessarily prove the DB theory on which it is based, but the results are nonetheless encouraging for the kind of approach to deformation theory.

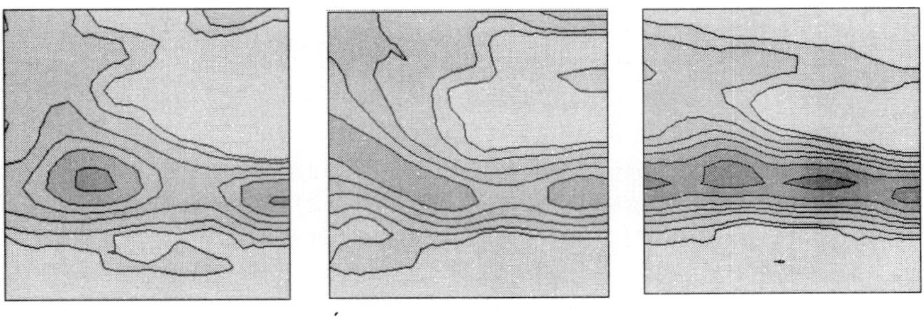

(a) 75% WR + Anneal 7 min at 710°C. Max intensity 11R

(b) WR 75% + Anneal 7 min at 710°C+CR 20%. Maximum intensity 10.55R.

(c) WR75% + Anneal 7 min at 710°C+ CR 20% + Ann 5 min at 710°C. Maximum intensity 15.72R

Fig. 12: $\varphi_2 = 45°$ ODF sections of the sample further cold rolled and annealed after warm rolling at 700°C.

CONCLUSIONS

1. The rolling and annealing textures of homogeneous cold rolling and single pass (inhomogeneous) warm rolling are very similar.
2. Extremely fine in-grain shear bands developed in warm rolled IF steel of 70 μm grain size have a number of similarities with microbands found in the material. The shear bands do not play a significant role in the recrystallization process because of the small misorientation between them and the matrix material.
3. Deformation bands parallel to the rolling direction, most easily visible in the rolling plane section, are associated with both sharp high angle boundaries and more gentle lattice curvature and play a major role in the recrystallization process in both the cold and warm rolling material.
4. DB theory is an advance on other Taylor models of plastic deformation in that individual grains do not give single end orientation, but two. Observation shows that is close to reality than other versions.
5. Using the warm rolled and annealed material as a starting material for conventional processing, produces excellent textures for drawing processes. The theory of DB is not proved by this experiment, but the results are extremely encouraging.

Acknowledgement: It is a pleasure to acknowledge the support of this work by the grant numbers CERG/HKU 7067/97E and CERG/HKU 7323/98E given by the RGC of Hong Kong Special Administration Region.

REFERENCES

Bailey, J.E. and Hirsch, P.B. (1962). "The recrystallisation process in some polycrystalline metals". Proc. Roy. Soc. London, **A 26711**.

Barnett, M.R. and Jonas, J.J. (1997). *ISIJ International*, **37:** 697.

Barnett, M.R. and Jonas, J.J. (1997). *ISIJ International*, **37:** 706.

Bitter, F. (1937). Introduction to ferromagnetism.

Bunge, H.J. (1969). Mathmatische Methoden der Texturanalyse, Akademie-Verlag, Berline.

DeArdo, A.J. (1997). International conference of thermomechanical processing of steels and other materials. The Minerals, Metals & Materials Society. 13.

Dillamore, I.L., Morris, P.L., Smith, C.J.E. and Hutchinson, W.B. (1972). *Proc. R. Soc Lond.*, **329:** 405.

Doherty, R.D. (1978). *In:* Recrystallization of Metallic Materials. Editor. F. Haessner, Riederer Verlag GmbH, Stuttgart, 23.

Duggan, B.J., Tse, Y.Y., Ning, H. and Quadir, M.Z. (2002). *Materials Science Forum*, **408-412:** 1151.

Hutchinson, W.B. (1994). *Mat. Sci. Forum*, **157-162:** 1917.

Lee, C.S. and Duggan, B.J. (1990). *Scripta Metallurgica*, **24:** 757.

Liu, G.L. and Duggan, B. (2001). *J. Met. and Mat. Trans A*, **32A:** 125.

Rajmohan, N., Hayakawa, Y., Szpunar, J.A. and Root, J.H. (1997). *Acta Mater*, **45:** 2485.

Roe, R. J. (1965). *J. Appl. Phys*. **6:** 2024.

Schläfer, D. and Bunge, H. J. (1974). *Texture*, **1:** 157.

Takechi, H., Katoh, H. and Nagishima, S. (1968). *Trans. AIME*, **242:** 56.

Taylor, G.I. (1938). *J. Inst. Met*, **62:** 307.

Tse, Y.Y., Liu, G.L. and Duggan, B.J. (2000). *Scripta Matarialia*, **42:** 25.

Van Houtte, P. (1982). *Mater. Sci. Eng*., **55:** 69.

Vigin, A.S. (1960). A Quantitative Measure of the Texture of a Polycrystalline: A Material-texture function.

Wilson, D.V. (1966). *J. Inst. Met*. **84**.

Zener, C. and Smith, C.S. (1948). *Trans. Metall. Soc. A.I.M.E*., **175:** 15.

Texture Determination of Zircaloy Using X-ray and EBSD Diffraction

H.S. Ubhi

Multifunctional Materials
QinetiQ Ltd
Farnborough, Hampshire, GU14 OLX. UK

INTRODUCTION

Crystallographic texture in materials is an important material property that can have a significant influence on mechanical anisotropy and consequent component performance e.g. in-service dimensional changes of structures made from hcp materials like Zircaloy.

X-ray diffraction (XRD) is a standard technique for texture measurements of materials and has been used for a number of years at QinetiQ to determine textures in Zircaloy. However, with the development and advances of the EBSD (electron back-scattered diffraction) technique, it is now possible to acquire texture data using this technique. An initial limited study that included a Zircaloy sample (Menhert et al., 1999) for the evaluation of this technique to determine textures showed good correlation with XRD measurements.

Quantification of the measured textures in Zircaloy is widely done by a simple method proposed by Kearns (1965). The parameter determined by this method is known as the *f*-factor which represents the effective resolved fraction of poles lying along a reference direction. Usually an orthogonal set of axes, conveniently the rolling, transverse and short transverse directions are considered and the *f*-factors are calculated in each of these directions. This *f*-factor is defined by the equation:

$$f_i = \int_0^{\pi/2} I_\rho \sin\varphi \cos^2\varphi \, d\varphi \qquad (1)$$

where i can be the normal (N), longitudinal (L) or transverse (T) direction, ϕ is the angle between the [00.1] and reference directions and I_ρ is the average pole density. Values of f vary from 0 to 1.

It can be shown that the sum of the *f*-factors for these three directions must be unity and that $f = 0.333$ will represent the isotropic (random) case. Kearns (1965) only considered the [001] direction, but subsequent researchers have extended this to include prism poles, with the result that Kelly and Watson (1972) proposed that the empirical Eq. (2) should hold

$$\sum f_s = f_{00.2} + 2f_{10.0} \approx 1 \tag{2}$$

However, if a convenient crystallographic orthogonal axis is chosen, viz., [00.1], [10.0] and [11.0], then it has been argued (Bowen, 1993) that the sum of the *f*-factors in these three directions must be unity i.e.

$$\sum f_s = f_{00.2} + f_{10.0} + f_{11.0} \equiv 1 \tag{3}$$

Following the above argument it has been now shown (Bowen, 1993; Mehnert) that

$$\sum f_s = f_{00.2} + f_{10.0} + f_{11.0} = 1 \tag{4}$$

And also that

$$f_{10.0} = f_{11.0} \tag{5}$$

Therefore, at QinetiQ this more rigorous equation has been used for quantification.

In addition, since measured X-ray textures (Juul Jensen and Lyttle, 1999) can show scatter between repeated measurements on the same sample especially over time and also due to counting statistics, experimental set up e.g. change of alignment or change of anode, etc. Therefore, whenever texture data has been collected on Zircaloy samples, repeatability measurements have also been made on two "standard" samples. The results from these samples have then been used as criteria for acceptance of data from the various batches of samples tested over a period of time. Results from these repeatability tests are presented and the problems encountered e.g. diffractometer set up, sample height, calibration, sample preparation, and solutions to overcome these problems, have been discussed.

With the development and advances of the EBSD technique (Schwarzer, 1997; Baba-Kishi, 2002) it is now possible to acquire texture data using this technique. An initial limited study that included a Zircaloy sample (Kearns, 1965), for the evaluation of this technique to the determination of texture showed good correlation with XRD measurements. The advantages and limitations of the EBSD technique over X-ray texture measurements are briefly presented for this part of the paper. Data from various EBSD acquisition and post-processing routines has been discussed and data from several Zircaloy samples presented. Comparison of this data with that from X-ray measurements has also been made in terms of pole figure, pole figure intensity and Kearn's *f*-factors correlations.

X-RAY DIFFRACTION

Experimental Procedures

All measurements were carried out on an automated Siemens D500 Texture Goniometer, which has a Bragg-Brentano geometry and a graphite monochromator in the diffracted beam. Before, during and after texture measurements the alignment of the gonimeter was checked, using a (111) cut silicon single crystal. In order to ensure that the Bragg angle at tilt angles of $\varphi = 0$ and 70.5° deviated from the absolute angle by less than 0.02° 2θ and that the intensity decrease due to defocussing was minimised and essentially constant around the $\varphi = 70.5^{\circ}$ circle. Typical peak positions and intensities are shown in Table 1 and the 70.5° circle is illustrated in Fig. 1.

Table 1: Typical peak positions and intensities from alignment checks with (111) cut single crystal

Tilt	*Peak position (deg)*	*Intensity (cps)*
$\varphi = 0^{\circ}$	28.419	151689
$\varphi = 70.5^{\circ}$ A	28.438	109451
$\varphi = 70.5^{\circ}$ B	28.441	109518
$\varphi = 70.5^{\circ}$ C	28.444	110034

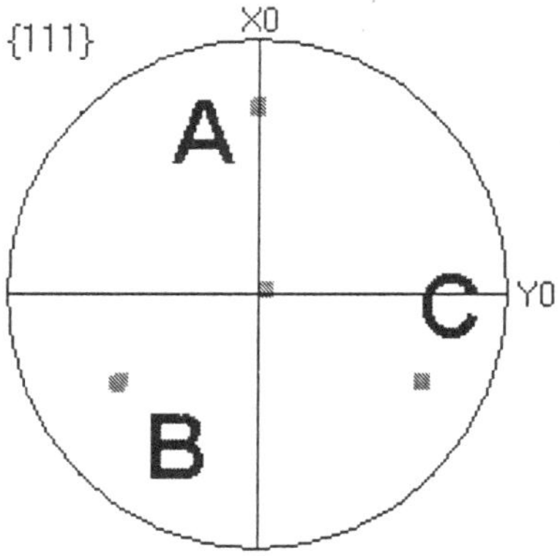

Fig. 1: (111) pole figure of the single crystal silicon crystal showing positions A, B and C to check goniometer alignment and defocus intensities.

Defocus Checks

Defocus checks to determine the well-known loss in intensity that occurs as the tilt angle on a goniometer is increased from 0 to 90° were made on a HIPed Zircaloy powder sample. A typical set of (00.2), (10.0) and (11.0) defocus curves are shown in Fig. 2. The data generated from these measurements was used to correct the intensities in all pole figures for this defocusing effect. The check was deemed satisfactory if the ratio of the

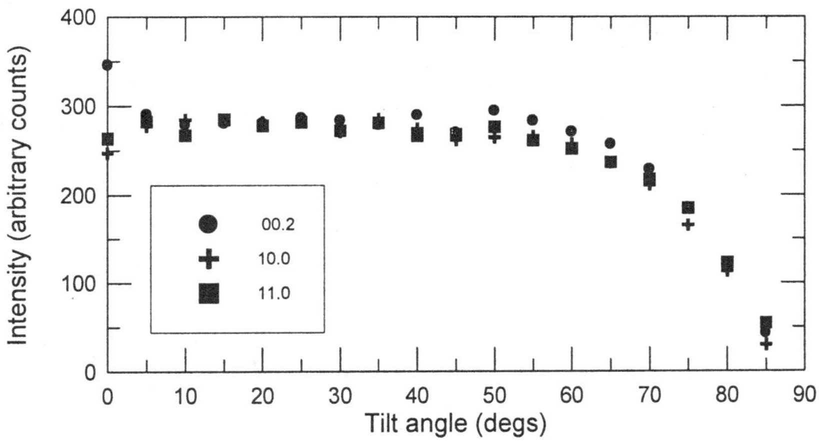

Fig. 2: Typical set of defocus curves from a HIPed Zircaloy powder sample.

mean intensity from 0-50° tilt and the intensity at 70° tilt was greater that 65%.

Data Collection

The Zircaloy samples, approximately 30 × 28 mm in size, in the mechanically polished condition were used for pole figure determinations. Four pole figures i.e. (10-10), (0002), (10-12) and (11-20) were measured for each sample without removal from the holder. Specimen background measurements were made at 30° and 80° 2θ. The data were collected in the reflected mode by continuous scanning at 5° step sizes of tilt and rotated up to a maximum tilt of 85°.

The collected data points were corrected by subtracting the background, adjusting for defocusing (using the random HIPed sample of Zircaloy) and then normalising to obtain times random intensity levels.

f-factor Calculation

To calculate the *f*-factors, the pole figures were processed in the following way:

To ensure that the integration of the intensities within the pole figure totals 100% (which can be ensured for a complete pole figure), the incomplete pole figures and the defocus curves were adjusted for the missing intensity by copying the 85° data to 90°. This has been found to give the best approximation to a complete pole figure (Bowen) without resorting to the very lengthy procedure of collecting data from a transmission specimen and then matching the reflection and transmission data.

Although all the pole figures showed four-fold symmetry, the very small variations in intensity at the equivalent positions in each quadrant of the pole figure need to be allowed for. This was carried out by averaging the intensities at equivalent positions in each quadrant. This has the effect of symmetrising each pole figure.

Repeatability Criteria

Repeatability checks were performed on standard samples S1 and S2. These were done first at the start of measurements of a batch of 10 samples and then at the end of the measurements. If the *f*-factors determined for the standard samples were within an acceptable range, the measured data was accepted. If not, the data was rejected and the whole alignment and defocus checks repeated and the goniometer adjusted until data was in the acceptable limits. A typical scatter band of *f*-factors for the two standard samples are given in Table 2.

Table 2: Scatter of *f*-factor for acceptable criteria for repeatability tests.

Sample S1	*Normal*	*Longitudinal*	*Transverse*
Lower limit	0.417	0.155	0.378
Upper limit	0.449	0.187	0.414
mean	0.433	0.171	0.396
Sample S2	*Normal*	*Longitudinal*	*Transverse*
Lower limit	0.599	0.104	0.231
Upper limit	0.657	0.158	0.251
mean	0.628	0.131	0.241

Results and Discussion

Data was routinely collected using the above procedure until a time when it became impossible to meet either the intensity criteria from the single crystal or the defocus check criteria. Renewal of the X-ray tube, replacement of the eulerian cradle, or re-alignment of the graphite monochromator did not result in satisfying either the intensity or defocus criteria. These problems started when the silicon single crystal fell out of its holder in which it was permanently mounted using palsticine. The plasticine had hardened and become brittle over time and cracked and allowed the sample to come off the holder. The sample was remounted using fresh plasticine and it was noticed that after alignment with the silicon single crystal there was a shift in Zircaloy lines. However, experience has shown that the alignment of diffractometer is robust and does not deteriorate over hours. Questions regarding the use of a monochromator were raised, as was the wisdom for the use of a single crystals for alignment. Monochromators are essentially designed to operate at zero or near zero tilt. Single crystal produce intensely sharp peaks and the slightest change in height necessitates re-alignment. It was then decided to remove the monochromator and use poly cystalline silicon and check 2θ for (111), (311) and (331) lines. This immediately resulted in meeting both the intensity and defocus criteria. However, a considerable scatter in the defocus curve was observed as shown in Fig. 3.

It is interesting to note that the intensity of this defocus curve is much higher than that in Fig. 2, produced using the monochromator. This curve also shows a spurious intensity at 0° tilt. The scatter in the defocus curve was thought to occur from the slight height changes that may occur during data acquisition. The samples are mounted using plasticine, which is pliable at ambient temperature. However, the sample holder in the texture gonimeter sits on top of a motor that oscillates and tilts the sample. The motor over a time generates enough heat to warm the sample holder such that it is conceivable that the plasticine at higher than ambient temperature flows under the weight of the Zircaloy sample especially when simultaneously tilted and oscillated, resulting in sample height change. Therefore, a new holder was designed obviating the need for plasticine which ensured no change in sample height during data acquisition. The resulting defocus curve from this new holder is shown in Fig. 4 and shows considerable reduction in scatter.

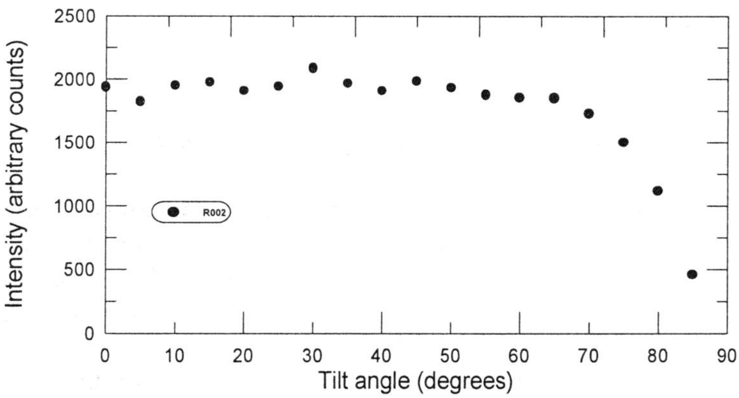

Fig. 3: Defocus curve without monochromator, but sample mounted on plasticine.

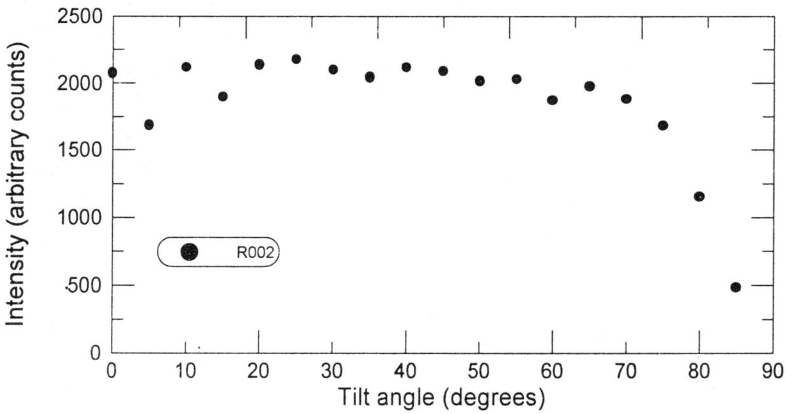

Fig. 4: Defocus curve with new holder.

The removal of the monochromator had little effect on the appearance of the pole figures of the two standard samples as shown in Figs. 5 and 6; however, there was a considerable change in the determined *f*-factors as illustrated in Tables 3 and 4. It is also clear from these tables that *f*-factors without the monochromator show an improvement in the sum *f*-1 parameter, i.e. {*f*(002) + *f*(100) + *f*(110)} −1 and more importantly there is a significantly better agreement between the *f*(100) and *f*(110) parameters for both the standard samples. Following these results some tests were made to investigate the effect of sample height on the intensity of defocus curve intensity ratios and also on *f*-factors. These results are shown in Figs. 7, 8 and 9, respectively.

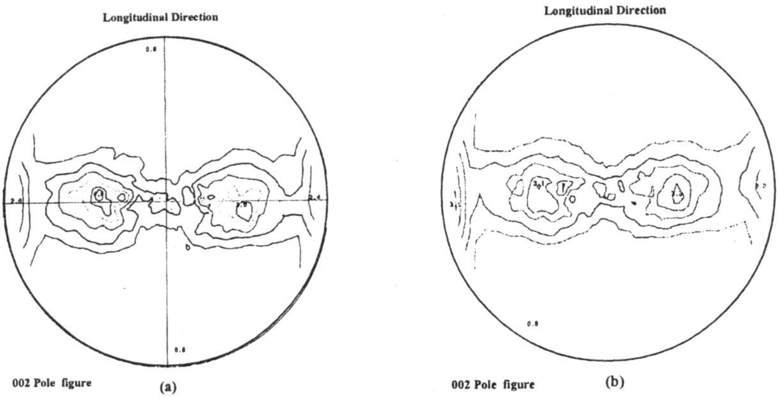

Fig. 5: (0001) Pole figures from standard sample S1.
(a) with monochromator and (b) without monochromator.

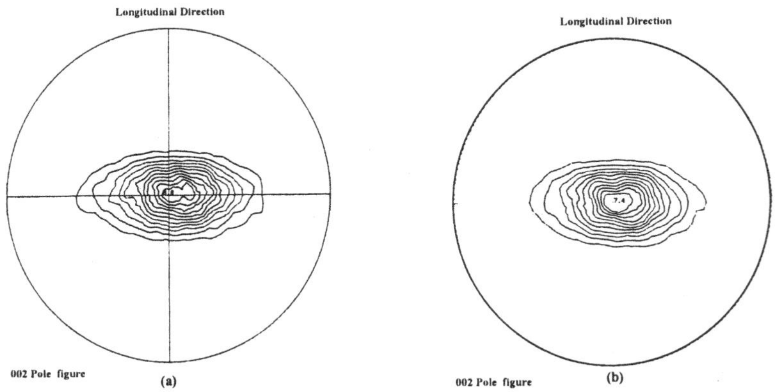

Fig. 6: (0001) pole figures from standard sample S2.
(a) with monochromator and (b) without monochromator.

A significant effect is observed in the intensity ratio for the defocus curves over a change in height of about 100 μm. Similarly, significant changes are observed in the *f*-factors for both the standard samples. On the y axes of the plots in Figs. 7, 8 and 9 are also indicated the scatter bands of *f*-factors for the acceptance criteria given in Table 2. It is clear from these

Table 3: *f*-factors for sample S1.
(a) with no monochromator, and (b) with monochromator.

(a) Samples S1	*f002*	*f100*	*f110*	*f100–f110*	*sum f–1*
N	0.411	0.257	0.259	–0.002	–0.073
L	0.169	0.444	0.447	–0.003	0.060
T	0.420	0.298	0.295	0.003	0.013
(b) Sample S1	*f002*	*f100*	*f110*	*f100-f110*	*sum f-1*
N	0.428	0.252	0.292	–0.040	–0.028
L	0.166	0.408	0.387	0.021	–0.039
T	0.406	0.341	0.321	0.020	0.068

Table 4: *f*-factors for sample S2.
(a) with no monochromator, and (b) with monochromator.

(a) Sample S2	*f002*	*f100*	*f110*	*f100–f110*	*sum f–1*
N	0.654	0.156	0.151	0.005	–0.039
L	0.100	0.453	0.461	–0.008	0.014
T	0.246	0.391	0.388	0.003	0.025
(b) Sample S2	*f002*	*f100*	*f110*	*f100–f110*	*sum f–1*
N	0.637	0.168	0.237	–0.069	0.042
L	0.122	0.445	0.403	0.042	–0.030
T	0.241	0.387	0.361	0.026	–0.011

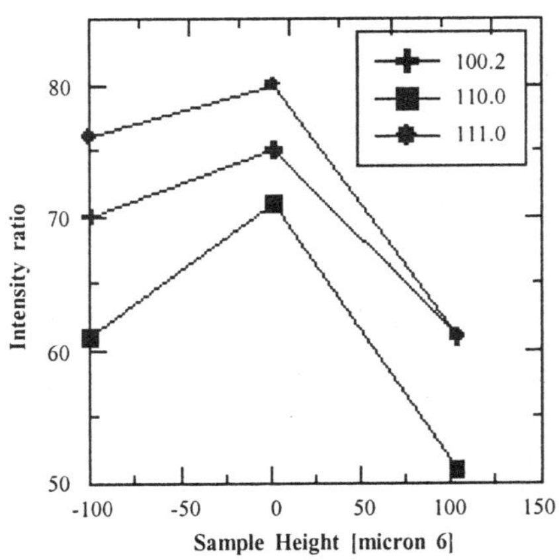

Fig. 7: Plot of intensity ratio versus sample height.

data that a significant amount of the scatter may be due to sample height changes. This could be due to a combination of either mounting practice or in-situ changes from the softening of the plasticine. From these data it can be inferred that sample height can have a significant affect on *f*-factors and care must be taken to keep tight control of specimen height during data acquisition.

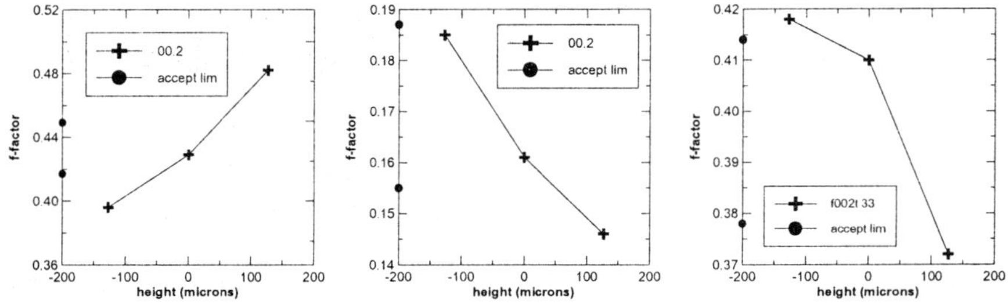

(a) (b) (c)

Fig. 8: Plots of (a) normal, (b) longitudinal and (c) transverse *f*-factors versus sample height for standard sample S1.

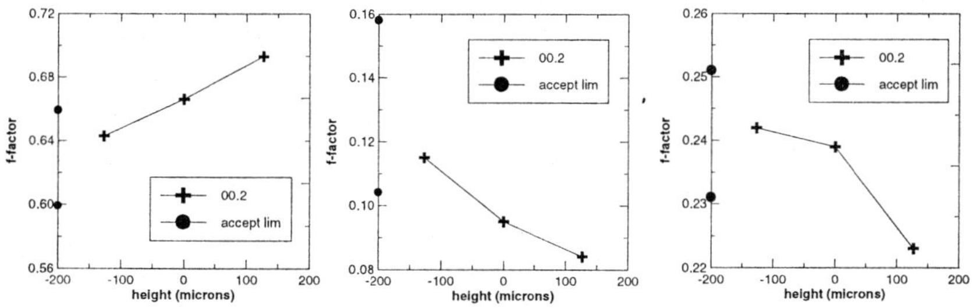

(a) (b) (c)

Fig. 9: Plots of (a) normal, (b) longitudinal and (c) transverse *f*-factors versus sample height for standard sample S2.

The effect of sample preparation on *f*-factors was also investigated at the same time as the above experiments. Pole figures from sample R, that was first mechanically polished and then electro-polished, are shown in Figs. 10 a and b, respectively. There is significant difference in these pole figures and this is clearly reflected in the *f*-factors as illustrated in Table 5. It is clear from these results that mechanical polishing leaves a considerable

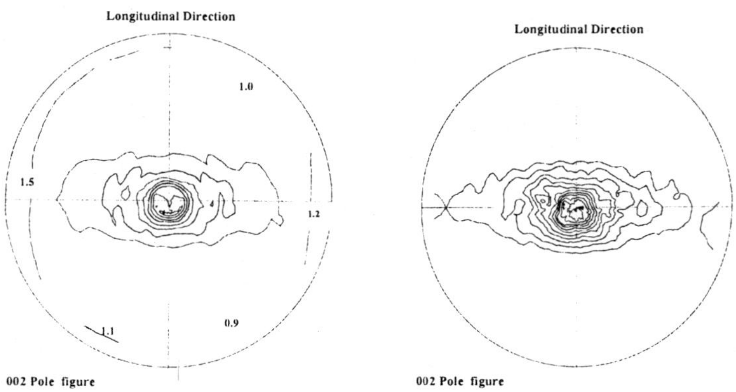

(a) (b)

Fig. 10: (0001) pole figures from (a) mechanically polished sample and (b) from the same sample, but after electro-polishing.

amount of residual surface deformation. As the penetration of CuKα radiation in Zircaloy is not very deep as illustrated in Fig. 11, a significant contribution to the measured pole figure comes from residual surface deformation. These results indicate the importance to eliminate surface deformation from sample preparation in order to obtain true textures and *f*-factors.

Table 5: *f*-factors from sample R (a) after mechanical polishing and (b) after electro-polishing.

(a) Sample R	*f002*	*f100*	*f110*	*f100–f110*	*sum f–1*
N	0.467	0.239	0.260	–0.021	–0.034
L	0.219	0.402	0.395	0.007	0.016
T	0.314	0.358	0.345	0.013	0.017
(b) Sample R	*f002*	*f100*	*f110*	*f100–f110*	*sum f–1*
N	0.634	0.170	0.164	0.006	-0.032
L	0.103	0.456	0.464	-0.008	0.023
T	0.264	0.375	0.372	0.003	0.011

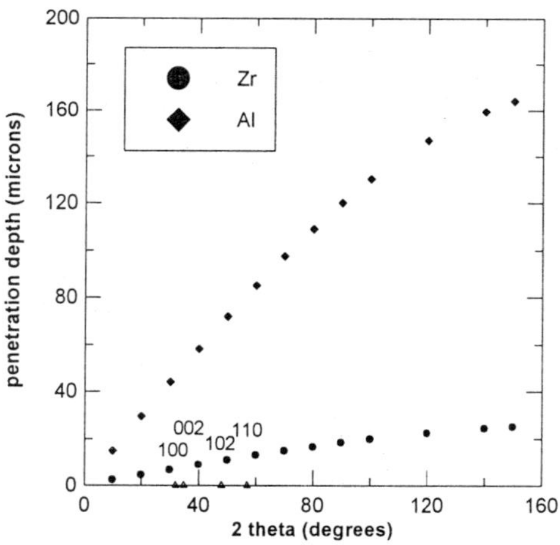

Fig. 11: Plot of penetration depth versus 2θ for aluminium and Zircaloy.

Data from repeatability measurements using the new set up from the two standard samples S1 and S2 are shown in Fig. 12. These data show that a scatter of about ±5% which is well within that reported in the literature (Juul Jensen and Lyttle, 1999).

In summary, to obtain *f*-factor parameters with a high degree of confidence from XRD, it is important to ensure appropriate set up conditions and have control of specimen height. Contributions from residual surface deformation should be minimised by using the correct sample preparation methods. In order to ensure repeatability and to monitor the performance of the diffractometer use of control samples must be incorporated in data acquisition runs.

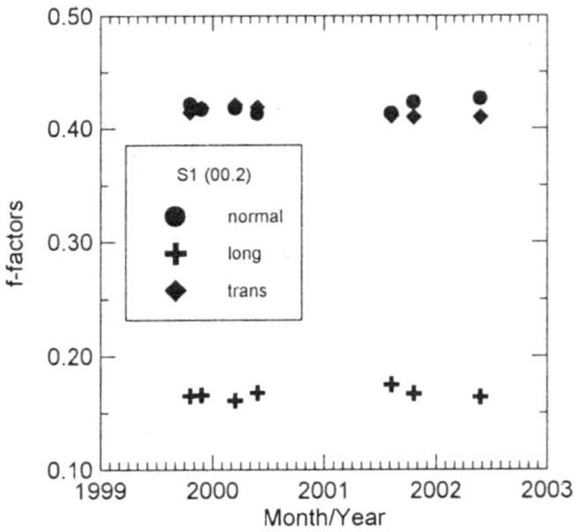

(a)

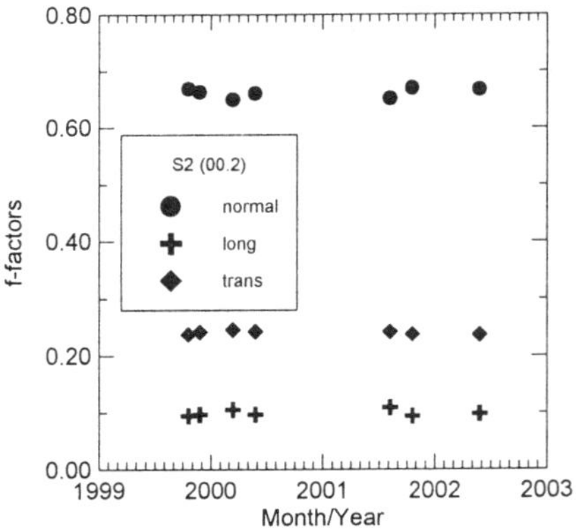

(b)

Fig. 12: (a) and (b) Plots of repeatability measurements for samples S1 and S2 respectively.

EBSD

An EBSD facility was installed at QinetiQ, Farnborough in early 1990. This was a manual system and allowed data acquisition rates at about a pattern per minute. This rate is too slow for producing average texture poles figures comparable with those determined by XRD which comprise data from tens of thousands of grains. However, with the introduction of automatic pattern recognition (Schwarzer, 1997) and improved indexing routines, integration of SEM beam and stage control into EBSD software, orientations from tens of grains per second are now possible. At these speeds it is therefore possible to acquire data that can provide average pole figures like those produced by XRD.

EBSD texture measurements has the advantage over XRD to generate complete pole figures since the technique suffers from no defocusing problems. As the measurements in EBSD are discrete, any desired set of pole figures can be derived from one set of measurements. Measurements do not have to be confined to a particular sample section, as the data can be easily manipulated to give pole figures in the say normal plane, from measurements made in the longitudinal or transverse sections. It is also possible to measure textures from different phases in a given sample. This is a particular advantage when the volume fraction of second phases may be small. Since there is no reliance on pole figure, intensity for quantification and calculation of *f*-factors conform with Equations (4) and (5) given in the introduction.

However, for successful EBSD data acquisition sample surfaces need to be free from residual deformation from polishing routines (Katrakova and Mucklich, 2001). Samples with high intrinsic dislocation densities provide poor quality patterns and may be difficult to index. There are also some instances where EBSD is prone to systematic indexing errors (Schwarzer, 1997).

Experiments

EBSD data for the present work was acquired from electro-polished Zircaloy samples used for X-ray texture measurements, i.e. data was collected from the same areas sampled by X-ray pole figure determinations. The EBSD system comprised a CCD camera, HKL Technology Channel 5 software plus beam and stage control. This is atttached to a CamScan S4 FEGSEM. EBSD patterns were collected using an operating voltage of 20 kV, a specimen current of about 4 mA, specimen tilt of 70° and a working distance of 20 mm. A (001) cut silicon crystal standard was used to initially calibrate the EBSD system. A zirconium match unit having values a = 0.3231 nm and c = 0.5148 nm and 6/mmm Laue group was used to index EBSD patterns.

EBSD texture data were acquired using the stage scan – safe sampling mode. The safe sampling mode employs a specified stage scan step size and at each step three patterns are acquired by moving the beam a specified distance, from its rest position. A distance of 0.3 μm was used in the present case. Data was accepted if at least two patterns of the three acquired were within 2° of each other. This routine has been specially devised to eliminate indexing mistakes that cannot be systematically removed by methods used to remove mis-indexed data from EBSD orientation maps.

EBSD Data Acquisition Durations and Post Processing

EBSD data acquisition durations vary largely according to the scan step size for a given system set up. Small step sizes for large grain materials will not only over sample data, but also considerably increase acquisition times. A step size larger than the grains in the material will lead to loss of data from small grains, while considerably reducing acquisition times. In addition EBSD scans seldom produce 100% indexing, due to poor pattern quality especially at grain boundaries, defects, etc. Therefore, some post processing of data is usually carried out. This combined with over or under sampling almost invariably has an effect on *f*-factors. In order to arrive at optimum levels of acquisition durations and sampling strategy, the effects of different acquisition modes, i.e. beam scan and stage scan plus post processing on *f*-factors are examined below, considering data acquired from the same area in sample B using beam scan and stage scan modes. Details of the scan modes and effects of post processing are given in Tables 6 and 7. The annotations used in these tables and throughout the report are:

1. bs-stc = beam scan – stitched data, i.e. data combined from two or more maps,
2. ss sf = stage scan – safe sampling mode
3. raw = data with no processing

4. spike removal = data with spike removal, spikes are pixels in the same grain that do not share any of the neighbouring orientations and are deemed indexing mistakes
5. extra-pol = extrapolated data, empty pixels are filled with the neighbouring orientation to remove all pixels with no data
6. N, L and T are normal, longitudinal and transverse directions.

Table 6: Details of scans from sample B. bs-stc = stitched beam scan data from four maps and ss2-sf2 and ss1 sf1 are stage scan data

Data	*step size/μm*	*Grid size*	*Maps*	*% index*	*acquistion time/hr*
bs-stc	1	125 × 125	2 × 2	70.0	6:08:24
ss2 sf2	5	100 × 100	1	53.2	4:26:49
ss1 sf1	10	50 × 50	1	46.8	1:07:07

Table 7: Details of effect on *f*-factors after post processing on scans

bs -stitch	Raw			raw + spike removal			raw + spike removal +extra-pol		
	{0001}	{10-10}	{11-20}	{0001}	{10-10}	{11-20}	{0001}	{10-10}	{11-20}
N	0.637	0.181	0.181	0.645	0.178	0.178	0.624	0.188	0.188
L	0.082	0.459	0.459	0.073	0.464	0.464	0.075	0.463	0.463
T	0.281	0.360	0.360	0.283	0.359	0.359	0.301	0.350	0.350

ss sf2	Raw			raw +extra-pol		
N	0.654	0.173	0.173	0.628	0.186	0.186
L	0.066	0.467	0.467	0.072	0.464	0.464
T	0.281	0.360	0.360	0.300	0.350	0.350

ss sf1	Raw			raw +extra-pol		
N	0.646	0.177	0.177	0.623	0.189	0.189
L	0.068	0.466	0.466	0.075	0.462	0.462
T	0.285	0.357	0.357	0.302	0.349	0.349

It is clear from Table 6 that considerable time saving can be achieved by employing stage scan modes. This, however, does result in lowering of indexing rate and the total sampled data. The grain size in the sample was found to be about 18 μm diameter and the step size used in the beam scan ensures most grains in the chosen area have been sampled and a true *f*-factor number should result from this data set. The raw data from the beam scan shown in Table 7 gives an *f*-factor value of *f*(0001) = 0.637 in the N direction. This data contains some mistakes and is 70% indexed. Removal of spikes i.e. indexing mistakes, results in a value of *f*(0001) = 0.645, giving an increase in the *f*(value). Indexing mistakes can result in an increase in *f*-values if the (0001) planes are close to the sample normal (N) or a decrease if the data are close to the samples transverse (T) or longitudinal (L) planes.

Extrapolation of the data to 100% has the result of producing an *f*-factor value of *f*(0001) = 0.624. This indicates that the majority of the extrapolated orientations lie away from the N plane and towards the T plane. The raw *f*-factor values for both the stage scans, also given in Table 7, are higher than those for the beam scans. No spike removal is applied to this data as the safe sampling mode ensures that there are no indexing mistakes and hence also gives a lower % indexed number. Extrapolation of both the stage scan data has had the effect to produce *f*-factor values to almost match those in the final beam scan processed data. The above data, therefore, shows that both beam scans and stage scans with safe sampling, when used to sample the same set of grains yield consistent *f*-factors after post processing. Similar consistency was observed in data acquired from five separate areas from sample B illustrated schematically in Fig. 13. This data is shown in Table 8 together with the safe sampled data ss sf1 and ss ss2 given in Table 7. From these sets of observations it was decided to acquire EBSD data using stage scanning in the safe sampling mode. A step size of 20 μm was used to cover a total of 4 mm^2 i.e a grid size of 50 × 50 for scans 1-4 and 100 × 100 for scan 5, to cover an area of 1 mm^2 and 2 mm^2, respectively of the sample from five separate locations in the area sampled by X-rays as illustrated in Fig. 13.

Table 8: Details of effect on *f*-factors after post processing on data from sample B

ss sf2	Raw (5 μm step)			raw + extra-pol		
N	0.654	0.173	0.173	0.628	0.186	0.186
L	0.066	0.467	0.467	0.072	0.464	0.464
T	0.281	0.360	0.360	0.300	0.350	0.350

ss sf1	Raw (10 μm step)			raw + extra-pol		
N	0.646	0.177	0.177	0.623	0.189	0.189
L	0.068	0.466	0.466	0.075	0.462	0.462
T	0.285	0.357	0.357	0.302	0.349	0.349

ss-sf				raw +extra-pol		
	5 maps 1 2 × 2mm			0.626	0.187	0.187
	4 1 × 1mm			0.071	0.465	0.465
	20 μm step			0.303	0.349	0.349

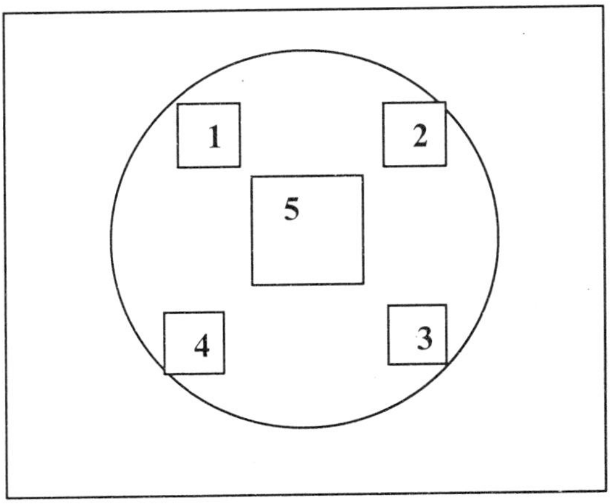

Fig. 13: Schematic figure showing the location of the five areas from where EBSD texture data was acquired area 5 = 2 mm^2, areas 1– 4=1 mm^2, area 1-5 are set in a circle of 10 mm diameter, approximately the area covered by X-rays (*not to scale*).

Results and Discussion

Results of EBSD pole figure measurements from four samples A, B, C and D given in Table 9 are shown in Figs. 14a, 15a, 16a and 17a, respectively. Sets of (0001), (10-10) and (11-20) pole figures are shown for sample B, while only the (0001) pole figures are shown for samples A, C and D.

Table 9: X-ray *f*-factors and grain sizes of samples A, B, C and D used for EBSD measurements

Sample	*X-Ray f002*	*Grain size μm*
A	0.494	40
B	0.580	18
C	0.587	11
D	0.666	33

Equivalent sets of pole figures from X-ray measurements are also shown in Figs. 14b, 15b, 16b and 17b, respectively. It is clear from these pole figures that the respective techniques produce similar texture details. Moreover, plots of EBSD and X-ray pole figure intensities for a set of ten samples (including six additonal samples not indicated in Table 5) shown in Fig. 18, indicate that intensities from the two types of pole figures are linearly correlated.

A comparison of the EBSD and X-ray *f*-factors for samples A, B, C and D are given in Table 10. The *f*-factor data for prism or secondary prism pole figures (10-10) and (11-20) show far better correlation than for (0001) *f*-factors, except for sample A which is in complete agreement. The better

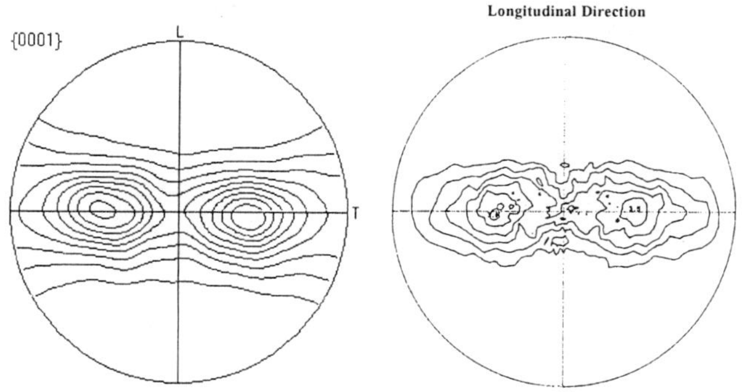

(a) **(b)**

Fig. 14: (a) EBSD and (b) X-ray (0001) pole figures from sample A, respectively.

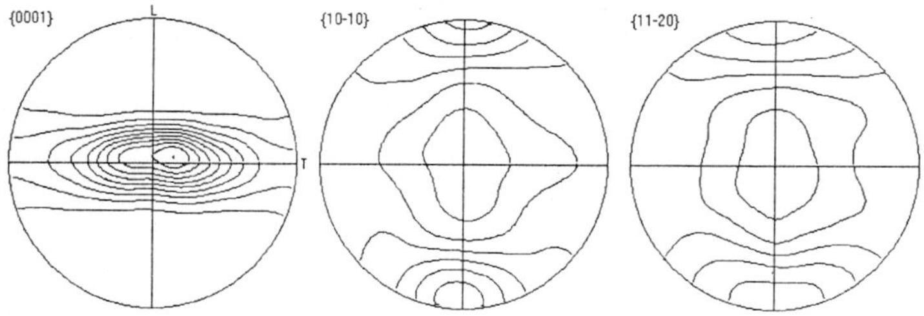

(a)

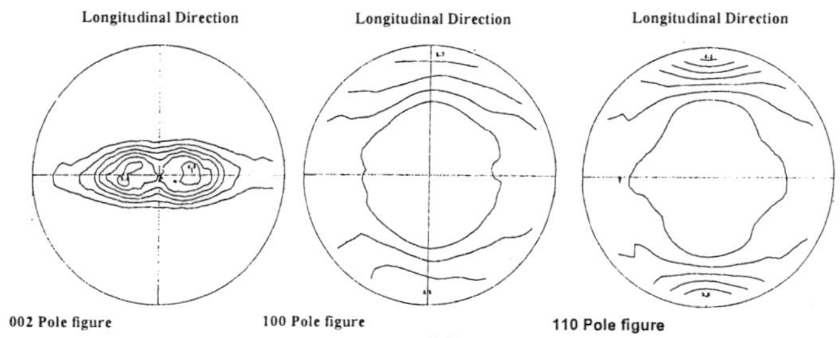

(b)

Fig. 15: (a) EBSD and (b) X-ray (0001), (10-10) and (11-20) pole figures for sample B, respectively.

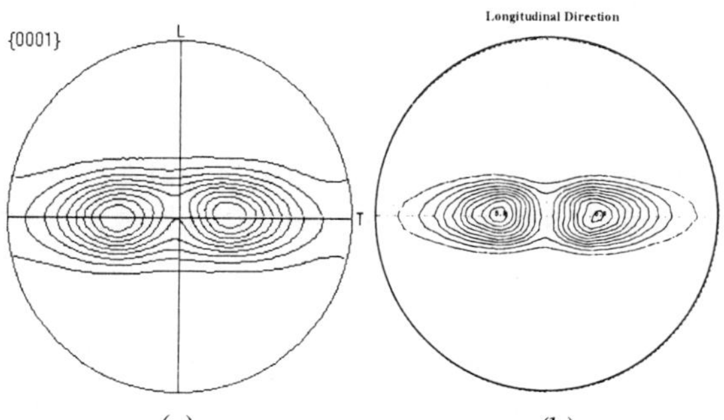

(a) (b)

Fig. 16 : (a) EBSD and (b) X-ray (0001) pole figures from sample C, respectively.

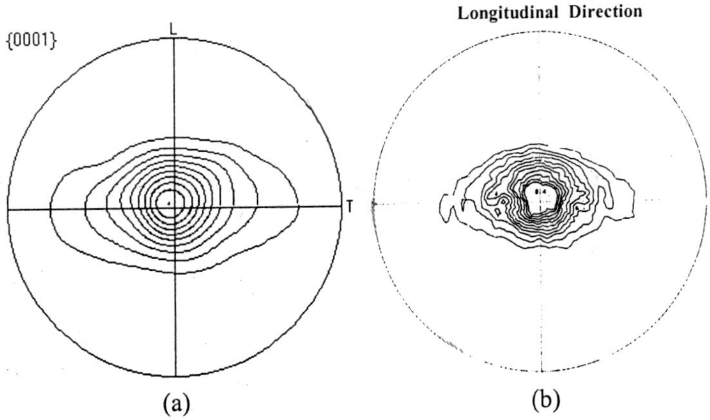

Fig. 17: (a) EBSD and (b) X-ray (0001) pole figures for sample D, respectively.

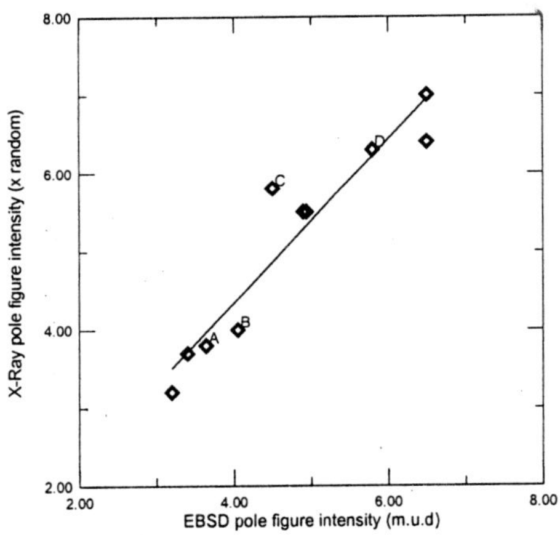

Fig. 18: Plot of X-ray versus EBSD pole figure intensities.

agreement between the prism or secondary *f*-factors may be due to the fact that X-ray *f*-factors are calculated from pole figure intensities, where the N, L and T factors are normalised, but the sum of basal and prism *f*-factors is not always unity. It is therefore possible that a systematic error biases the basal *f*-factors and the secondary *f*-factors are more reliable and compare favourably with EBSD prism *f*-factors.

Plots for EBSD versus X-ray *f*-factors for normal, transverse and longitudinal directions are shown in Figs. 19a, b and c, respectively. These plots again include the data from the four samples given in Table 5 plus data sets from six other samples. Table 11 shows a summary of the slopes, constants and correlation coefficients for the linear fits for these plots. This table confirms that the best correlation is between the secondary prism factors in the normal and transverse direction. The worst correlation is found for the L *f*-factors, possibly due to these values being scattered closely about similar values, Fig. 19c, for all samples measured.

Table 10: *f*-factors from samples A, B, C and D from EBSD and X-ray measurements

Sample	*EBSD*	*EBSD*	*X-ray*	*X-ray*	*X-ray*
fN	{0001}	{10-10}	{0001}	{10-10}	{11-20}
A	0.494	0.253	0.494	0.251	0.246
B	0.626	0.187	0.580	0.184	0.181
C	0.591	0.204	0.580	0.204	0.196
D	0.708	0.146	0.666	0.165	0.153
fL	{0001}	{10-10}	{0001}	{10-10}	{11-20}
A	0.077	0.461	0.108	0.461	0.467
B	0.071	0.465	0.102	0.456	0.467
C	0.051	0.474	0.067	0.415	0.446
D	0.089	0.455	0.108	0.436	0.454
fT	{0001}	{10-10}	{0001}	{10-10}	{11-20}
A	0.429	0.286	0.399	0.288	0.288
B	0.303	0.349	0.318	0.360	0.352
C	0.357	0.321	0.353	0.381	0.358
D	0.203	0.398	0.226	0.399	0.392

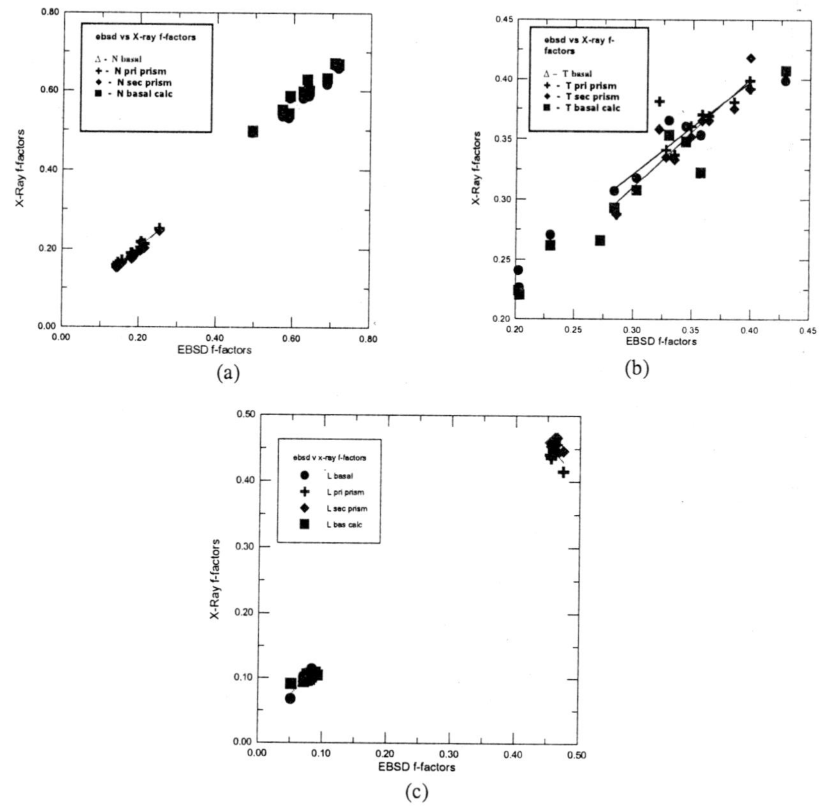

Fig. 19: (a), (b) and (c). Plots of EBSD versus X-ray *f*-factors for normal, transverse and longitudinal directions.

Table 11: Summary of slopes, constants and correlation coefficients for linear fits for plots in Fig. 19.

N	*slope B*	*constant A*	*corr coef*
basal'0001	0.779	0.100	0.888
pri prism'101-0	0.810	0.043	0.968
sec prism'11-20	0.836	0.030	0.950
L	*slope B*	*constant A*	*corr coef*
basal'0001	0.987	0.025	0.751
pri prism'101-0	–1.040	0.923	0.225
sec prism'11-20	–0.425	0.654	0.130
T	*slope B*	*constant A*	*corr coef*
basal'0001	0.776	0.081	0.929
pri prism'101-0	0.772	0.090	0.731
sec prism'11-20	0.908	0.038	0.864

Linear fit Y=B*X+A X=EBSD *f*-factor Y=X-Ray *f*-factor

CONCLUDING REMARKS

X-ray Diffraction

Experience at QinetiQ on collecting X-ray texture determinations from Zircaloy shows that reliable measurements of *f*-factors depends on ensuring appropriate texture goniometer set up conditions, strict control of sample height and specimens free of residual deformation from polishing routines. It is strongly recommended to incorporate a long-term control sample to monitor repeatability of the texture goniometer over time. This can show up any deviations arising from normal set up conditions and can potentially avoid expensive repeat tests.

EBSD Texture Measurements

The EBSD technique is becoming indispensable for materials characterisation, and the data from the present study shows that EBSD can provide texture information as good as that obtained from XRD, in terms of pole figures and quantification e.g. Kearns *f*-factors.

Good sample preparation practice is important to ensure reliable data, in addition to taking care in selection of data sampling strategy, e.g. total area measured, use of beam or stage scan, step sizes, etc.

Acknowledgements: The author would like to thank Dr. E.C. Darby and Mr. R Gifford (Rolls-Royce plc) for permission to release texture data for this report.

REFERENCES

Baba-Kishi, K. Z. (2002). *Journal of Materials Science*, **37:** 1715-1746.

Bowen, A. W. unpublished work.

Bowen, A.W. (1993). Titanium '92: Science and Technology. Froes, F.H. and Caplan, I. (Eds). The Minerals, Metals and Materials Society, Warrendale, PA.

Juul Jensen, D. and Lyttle, M.T. (1999). "Reproducibility of Experimental Texture Measurement", Proceedings of the Twelfth International Conference on Textures of Materials, ed. Spunzar, J.A., **1:** 46-51, NRC Research Press, Ottawa.

Katrakova, D. and Mucklich, F. (2001). *Prakt. Metallogr.* **38:** 10.

Kearns, J.J. (1965). Westinghouse Elec Corp Report WAPD-TM-472.

Kearns, J.J. (2001). *Journal of Nuclear Materials*, **2991:** 71-174.

Kelly P.M. and Watson, K.G. (1972). *Journal of Nuclear Materials*, **44**: 71.

Mehnert, K. unpublished work.

Mehnert, K, Ubhi, H.S. and Day, A.P. (1999). "Comparison of Texture Data Measured by EBSD and Conventional X-ray Diffraction", Proceedings of the Twelfth International Conference on Textures of Materials, **1:** 217-222, J.A. Spunzar (ed) NRC Research Press, Ottawa.

Schwarzer, R. (1997). *Micron*, **28:** 249-265.

Texture Analysis Using General Area Diffraction Detector System

Srinivas Sathiraju and Paul N. Barnes

Air Force Research Laboratory
Wright-Patterson Air Force Base, OH 45433-7919, USA

INTRODUCTION

Ever since the discovery of high temperature superconductor (HTS) materials, significant effort has occurred to develop the material into HTS wires for employment in more efficient and powerful energy applications (Watanabe et al., 2003; Rosner, 2001; Hassenzahl, 2001). Although difficult but not insurmountable hurdles remain, today we are close to seeing practical electric power devices and energy systems based on HTS technology. The coming years may witness the implementation of a variety of superconducting manufacturing techniques and the introduction of additional products incorporating HTS technology; indeed products are currently being offered (American Superconductor Corporation, USA; Sumitomo Electric Industries, Japan and THEVA, Germany).

The primary commercial demand in the future for HTS wire will be electric power utility equipment. Complete development of the technology will lead to industrial commercialization of the HTS conductor in electric power applications such as transformers, transmission cables, motors, fault current limiters, synchronous condensers and generators. What drives the introduction of these new HTS devices is the significant reduction in size and weight with the associated increase in power density along with a large reduction in energy usage due to low electrical losses. An added benefit can be the noise reduction in HTS machinery and transformers. Specific advantages can include for the different power systems: HTS cable with 3-5 times more power capacity than the conventional counterpart, elimination of fire and environmental hazards resulting from the use of cooling oil in conventional transformers, and a 4-6 fold reduction in fault current levels with HTS fault current limiters with no capacitive correction. For high horsepower HTS motors a reduction in production cost is also possible.

Even future weapon systems in the military will depend heavily on high electrical power input at the multi-megawatt level (Tolliver et al., 2003; Barnes et al., 2004). Examples of these weapon systems include electrically driven directed energy weapons (DEW), such as high power microwaves and solid-state lasers. Not only do these innovative weapon systems require significant amounts of reliable electrical power, they often must be packaged in a limited space and within strict weight limits on either ground-mobile or airborne platforms. As such, space and weight are at a premium. Currently no multi-megawatt class power generation system is flight worthy. A new class of generators made of high temperature superconductors (HTS) will enable megawatt-class power systems to take flight. These airborne applications may come true if we can make long length HTS conductors.

Even though HTS superconductors are generally oxide ceramics and as such brittle, recent progress by Japan, Europe and the United States have made conductor availability quite feasible. Currently available HTS conductor is made of bismuth strontium calcium copper oxide (BSCCO), the first generation HTS conductor. Yttrium barium copper oxide (YBCO) is accepted worldwide as the future second generation HTS conductor. An advantage of YBCO superconductors is its use at liquid nitrogen temperatures (77 K) for reduced cooling requirements compared to 20-30 K for BSCCO. For applications subject to magnetic fields, such as gėnerators, YBCO can maintain current densities $>10^5$ A/cm^2 in Tesla level magnetic fields whereas BSCCO cannot.

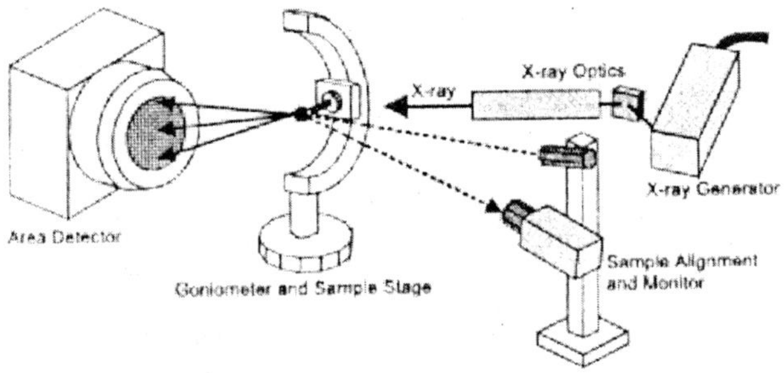

Fig. 1: Typical GADDS X-ray system block diagram.

Advances in the YBCO coated conductors have resulted in dramatic improvements in the quality and length of wire available—from 10^5 A/cm^2 at cm lengths about three years ago to lengths of several metres with 10^6 A/cm^2 performances (Selvamanickam et al., 2003; Verebelyi et al., 2003; Kakimoto et al., 2003; Usoskin et al., 2003). These successes have been achieved by improved deposition methods for the HTS layer, a better buffer layer architecture, and fundamental control of the biaxial alignment of the initial buffer layer or underlying metallic substrate. Even so, development must continue on the basic YBCO conductor to extend the lengths in excess of 100 metre lengths and ultimately to kilometre lengths. Now that reel-to-reel processing has been established, *in situ* or *ex situ* characterization tools are paramount to establish process control and better understand the structure-property relationship of these HTS materials on metallic tapes. Today industrial researchers and national labs all over the world are using general area diffraction detector system (GADDS) attached with X-ray equipment which is a powerful tool to understand the texture analysis of these coated conductor tapes. The GADDS X-ray system offers a quick ex situ measurement of these HTS tapes (GADDS XRD manual and elements of X-ray diffraction by Cullity (1978)). In this paper, we will discuss the features of the GADDS system and present some results obtained on coated conductors using this equipment.

GENERAL AREA DIFFRACTION DETECTOR SYSTEM (GADDS)

Figure 2 shows the typical block diagram of the GADDS system configuration, which normally has five major units:

1. The X-ray source,
2. Optics for conditioning of the primary X-ray beam,
3. The goniometry and sample stage to establish and maneuver the geometric relationship between primary beam, sample, and detector,
4. A sample alignment and monitor to assist users in positioning the sample into the instrument centre and in monitoring the sample state and position, and
5. A Hi Star area detector system to intercept and record the scattering X-rays from a sample and to save and display the diffraction pattern in to a two-dimensional image frame.

Fig. 2: General Area Detector Diffraction System.

In addition to the basic configuration components, accessories can be used such as a low temperature stage, a high temperature stage, a helium (or vacuum) beam path for small-angle X-ray scattering (SAXS), a beam stop, and alignment and calibration fixtures. Figure 1 shows the picture of Bruker

GADDS system. A computer that uses software specific to the GADDS controls the whole system. This system is designed for speed, precision, flexibility, versatility and reliability as part of a new generation of GADDS products. Area detector can use a variety of X-ray sources from a sealed tube generator to a rotating anode generator (RAG) to a synchrotron anode generator (SAG) to synchrotron radiation (with CCD detector). The sealed tube generator is the most commonly used X-ray source in the GADDS system. The incorporated targets consist of commonly used target materials that are generally used for ordinary point detector XRD systems such as Ag, Mo, Cu, Co, Fe and Cr. Table 1 lists the typical applications for the given target materials.

Table 1: Applications of target materials

Target	*Typical applications*
Ag, Mo	Low absorption: single crystal, transmission diffraction (with CCD Detector)
Cu	Most powder diffraction, stress, texture, thin films, polymer, SAXS single crystal
Co, Fe	Used for ferrous alloys (steels) to reduce Fe fluorescence, ideal for residual stress
Cr	Ideal for materials with large unit cell, ideal for residual stress with high resolution

The focal spot of the system is typically rectangular with a length to width ratio of 10 to 1. The projection along the length of the focal spot at a takeoff angle from the anode surface is called the spot focus, the square focus, or the point focus. The projection of the focal spot perpendicular to its length is called line focus. Thus, line focus and spot focus are separated by an angle of 90° around the tube cylinder. The line focus is commonly used for the conventional diffractometer with a point detector or PSD. A standard GADDS system uses the spot focus. The take off angle can be set from 3° to 7° (6° for most systems). Table 2 lists focal spot size, line focus size and spot focus size at a 6° take off angle for typical X-ray tubes used with GADDS systems. The micro focus sealed tubes have the brightest focal spot of all sealed tubes. Rotating anode generators have a very high brilliance compared with sealed tubes. The intensity over the focal spot is not evenly distributed, however.

Table 2: Some typical X-ray tubes

Tube type	*Focal spot size at anode (mm^2)*	*Line focus size (mm^2)*	*Spot focus size (mm^2)*
Normal Focus	1 × 10	0.1 × 10	1 × 1
Fine focus	0.4 × 8	0.04 × 8	0.4 × 0.8
Long fine focus	0.4 × 12	0.04 × 12	0.4 × 1.2
Micro focus	0.15 × 8	0.015 × 8	0.15 × 0.8

The focal spot profile is the intensity distribution over the area of the beam cross-section and is eventually translated to the beam profile. The beam profile is sometimes very important to the diffraction result. The focal spot profile across the beam from the fine focus and the long fine focus sealed tubes are typically saddled in the centre with the maximum near the edge. The intensity at the centre can be as low as 50% of the maximum. The focal spot profile for RAG is normally more evenly distributed, like a flat topped Gaussian distribution. The focal profile from a fine focus or long fine focus sealed tube can satisfy most GADDS applications. The micro focus sealed tube and RAG may be necessary for some applications.

GAADS is a very powerful tool to analyze the chemical composition in powder samples. Because of its capability to collect the diffracted intensity from a large angular range, its GADDS detector has strong advantages compared to a conventional point detector system. The first advantage is its speed. The large area of the GADDS detector allows for a large 2θ range to be analyzed without any movement of sample and detector. This results in a huge speed advantage over conventional systems. Figure 3 shows the comparison between a conventional point detector and area detector.

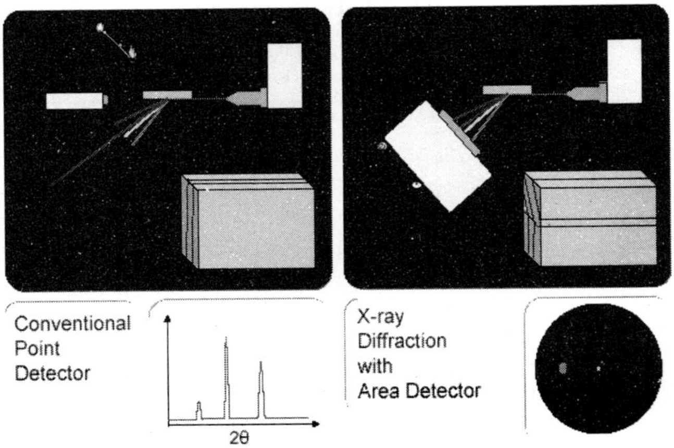

Fig. 3: Comparison of point detector and area detector.

Phase identification can be done by integration over a selected range of 2θ and χ angles. With a direct link to the International Centre for Diffraction Data (ICDD) database, profile fitting with conventional peak shapes and fundamental parameters, quantification of phases, and lattice parameter indexing and refinement make powder diffraction analysis easy and fast. Due to the integration along the Debye rings the integrated data gives better intensity and statistics for phase ID and quantitative analysis, especially for those samples with texture, large grain size, or small quantity. Figure 4 shows the typical example of the phase identification.

Texture measurement using an area detector is extremely fast compared to measurement using a scintillation counter or a linear position detector

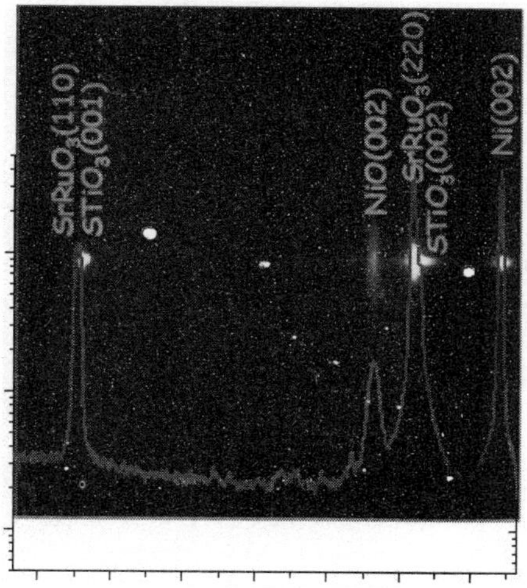

Fig. 4: Typical example of Phase ID from GAADS X-ray diffraction.

PSD. The area detector collects texture data and background values simultaneously for multiple poles and multiple directions. Due to the high measurement speed, GADDS can measure pole figures at very fine steps allowing detection of very sharp textures. GADDS is the best tool for quantitative texture analysis. Classic examples of materials that have been examined by texture analysis are geologic samples, rolled metal sheets and polymer fibres. New materials that are examined include thin film layers on silicon and high temperature superconducting thin films on metal substrates, bi-axially textured metallic substrates, and ion beam assisted deposition of thin films of various materials for texture analysis. Crystallographic texture is an important property of materials, which is defined by specific properties such as position, crystallite size, grain boundaries, shape and orientation of the individual crystallites. The meaning of orientation becomes obvious when looking at macroscopic properties that are anisotropic for single crystals. As more materials are formulated at a molecular level, texture must be specified and controlled to ensure proper product performance. Texture analysis is the key to understand material properties like: mechanical strength and elasticity, electrical resistance and capacitance, thermal conductivity, magnetic and optical properties. Figure 5 shows the pole figure analysis of a solution based deposited buffer layer.

Stress measurements using the two-dimensional area detector are based on a new 2-D stress algorithm, which gives a direct relationship between the stress tensor and the diffraction cone distortion. Since the whole or a part of the Debye ring is used for the stress calculation, GADDS can measure stress with high sensitivity, high speed, and high accuracy. It is especially suitable for samples with large crystals and textures.

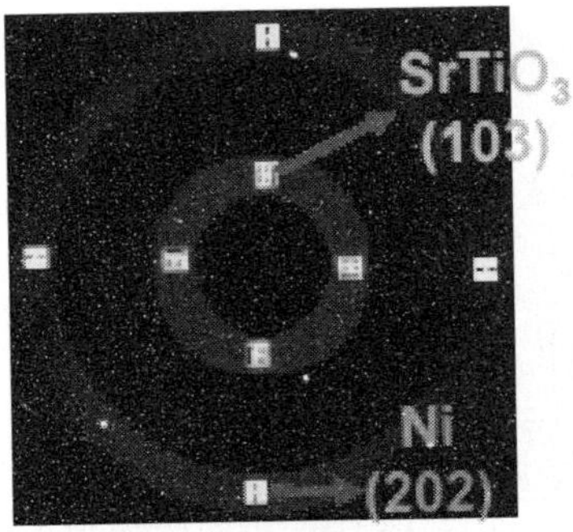

Single domain with (001) normal·

Fig. 5: Texture analysis 103 peak of $SrTiO_3$ buffer layer deposited on textured RABITS Ni.

Simultaneous measurement of stress and texture is also possible since 2-D data consists of both the stress and texture information. The percentage of crystallization can be measured faster and more accurate with the data analysis over the 2-D frames, especially for samples with an anisotropic distribution of the crystalline orientation. An amorphous region can be defined with the crystalline region included when the crystalline region and the amorphous region overlap. GADDS can also calculate and display the Compton scattering so the Compton affect can be excluded from the amorphous result. The rolling ball algorithm calculates the percent of crystallization by extracting an amorphous background frame.

CONCLUSIONS

The GADDS X-ray system is a powerful tool for analysis of a variety of specimens. Thin film samples with a mixture of single crystal, random polycrystalline layers, and highly textured layers can be measured with all the features appearing simultaneously in diffraction frames. Stress and texture can be measured with speed or even simultaneously with the new stress and texture approach developed as a pole figure or fibre plot. Any weak and spotty diffraction pattern can generally be compensated by integration over the 2-D diffraction pattern. The GADDS measurement technique is a very versatile tool for analyzing crystalline texture, chemical composition and stress in samples.

Acknowledgements: Sathiraju acknowledges National Research Council, Air Force Office of Scientific Research and Propulsion Research Directorate for the research support. Sathiraju is thankful to researchers from Los Alamos National Laboratory for the GADDS data.

REFERENCES

American Superconductor Corporation, United States.

Barnes, P.N., Rhoads, G.L., Tolliver, J.C., Sumption, M.D. and Schmaeman, K.W. (2004). IEEE Proceedings of the 12th Symp. on Electromagnetic Launch Technology.

Cullity, B.D. (1978). Elements of X-ray diffraction. Addison-Wesley, Massachusetts, USA.

GADDs XRD manual from Bruker X-ray systems Inc.

Hassenzahl, W.V. (2001). IEEE *Trans. on Appl. Supercond.*, **11:** 1447.

Kakimoto, K., Iijima, Y. and Saitoh, T. (2003). *Physica*, **C392:** 783.

Rosner, C.H. (2001). IEEE *Trans. on Appl. Supercond.*, **11:** 39.

Selvamanickam, V., Lee, H.G., Li, Y., Xiong, X., Qiao, Y., Reeves, J., Xie, Y., Knoll, A. and Lenseth, K. (2003). *Physica*, **C 859:** 392.

Sumitomo Electric Industries, Japan.

THEVA, Germany, htttp://www.theva.com

Tolliver, J.C., Rhoads, G.L., Barnes, P.N., Adams, S.F. and Oberly, C.E. (2003). Superconducting Generators: Enabling Airborne Directed Energy Weapons. AIAA Paper 2003-5917, International Energy Conversion Engineering Conference.

Usoskin, A., Frehardt, H.C., Issaev, A., Dzick, J., Knoke, J., Oomen, M.P., Leghissa, M. and Neumueller, H.W. (2003). IEEE *Trans. on Appl. Supercond*, **13:** 2452.

Verebelyi, D.T., Schoop, U., Thieme, C., Li, X., Zhang, W., Kodenkandath, T., Malozemoff, A.P., Nguyen, N., Siegal, E., Buczek, D., Lynch, J., Scudiere, J., Rupich, M., Goyal, A., Specht, E.D., Martin, P. and Paranthaman, M. (2003). *Supercond. Sci. Technol*, **16:** L19.

Watanabe, T., Shiohara, Y. and Izumi, T. (2003). IEEE *Trans. on Appl. Supercond* **13:** 2445.

Determination of Preferred Orientation Factors of Zirconium Alloy Components Using XRD Patterns

D.N. Sah, Sunil Kumar, Suparna Banerjee, S.M. Sathe, K.B. Khan[1], A. K. Tyagi[2] and B. Ravi Kumar[3]

Post Irradiation Examination Division
[1]Radiometallurgy Division
[2]Applied Chemistry Division
BARC, Trombay, Mumbai, India
[3]National Metallurgical Laboratory, Jamshedpur, India

INTRODUCTION

Pressure tube, calandria tube and fuel cladding tube are critical core components in Pressurised Heavy Water Reactors (PHWRs). They are made from zirconium alloys because of their low neutron absorption cross section, good corrosion resistance and mechanical strength. However, α zirconium has hexagonal closed packed (hcp) crystal structure (Fig. 1) and exhibits anisotropy. Zr-alloy components develop preferred orientation (or texture)

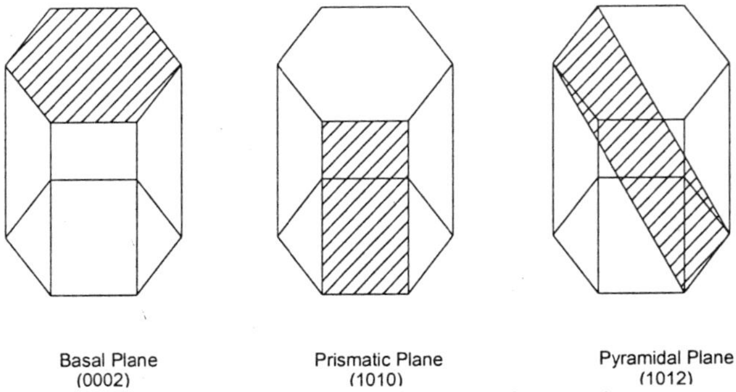

Fig. 1: Crystal structure of alpha Zr and some important crystallographic planes.

during fabrication involving thermo-mechanical treatments. Crystallographic texture has influence on mechanical properties, creep, irradiation growth, orientation of hydride precipitates, delayed hydride cracking (DHC) and stress corrosion cracking (SCC) behaviour of component during service (Lingamurthy, 1991; Simpson, 1999; Hwang, 1991; Savino, 1980). Properties of the components in different directions are directly related to the texture in the components. Characterisation of texture in these components is, therefore, essential. Texture coefficients of relevant planes in various zircaloy components of PHWR were measured and reported earlier (Sah et al., 2002). Preferred Orientation factors have been determined for those samples and some more samples in the present work. Texture parameters can be measured by severai techniques (Lewis et al., 1982; Kearns, 1965; Kearns, 2001). Measurement can be made using conventional pole figure method but it is necessary to obtain a complete pole figure for this purpose. In the present work, 2θ X-ray diffraction (XRD) scans have been used to determine the preferred orientation factors, f_r , f_t and f_n which represent the fraction of grains with their basal poles oriented in rolling direction, transverse direction and normal (radial) direction of the tube, respectively.

EXPERIMENTAL PROCEDURE

Specimen Preparation

Specimens for XRD analysis were taken from three well-defined directions in the tubes e.g., radial (normal) direction (ND), longitudinal or rolling direction (RD) and transverse direction (TD) in such a way that the reflecting surface was normal to the selected direction (Fig. 2). As the wall thickness of the tubes was small (0.4 mm – 4 mm approx.) composite samples were made in case of RD and TD to provide a surface area of more than 1 cm × 1 cm for exposure to X-ray beam. The sample surface was ground, polished and chemically etched to make it flat and to remove the deformed layer. The chemical etching of the sample surface was carried out using a solution containing 10 vol.% HF, 45 vol.%HNO_3 and 40 vol.% H_2O.

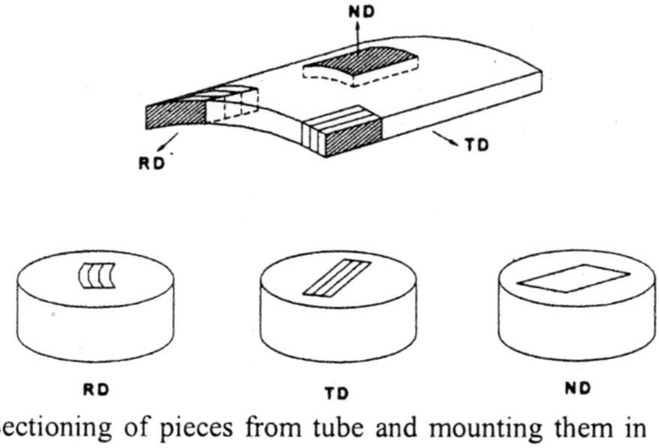

Fig. 2: Sectioning of pieces from tube and mounting them in araldite for XRD study. Shaded area is taken as the reflecting surface.

X-Ray Diffraction

The XRD patterns were recorded on the samples, in the 2θ range of 25° – 125°, on a Philips X-ray diffractometer (Model PW 1710) with monochromatized CuKα radiation. Silicon was used as an external standard. An XRD pattern of zirconium powder sample was also recorded to represent a random sample. XRD patterns on samples of Zr-2.5Nb (wt.%) pressure tube were additionally recorded at two other laboratories: one using ITAL structure XRD system and other using Siefert XRD system. Each laboratory used separate powder samples. CuKα radiation was used at two laboratories and CoKα radiation was used at one laboratory. The intensities were corrected for background. Intensity ratio ($I_{(hkil)}/I_{0(hkil)}$, where $I_{(hkil)}$ is the intensity of (hkil) reflection in the sample and $I_{0(hkil)}$ is the intensity of the same reflection in the powder sample) was determined for each reflection in each sample from the XRD pattern. This ratio, which represents the intensity of reflection in the unit of times random, was used for calculation of orientation parameters.

Calculation of Orientation Parameters

The orientation factor, which represents the effective fraction of basal poles aligned along a particular reference direction is given by (Kearns, 1965).

$$f = \frac{\int_0^{\pi/2} I_\Phi \sin\Phi \cos^2\Phi \, d\Phi}{\int_0^{\pi/2} I_\Phi \sin\Phi \, d\Phi} \tag{1}$$

where, I_Φ is average intensity (in units of times random) at an angle of tilt Φ from a reference direction. The f can have a value from 0 to 1 indicating perfect alignment of basal poles perpendicular or parallel to the reference direction, respectively. When the basal poles are randomly distributed then the value of f is 1/3 in each of the orthogonal directions of the sample. The sum of f values in three directions equals unity.

Preferred orientation factors, f_r, f_t and f_n , also known as Kearn's factors, were estimated from the intensity data of 2θ XRD scan of RD, TD and ND samples. The orientation factors f_r, f_t and f_n represent the effective fraction of basal poles in rolling direction, transverse direction and normal direction, respectively of the component. A computer programme was developed and used to calculate orientation factor f using equation (1). The average intensity (I_Φ) for the range, Φ = 0°–90°, at intervals of 10°, was obtained from the intensity of reflections of (10i*l*), (11i*l*) and (21i*l*) set of planes in 2θ XRD scan using method described elsewhere (Kearns, 1965). Intensity at the mid points of the interval (5°, 15°, 25°, 35°, 45°, 55°, 65°, 75° and 85°) was used for integration in equation (1).

RESULTS

Zr-2.5Nb Pressure Tube

The XRD patterns on the ND, TD and RD samples of Zr-2.5Nb pressure tube are shown in Fig. 3. The presence of texture or preferred orientation is evident from the comparison of XRD patterns on the three samples. The plot of average intensity (I_{Φ}) vs tilt angle for the ND, RD and TD samples are shown in Fig. 4. XRD patterns and intensity vs tilt angle plots obtained on the samples analyzed at other two laboratories were similar. The orientation factors determined from the XRD data of the three laboratories are given in Table 1. The values determined at the three laboratories are found to be close to each other. The average of the three values is also given in the Table 1. It is observed that about 50% of the grains are oriented with their c-axis (basal pole) aligned with the tangential (circumferential) direction and about 40% of grains are oriented with their basal poles along the radial (normal) direction.

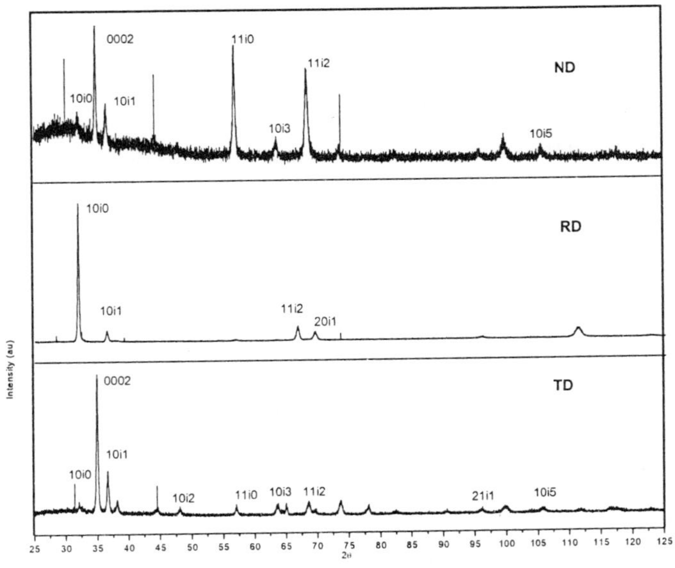

Fig. 3: XRD pattern of Zr-2.5Nb samples.

Zircaloy Cladding Tube

Samples taken from PHWR fuel cladding (0.38 mm thick, 15.3 mm OD) and TAPS (Tarapur Atomic Power Station) fuel cladding tube (0.8 mm thick, 12.5 mm OD) were analysed. The XRD patterns of the three samples of TAPS fuel cladding tube are shown in Fig. 5. It may be seen that (0002) shows a very intense peak in ND sample but it is either absent or insignificant in the other two samples. The XRD pattern of PHWR cladding was almost similar to TAPS fuel cladding.

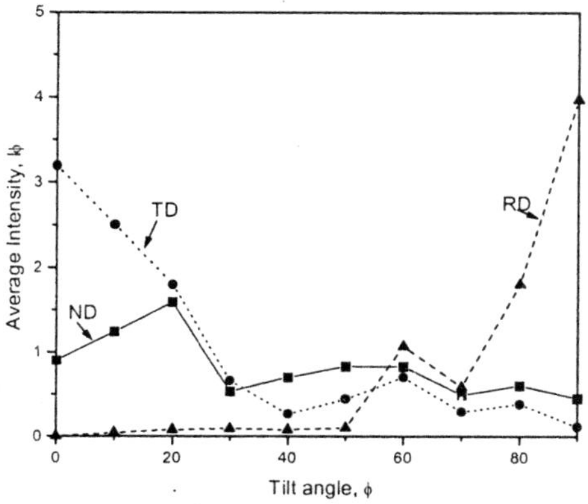

Fig. 4: Plot of average intensity vs tilt angle (ϕ) for specimens of Zr-2.5Nb pressure tube.

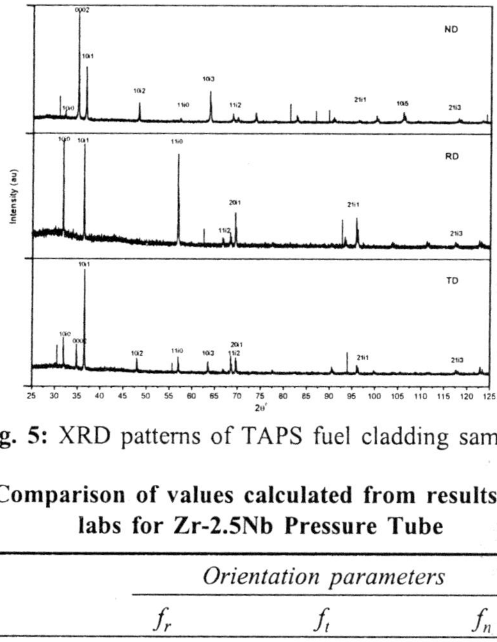

Fig. 5: XRD patterns of TAPS fuel cladding samples.

Table 1: Comparison of values calculated from results of different labs for Zr-2.5Nb Pressure Tube

Source	*Orientation parameters*			Σf
	f_r	f_t	f_n	
APCD, BARC	0.09	0.53	0.40	1.02
PIED, BARC	0.04	0.46	0.42	0.92
NML, Jamshedpur	0.04	0.54	0.42	1.00
Average	0.06	0.51	0.41	0.98

Calculated orientation parameters for TAPS fuel cladding and PHWR fuel cladding are given in Table 2. The PHWR cladding shows a strong radial basal pole texture. 67% of grains in the cladding are oriented with their basal poles in the radial or normal direction of the tube whereas about 29% are oriented with their basal poles in the tangential direction of the tube. In case of TAPS fuel cladding, the percentage of grains with their basal poles in the radial direction is 58.

Table 2: Preferred Orientation factors of Zr alloy component

Component	*Orientation factors*			Σf
	f_r	f_t	f_n	
Zircaloy-2 calandria tube	0.17	0.29	0.57	1.03
PHWR fuel cladding tube	0.11	0.24	0.67	1.02
TAPS fuel cladding tube	0.13	0.29	0.58	1.00

Zircaloy Calandria Tube

The calculated orientation factors of calandria tube are given in Table 2. The orientation parameters for calandaria tube are found to be almost similar to TAPS fuel cladding tube. 57% of grains are oriented with their basal pole in the radial direction whereas 29% are oriented with their basal poles in tangential direction.

DISCUSSION

Measurement of preferred orientation parameters in the sample of Zr-2.5Nb pressure tube was made in three different laboratories using three different XRD systems. The results from these measurements show that the values obtained at the three laboratories are comparable. Results show that the basal pole fraction in tangential direction, f_t is higher than the basal pole fraction in radial (normal) direction, f_n. The average values of f_r, f_t and f_n determined from the results of three laboratories for Zr-2.5Nb pressure tube are found to be 0.06, 0.51 and 0.41, respectively. These values are similar to the values reported for zircaloy-2 pressure tubes (Holt, 1979).

The texture of the finished component depends on the fabrication process and the strain path followed. The cladding tubes and pressure tubes are manufactured by cold pilgering method. A characteristic of the pilgering process is that a reduction in cross section is achieved by reduction in the wall thickness (t) and the diameter (D). During pilgering the tube experiences compressive stress in the thickness direction, which orients the basal poles in the radial direction. The final stable position of the basal poles can be varied according to the Q factor, which is defined as the ratio of thickness reduction to diameter reduction. For high Q values the basal poles get aligned to the radial direction because basal poles tend to align parallel to the compressive force during cold rolling.

The values of f_r, f_t and f_n in PHWR cladding are found to be 0.11, 0.24 and 0.67, respectively showing a strong radial basal pole texture. Among all the components, PHWR cladding has shown highest f_n value. Strong radial basal pole texture is produced due to high Q value used in the final pass of cold pilgering of PHWR cladding tube. Radial basal pole texture is a desirable feature for fuel pin cladding as it provides high ductility and also ensures circumferential orientation of hydride precipitates. High radial basal pole texture ensures predominantly circumferential hydrides in the cladding because hydrides precipitate on a plane, which is almost parallel to the basal plane. The fraction of radial basal pole in TAPS fuel cladding is also high (0.57) but significantly lower than in PHWR cladding.

The f_r, f_t and f_n in calandria tube are found to be 0.17, 0.29 and 0.57, respectively. The seam welded calandria tube, from which samples had been prepared, was made from cold rolled and annealed zircaloy sheets. During sheet rolling process the sheet is compressed in the normal direction, which causes it to lengthen in the rolling direction while showing little change in transverse direction (or the width). Since the basal poles preferentially align parallel to the compressive force during cold working, basal poles tend to align parallel to the normal direction in the sheet.

CONCLUSIONS

1. Preferred orientation factors of Zr-alloy components of PHWR namely Zr-2.5Nb pressure tube, zircaloy fuel cladding tubes and zircaloy calandria tube have been determined using XRD patterns.
2. Zr-2.5Nb pressure tube samples were analysed at three different laboratories using three different XRD systems. The orientation factors determined from data of the three laboratories were comparable to each other. The values of preferred orientation factors f_r, f_t and f_n were found to be 0.06, 0.51 and 0.41, respectively.

Acknowledgement: The authors would like to express their thanks to Shri H. S. Kamath, Director, Nuclear Fuels Group, BARC and Shri K. C. Sahoo, Head, PIE Division, BARC for their interest in this work.

REFERENCES

Holt, R.A. (1979). *Journal of Nuclear Materials*, **82:** 419.

Hwang, H.K. (1991). *Metallurgical Transactions*, **22A:** 2247.

Kearns, J.J. (1965). "Thermal expansion and preferred orientation in Zircaloy" Report. WAPD-TM-472.

Kearns, J.J. (2001). *Journal of Nuclear Materials*, **299:** 171.

Lewis, J.E., Schoenberger, G. and Adamson, R.B. (1982). Round Robin Study in Zirconium in Nuclear Industry. ASTM STP 754, ASTM International, West Consohoeken, PA, 39.

Lingamurthy, K. (1991). Proc. of Symp. on Zirconium alloys for Reactor components (ZARC-91), 1.

Sah, D.N. Sunil Kumar, Sathe, S.M. and Tyagi, A.K. (2002). *In:* Proc of Symposium on Zirconium (ZIRC-2002), BARC Mumbai, 326.

Savino, E.J. (1980). *Journal of Nuclear Materials*, **98:** 98.

Simpson, L.A. (1999). *Journal of Nuclear Material*, **273:** 52.

Characterization of Microstructure, Texture and Residual Stress of PHWR Fuel Clad Material Using X-ray Diffraction

K. Kapoor,* S.V.R. Rao, D. Lahiri and T. Sanyal

Advanced Materials Characterization Laboratory
Nuclear Fuel Complex, Hyderabad-700062, India

INTRODUCTION

Fuel clad tubes, due to their critical applications, has stringent specification for mechanical, physical anf chemical properties. Some of these, like residual stress, could be directly evaluated by X-ray diffraction (XRD), whereas others like resistance to hydride cracking and stress corrosion cracking, fracture toughness and mechanical strength could be predicted with confidence by other applications of XRD like line profile analysis (LPA) and texture analysis. Out of the several applications of XRD, some have been used in the present study (Fig. 1) for characterisation of properties.

LPA is the analysis of the shape XRD peaks and their broadening resulting with different thermo-mechanical treatments. The peak shape is analyzed using Fourier transformation and information about coherent domain size, micro-strain and stacking fault probabilities is determined from the Fourier coefficients of the peak-curve. Analysis of broadening of the peaks by integral breadth method also gives idea about the coherent domain size, dislocation density and micro-strain present in the material.

Texture analysis by XRD gives the idea about the orientation of the grains. There are two methods for quantification of texture used here. The quantitative texture (Kearn's 'f') parameter gives an idea of the distribution of the basal poles in terms of fractions in sample co-ordinate system, i.e., rolling direction (RD), transverse direction (TD) and normal direction (ND). This parameter is useful for knowing basic idea of the texture in the material. But for exact texture measurement an X-ray goniometer is used to obtain the pole figures, which are the two dimensional representation of orientations. Further, three-dimensional orientation distribution function (ODF) is calculated from these pole figures by mathematical inversion techniques. From the ODF, the volume fraction of the prominent orientations is determined which correlate with important properties like mechanical, creep, irradiation growth, delayed hydride cracking and stress corrosion cracking.

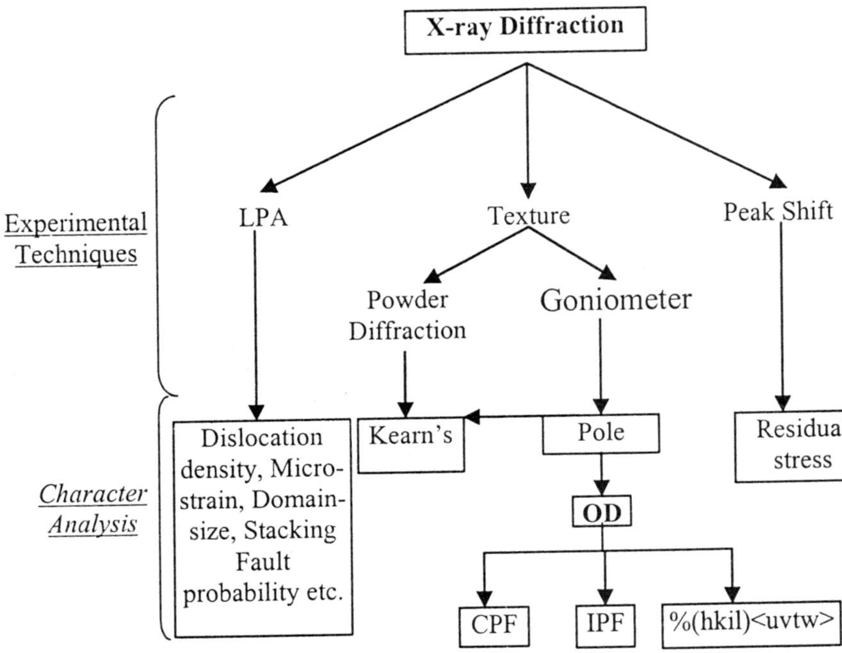

Fig. 1: Application of XRD Technique in Analysis of Properties of Fuel-Clad Tubes.

Shift in the peaks of XRD in a material in comparison with annealed powders (ideally stress free material) is used in calculation of residual stress of the material. The residual stresses determine the dimensional stability, hydride reorientation and stress corrosion cracking of the material.

The X-ray analytical methods, mentioned above, have been effectively applied to Zircaloy-4 fuel clad tube in the present study and the results on the characterisation of this material are compiled here. The process flow sheet for Zircaloy-4 fuel clad tube is given in Fig. 2. The critical processing steps in processing of Zircaloy-4 tubes include the heat-treatment at the billet stage, hot and cold working stages. It may be observed that there are a number of steps involved in the processing to make the thin walled (0.52 mm) tube starting from 300 mm dia solid round ingot.

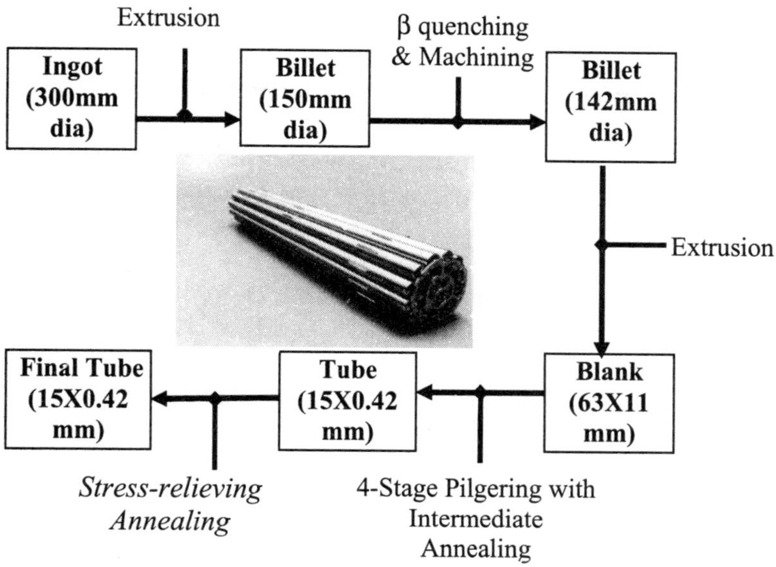

Fig. 2: Process Route of Zircaloy-4 Fuel Cladding tube, with fuel bundle in insert.

XRD TECHNIQUES FOR CHARACTERISATION OF ZIRCALOY-4 CLAD MATERIAL

Microstructure Analysis Using XRD Line Profile Analysis

The diffraction pattern obtained from a metallic sample gives information about imperfections in the material i.e. dislocations, small crystallite (domain) size, microstrains within the grains due to dislocations and stacking faults. The X-ray peak is broadened due to small crystallite size, strain due to dislocations and stacking faults. The analysis of the shape of the peaks gives information about the material and is referred as Line Profile Analysis.

Dislocation density (ρ) is an important material property, which gives the length of the dislocations present per unit volume (m/m^3) in the material. It has a strong influence on out-of-pile and in-pile properties of critical clad and structural materials for PHWR. X-ray line broadening technique has been used widely to estimate the dislocation density. But use of this technique requires careful experimentation, sample preparation and calculations.

The line profiles of the reflections of various planes during XRD are characteristic of the state of the material. The shapes of line profiles are also affected by instrument and sample shape, which is referred as instrumental broadening. This instrumental broadening needs to be eliminated to obtain broadening due to metallurgical effects. Once such data is obtained, estimation of coherent domain size, micro strains within these domains, density of dislocation, stacking fault probability etc. can be done. Coherent domain size (*D*) is the size of coherently diffracting region within a grain, representing fault free region between dislocations (of the order of 100–1000 Å) higher the ρ lower the D. Within this region, due to dislocations micro strain (ε) are present between the atomic planes (of the order of 10^{-3}). The observed broadening in X-ray line profiles is due to the domain size (D) and micro strain (ε) present in the material. Once the D and ε are determined from the broadening, the dislocation density can be estimated. There are three basic techniques of X-ray line profile analysis namely Fourier analysis (Warren, 1959), Integral Breadth (Warren, 1972) and Variance analysis (Wilson, 1962). It has been shown that each of the above techniques leads to similar results for domain size, dislocation density etc. In the integral breadth method the peaks that are affected by stacking fault are also included to give overall dislocation density including the stacking fault concentration. On the other hand, Fourier analysis gives separate estimation of stacking fault concentration in addition to ρ, D and ε. The correction for instrumental broadening (b) is most important step in the estimation of material properties from line profile analysis. The Fourier analysis technique has been followed here for line profile analysis.

Fourier Analysis

In the Fourier analysis the line breadth is corrected for instrumental broadening using Stoke's correction (Stokes, 1948). The profile of the peak corrected for instrumental broadening can be expressed in terms of Fourier series (Chaterjee and Sengupta Jr., 1974). By considering each reflection of closed packed hexagonal (cph) structure ($HK.L_0$) as a (00l') reflection in terms of orthorhombic axes, the Fourier series in terms of a position in reciprocal space is expressed as,

$$P'_{2\theta} = k'\Sigma\, A_n^S A_n^D \cos 2\pi\, n\, (h_3' - l') \tag{1}$$

where $h_3' - l' = 2\,|a_3'|\,(\sin\theta - \sin\theta_0)/\lambda$ and $|\,a_3'\,|$ is a fictitious distance chosen to correspond to the sin θ interval within which the peak is expressed as a Fourier series. A_n^S and A_n^D represents the size and distortion coefficients, respectively and A_n^S includes the effect of both domain size and faulting. The distortion coefficient can be expressed as,

$$A_L^D = \exp\,(-2\pi^2 L^2 < \varepsilon_L^2 > /d^2) \tag{2}$$

where $L = n\,|\,a_3'\,|$ which has a significance normal to the reflecting planes, $<\varepsilon_L^2>$ is the mean square strain and d is the interplanar spacing of the reflecting planes. The separation of A_L^D and A_L^S is done by log plot of $A_L = A_L^D . A_L^S$ against $1/d^2$ using the Eq. 2. The intercept yields A_L^S and the slope gives $<\varepsilon_L^2>$. Slope of the plots A_L^S vs. L gives the domain size (D) normal to the reflecting planes, deformation and growth fault probabilities (α and β, respectively) by the following relations,

$$-\left(\frac{dA_L^S}{dL}\right)_0 = \frac{1}{D} \quad \text{for} \quad H-K = 3N \tag{3}$$

$$-\left(\frac{dA_L^S}{dL}\right)_0 = \frac{1}{D} + \frac{\{L_0 \mid d(3\alpha + 3\beta)\}}{C^2} \quad \text{for } H\text{–}K = 3\text{N} \pm 1,\ L_0 \text{ even} \tag{4}$$

$$-\left(\frac{dA_L^S}{dL}\right)_0 = \frac{1}{D} + \frac{\{L_0 \mid d(3\alpha + \beta)\}}{C^2} \quad \text{for } H\text{-}K = 3N \pm 1,\ L_0 \text{ odd} \tag{5}$$

where $C = 2d_{00.2}$.

Dislocation density can be calculated from D and ε_L using following formulae:

$$\rho_D = \frac{3\eta}{D^2} \qquad \eta = 1 \tag{6}$$

$$\rho\varepsilon = \frac{2\,k\varepsilon_L^2}{b^2} \qquad k = 10 \tag{7}$$

where b = burges vector (a) = $\frac{1}{3}$ $[11\,\bar{2}\,0]$. (8)

Dislocation density (ρ) is obtained from ρ_D and $\rho\varepsilon$ as follows:

$$\rho = (\rho_D \cdot \rho\varepsilon)^{1/2} \tag{9}$$

Determination of Texture by Kearn's 'f' Parameter Using XRD

Texture determination has been performed both by powder XRD technique (X-ray line scan) for calculating Kearn's 'f' parameter and X-ray goniometer technique for obtaining pole figures and ODF.

Definition of Kearn's 'f' Parameter

The Kearn's 'f' parameter is the fraction of basal poles oriented in the three principal directions i.e. Longitudinal, Transverse and Normal (f_L, f_T and f_N) in the sample.

Such that $f_L + f_T + f_N = 1$ (10)

The number for any reference direction, say f_L, is arrived at by summation of the volume fraction (V_i) of crystals having basal poles inclined at angle (ϕ_i) to longitudinal direction, multiplied by cosine^2 (ϕ_i) as projection of this quantity (V_i) in the reference longitudinal direction.

Thus
$$f = \sum_{\phi_i = 0}^{\phi_i = \pi/2} V_i \cos^2 \phi_i \tag{11}$$

The volume fraction of crystals having inclination ϕ_1 to ϕ_2 with the reference direction is given by:

$$V_i = \frac{\int_{\phi_1}^{\phi_2} I(\phi)\sin(\phi)\,d\phi}{\int_0^{\pi/2} I(\phi)\sin(\phi)\,d\phi} \tag{12}$$

where $I(\phi)$ is the average intensity of poles at an angle ϕ,

$$I(\phi) = \int_0^{2\pi} I(\phi, \beta)\cdot d\beta \tag{13}$$

From (12) and (13) we get,

$$f = \frac{\int_0^{\pi/2} I(\phi)\cdot \sin\phi \cdot \cos^2\phi \cdot d\phi}{\int_0^{\pi/2} I(\phi)\cdot \sin(\phi)\cdot d\phi} \tag{14}$$

Fraction of poles in the radial, transverse and normal directions are denoted as f_R, f_T and f_N, respectively.

Measurement of Kearn's 'f' Parameter using Powder Diffraction Data

The samples for ND, TD and LD represent the normal direction, transverse direction and longitudinal direction of the tube respectively. The samples were prepared with cut pieces stacked together as shown in Fig. 3. In ND samples the LD-TD plane of the sample was mounted as surface of the

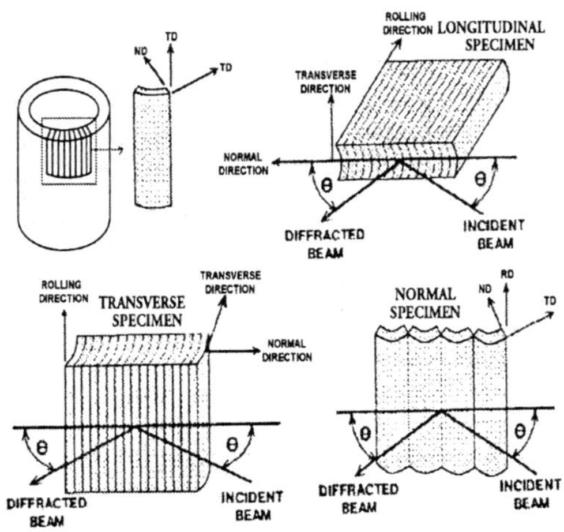

Fig. 3: Procedure for specimen preparation from tube sample for texture analysis.

specimen. In the TD sample, ND-RD surface was facing sample surface and in RD sample, ND-TD was the mounted surface.

Calculation of Kearn's 'f' Parameter

From the diffraction data given in Fig. 4 and with 18 peaks of sample taken for a reference direction, the texture coefficient $TC_{(hkil)}$ for (hkil) plane was calculated using the following formula (Mahmood and Linga Murty, 1991).

$$TC_{(hkil)} = \frac{\dfrac{I_{(hkil)}}{I_{o(hkil)}}}{\left[\dfrac{1}{N}\right]\sum\left(\dfrac{I_{(hkil)}}{I_{o(hkil)}}\right)} \tag{15}$$

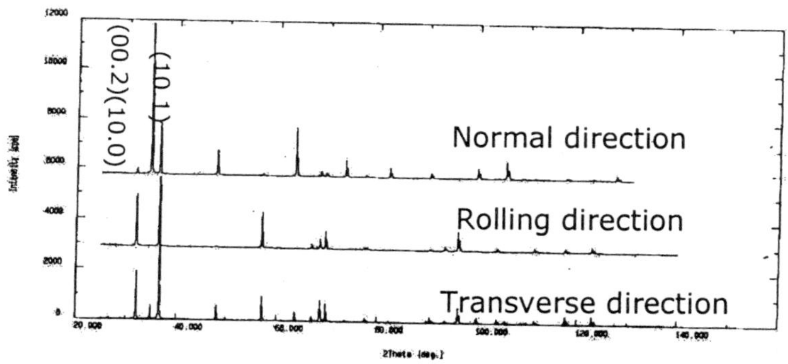

Fig. 4: Diffraction data for samples in three principal directions RD, TD and ND.

The inverse pole figure was then drawn along RD, ND and TD using these texture coefficients. Subsequently, the intensity I was determined as a function of ϕ, the tilt angle for the three sets of planes $(10\bar{1}l)$, $(21\bar{3}l)$ and $(11\bar{2}l)$, where ϕ is varying from 0 to $\pi/2$ for each set. The average intensity $I_{av}(\phi)$ is calculated using the following formula:

$$I(\phi)_{av} = \frac{I}{30}\left[19.1\left(\frac{I_1 + I_2}{2}\right) + 10.9\left(\frac{I_{21} + I_3}{2}\right)\right] \tag{16}$$

where I_1, I_2 and I_3 are the intensities at a given value of ϕ for the three sets of lines $(10\bar{1}l)$, $(21\bar{3}l)$ and $(11\bar{2}l)$, respectively. Finally 'f' is obtained from equations 11 to 14.

Determination of Kearn's f Parameter from the Full Pole Figure

Experimental Full Pole Figure Determination

Schultz reflection in combination with transmission technique (Schultz, Jr., 1949; Decker, 1948) was used to get the complete pole figure. In reflection technique, the source and the detector were kept such that both were facing the same side of the specimen surface. A schematic sketch is shown in Fig. 5. The X-ray beam in this setup is point focused. It uses an optical focusing system and axis A is orthogonal to the goniometer axis. As seen in the figure, α rotation refers to rotation of specimen around axis A, direction of arrow indicates the +α rotation direction; β rotation represents the rotation around axis B; γ oscillation represents the reciprocating motion in the specimen plane 45° with respect to axis A.

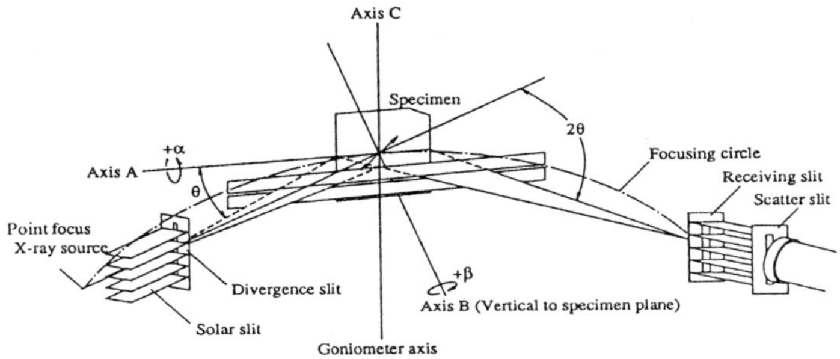

Fig. 5: Functional diagram of Schultz.

In transmission technique the specimen in the form of a thin foil was placed between the source and detector. A schematic sketch is shown in Fig. 6. The X-ray beam in this setup is line focused. α, β and γ refers to similar notation as explained above. At lower α angles (0°-20°) transmission technique was used. Reflection technique is not possible at this angular range because these angles were grazing angles for the X-ray beam, which results in the defocusing. The X-ray diffraction data were collected as the specimen was tilted from 0-90° in 5° steps (0-20° in transmission mode and 20-90° in reflection mode) about rolling direction. At each step it was rotated 0-360° about tube normal direction. The detector position with respect to source was such that the Bragg diffraction condition for (00.2) plane was always maintained. The rotation and tilt angles were coupled in such a way that for every 360° rotation about tube normal direction a tilt of 5° about tube rolling direction was given to the sample.

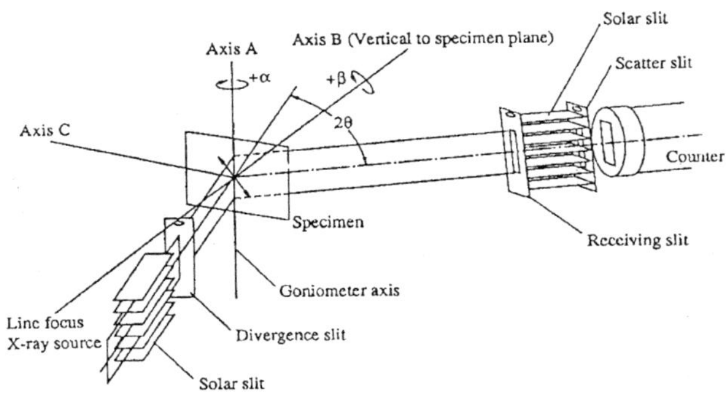

Fig. 6: Functional diagram of transmission method.

Determination of 2θ, BG1, BG2 Peak Positions

During the preliminary setting this data is most important and it determines the accuracy of the results. Initially using an approximate 2θ, beta angle is determined where maximum intensity of peak is observed. Then fixing the sample at this beta position a 2θ scan is carried out to get maximum intensity of the peak. The 2θ, BG1 (lower background), and BG2 (higher background) peak positions are determined. The raw data were processed for smoothening, background elimination, and normalized with respect to random (powder) specimen. Also the necessary slits sizes (divergence, scattering and receiving) for this system to get optimum X-ray intensity were optimized. The processed data was plotted as stereographic projection, called pole figure, showing the preferred orientation of a given plane in the tube rolling and tangential direction. The results of the data processing for obtaining the full pole figure is shown in Fig. 7.

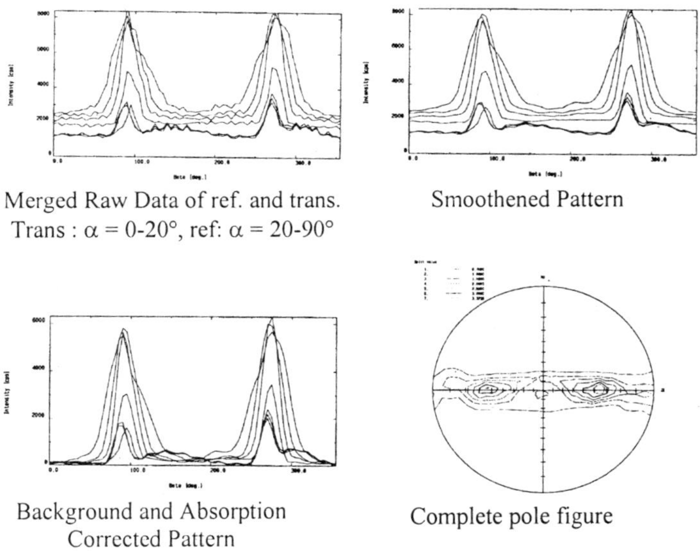

Merged Raw Data of ref. and trans.
Trans : α = 0-20°, ref: α = 20-90°

Smoothened Pattern

Background and Absorption Corrected Pattern

Complete pole figure

Fig. 7: Direct Complete Pole Figure Method (Schultz reflection + transmission).

TEXTURE CHARACTERISATION BY XRD THROUGH ODF

Calculation of ODF from Pole Figures

ODF represents the volume fraction of the crystallites in a polycrystalline material having a certain crystallographic orientation (g) with respect to the sample fixed co-ordinate system,

$$\frac{dV(g)}{V} = f(g)\ d(g) \tag{17}$$

Pole figure is a two-dimensional distribution function depending only on the orientation of one specific crystal direction and is independent of the rotation of crystal about that direction. An orientation in three-dimensional space is represented by Euler angles as φ_1^{hkl}, ϕ^{hkl}, φ_2^{hkl}, which can be expressed then in terms of pole density function as,

$$P_{hkl}(\alpha, \beta) = \frac{1}{2\pi}\int f(\varphi_1^{hkl}, \varphi^{hkl}, \varphi_2^{hkl})\, d\varphi_2^{hkl} \tag{18}$$

where $P_{hkl}(\alpha, \beta)$ is the pole density function and $f(\varphi_1^{hkl}, \phi^{hkl}, \varphi_2^{hkl})$ is the orientation distribution function. When several pole figures belonging to different lattice planes (hkl) are considered, the Euler angles are transformed from specific orientation φ_1^{hkl}, ϕ^{hkl}, φ_2^{hkl} to general angles φ_1, ϕ, φ_2. The calculation of three dimensional function $f(g)$ from several two-dimensional projections $P_{hkl}(\alpha, \beta)$ is called the pole figure inversion. The pole figure inversion has been done using discretization method (Bunge and Puch, 1984) to obtain ODF plots.

RESIDUAL STRESS MEASUREMENT BY XRD

Modified Procedure for Textured Material

The surface residual stresses are usually measured using the multi-exposure side-inclination technique. In this technique the peak mid position is measured by curve fitting (Gaussian/Pearson IV). The shift in the peak position is determined as a function of change in inclination angles. The residual stress is determined using the following relationship:

$$\sigma\phi = -\left\{\frac{E}{2*(1+\nu)}\right\} * \cot\theta * \frac{d(\Delta 2\theta)}{d\sin^2\Psi} \tag{19}$$

where $\sigma\phi$ is the stress in the direction ϕ, E is elastic modulus and ν is poison's ratio, Ψ is the inclination angle measured in direction ϕ and θ is the Bragg's diffraction angle and $\Delta 2\theta$ is the shift in the Bragg's angle due to change in lattice spacing measured at given Ψ and ϕ. These stresses are then measured in the three rotation angles using bulk elastic constants and then principal stresses are determined.

Figure 8a gives the functional diagram of the measurement principle. As seen (case 'A') in stress free sample, with variation in inclination angle (ψ) the d spacing will be constant. When stress is present (case 'B') the 'd' spacing varies as d_i in the figure. The relationship between the 'd' and $\sin^2\psi$ will be linear in case of isotropic materials as shown in Fig. 8b and will be generally nonlinear in case of textured materials as shown in Figs. 8c and 8d. It has been observed in the present case that there is also some non-linearity in the 'd' vs $\sin^2\psi$ relationship. This non-linearity has been shown to be purely due to elastic anisotropy. For this a plot of the average elastic compliance r^{1111}, calculated from the single crystal elastic compliance and the texture of the material, has been plotted as a function of $\sin^2 \psi$ for two measurement directions (Fig. 9). Its curvature has been found to be similar to that of the '*d*' vs $\sin^2 \psi$ in the two measuring direction, indicating non-linearity to be purely due to elastic anisotropy.

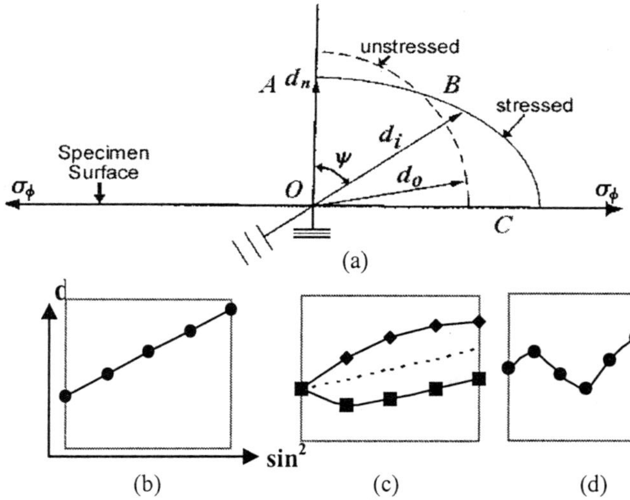

Fig. 8: (a) Vector diagram indicating variation of lattice spacing *d* with change in inclination angle for stressed and stress-free conditions, (b) showing linear relationship between '*d*' vs $\sin^2 \psi$ for un-textured material, (c) and (d) non-linearity observed in textured materials.

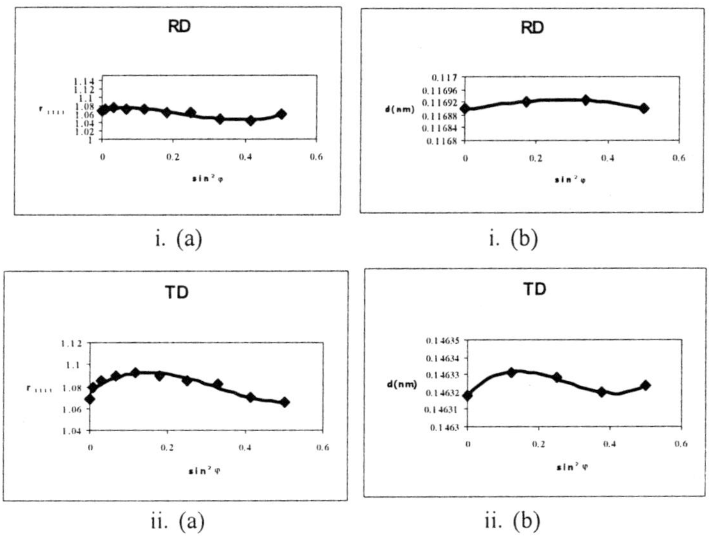

Fig. 9: (a) The plot of calculated r_{1111} vs. $\sin^2\psi$ and (b) The plot of experimentally determined '*d*' vs. '$\sin^2\psi$' for cold worked and stress relieved material in (i) rolling direction, (ii) transverse direction.

In order to reduce the anisotropic effect it is required to obtain some specific combination of ϕ and ψ values, which give a texture independent path during the stress measurement. Secondly include the effect of texture on the elastic constants. This is implemented in the following way:

(a) Find specific combination of ϕ's and ψ's for which the height and nature of the intensity peaks are similar in both the detectors. Disable inclination angles where variations are pronounced.

(b) Calculate the directional elastic constant used for the bulk polycrystalline samples using the Constant Stress Model (CSM) (Zarkades and Larson, 1970).

The intensity of the diffracting planes used for lattice strain measurement were indexed on a Wulff net. For the interpretation of the effect of crystallographic texture on the peak intensity, the corresponding pole figure of the plane was superimposed with peak intensity during stress measurement. This is presented in Fig. 10. It is seen from this that the φ directions selected for the measurement of stress are nearly texture independent. Hence the error in the reported stress values is minimum.

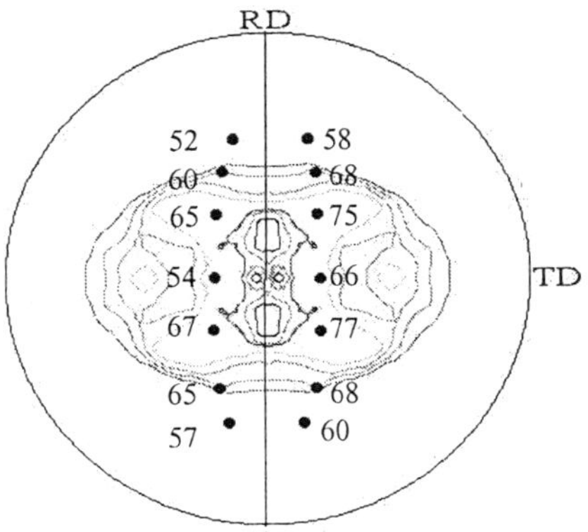

Fig. 10: Intensity recorded by the two detectors in texture free directions (inclination and rotation).

EXPERIMENTAL SETUP

Fourier Analysis

The peak data is obtained using following settings (for sample and instrumental broadening)

Radiation	:	CuKα
Scan rate	:	0.1° per min.
Step size	:	0.002°
Kα/Kβ separation	:	Ni filter (for single wave length)
Monochromator	:	Curved crystal type

Background connection	:	Averaging at the end (Sonnevelt – Visser's method)
Smothering	:	Savitzkey–Golay's (67 points)
LP correction	:	Using software

Kearn's 'f' from X-ray Line Scan

This method uses the diffraction data for the samples taken from the tube in ND, TD and RD directions. The specimens were mechanically polished and etched with 40% H_2O, 45% HNO_3 and 10% HF. A programme has been developed which requires diffraction data as input and directly gives the Kearn's parameter. Rigaku Dmax 2000 with CuKα radiation was used as the X-ray source (40 kV and 30 mA with Ni filter). Monochromatic Kα_1 was obtained for diffraction using crystal monochromator for Kβ separation and Kα_2 stripping by software. Following experimental and processing parameters

1. Scan speed: 1 degree/minute
2. Scan Step: 0.01 degree/step
3. Smoothening: Savitzkey Golay's method
4. Kα_2 Stripping intensity ratio: 0.5

Pole Figure

For texture measurement Rigaku Dmax 2000 X-ray diffractometer fitted with a horizontal goniometer for texture measurement was used. For sample preparation, two specimens with dimensions 12 mm (width) X 25 mm (length) were taken from a pressure tube. The length direction was parallel to the tube rolling direction and the tube normal (thickness) direction was parallel to the sample surface normal. To prepare the specimen for reflection geometry, the outer surface of the tube was ground flat, polished mechanically and etched. The sample was irradiated with X-rays falling on this surface. In such a configuration, the tube normal direction is the centre of the pole figure. In the case of transmission technique, foil specimen of 0.05 mm uniform thickness was used. To prepare a thin foil from the tube, initially mechanical removal was carried out to reduce the thickness to 0.1 mm. This was followed by a further reduction in the thickness to 0.05 mm by chemical dissolution. For uniform removal of material by chemical dissolution, an in-situ metallography polisher having rotating strub fixed with a micro cloth soaked with pickling solution (40% H_2O, 45% HNO_3 and 10% HF) was used. In transmission technique the specimen in the form of a thin foil was placed between the source and detector.

Residual Stress

The residual stress was measured using the Stress Tech X-Stress 3000 equipment. The equipment consists of a goniometer with an X-ray source and a compact high voltage cum cooling system. The probe used was TiKα X-ray radiation, with two position sensitive detectors for obtaining the full XRD peak in one exposure. The goniometer is capable of giving inclination, rotation and vertical movements. The stress measuring directions were 0, 45 and 90° to the tube axial direction. In each direction, measurements were done in five inclination angles in either side of inclination (positive and negative). From the three measurements the principal stresses and their orientation was determined. Also from the three measured stresses the stress tensor was determined using the Hauk-Dolle method.

RESULTS

Microstructure Analysis

Line profile analysis for the tube in final condition has been performed using the Fourier analysis technique. The total peak shape analysis involved peaks from nine planes, namely, (11.0), (11.4), (30.2), (10.2), (10.4), (21.2), (10.3), (20.3) and (21.3). The peak shape after instrumental correction is shown in Fig. 11. This peak is obtained by inverse fourier transform after stokes correction.

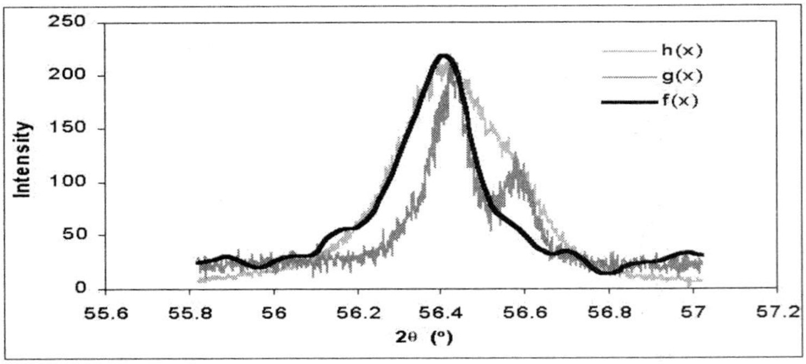

Fig. 11: XRD peak of (11.0) plane for pilgered + stress relieved sample, as obtained [$h(x)$], with only instrumental broadening [$g(x)$] and the derived after stokes correction i.e. without instrumental broadening [$f(x)$].

The Fourier coefficient A_L vs. L plot for the planes are given in Figs. 12a, b and c for $H–K = 3N$, $H–K = 3N \pm 1(L_0$ even) and $H–K = 3N \pm 1(L_0$ odd) planes, respectively. The ln A_L vs. i/d^2 plots with different L values for fault unaffected reflections (Fig. 13a) have been used for calculating the values of strain (ε_L). Coherent domain size (D) has been calculated from the slope of the A_L^S vs. L plot for the same reflections shown in Fig. 13b. These parameters along with the dislocation density (ρ) have been reported in

Table 1. The A_L^S vs. L plots for fault-affected reflections are presented in Fig. 14. The effective domain size ($D_{eff.}$) for fault-affected reflections and the deformation (α) and growth (β) fault probabilities are reported in Table 2. It has been observed that the coherent domain size is small in the final tube with high strain values and dislocation densities. The fault probabilities are also quite low.

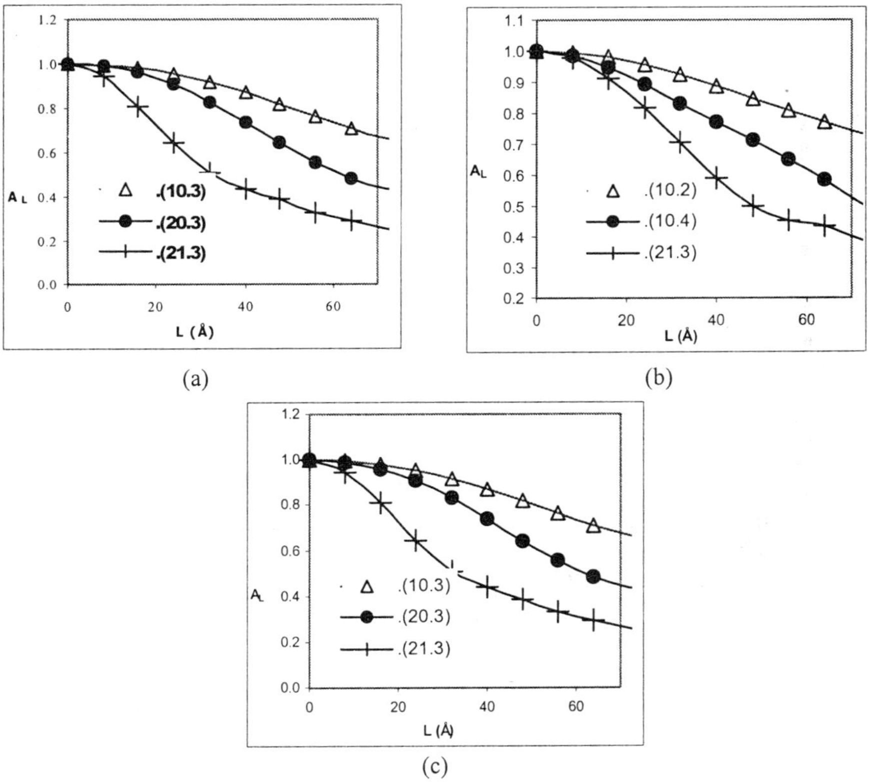

Fig. 12: Fourier Coefficient A_L vs. L Plots for Zircaloy-4 Fuel Cladding Tube in Final Condition for, (a) $H–K = 3N$ Reflections; (b) $H–K = 3N \pm 1$; L_0 = even, Reflections; (c) $H–K$ = 3N ±1; L_0 = odd, Reflections.

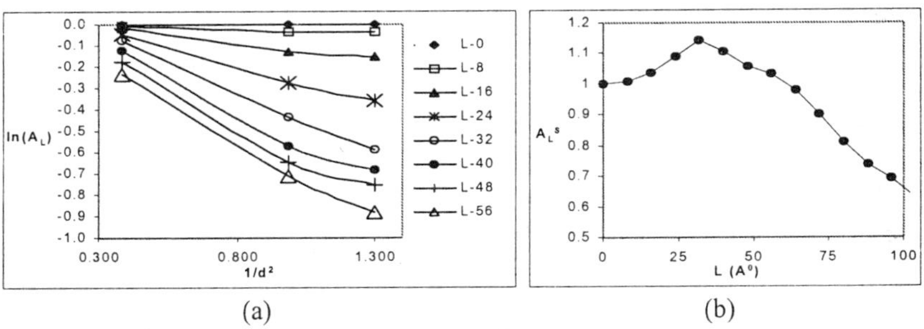

Fig. 13: (a) Fourier coefficient A_L vs. $1/d^2$ Plots for Zircaloy-4 fuel cladding tube in final condition (b) Size coefficient A_L^S vs. L plots of Zircaloy-4 fuel cladding tube in final condition for $H–K = 3N$ reflections.

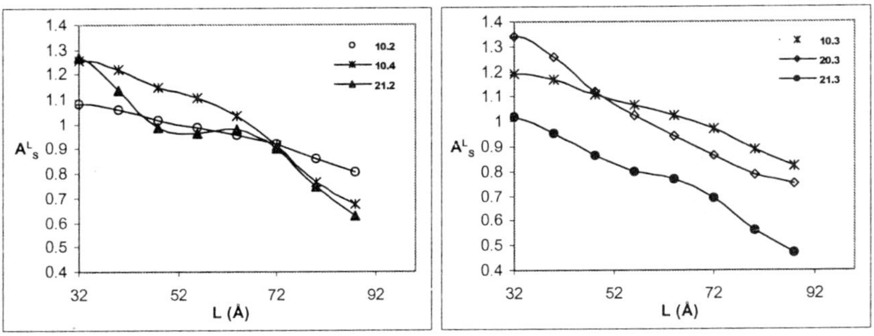

(a) (b)

Fig. 14: Size coefficient A_L^S vs. L plots of Zircaloy-4 fuel cladding tube in final condition for (a) H–K = 3N ± 1; L_0 = even, Reflections; (b) H–K = 3N ± 1; L_0 = odd, reflections.

Table 1: Average domain size *D*, r.m.s strain ε_L and dislocation density ρ in Zircaloy-4 fuel tube material from fault-unaffected reflections

Reflections: H–K = 3N	*D (A°)*	$(\varepsilon_L^2)^{1/2}_{L=56Å}$	*ρ (m/m³)*
(11.0), (11.4), (30.2)	212	0.00341	50 × 10^{14}

Table 2: Effective particle size D_e, and fault probabilities α and β in Zircaloy-4 fuel tube material from fault-affected reflections

Reflections: H–K =3N ± 1		D_e *(Å)*	$3\alpha + 3\beta$	$3\alpha + \beta$	α	β
	(10.2)	275	–0.00026			
L_0 even	(10.4)	174	0.00041			
	(21.2)	294	–0.00062		− 9.28 × 10^{-4}	8.76 × 10^{-4}
	(10.3)	189		0.000183		
L_0 odd	(20.3)	83		0.003121		
	(21.3)	122		0.001803		

Determination of Quantitative Texture by Kearn's 'f' Parameter

Kearn's 'f' parameter calculated for extruded blank and final tube are reported in Table 3. It is observed that in the extruded blank, the fraction of basal pole is same in both normal, radial and transverse directions. But, during cold working by pilgering operation large orientation change takes place and in the final tube, the fraction of basal pole in normal direction is double of that in the transverse direction. In both the conditions, the fraction in longitudinal direction is negligible. For comparison purpose the texture of other components i.e. calandria tubes and the pressure tube is given.

Table 3: Kearn's 'f' parameter for longitudinal, normal and transverse directions for extruded blank and final tube

Material	*Fuel Tube (average of 10)*		*Seamless Calanderia Tube (average of 20)*	*Pressure Tube (average of 20)*
Kearn's parameter	*Extruded blank*	*Final Tube*		
f_L	0.08	0.05	0.11	0.03
f_N	0.46	0.72	0.65	0.35
f_T	0.41	0.22	0.25	0.62

Determination of Texture from Pole Figure and ODF

The pole figures for the extruded final tube material, determined after symmetrisation, is shown in Fig. 15 for (00.2), (10.1) and (11.0) planes. The basal pole figures show clearly that the basal pole maxima is shifted more towards the normal direction in the final condition in comparison to the extruded sample. For the (11.0) planes the maxima lies in rolling direction (RD) both in the extruded and the final tube. This indicated that there is no rotation of the crystallite along the 'c' axis during the thermo-mechanical treatment from the extruded to the final stage. This is further confirmed by the (10.1) pole figures, where the maxima lies between the ND-RD in both the cases. The (00.1) inverse pole figures shown in Fig. 16 (taken in ND) also confirms the fact that the maxima of basal poles shifts towards the normal direction in case of final material with noticeable increase in the pole density on themo-mechanical processing.

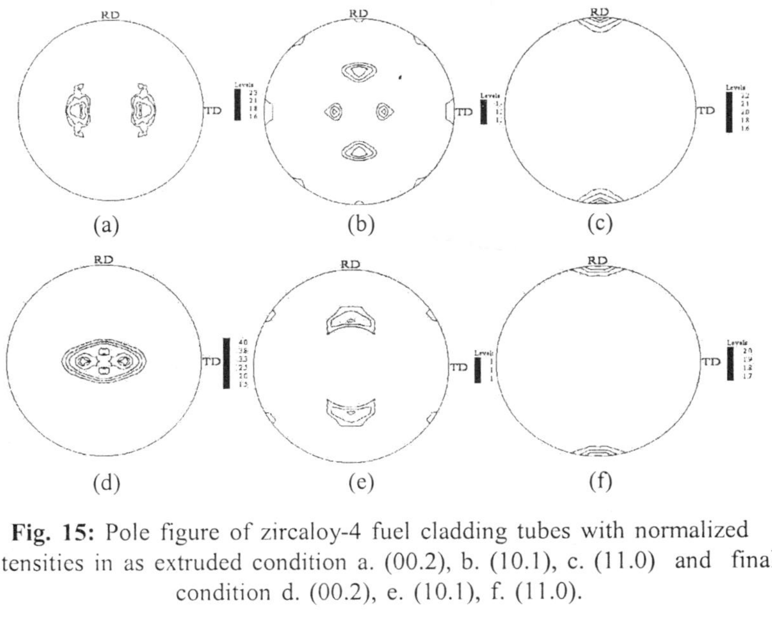

Fig. 15: Pole figure of zircaloy-4 fuel cladding tubes with normalized intensities in as extruded condition a. (00.2), b. (10.1), c. (11.0) and final condition d. (00.2), e. (10.1), f. (11.0).

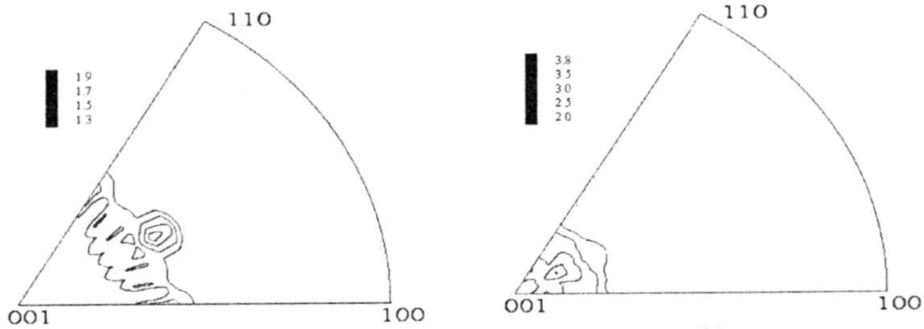

(a) (b)

Fig. 16: (00.1) Inverse pole figure of zircaloy-4 fuel cladding tubes with normalized intensities in (a) as extruded condition and (b) final condition.

The ODF in ϕ_2 sections at 5° intervals for the Zircaloy-4 fuel cladding tube in extruded final condition has been shown in Figs. 17a and b, respectively. The prominent components along with their relative intensity values expressed as ODF number '*f*(*g*)' and volume fraction of the same orientation in the material has been reported in details in Table 4. A fibre parallel to φ_2 is observed with a spread of around 30° (15°- 45°) in case of extruded material with the main texture component being $(0\ 1\ \bar{1}\ 2)<2\ 1\ \bar{1}\ 0>$ along with two other components with same intensity but of much lesser volume fraction. In case of final tube material, the fibre parallel to φ_2 is having a spread of 20° (50°-10°), lesser than the extruded one, but is of much higher intensity. Two main components of texture present in the final tube are $(0\ 1\ \bar{1}\ 5)<2\ 1\ \bar{1}\ 0>$ and $(0\ 1\ \bar{1}\ 7)<4\ \bar{6}\ 1\ 1>$ with much higher intensity (in comparison with the components in the extruded sample) and with huge volume fraction. This clearly indicates the immense increase in the degree of anisotropy for the final tube material i.e, the formation of a strong texture in the material with fabrication. Shift of the maximum intensity position towards lower ϕ regions in the final sample ODF can be corresponded to the shift of basal pole maxima towards ND.

Determination of Residual Stress

Firstly the directional elastic constants were determined using the constant stress model. These are reported for the two measurement directions in Table 5. Using these elastic constants the residual stresses were determined. The stresses were found to be compressive in both the longitudinal and the transverse directions. Also the principal stresses and their orientation determined is reported in Table 6. The stress tensor was determined using the Hauk-Dolle method and the results are reported in Table 6. It is observed that the stresses are fully compressive which is beneficial for the good performance of the component.

Table 3: Kearn's 'f' parameter for longitudinal, normal and transverse directions for extruded blank and final tube

Material	*Fuel Tube (average of 10)*		*Seamless Calanderia Tube*	*Pressure Tube*
Kearn's parameter	*Extruded blank*	*Final Tube*	*(average of 20)*	*(average of 20)*
f_L	0.08	0.05	0.11	0.03
f_N	0.46	0.72	0.65	0.35
f_T	0.41	0.22	0.25	0.62

Determination of Texture from Pole Figure and ODF

The pole figures for the extruded final tube material, determined after symmetrisation, is shown in Fig. 15 for (00.2), (10.1) and (11.0) planes. The basal pole figures show clearly that the basal pole maxima is shifted more towards the normal direction in the final condition in comparison to the extruded sample. For the (11.0) planes the maxima lies in rolling direction (RD) both in the extruded and the final tube. This indicated that there is no rotation of the crystallite along the 'c' axis during the thermo-mechanical treatment from the extruded to the final stage. This is further confirmed by the (10.1) pole figures, where the maxima lies between the ND-RD in both the cases. The (00.1) inverse pole figures shown in Fig. 16 (taken in ND) also confirms the fact that the maxima of basal poles shifts towards the normal direction in case of final material with noticeable increase in the pole density on themo-mechanical processing.

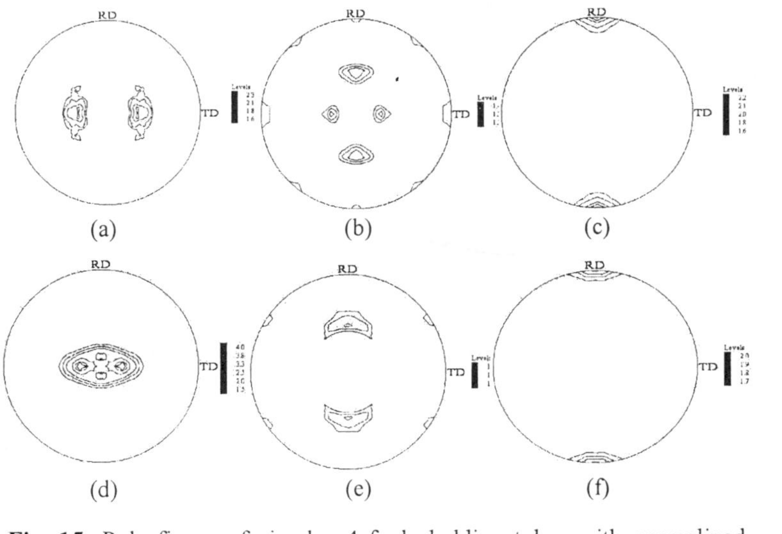

Fig. 15: Pole figure of zircaloy-4 fuel cladding tubes with normalized intensities in as extruded condition a. (00.2), b. (10.1), c. (11.0) and final condition d. (00.2), e. (10.1), f. (11.0).

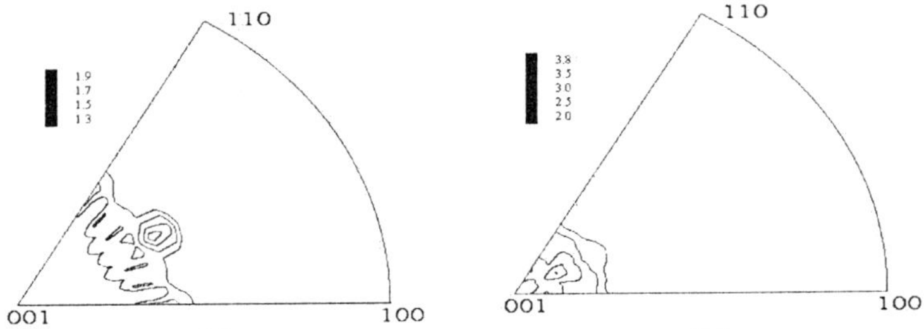

(a) (b)

Fig. 16: (00.1) Inverse pole figure of zircaloy-4 fuel cladding tubes with normalized intensities in (a) as extruded condition and (b) final condition.

The ODF in ϕ_2 sections at 5° intervals for the Zircaloy-4 fuel cladding tube in extruded final condition has been shown in Figs. 17a and b, respectively. The prominent components along with their relative intensity values expressed as ODF number '*f*(*g*)' and volume fraction of the same orientation in the material has been reported in details in Table 4. A fibre parallel to φ_2 is observed with a spread of around 30° (15°- 45°) in case of extruded material with the main texture component being (0 1 $\bar{1}$ 2)<2 1 $\bar{1}$ 0> along with two other components with same intensity but of much lesser volume fraction. In case of final tube material, the fibre parallel to φ_2 is having a spread of 20° (50°-10°), lesser than the extruded one, but is of much higher intensity. Two main components of texture present in the final tube are (0 1 $\bar{1}$ 5)<2 1 $\bar{1}$ 0> and (0 1 $\bar{1}$ 7) <4 $\bar{6}$ 1 1> with much higher intensity (in comparison with the components in the extruded sample) and with huge volume fraction. This clearly indicates the immense increase in the degree of anisotropy for the final tube material i.e, the formation of a strong texture in the material with fabrication. Shift of the maximum intensity position towards lower ϕ regions in the final sample ODF can be corresponded to the shift of basal pole maxima towards ND.

Determination of Residual Stress

Firstly the directional elastic constants were determined using the constant stress model. These are reported for the two measurement directions in Table 5. Using these elastic constants the residual stresses were determined. The stresses were found to be compressive in both the longitudinal and the transverse directions. Also the principal stresses and their orientation determined is reported in Table 6. The stress tensor was determined using the Hauk-Dolle method and the results are reported in Table 6. It is observed that the stresses are fully compressive which is beneficial for the good performance of the component.

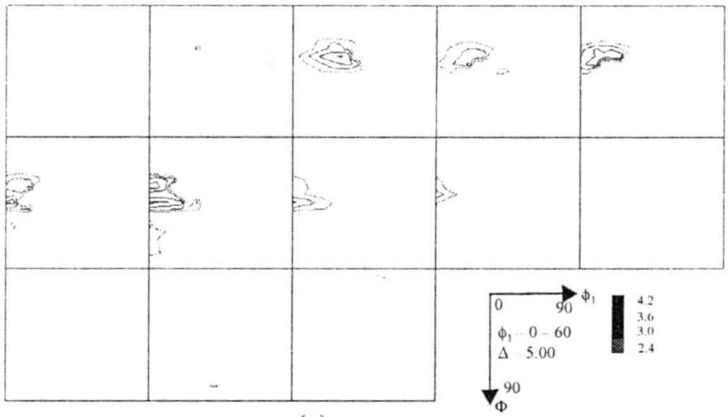

(a)

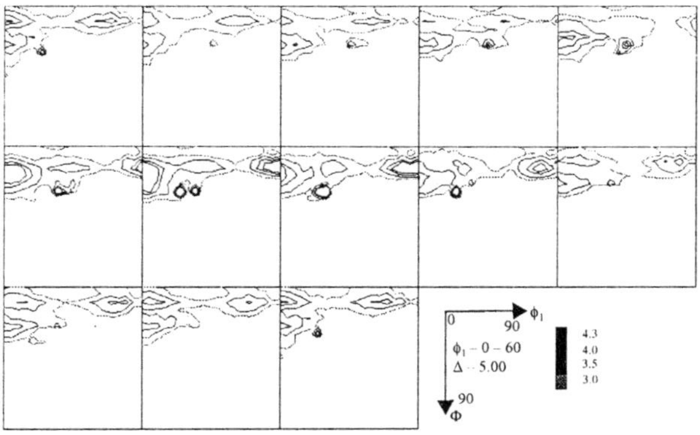

(b)

Fig. 17: φ_2 sections of the orientation distribution function at 5° interval for zircaloy-4 fuel cladding tubes with normalized intensities in (a) as extruded condition and (b) final condition.

Table 4: Volume fraction and ODF Value *f*(*g*) of different prominent orientation components for Zircaloy-4 fuel cladding tubes in extruded and final condition

Extruded condition			*Final condition*		
Orientation component	*Volume fraction*	*f(g)*	*Orientation component*	*Volume fraction*	*f(g)*
$(0\ 1\ \bar{1}\ 2)\langle 2\ 1\ \bar{1}\ 0\rangle$	4.61	4.30	$(0\ 1\ \bar{1}\ 5)\langle 2\ 1\ \bar{1}\ 0\rangle$	16.16	5.25
$(0\ 1\ \bar{1}\ 3)\langle 2\ 1\ \bar{1}\ 0\rangle$	1.55	4.35	$(0\ 1\ \bar{1}\ 7)\langle 4\ \bar{6}\ 1\ 1\rangle$	10.27	5.11
$(2\ 1\ \bar{3}\ 6)\langle 4\ \bar{9}\ 5\ 1\rangle$	1.83	4.30			

CONCLUSIONS

1. X-ray line profile analysis of cold worked–stress relieved Zircaloy-4 fuel tube shows that in cold worked–stress relieved condition material has low coherrent domain size (of the order of 200 Å), high microstrain (3.4×10^{-3}) and low stacking fault probabilities. The dislocation density measured is of the order of 10^{15}m/m^3.

Table 5 : Directional elastic constants and calculated residual stresses

Property → / *Direction ↓*	*Elastic Constants*		*Residual stress (MPa)*
	Young's modulus, E (Gpa)	*Poisson's ratio (μ)*	
Rolling direction (RD)	98.8	0.36	−103.8 ± 17.2
Transverse direction (TD)	91.2	0.33	−87.5 ± 19.4

Table 6: Calculated principle stresses and stress tensor form the measured stresses

Principle stress		*Stress tensor*			
Representation	*Value*	*Represen-tation*	*Stress (MPa)*	*Represen-tation*	*Stress (MPa)*
σ_1 (MPa)	−84.4	σ_{11}	−103.8 ± 16.7	σ_{12}	7.7 ± 9.0
σ_2 (MPa)	−106.8	σ_{22}	−87.5 ± 18.8	σ_{13}	−2. ± 3.6
ϕ(°)	21.0	σ_{33}	0.0	σ_{23}	1.0 ± 3.7

2. The parameter '*f*' has been used to quantify the texture using modified methods. The parameter can be used for industrial quantification of the texture for quality control. In case of Zircaloy-4 fuel tube high radial basal pole intensity ($f_N = 0.72$) is observed.
3. The volume fraction of major texture components has been calculated using ODF. Strong orientations like (h k i l)<2 $\bar{1}$ $\bar{1}$ 0> results during cold working.
4. Measurement of residual stress has been done using a modified XRD technique to take care of the texture present in the material. It has been shown that the non-linearity in '*d*' vs. '$\sin^2 \psi$' plot is purely due to texture. The residual stress measured in Zircaloy-4 fuel tube is compressive in both the longitudinal and transverse directions.

REFERENCES

Bunge, H.J. and Puch, K.H. (1984). *Z. Metlkd.*, **75:** 124.

Chaterjee, S.K. and Sengupta, Jr. S.P. (1974). *Mat. Sci.*, **9:** 953.

Decker, B.F., Asp, E.T. and Harker, Jr. D. (1948). *Appl. Phys.*, **19:** 388.

Mahmood, S.T. and Linga Murty, K. (1991). Effect of tensile deformation on crystallographic texture in Zircaloy sheet. ASTM, STP, **1132:** 119.

Schultz, Jr. L.G. (1949). *Appl. Phys.*, **20:** 741.

Stokes, A.R. (1948). *Proc. Phys. Soc.* (London), **61:** 382.

Warren, B.E. (1959). *Prog. Metal Phys.*, **8:** 147.

Warren, B.E. (1972). "X-ray Diffraction". Addison-Wesley, Massachusetts, USA.

Wilson, A.J.C. (1962). *Proc. Phys. Soc.* (London), **8:** 286.

Zarkades, A. and Larson, F.R. (1970). The Science Technology and Application of Titanium, (ed.) R.I. Jaffee and N.E. Promisel. Pergamon Press, 933.

Effect of Iridium Buffer Layer on YNi_2B_2C Thin Film Growth

K. Subba Rao, G.H. Cao, R. Tamm, C.-G. Oertel, W. Skrotzki, S.C. Wimbush[1] and B. Holzapfel[1]

Institute of Structural Physics, Division of Metal Physics
Dresden University of Technology, D-01062 Dresden, Germany
[1]IFW Dresden, Helmholtzstr. 20, D-01069 Dresden, Germany

INTRODUCTION

The class of rare-earth (R) transition metal borocarbides (Müller and Narozhnyi, 2001) of the type R-Ni_2B_2C attracted the interest of many groups because of their wide variety of physical properties, such as the coexistence between superconductivity and magnetism. Initial characterization of their superconducting parameters has revealed that a moderately high density of states at the Fermi level, comparable to that of A15 compounds, is part of the reason for the high superconducting transition temperatures T_c of up to 16K. The structure of the R-Ni_2B_2C phase (Siegrist et al., 1994) is body centred tetragonal (bct, space group I_4/mmm), but with an extra constituent (carbon) at the 2b sites in the basal plane.

Although many detailed studies have been reported on these compounds in polycrystalline bulk form or in single crystals (Hilscher and Michor, 1998), high quality films of R-Ni_2B_2C would certainly be of great importance to perform a series of key experiments, like tunnelling spectroscopy, Josephson effect in planar junctions, high sensitivity microwave measurements, etc., both to understand the nature of superconductivity in this class of compounds and to explore possible applications.

The first YNi_2B_2C films were synthesized by magnetron sputtering on MgO substrates maintained at room temperature (Arisawa et al., 1994). By post annealing at 1050 °C, the films were found to transform from the amorphous to the crystalline YNi_2B_2C phase. The films were highly c-axis oriented and showed a superconducting transition at 15K. Similar results were obtained during in-situ growth at 800 °C with T_c up to 15.3K (Andreone et al., 1996; Arisawa et al., 1998). A key role in obtaining high quality films was played by the substrate temperature. Changing the deposition technique to pulsed laser deposition (PLD) (Häse et al., 1997) showed that films deposited at room temperature were completely amorphous and, for substrate temperatures below 800 °C, the YNi_2B_2C phase was not detectable in the films, neither for post-annealed nor for in-situ grown films. PLD experiments by Cimberle et al. (1997) and Wimbush et al. (2000) succeeded in YNi_2B_2C film growth on MgO even at deposition temperatures as low as 720 °C and 600 °C, respectively, by optimising the deposition parameters.

A steady increase in T_c with increasing deposition temperature was observed as the formation of the YNi_2B_2C phase improved. Towards the higher temperatures, the T_c of the films approached the bulk value, although the ultimate deposition temperature was limited by the tendency for films deposited at temperatures higher than 775 °C to peel off from the substrate. For PLD as well as for magnetron sputtering X-ray diffraction (XRD) demonstrates the simultaneous formation of Y_2O_3 (Arisawa et al., 1998; Häse et al., 1997 and Wimbush et al., 2000). Transmission electron microscopy reveals that this oxide exists as a continuous interlayer between MgO substrate and YNi_2B_2C film having a clear orientation relationship to MgO and YNi_2B_2C (Reibold et al., 2002; Cao et al., 2004). Consequently, textures develop which around 650 °C and 750 °C are a- and c-axis oriented, respectively (Wimbush et al., 2000). The a-axis oriented film is seen to grow in the two equivalent orientations on the substrate $(100)[001]_{YNi2B2C} \parallel (001)[100]/[010]_{MgO}$, while the c-axis oriented film grows in a single orientation rotated by 45° with respect to the cube-on-cube texture of the a-axis: $(001)[110]_{YNi2B2C} \parallel (001)[100]_{MgO}$. Moreover, oxide formation leads to deviations from stoichiometry, which are balanced by the formation of other phases at the interface, such as YNi_4B, Ni_3B, etc. (Reibold et al., 2002; Cao et al., 2004).

To prevent the formation of an oxide layer at the interface, it was the aim of this preliminary investigation to study YNi_2B_2C film growth on single crystal MgO (100) substrates with an Ir buffer layer. Among the various metallic materials suitable for use as a buffer, Ir is a good candidate, because it is stable at high deposition temperatures and has a lattice constant of 0.383 nm compared to YNi_2B_2C (a ≈ 0.352 nm and c ≈ 1.09 nm) and to MgO (a = 0.421 nm). Thus, the lattice parameter of Ir lies between that of MgO and YNi_2B_2C. Consequently, a good epitaxial growth of Ir on MgO and YNi_2B_2C on the Ir buffer layer may be expected.

EXPERIMENTAL PROCEDURE

The YNi_2B_2C films were prepared by PLD as described by Wimbush et al. (2000). Thin films were grown on MgO (100) substrates with Ir buffer layer. The deposition temperature was kept at 700 °C and 750 °C during deposition of the Ir and Ir/YNi_2B_2C film, respectively. Phase analysis of the films was done by XRD using CoKα radiation. The microstructure of the films was investigated by TEM in a Philips CM 200 microscope equipped with a LaB_6 electron source and operated at 200 kV. Cross-sectional TEM foils were prepared by a standard technique involving mechanical grinding, dimpling and ion-milling. The chemical composition of the borocarbide film was determined by energy dispersive X-ray analysis (EDX).

To determine the texture, incomplete pole figures were measured by XRD using CuKα radiation and a secondary monochromator. These were (111) for the Ir buffer layer (rotation between 0° and 360°, tilt from 35° to 85°, increment steps of 0.5°) and (112) and (004) for the YNi_2B_2C film (rotation between –6° and 96°, tilt from 0° to 70°, increment steps of 2°).

RESULTS AND DISCUSSION

Figure 1a shows the XRD pattern of an iridium layer deposited on a MgO(100) substrate. The orientation relationship found is $(100)_{Ir}$ // $(100)_{MgO}$. The same is shown in Fig. 1b for an YNi_2B_2C thin film deposited on the MgO (100) substrate with Ir buffer layer. The YNi_2B_2C film is predominantly [001] oriented although a (112) peak is also visible indicating the presence of a small randomly oriented volume fraction. After deposition of the YNi_2B_2C film the Ir(200) peak is absent in the XRD pattern. During the high-temperature deposition of YNi_2B_2C the Ir has diffused into the YNi_2B_2C layer. This has been confirmed by an EDX line scan across the interface. Ir was not detected in the MgO substrate.

The bright field TEM cross-sectional image of the YNi_2B_2C film deposited on an Ir buffer layer is shown in Fig. 2a. There is a sharp interface between MgO and YNi_2B_2C. Figures 2b and c are selected area electron diffraction patterns of the MgO substrate and the YNi_2B_2C film along the [100] zone

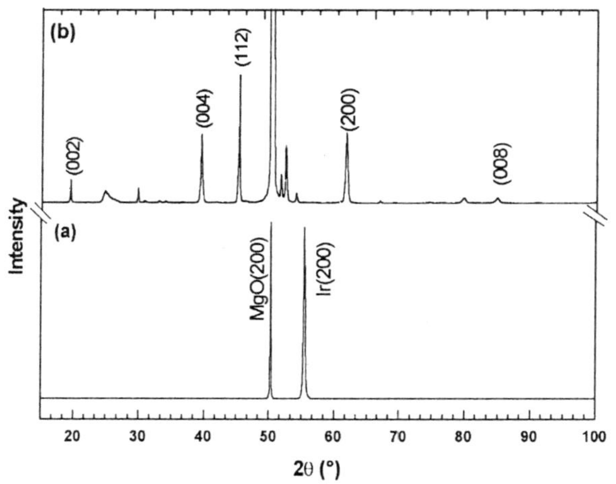

Fig. 1: (a) XRD θ-2θ scan of an Ir layer on MgO substrate, (b) XRD θ-2θ scan for an YNi_2B_2C film deposited on Ir/MgO substrate.

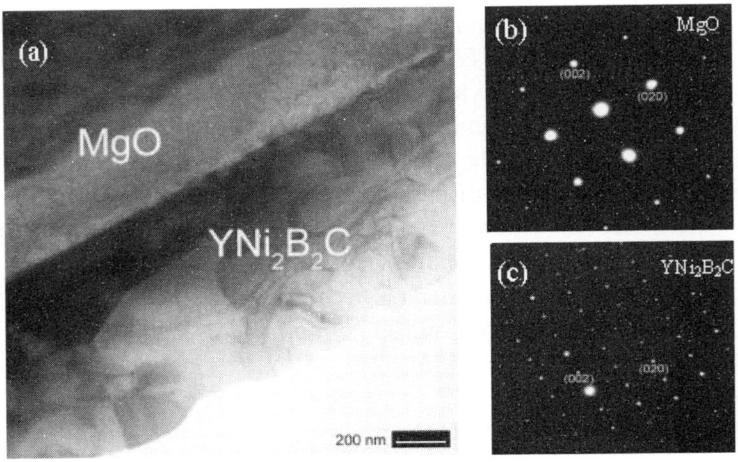

Fig. 2: (a) Cross-section TEM bright field image of an YNi_2B_2C film on an Ir buffer layer. (b) SADP of MgO along the [100] zone axis, (c) SADP of YNi_2B_2C along the [100] zone axis.

axis yielding the orientation relationship $(001)[100]_{MgO}//(001)[100]_{YNi2B2C}$. Electron diffraction proves that the white strip is MgO and therefore belongs to the MgO substrate. It may have been produced by the bombardment of Ar ions during ion milling. From electron diffraction it is clear that the fine grains in the YNi_2B_2C film are preferentially aligned with their c-axis normal to the substrate. The composition ratio of Y to Ni was confirmed to be 1:2 within the error of EDX analysis. Because of the small atomic weights of B and C their concentrations could not be determined.

The Ir buffer layer seen in Fig. 3a has a strong cube-on-cube texture, while in Figs. 3b and c, the YNi_2B_2C film grows with the c-axis preferentially aligned perpendicular to the substrate surface. However, the texture is not a simple c-axis fibre texture, but it consists of four components rotated about the c-axis by 15° (Fig. 3c). The orientation relation given by the electron diffraction patterns of Figs. 2b and c belongs to one of the four components found by texture measurements. The texture observed for borocarbide films deposited on an Ir buffer layer is different from that reported in (Wimbush et al., 2000) for direct growth on MgO. However, it should be noted that the in-plane alignment is quite weak. Further work is necessary to improve the borocarbide texture in the direction of a desired single texture component.

Figure 4 shows the temperature dependence of the resistance of YNi_2B_2C deposited on an Ir buffer layer at 900 °C. The critical temperature T_c^{90} is 13 K with a broad transition to superconductivity of about $\Delta T_c = 4$ K, where T_c^{90} is the temperature at which the resistance has dropped to 90% and the transition width $\Delta T_c = (T_c^{90} - T_c^{10})$. Ir probably substitutes Ni leading to deviations from stoichiometry. According to Behr et al. (1999) and Ledig

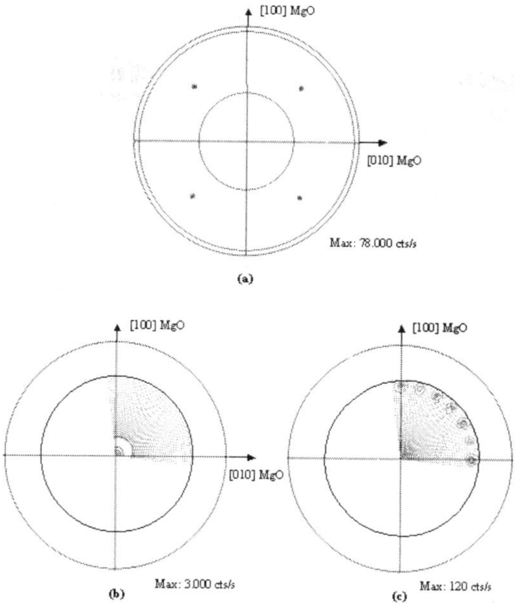

Fig. 3: X-ray texture measurements: (a) (111) pole figure of an Ir film deposited on MgO, (b) (004) and (c) (112) pole figures of an YNi_2B_2C film deposited on Ir/MgO.

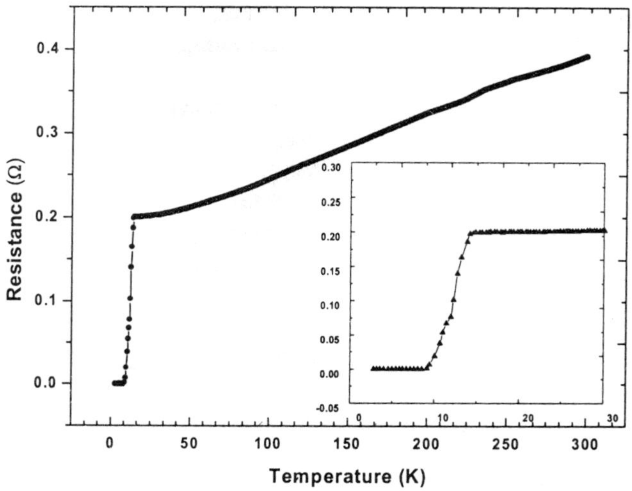

Fig. 4: Temperature dependence of the resistance of YNi_2B_2C on Ir/MgO. The inset shows in detail the superconducting transition.

et al. (1999), these deviations may be the reason for the low T_c measured. Moreover, the large transition width may result from the diffusion profile. Large deviations may also lead to the formation of other phases. Indeed, occasionally YNi_4B has been found by TEM in the interface region.

CONCLUSIONS

An Ir buffer layer has been successfully used to prevent the formation of an Y_2O_3 layer during PLD of YNi_2B_2C on MgO. However, as Ir diffuses into the YNi_2B_2C layer the superconducting properties of the film become deteriorated. Future work should concentrate on lowering the deposition temperature in order to decrease the diffusivity. However, this should not be allowed to conflict with optimal texture formation of the borocarbide films.

Acknowledgement: This work was supported by the Deutsche Forschungsgemeinschaft as part of SFB463, "Rare earth transition metal compounds: Structure, magnetism and transport".

REFERENCES

Andreone, A., Iavarone, M., Vaglio, R., Manini, P. and Cogliati, E. (1996). *Appl. Phys. Lett.*, **69:** 118.

Arisawa, A., Hatano, T., Hirata, K., Mochiku, T., Kitaguchi, H., Fujii, H., Kumakura, H., Kadowaki, K., Naamura, K. and Togano, K. (1994). *Appl. Phys. Lett.*, **65:** 1299.

Arisawa, S., Hatano, T., Nakamura, K. and Togano, K. (1998). *Physica*, C **308:** 67.

Behr, G., Löser, W., Graw, G., Nenkov, K., Krämer, U., Belger, A. and Wehner, B. (1999). *J. Mater. Res.*, **14:** 16.

Cao, G.H., Simon, P. and Skrotzki, W. (2004). *J. Mater. Res.*, in press.

Cao, G.H., Simon, P., Krämer, U., Wimbush, S.C. and Holzapfel, B. (2004). *Chem. Mater.*, in press.

Cimberle, M.R., Ferdeghini, C., Guasconi, P., Mare, D., Putti, M., Siri, A.S., Canepa, F., Manfrinetti, P. and Palenzona, A. (1997). *Physica*, C **282-287:** 573.

Häse, K., Holzapfel, B. and Schultz, L. (1997). *Physica*, C **288:** 28.

Hilscher, G. and Michor, H. (1998). Studies of High Temperature Superconductors, Vol. 26, ed. Narlikar, A.V., Nova Science Publishers, New York.

Ledig, L., Hough, D. and Oertel, C.-G. (1999). J. Eckert and W. Skrotzki, *J. Alloys Comp.*, **285:** 27.

Müller, K.H. and Narozhnyi, V. (2001). *Rep. Prog. Phys.*, **64:** 943.

Reibold, M., Wimbush, S.C., Holzapfel, B. and Krämer, U. (2002). *J. Alloys Comp.*, **347:** 24.

Siegrist, T., Zandbergen, H.W., Cava, R.J., Krajewski, J.J. and Peck, Jr. W.F. (1994). *Nature*, **367:** 254.

Wimbush, S.C., Häse, K., Schultz, L. and Holzapfel, B. (2000). *J. Phys. Condens. Matter*, **13:** L355.

X-ray Diffraction and Absorption Tomography for Three Dimensional Measurements in Bulk Material

S.F. Nielsen

Materials Research Department
Risø National Laboratory, DK-4000 Roskilde, Denmark

INTRODUCTION

Thermomechanical processing of materials generally introduce a change in microstructure and the properties of the thermomechanically processed material depend on this microstructure. Characterisation of the microstructure is thus essential in order to understand the thermomechanical processes, but the characterisation is complicated by the fact that the available experimental techniques in general cannot provide the complete description of the microstructure. Optical and electron microscopy gives a detailed two dimensional picture of the microstructure, which relatively easily may be combined with diffraction methods for determination of crystallographic orientations for an even more complete description of the microstructure. Examples of such methods for additional orientation determination are analysis of Kikuchi patterns (Liu, 1994) and electron back-scattering patterns, EBSP (Krieger Lassen et al., 1992; Wright and Adams, 1992). The two-dimensional microscope techniques are ideal for static characterisation at discrete processing intervals (*post mortem analysis*), but not for *in-situ* characterisation of the dynamics of structural development.

Three-dimensional X-ray diffraction (3DXRD) microscopy (Poulsen et al., 1997; Juul Jensen and Poulsen, 2000) is a new technique, which offers the possibility of time resolved three-dimensional mapping of structures to the micrometer scale. The technique is thus complementary to more traditional microscopy and enables acquisition of experimental data on scientific topics, which earlier could only be analysed theoretically. The aim of the present paper is to give an overview of 3DXRD microscopy and its potentials for *in-situ* characterisation of thermomechanical processes. For more in-depth description of the method and the obtained scientific results, the reader is referred to the relevant papers in the reference list, or the web pages www.metals4d.dk, www.risoe.dk/afm/synch and www.3dstrain.dk.

THE 3D X-RAY DIFFRACTION MICROSCOPE

The three dimensional X-ray diffraction microscope (3DXRD) are installed at the Materials Science beamline (ID11) at the European Synchrotron Radiation Facility (ESRF) in France. The microscope was developed jointly by the Materials Research Department at Risø and ESRF and it was commissioned during the summer of 1999. The microscope allows two-dimensional focusing of hard X-rays (45-90 keV), by using a bent Laue crystal and a bent multilayer as focusing elements (see Fig. 1). The technical details of the same is described elsewhere (Lienert et al., 1999; Wright and Adams 1992). Focal spot sizes are achieved down to $5 \times 5\ \mu m^2$, and the divergence of the monochromatic beam is approximately 0.1-1 mrad.

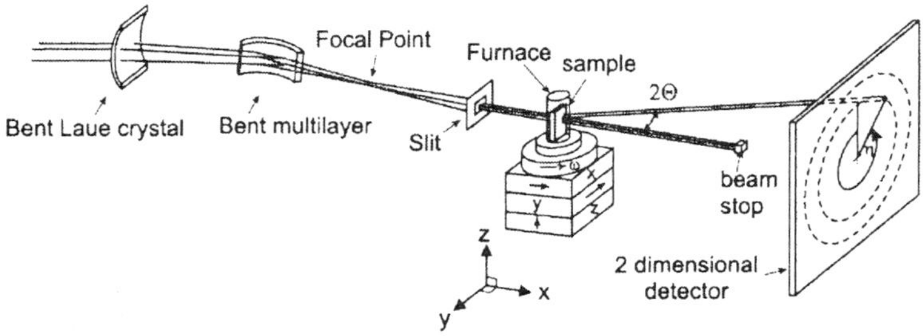

Fig. 1: An illustration of the 3DXRD microscope. The diffracted beam exits the sample in an angle of 2θ and is defined on the two dimensional detector by the azimuthal angle η on the Debye-Scherrer ring. ω is the rotation of the sample (Juul Jensen and Poulsen, 2000).

The ESRF synchrotron provides a large photon flux of about 10^{11} counts per second and the high energy ensures a high penetration power. When the energy of the X-ray beam is 80 keV the penetration depth is 5 mm in steel and 4 cm in aluminium. The penetration depth is normally defined as the depth where the intensity of the transmitted beam is reduced to 10%. The combination of these features makes the X-ray diffraction (XRD) microscope ideal for non-destructive characterisation of the microstructure, in the μm-scale range, within the bulk of crystalline materials.

The microscope is a two-axis diffractometer and consists basically of a detector arm and a sample tower. The sample tower can be translated along three axes in an orthogonal co-ordinate system (x, y, z). A rotation unit is mounted above the tower and a sample stage with an extra set of x and y translations to be used for alignment of the sample. The sample tower is designed to carry loads up to 200 kg making it possible to mount a stress rig or a furnace at the sample position for *in-situ* measurements. Even with a heavy load the sample tower can be rotated and translated with an absolute accuracy of a few microns. All translations and rotations are motorised and controlled through the SPEC program.

Figure 2 is an image of the 3DXRD microscope. The white synchrotron beam enters the beamline through a hole in the wall, and a monochromatic beam is singled out in the optics box. If a sample is placed in the monochromatic beam the resulting diffracted beam can be studied by positioning a two dimensional detector in its path by scanning the detector arm. Presently three different CCD's are available at the beamline; one wide-range with a pixel size of 120 μm (Frelon), one intermediate-range with a pixel size of 67 μm (Metoptics), and one semi-transparent narrow-range detector with a pixel size of 5 μm (Quantics).

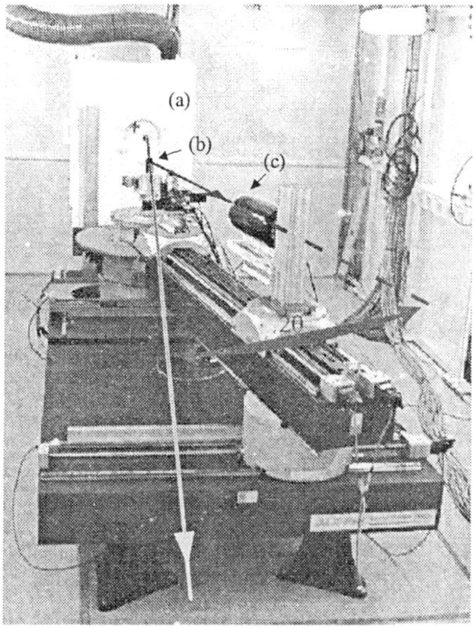

Fig. 2: Image of the 3D X-ray diffraction microscope at beamline ID11. (a) is the optics box with the monochromator, (b) is the sample and (c) is a two dimensional detector. The monochromatic and the diffracted beams are indicated by arrows.

Even though the 3DXRD microscope has a very simple set-up it can be used for depth-resolution studies. The main problem in achieving 3D information in all XRD experiments is to determine the origin of the diffraction in the direction of the incoming beam. Three different methods to obtain this longitudinal resolution can be applied: conical slits (Nielsen et al., 2000), X-ray tracking with line focus (Poulsen et al., 2001) or point focus and focusing analyzer optics (Lienert et al., 1999). The former two will be described in the following in greater detail.

SLITS AND LONGITUDINAL RESOLUTION

In most XRD experiments a sample is placed in the X-ray beam and the resulting diffraction is studied. It is not obvious, however, from exactly where inside the sample the diffraction originates. The longitudinal resolution is missing. A gauge volume within the sample can be defined by using conventional slits to confine both the incoming and the diffracted beam (see Fig. 3). By back-tracing the edges of the slits, a volume element in the sample is found for both the incoming and the diffracted beam. The crossing of the two volume elements is the gauge volume within the sample where the diffraction must have originated. Alternatively, focusing instead of slits could be used to confine the incoming beam.

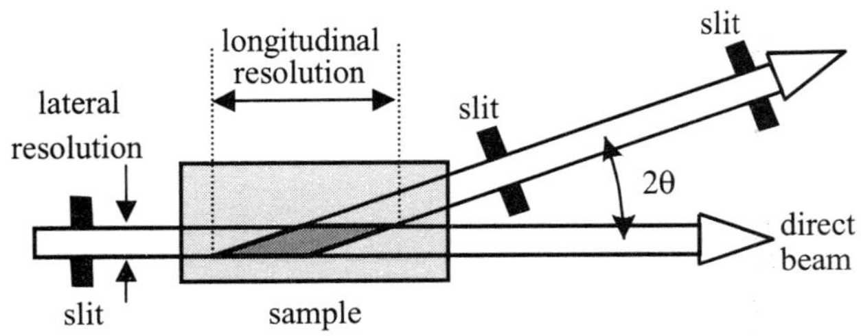

Fig. 3: Two dimensional illustration of the cross-beam technique where slits are used to confine both the incoming and diffracted beam. The resulting gauge volume is shaped as a needle.

A disadvantage of this technique is that only one gauge volume, one reflection and one orientation is measured at a time. This problem can be overcome, however, by using a conical slit (CS) with openings along the Debye-Scherrer cones. When a CS is positioned in the diffracted beam it will define a three-dimensional gauge volume in the sample. If a two-dimensional detector is used it will be possible to record a full Debye-Scherrer ring and thereby to get complete information on the texture from one particular gauge volume in only one image. By using a detector far away from the origin of the diffraction, to ensure a high angular resolution, information of the strain-field could in principle be obtained simultaneously. Another advantage is that one diffraction ring consists of several diffraction spots yielding information from many different reflecting grains in one image. A CS provides a considerable faster way to obtain longitudinal resolution than traditional slits.

Figure 4 illustrates the experimental set-up when using a CS. A sample is positioned in a monochromatic beam that is defined in two dimensions either by a slit or by focusing. The CS is mounted after the sample in the diffracted beam. Further downstream a two-dimensional detector can record the diffraction that has been allowed through the conical slit.

By varying the material thickness of the conical device it can be designed to function as a slit or as a collimator. As a slit, the strain profile of the

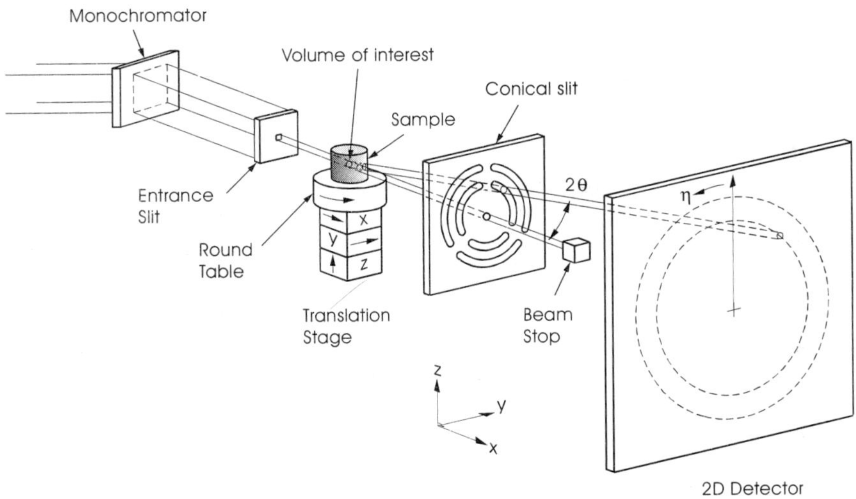

Fig. 4: An illustration of the experimental set-up. The CS selects the diffraction from a specific gauge volume in the sample.

sample can be directly monitored in the images, provided the 2D detector has sufficient resolution. Moreover, for coarse-grained samples the integrated intensity of a single grain can be deduced from a single exposure by summing over the strain profile. For fine-grained samples, the option of summing over the strain profile implies that texture components are directly observable. As a collimator, only a part of the strain profile is allowed to pass. Hence, neither strains, nor integrated intensities are directly observable. Instead it is necessary either to scan the incoming energy or to deconvolute the spatial degrees of freedom (Lienert et al., 2000). Intensities will also be substantially smaller in this case. On the other hand, a collimator defines the gauge volume better than a slit. In the limit of an infinitesimally thin slit the conical openings will be rings. From here on the conical slit openings will be referred to as rings.

DESIGN AND MANUFACTURING OF THE CONICAL SLIT

The manufactured CS has six slit openings, each with a gap of 25 μm. The size of the gaps was chosen as a compromise between achieving a high resolution and still having a reasonable intensity in the Debye-Scherrer rings on the detector. The CS is 4 mm thick, made of tungsten carbide and designed to be used at X-ray energies just above the tungsten *K*-edge for the investigation of fcc materials. The six fcc powder rings that are allowed through the CS are the {111}, {200}, {220}, {222}, {331} and {422} reflections. The CS also has a central hole to allow the direct beam to pass through the slit. The heat-load from the direct beam can then be shifted to an appropriate beam-stop after the CS.

In order for diffraction to occur from a crystalline sample, the Bragg equation has to be fulfilled. The conical slit openings in the CS had to be manufactured with an angle 2θ so the Bragg diffraction at one particular energy from a given fcc material with a certain set of *d*-spacings would be allowed through the slit. The slit openings in the CS is designed to fit the calculated Debye-Scherrer cones from an Ag powder with a lattice constant a_{Ag}= 0.40862 nm, a distance of 10 mm between the powder and the CS, and an X-ray energy of 69.263 keV. Other fcc materials can be investigated by changing the energy. The X-ray energy that is necessary to make the diffraction from a particular fcc material fit the slit openings in the CS can be calculated by the relation $2d \sin \theta = 12.398/E$, where d is the lattice distance for the fcc material, and 2θ is the fixed opening angle in the CS.

The CS was constructed at Institute für Mikrotechnik Mainz, Germany by wire electrodischarge machining (wire-EDM). This is a thermal method where material in the working zone is melted and removed through a surrounding dielectric liquid. Various EDM techniques have been established, for example in the tool making industry, for the generation of moulds or dies for punching purposes. The EDM technique offers a high degree of freedom in precisely manufacturing small components. Unfortunately this method can only be used with wires down to 30 μm in diameters. The gaps in the CS were designed to be 25 μm and it was, therefore, impossible to produce the whole CS from one single piece of tungsten. Instead seven conical parts were manufactured individually. When assembled the seven pieces form the CS (Fig. 5a,b). The conical parts were manufactured with small uniform taps to ensure a constant slit gap (Fig. 5c). The taps also have the advantage that they make the conical parts self-aligning when they are assembled. Machining by wire-EDM heats the surroundings of the cutting zone. Unfortunately heating can introduce microcracks in the material. By lowering the discharged energy the heat effected layers can be removed furnishing a very smooth surface finish (Fig. 5d). The main problem in the fabrication process was the formation of elliptically shaped conical parts instead of the intended circular cross section. The inner and outer diameters on both sides of each individual part were measured in a scanning electron microscope and the absolute error on the diameters was found to be at worst 13 μm and in average 4.6 μm.

EXPERIMENTAL AND CALCULATED DEPTH RESOLUTION

A test of the CS was performed at the synchrotron beamline ID15, ESRF. The aim was to measure the experimental depth resolution. The CS only allows diffraction from a specific gauge volume to pass through the slit, or seen from the other direction the CS so to speak focuses on the gauge volume. By scanning the CS along the x-axis the CS will focus on different gauge volumes within the thickness of the sample (see Fig. 6). When focusing

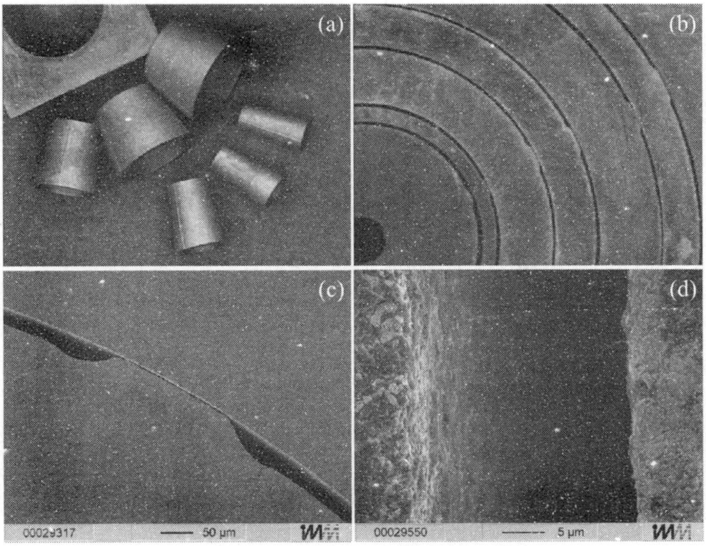

Fig. 5: (a) The disassembled CS consists of seven parts, (b) top view of the assembled CS, (c) one tab that provide a constant silt opening of 25 µm, and (d) detailed view of gap between two parts.

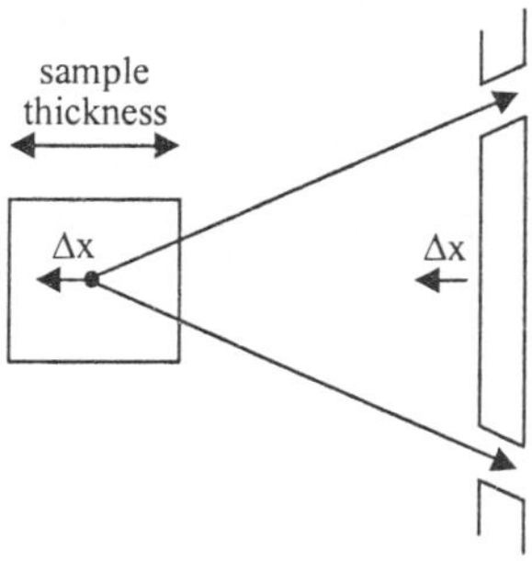

Fig. 6: The figure illustrates that a movement Δx of the CS will cause the gauge volume to move the same amount. The depth resolution of the CS can then be directly measured if the thickness of the sample is known by scanning the gauge volume through the sample.

outside the sample no intensity, should in principle, be measured. If the thickness of the sample is known, a scan along the x-axis will provide a direct measure of the longitudinal resolution of the CS.

The sample used for this test was a 100 µm thick layer of nickel (Ni) powder, packed between two glass plates. The Ni powder was provided by P. Suortti and he has earlier described the powder as well annealed and with virtually no strain (Suortti, 1977). The incoming beam was defined by a square aperture to 22×24 µm^2, and the beam profile was found to be almost Gaussian in both directions. To acquire the images a two-dimensional image intensifier was placed 30 cm from the Ni sample. The aligning of the CS was performed as described in the previous section. However, assemblage errors of the CS made it impossible to optimise the intensity in all the rings at the same time. The CS was scanned in a three-dimensional mesh by translating in the x, y and z directions. An exposure was acquired at each position. The global assemblage errors can be found by evaluating the azimuthal variation of the intensity in the six rings in all the images and comparing the x, y and z settings where the individual rings are aligned. During the experiment, no attempt was made to find the optimal settings for the aligning of all the rings that take the assemblage errors into account. Instead we chose to align the CS with respect to the {200} ring. Figure 7 shows a typical image of the powder diffraction that is transmitted through the CS. The figure clearly shows that for example the {422} ring is not completely aligned (some intensity is missing in the lower left part of the diffraction ring).

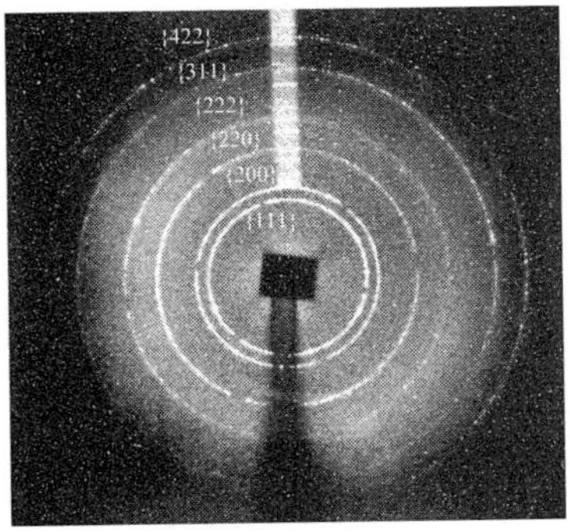

Fig. 7: A typical recorded image of powder diffraction when using the CS. The {hkl} indexes of the individual rings are written on top of the image.

To find the depth resolution the CS was scanned in the x-direction through the {200} alignment centre. Thirteen images were recorded with equidistant steps of 100 μm along the x-axis. For each ring the intensity was integrated radially as well as azimuthally, by using the image-processing programme *FIT2D* (Hammersley et al., 1996). The normalised integrated intensity distributions as function of the x-position of the CS are plotted in Fig. 8 for the four innermost rings. Due to a lack of recorded intensity the two outermost rings are not included. The variation of the integrated intensity with x was fitted to a Gaussian. The resulting fitted centre positions and widths are listed in Table 1. The table also contains calculated values for the widths of a perfect CS, assuming a Gaussian profile of the incoming beam.

As seen from Table 1 the measured and calculated depth resolutions are in reasonable agreement with better resolution for higher-order reflections as expected. In general the measured resolutions are smaller than the calculated. This is caused by the combined effect of the assembly and aligning errors that make the slits appear more narrow when seen from the

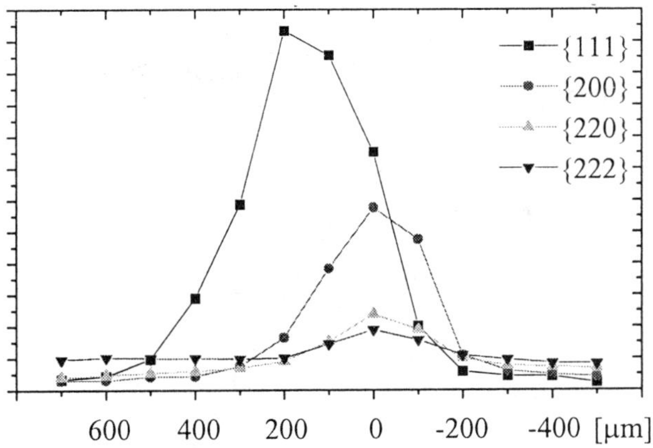

Fig. 8: Normalised plot of the integrated intensity as function of the x-position of the CS. $x = 0$ corresponds to the position where the CS is focusing on the middle of the 100 μm thick Ni powder sample.

diffracted beam. The centre positions are constant within the errors on the measurements, with exception of the {111} reflection, which is offset by 153 μm. The magnitude, seen in relation to the measured width, implies that assembly errors have caused the focal point of the {111} reflection to be shifted. The {111} slit in the CS is so to speak cross-eyed compared to the other slit openings. In general we estimate the average machining and assemblage errors on the diameter of the CS rings to be both of the order of 5 μm.

Table 1: The calculated and measured FWHM of the four reflections. The measured FWHM are decreasing for higher reflections. The centre positions of the measured peaks are in good agreement, with exception of the {111} peak. This indicates that the {111} slit opening is cross-eyed compared to the other slit openings.

Reflection	*Calculated FWHM [μm]*	*Measured FWHM [μm]*	*Measured centre position [μm]*
{111}	366	275 ± 12	153 ± 5
{200}	315	222 ± 12	0 ± 6
{220}	239	217 ± 16	10 ± 7
{222}	181	192 ± 15	4 ± 7

GRAIN BOUNDARY MAPPING

The main use of the CS is expected to be local strain and stress measurements (Lienert et al., 2000). The CS can also be applied for mapping grain boundaries and grain orientations in polycrystalline samples. Poulsen et al. (1997) have tested a CS in a grain mapping experiment at beamline ID11, ESRF. The aim of this experiment was to map the grain boundaries and grain orientations within an embedded layer of a coarse-grained 4.5×4.5×4.5

mm^3-sized Cu polycrystal. Afterwards the sample was sectioned and the same layer that was mapped in the synchrotron experiment was investigated by electron microscopy (EBSP). The result from the two different techniques was then compared.

The set-up was identical to that presented in Fig. 4, with the incoming beam being defined by a 50 × 50 μm^2 aperture. Results shown here refer to one specific layer and the (200) data only. To find the orientation of a specific grain the sample was first rotated from –90° to 90° in steps of 10° while oscillating by ±5° at each step. By identifying at least three major spots relating to lattice planes that are perpendicular to each other, a first fit was obtained. The exact orientation was determined by acquiring exposures at constant ω and 'scanning' ω with an accuracy of $\Delta\omega = 0.1°$. To map a specific grain the variation in total intensity of a selected reflection was monitored while scanning the gauge volume in x and y. The boundary, defined at the half-intensity point, was determined by interpolation.

Resulting synchrotron and EBSP data are shown in Fig. 9. The map provided by synchrotron diffraction is not complete, due to an inadequate amount of beamtime (about 12 hours). Grain positions measured by the two

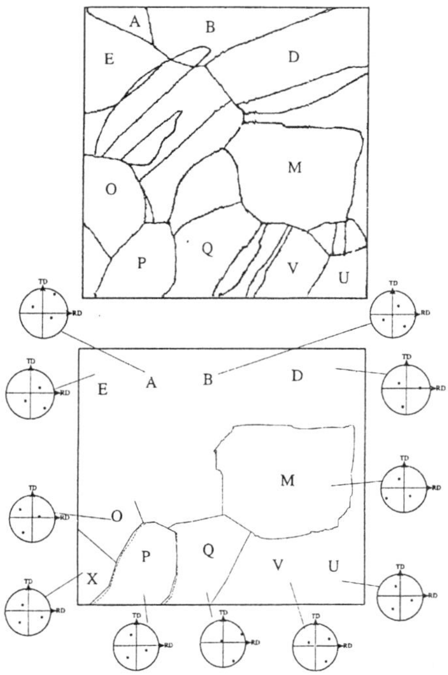

Fig. 9: Grain mapping of the top layer in a Cu poly-crystal by electron microscopy (EBSP) above, and using the hard X-ray conical slit set-up illustrated in Fig. 4, below. Also shown are the determined <200> pole figures for the individual grains. The boundary of grain P was measured using a (200) reflection from grain P (solid line) as well as (200) reflections from neighbouring grains X, O and Q (dashed lines).

methods correspond within 150 μm and orientations within 2°. There is a large potential for substantial improvements of these numbers, especially the spatial accuracy. Firstly, the impinging beamsize can be reduced to 5x5 μm^2 by focusing the monochromatic beam. Secondly, measuring several reflections from the same grain, corresponding to approximately 90° sample rotations, can diminish the effect of the poor resolution in the direction along the beam. Thirdly, slight misalignments during synchrotron measurement implied that the measured layer is not identical to the surface measured by EBSP.

The work with the CS has been published (Nielsen et al., 2000) and it is probably the first attempt to use a CS for depth profiling, three-dimensional mapping of grain boundaries and grain orientations, in an XRD experiment. Earlier a conical device has been used in the diffracted beam (Häuserman and Itié, 1992), but only for 'beam cleaning' and energy determination in high-pressure research. The conical device was in their case a conical collimator with only one ring.

The basic principle can naturally be applied to X-rays with lower energies or to neutrons, but in these cases flat conical devices and flat 2D detectors cannot be applied due to the larger scattering angles. The main drawback by using a CS is that it will only work for samples that belong to one specific symmetry group and only at one specific energy. The CS described in this paper was designed for fcc materials, which of course limits the applicability of the CS. The conical parts all suffer from some degree of eccentricity, even though the most modern wire-EDM machining was used to produce the conical parts. When assembled, the achieved slit width is approximately 25 ± 5 μm. This uncertainty influences the resolutions of the CS and makes it difficult to compare intensities through different rings. The assembly errors could have been avoided if it had been possible to manufacture the six slits in one piece of tungsten. When the slit is assembled from seven conical pieces the combined machining and assembly errors could be determined by performing a calibration of the CS. However, this would complicate the data analysis substantially.

The tab system that ensures self-alignment of the cones needs to be modified. Structuring the inside of the cones and at the same time increasing the height of tabs on the outside would definitely lead to an improvement. Alignment and tilting errors between the cones due to the geometrical errors of tabs over the entire height could be minimised. The cones would be secured against each other. Therefore, distortion could be excluded as well. A better self aligning system for the cones in the CS would ensure that all slit openings focus on the same gauge volume, and cross-eyed slits like the (111) ring would be avoided. A straightforward way to reduce the errors is to make the conical device thinner. This is feasible, as the penetration depth at 70 keV is 62 μm in tungsten. Hence even at 1 mm thickness, the transmitted fraction is 10^{-7}. A thinner device would also constitute a more ideal slit.

The aligning of the present CS is a complicated task because an error in the position of the CS and the related angular settings has the same effect. The misalignment of each of these "pairs of parameters" can be recognised from the appearance of the intensity transmitted through the CS. With an appropriate beam and an appropriate reference sample, the position and the orientation of the CS can be adjusted.

As illustrated earlier the CS can be used to map grain boundaries and grain orientations and compared to conventional slits the acquisition time is short. However, if a large ensemble of grain were to be mapped, it would not be possible with the CS within a reasonable time. The CS has a large potential within stress and strain measurements from a particular gauge volume, but when speed is required in order to map grain boundaries and grain orientations in a large volume element it would be better to use the so-called X-ray tracking technique.

X-RAY TRACKING

The microstructure in polycrystalline materials has mostly been studied in planar sections by microscopy techniques like SEM on sample surfaces, or TEM on thin foils. The 3DXRD microscope at ESRF provides a new and fast non-destructive technique for studying planar sections in the bulk of samples by using a so-called tracking technique. All essential features like the position, volume, orientation, stress-state of the grains can be determined, including the morphology of the grain boundaries.

The tracking algorithm is inspired by the use of three-dimensional detectors in high-energy physics and is based on a monochromatic high energy X-ray beam focused into a line and a two-dimensional detector. The principle is sketched in Fig. 10. When the incoming beam is focused into a line it defines a layer within the sample.

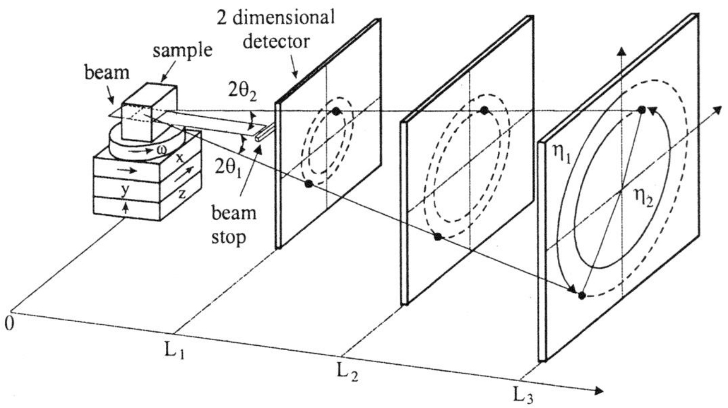

Fig. 10: Illustration of the tracking technique. When the detector is translated away from the sample the diffraction spots will move outwards on the detector. The positions of the diffracting grains in the sample are determined by linear fits. The fits also provide the 2θ and η values.

All grains that happen to fulfil the Bragg condition in this layer give rise to diffraction spots on the detector. The diffraction spots are the projected images of the corresponding diffracting grains in the observed layer and the projection angle is a combination of the scattering angle 2θ and the azimuthal angle η on the detector.

The position of each diffraction spot on the detector is measured and the intensity-weighted centre-of-mass (CM) of each diffraction spot is computed. This procedure is repeated at several detector distances corresponding to different L values in Fig. 10. Linear fits through corresponding CM points extrapolate to the CM of the diffracting grains and provide the angles 2θ and η. To obtain the cross-sectional grain shape, the periphery of the diffraction spot in the image acquired at the closest distance is projected into the illuminated sample plane along the direction determined by the fit.

By rotating the sample around an axis ω, perpendicular to the illuminated plane, all grains will come to fulfil the Bragg condition and a complete map of all grain boundaries and grain orientations in the plane is thus produced. A three-dimensional map can be obtained simply by translating the sample in z and repeating the procedure for several layers. Examples of actual images acquired with different detector-to-sample distances are given in Fig. 11. The diffraction spots are clearly seen to move outwards on the detector when the detector is translated away from the sample. The detector is in this case a CCD-plate, which is built into a detector system that consists of two fluorescence screens 200 μm apart. This allows the detector to be aligned in such a way that the direct beam goes between the two fluorescence screens. Only the diffracted beam hits the fluorescence screens and is transferred by optics to the CCD. The direct beam has a two-dimensional Gaussian shape and some of the tails will be detected by the part of the fluorescence screens that is closest to the gap between them. This effect is seen in Fig. 11 as two bright rectangles in the centre of the images. All other spots are diffraction spots from one layer in an aluminium sample.

Fig. 11: Three images of the diffraction pattern from an Al sample. The distance from the sample to the detector in (a), (b) and (c) are 7.6 mm, 10.3 mm and 12.9 mm, respectively. The two rectangles in the middle of the images correspond to the tails of the direct beam, which are detected by the edges of the two fluorescence screens in the detector.

THE PERIPHERY OF A DIFFRACTION SPOT

The incoming monochromatic beam can be considered as a series of parallel single X-rays in the x-y plane. Each ray will be either diffracted or not diffracted in a particular grain. The resulting intensity distributions across a diffraction spot should therefore in theory be a step-function, but this will only be the case for a perfect crystal and a perfect detector. Fig. 12 is an example of the intensity distribution across two diffraction spots. In Fig. 12a the intensity distribution is nearly flat at the centre, with some tails reflecting the instrumental resolution and the mosaic spread of the grain. If the grain has a large mosaicity, certain parts of the grain will diffract at a slightly different ω rotation of the sample. Some parts of the intensity will be missing, and the intensity distribution will have local minima instead of a flat plateau at the centre (Fig. 12b).

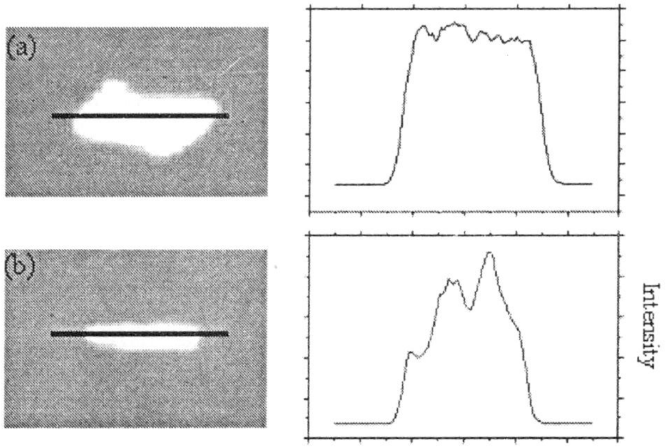

Fig. 12: Two diffraction spots (a) and (b) with low and high mosaicity, respectively. The intensity along the black lines was measured and the intensity distributions are plotted next to the diffraction spots.

The variation in the shape of the intensity distributions across the diffraction spots makes it difficult to determine the periphery of the spots. As a first approximation the periphery of the spots can be determined by setting a fixed intensity threshold. The diffraction is stronger from some (hkl) reflections than from others. This can be taken into account by letting the fixed intensity threshold be a function of the maximum intensity of the spot under investigation.

Another and more consistent way to define the periphery of a diffraction spot is to approach the spot from all directions on separate lines. The edges of a diffraction spot can then be defined in these intensity profiles as the points where the profiles have the steepest slopes. By combining all the edge points we get the periphery of the spot. When using a fixed threshold on a diffraction spot where some intensity is missing due to mosaicity, a local minimum close to the edges of the spot would cause the spot to appear smaller than it actually is. This would be the case for the diffraction spot in Fig. 12b if the threshold were fixed at 40%. When the steepest slope approach is used on Fig. 12b it is much more likely to determine the correct periphery of the diffraction spot, but it should be underlined that edge finding always will be a matter of definitions. All the reconstructed grain boundaries in this paper have been determined by using the steepest slope approach because it gives boundaries with the smallest misfit compared to the boundaries in the EBSP images.

THE GRAIN BOUNDARY POSITION

The tracking algorithm is based on acquiring a series of images of the same diffraction pattern while the detector is translated away from the sample (see Fig. 10). If individual points on the peripheries of the same diffraction spot in this series could be identified, then corresponding points could be fitted and extrapolated into the sample plane. The identification of individual points is unfortunately not possible. Instead it was decided to determine the peripheries of corresponding diffraction spots in the series and calculate the intensity-weighted centre of mass (CM) of all the pixels within these peripheries. The CM values were then fitted by a straight line and the determined angles 2θ and η were used to project all points on the periphery of the diffraction spot recorded at the closest distance to the sample into the sample plane. The shortest possible distance between the sample and the detector ensures the lowest effect of any divergence in the diffracted beam on the diffraction spot.

When the sample is rotated, several different reflections will be observed from the same grain as different diffraction spots. When the periphery of these spots have been determined they can be projected into the sample plane by their respective directions and they should correspond to exactly the same grain boundary in the sample plane. This will only be the case if the sample is perfectly aligned in the centre of the ω rotation. If it is not perfectly aligned then the vertical axis of the sample will not be identical with the rotation axis and the rotation will make the sample swing around in the shape of a cone (see Fig. 13). Even if the sample is very carefully aligned there will always be some sphere of confusion due to the precision of the translation and rotation stages. This error can be minimised by using several reflections from the same grain and minimising the misfit between their respective projected peripheries in the sample plane.

The parameters that have to be fitted are the ω rotation, the sample translations x and y, the sample-to-detector distances and the position of the direct beam on the detector. This fit was optimised by visual inspection of the misfit between the projected peripheries of different reflections from the same grain. The most important parameter was found to be the position of

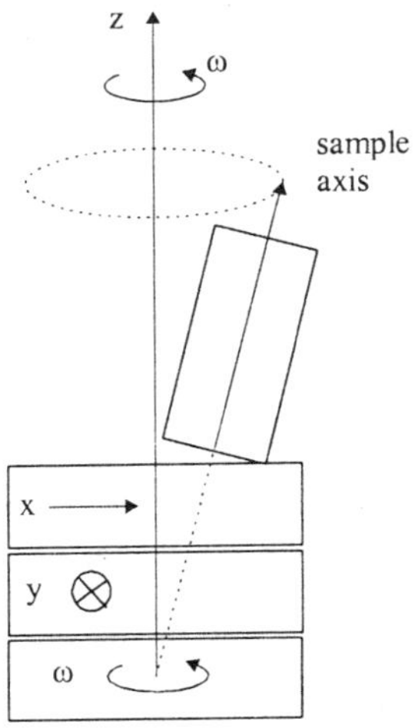

Fig. 13: If the sample is not perfectly aligned then the vertical sample axis will turn around like a cone when the sample is rotated in ω. In the projection of the periphery of the diffraction spots it is assumed that the sample axis is aligned with the rotation axis. The periphery of the diffraction spot will therefore be projected back to a wrong position in the sample plane that is not identical with the boundary of the actual diffracting grain.

the sample relative to the ω rotation axis, e.g. the sample translations in x and y. Visual inspection is not the best way to determine a global minimum, but a large improvement was achieved by this fit even though it was not possible to get all the projected peripheries of different reflections from the same grain to be identical in the sample plane. In Fig. 14, the determined grain boundaries from three different reflections from the same grain have been superimposed as white lines on an EBSP images of the same Al sample. The tracking was performed 10 μm below the sample surface.

The variations of the shape of the determined boundaries in the sample plane are partly caused by the fit. Only a local minimum has been determined. Another reason for the variation is that different peripheries are projected in different directions. Depending on the η angle, each projected periphery will have a large uncertainty in one direction and a small uncertainty in the perpendicular direction. For $\eta = 0$ the uncertainty of the determined boundary will be large in the x-direction and small in the y-direction.

It was attempted to take the dependence of the projection direction into account when comparing the boundaries with the EBSP image by averaging the determined boundaries from different reflections from the same grain. This was done by selecting a centre point of the grain and rotating a straight line around the centre point. The mean grain boundary was then determined by averaging the points where the straight line intersects the boundaries from different reflections. The proper way of weighting the uncertainties in this averaging was unfortunately not determined. Figure 14 shows the calculated mean of three different boundaries of the same grain where the uncertainties have not been taken into account. By comparing the misfit between the determined boundaries and the EBSP image we see that the boundary from the (331) reflection has the smallest misfit. It was therefore decided to use single reflections in all plots of determined boundaries instead of averaging between boundaries from different reflections that in principle should improve the result.

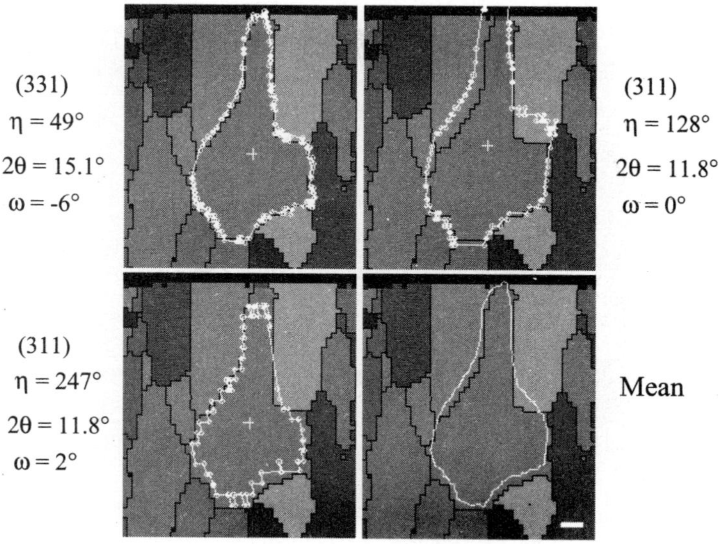

Fig. 14: The boundary of one grain determined from three different reflections. The image in the lower right corner is the calculated mean of the three boundaries. The white scale bar in the right corner is 100 μm.

The tracking technique is a non-destructive technique for bulk studies, but due to the novelty of the technique it was necessary to carry out the experiments in such a way that it is possible to compare the results with standard techniques. As a validation test of the technique, a coarse-grained 99.996% pure aluminium sample was investigated. The sample was first annealed for 12 hours at 500 °C and slowly cooled to minimise the mosaic spread in the grains. One sample surface was polished and the grains at this surface were mapped by electron microscopy. The EBSP image was acquired with a step size of 20 μm. The range of orientation variations within the grains was found to be less than 1°. Next, the sample was aligned with the same surface parallel to the beam at the 3DXRD microscope and the tracking procedure was performed with a line focused X-ray beam (800×5 μm^2) incident 10 μm below the surface. The sample dimensions were 2.5×2.5 mm^2, making it necessary to acquire information from three stripes across the sample. With one second exposure time, 22 ω-settings with $\Delta\omega = 2°$ and L = 7.5, 10.3 and 13 mm, the total data acquisition time was less than two minutes. The speed of this technique should be compared with conventional diffraction set-ups for depth resolved studies using slits formed as pinholes, grids (Wroblewski et al., 1999) or cones (Nielsen et al., 2000) before and/ or after the sample, which lead to very slow data acquisitions.

The resulting grain boundaries determined by the two different techniques are superposed in Fig. 15. The black lines in the figure indicate the grain boundaries as determined by electron microscopy (EBSP), and the white lines indicate the grain boundaries resulting from the synchrotron X-ray

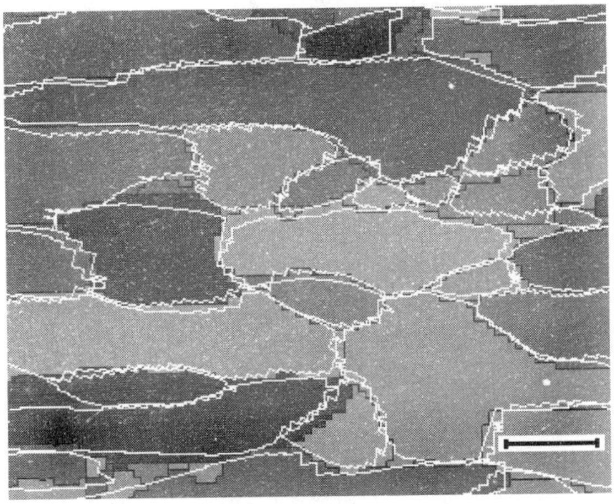

Fig. 15: The determined grain boundaries from the tracking experiment are superimposed on an EBSP image of the sample surface. The grey scale of the grains in the EBSP image indicate the macroscopic orientation of the grains. The scalebar in the image is 400 μm.

tracking experiment. The tracking boundaries are raw data from single reflections with no interpolation or averaging between reflections from the same or neighbouring grains. The misfit between the tracking and the EBSP boundaries have been found by linear intercept to be 26 μm in average with a maximum of 40 μm. Maps of the grain boundary structure determined by the tracking technique will thus be of sufficient quality for many applications. For example, in recrystallisation studies the evolution of separate nuclei can be followed during annealing (Lauridsen et al., 2000), and in estimates of grain-to-grain interactions where centroid descriptions often are sufficient. To increase the quality of the maps, software is presently being developed that makes use of the inherent crystallographic features. This includes automatic procedures for interpolation between reflections from the same and neighbouring grains. All tracking results presented in this paper have been calculated by hand with the exception of the orientations of the grains.

The grey scale of the grains in Fig. 15 indicate different grain orientations from the EBSP measurement. The EBSP was acquired on a standard JEOL840 SEM microscope where it is not possible to align the sample relative to the microscope axis with a sufficient precision. The tilts of the sample are normally corrected by eye. It is therefore not possible to compare the determined macroscopic orientation from the EBSP measurement and the tracking experiment. Relative changes in the orientation like the misorientation across a grain boundary can, on the other hand, easily be compared.

The misorientations across the grain boundaries in Fig. 15 have been calculated from the tracking data and found to be equal to the EBSP measurement with an uncertainty of less than 1.2°. This misfit between the calculated misorientations from the two techniques is partly due to the combined uncertainties of the measurements in the set-ups and partly due to the way the misorientations were calculated. The misorientations were found from the EBSP data by calculating the misorientation between the average of 10 measured points on each side of the grain boundaries in contrast to the tracking technique where all points within the grain area automatically are averaged when the orientation of a grain is determined. In the calculation of the orientation of a grain the 2θ and η values are used to define the reflection. These angles are determined by the linear fit through the calculated intensity weighted CM values of the corresponding diffraction spots. In case of mosaicity the CM values of the diffraction spots will be wrong resulting in not only a wrong position and shape of the reconstructed grain boundary but also in a wrong calculated orientation of the grains. In the present experiment the mosaicity was 1° and the sets of reflections from the same grains only had a variation of less than 0.1° relative to their theoretical calculated G-vectors.

GRAIN ORIENTATIONS

In the continuing development of the 3DXRD-microscopy technique focus is presently on establishing software for on-line data analysis. This is a non-trivial task, due to the magnitude of data (presently of order 10 GB/hour), and the dimensionality of the data, spatial position (x, y, z) and crystallographic orientation (φ_1, Φ, φ_2), which needs to be reconstructed. A reconstruction algorithm suitable for this task, named GRAINDEX, (Lauridsen et al., 2001) is already available.

A full data set obtained by the tracking technique typically consists of a set of some hundred images acquired at different detector-sample positions and different settings of the ω rotation stage (see Fig. 1). The total number of spots are of order 100,000. Graindex associates the spots in the images with reflections and sorts these according to which grain they originate from. The sorting is presently based on two criteria: the spatial origin of the reflection and the crystallography. For small grains the instrumental resolution of order 25 μm implies that emphasis must be on crystallographic sorting.

At first glance, it seems the crystallographic sorting could be based on comparing the angles between all pairs of reflections to a list of theoretical angles, dictated by the space-group symmetry. However, as n reflections give rise to 2^n possible groups, the speed of analysis becomes prohibitive for large n. Instead, a novel approach is applied where the whole symmetry of the space-group is used. Basically, a discrete set of crystallographic orientations, resembling all possible orientations of a grain, is tested and compared to the recorded images.

Once the grains have been sorted, the positions, volumes and crystallographic orientations of the grains can be fitted. Alternatively, the relevant part of the data may be reinvestigated for a stress analysis. Hence, the Graindex algorithm has been indispensable for initial analysis of the various data sets. In addition to the two mentioned criteria, the integrated intensity could be used for sorting provided the stoichiometry of the grains is fixed. Alternatively, single crystal refinements can be performed to elucidate the structure factors (and thereby the stoichiometry) of the grains (Schmidt et al., 2003).

GROWTH KINETICS OF INDIVIDUAL GRAINS

Recrystallization is the combined process of nucleation and growth (Humphreys and Hatherley, 1995; Hansen et al., 2000; Sakai and Suzuki, 1999). Several experimental techniques like electron microscopy and neutron diffraction can be used for *in-situ* characterisation during annealing. However, microscopic techniques are limited to observations on sample surfaces and such surface investigations of individual grain growth have always been questionable because it cannot be directly proved that the surface observations are truly representative of the bulk material. In the surface, special conditions may dominate, for example, due to different diffusion conditions, grain boundary geometries and thermal grooving (Humphreys and Hatherley, 1995; Rabkin and Klinger, 2001). Neutron diffraction is capable of probing the bulk of a sample, but the spatial resolution of this technique is typically too low to resolve individual grains. Neutron diffraction can only, in the case of mm-size grains, be used to study the growth of individual grains (Branger et al., 2000). Most experimental techniques for studying recrystallization phenomena have been used for static descriptions of the recrystallization in a series of samples. Consequently the recrystallization kinetics can only be studied by combining several static descriptions of different samples, resulting in average kinetic descriptions, which is the basis for present day understanding.

The 3DXRD microscope provides a unique combination of high penetration power, high intensity and a spatial resolution comparable to the size of a typical nucleus (Bay and Hansen, 1979). These properties make the microscope ideal for studying the nucleation and growth of individual grains within bulk material (Lauridsen et al., 2000). An as-deformed specimen can be mounted in the 3DXRD microscope, which can be set up to study the recrystallisation inside a specimen gauge volume of typically 200 × 200 ×

1000 μm^3. The specimen can be rotated around an axis ω parallel to the z axis and images can be acquired for each ω setting. The specimen is typically oscillated by Δω = 0.5° during each exposure in order to obtain a complete integration of the intensity in the diffraction spots. While the specimen is annealed *in-situ* in an argon atmosphere, a series of Debye-Scherrer diffraction patterns can be acquired by a two-dimensional detector. Each diffraction spot in the Debye-Scherrer rings corresponds to a specific grain and as the grain grows during the annealing the intensity increases in the corresponding diffraction spot.

Therefore, growth characteristics of individual grains can be determined by measuring the integrated spot intensities as a function of annealing time. In order to make sure that the integrated spot intensity is proportional to the volume of the grain, the Bragg reflection has to lie completely within the ω-range measured and the grain has to lie completely within the gauge volume defined by the slits and the thickness of the specimen. The first criterion can be validated by using consecutive ω-settings. The first and last ω-setting can only be used for validation purposes; hence a minimum of three consecutive ω-settings are required. The integration of a Bragg reflection is only valid if there is no intensity of the reflection left at the two neighbouring validation settings. The number of ω-settings applied is a trade-off between grain statistics and acquisition speed. By using five consecutive ω-settings a time resolution of about 2 min in the growth curves can be obtained. The second criterion was validated by opening up the slits, and hence illuminating a wider channel through the sample while keeping the oscillation Δω constant. If the intensity in a diffraction spot increases while opening up the slits, the corresponding grain is not fully embedded in the gauge volume and it is disregarded in the further analysis.

Figure 16 is an example of growth curves observed in a 90% cold-rolled commercial aluminium alloy AA1050 during annealing at 270 °C (Lauridsen et al., 2003).

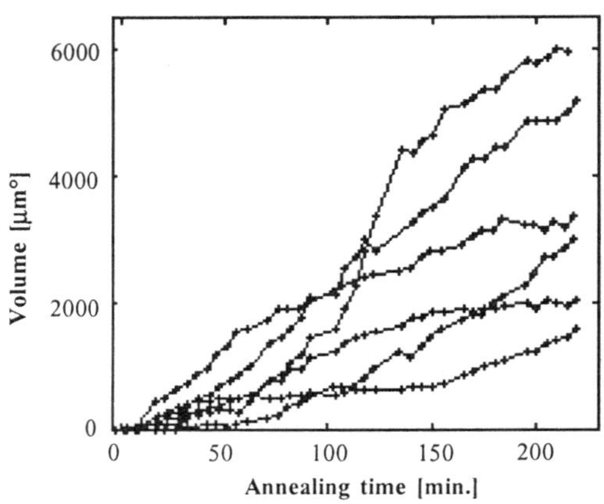

Fig. 16: Growth curves that show large variations in the growth kinetics of the individual grains.

LATTICE ROTATIONS DURING DEFORMATION

Electron microscopy gives detailed information on the deformation microstructure and local distribution of crystallographic orientations, which may be related to active slip systems. However, these methods do not allow measurements of the evolution in structure or orientation for specific individual grains other than those in the surface.

Experimental observations of the rotation pathways of individual grains within the bulk of polycrystalline metals as a function of tensile strain have been performed using the 3DXRD microscope (Margulies et al., 2001). An Instron tensile stress rig is mounted on the sample stage, and a CS is aligned between the sample and detector such that its focal point coincides with the rotation axis of the sample stage and the focal point of the X-ray focussing optics. This acts to define a gauge volume within the bulk of the sample, which will give rise to diffraction spots on the detector. Diffraction from volumes outside this gauge volume will be rejected by the slit. In this way we can provide a three dimensionally resolved reference volume within the sample, and limit the spot overlap problem, which would otherwise be faced when examining thick or fine-grained samples. The conical slit is used in conjunction with a point focused beam which leads to a gauge volume of approximately 5 × 5 × 250 μm^3. By assuring that the gauge volume is aligned on the centre of rotation, and that the sample is probed far from the surface, we can increase the number of valid grains measured at each position within the sample. The type of measurements that have been made, are illustrated here with the results from 95 grains with nearly random initial orientation. Measurements are made for strains up to 6% in steps of 2%. The rotations of the tensile axis for the 95 grains are presented in Fig. 17. All

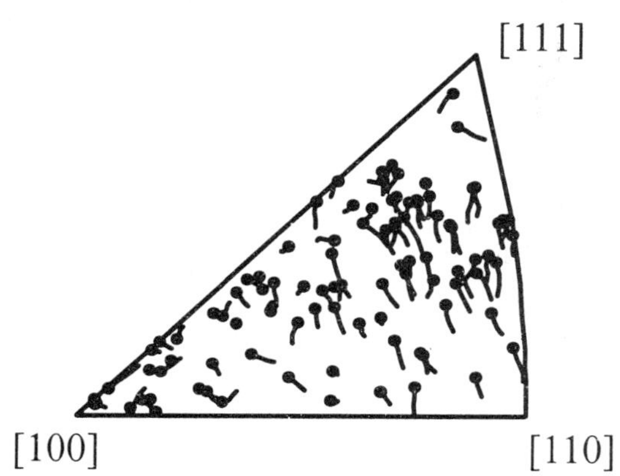

Fig. 17: The rotation of the tensile axis of 95 embedded aluminium grains during tensile deformation (0 to 6%) plotted in the stenographic triangle. The curves are the observed paths for the average orientation of each grain. The final orientation of the tensile axis is marked with a filled circle. (Poulsen et al., 2003)

the grains undergo plastic deformation, which is also evidenced by spot broadening. The rotations of the individual grains at 6% elongation vary between 0.2 and 5.5° and the rotation paths are as seen on the figure approximately straight.

PLASTIC DEFORMATION

When studying plastic deformation of materials it is often necessary to know the deformation in local areas on the surface or in the bulk. Depending on the deformation process and the material there can be local plastic strain variations down to the μm-scale, which are important for the mechanical properties of the material.

Spatial variations in plastic deformation can be studied experimentally by observing scratchs, etch patterns or grids placed on the surface (Bartholomeusz and Wert, 1994; Thorsen and Bilde-Sørensen, 1999; Liu and Fischer, 1997; Berka et al., 1998; Soppa et al., 1999; Crostack et al., 2001). Figure 18 is an illustration of a Cu sample with an aluminium oxide grid on the surface. The spatial resolution of these methods depend on the size and the distance between recognizable features before and after a given deformation increment.

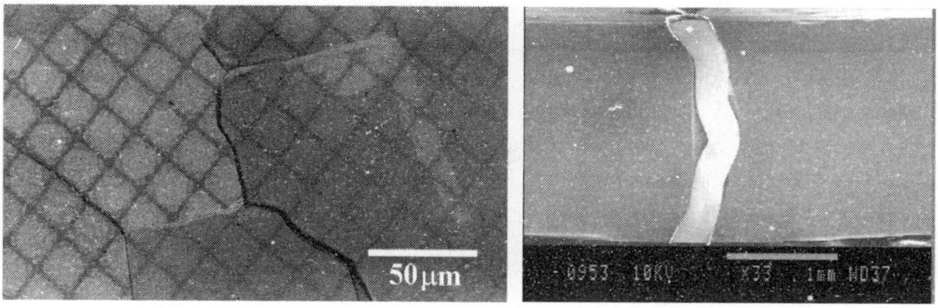

Fig. 18: Left: Cu sample with an aluminium oxide grid on the surface, which show that the grid points are displaced with respect to each other in this case due to diffusional creep (Krieger Lassen, 1992). Right: A gold wire inserted in an aluminium sample that have been roled (30%) and sectioned at the gold wire. The SEM image with the deformed gold wire show the bulk have been deformed (Wert, 2002).

Corresponding measurements within the material are experimentally even more complicated. Two-dimensional grids or marker wires within samples have in the past been used to study deformation processes within bulk material (Ya et al., 1965; Kak and Slaney, 1988). The spatial resolution of these methods is in order of 0.1 mm, which is insufficient to study local deformation. At the same time these methods demand drilling holes or welding material together around a grid, which most likely will change the material properties, and the application of these methods is therefore limited.

A novel method to determine the plastic deformation in bulk material with a spatial resolution in the μm- range is presented in this paper. X-rays from a synchrotron are used to perform absorption tomography on a sample that contains marker particles with high absorption contrast. The displacements of individual particles are identified as function of imposed strain and the local plastic deformation gradient components are determined in the subsequent analysis.

A sample was made by compacting aluminium powder mixed with 1 vol% tungsten powder. The size of both types of powder particles was in the range of 1-10 μm. The powder mixture was cold compressed (30 MPa) and hot compressed (60 MPa) at 825 K in five minutes. The result was a compact sample with a relative homogenious distribution of W particles (Fig. 19). A cylinder with a diameter of 1 mm and a height of 2 mm was cut from the material and investigated tomographically before and after compression to macroscopic strains of 2.7, 6.2 or 9.5%.

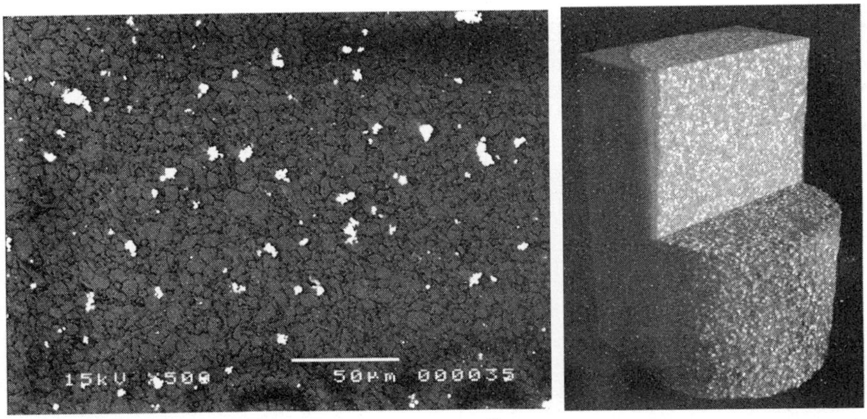

Fig. 19: Left: SEM image of the W particles (white) and the Al grains (grey). Right: Tomographic reconstruction of the cylindrical specimen.

The tomographic technique was developed in the late 1970s (Kak and Slaney, 1998) and is based on recording images of the absorption contrast, so called radiograms, while the sample is rotated around an axis perpendicular to the X-ray beam. If the absorption contrast is recorded on a 2D detector, the image will be a projection of the sample absorption from a given angle. By combining projections of the absorption from different angles, the absorption can be calculated in the bulk of the sample. This is called a tomographic reconstruction. Advantages of using synchrotron radiation for tomographic investigations is that the divergence of the beam is small and the intensity high. This combination makes it possible to do micro-tomography with a spatial resolution of 1-2 μm.

The experiment described here was performed at HASYLAB (Beckmann et al., 1999) with an X-ray energy of 24 keV. The spatial resolution in the reconstructed volumes is 2 μm. The sample was deformed step-wise to enable identification of the marker particles both before and after a deformation step from the distance they are displaced. This approach is possible as long as the particles are displaced less than the average distance between the particles in one deformation step. When the positions of the marker particles are known both before and after the deformation, then the displacement field can be determined. Figure 20 shows a 3D image of the displacement of all the particles in the sample after compression to 9.5%.

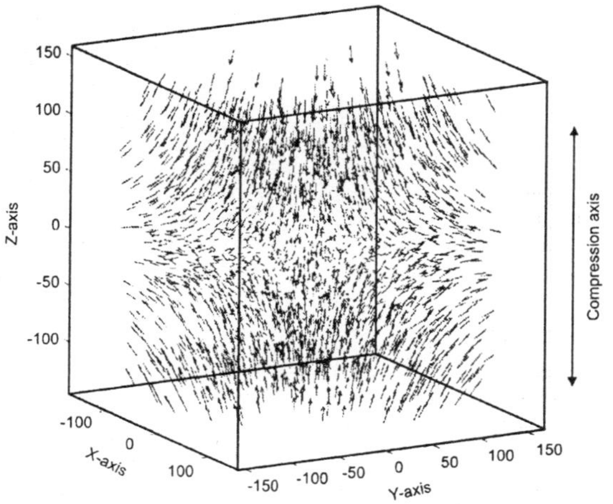

Fig. 20: A 3D image of the displacement of the marker particles after compression to 9.5%.

The nine components of the displacement gradient tensor is defined by:

$$e_{ij} = \left(\frac{\partial u_i}{\partial x_j}\right) \forall \, i, j \tag{1}$$

where u_i are the particle displacement components and x_j are the position components.

The displacement gradient tensor was determined for each particle by a least-squares fit to the relative displacement of the eight nearest neighbours. The nine e_{ij}'s were in this way determined at the position (x, y, z) of all the marker particles. This makes the displacement gradient tensor hard to visualize in three dimensions. One useful visualization method is to interpolate each *e*-component on a regular grid and to assign each grid point the average value of that *e*-component for all the particles that lies within a radius of half the distance between the grid points. The nine *e*-components can now be visualized by plotting them layer by layer (see Fig. 21). Figure 21 shows a very homogeneous straining which is as expected for a powder metallurgical sample of aluminium grains with approximately the same size as the marker particles. Never-the-less there do exists small variations in the straining (e.g. see Fig. 21 near x = 200 μm, y = 400 μm and z = 120 μm).

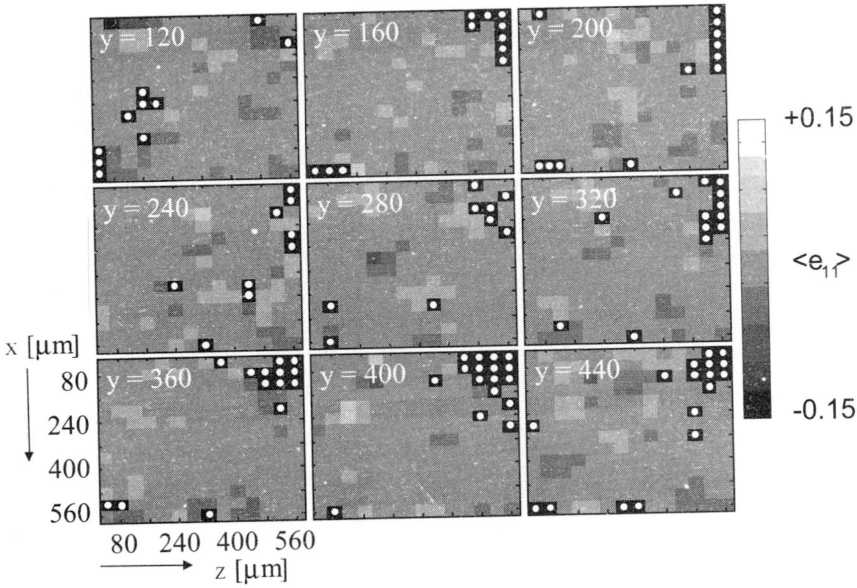

Fig. 21: Strain contours in 3D for the displacement gradient tensor component e_{11}. The black voxels with the white dots are created during the interpolation on the regular grid where no particles were found to be within 15 μm from the grid point.

The described method for non-destructive determination of the local plastic deformation in three dimensions in bulk material can be applied to study most cases of conventional deformation processes like rolling, tension and torsion. The method can be applied to study heavy deformation as long as the marker particles do not fragment and as long as the deformation is performed step-wise, in such a way that the particles can be re-identified after each strain increment.

During deformation of a metal that contains particles, which are harder than the matrix material, local compatibility stresses induce a plastic strain gradient in the matrix near each particle. This local deformation zone extends typically one particle diameter into the matrix material from the surface of the particle (Humphreys, 1985). The described method uses the displacement of the particles to determine the local strain. Since the spacing of the particles is much larger than the deformation zone dimension, the deformation zone local to each particle is displaced with the particle and does not influence displacement of neighbouring particles. Thus, to a first approximation the measurements are not perturbed by marker particles. The lower limit for the size of the marker particles are determined by the spatial resolution of the applied X-ray detector. At present the best X-ray detectors for tomography experiments can detect particles with a diameter above 0.7 μm.

CONCLUSIONS

Two different non-destructive synchrotron techniques capable of mapping grain boundaries and the crystallographic orientation of grains in the bulk of polycrystalline samples have been demonstrated in this paper. The conical slit is a useful tool for depth profiling that can be used on even highly deformed samples. One Debye-Sherrer ring consists of several diffraction spots and a conical slit therefore provides a considerable faster way to obtain longitudinal resolution than traditional slits where each reflection has to be scanned individually. The X-ray tracking technique is an even faster technique and therefore more suitable for three-dimensional mapping. A complete map of a 2.5×2.5 mm^2 layer can be obtained in less than 2 min. As demonstrated the 3DXRD allow us to study the growth and rotational history of individual grains in a polycrystalline sample.

A novel method for non-destructive characterization of local plastic deformation in bulk material was also presented. The method can be applied on all types of materials, crystalline as well as amorphous, which contain marker particles. The described method can be applied to study most cases of conventional deformation processes like rolling, tension, torsion and compression (Nielsen et al., 2003). The method can also be applied to study heavy deformation as long as the marker particles do not fragment and as long as the deformation is performed step-wise, in such a way that the particles can be re-identified after each strain increment.

The lower limit for the size of the marker particles is determined by the spatial resolution of the applied X-ray detector. At present, the best X-ray detectors for tomography experiments can detect particles of a diameter exceeding 0.7 μm.

REFERENCES

Bartholomeusz, M.F. and Wert, J.A. (1994). *Mater Charact.*, **33:** 377.

Bay, B. and Hansen, N. (1979). *Metallurgical Transactions*, A **10:** 279.

Beckmann, F., Bonse, U. and Biermann, T. (1999). Proc. SPIE Conf. Developments in X-ray Tomography II, Denver Colorado, **3772:** 179.

Berka, L., Sova, M. and Fischer, G. (1998). *Experimental Techniques*, **22:** 22.

Branger, V., Mathon, M.H., Baudin, T. and Penelle, R. (2000). *Scripta Mater.*, **43:** 325.

Crostack, H.A., Fischer, G., Soppa, E., Schmauder, S. and Liu, Y.L. (2001). *J. Microscopy* – Oxford, **201:** 171.

Hammersley, A.P., Svensson, S.O., Hanfland, M., Fitch, A.N. and Häusermann, D. (1996). *High Press. Res.*, **14:** 235.

Hansen N. et al. (2000). ed. Proceedings of the 21th Risø International Symposium on Materials Science "Recrystallization - Fundamental Aspects and Relations to Deformation Microstructure", Risø.

Häuserman, D. and Itié, J.P. (1992). *Rev. Sci. Instrum.*, **63:** 1080.

Humphreys, F.J. and Hatherley, M. (1995). Recrystallization and Related Annealing Phenomena. Pergamon Press, New York.

Humphreys, F.J. (1985). *In*: Dislocations and Properties of Real Materials, ed. M.H. Loretto, Institute of Metals, London, 175.

Juul Jensen, D. and Poulsen, H.F. (2000). Recrystallization in 3D. *In*: Proceedings of the 21th international symposium on materials science: Recrystallization – Fundamental aspects and relations to deformation microstructure. Edited by N. Hansen et al. (Risø National Laboratory, Roskilde), pp.103-124.

Juul Jensen, D., Kvick, Å., Lauridsen, E.M., Margulies, L., Nielsen, S.F. and Poulsen, H.F. (2000). *Mat. Res. Soc. Symp. Proc.*, **590:** 227.

Kak, A.C. and Slaney, M. (1988) Principles of computerized tomographic imaging. New York: IEEE Press.

Krieger Lassen, N.C., Juul Jensen, D. and Conradsen, K. (1992). *Scanning Microscopy*, **6:** 115.

Lauridsen, E.M., Juul Jensen, D., Poulsen, H.F. and Lienert, U. (2000). *Scripta Mater.*, **43:** 561.

Lauridsen, E.M., Poulsen, H.F., Nielsen, S.F. and Juul Jensen, D. (2003). *Acta Mater.*, **51:** 4423.

Lauridsen, E.M., Schmidt, S., Suter, R.M. and Poulsen, H.F. (2001). *J. Appl. Cryst.*, **34:** 744.

Lienert, U., Martins, R., Grigull, S., Pinkerton, M., Poulsen H.F. and Kvick, Å. (2000). *Mat. Res. Soc. Symp. Proc.*, **590:** 241.

Lienert, U., Poulsen, H.F., Honkimaki, V., Schulze, C. and Hignette, O. (1999). *J. Synchrotron Rad.*, **6:** 979.

Lienert, U., Poulsen, H.F., Martins, R.V. and Kvick, Å. (2000). *Mater. Sci. Forum* **347:** 95.

Liu, Q. (1994). *J. Appl. Cryst.*, **27:** 755.

Liu, Y.L. and Fischer, G. (1997). *Scripta Mater.*, **36:** 1187.

Margulies, L., Winther, G. and Poulsen, H.F. (2001). *Science*, **291:** 2392.

Nielsen, S.F., Lauridsen, E.M., Juul Jensen, D. and Poulsen, H.F. (2001). *Mat. Sci. Eng.*, **A319-321:** 179.

Nielsen, S.F., Poulsen, H.F., Beckmann, F., Thorning, C. and Wert, J.A. (2003). *Acta. Mater.*, **51:** 2407.

Nielsen, S.F., Wolf, A., Poulsen, H.F., Ohler, M., Lienert, U. and Owen, R.A. (2000). *J. Synch. Rad.*, **7:** 103.

Poulsen, H.F., Garbe, S., Lorentzen, T., Juul Jensen, D., Poulsen, F.W., Andersen, N.H., Frello, T., Feidenhans'l, R. and Graafsma, H. (1997). *J. Synch. Rad.*, **4:** 147.

Poulsen, H.F., Margulies, L., Schmidt, S. and Winther, G. (2003). *Acta Mater.*, **51:** 3821.

Poulsen, H.F., Nielsen, S.F., Lauridsen, E.M., Schmidt, S., Suter, R.M., Lienert, U., Margulies, L., Lorentzen, T. and Juul Jensen, D. (2001). *J. Appl. Cryst.*, **34:** 751-756.

Rabkin, E. and Klinger, L. (2001). Revue de Metallurgie-Cahiers de *Informations Techniques*, **12:** 1059.

Sakai, T. and Suzuki, H.G. (ed.) (1999). Proceedings of ReX'99 "The fourth International Conference on Recrystallization and Related Phenomena".

Schmidt, S., Poulsen, H.F. and Vaughan, G.B.M. (2003). *J. Appl. Cryst.*, **36**: 326.

Soppa, E., Schmauder, S., Fischer, G., Thesing, J. and Ritter, R. (1999). *Computational Materials Science*, **16:** 323.

Suortti, P. (1977). *P. Acta Cryst.*, **A33:** 1012.

Thorsen, P.A. and Bilde-Sørensen, J.B. (1999). *Materials Sci. and Eng.*, **A265:** 140.

Wert, J.A. (2002). *Acta Mater.*, **50:** 3125.

Wright, S.I. and Adams, B.L. (1992). *Met Trans.*, **A23:** 759.

Wroblewski, T., Clauss, O., Crostack, H.A., Ertel, A., Fandrich, F., Genzel, C., Hradil, K., Ternes, W. and Woldt, E. (1999). *Nucl. Inst. Meth.*, **A428:** 570.

Ya, I., Tarnovskii, A., Pozdeyev, A. and Lyashkov, V.B. (1965). Deformation of Metals during Rolling. Oxford: Pergamon Press.

Structural Characterization of Ultra Thin Films Using Specular X-ray Reflectivity Technique

S. Banerjee

Saha Institute of Nuclear Physics
1/AF Bidhan Nagar
Calcutta 700 064, India

INTRODUCTION

Ultra thin films are very important technologically for device application. Structural characterizations of these films are necessary to design the films with required physical properties and in understanding its performance. Grazing incidence X-ray reflectivity (GIXR) technique is recently becoming very popular for structural characterization of ultra thin inorganic and organic films. There are various destructive and nondestructive tools to characterise ultra thin films. GIXR is a non-destructive tool for structural characterization of ultra thin films (Parratt, 1954). In GIXR technique, the X-ray beam is incident on the sample at a grazing angle and the interfered reflected beam is collected by an X-ray counter. This is shown schematically in Fig. 1. The interference occurs due to the presence of surface/film interface and the film/substrate interface in the sample. The interference oscillation also gets modified if density variation exists along its depth or is of a multi-layer in nature. The interference periods depend on the thickness of the layer and the amplitude of the interference oscillations depends on the contrast of the electron density between the layers and the interfacial roughness. Thus the GIXR data contains structural information of the film such as thickness, densities and roughness of the film and its individual layers. But one has to note that from specular GIXR measurement one obtains the densities along the depth of the film and not along the plane of the surface. We will discuss here only the X-ray reflectivity study of a single layer films deposited on a substrate and try to obtain the density variation as a function of its depth.

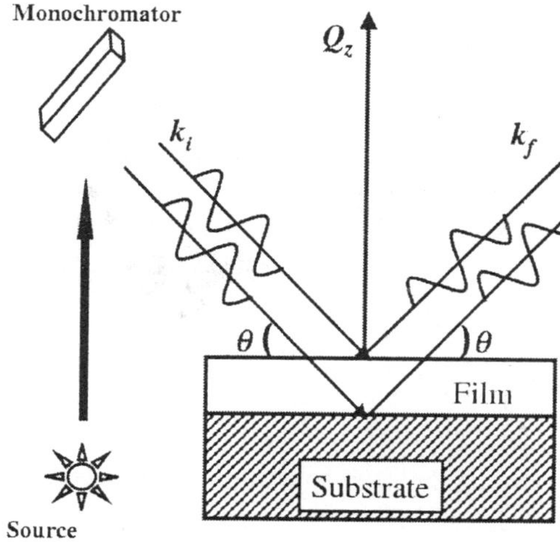

Fig. 1: Schematic of the X-ray reflectivity set up showing the positions of X-ray source, monochromator and the detector with respect to the sample. The incident and the scattered wave vector are also shown in the diagram.

We will first describe the properties of X-ray in the case of specular GIXR and then would like to discuss here various analysis schemes to analyse the X-ray reflectivity data which are existing now. But depending on the system to be studied one has to choose from the available schemes the one which can generate truthfully the parameters having real physical values and simultaneously generates the experimental reflectivity data accurately or simultaneously one can use all the schemes to get proper model of the film and can be confident about the result and interpretation. There are now mainly three schemes to analyze the X-ray reflectivity for ultra thin films and they are (1) a model dependent scheme based on dynamical multiple scattering leading to a recursive formula (Parratt, 1954; Born and Wolf, 1980; Vidal and Vincent, 1984; Gibaud, 1999), (2) model independent method based on distorted wave Born approximation (DWBA) (Sanyal et al., 1993; Zhou and Chen, 1995; Banerjee et al., 1996; Sanyal et al., 1996; Banerjee et al., 2002) and (3) a Fourier inversion technique based on Born approximation (Ming et al., 1996; Banerjee et al., 1999). As a case study, we will demonstrate the above three formalisms on two ultra thin film systems : (1) an ultra thin film of TiN grown by chemical vapour deposition (CVD) technique and (2) an ultra thin film of oxynitride both grown on Si(001) substrate.

SPECULAR X-RAY REFLECTIVITY FORMALISMS

Properties of X-ray

An X-ray is an electromagnetic plane wave and can be described by its electric field vector $\overline{E}(r) = \overline{E}_0\, e^{\text{in(r)k.r}}$ where $k = \dfrac{2\pi}{\lambda}$ is the scattering wave vector and $n(r)$ is the refractive index of the medium. In a medium, this electromagnetic plane wave obeys the Helmholtz equation

$$\nabla^2 \bar{E}(r) + k^2 n^2(r) \bar{E}(r) = 0 \tag{1}$$

If we consider there are effective f_j electrons which determine the forced oscillation strength with frequency ω_j, damping γ_j in each atoms and N atoms per unit volume, then refractive index can be expressed as (Born and Wolf, 1980)

$$n^2(r) = 1 + \frac{Ne^2}{\varepsilon_0 m} \sum_j \frac{f_j}{\omega_j^2 - \omega^2 - 2i\omega\gamma_j} \tag{2}$$

where ε_0 is the permittivity of the free space, e and m are the charge and mass of the electron, respectively. Also, we should note that f_j is a complex number and is expressed as $f_j = f_j^0 + f_j'(E) + if_j''(E)$ where $f_j'(E)$ and $f_j''(E)$ are the dispersion and absorption corrections depending on radiation frequency ω and are well tabulated for each element. For X-rays we know $\omega > \omega_j$ then Eq. (2) can be expressed as

$$n(r) = 1 - \delta(r) + i\beta(r) \tag{3}$$

where the dispersion and the absorption terms are given by

$$\delta(r) = \frac{\lambda^2}{2\pi} r_e \rho(r) \sum_j \frac{f_j^0 + f_j'(E)}{Z} \tag{4}$$

and

$$\beta(r) = \frac{\lambda^2}{2\pi} r_e \rho(r) \sum_j \frac{f_j''(E)}{Z} \tag{5}$$

where $r_e = \dfrac{e^2}{4\pi\varepsilon_o mc^2} = 2.814 \times 10^{-5}$ Å is the classical electron radius, $\rho(r)$ is the electron density, $Z = \sum_j Z_j$ where Z_j denotes the number of electrons for each element in the medium. It should be pointed out that $\delta(r)$ is always a positive quantity and for X-rays it is generally one or two order of magnitude larger than $\beta(r)$ and hence refractive index for the case of X-rays in any material is always less than value 1 i.e., its value is less than in vacuum. And hence, as we know that for the case of visible light one can obtain total internal reflection if the light is incident from a denser medium onto a rarer medium below a certain critical angle from the plane of the interface. A similar phenomenon, but in opposite manner, is observed for the case of X-ray when the X-ray is incident from denser medium (vacuum or air) onto a rarer medium (any material) and then we observe total external reflection below a certain critical angle.

Specular Reflection from a Surface of a Substrate

If we consider X-ray plane wave incident on a surface of a substrate as $e^{ik_0 \cdot r}$, then the reflected wave can be represented by $re^{ik_0 \cdot r}$ and the transmitted wave into the substrate as $te^{ik_s \cdot r}$, where r and t are the reflectance and transmittance coefficient, respectively and k_0 and k_s are the wave vector in air and in substrate, respectively. This is revealed schematically in Fig. 2(a). If the surface is represented by $z = 0$ as shown in Fig. 2 then the X-ray will obey one dimensional Helmholtz equation in both air and substrate as given below

$$\nabla^2 u(z) + k_0^2 u(z) = 0 \quad \forall \ z \leq 0 \tag{6}$$

and

$$\nabla^2 u(z) + k_0^2 u(z) = 0 \quad \forall \ z \geq 0 \tag{7}$$

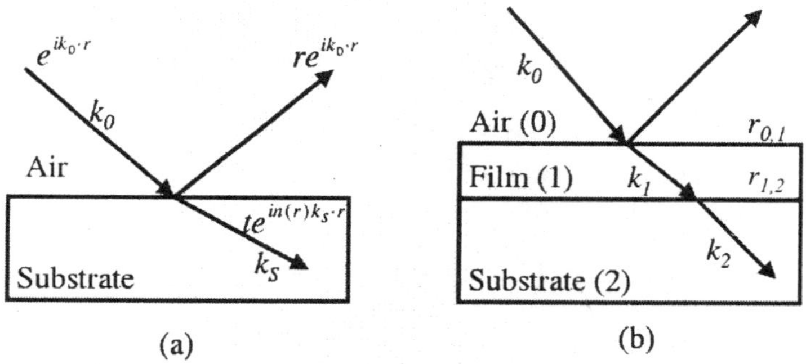

Fig. 2: The schematic showing the incident, scattered and the transmitted wave vectors for (a) substrate and (b) with a thin film on a substrate, $r_{0,1}$ and $r_{1,2}$ are the reflectance coefficient at the interface of air/film and film/substrate, respectively.

where $k_S^2 = k_0^2 - 4\pi\rho_S$ and ρ_S is the electron density of the substrate while $u(z)$ in both the regions can be expressed as

$$u(z) = e^{ik_0 z} + re^{-ik_0 z} \quad \forall \ z \leq 0 \tag{8}$$

and

$$u(z) = te^{-ik_S Z} \quad \forall \ z \geq 0 \tag{9}$$

where Eqns. (8) and (9) satisfy Eqns. (6) and (7), respectively. Using boundary conditions i.e., $u(z^+)_{z=0} = u(z^-)_{z=0}$ and $u/(z^+)_{z=0} = u/(z)_{z=0}$, we then obtain

$$1 + r = t \tag{10}$$

and

$$r = \frac{k_0 - k_S}{k_0 + k_S} \tag{11}$$

Here, in this manuscript, we will not bother with the expression of transmission coefficient t as we will not be measuring this coefficient experimentally. The reflectivity R from the surface will be

$$R = rr^* \tag{12}$$

where r^* is the complex conjugate of the above reflectance coefficient r (Eq. (11)).

Specular Reflection from a Single Layer of Thin Film on a Substrate

Now one can label air, film and substrate as medium 0, 1 and 2, respectively and are shown in Fig. 2(b). One can obtain the reflectance coefficient from a single layer film deposited on a substrate by again constructing the Helmholtz equations for each of the layers and then matching the solutions and its derivatives at each of the boundaries (air/film and film/substrate interface in this case of single layer on substrate) one can get the reflectance coefficient as

$$r = \frac{r_{0,1} + r_{1,2}\, e^{i2k_{z,1}d}}{1 + r_{0,1}\, r_{1,2}\, e^{i2k_{z,1}d}} \tag{13}$$

where $r_{0,1}$ and $r_{1,2}$ are the reflectance coefficient of the air/film and film/substrate interface. By taking its product with its complex conjugate one then obtains the reflectivity as

$$R = \frac{r_{0,1}^2 + r_{1,2}^2 + 2r_{0,1}\, r_{1,2} \cos 2k_{z,1}d}{1 + r_{0,1}^2\, r_{1,2}^2 + 2r_{0,1}\, r_{1,2} \cos 2k_{z,1}d} \tag{14}$$

thus, one observes oscillatory behaviour in the reflectivity profile as shown in Fig.1(b) and the maximum occurs at $k_{z,1}d = n\pi$ where $k_{z,1} = \frac{2\pi}{\lambda} \sqrt{\sin^2\theta - \sin^2\theta_c}$ and θ_c is the critical angle for the total external reflection.

Recursive Formalism

One can also consider a single film to be of a multi-layer type if the density of the film fluctuates across its depth. To analyse the X-ray reflectivity data for a multilayer system one generally uses a recursive formalism similar to Parratt's formalism (Parratt, 1954) or the matrix method (Born and Wolf, 1980; Vidal and Vincent, 1984; Gibaud, 1999) to obtain the structural parameters of the film, such as the thickness, the interfacial roughness and the electron density of each of the layers. In the above formalism the total film is considered to be consisting of n+1 layers and the X-ray reflectivity as a function of scattering wave vector k is expressed by a recursive formula:

$$r_{n-1,\,n}(k) = \frac{r'_{n-1,n} + r'_{n,\,n+1} \exp(2id_n k_n)}{1 + r'_{n-1,\,n}\, r'_{n,\,n+1} \exp(2id_n\, k_n)} \tag{15}$$

where $r'_{n-1,n}$ is Fresnel reflectance coefficient for the interface of layer n and layer n–1 and is equal to $(k_{n-1} - k_n)/(k_{n-1} + k_n)$ as described earlier. The label $n + 1$ is for the substrate and $n = 0$ is the label for the air. k_n is the scattering wave vector in the nth layer and $k_n = (k_0^2 - k_{c,n}^2)^{1/2}$ where $k_0 = (2\pi/\lambda)$ sin θ and $k_{c,n} = (2\pi/\lambda)\sin\theta_{c,n}$, θ and $\theta_{c,n}$ are the angle of incidence and critical angle for the total external reflection, respectively. Electron density of the layer is related to $\theta_{c,n}$ $(= \lambda\ (r_e\rho_e/\pi)^{1/2})$, where λ is the incident wavelength of the X-ray beam, ρ_e is the electron density of the nth layer and is related to mass density of that layer (Ashcroft and Mermin, 1976) and r_e is the electron radius (2.8 × 10^{-5} Å). The reflectance coefficient obtained at the detector is $r_{0,1}$. $r_{n-1,n}$ is used to calculate the reflectance for the layer above the substrate and recursively one can obtain $r_{0,1}$. To take into account the interfacial roughness, one has to multiply each of the reflectance coefficients by $\exp(-2k_n^2\sigma_n^2)$ (Nevot and Croce, 1980), where σ_n is the interfacial roughness of the nth interface. We use the above expression (Eq. 15) to fit the reflectivity data. In the recursive formalism one has to sometime use a user-defined model for the structure of the film across its depth, which consist of a number of virtual layers (boxes) describing the film. The smallest box size is determined by the maximum value of the scattered angle measured. If one considers the virtual layers of equal small thickness, determined by the large value of the scattering vector k_z then one can get various non-unique solutions for the fit parameter because of the nature of non-linear least square fitting and discussion about this is beyond the scope of this manuscript. By fitting the reflectivity data using Eq. (15) one obtains the thickness d_n, electron density (obtained from $k_{c,n}$) and interfacial roughness (σ_n) of the layers as the fit parameters.

Distorted Wave Born Approximation (DWBA) Formalism

This is to certain extent a model independent scheme based on DWBA as shall be discussed below. This scheme works well for only ultra thin films for thickness < 400 Å having small variation in the density of the film across its depth. In this scheme the film is considered to be composed of number of thin equal thickness of virtual slices or boxes of electron density ρ_i of the ith slice/box. If one can evaluate the density of each box then one obtains the density profile of the film as a function of depth. The reflectivity of the film using DWBA formalism is expressed as (Sanyal et al., 1993; Zhou and Chen, 1995; Banerjee et al., 1996; Sanyal et al., 1996; Banerjee et al., 2002)

$$R(k) = \left| ir_0(k) + \frac{(2\pi r_e)}{k}\left(a^2(k)\,\Delta\rho(q_z^f) + b^2(k)\,\Delta\rho^*(q_z^f)\right)\right|^2 \qquad (16)$$

The first term in Eq. (16) containing $r_0(k)$ is the specular reflectance coefficient of the film with an average electron density (AED) of ρ_0. The second term in Eq. (16) is for the small fluctuation of the electron density $\Delta\rho(z)$ over ρ_0 across the film where $a(k)$ and $b(k)$ are the transmission and reflection coefficient in the film, respectively and r_e is the electron radius. $q_z = 2k_z = 4(\pi/\lambda)\sin\theta$, q_z^f [$= (q_z^2 - q_c^f)^{1/2}$] is the wave vector in the film, the subscript c indicates the critical value of the wave vector having a critical angle (θ_c) up to which the total external reflection occurs and the superscript f indicates the value of the wave vector in the film. The $\Delta\rho(q_z^f)$ is the Fourier transform of variation of electron density $\Delta\rho(z) = \rho(z) - \rho_0$, where z is the depth of the film and is expressed as

$$\Delta\rho(q_z^f) = \int_{-\infty}^{\infty} \Delta\rho(z) e^{iq_z^f} dz \tag{17}$$

and $\Delta\rho(q_z^f)$ can be written in summation form in terms of $\Delta\rho_i$ ($= \rho_i - \rho_0$) of thickness d of the ith box as

$$\Delta\rho(q_z^f) = \frac{i}{q_z^f}\left[\left(\sum_{j=2}^{j=N} (\Delta\rho_j - \Delta\rho_{j-1})\, e^{iq_z^f(j-1)d} + \Delta\rho_1 - \Delta\rho_N\, e^{iq_z^f Nd}\right)\right] \tag{18}$$

where N is the total number of boxes used in the calculation. By selecting the appropriate number of boxes and ρ_0 of the film, we fit Eq. (16) with $\Delta\rho_i$ as the fit parameters after convoluting the data with Gaussian instrumental resolution function. The initial value of $\Delta\rho_i$'s are set equal to zero for the fitting. Here, since we start the fitting with an average electron density of the film as ρ_0, the fitting converges to the solution much quicker because ρ_0 act as an anchor to the fit parameters $\Delta\rho_i$ such that they do not wander away in the parameter space to another local minimum. Thus this scheme works well for ultra thin films having small fluctuations in electron density as a function of depth.

Iterative Scheme Using Born Approximation (BA) formalism

If we consider scattering wave vector greater than the critical value q_c i.e., angle greater than θ_c the critical angle for total external reflection, then the X-ray reflectivity can be expressed as the Fourier transformation of the derivative of the electron density profile ($\rho'(z) = d\rho(z)/dz$) as (Als-Neilsen et al., 1982; Als-Neilsen, 1984):

$$R(q) = R_F(q)\left|\frac{1}{\rho_\infty}\int_{-\infty}^{\infty} \rho'(z)\exp(iqz)\right|^2 \tag{19}$$

where ρ_∞ and $R_F(q)$ are the electron density and Fresnel reflectivity of the substrate, respectively. From Eq. (5) we can easily obtain the auto-correlation function (ACF) of the derivative of the density profile by taking Fourier transform of the ratio of reflectivity data and $R_F(q)$ i.e.,

$$\text{ACF}\ [\rho'(z)] = \int_{-\infty}^{\infty} \rho'(t)\ \rho'(t-z)dt = \text{const} \int_{-\infty}^{\infty} \frac{R(q)}{R_F(q)} \exp(-iqz)dz \qquad (20)$$

The peaks in the Fourier spectrum indicate the layer thickness involved. In this scheme, one tries to obtain the actual electron density profile (EDP) as a function of depth normal to the film from the initial guess model. The guess is made using the layer thickness of the film obtained from the Fourier spectrum and the average electron density value of the film from the value of q_c. From Eq. (19), one can obtain the reflectivity profile $R_m(q)$ from the derivative of the EDP of the model ($\rho'_m(z)$). The experimental data we denote by $R_e(q)$. By taking its ratio with $R_m(q)$ one can express the actual derivative of EDP ($\rho_e'(z)$) as (Ming et al., 1996; Banerjee et al., 1999)

$$\rho_e'(z) = F^{-1}\left[\sqrt{\frac{R_e(q)}{R_m(q)}}\, F\left[\rho_m'(z)\right]\right] \qquad (21)$$

In the above expression F and F^{-1} are the forward and inverse Fourier transform pair. In analyzing the reflectivity data, an iterative procedure is carried out with Eq. (7) where we initially use a model profile $\rho_m(z)$ to generate a profile $R_m(q)$ and hence obtain the $\rho'_e(z)$. The model reflectivity profile $R_m(q)$ for the next iteration is now calculated by setting $\rho_m(z)$ equal to $\rho_e(z)$ [= $\cap dz\ \rho'_e(z)$] of the previous iteration. Continuing the iteration, the model EDP will slowly converge to the actual EDP. It converges because the information regarding the actual experimental peak position of $\rho'_e(z)$ is contained in $F^{-1}[\sqrt{R_e(q)}]$ which is convoluted with two other function $F^{-1}[1/\sqrt{R_m(q)}]$ and $\rho'_m(z)$. With successive iteration, $R_m(q)$ should tend towards $R_e(q)$ since $\rho_m(z)$ gets modified and the procedure should converge to the actual peak position of $\rho'_e(z)$, provided the model is within the radius of convergence of the experimental profile. In practice, the refinement of the density profile is obtained by selecting a window of the size of the film thickness that is of our interest, and in all iteration only this window size is considered.

SYSTEMS STUDIED

Ultra Thin Film of TiN

TiN thin films are commonly used as diffusion barrier and adhesion layer in integrated circuit devices (Danek et al., 1996). Ultra thin films of TiN have been deposited by metal-organic CVD from tetrakis-dimethylamino-titanium (TDMAT) precursor on silicon wafers covered by native oxide. The deposition time was 10 seconds and a thickness of 100 Å is measured by ellipsometry. The as-deposited samples were then plasma treated in N_2/H_2 atmosphere at RF = 750 W for 10, 20 and 40 seconds and labeled as sample A, B and C, respectively. Specular and off-specular X-ray reflectivity data were collected using laboratory X-ray source (Philips diffractometer) of wavelength 1.54 Å. The off-specular intensity from all samples was found to be very small having values close to background count. In Fig. 3, we have shown the X-ray reflectivity data for the plasma treated samples. The best reflectivity fit is shown as a solid line and is labeled as R for the fit obtained using three boxes by recursive method and labeled as D for the fit obtained by the DWBA formalism [Note: The number of boxes considered

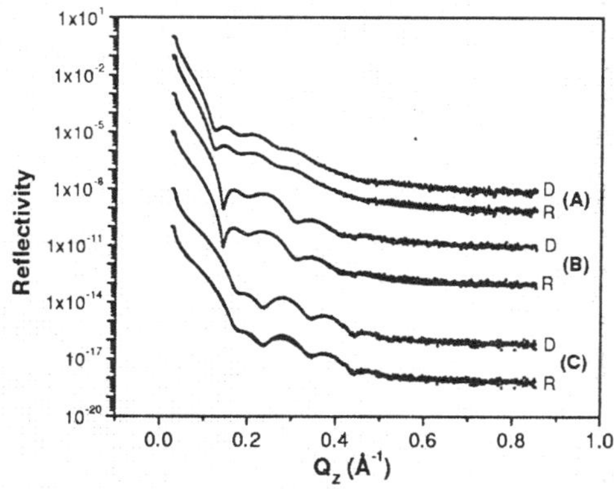

Fig. 3: Specular X-ray reflectivity data of TiN thin film (•) for sample A,B and C, the solid lines marked D and R are the curves obtained from DWBA and recursive formalism respectively.

in the recursive method does not mean that the film consists of that many layers. It only provides a model to obtain the EDP of the film]. In the inset of Fig. 4, we show the electron density values for each of the layers as histograms. The EDP is easily derived from the fit parameters. Since $d\rho/dz$ at each interface was considered to have a Gaussian form, the EDP $\rho(z)$ is represented by a series of error functions shown as smooth lines in the inset of Fig. 4. In the case of DWBA method, the thickness of each layer was fixed to 4Å and 24, 20 and 19 numbers of layers were considered resulting to 96, 80 and 76Å for sample A, B and C, respectively as the total range of analysis. One generally takes in the DWBA scheme the range larger than the nominal total thickness of the film to make sure that some portion of the substrate is included in the analysis. The EDP obtained in this calculation does not depend on *a priori* model of the film and is therefore model independent. The EDP obtained by this formalism is shown as symbols in Fig. 4. We observe from Fig. 4 that the EDP obtained from the recursive formalism and that from the DWBA formalism more or less fall on each other.

At a closer look in the inset of Fig. 4 one can see from the histograms that the samples A and B exhibit larger values of the electron density for the layer adjacent to the substrate. Convolution with an error functional form of interfacial roughness whose width is determined by the value of σ,

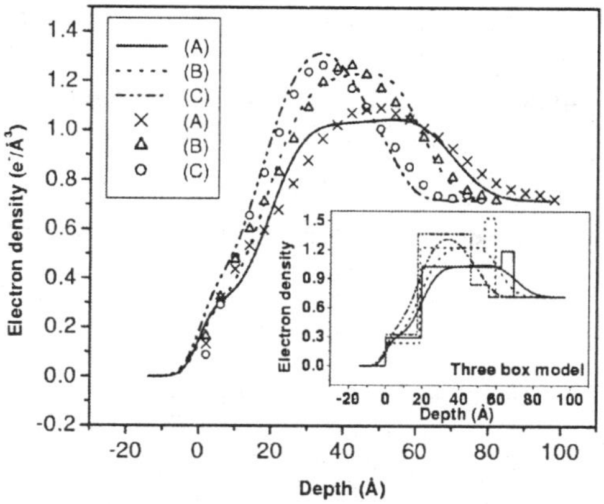

Fig. 4: EDP of the TiN thin film obtained from the DWBA (shown as symbols) and the recursive formalism (shown as lines) for sample A (cross), B (triangles) and C (circles). Inset: It shows the same EDP along with the electron density value as histograms to compare the EDP with the actual electron density value obtained from the recursive formalism for sample A (solid), B (dash) and C (dash dot dot).

and due to its larger value results in to lowering of the electron density. This is a direct consequence of a large value of σ obtained from the fit. We would like to stress here that the EDP obtained after the convolution is a more meaningful quantity than just the electron density value and its respective thickness for each of the layers. It is interesting to point out that σ values can be very large and its value can be close to that of layer thickness as was shown earlier (Banerjee et al., 2002). This is so because in Eq. (15) k_n appears as a product with thickness d_n and the interfacial roughness σ_n and for fitting this cannot be separated out. Also when one use non-linear least square fitting technique to obtain the fit parameters one generally obtains non-unique solutions and one has to choose from it the best fit parameters which appears more physical. Whereas, if one compares the EDP obtained from the recursive formalism with the parameters obtained from DWBA and if both matches well then one can to a certain extent have more confident about the EDP. Even though, we know one cannot be absolutely certain about the EDP obtained from the above two formalism as we loose the phase information during the measurement of reflectivity. As mentioned above, in DWBA one starts the fitting with $\Delta\rho_i = 0$ and for small variation in the electron density value from the average value of ρ_o the fit parameters '$\Delta\rho_i$' converges to solution quickly as all the values are of the same order and one need not bother about the interfacial roughness to get the correct EDP in contrast to recursive formalism. From Fig. 4 we observe that the film thickness on plasma treatment reduces to ~ 80 Å, ~ 70 Å, and ~ 60 Å for sample A, B and C, respectively. These values are obtained from the position of the depth where the value of the electron density approaches the value of the Si substrate. The air/film and film/substrate interfaces are not very sharp and the interface can be considered to be diffuse. We also observe that the electron density of the film increases with the plasma processing. The reduction of the film thickness as a function of plasma processing is due to the plasma etching. The carbon present in CVD film gives rise to complex polymeric structures of low density with embedded TiN clusters. The removal of carbon during plasma treatment causes TiN clusters to form a compact TiN film resulting in an increase in the electron density.

Ultra Thin Oxynitride Films

Dielectric films of oxynitride have replaced pure silicon oxide films as gate and tunnel oxide films in the ultra large scale integrated (ULSI) technology because of its superior properties in terms of the boron barrier, resistance to electrical stress and higher dielectric strength. The exact mechanism of the role of nitrogen in inhibiting boron diffusion is not completely clarified. The most interesting aspect is that the nitrogen places itself at the interface of SiO_2/Si independent of how it has been introduced by either NO or N_2O, and also independent on the condition of preparation. Characterization of nitrogen distribution in oxynitride thin films is becoming a very challenging task, given the extreme low thickness used in today's technology. The structural and chemical characterization of the oxynitride film has been carried out using the reflectivity technique. Here we would describe X-ray reflectivity analysis of an ultra thin oxynitride film, grown from 3.5 nm wet oxides, by nitridation in NO atmosphere. Similarly as in the above case the longitudinal specular and off-specular data were collected for GIXR measurement using laboratory X-ray source of wavelength 1.54Å, the off-specular intensity was found to be very small having the value close to background counts. Here we would like to compare the Born approximation (BA) with DWBA and recursive formalism.

In Fig. 5, we show specular X-ray reflectivity for an ultra thin oxynitride film. First, we shall analyse the reflectivity data using the recursive formalism to obtain the electron density profile (EDP) of the film as a function of depth. We assumed that the total film on the substrate consisted of mainly three layers i.e., the top surface layer, the bulk oxide layer and a transition layer at the film/substrate interface. Considering a three-layer model, we obtain the EDP from the recursive formalism and is shown in Fig. 6. In Fig. 6, we have shown the EDP as a smooth solid line. We observe an increase of electron density in the transition layer at the interface of film/substrate. This increase of electron density at the interface of SiO_2/Si has been attributed due to the accumulation of N as this was observed using secondary ion mass spectrometry (SIMS) analysis described elsewhere (Banerjee et al., 2002).

For the above sample we have also carried out the X-ray reflectivity analysis using a model independent approach in the framework of DWBA

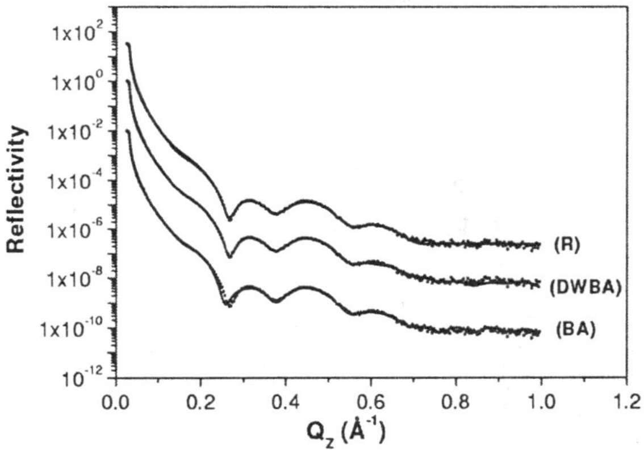

Fig. 5: Specular X-ray reflectivity data of oxynitride film (•) of a SiO_2 film nitraded in NO atmosphere to form an oxynitride film, the solid lines are the fit obtained by recursive formalism (R), using DWBA formalism (DWBA) and Fourier inversion technique based on Born approximation (BA).

and Fourier inverse technique based on BA. Based on the above two techniques the reflectivity curves obtained are shown in Fig. 5. All the three reflectivity curves obtained by the above three formalism fits the experimental data quite well. The EDP obtained from the above two formalisms are shown in Fig. 6 as symbols. Both formalisms give the same result i.e., the EDP falls on each other. In Fig. 6 we have also compared the EDP's obtained from recursive formalism with that from DWBA and BA. The shoulder obtained from the recursive formalism at the surface of the film marked by an arrow was also obtained from the DWBA and BA formalism. As earlier mentioned, we can have better confidence about the EDP if all the above formalism gives similar EDP. Thus, we could say that one can reach to a realistic solution for an ultra thin film containing small variation in the electron density profile if one can compare the EDP with each of the method.

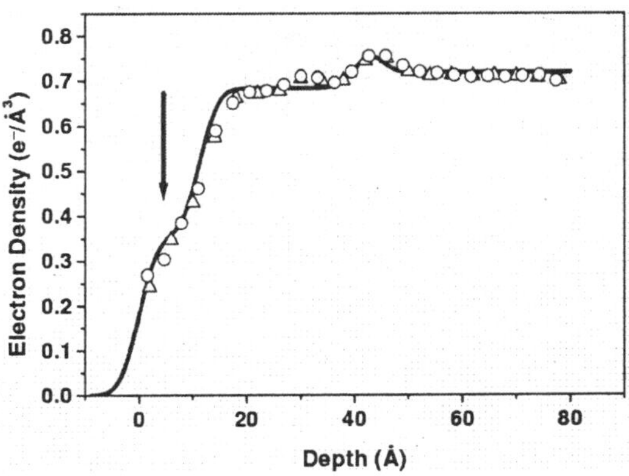

Fig. 6: Electron density profile of an ultra thin oxynitride film obtained using recursive formalism (solid line), DWBA (triangles) and BA (circles).

CONCLUSIONS

We have shown that if an ultra thin film consists of gradual variation in electron density as a function of depth then one can fit the reflectivity profile using recursive formalism considering several layers and evaluating the EDP from the fit parameters. Since one can obtain various non-unique solutions that can fit the reflectivity data quite well, then it becomes very important to obtain the EDP using the DWBA formalism and the BA. The DWBA formalism is based on non-linear least square fitting whereas BA is based on an iterative method. By considering and comparing the EDP obtained from all the three methods one can guess the structure of the system as a function of depth with a better confidence.

Acknowledgements: We would like to acknowledge S. Ferrari, D. Chateigner, A.Gibaud and ST Microelectronics (Agrate, Milan, Italy) for providing TiN and oxynitride thin films.

REFERENCES

Als-Neilsen, J. (1984). *Physica* B + C, **126:** 145.

Als-Neilsen, J., Christensen, F. and Pershan, P. S. (1982). *Phys. Rev. Letts.*, **48:** 110.

Ashcroft, N.W. and Mermin, N. D. (1976). Saunders College Publishing, 4.

Banerjee, S., Gibaud, A., Chateigner, D., Ferrari, S., Wiemer, C. and Dekadjevi, D. T. (2002). *Appl. Phys. Letts.*, **80:** 512.

Banerjee, S., Chakraborty, S. and Lai, P. T. (2002). *Appl. Phys. Lett.*, **80:** 3075.

Banerjee, S., Gibaud, A., Chateigner, D., Ferrari, S. and Fanciulli, M. (2002). *J. Appl. Phys*, **91:** 540.

Banerjee, S., Raghavan, G. and Sanyal, M.K. (1999). *J. Appl Phys.*, **85:** 7135.

Banerjee, S., Sanyal, M.K., Datta, A., Kanakaraju, S. and Mohan, S. (1996). *Phys. Rev.*, B **54:** 16377.

Born, M. and Wolf, E. (1980). Principles of Optics. Pergamon, London, 6th Edition.

Danek, M., Liao, M., Tseng, J., Littau, K., Saigal, D., Zhang, H., Mosely, R. and Eizenberg, M. (1996). *Appl. Phys. Lett.*, **68:** 1015.

Gibaud (1999). X-ray and Neutron Reflectivity: Principles and Applications, **3:** 87.

Ming Li, Moller, M.O. and Landwehr, G. (1996). *J. Appl Phys.*, **80:** 2788-2790.

Nevot, L. and Croce, P. (1980). *Rev. Phys. Appl.* **15:** 761.

Parratt, L.G. (1954). *Phys. Rev.*, **95:** 359.

Russell, T. P. (1990). *Material Science Reports*, **5:** 171.

Sanyal, M. K., Basu, J. K., Datta, A. and Banerjee, S. (1996) *Europhys. Lett.*, **36:** 265.

Sanyal, M. K., Sinha, S. K., Gibaud, A., Huang, K. G., Carvalho, B. L., Rafailovich, M., Sokolov, J., Zhao, X. and Zhao, W. (1993). *Europhys. Lett.*, **21:** 691.

Vidal, B. and Vincent, P. (1984). *Applied Optics*, **23:** 1794.

Zhou, X.-L. and Chen, S.H. (1995). *Phys. Rep.*, **257:** 233.

Recent Progress in Instrumentation, Evaluation Techniques and Applications of Small-Angle X-ray Scattering

Otto Glatter

Institute of Chemistry, University of Graz, Austria

INTRODUCTION TO SAXS AND DATA EVALUATIONS SCHEMES

SAXS and Complementary Techniques

The purpose of this first part is to introduce small-angle X-ray scattering (SAXS) as a method for investigation of nano-structured matter like colloidal systems, complex fluids and other non-periodic systems. Only a few important points will be explained here, for details the reader is referred to the original literature or to textbooks (Guinier and Fournet, 1955; Brumberger, 1967; Glatter and Kratky, 1982; Feigin and Svergun, 1987; Brumberger, 1995; Lindner and Zemb, 2002).

The physical principles of scattering are the same for (wide-angle) X-ray diffraction (XRD) and SAXS. The incoming wave induces dipole oscillations at every electron and so the generation of secondary waves which add at large distances (far field approach) to the overall scattering amplitude. All secondary waves have the same frequency but will have different phases caused by the different path lengths. Due to the high frequency it is only possible to detect the scattering intensity—the complex square of the scattering amplitude—and its dependence on the scattering angle. The situation is nearly the same for scattering of X-rays (SAXS) and neutrons (SANS). The interaction with matter is different, but the problem of interfering secondary waves is the same. Instead of the electron density for X-rays we have to deal with the *scattering length density* for neutrons. The essential fact in neutron scattering is the pronounced difference in the scattering length between hydrogen and deuterium, which is important for the variation of the contrast between the particles and the matrix. Another important feature of neutron scattering is the high incoherent scattering cross section of hydrogen. For the following, however, we focus on SAXS and we use the term electron density instead of scattering length density.

The Scattering Problem

What type of investigations can be performed with SAXS? It is possible to study monodisperse and polydisperse systems. In the case of monodisperse systems it is possible to determine size, shape and under certain conditions the internal structure. Monodispersity cannot be deduced from SAXS data and must therefore be assumed or checked by independent methods. For polydisperse systems a size distribution can be evaluated under the assumption of a certain shape of the particles (*particle sizing*).

Dilute Monodisperse Systems

We start with the simplest possible case, a very dilute system of scatterers. This assumption will not only guarantee that the condition of single scattering is fulfilled automatically, but also leads to a situation that all scatterers are *uncorrelated*, i.e. every particle of the system can be regarded as independent in respect of translation and orientation and that there exists no long range order. This means that we assume that there exists no *interparticle interaction* at all. This condition can often be met by colloidal systems of globular particles with a low volume fraction (typically below one per cent), but we have to keep in mind that high particle charges or long thin cylinders may necessitate much higher dilutions. The condition of a very dilute system leads also to the situation that $\bar{\rho}$ is equal to the density of the solvent and $\Delta\rho$ is just the difference between the density of the particle and the density of the solvent (or surrounding medium). If all particles are identical we get:

$$I_s(q) = N < |F(\mathbf{q})|^2 > = NI(q) \tag{1}$$

with $q = (4\pi/\lambda)\sin(\theta/2)$. We have introduced the particle scattering amplitude $F(\mathbf{q})$ which is the scattered field resulting from integration over the particle volume only. Averaging and calculation of the complex square finally leads to

$$I(q) = 4\pi\int_0^\infty p(r)\frac{\sin(qr)}{qr}dr \tag{2}$$

where we have introduced the *pair distance distribution function* (PDDF) $p(r)$ with

$$p(r) = \Delta\tilde{\rho}^2(r)\cdot r^2 \tag{3}$$

with $$\Delta\tilde{\rho}^2(r) = <\Delta\tilde{\rho}^2(\mathbf{r})> = <\int_V \Delta\rho(\mathbf{r}_1)\,\Delta\rho(\mathbf{r}_1-\mathbf{r})/d\mathbf{r}_1> \tag{4}$$

We can relate the meaning of a distance histogram to the PDDF $p(r)$ if the particles are homogeneous. The height of $p(r)$ is proportional to the number of distances that can be found inside the particle within the interval r and $r + dr$ (see Fig. 1).

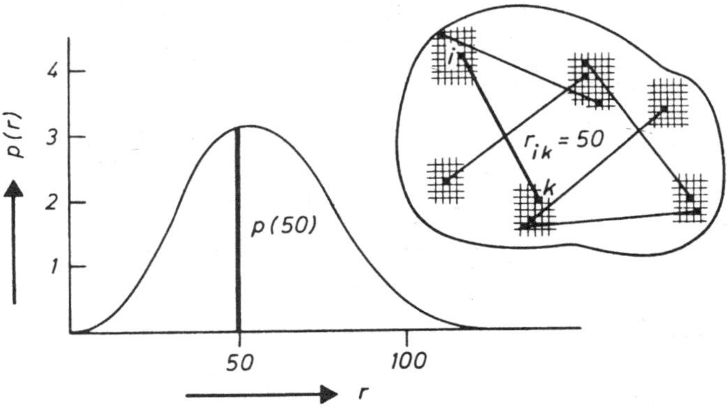

Fig. 1: Definition of the PDDF $p(r)$: The height of this function is proportional to the number of possible connection lines with a length between r and $r + dr$.

If we now leave the assumption of homogeneous densities in the particles we come to the (difference-density) pair-distance distribution, i.e. $p(r)$ is proportional to the difference-density length pairs in the interval r and $r + dr$. Still, the orientation of all connection lines does not matter due to the spatial averaging process. We can understand the PDDF for inhomogeneous particles as follows: We subdivide the particle into many small volume elements (small compared to the local density fluctuations). The $p(r)$ function is proportional to the product of the difference scattering lengths $n_i n_k$ [$n_i = \Delta\rho(\mathbf{r}_i)\, dV(\mathbf{r}_i)$] of two volume elements i and k with a centre-to-centre distance between r and $r + dr$ and we sum over all pairs with this distance (see Fig. 2).

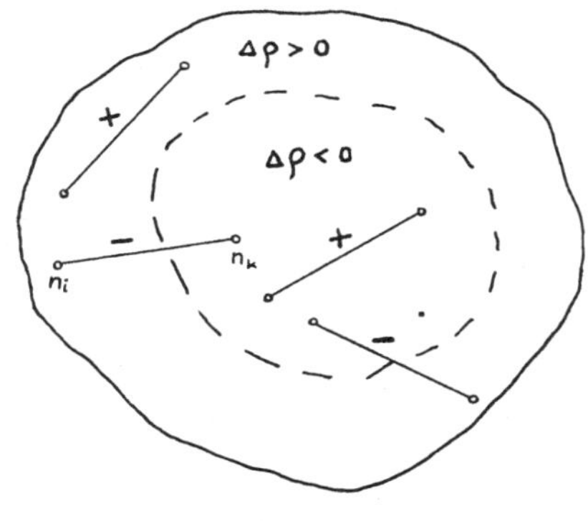

Fig. 2: Definition of the PDDF $p(r)$ of inhomogeneous particles. Each line must be weighted by $n_i . n_k$.

The difference density may be either positive or negative. Whenever we combine volume elements with opposite sign of the density we get negative contributions to $p(r)$ leading to minima and even to negative values of the PDDF at this distance in some cases. We see that the PDDF allows a clear interpretation for homogeneous and inhomogeneous system. This is of great importance for the evaluation of scattering experiments.

The scattering problem and the inverse scatting problem are illustrated in Fig. 3. The calculation of the scattering functions, i.e., the calculation of the scattering curve and of the PDDF for a given structure is called the scattering problem. We have seen that the particle, after spatial averaging, is represented by its correlation function or by the PDDF [Eqs. (4) and (3)]. The Fourier transform of this PDDF $p(r)$ is the scattering function $I(q)$ [Eq. (2)].

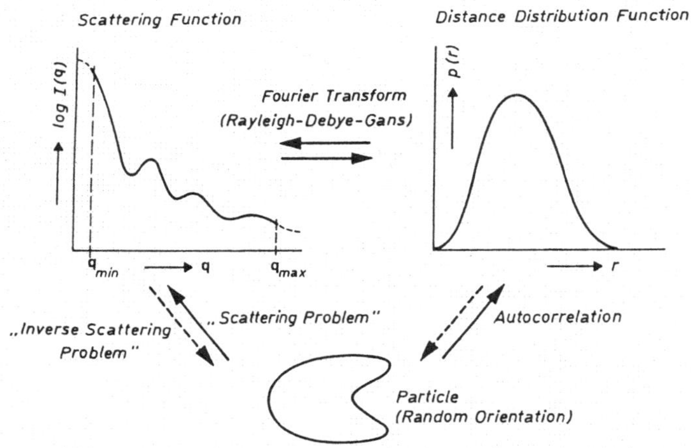

Fig. 3: The scattering problem and the inverse scattering problem.

The inverse scattering problem is the problem to find the structure of the particle from its scattering function. This task is not trivial. The scattering curve is measured only in a limited q-regime from q_{min} to q_{max}. This led to the development of the so-called *indirect Fourier transformation* (IFT) method. In this method the function $p(r)$ is calculated from $I(q)$ with minimized termination effects including the correction for instrumental broadening (Glatter, 1977 and 1982).

Polydisperse Systems

Until now we have discussed all problems under the assumptions that all particles have the same size and shape, which can be determined from $I(q)$ and especially from $p(r)$ (Glatter, 1979). We now want to skip this assumption of monodispersity.

It is in principle impossible to determine simultaneously the shape and the size distribution of a polydisperse system from its scattering data. It can be shown that the scattering curve of homogeneous ellipsoid can be simulated exactly by a corresponding distribution of homogeneous spheres (Mittelbach and Porod, 1965; Mittelbach, 1965).

Polydispersity is a common problem in practical applications. In these cases we have to know or to assume the shape and can then determine the size distribution. The most important case is the spherical one because many systems can be approximated by spheres and one is often interested in a rough information on the size and size distribution (monomodal, bimodal etc.). We can define three different types of distributions, the so-called *intensity weighted distribution* $D_i(R)$, the *volume* or *mass distribution* $D_v(R)$ and the *number distribution* $D_n(R)$. The intensity distribution gives the relative contribution of the different sizes to the total scattering intensity and is free from model assumptions. The volume or mass distribution states the relative amount of volume in the different size classes and the number distribution simply defines the relative number of particles in the different size classes. The later two distributions need the relation between volume and intensity or the size dependence of the intensity for one particle. These relations are quite simple for spheres :

$$D_i(R) = R^3 D_v(R) \tag{5}$$

and

$$D_i(R) = R^6 D_v(R) \tag{6}$$

The total scattering intensity is just the sum over the contributions from the different size classes, or for continuous distribution the corresponding integral

$$I(q) = c_i \int_0^\infty D_i\,(R) \cdot P_0(q,\,R)\;dR$$

or

$$I(q) = c_v \int_0^\infty D_v\;(R) \cdot R^3 \cdot P_0(q,\,R)\;dR \tag{7}$$

or

$$I(q) = c_n \int_0^\infty D_n\,(R)\;R^6 \cdot P_0(q,\,R)dR$$

where the coefficients c_i, c_v and c_n are normalization constants and $P_0(q,R)$ is the normalized form factor of a sphere with radius R. Eq. (7) is very similar to Eq. (2) for the monodisperse case, the PDDF of the particle is replaced by the size distribution $D(R)$ and the sin x/x terms are replaced by the form factor of the particles. The inversion of Eq. (7) is also implemented into the indirect Fourier transformation (Glatter, 1980). The effect of the polydispersity on the scattering curves is illustrated in Fig. 4 for the case of a Gaussian size distribution of spheres with a varying width. We can see clearly the increase of the smearing effect with increasing width and increasing q-value.

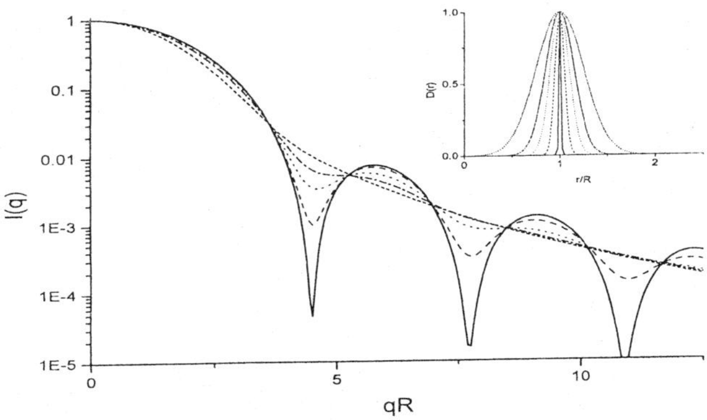

Fig. 4: Scattering curves of Gausian size distributions of spheres with varying width.

Concentrated Systems

All the above statements are strictly true for highly diluted systems where the interparticle distances are much larger than the particle dimensions. In the case of semi-dilute systems the result of a SAXS experiment is influenced as well by the structure of the particles as by their spatial arrangement. In the case of globular particles the scattering curve is the product of the particle scattering function $P(q)$ (*form factor*) and of the interparticle interference function $S(q)$ (*structure factor*).

$$I(q) = nP(q)\ S(q) \tag{8}$$

The first part of this equation $nP(q)$ describes scattering from n spheres, where n is the particle number density, and the second part is the *structure factor* $S(q)$, which describes the effect of positional correlations.

In order to show similarities and dependencies we summarize the corresponding equations in the following Table 1.

Table 1: Summary of the Fourier transformation equations for interacting spherical systems.

$$I(q) = n\ P(q)\ S(q) \tag{8}$$

$$P(q) = 4\pi \int p(r) \frac{\sin qr}{qr} dr \tag{2}$$

with
$$p(r) = \gamma(r)\ .\ r^2 \tag{9}$$

$$S(q) - 1 = 4\pi n \int s(r) \frac{\sin qr}{qr} dr \tag{10}$$

with
$$s(r) = h(r)r^2 = [g(r) - 1]r^2 \tag{11}$$

where $g(r)$ is the pair correlation function and for analogy, we define

$$I(q) = 4\pi \int i(r) \frac{\sin qr}{qr} dr \tag{12}$$

For non-interacting, infinitely diluted systems we get $S(q) = 1$ and $i(r) \equiv p(r)$. We shall use the function $i(r)$ as the Fourier transform of $I(q)$ only where it is essential to differentiate between the transforms of $I(q)$ and $P(q)$. In all other cases we shall simply use $p(r)$. The reciprocal space functions $P(q)$, $S(q)$ and $I(q)$ for a system of hard spheres are shown in Fig. 5 while the corresponding real space functions $p(r)$, $h(r)$, and $i(r)$ are depicted in Fig. 6. The multiplication of the form factor $P(q)$ with the structure factor $S(q)$ causes mainly changes in the low-q regime while the real space function $p(r)$ is changed in the whole r-space as can be seen from the comparison of $p(r)$ with $i(r)$. Obviously, the function $i(r)$ is no longer helpful for the determination of the shape of the particle as $p(r)$ was.

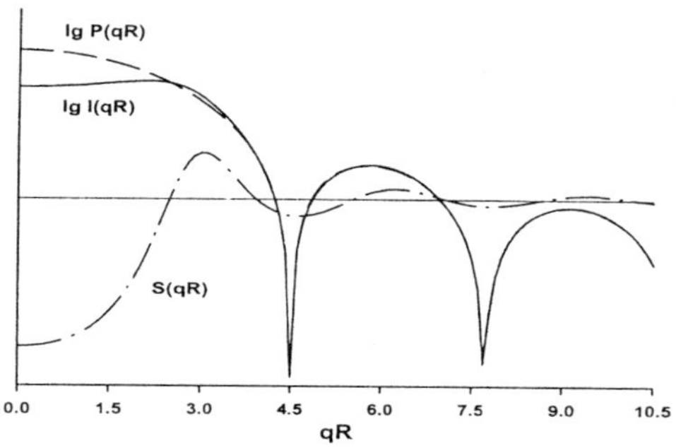

Fig. 5: Reciprocal space functions for spheres with hard sphere interaction.

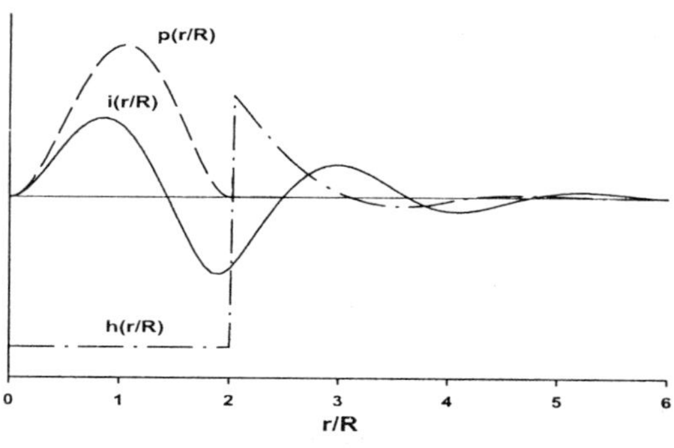

Fig. 6: Real space functions for spheres with hard sphere interaction.

A generalized version of the IFT method, the *Generalized Indirect Fourier Transformation* (GIFT) makes use of the factorization in Eq. (8) while keeping the ideas of the IFT method. This recently developed technique (Brunner-Popela and Glatter, 1997; Weyerich et al., 1999; Brunner-Popela et al., 1999; Bergmann et al., 2000; Fritz et al., 2000) can, however, not be described in detail here.

NEW INSTRUMENTATION

A general review about small angle scattering instruments was given by Pedersen (1995). Laboratory X-ray sources emit a highly divergent beam. The Kratky compact camera (Kratky, 1954; Kratky and Skala, 1958; Stabinger and Kratky, 1979; Kratky and Stabinger, 1984) was constructed to maximize the intensity in the sample using a slit collimation system. The performance of this camera can be further increased if the primary beam is collimated from a divergent into a parallel or convergent beam. A recently developed device for this purpose is the so-called "Göbel mirror". This mirror is made of parabolically or elliptically bent multilayers designed to collimate divergent X-rays from laboratory X-ray sources into a parallel and monochromatic beam of high brilliance (Schuster and Göbel, 1995; Schuster and Göbel, 1996). Modification of the block collimation system in combination with a Göbel mirror are leading to a different beam geometry resulting in an intensity increase by a factor of about 10 to 20. The gain in intensity implicates the use of imaging plate detectors which have a wide linear range in intensity than the gas-filled electronic position sensitive detectors and allow the full use of the increased intensity. Hence the quality of the SAXS data is improved by the higher intensity primary beam, the much lower background due to

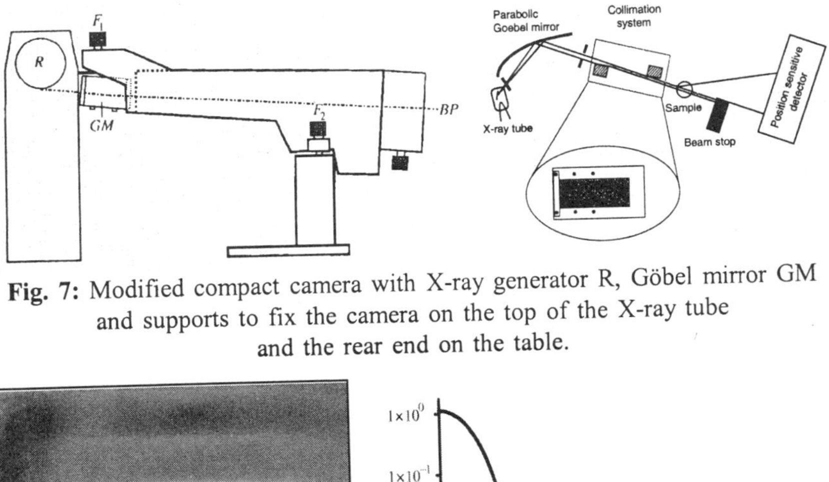

Fig. 7: Modified compact camera with X-ray generator R, Göbel mirror GM and supports to fix the camera on the top of the X-ray tube and the rear end on the table.

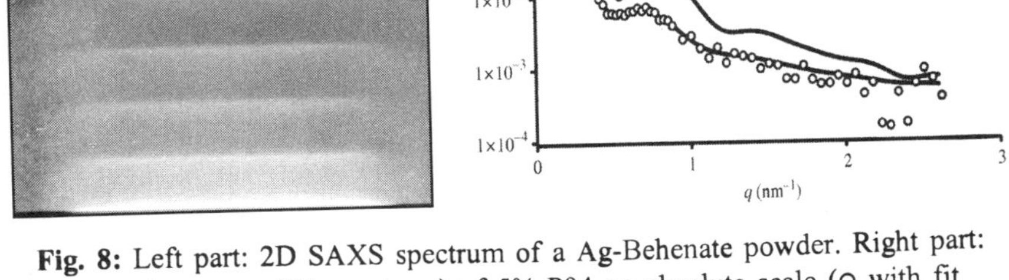

Fig. 8: Left part: 2D SAXS spectrum of a Ag-Behenate powder. Right part: Scattering curve (1D spectrum) of 5% P94 on absolute scale (o with fit function) and corresponding desmeared scattering curve (full line).

the exclusive use of CuKα radiation, and a detection unit which is linear in the measured intensity regime. These advantages such as intensity gain, lower background, better quality of the data have been demonstrated recently (Bergmann et al., 2000) with some selected experimental results. Based on these developments the new *SAXSess* system has been designed by Anton Paar, Graz, Austria. This camera is distributed by PANalytical worldwide.

CALIBRATION TO ABSOLUTE SCALE

For both, SAXS and SANS, the importance of the absolute calibration has been recognized since the beginnings of the technique. Here we focus on SAXS measurements using slit collimation systems (like it applies for the *SAXSess* system). In former days only molecular weights or scattering per particles were determined, but today absolute calibration implies to put the scattering curve on the unit of [cm^{-1}]. It is necessary to measure the so-called absolute intensity which is the ratio of the scattering intensity to the primary intensity P_0. Basically there are two possible ways to determine the absolute intensity. The first one is the direct method which means the mechanical attenuation of the primary beam by a rotating disc or a moving slit. The second one is the indirect method by using secondary standards. For this purpose water is well suited as calibration standard because of the angle independent scattering. The essential advantage is that the scattering of water only depends on the physical property of the isothermal compressibility. With the slit collimation system the scattering curve of water can be measured with high enough statistical accuracy (Orthaber et al., 2000).

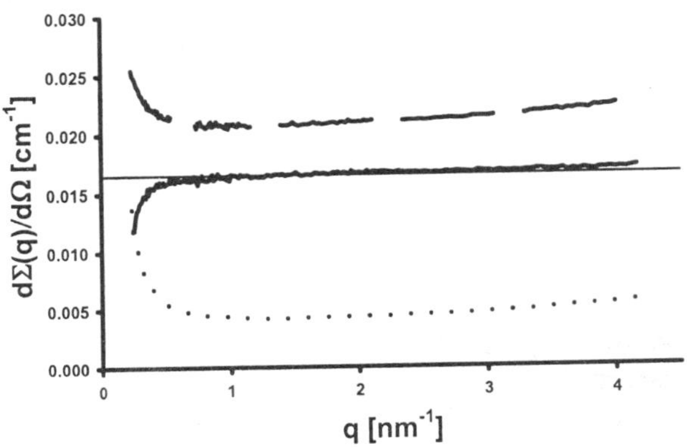

Fig. 9: Scattering curves for calibration. Broken line : water scattering, raw data. Dotted line : blank scattering of the empty capillary. Thick full line: scattering curve of water after subtraction of the blank scattering. Thin full line: horizontal fit.

As an example the scattering curve of the protein lysozyme on absolute scale (Orthaber et al., 2000) is shown below.

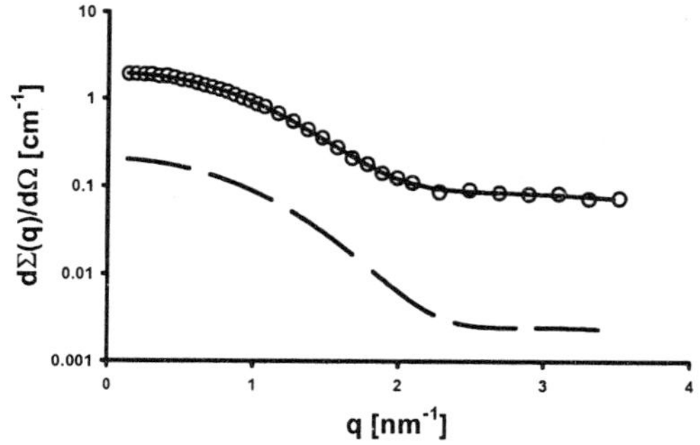

Fig. 10: Scattering curves of 2% lysozyme in 0.15 *M* NaCl solution. Circles: experimental data [a.u.]. Full line: fit to the data [a.u.]. Dashed line: desmeared scattering curve on absolute scale.

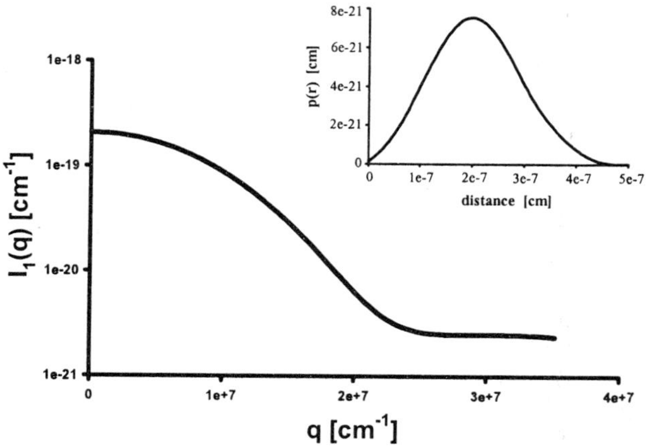

Fig. 11: Particle scattering curve $I_1(q)$ of lysozyme. Inset: corresponding PDDF of one lysozyme particle.

APPLICATIONS

SAXS has many fields of applications. By far not all of them can be even mentioned here. In principle, we can divide the applications into two main fields: determination of size, shape and internal structure in monodisperse systems. In these systems we can also study interparticle interactions. The other field is characterization of polydisperse systems. Here one is mostly interested in particle sizing, i.e., to determine the size distribution in the polydisperse system. Other applications would not even demand for a particulate system. So we can determine the specific surface, i.e., surface per volume even in the case of perculating infinitely large systems (Guinier and Fournet, 1955; Brumberger, 1967; Glatter and Kratky, 1982).

Historically, SAXS was in the beginning used a lot to characterize monodisperse systems. Most important was may be the application in protein research before the introduction of protein crystallography. Proteins and other biological macromolecules are reproduced perfectly well in nature, so it was just a question of preparation to end up with a monodispese system. These types of investigations are now easily done on absolute scale (see example in the previous section) and are still important. Not every macromolecule can easily be crystallized and it is also often important to know the state of the molecule in aqueous solution, i.e., to know if a protein is in monomeric or oligomeric state when dissolved in an aqueous medium.

Particle sizing in the nanometer regime, as it is possible by SAXS is of increasing importance due to the rapid developments in nano-technology. A typical example will be given below.

Finally, SAXS has found a wide field of applications in the study of self-assembling systems (Robinson, 2003). Here amphiphilic molecules arrange in thermodynamic equilibrium into a large variety of structures (Israelachvili, 1991). Those are quite monodisperse and allow to use the standard evaluation routines developed long ago for such systems (Glatter, 1991). However, the structure of such systems is quite often a function of concentration, so simple dilution series cannot be applied and particle interactions have to be taken into account in the evaluation procedure like it is done with minimum *a priori* information in the GIFT method (see also examples below).

Nano Gold Colloids

SAXS data from small colloidal gold particles are shown. Three different samples had been measured: One preparation containing particles with a radius of about 3 nm, a second one with a mean radius of about 7 nm and finally a mixture of the first two samples. The scattering data are presented in Fig. 12, the raw data after background subtraction with fit curves in the left part, the desmeared scattering curves (corrected for instrumental broadening) in the right part.

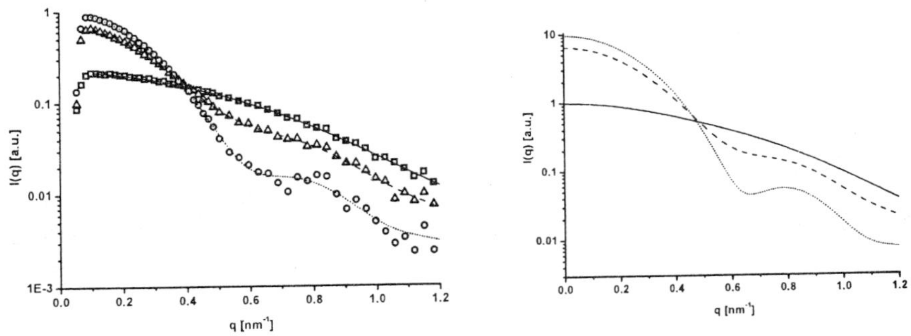

Fig. 12: Left part: Measured scattering data after subtraction of background with fit curves from the IFT algorithm. Squares and full line: small particles, circles and dotted line: large particles, triangles and dashed line: mixture. Right part: desmeared scattering curves (corrected for instrumental broadening) for the same samples as shown in the left part.

From these results we see, that rather narrow monomodal and bi- (or multi-) modal size distributions can be resolved from such experiments.

Concentrated P94 Solutions

In the following we show SAXS data (Figs. 14 and 15) from micelles formed by self-assembly of the triblock copolymer P94 in aqueous solution at 40 °C at different concentrations between 1% and 15% (w/w). What do we learn from these SAXS results? The measured scattering curves show a concentration independent high-q regime, the intensity increases with concentration, the micelles are close to spherical symmetry (pronounced minima) and we find a correlation peak below $q = 0.5\ \mathrm{nm}^{-1}$ for the highest concentrations. The form factor, represented in real space by the $p(r)/c$ function, exhibits no changes at all with concentration. This means that there are just more of the same micelles with increasing concentration and the micelles have a diameter of about 15 nm. The structure factor in the right part of Fig. 15 shows the increasing effect of micellar interaction with concentration. The interaction can be modeled by PY hard sphere interaction with a mean hard sphere radius $R_{HS} = 7.4$ nm and low polydispersity. The effective volume fraction is, however, more than twice the weight fraction $\Phi_{eff} \approx 2.1 \times conc$! These results can be understood with a relatively open shell structure of highly hydrated PEO making this shell relatively stiff. This stiffness is indicated by a hard sphere radius and a diameter of the micelles being practically constant over the wide concentration range measured.

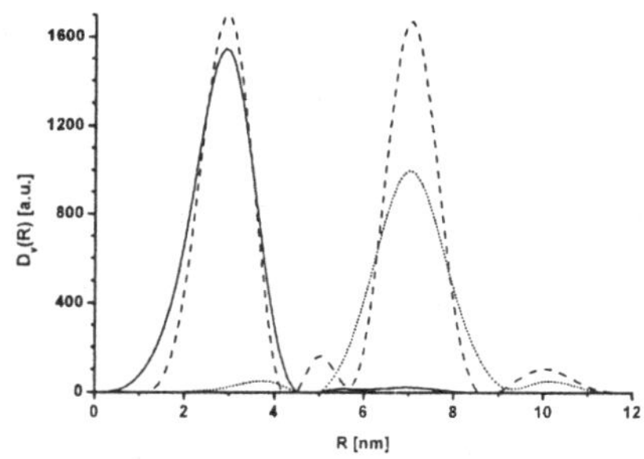

Fig. 13: Size distributions (volume or mass) calculated from the scattering data shown in Fig. 12.

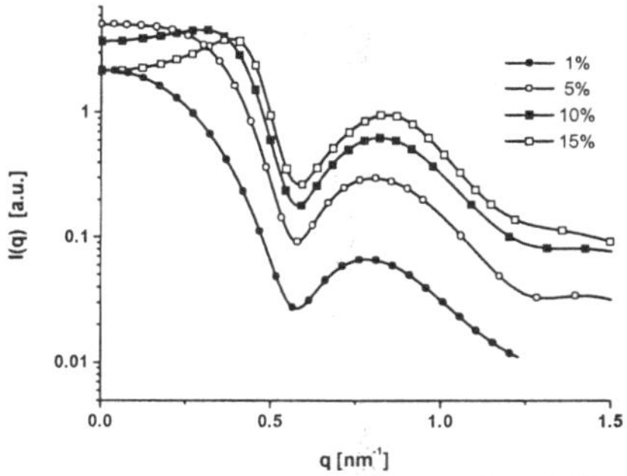

Fig.14: SAXS curves of P94 for different concentrations at 40 °C. The desmeared intensity is plotted in logarithmic scale vs. scattering vector. The curves are scaled in height by arbitrary factors for better visibility.

Multi-lamellar Stacks, DPPC

Multi-lamellar phases can be identified and characterized by small-angle scattering of X-rays (SAXS) or neutrons (SANS). Equidistant peaks are the typical signature and their spacing allows the fast determination of the repeat

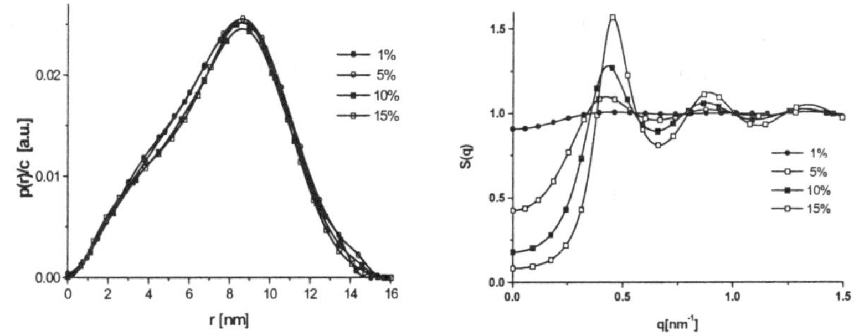

Fig. 15: Left part: PDDFs, divided by the concentration, for P94 at 40°C for different concentrations. Right part: Structure factors $S(q)$ for P94 at 40°C for different concentrations.

distance, i.e., the mean distance between the mid-plane of neighbouring bilayers. The scattering function can be described as the product of a structure factor and a form factor. The structure factor is related to the ordering of the bilayers and is responsible for the typical equidistant peaks, but it also contains information about the bilayer flexibility and the number of coherently scattering bilayers. The form factor depends on the thickness and the internal

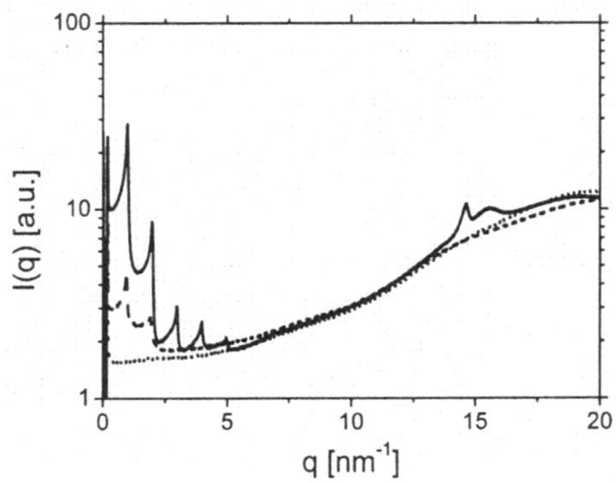

Fig. 16: SAXS and WAXS data of DPPC at 2 °C (full line) and 70 °C (dashed line) and water (dotted line).

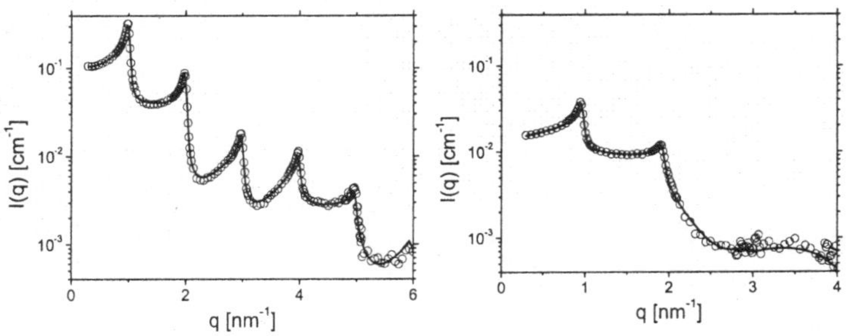

Fig. 17: Scattering curve after subtraction of the water file (open symbols) and fit function (full line) of DPPC at 2 °C (left) and 70 °C (right), respectively.

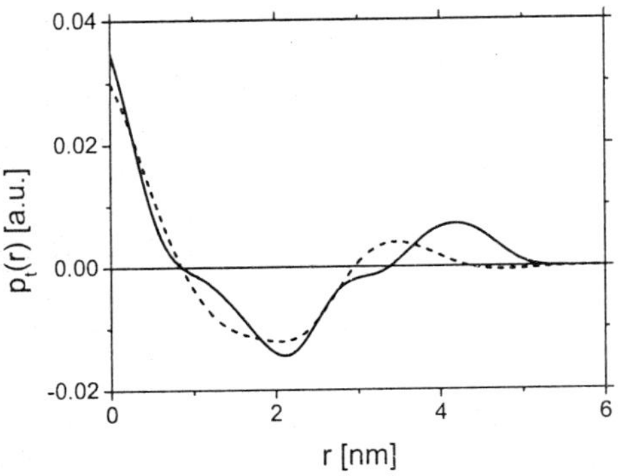

Fig. 18 Thickness distance distribution function of DPPC at 2°C (full line) and 70°C (dashed line).

structure (scattering length density distribution) of a single bilayer. We extend the recently developed generalized indirect Fourier transformation (GIFT) method to such systems (Frühwirth et al., 2004). This method allows the simultaneous determination of the structure factor and the form factor of the system, including the correction of instrumental broadening effects. A few-parameter model is used for the structure factor while the form factor can be determined completely model free. The method is tested successfully with simulated scattering data and by application to experimental SAXS data from some demonstration experiments.

SUMMARY

This article is based on my lecture I gave at the XTRI 2003 at the DMRL in Hyderabad, India. It should serve as an overview over the state of the art of SAXS applied to soft matter. Not all fields of application could be described, neither does it represent a review over the existing literature. It is more something like a partial status report from my very personal point of view.

Acknowledgements: It is impossible to write such a contribution without the direct or indirect help of most of the actual and of many former members of my research group. I would like to thank all of them for supporting me with printable versions of the figures and all the other little helps. This includes, in alphabetic order Alexander Bergmann, Judith Brunner-Popela, Norbert Freiberger, Gerhard Fritz, Thomas Frühwirth, Christian Moitzi, Doris Orthaber, Günther Scherf and Heimo Schnablegger. It was also prepared under quite some time pressure, even though the organizers, especially editor Dr. Ashok Singh allowed me some overtime to finish.

REFERENCES

Bergmann, A., Fritz, G. and Glatter, O. (2000). *J. Appl. Cryst.*, **33:** 1212.

Bergmann, A., Orthaber, D., Scherf, G. and Glatter, O. (2000). *J. Appl. Cryst.*, **33:** 869.

Brumberger, H. (ed.) (1967). Small-Angle X-ray Scattering. Gordon and Breach, New York.

Brumberger, H. (ed.) (1995). Modern Aspects of Small-Angle Scattering. Kluwer Academic Publishers, Dordrecht.

Brunner-Popela, J. and Glatter, O. (1997). *J. Applied Cryst.*, **30:** 431.

Brunner-Popela, J., Mittelbach, R., Strey, R., Schubert, K.V., Kaler, E.W. and Glatter, O. (1999). *J. Chem. Phys.*, **21:** 10623.

Feigin, L.A. and Svergun, D.I. (1987). Structure Analysis by Small-Angle X-ray and Neutron Scattering. Plenum Press, New York.

Fritz, G., Bergmann, A. and Glatter, O. (2000). *J. Chem. Phys.*, **113:** 9733.

Frühwirth, Th., Fritz, G., Freiberger, N. and Glatter, O. (2004). *J. Appl. Cryst.*, submitted.

Glatter, O. (1977). *J. Appl. Cryst.* **10:** 415.

Glatter, O. (1979). *J. Appl. Cryst.* **12:** 166.

Glatter, O. (1980). *J. Appl. Cryst.* **13:** 7.

Glatter, O. (1982). *In:* Small Angle X-ray Scattering, (ed.) Glatter, O. and Kratky, O., Academic Press, London, 119.

Glatter, O. (1991). *Progr. Colloid Polym. Sci.*, **84:** 46.

Glatter, O. and Kratky, O. (ed.) (1982). Small Angle X-ray Scattering. Academic Press, London.

Guinier, A. and Fournet, G. (1955). Small Angle Scattering of X-rays. Wiley, New York.

Israelachvili, J. (1991). Intermolecular & Surface Forces. Academic Press, San Diego, 2nd ed.

Kratky, O. (1954). *Z. Elektrochem.*, **58:** 49.

Kratky, O. and Skala, Z. (1958). *Z. Elektrochem.*, **62:** 73.

Kratky, O. and Stabinger, H. (1984). *Colloid & Polymer Sci.*, **262:** 345.

Lindner, P. and Zemb, Th. (ed.) (2002). Neutrons, X-rays and Light, Scattering Methods Applied to Soft Condensed Matter. Elsevier, Amsterdam.

Mittelbach, P. (1965). *Kolloid-Z. u. Z. Polymere*, **206:** 152.

Mittelbach, P. and Porod, G. (1965). *Kolloid-Z. u. Z. Polymere*, **202:** 40-49.

Orthaber, D., Bergmann, A. and Glatter, O. (2000). *J. Appl. Cryst.*, **33:** 218.

Pedersen, J.S. (1995). *In:* Modern Aspects of Small Angle Scattering, (ed.) Brumberger, H. Kluwer Academic Publishers, Dordrecht, 57.

Robinson, B.H. (ed.) (2003). Self-Assembly. IOS Press, Amsterdam.

Schuster, M. and Göbel, H. (1995). *J. Phys. D.: Appl. Phys.*, **28:** A270.

Schuster, M. and Göbel, H. (1996). *J. Phys. D.: Appl. Phys.*, **29:** 1677.

Stabinger, H. and Kratky, O. (1979). *Makromol. Chem.*, **180:** 2995.

Weyerich, B., Brunner-Popela, J. and Glatter, O. (1999). *J. Appl. Cryst.* **32:** 197.

Powder and Glancing Incidence X-ray Diffraction Techniques: Some Applications in Research and Industry

V. Ravichandran

Materials Science Center, Department of Nuclear Physics
University of Madras, Chennai 600 025, India

INTRODUCTION

It is well recognized that the discovery of X-rays has caused a revolution in medical diagnostics as the first ever tool to probe into the human body. Perhaps an even greater revolution it has turned out to be as a probe to explore matter at the atomic scale. This has become possible as X-rays having wavelength of about 0.1 nm are diffracted by the atoms periodically arranged at comparable interatomic distances in the crystalline solid state. Though other spectroscopic phenomena by X-rays are also useful in probing matter, X-ray diffraction (XRD) has attracted immense popularity as a technique to elucidate the structure of any material, biological or otherwise. The molecular structure of a material is of fundamental importance to the understanding of its properties and capabilities. In spite of the limitation that the specimen should be single crystalline for a successful determination of its molecular structure, the XRD technique has contributed enormously to the growth of modern biology and materials science and technology. In recent years, the application of XRD to polycrystalline form of materials is also evolving at a rapid pace to mature into a routine technique in a variety of areas including structure determination. The purpose of the present article is to bring out the variety and exclusivity of XRD powder technique through real-life examples of a cross-section of the multitude of application areas in the industrial and scientific arena.

POWDER DIFFRACTION APPLICATIONS TO INDUSTRY

A wide variety of industrial activities involve material production and the materials encountered are polycrystalline solids in most cases. The structures of these materials are the key determinants of their solid-state properties. Therefore, powder XRD has established itself as an important analytical tool. Each material is characterized by its unique diffraction pattern. There are three types of information in any diffraction pattern—geometrical, structural and physical. In the case of powder diffraction, these three types are obtained from the angular position, integrated intensity and the profile of the diffraction peak, respectively. The diffraction pattern of the material is thus an invaluable fingerprint of its properties and hence databases (Jenkins and Smith, 1987; Allen and Kennard, 1987; Mighell et al., 1987) of these patterns with nearly half-a-million entries are maintained for the benefit of the scientific and industrial users. These patterns are useful primarily in the identification and quantitative characterization of the crystalline phase(s) constituent of a solid. Even if the chemical nature of the solid is known, such a phase analysis is crucial to several applications as understood from a few select examples to follow.

1. While potassium sulfate is a good fertilizer chemical, its chemical analogue potassium pyrosulfate is highly acidic and detrimental to crops. Even though routine chemical compositional analysis may not distinguish between the two, powder diffraction patterns can be routinely employed to identify and estimate the bad pyrosulfate (Chung, 1981).
2. Gypsum ($CaSO_4$. $2H_2O$), an important component of cement, gets partially dehydrated into bassanite ($CaSO_4$. 0.5-0.8 H_2O) and anhydrite ($CaSO_4$) in the manufacturing process due to the elevated mill temperature. The ratio of these two phases determines the performance of the cement.
3. Titanium dioxide, the best white pigment material in paint industry, has three polymorphs. The rutile phase is the useful pigment while the anatase phase causes chalking of paint. Use of the raw material, containing more than the permissible 0.2% of anatase, may render the paint chalky. Powder diffraction analysis is therefore an essential requirement in the manufacturing line of this product.
4. A green painted surface was noticed to turn blue after a brief period. Only XRD could solve the mystery. The paint originally made of lead chromate (for yellow) and iron (for blue) mixed in proportion to produce the green-blue contained lead chloride (white) in the sample from the affected surface, as seen by a powder diffraction analysis. It was concluded that environmental attack by fumes of HCl on the surface has converted the chromate to chloride (Scott, 1975).

The application of powder diffraction technique for phase analysis is very wide spread and the list of user industries is as diverse as to include

Aerospace, Automotive, Biotechnology, Cement, Ceramic, Chemical, Detergents, Fertilizer, Forensic, Hospitals, Jewellery, Lighting, Metallurgy, Mining, Paint, Petroleum, Pharmaceuticals, Photography, Polymer and Steel. Some industries extensively depend on XRD for the process quality control. The hydrometallurgical process in extractive metallurgical industry is a case in point.

Though, qualitative and quantitative phase analysis together form a major share of XRD applications in industries, other characteristics such as crystallite (grain) size, lattice strain, sheet texture and crystallinity are also extractable from the diffraction patterns. For instance, in pharmaceutical industry, polymorphism in the structure of many drug molecules and influence of degree of crystallinity in its bio-availability are important. An example of the pattern for a crystalline form of the antibiotic ceftriaxone with an included amorphous component is shown in Fig.1. This is taken from one of our studies on the effect of crystallinity on the efficacy of the drug. The various powder diffraction techniques and their representative application areas are given in Table 1.

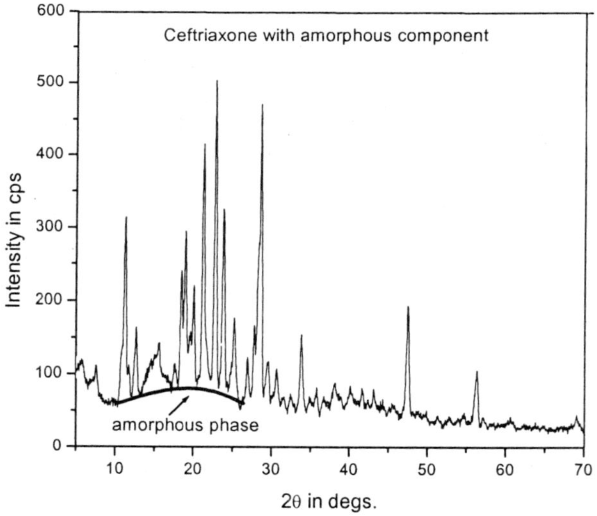

Fig. 1: XRD pattern of crystalline ceftriaxone with included amorphous phase.

STRUCTURAL PROBE OF PARACRYSTALLINE ORDERS

While XRD is undoubtedly the choice technique to arrive at the complete structure of materials possessing crystalline long-range order, it is equally powerful in probing lesser degrees of order. A good illustration of this is its utility in characterizing the mesoporous silicate network structures. These materials are synthesized by forming a silicate network layer over tubules of hexagonal liquid crystal templates assembled from rod-shaped micelles.

Table 1. Some XRD Applications

Techniques	*Application in*
Compositional	Abrasives
Crystallinity	Pharmaceuticals
Crystallite size	Cements
Crystal orientation	Clays
Lattice parameters	Corrosion
Low-angle diffraction	Drugs
Microdiffraction	Environmental
Phase identification	Mining etc.
Rocking curve	Phosphors
Macro-stress/strain	Metallurgical
Micro-stress/strain	Semiconductors
Texture/poll figures	Zeolites

The tubules bundle into higher organizations as shown in Fig. 2. When the templates are removed the stable silicate networks remain. The resultant structures have well defined hollow hexagonal motifs with uniform pore size for their cross sections but they do not possess a three-dimensional crystalline order and the lengths of tubules are variable. Mesoporous channels

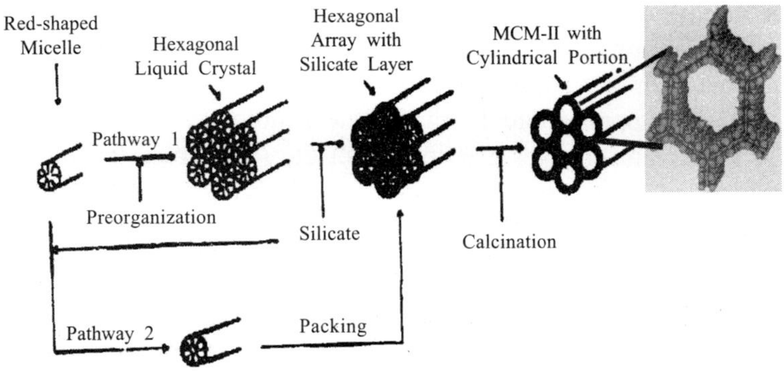

Fig. 2: Schematic showing the assembly of templated MCM structure (after Beck at al., 1992)

are synthesized for a variety of applications, the primary one being of a molecular sieve. XRD is invariably first performed on the synthesized mesoporous material to characterize the integrity of the framework structure and evaluate the pore size. Figure 3 shows the diffraction pattern of a mesoporous MCM41 sodium silicate structure synthesized in our laboratory based on the standard procedure (Beck et al., 1992).

The XRD pattern obtained is typical of these structures. It shows a few (hk.0) peaks of a hexagonal lattice and the dimension of the hexagon is readily obtained from the *d* value of the first peak. The nature of these diffraction patterns with only a few (hk0) peaks indicates that the material

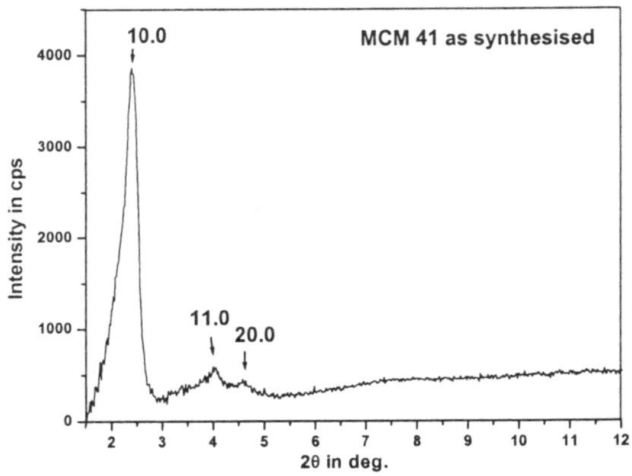

Fig. 3: XRD pattern for the as-synthesised Si-MCM-41 showing the peaks corresponding to a hexagonal cell of a = 4.25 mm

possesses very little order along the c axis. Higher order peaks with 'h' or 'k' indices are not generally seen, indicating the hollow nature of the structure. Notwithstanding the poorly crystalline nature of these materials, the hexagonal nature of the framework is exhibited very well. In fact, the profile shape and intensity of the (10.0) peak can be used to draw inferences about the framework integrity and bundle size in the synthesized material. Figure 4 shows the subtle changes in the intensity profiles as a consequence of chemically substituting some of the Si atoms by Al, even though no specific correlation between Si/Al ratio and profile characteristics is exhibited. When Cd^+ ions were incorporated into the channels and subsequently were converted into CdS, significant changes are noticed in the (10.0) peak profiles. The changes in the intensity are perhaps due to modifications occurring in the framework structure. The small peak shift is attributable to strains induced by the inclusion.

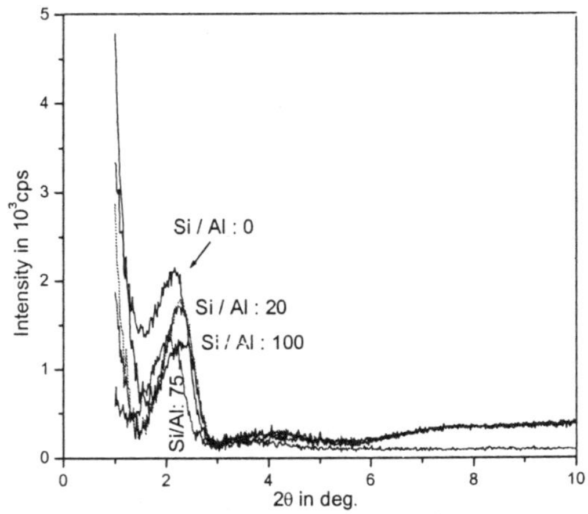

Fig. 4: XRD patterns showing the intensity variations and small peak shifts in the main (1 0 0) peak for various ratios of Si/Al in Al-MCM41.

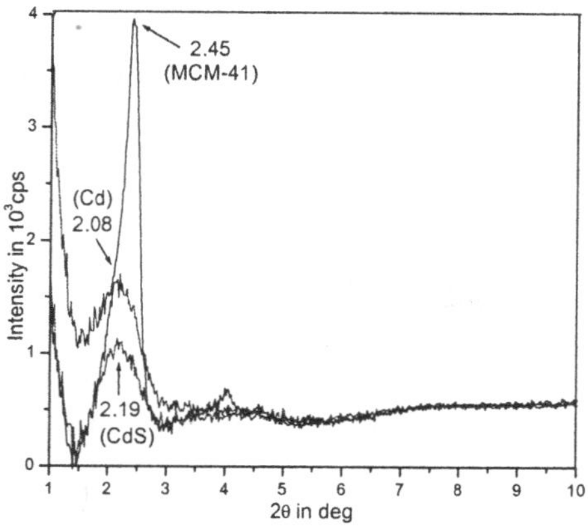

Fig. 5: XRD patterns showing changes in (10.0) peak profiles on incorporation of Cd ions and conversion into CdS nanoclusters.

STRUCTURAL PROBE OF ION IMPLANTATION EFFECTS

An energetic charged particle injected into a solid is slowed down as it transfers energy kinematically to the host atoms by a cascade of collisions resulting in energy loss. It propagates almost as a neutral atom to ultimately stop in the solid. In the low energy (keV per nucleon) regime, ions elastically collide with the nuclei dislodging them and get implanted. At higher energies (MeV per nucleon), energy is primarily lost to electrons through inelastic scattering, thereby exciting them or even knocking them out. This energy deposited in any case can finally result in lattice damage, the nature of damage/defect depending on the physical properties of the material. However, controlled defects of desired type can be produced at the desired locations opening up possibilities of engineering new materials. Even though energies involved are very high compared to thermal processes, the highly non-equilibrium states obtained here, unlike under thermal equilibrium conditions, potentially can result in unusual processes leading to novel materials.

Here are some examples of how XRD has revealed the effects ion implantation can produce in the structural characteristics of a material at the atomic scale. Low energy ions have a small range and get implanted less than a hundred nanometers below surface. Conventional XRD provides information averaged over much deeper layers also, determined by the penetration depth in the material. This precludes a study of only the implantation-affected region. This limitation can be overcome by allowing X-ray to be incident at very small angles grazing the surface.

When X-ray incident on the material is at grazing angles, its diffraction provides structural information mainly from surface layers since the penetration depth of X-rays is reduced by three orders of magnitude from 1-10 μm for a normal diffraction to 1-10 nm. This is because, in this scattering geometry, the Bragg condition is realized by X-rays that are totally externally reflected by the surface layers.

The geometry of grazing incidence X-ray diffraction (GIXD) is shown in Fig. 6. Let E_0 represent a wave incident on the surface making a small angle Φ_0. This wave produces the specularly reflected wave *Es*. Now, let us rotate the sample about a normal to its surface, as shown in Fig. 6, but preserving the small angle Φ_0. As we rotate, the wave E_0 may meet some atomic plane perpendicular to the surface at the Bragg angle Θ_B for that plane and thereby diffraction may occur. Though this diffracted wave will be directed inwards, the reciprocal lattice vector being parallel to the surface, its specularly reflected counterpart E_h occurs taking off the crystal at a small angle Φ_h. The wave E_h may also be treated as the Bragg diffraction of specularly reflected wave *Es*. In any case, E_h contains structural information of a very thin surface layer. The angle Φ_0 or Φ_h or both should be less than the critical angle Φ_c for total external reflection. Thus, the illuminated depth should be as small as a few nanometres in general, and the diffracted intensity would yield from this small depth. The penetration depths of CuKα in GaAs at grazing incidence angles are shown in Fig. 7, as a typical example.

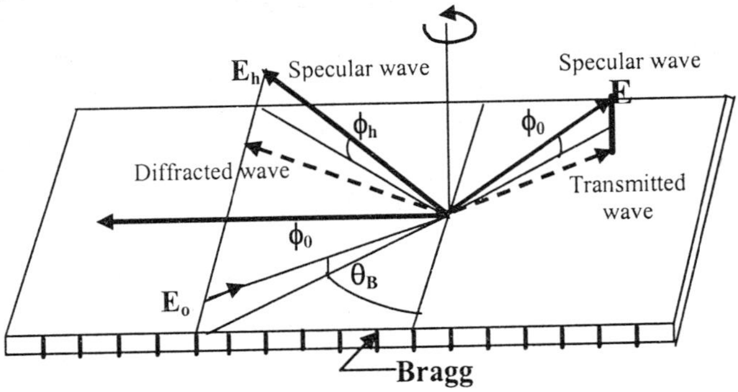

Fig. 6: Schematic of the GIXD phenomenon and geometry (Golovin et al., 1984).

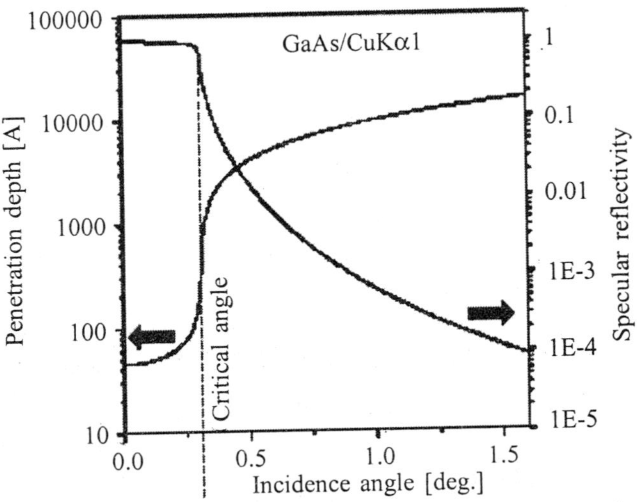

Fig. 7: To show how grazing incidence probes only near-surface layers.

The power of GIXD is in its ability to probe structure at very low penetrations. This technique is thus suitable for selectively studying thin films coated on a substrate without the interference of any contribution from substrate. Further, by gradually increasing the probing depth, it is possible to follow structural features at deeper layers as a function of distance from surface. In this context, GIXD is a very valuable tool for analyzing the structural changes obtained in ion implantation experiments. The following few examples are intended to illustrate this.

ION IMPLANTATION INTO Ti THIN FILMS

Analysis of Phase Transformation

This experiment was done on basal plane textured Ti thin films of thickness 150 nm obtained by sputter deposition on thoroughly cleaned float glass substrates. Implantation of 60 keV N^+ ions was carried out on these films under UHV conditions (1×10^{-5} Pa), at normal incidence and at room temperature. Samples were implanted to doses of 8.8×10^{14}, 1.8×10^{15}, 4×10^{16}, 7×10^{16}, 1×10^{17}, 2×10^{17}, 3×10^{17} and 5×10^{17} ions/cm^2. GIXD was performed using CuKα radiation incident at a fixed angle of $\alpha = 1°$, so that a large volume fraction of the film is covered with a minimum glassy background (X-ray penetration depth 177.5 nm). The mean projected range of 60 keV N^+ ions was computed using TRIM code (Biersack and Haggmark, 1980) and was found to be $\approx$ 93 nm.

Figure 8 shows the diffraction patterns obtained mainly over the implanted region (Rajan et al., 2000). The patterns from the samples implanted at various doses are stacked up in the increasing order of doses. The gradual transformation of hexagonal Ti to cubic δ-TiN is vividly seen in the systematic evolution of TiN diffraction peaks from those of Ti. In particular, the development of uniform and non-uniform strain in the atomic layers of the texture plane (0002) of Ti resulting in a stacking reorder corresponding to the close packed diagonal plane (111) of TiN, is brought out clearly. This phenomenon is more dramatic in the expanded view of Ti(0002) profile shown in Fig. 8b.

Formation of fcc-Ti Phase by Implantation

The GIXD pattern of the implantation product in the above experiment, on closer scrutiny, also revealed a weak peak at $2\theta = 44.1°$ (Fig. 8a) even for an implantation dose as low as 8.8×10^{14} ions/cm^2. The presence of this peak in the diffraction pattern is very much evident when the data was recollected with better statistics as shown in Fig. 9. This peak ($d = 0.2052$ nm) does not belong to any of TiN peaks and, further, compound formation at such a low dose is not in any case expected. Thus, it was suspected that the energy transfer from the bombarding ions has caused the formation of

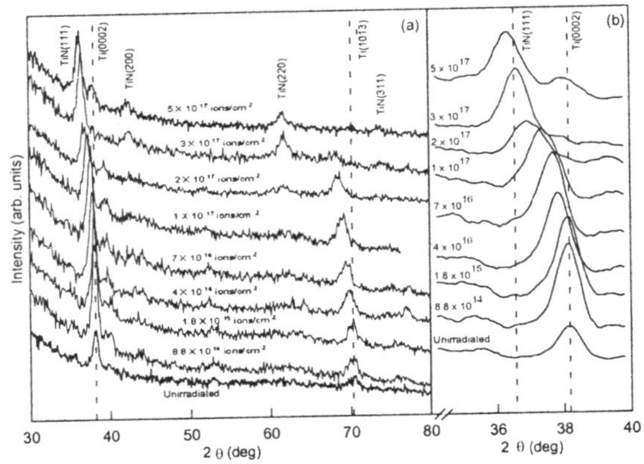

Fig. 8: (a) GIXD scans of N^+ implanted (0002)Ti film (X-ray incident angle = 1°) showing the Ti(0002) → TiN(111) evolution. Doses above 2×10^{17} ions/cm^2 show all the δ-TiN reflections possible in the given range. (b): Expanded view of smoothened profiles.

a new phase of Ti. In order to verify this hypothesis and preclude any remote possibility of compound formation, it was decided that ions of inert species could be injected into the Ti film instead of N^+ and analysed (Rajan, 2003).

A (0002) textured Ti film of thickness 150 nm was therefore irradiated with 90 keV He^+ ions. The range of He^+ ions being 560 nm, the ions will almost entirely be transmitted through the film and the results of the experiment could be associated solely with irradiation-induced damage. The GIXD pattern in Fig. 10 reveals the formation of a new phase on He^+ irradiation to a dose of 1×10^{17} ions/cm^2. The new phase was characterized by the couple of distinct peaks and a left shoulder to the (0002) peak of Ti. These peaks, including the shoulder as extracted by deconvolution (Fig. 11)

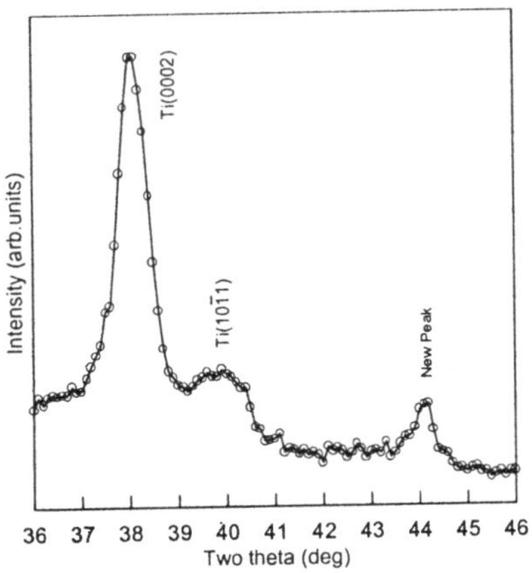

Fig. 9: A new peak for N^+ implanted Ti.

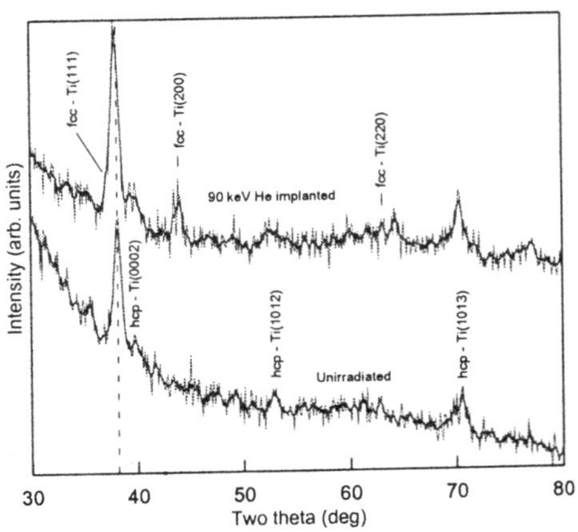

Fig. 10: GIXD of He^+ implanted Ti film and blank.

of the peak at (0002) Ti position, correspond to the fcc-Ti. Implantation of the bulky Ar^+ ions at 100 keV with a range of 75 nm produced the same phase more effectively as seen in Fig. 12. However, the fcc-Ti phase produced by irradiation was a metastable phase and its signature peaks disappeared on annealing at 523 K.

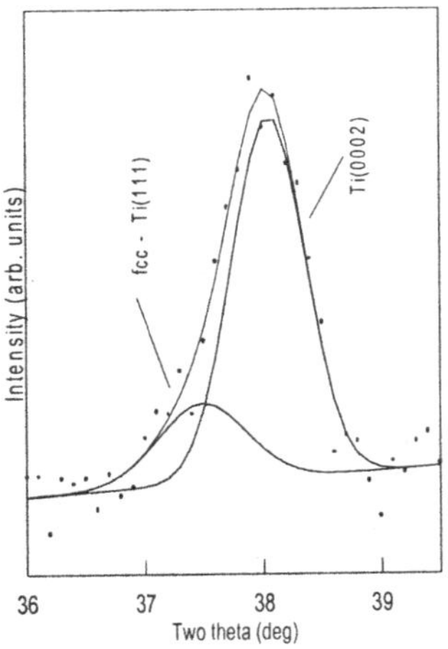

Fig. 11: Deconvolued Ti (0002) showing fcc-Ti peak.

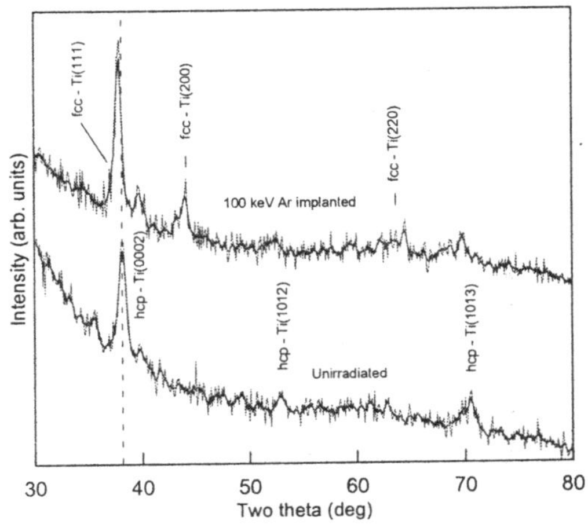

Fig. 12 Formation of fcc-Ti in Ar^+ implanted sample.

ION IMPLANTATION INTO InP CRYSTALS

Structural Effect of N^+ Implantation

Understanding of the effects of ion implantations into semiconductors is important since implantation is used as a semiconductor processing technique in device making. III-V compound semiconductors have attracted considerable attention in recent years due to their application potentials. Indium phosphide is one such candidate for optoelectronic and high-frequency devices. Implantation of N^+ ions of 150 keV was carried out on a semi-insulating (Fe-doped) InP single crystal substrates of (1 0 0) orientation at room temperature. GIXD was performed on the implanted samples using $CuK\alpha_1$ radiation at an incident angle of 0.3 in order to analyse the structural effect of implantation. The implanted sample was later annealed in a quartz tube furnace at 673 K for 30 min in an argon atmosphere by keeping it encapsulated between InP wafers. Figure 13 shows the GIXD patterns obtained for various implantation doses. The appearance of a broad hump corresponding to InP(111) peak indicates that implantation induced damages have produced orientational disorders in the close packed (111) layers. The significant growth of the InP(311) peak on increasing the implantation dose with a shift to the low angle suggests the expansion of InP lattice on incorporation of nitrogen and subsequent formation of an alloy of InP_xN_{1-x}. The diffraction pattern of the annealed sample (Fig. 14) shows annealing out of the

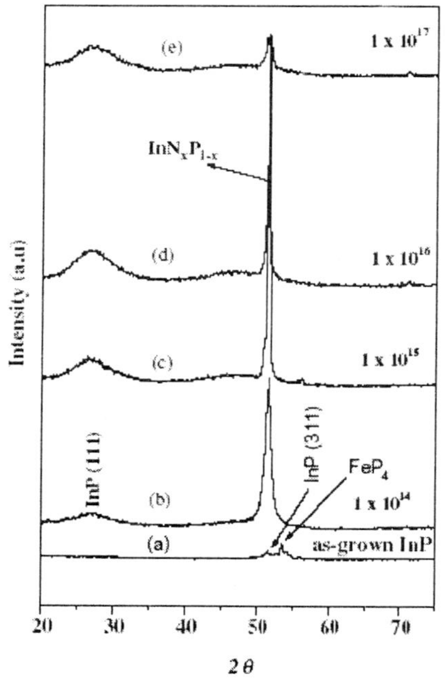

Fig. 13: Implantation-induced damages in InP as a function of dose as revealed by GIXD.

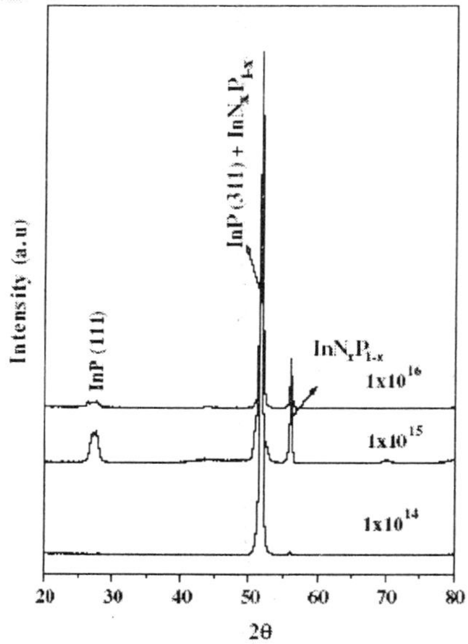

Fig. 14: Annealing of the induced damages and compound formation shown by GIXD.

misorientations of (111) layers to a large extent and consolidation of the alloy peaks by crystal growth (Santhakumar et al., 2003).

Synthesis of InN by Implantation Route

Indium nitride has generated lot of interest due to its direct band gap in the visible range (Tansley and Foley, 1986) for the production of InN/Si tandem solar cells. The large drift velocity for carriers in InN is another advantage. But growth of InN crystals is hampered by its low dissociation temperature and nitridation of III-V compounds is generally difficult due to the small sticking probability of nitrogen on the surface. Studies on InN have been performed mostly on epitaxially grown samples (Wu et al., 2002). Ion implantation is recognized as an important technique to synthesise of even difficult-to-form materials in pure form since it allows introduction of hosts far beyond the normal solid solubility. Therefore, the nitrogen implantation experiments on InP were extended with an aim to achieve the formation of InN by implantation.

It was possible to produce oriented single crystalline regions of InN by implantation of N^+ ions into an InP wafer at an elevated temperature (Santhakumar et al., 2003). Implantation of a lower energy ions at only 50 KeV was performed at a substrate temperature of less than 673 K much below the dissociation temperature of 873 K for InN. The GIXD of the implanted sample at an incidence angle of 0.3° corresponding to an X-ray penetration of 150 nm is shown in Fig. 15 for the doses of 1 × 10^{17} and 4 × 10^{17} ions/cm^2.

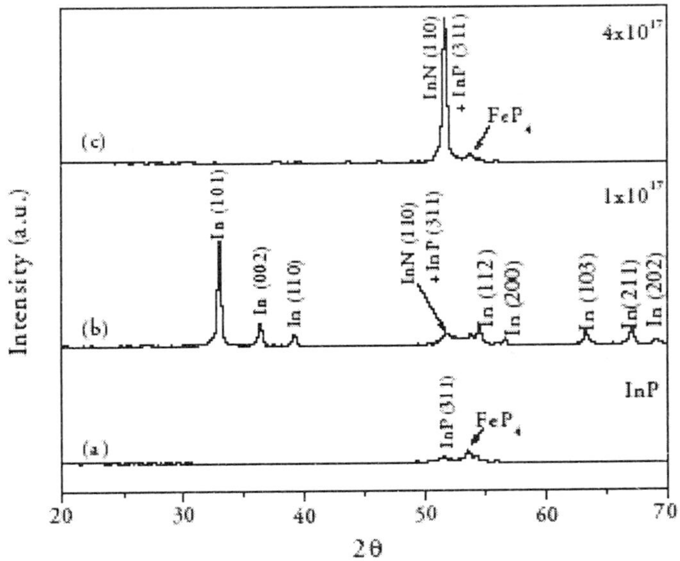

Fig. 15: Transformation of InP → In → InN by N^+ implantation into InP as revealed by GIXD

It is seen that at the lower dose no indications of formation of InN are visible but the diffraction pattern matches very well for metallic Indium suggesting that the combined effect of substrate temperature and implantation at this dose is one of removal of phosphorous species. Subsequent implantation at a higher dose has provided sufficient nitrogen concentration to form crystals of InN. The GIXD of the 4 x 10^{17} ions/cm^2 implantation shows a single sharp peak corresponding to the InN(110) indicating the presence of a highly oriented hexagonal InN crystals in the sampled region.

REFERENCES

Allen, F.H. and Kennard, O. (1987). Cambridge Structural Database. *In: Crystallographic Databases,* Eds. Allen, F. H., Bergerhoff, G. and Sievers, R. International Union of Crystallography, Chester, UK, 31.

Beck, J.S., Vertali, C., Roth, W.J., Leonowetez, M. E., Kresge, C.T., Schmitt, K.D., Chu, C.T.W., Olson, D.H., Sheppard, E.W., Mecullen, S.B., Higgins, J.B. and Schlenkel, J.L. (1992). *J. Am. Chem. Soc.*, **114:** 10834.

Biersack, J.P. and Haggmark, L.G. (1980). *Nucl. Instr. Meth.* **174:** 257.

Chung, F.H. (1981). *In:* Analytical Measurements and Instrumentation for Process and Pollution Control. Ed. Cheremisinoff, P.N. and Perlis, H.J. Ann Arbor, MI, Ann Arbor Science, 151.

Golovin, A.L., Imamov, R.M. and Stepanov, S.A. (1984). *Acta Cryst.* A, **40:** 225.

Jenkins, R. and Smith, D.K. (1987). Powder Diffraction File. *In: Crystallographic Databases.* Eds. Allen, F.H., Bergerhoff, G. and Sievers, R. International Union of Crystallography, Chester, UK, 158-197.

Mighell, A.D., Stalick, J.K. and Himes, V.L. (1987). NBS Crystal Data : Database description and applications. *In: Crystallographic Databases,* Eds. Allen, F.H., Bergerhoff, G. and Sievers, R. International Union of Crystallography, Chester, UK, 133.

Rajan, N. (2003). In the Ph.D. thesis submitted to the University of Madras, 110.

Rajan, N., Nair, K.G.M., Rao, G.V.N., Vaidhyanathan, L.S., Sastry, V.S., Magudapathy, P., Krishnan K. and Sampathkumar, T.S. (2000). *Appl. Phys.,* **A70:** 65.

Santhakumar, K., Jayavel, P., Reddy, G.L.N., Sastry, V.S., Nair K.G.M. and Ravichandran, V. (2003). *Nucl. Instr. and Meth.* **B212** : 197.

Santhakumar, K., Kesavamoorthy, R., Nair, K.G.M., Jayavel, P., Kanjilal, D., Sankara Sastry, V. and Ravichandran, V. (2003). *Nucl. Instr.Meth. Res.,* **B 212:** 521.

Scott, R. W. (1975). *In:* Characterisation of Coatings: Physical Techniques, Part II. Ed. Myers, R.R., Marcel Dekker, New York.

Tansley, T. L. and Foley, C. P. (1986). *J. Appl. Phys.,* **59:** 3241.

Wu, W., Walukiewicz, W., Shan, K.M., Yu, J. W., Ager III, Haller, E.E., Hai Lu, and William J. Schaff (2002). *Phys. Rev.,* **B66:** 201403(R).

Small-Angle X-ray Scattering Activities at Trombay

S. Mazumder, D. Sen and A.K. Patra

Solid State Physics Division
Bhabha Atomic Research Centre
Trombay, Mumbai- 400 085

INTRODUCTION

Small-angle scattering (SAS) is a powerful technique to investigate the structural features of the inhomogeneities i.e., density fluctuations in condensed matter of length scale ranging from nanometer to micron (Glatter and Kratky, 1982; Guinier and Fournet, 1955). In a unique way, SAS opens the door to the "nanocosmos" in various domains of science and technology. These include classical metallurgy (precipitate hardening), materials research (nano-composites), soft matter (gels, colloids, polymers), chemistry (catalysis) and engineering (porous filters, structural and functional ceramics). The lower limit of the length scale resolved by SAS overlaps with that of crystallography while the upper limit coincides with length scale accessed by microscope. SAS phenomenon is distinct from Bragg scattering in the sense that it corresponds to the scattering at momentum transfer region small compared to that of the smallest reciprocal lattice vector in crystalline substances. Further, it does not call for either short or long-range order of atomic arrangement in materials.

Rotating anode sources and currently high flux synchrotron sources are being used for Small-Angle X-ray Scattering (SAXS) investigations. Small-Angle Neutron Scattering (SANS) needs a high flux neutron source like nuclear reactor.

At Trombay, we are involved in SAXS and SANS investigations on the mesoscopic structure of various technologically relevant materials. In this country, we first started SANS activity way back in 1980 using neutron beam from CIRUS research reactor at Trombay (Desa et al., 1985). Mounting a SAXS goniometer on a 12 KW rotating anode source, we have investigated porous silicon (Mazumder et al., 1998), porous polymeric membranes (Sen et al., 2000; Sen et al., 1999; Pandey et al., 2003), nano-particles (Sen et al., 2000; Deb et al., 2002; Sastry et al., 2003; Vogel et al., 2002; Jacob et al., 1999), coal (Sen et al., 2001; Sastry et al., 2000) etc. We have used SAXS facility (Amenitsch et al., 1997) at ELETTRA synchrotron, Italy, for investigating the phase separation and precipitation in some important nuclear materials like maraging steel (Mazumder et al., 1999; Tewari et al., 2000), Zircaloy-2 fuel cladding tube (Sen et al., 2000) etc. Currently we are involved in the development of a SAXS facility at India's forthcoming synchrotron radiation source at Centre of Advanced Technology (CAT), Indore. We have also developed multiple small-angle scattering formalism to analyze data from strongly scattering systems (Mazumder et al., 1998; Mazumder et al., 1995; Mazumder, 1993; Mazumder and Sequeira, 1992, 1991, 1990, 1989, 1988, Swain et al., 2001; Mazumdar et al., 2003). We have built a double crystal based SANS instrument at guide tube laboratory of Dhruva reactor, Trombay (Mazumder et al., 2001; Mazumder et al., 2002). We have also characterized materials using SANS (Sen et al., 2002; Sen et al., 2003; Mazumder et al., 2001).

In this paper, we present our SAXS activities and also give a brief sketch of our developmental activities of setting up a SAXS facility using a beam line of Indus-2 at CAT, Indore.

INVESTIGATIONS USING ROTATING ANODE X-RAY SOURCE

Pore Morphology in Porous Silicon

Silicon is the heart of the microelectronics revolution. However, it is an extremely inefficient light emitter, and for this reason has not enjoyed the same level of dominance in optical applications. However, porous silicon (PS) shows the visible photo-luminescence. To describe, porous silicon is a network of nano meter-sized silicon regions surrounded by pore space and has found potential application as material for opto-electronic integration. Detailed information on pore structure is relevant for understanding its light emitting characteristics.

The pertinent structures in porous silicon can be accessed by SAXS and there are a few reports (Vezin et al., 1991; Vezin et al., 1992; Goudeau et al., 1989; Naudon et al. 1994; Binder et al., 1996) of SAXS investigations on PS. However, in all the activities mentioned, the conclusions have been drawn using single scattering approximation (Mazumder and Sequeira, 1992). Since the pores in the silicon matrix are strongly scattering (scattering length density contrast being 1.97×10^{-3} nm^{-2}, with some cylindrical pores of lengths of the order of thickness of the porous layer, and the typical porosity in the range of 30-85 %), we considered it important to check the validity of the single scattering approximation by carrying out measurements at two different wave lengths on the same sample. Here we present the results

(Mazumdar et al., 1998) obtained on a lightly doped p-type porous silicon by SAXS experiments carried out at 0.154 and 0.071 nm, respectively with CuKα and MoKα X-rays.

The raw measured and the corrected data using CuKα and MoKα are depicted in Fig. 1. Scattering data recorded using MoKα shows a distinct shouldering while the data recorded with CuKα does not have that feature. Further the two profiles have distinctly different functionality. The inset in the Fig. 1 depicts the wavelength dependence of the scattering profiles in normalized scale to appreciate the distinct functionalities of the profiles. It is important to note that in conventional SAS, where the single scattering approximation holds good, variation of sample thickness or wavelength does not change the functionality of the profile i.e. profiles are different with respect to a scale factor only.

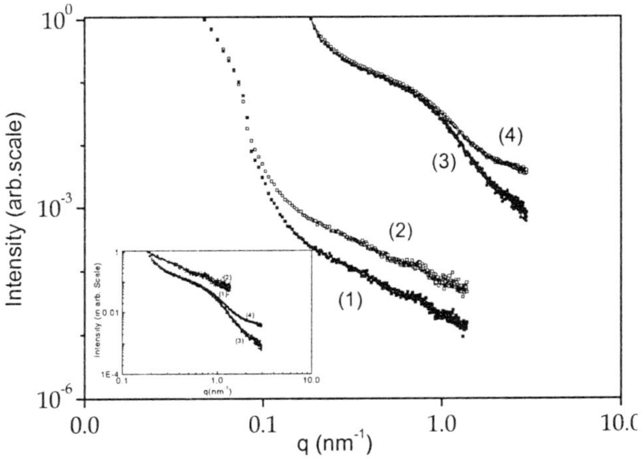

Fig. 1: Raw and corrected, collimation and transmission, data on 13.5 μm thick porous layer sample using MoKα and CuKα. Curves (2) and (4) correspond raw data while curves (1) and (3) correspond the corrected data. CuKα has been used to record data set (2) while MoKα has been used to record data set (4). Over the q range 2.615-2.986 nm^{-1}, curve (3) and (4) follow power laws $\sim q^{-3.022}$ and $\sim q^{-1.394}$, respectively. In the q range 1.214-1.385 nm^{-1}, curve (1) and (2) follow power laws $\sim q^{-1.150}$ and $\sim q^{-0.196}$, respectively.

From the very first observation of the dependence of the functionality of the scattering profile with wavelength, it was concluded that the pores could be cylindrical in shape with axis of the cylinder aligned with the incident beam direction. In such a situation, the scattering mean free path L of radiation takes the form $L = \frac{1}{\varphi l \lambda^2 D^2}$, where λ is the wavelength of the probing radiation, D is the scattering length density contrast, l is the length of the capillary pores and φ is the packing fraction of the pores in the medium. Since, the length of the capillary pores could be of the order of μm, we consider that the multiple scattering effects will be quite significant. To give an example, for 50 % porous sample and cylindrical pores with 1.0 μm length L = 21.73 μm for CuKα (λ = 0.154 nm) and L = 102.23 μm for MoKα (λ = 0.071 nm).

So it is evident that for sample with porous layer thickness 13.5 μm, it will be grossly mistaken to treat the data under single scattering approximation which is valid only for thickness ≤ L/10 (Mazumder and Sequeira, 1992). It is important to note that for spherical inhomogeneity of radius R, L is given by $L = \frac{2}{3\varphi R\lambda^2 D^2}$, so if one is to observe appreciable effect of multiple scattering, R has to ≥ 1.0 μm.

To verify our assessment that cylinders are oriented perpendicular to surface, we have tilted one of the sample and observed profile changes, shown in Fig. 2, with tilting. However, rotation of the sample in its own plane about the incident beam axis does not change the profile to any significant extent. We have estimated the scattering mean free path L of both the probing radiation in the sample. The estimated L values are L = 38.352 (±2.685) μm for MoKα and L= 8.157(±0.571) μm for CuKα radiation.

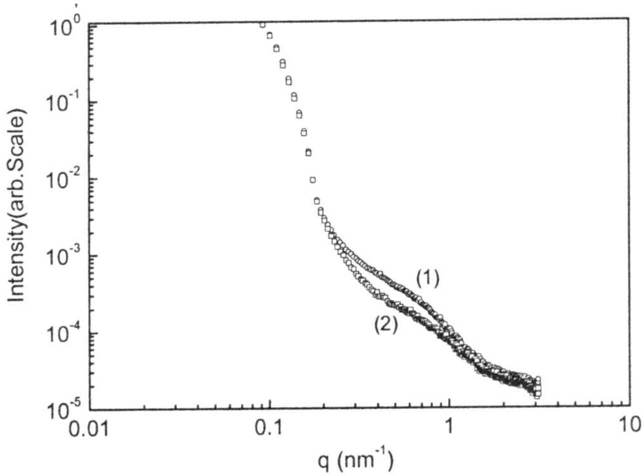

Fig. 2: Scattering profiles of a sample with 6.75 μm thick porous layer recorded using MoKα before and after tilting the sample by about 7 degree. Curve (1) corresponds to the raw data recorded without tilting and curve (2) corresponds to that recorded with tilting.

For 50 % porosity, as measured by the gravimetric method, we estimate the length l of the pores to be 2.664 (±0.186) μm when the system is monodisperse. For a polydisperse system, 2.664 μm would correspond to $< l^2 >/< l >$ assuming all the cylindrical pores are of uniform cross-section and averaging is done with respect to population distribution. However, for a system of mixture of cylindrical and spherical pores, 2.664 μm would correspond to $g\frac{\langle l^2\rangle}{\langle l\rangle}+\frac{3}{2}(1-g)\frac{\langle R^4\rangle}{\langle R^3\rangle}$ where g denotes the population fraction of cylindrical pores.

The extracted single scattering profile is shown in Fig. 3. The profile has shoulders at its tail due to either shape dispersion or the interference effects between the pores. As the system is very dense, such interference effects occur at relatively higher value of q. In the q range 2.26-3.06 nm^{-1}, the scattering profile follows a power law $\sim q^{-2.934}$ which is close to $\sim q^{-3.022}$ as observed in the case of corrected profile of the sample recorded using MoKα. It is noteworthy that for cylindrical pores with axis of the pores aligned with the incident beam, the expected power law is $\sim q^{-3}$ (Porod, 1982).

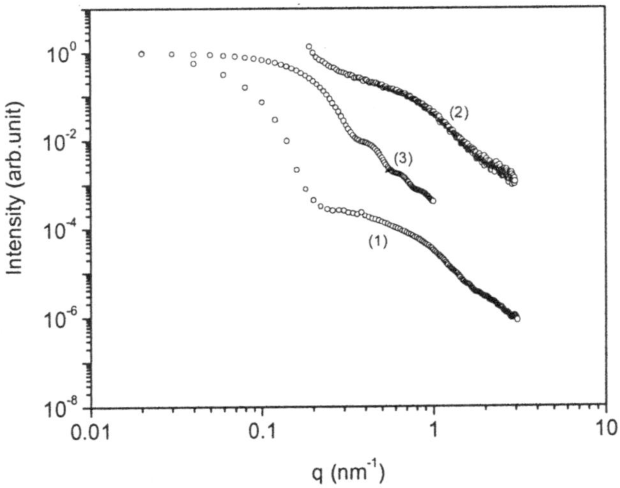

Fig. 3: Estimated single scattering profile, curve (1), vis-a-vis. the recorded, using MoK_α, scattering profile, curve (2), which is affected by multiple scattering.

A mixture of different shaped objects can explain the form of the single scattering profile. We did model calculations with a mixture of spherical and cylindrical pores, which also has the same features as are shown by curve (3) in Fig. 3. Curve 3 in Fig. 3 is the outcome of a model calculation with a mixture of cylindrical particles of radius 1.0 nm, length 100.0 nm and dispersion of spherical particles with radii ranging between 5 and 15 nm following a square-wave distribution. It is not found necessary to invoke fractal structure to explain the data. The single scattering profile has two distinct Guinier slopes, viz., -264.88 nm^2 in the wave-vector range ≤ 0.16 nm^{-1} and -2.96 nm^2 in the wave vector range 0.4-0.7 nm^{-1}. The first slope is interpreted as due to spherical particles and the second one is due to the radius of the cylindrical pores, which is estimated to be 3.44 nm. If the physical system is modeled as mixture of polydisperse spherical particles and polydisperse cylindrical particles with dispersion only in length, then from the slope at lower q values, we have $\langle R^8\rangle/\langle R^6\rangle$=1324.23 nm^2.

The estimated single scattering profile is best interpreted with a mixture of polydisperse spherical pores, curve (1) in Fig. 4, and polydisperse cylindrical pores, curve (2) in Fig. 4, with fixed cross-sectional radius of 3.44 nm. The fit used the constraints, $\dfrac{\langle R^8\rangle}{\langle R^6\rangle} = 1324.23$ nm^2 and $g\dfrac{\langle l^2\rangle}{\langle l\rangle} + \dfrac{3}{2}(1-g)\dfrac{\langle R^4\rangle}{\langle R^3\rangle} = 2.664$ μm with g found to be 0.845.

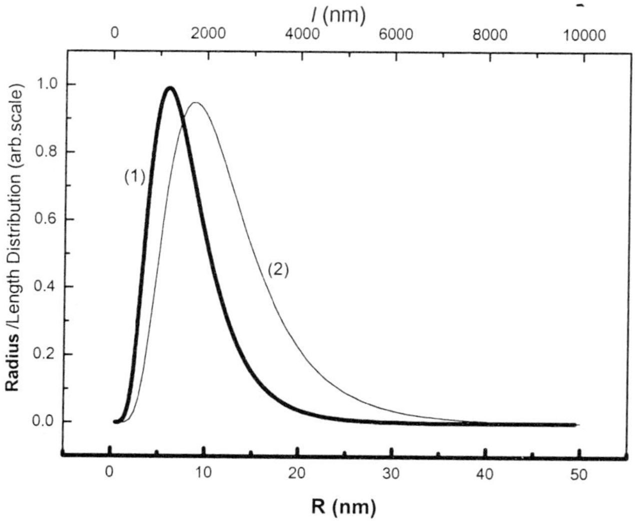

Fig. 4: Curve (1) is the estimated radius distribution function depicted by the thick line. The distribution has the following characteristics: $\langle R\rangle$ = 8.339nm, $\langle R^2\rangle$ = 87.247 nm^2 and peak position =5.930 nm. Curve (2) is the estimated length distribution function. The scale of the x-axis relevant to this curve is indicated at the top of the frame. The distribution has the following characteristics $\langle l\rangle$ = 2.487 μm, $\langle l^2\rangle$ = 784 μm^2 and peak position = 1.745 μm.

We concluded that the pores in lightly doped porous silicon have mixed shapes. The observed feature of functionality varying with wavelength is indicative of the presence of the cylindrical pores. Cylindrical pores have a broad distribution with peak at a fraction of porous layer thickness. However, some pores could be as long as half of porous layer thickness. Multiple scattering has to be accounted for explaining structural features of porous silicon by SAXS experiments. We did not need the assumption of fractal surfaces to explain the data.

INVESTIGATIONS USING SYNCHROTRON BASED X-RAY SOURCE

Phase Separation in 350-grade Maraging Steel

The kinetics of phase separation from a homogeneous phase into a two-phase region is a subject of continuing fascination for the last several decades. In this regard, the dynamical scaling behaviour during phase separation is of considerable importance towards understanding the phase separation of an alloy system. Experiments (Langer et al., 1975; Komura, et al., 1984; Katano and Iizumi, 1984; Komura et al., 1985; Furusaka et al., 1985) on binary alloys show that the scaling phenomenon holds only at the late stages of phase separation. The present section deals with the investigation (Mazumder et al., 1999) on phase separation behaviour of 350-grade maraging steel, a material of strategic importance. The current study shows that the dynamical scaling law holds, in contrast to its validity during late stages of phase separation in binary alloys (Langer et al., 1975; Komura, et al., 1984; Katano and Iizumi, 1984; Komura et al., 1985; Furusaka et al., 1985) during the early stages of phase separation in the multicomponent alloy maraging steel. However, the scaling phenomenon is valid only at the early stage when the interface between the secondary phase and matrix is diffused. At late stages the interface becomes sharp but the dynamical scaling law does not hold good.

Maraging steel well known for its high strength-to-weight ratio, high toughness and easy machining in solution-annealed condition is extensively used in strategic applications. At temperatures around 450 °C, ω related phases are known to form in 18 wt% Ni steels. The formation of ω through lattice collapse mechanism takes place at a large number of sites (approximately 10^{19} sites per cm^3) in the matrix and cold work can influence this density.

Two specific temperatures, viz. 430 °C and 510 °C, have been suitably chosen in order to study the competition between the formation of the ordered ω phase and the inter-metallic precipitates, the former known to occur at lower temperatures. The selection of higher aging temperature is governed by the fact that the steel chosen, (composition: 18 wt % Ni, 4.2 wt % Mo, 12.5 wt % Co, 1.7 wt % Ti, 0.1 wt % Al, 0.02 wt % Mn, 0.004 wt % C, and rest Fe) is conventionally hardened by vacuum aging at 510 °C. At each temperature, the specimens have been aged for different durations of time. At 510 °C, the aging time ranges from 5 min to 18 hrs, while at 430 °C, time ranges between 30 min and 72 hrs. Simultaneous small-angle and wide-angle X-ray scattering measurements have been carried out at synchrotron source ELETTRA, Trieste, Italy using 16 KeV (wavelength λ = 0.077 nm) branch of SAXS beamline (Amenitsch et al., 1997).

Figures 5 and 6 show the time-dependent scattering function $S(q, t)$ [$q = 4\pi \sin(\theta)/\lambda$ where 2θ is the scattering angle] at 430 and 510 °C, respectively.

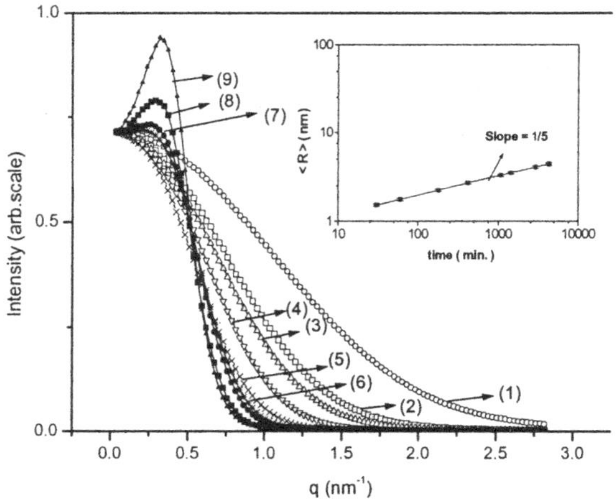

Fig. 5: Time evolution of the scattering function $S(q, t)$ at 430 °C for different durations of time as mentioned below against each profile. (1) re-crystallized, (2) 30 min, (3) 1 hr, (4) 3 hrs, (5) 7 hrs, (6) 18 hrs, (7) 24 hrs, (8) 48 hrs, (9) 72 hrs. The inset shows the time evolution of average precipitate radius showing $t^{1/5}$ law.

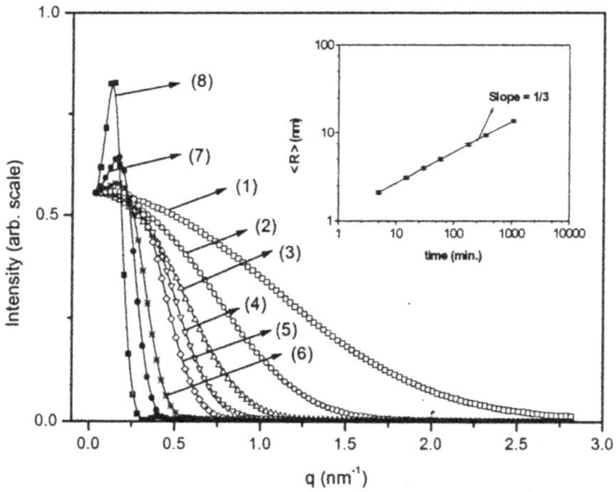

Fig. 6: Time evolution of the scattering function $S(q, t)$ at 510 °C for different durations of time as mentioned below against each profile. (1) recrystallized, (2) 5 min, (3) 15 min, (4) 30 min, (5) 1 hr, (6) 3 hrs, (7) 6 hrs, (8) 18 hrs. The inset shows time evolution of average precipitate radius showing $t^{1/3}$ law.

As evident from Figs. 5 and 6, with increasing aging time the scattering function becomes sharp with the position of the maxima shifting towards smaller wave vectors and with the peak intensity increasing quite sharply.

These observations indicate that both growth and the spatial correlation of the secondary phase intensify with aging. It was also observed that cold work facilitates the growth of the precipitates and their spatial correlation.

Porod exponents, as estimated from log [intensity] vs. log (q), in the q range 1.8 – 2.86 nm^{-1} for various specimens lay between ~4.0-5.3.

The deviation of the Porod exponent value from 4 was more pronounced for specimens aged at 430 °C. The proximity of the values to 4 for specimens aged for long time are indicative of well defined interface. The positive deviation of the Porod exponents of the profiles from 4 is indicative of the diffuse interface between the precipitates and the matrix, which arises essentially due to the presence of composition modulation in the matrix. This is expected as the formation of both the ordered ω and Ni_3(Ti, Mo) phases involve development and amplification of concentration waves of appropriate wavelength in the pre-nucleation stage.

While in the former case a displacement wave interferes with the concentration wave in the bcc lattice, the latter involves a structural transition into the fcc lattice and a growth of a concentration wave therein. The concentration modulation gives rise to the side peak observed in the WAXS profiles, as shown in Figs. 7 and 8, with a characteristic wavelength (~ 0.21 nm) in the system and can also explain the diffuse interface of the precipitates. With aging this modulation dies down. At 510 °C, this modulation dies down faster than that at 430 °C. At the beginning of this precipitation process, the interface is more diffused but as the aging proceeds, the precipitate-matrix boundaries become sharper. For cold-rolled specimens

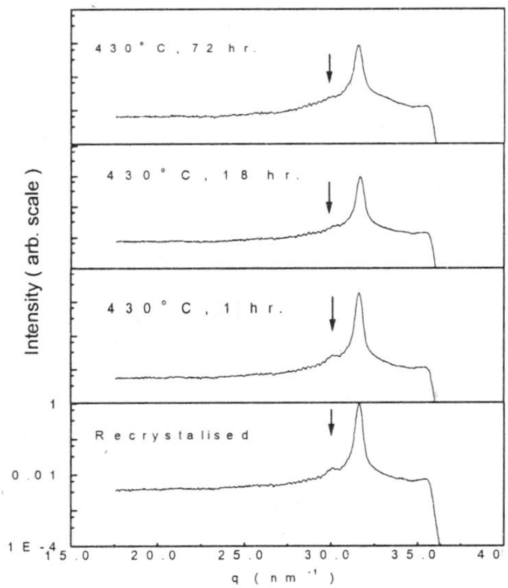

Fig. 7: WAXS profiles of specimens aged at 430 °C for some specified duration of time. The side peak is indicated by the arrow. The strong peak is due to reflection from (110) of the martensite bcc matrix.

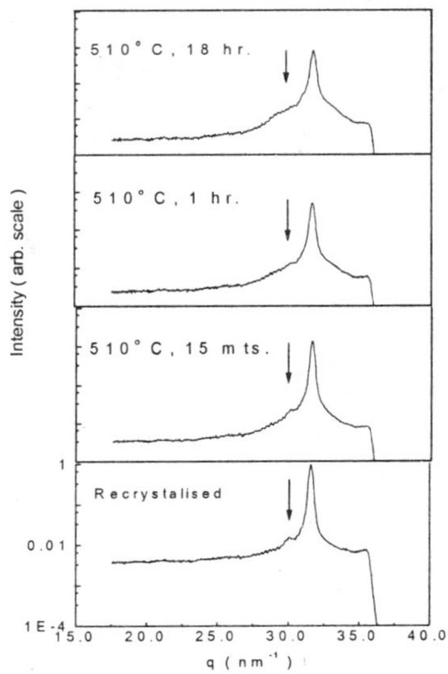

Fig. 8: WAXS profiles of specimens aged at 510 °C for some specified duration of time. The side peak is indicated by the arrow.

the side peak is weak. A clear evidence of formation of ω phase is also obtained in the electron diffraction patterns (not shown here).

At temperatures above 450 °C, the nucleation of A_3B type of precipitates occurs through concentration fluctuations rather than through the collapse of the lattice. In this case, the conversion of an embryo into a nucleus through a concentration fluctuation results in sharp interfaces relatively quickly with well-defined composition of the nucleus. These studies hint at the existence of two distinct time-temperature transformation (TTT) curves for precipitation in maraging steels, one below and the other above 450 °C.

$<R>$ follows a scaling law with aging time with exponent 0.2152, somewhat closer to 1/5, at 430 °C and with exponent 0.35, close to 1/3, at 510 °C. The normalized scaling function $F[q/q_1(t)] = [q_1(t)]^3 S(q,t)/\Sigma q^2 S(q,t)\,\delta q$ (where $q_1(t)$ and δq are the first moment of the scattering function $S(q,t)$ and experimental q increment, respectively) were calculated.

The plots of $F[q/q_1(t)]$ for specimens with different aging time at 430 and 510 °C are shown in Figs. 9 and 10, respectively. At 430 °C, the scaling function $F(q/q_1(t))$ is independent of time up to an aging time of 7 hrs.

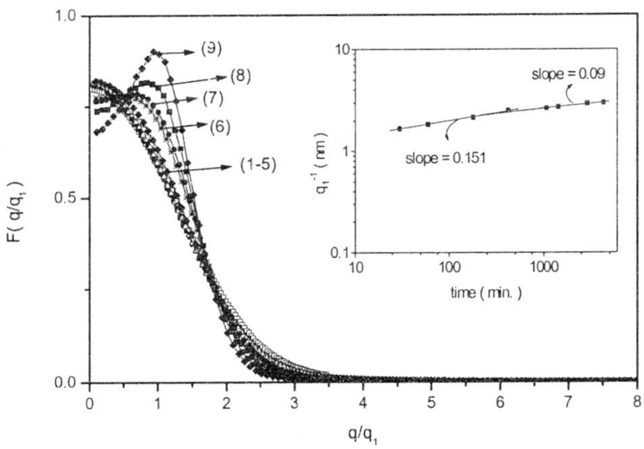

Fig. 9: Scaled function at 430 °C for different durations of time. Inset shows the variation of the reciprocal of $q_1(t)$ with time plotted on a logarithmic scale.

At this temperature, $F(q/q_1(t))$ scales as $(q/q_1)^{-4.43 \pm 0.19}$ for large (q/q_1). At 510 °C the time independent feature of scaling function is observed up to 15 min at 510 °C. $F(q/q_1(t))$ scales as $(q/q_1)^{-4.21 \pm 0.16}$ for large (q/q_1) indicating diffused interfaces between precipitates and the matrix.

Beyond these points, the change of the scaling function takes place in a rather broad time range at both the temperatures. Irrespective of aging temperature, phase separation broadly follows two distinct patterns. In the initial stage, dynamical scaling holds for the scattering function $S(q,t)$ which exhibits $q^{-(4+\delta)}$ (where $\delta > 0$) for high q values. No q^{-2} law was observed. However, at late stage dynamical scaling does not hold good but q^{-4} law

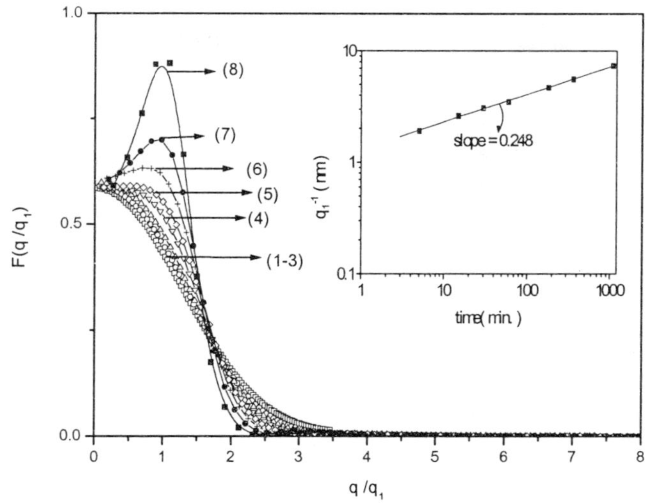

Fig. 10: Scaled scattering function $F(q/q_1)$ at 510 °C for different durations of time. Inset shows the variation of the reciprocal of q_1(t) with time plotted on a logarithmic scale.

holds. These observations are in sharp contrast with those obtained in past with AlZn and FeCr alloys. In AlZn and FeCr alloys, at early stage dynamical scaling does not hold good for the scattering function $S(q, t)$ which exhibits q^{-2} power law for high q values as predicted by Langer, Baron and Miller (1975). At late stage, both q^{-4} law and dynamical scaling hold good.

We concluded that the nature of the final precipitates and the routes through which these transformations occur appear to be different at the two temperatures namely 430 and 510 °C, respectively. It could be due to different thermodynamic stability criterion at the two temperatures. The system exhibits scaling phenomena but only at the early stage of phase separation. At the late stage, the scaling phenomenon does not hold good. Possibility exists for non-unique characteristic length. Porod exponent being close to 5 is not indicative of the existence of multicomponent order parameter. TEM investigations also clearly indicate that at each aging temperature, the system exhibits only one order parameter as separation of only one phase takes place.

Precipitation in Zircaloy-2 Fuel Cladding Tube

Morphology of precipitates and kinetics of phase separation of intermetallic precipitates in the transition metal based alloys play an important role in their corrosion behaviour. Zirconium based alloys are of practical importance for their extensive applications in the nuclear industries. Zircaloy-2 is one of the most commonly used materials in boiling water reactors and pressurized water reactors permitting the attainment of a high level of integrity and reliability. However, in the high temperature water/steam environment inside a nuclear reactor, corrosion occurs in Zircaloy and is manifested by the formation of a non-protective type of oxide layer, limiting the life of a fuel cladding tube. In addition to uniform corrosion, another type of corrosion occurs in Zircaloy due to the formation of local nodules of unprotected oxide particles and is known as nodular corrosion. The nodules grow in size with increasing exposure and ultimately form a closed thick layer. The corrosion behaviour of this alloy is largely dependent on size distribution, morphology and also on the chemical compositions of the precipitates. Thus for increased discharge burn-ups, different strategies are necessary to optimize corrosion by controlling the microstructure of the precipitates. In order to reduce uniform corrosion, a minimal size of the precipitates has to be ensured while to avoid nodular corrosion, a maximal size should not be exceeded as the probability of nodular corrosion decreases with a decrease in the average precipitate size.

In this study (Sen et al., 2000), the precipitate morphology and the precipitation kinetics were investigated in as-received (virgin) and β quenched Zircaloy-2 as well as in specimens aged at 823 K for 1 hr, 2 hrs and 4 hrs, respectively following a β quenching. As-received (virgin) samples were subjected to stress relief anneal after pilgering (cumulative annealing parameter, ΣA, being 3.3306×10^{-14} hr) a Zircaloy-2 tube (composition: 1.50 wt.% Sn, 0.14 wt.% Fe, 0.12 wt.% Cr, 0.12 wt.% Ni, 1200 wppm oxygen and balance Zr) with a wall thickness of 0.4 mm. The further thinned specimens in the sealed condition were soaked at a temperature of 1323 K for 10 minutes and then quenched in water at room temperature. Subsequently, the specimens were aged at 823 K for 1 hr, 2 hrs and 4 hrs, respectively after β quenching.

The experimental wavelength of the X-ray radiation was chosen purposely at 0.077 nm as the absorption edge of zirconium is slightly below the above experimental wavelength. Hence, due to the anomalous scattering effect near this wavelength, the intensity of the peaks (in WAXS) due to the precipitates phases in Zircaloy becomes significant, though the weight percent of the other alloying elements are small compared to that of the zirconium in the alloy.

Figure 11 shows the time dependent small angle scattering function $S(q, t)$ for the as-received, β quenched and aged specimens. It is evident from the Fig. 11 that with increasing aging time, the scattering function becomes sharp while the sharpness is maximum for the as-received sample.

Porod exponents as estimated from log (intensity) vs. log (q), in the q-range >2.0 nm^{-1} for various specimens deviates from the value of 4 for all the specimens and that its value lay between ~ 2.4 to 2.7. It is noteworthy that the characteristic Porod exponent for spherical particles with sharp interfaces is 4 and that for particles deviating from sphericity, is less than 4 (e.g. 2 for a plate-like particles and 1 for a rod like particles in the

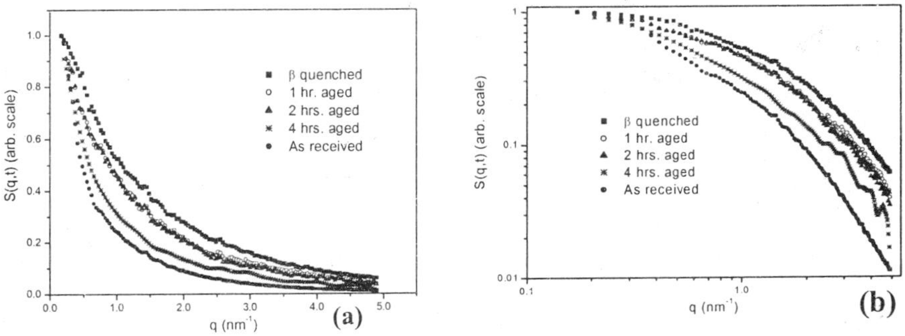

Fig. 11: Time dependent scattering function $S(q, t)$ of Zircaloy-2 specimens. The profiles are normalized to unity (a) linear scale (b) double logarithmic scale.

intermediate q range). Hence a suitable mixture of spherical and non-spherical particles can give rise to a Porod exponent, which is less than 4.

From TEM micrograph (not shown here), it was seen that the rod-like particles had not formed but spherical and plate-like particles were present. Hence, we tried to explain the SAXS data by modeling the system in terms of a poly-disperse mixture of spherical and plate-like particles. The estimated volume square weighted size distributions for spherical and plate-like particles are shown in Figs. 12 and 13, respectively. The plots of $F[q/q_l(t)]$ for specimens with different aging time are shown in Fig. 14. It is evident that the scaling function $F[q/q_l(t)]$ for the samples superimpose one over another and hence indicates the time independence of the same. The wide-angle X-ray scattering (WAXS) data for the specimens are shown in Fig. 15.

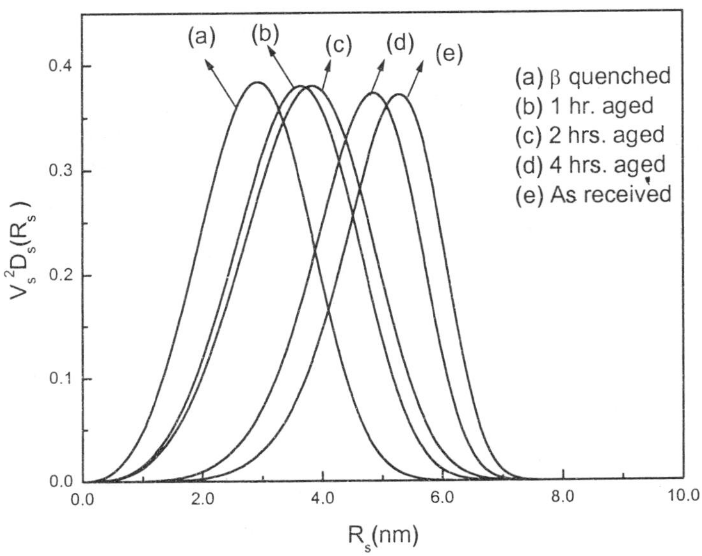

Fig. 12: The volume square weighted radius distribution of the spherical precipitates for different specimens.

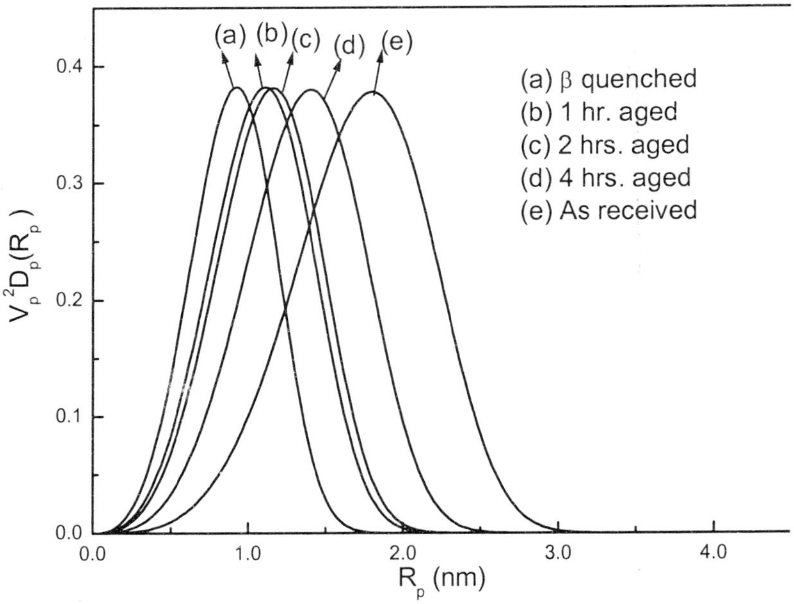

Fig. 13: The volume square weighted radius distribution of the plate-like precipitates for different specimens.

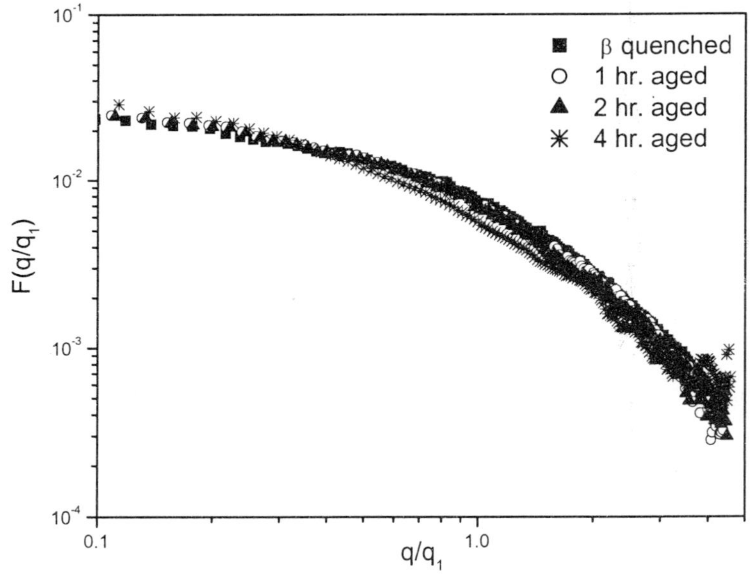

Fig. 14: Scaled structure factor $F(q/q_1)$ for different aging times for Zircaloy-2.

Aging at 823 K has not altered the microstructure substantially. However, the presence of small precipitates with size of few nanometers has been observed on aging. They are found to be distributed uniformly inside the grain as well as at the grain boundaries. In order to identify their

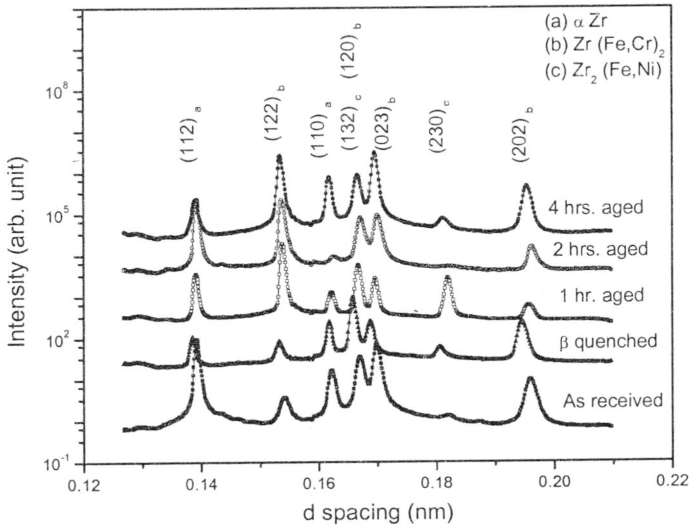

Fig. 15: WAXS profiles of the Zircaloy-2 fuel cladding specimens.

crystallographic structures, selected area diffraction (SAD) patterns were obtained. From these SAD patterns, the presence of two types of intermetallic precipitates has been identified as Zr_2(Fe, Ni) with tetragonal [$CuAl_2$ type, with $a_0 = 0.66$ nm and $c_0 = 0.55$ nm] and Zr(Fe, Cr)$_2$ with hexagonal [$MgZn_2$ type, $a = 0.51$ nm and $c = 0.82$ nm] crystal structure. These observations are further supported by the WAXS data.

Presence of different types of precipitates can occur in a multi-component alloy like Zircaloy. In fact WAXS and TEM results also support the above fact by confirming the existence of two types of precipitates, Zr_2(Fe, Ni) and Zr(Fe, Cr)$_2$ and has reasonably large lattice mismatch with the α zirconium matrix. It is to be noted that the precipitates, formed after quenching from a high temperature, are generally spherical when the lattice mismatch between the matrix and the precipitates is small while the non-sphericity of the precipitates increases with the increase in the lattice mismatch. It is noteworthy that lattice mismatch between the α zirconium [with $a = 0.32$ nm and $c = 0.51$ nm] matrix and the Zr(Fe, Cr)$_2$ precipitates is reasonably high and the formation of non spherical e.g. plate-like precipitates in addition to the spherical precipitate is expected.

From the present investigations, we conclude that two types of small precipitates (< 10 nm), spherical and plate-like, are present in the Zircaloy-2. The presence of two different shapes are responsible for the observation of a Porod exponent ~ 2.5 in SAXS profiles. The presence of non-spherical e.g. plate-like precipitates are expected due to the reasonably large lattice mismatch between the α zirconium matrix and the precipitates. β quenching helps in reducing the precipitate size, which is an essential criterion to reduce the nodular corrosion. With increase in aging period the precipitate starts growing in size. The relatively large precipitate size in the as-received specimen can be attributed to the enhancement of the nucleation and growth due to the number of quenched-in vacancies created at the time of tube fabrication.

TEM and WAXS investigations have indicated that the un-mixing occurs via the formation of precipitates with two types of crystallographic structures, tetragonal Zr_2(Fe, Ni) and hexagonal Zr(Fe, $Cr)_2$. Appearance of no new peak or no significant change in the peak positions in WAXS profiles indicates that no new phase appears in the course of heat treatment. The system exhibits scaling behaviour and the simple scaling laws also hold good for a multi-component alloy like Zircaloy-2.

Small-angle X-ray Scattering Facility at Indus-2

Considering the demand of SAXS facilities in our country and our experience with rotating anode based SAXS facility, synchrotron based SAXS facility at ELETTRA and Small-Angle Neutron Scattering (SANS) spectrometers at CIRUS (Desa et al., 1985) and DHRUVA (Mazumder et al., 2001; Mazumder et al., 2002) reactors, we propose to set-up a SAXS facility at INDUS-2. The facility will be equipped with asymmetric cut double crystal monochromator (DCM), toroidal mirror and two discrete wavelengths namely 0.154 and 0.077 nm. We propose to use a CCD detector for SAXS measurements and a gas-filled delay line Gabriel type linear position sensitive detector for WAXS measurements.

In a scattering experiment one has to define the direction and energy of the primary beam to know the initial momentum. The direction of the beam is defined by the slits, which simply cut out a narrow solid angle of the radiation emerging from the source. This property is also shared by mirrors and monochromators, which, because of their finite size, accept only a small part of the radiation.

The elements defining the shape and size of the incident beam on the specimen depend critically upon the size of the specimen to be investigated and the desired spatial resolution. For most applications at synchrotron sources, the use of a convergent geometry defined by a focusing mirror and monochromator is suitable.

Monochromator selects specific wavelength range of interest. In our case, we intend to tune the wavelength in the range 0.154-0.077 nm. When only one monochromator is used, the direction of the emerging monochromatic beam will depend on the wavelength according to Bragg's law. The instrument has to rotate around the monochromator whenever a change in wavelength is desired. Though the mechanical problems can be solved for medium-size instruments, the rotation of a very long instrument like SAXS causes many problems. Therefore, we propose double-crystal monochromator (DCM) which produces a constant beam position at all wavelength.

Focusing with monochromator crystals can be achieved by bending them either cylindrically or even spherically. But bending introduces defects which leads to a somewhat enlarged single crystal reflection curve compared to the ideal values. For focusing purpose, we propose the asymmetric cut Si-crystals as monochromators. Symmetric cut crystals accomplish a reduction in the size of the bundle of beam. We propose two pairs of asymmetrically cut flat Si (111) crystals as monochromators. Small grazing angle will result in a reduction of power density by more than one order of magnitude compared to the normal incidence and backside cooling will be sufficient to keep the crystal in shape.

We propose a toroidal mirror for further focusing the monochromatic beam. A toroidal mirror focuses a beam both in the horizontal and vertical direction. The surface roughness of the mirror is to be minimized particularly in terms of low frequency surface distortions, which tend to broaden the reflected beam. The focal distance of the mirror will be kept variable since the scattering geometry may change during the course of a series of experiments. To avoid the changes in the characteristics of the mirror with time, i.e., the radiation damage on the reflecting surface and plastic flow of the mirror, we propose to put the mirror after the double crystal monochromator assembly. However, producing quality mirrors with an acceptable level of imperfections is extremely difficult. The length of the mirror required further compounds the problem. The very small angle of incidence of the incident beam with a finite source size mandates that mirrors be approximately 60-100 cm in length.

To circumvent the difficult task of maintaining high tolerances on a large mirror length, we propose to put two smaller mirrors one after another. This arrangement will have the desirable features that tolerances are more accurately held over the shorter lengths of the individual mirrors and each element can be replaced independently. Scattering from the edges of the mirrors is circumvented via synchronous movements of the end of one mirror and the edge of the following mirror. Oxidation and deposition of the contaminants on the mirror surface degrade the performance of the mirror. So we propose to keep the mirror under vacuum or under a helium blanket.

To reduce the parasitic instrumental background and improve the resolution of the SAXS instrument, several slits cooled by water will be installed along the beam line. There are beam defining slits situated up stream of the monochromator; baffle slits (down stream of the monochromator); aperture slits (down stream of the mirror) and guard slits (before the sample holder). All slit blades will be made from tungsten carbide and will be located in vacuum.

A schematic layout of the proposed SAXS beamline is shown in Fig. 16. The beamline contains carbon filter and beryllium windows, which separate the clean HV of the beamline from the storage ring UHV. The filter will remove all photons with energies < 4 Kev. The approximate distances are

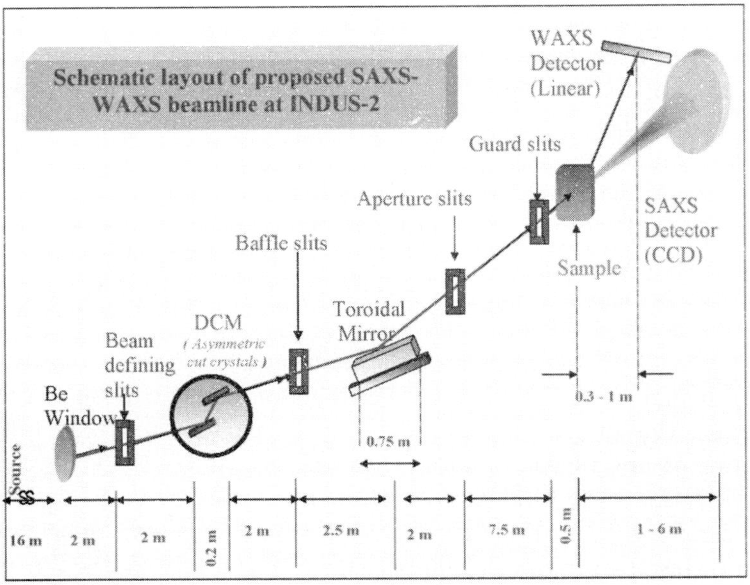

Fig. 16: The schematic diagram of the proposed SAXS beamline at Indus-II.

following: Beryllium window - 1st. Monochromator centre 18 metre; 1st Monochromator centre- 2nd Monochromator centre 4 metre. 2nd Monochromator centre - Toroidal mirror centre ~ 4.5 metre; Toroidal mirror centre Sample stage centre ~ 9.6 metre; Sample stage centre- SAXS detector centre (1-6 metre variable); Sample stage centre - WAXS detector centre (0.3-1.0 metre variable).

With this proposed instrument, the SAXS resolution in the scale of 1-100 nm in *d*-spacing at 0.154 nm could be achieved. Provision has been made for simultaneous wide angle diffraction measurement in the *d*-spacing range of 0.12-0.8 nm at 0.154 nm. With two different energies, we have provision for experimenting on optically thin and dense systems (like alloys of heavy elements). Multi-purpose sample stage, which include an IR laser and different sample cells optimized for temperature and pressure scans is proposed.

A beam line consists of source, front end, beam transport channel and experimental station. The source of radiation is proposed to be a bending magnet. The machine crew will provide front end. The proposed optical lay out and mechanical layout are shown in the Figs. 17 and 18, respectively.

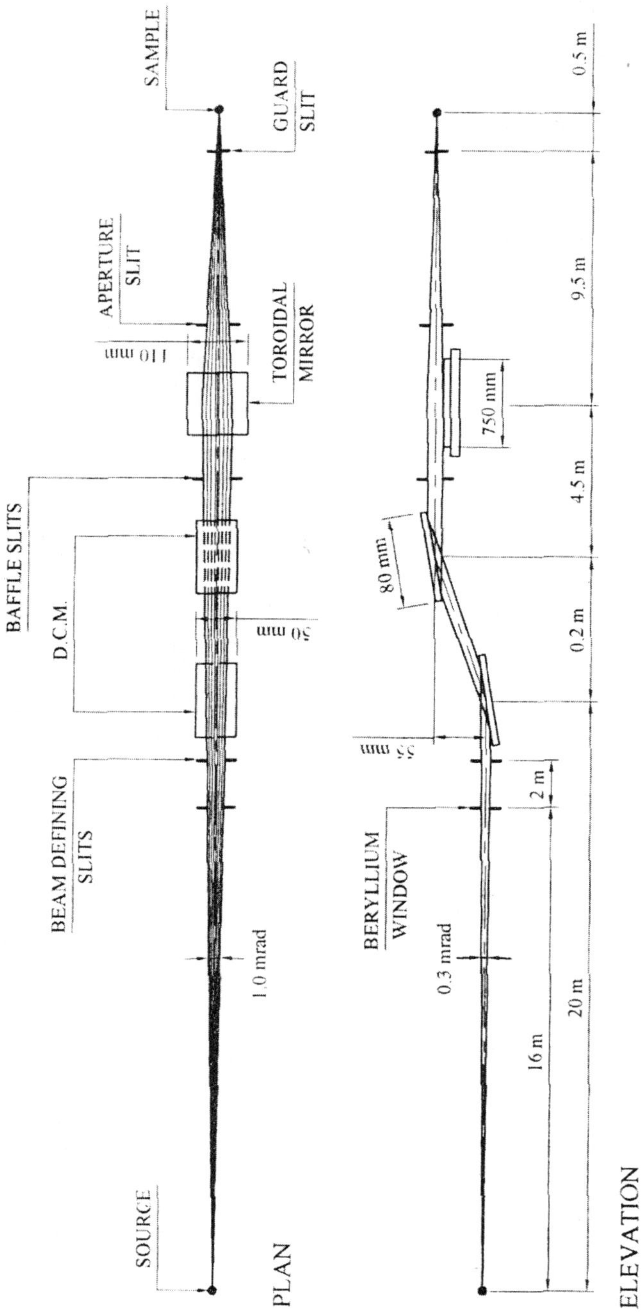

Fig. 17: The optical layout of the proposed SAXS beamline at Indus-II.

Item No.	Description	Qty
1	Beam Position Monitor	3
2	S.I.P.	6
3	T.M.P.	1
4	Bellows	21
5	Be-Window	5
6	Gate Valve	5
7	Beam Defining Slit	1
8	Fluarecent Screen	1
9	D.C.M.	1
10	Baffle Slit	1
11	Toroidal Mirror With Chamber	1
12	Aperture Slit	1
13	Guard Slit	1
14	Beam Shutter	1
15	Experimental Station	1
16	Fast Closing Valve	1

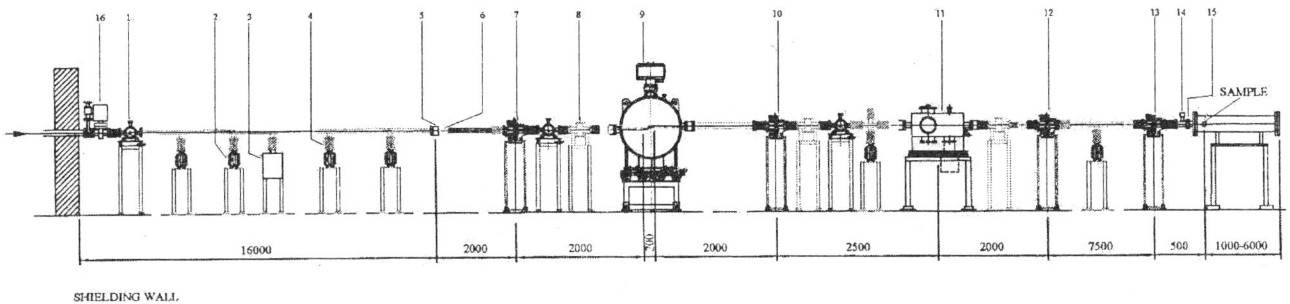

Fig. 18: The mechanical layout of the proposed SAXS beamline at Indus-II.

REFERENCES

Amenitsch, H., Bernstorff, S., Kriechbaum, M., Lombardo, D., Mio, H., Rappolt, M. and Lagner, P. (1997). Elettra Laboratory Beamline Hand Book, November (1996). *J. Appl. Crystallogr.*, **30:** 872.

Binder, M., Edelmann, T., Metzger and Peisl, J. (1996). *Solid State Commn*, **100:** 13.

Binder, M., Edelmann, T., Metzger, T.H., Mauckne, G., Goerigk, G. and Peisl, J. (1996). *Thin Solid Films*, **276:** 65.

Deb, P., Biswas, T., Sen, D., Basumallick, A. and Mazumder, S. (2002). *J. Nanoparticles Res.*, **4:** 91-97.

Desa, J.A.E., Mazumder, S., Sequeira, A. and Dasannacharya, B.A. (1985). *Solid State Physics Symposium*, **28C:** 318.

Furusaka, M., Ishikawa Y. and Mera, M. (1985). *Phys. Rev. Lett.*, **54:** 261.

Glatter, O. and Kratky, O. (1982). Small angle X-ray scattering. Academic London.

Goudeau, P. et al. (1989). *J. Appl. Phys.*, **66:** 625.

Guinier, A. and Fournet, G. (1955). Small angle scattering of X-rays. Wiley, New York.

Jacob, J., Kumar, A., Asokan, S., Sen, D., Chitra, R. and Mazumder, S. (1999). *Chem. Phys. Lett.*, **304:** 180-186.

Katano, S. and Iizumi, M. (1984). *Phys. Rev. lett.*, **52:** 835.

Komura, S., Osamura, K., Fujii, H. and Takeda, T. (1984). *Phys. Rev. B*, **30:** 2944.

Komura, S., Osamura, K., Fujii, H. and Takeda, T. (1985). *Phys. Rev. B*, **31:** 1278.

Langer, S., Baron, M. and Miller, H.D. (1975). *Phys. Rev.A*, **11:** 1417.

Mazumder, S., Bhagwat, K.V. and Sequeira, A. (1995). *J.Phys.*: Condens. Matter, 9737-46.

Mazumder, S. (1993). *J. De Physique IV*, 519-22.

Mazumder, S., Jayaswal, B. and Sequeira, A. (1998). *Physica* B, **241-243:** 1222.

Mazumder, S., Sen, D. and Patra, A.K. (2003). *J. Appl. Cryst.*, **36:** 840-844.

Mazumder, S., Sen, D., Saravanan, T. and Vijayaraghavan, P.R. (2001). *Neutron Research*, **9:** 39.

Mazumder, S., Sen, D., Saravanan, T. and Vijayaraghavan, P.R. (2001). *Current Science*, **81:** 101.

Mazumder, S., Sen, D., Saravanan, T. and Vijayaraghavan, P.R. (2002). *Neutron News*, **13:** 2.

Mazumder, S., Sen, D., Saravanan, T. and Vijayaraghavan, P.R. (2002). *Appl. Phys (A)*, **74:** S183-S185.

Mazumder, S., Sen, D., Batra, I.S., Tewari, R., Dey, G.K., Banerjee, S., Sequeira, A., Amenitsch, H. and Bernstorff, S. (1999). *Phys. Rev. B*, **60:** 822-830.

Mazumder, S., Sen, D., Roy, S.K., Hainbuchner, M., Baron, M. and Rauch, H. (2001). *J. Phys.*: Condens. Matter, **13:** 5089-5102.

Mazumder, S., Sen, D., Sastry, P.U.M., Chitra, R., Sequeira, A. and Chandrasekaran, K.S. (1998). *J. Phys.*: Condens. Matter, **10:** 9969-9974.

Mazumder, S. and Sequeira, A. (1988). *Mat. Sci. Forum*, 427-432.

Mazumder, S. and Sequeira, A. (1988). *Pramana-J. Phys.*, 557-568.

Mazumder, S. and Sequeira, A. (1989). *Phys.Rev.B*, **39:** 6370.

Mazumder, S. and Sequeira, A. (1990). *Phys.Rev. B*, **41:** 6272.

Mazumder, S. and Sequeira, A. (1991). *Physica B*, 200-205.

Mazumder, S. and Sequeira, A. (1992). *J. Appl. Cryst.*, 221-230.

Mazumder, S. and Sequeira, A. (1992). *Pramana- J. Phys.*, **38:** 95.

Mazumder, S. and Sequeira, A. Roy, S.K. and Biswas, A.R. (1993). *J. Appl. Cryst.*, 357-362.

Naudon, A., Goudeau, P., Halimaoui, A., Labbert, B. and Bomchil, G. (1994). *J. Appl. Phys*, **75:** 780.

Pandey, A.K., Goswami, A., Sen, D., Mazumder, S. and Childs, R.F. (2003). *J. Membrane Science*, **217:** 117-130.

Porod, G. (1982). *In:* "Small-angle X-ray Scattering", edited by O. Glatter and O. Kratky (Academic, London), 17.

Sastry, P.U., Sen, D., Mazumder, S. and Chandrasekaran, K.S. (2000). *Solid State Communication*, **114:** 329-333.

Sastry, P.U., Sen, D., Mazumder, S., Chavan, S.V. and Tyagi, A.K. (2003). *J. Solid State Chemistry*, **176:** 57-61.

Sen, D., Mazumder, S. and Tarafdar, S. (2002). *Appl. Phys* (*A*), **74:** S1049-1051.

Sen, D., Deb, P., Mazumder, S. and Basumallick, A. (2000). *Materials Research Bulletin*, **35:** 1243-1250.

Sen, D., Mahata, T., Patra, A.K., Mazumder, S. and Sharma, B.P. (2003). *J. Alloys and Compounds* (In press).

Sen, D., Mazumder, S. and Tarafdar, S. (2002). *J. Mat. Sci*, **37:** 941-947.

Sen, D., Mazumder, S., Chitra, R. and Chandrasekaran, K.S. (2001). *J. Mat. Sci.*, **36:** 909-912.

Sen, D., Mazumder, S., Sequeira, A., Ghosh, A.K., Ramachandran, V., Hanra, M.S. and Misra, B.M. (1999). *J.Macromol. Sci. - Phys*. B, **38:** 341-348.

Sen, D., Mazumder, S., Tewari, R., De, P.K., Amenitsch, H. and Bernstorff, S. (2000). *J. Alloys and Compounds*, **308:** 250-258.

Sen, D., Patra, A.K., Mazumder, S. and Ramanathan, S. (2002). *J. Alloys and Compounds*, **340:** 236-241.

Sen, D., Patra, A.K., Mazumder, S. and Ramanathan, S. (2003). *J. Alloys and Compounds*, **361:** 270-275.

Sen, D., Mazumder, S., Sengupta, P., Ghosh, A.K. and Ramachandran, V. (2000). *J. Macromol. Sci.- Phys.* B, **39:** 235-243.

Swain, A.K., Parida, J.K., Bisoyi, D.K., Mazumder, S. and Mohanty, A.K. (2001). *J. Appl. Cryst.*, **34:** 510-518.

Tewari, R., Mazumder, S., Batra, I.S., Dey, G.K. and Banerjee, S. (2000). *Acta Metall.*, **48:** 1187-2000.

Vezin, V., Goudeau P., Naudon, A., Herino, A. and Bomchil, G. (1991). *J. Appl. Cryst.*, **2:** 581.

Vezin, V., Goudeau P., Naudon, A., Herino, A. and Bomchil, G. (1992). *Appl. Phys. Lett.*, **60:** 2625.

Vogel, W., Dhayagude, D., Chitra, R., Sen, D., Mazumder, S. and Kulkarni, S.K. (2002). *J. Mat. Sci.*, **37:** 4545-4553.

Role of Combined SANS and SAXS Studies in Micellar Solutions

P.S. Goyal* and V.K. Aswal

Solid State Physics Division
(*IUC-DAEF, Mumbai Centre)
Bhabha Atomic Research Centre, Mumbai 400 085

INTRODUCTION

Micellar solutions are suspensions of aggregates of surfactant molecules in water (Chevalier and Zemb, 1990; Israelachvili, 1992). Surfactant molecule consists of a polar hydrophilic head group and a long hydrophobic chain connected to the head group. The coexistence of the two opposite types of behaviour (hydrophilic and hydrophobic) in the same molecule leads to the self-aggregation of the surfactant molecules when dissolved in water. The aggregates are called as micelles. The typical size of a micelle is about 50 Å and is made up of about 100 surfactant molecules. A schematic representation of a surfactant molecule and a spherical micelle is shown in Fig. 1. The micelles are formed by the delicate balance of opposing forces: the attractive tail-tail hydrophobic interaction provides the driving force for the aggregation of the surfactant molecules, while the electrostatic repulsion between the polar head groups limits the size that a micelle can attain. As a result, the characteristics of these aggregates are easily controlled by the small changes in chemical structure of the surfactant molecule and the solution conditions such as concentration, temperature and ionic strength. The aggregates formed are of various types, shapes and sizes such as spherical or ellipsoidal, cylindrical or thread-like micelle, disk-like micelle, membrane and vesicles (Chevalier and Zemb, 1990). The study of formation of these different structures is important as the surfactant solutions are widely used in various household, industrial and research applications.

Surfactant molecules such as cetyltrimethylammonium bromide (CTABr) ionize in aqueous solution and the corresponding micelles are aggregates of CTA^+ ions. The micelle is charged and is called an ionic micelle. The Br^- ions, known as counterions, tend to stay near the CTA^+ micellar surface.

The shape, size, fractional charge of the micelle and the intermicellar interaction depend on the distribution of these counterions around the micelle (Israelachvili, 1992). Since the works of Oosawa (1971) and Manning (1969), the concept of counterion condensation is widely accepted in the field of linear polyelectrolytes. It has been shown that when the charge density on an infinitely long cylinder is increased beyond a critical value, counterions condense around the cylinder so as to reduce the effective charge density to the critical value. Similar concepts have also been used in colloidal suspensions made of spherical charged colloids (Belloni, 1998). Study of counterion distribution around a micelle is of interest as it decides the effective charge on the micelle and hence the micellar structure and the interaction between the micelles (Chevalier and Zemb, 1990; Aswal et al., 1998).

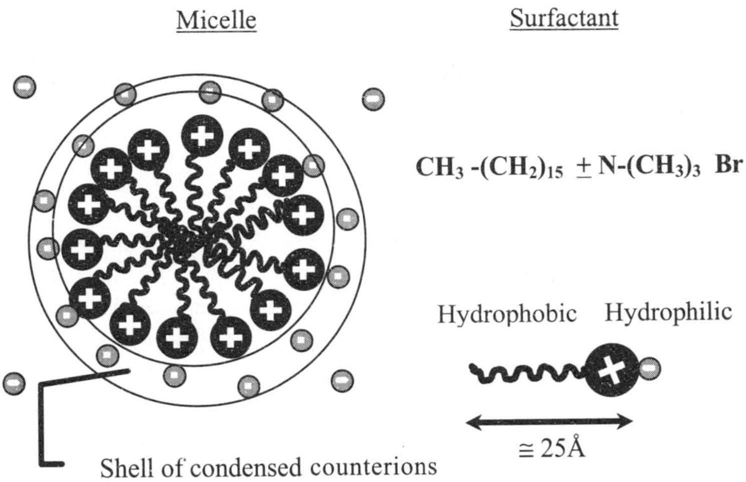

Fig. 1: Schematic of surfactant molecule and spherical micelle with counterion distribution around the micelle.

Small-angle neutron scattering (SANS) and small-angle X-ray scattering (SAXS) are excellent techniques for studying the structures on length scale of 10-1000 Å (Chen and Lin, 1987). These techniques are now routinely used for studying the sizes and shapes of "nano particles" (precipitates, voids, maromolecules and micelles) etc. dispersed in uniform medium. This paper deals with a combined SANS and SAXS study for micellar solutions. Neutron scattering in micellar solutions is from the core of the micelle. However, X-rays are largely scattered by counterions, especially when the counterion has a large atomic number (e.g. Br^-). The neutron scattering intensity from the counterion distribution is negligible in comparison to that from the core. Thus, neutrons see the core of the micelle and X-rays give information relating to the counterion condensation/distribution around the micelle (Wu et al., 1988).

SMALL-ANGLE SCATTERING

Small Angle Neutron Scattering (SANS) and Small Angle X-ray Scattering (SAXS) have been referred to as Small Angle Scattering. The experimental details and the data analysis techniques used in the two scatterings are similar and the only difference is in the radiation used. As it will be discussed below, because of differences in their interactions with matter the contrast factors and the scattering length densities will be different for neutrons and X-rays. The details of above techniques have been discussed in a book (Feigin and Sverguu, 1987). In the following, we confine to SANS. However, most of it applies to SAXS also.

SANS is a diffraction experiment. However, unlike conventional diffraction experiments, where the structure of a material is examined at atomic resolution (~2 Å), SANS is used for studying the structure of materials with a spatial resolution of ~100 Å (Chen and Lin, 1987; Wu et al., 1988; Feigin and Sverguu, 1987). SANS experiment involves scattering of a monochromatic beam of neutrons from the sample and measuring the scattered neutron intensity as a function of the scattering angle. The wave vector transfer Q (=$4\pi \sin\theta/\lambda$ where λ is the neutron wave length and 2θ is the scattering angle) in these experiments is small, typically in the range of 10^{-4} to 1.0 $Å^{-1}$. The wavelength of neutrons used for these experiments usually being ~4-10 Å, the smallest Q occurs at small scattering angles (~1^{O}). The technique is therefore called "Small Angle Neutron Scattering". An indigenously built SANS diffractometer (Aswal and Goyal, 2000) is operating at Dhruva reactor, B.A.R.C. and is being regularly used by BARC and university scientists.

The small-angle scattering intensity $I(Q)$ as a function of scattering vector Q for a micellar solution can be expressed as (Chen and Lin, 1987)

$$I(Q) = nP(Q)\ S(Q) \tag{1}$$

where n is the number density of the particles. $P(Q)$ is the intraparticle structure factor and depends on the shape and size of the particles. $S(Q)$ is the interparticle structure factor and is decided by the spatial distribution of the particles. $P(Q)$ is given by the integral

$$P(Q) = \left| \int \rho(r) - \rho_s)\exp(i\boldsymbol{Q}.\boldsymbol{r})dr \right|^2 \tag{2}$$

In the simplest case of a monodispersed system of homogeneous spherical particles with a radius R, $P(Q)$ is given by

$$P(Q) = (\rho - \rho_s)^2\ V^2 \left[\frac{3J_1(QR)}{QR} \right]^2 \tag{3}$$

where $V = (4/3)\pi R^3$, ρ_s is the scattering length density of the solvent and ρ is the mean scattering length density of the particle.

The expression for $S(Q)$ depends on the relative positions of the particles. In case of isotropic system, $S(Q)$ can be written as

$$S(Q) = 1 + 4\pi n \int (g(r)-1)\frac{\sin Qr}{Qr} r^2 dr \tag{4}$$

where $g(r)$ is the radial distribution function. $g(r)$ is the probability of finding another particle at a distance r from a reference particle centered at the origin. The details of $g(r)$ depend on the interaction potential $U(r)$ between the particles. For the results reported in this paper, $U(r)$ was assumed to be screened Coulomb potential and $S(Q)$ was calculated under mean spherical approximation (Haldar et al., 2001).

The term $(\rho - \rho_s)^2$ is referred as a contrast factor. The above equations are valid both for the SAXS and the SANS experiments. The contrast factor, however, depends on the radiation used (Windsor, 1988; Aswal et al., 2000). The values of ρ and ρ_s depend on the chemical composition of the micelle and the solvent and are different for neutrons and X-rays. The differences in ρ values for neutrons and X-rays arise from the fact that while neutrons are scattered by the nucleus of an atom, the X-rays are scattered by the electron clouds around the nucleus. It is seen that as one goes across the periodic table, the neutron scattering lengths vary in a random way and the X-ray scattering lengths increase with the atomic number of the atom. For example, unlike X-rays where $\rho_s (H_2O) = \rho_s (D_2O)$, the values of ρ_s changes significantly for neutrons when solvent is changed from H_2O to D_2O. X-rays are scattered more strongly from heavy elements (e.g. Cl^-, Br^- etc.) as compared to light elements such as C, H etc. That is why, X-rays are scattered by the counterions and neutrons by the core of the micelle.

RESULTS AND DISCUSSION

Comparison of SANS and SAXS Data from CTABr Micellar Solutions

Figure 2 shows the comparison of SANS and SAXS data on 100 mM CTABr micellar solution. Both these data show a correlation peak at $Q \sim 0.05$ Å^{-1}, which is due to a corresponding peak in the interparticle structure factor $S(Q)$ (Aswal et al., 2000; Aswal and Goyal, 2000). The fact that the average distance between the micelles mainly decides position of the correlation peak, it is independent of the radiation used. The peak usually occurs at $Q_m \sim 2\pi/d$, where d is the average distance between the micelles. The second peak in the SAXS data arises from scattering of shell-like structure of the condensed counterions around the micelles. The analysis of SANS data using Eq. (1) determines, as discussed in our earlier paper (Aswal and Goyal, 2000), the shape and size of the core of the micelle. It is found that the CTABr micelles are prolate ellipsoidal with the semimajor axis (a) = 40.2 Å and semiminor axis ($b = c$) = 24.0 Å, respectively. The fractional charge on the micelle was found to be 0.23 suggesting 77% of Br^- counterions reside near the micelle surface. The thickness of the layer of these counterions, referred to as condensed counterions, was obtained from SAXS data. This was in fact the only unknown parameter in analysis of SAXS data, as all other parameters were already known from SANS data. The fitted value of the thickness over which the Br^- counterions are condensed is 4.2 Å. The fact that ionic radius of bare Br^- ion is 1.95Å, the above suggests that most of condensed Br^- ions reside in a mono-layer of counterions around the micelle.

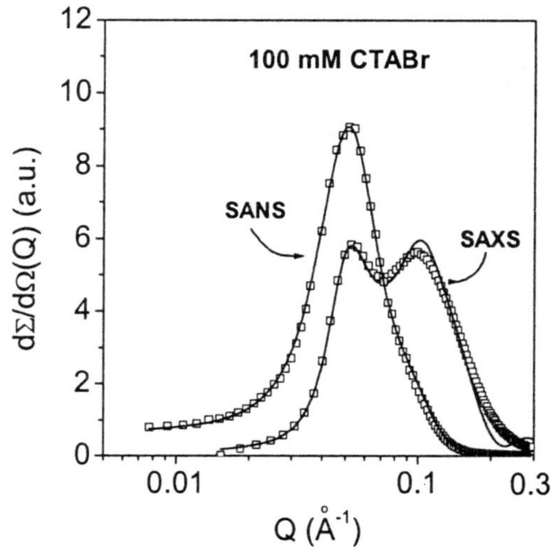

Fig. 2: Comparison of SANS and SAXS data for 100 mM CTABr.

Effect of Counterion Size on the Counterion Condensation

It has been known that the phase diagrams of the ionic surfactants depends on the size of the counterions. For example, while the micelles of ionic surfactant cetyltrimethylammonium bromide (CTABr) shows sphere to rod-like transition with the increase in the surfactant concentration at about 270 mM, the micelles of cetyltrimethylammonium chloride (CTACl) remain spherical even up to very high surfactant concentrations. It may be noted that sizes of $Br^-$and Cl^- counterions are 1.95 Å and 1.80 Å, respectively. To explain the above differences, we have carried out comparative SANS and SAXS studies for CTABr and CTACl micellar solutions. The results are shown in Fig. 3.

The detailed analysis of SANS data shows that micelle size is smaller and the micellar charge is larger for CTACl micelles as compared to those for CTABr micelles (Aswal and Goyal, 2000 and 2003). SAXS data show that the thickness of layer of Cl^- counterions around CTACl micelle is about 4.6Å, which is about 30% larger than the mono-layer thickness. This result along with the one reported in earlier section shows that Cl^- counterions in

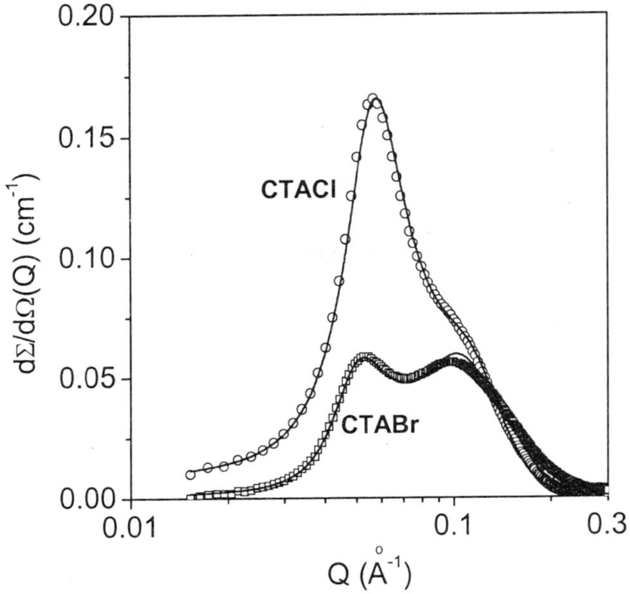

Fig. 3: Comparison of SANS and SAXS data for 100 mM CTABr and CTACl micellar solutions.

CTACl solutions are more dispersed in water as compared to Br^- counterions in CTABr solutions. To our knowledge, this is the first experimental observation of the above fact.

Selective Counterion Condensation in Ionic Micelles

The effect of addition of salts KBr and KCl to the ionic micellar solutions of cationic surfactant (e.g. CTABr or CTACl) is quite different (Aswal et al., 1995). We believe this to be connected with the differences in the condensation of Br^- and Cl^- ions around the charged micelles. To get a better insight, SANS and SAXS experiments were carried out on CTABr/KCl and CTACl/KBr solutions. Molar concentrations of the surfactant and the salt in these solutions were equal. The results are shown in Fig. 4. It may be mentioned that results of SANS studies have been reported earlier (Aswal and Goyal, 2003). It is interesting to note that SANS and SAXS data for CTABr/KCl and CTACl/KBr solutions are similar. These studies suggest that Br^- counterions preferentially condense around CTA^+ micelles both in CTABr/KCl and CTACl/KBr solutions. In short, above studies show that layer of Cl^- ions around CTACl micelles is replaced by Br^- ions, when one adds KBr to the solution. Such a replacement of counterions does not take place when KCl is added to CTABr solutions.

In summary, small-angle scattering is an ideal method for the study of the structural aspects of the micellar solutions. The use of SANS and SAXS provide complementary information on the micelles. The combined

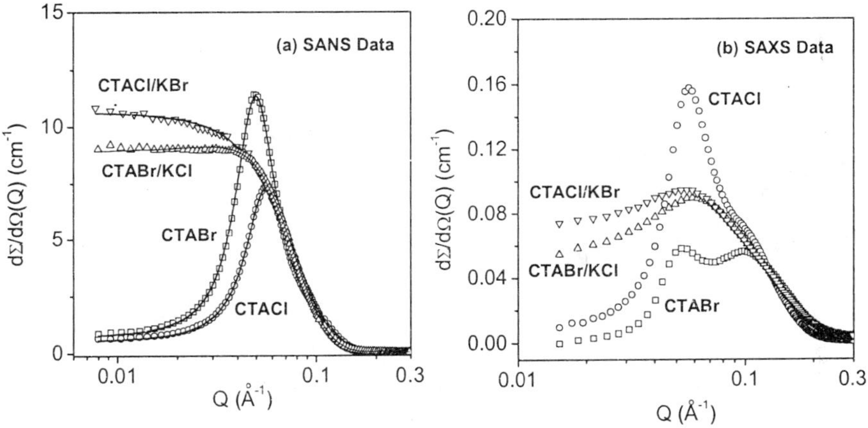

Fig. 4: (a) SANS and (b) SAXS data from equimolar surfactant to salt micellar solutions of CTABr/KCl and CTACl/KBr. For comparison the data from pure CTABr and CTACl micellar solutions are also shown.

experiments of these two techniques have been used to study the counterion condensation in ionic micelles. While neutron scattering in micellar solutions is from the core of the micelle, X-rays are largely scattered by counterions, especially when the counterion has a large atomic number (e.g. Br^-).

REFERENCES

Aswal, V.K. and Goyal, P.S. (2000). *Current Science*, **79:** 947.

Aswal, V.K. and Goyal, P.S. (2000). *Phys. Rev.*, **E61:** 2947.

Aswal, V.K. and Goyal, P.S. (2003). *Chem. Phys. Lett.*, **368:** 59.

Aswal, V.K. and Goyal, P.S. (2003). *Phys. Rev.*, **E67:** 051401.

Aswal, V.K., Goyal, P.S., De, S., Bhattacharya, S., Amenitsch, H. and Bernstorff, S. (2000). *Chem. Phys. Lett.*, **329:** 336.

Aswal, V.K., Goyal, P.S., Menon, S.V.G. and Dasannacharya, B.A. (1995). *Physica*, **B213:** 607.

Aswal, V.K., Goyal, P.S., Thiyagarajan, P. (1998). *J. Phys. Chem.*, **102:** 2469.

Belloni, L. (1998). *Coll. Surf.*, **A140:** 227.

Chen, S.H. and Lin, T.L. (1987). *In:* Methods of Experimental Physics. D.L. Price, K. Skold (Eds.) Academic Press, New York, **23B:** 489.

Chevalier, Y. and Zemb, T. (1990). *Rep. Prog. Phys.*, **53:** 279.

Feigin, L.A. and Sverguu, D.I. (1987). "Structural Analysis by Small Angle X-ray and Neutron Scattering" Plenum Press, New York.

Haldar, J., Aswal, V. K., Goyal, P. S. and Bhattacharya, S. (2001). *J. Phys. Chem.*, **B105:** 12803.

Israelachvili, J.N. (1992). *Intermolecular and Surface Forces*. Academic Press.

Manning, G.S: (1969). *J. Chem. Phys.*, **51:** 924.

Oosawa, F. (1971). *Polyelectrolytes*. Dekker, New York.

Windsor, C.G. (1988). *J. Appl. Cryst.*, **21:** 582.

Wu, C.F., Chen, S.H., Shih, L.B. and Lin, J.S. (1988). *Phys. Rev. Lett.*, **61:** 645.

Bulk Amorphous and Hybrid Nanocomposite Alloys of Novel High-Energy-Density Magnetic Properties and Applications

S. Ram

Materials Science Centre
Indian Institute of Technology
Kharagpur-721302, India

INTRODUCTION

Once upon a time it was thought that a reasonably high value of energy-density $(JH)_{max}$, of the order of a few kJ/m^3 to a few hundreds kJ/m^3, appears in certain ferromagnetic or ferrimagnetic solids of crystalline structure that owe a concomitantly large value of magnetocrystalline anisotropy H_a. Hexagonal ferrites and $R_2M_{14}B$, RM_5, or R_2M_{17} type intermetallics (with R as a rare-earth element and M a transition metal) are typical examples of such materials (Strnat et al., 1967; Sagawa et al., 1984; Croat et al., 1984; Ram, 1989, 1994, 1997, 2001; Schnitzke et al., 1990; Ram and Joubert, 1992; Ram et al., 1995; Ram et al., 1997; Kumar et al., 2002; Kumar et al., 2002; Kumar et al., 2003). In principle, a maximum value of $(JH)_{max}$ lies in an assembly of ideal single domain particles of critical size D_c (Buschow, 1986; Gong et al., 1991). In conventional hard magnetic materials of ceramics or intermetallics, the value of D_c often varies between 20 to 100 nm as controlled by the H_a value. Soft magnetic materials of pure Fe, Co, or Ni metal (crystalline) have a similar value of $D_c \sim 20$ nm (Gong et al., 1991). Irrespective to H_a value, both the soft and hard magnetic materials develop a superparamagnetic character at D below D_c while a multidomain character at D above D_c. Thus, in principle, coercivity H_c, which determines the $(JH)_{max}$ value, has a zero value in both the superparamagnetic as well as in the multidomain particles. Shape anisotropy H_b in particles is another important parameter, which controls an optimal value of H_c or $(JH)_{max}$ in D_c

particles (Ram, 1989; Buschow, 1986). A value of H_c as large as 50 kA/cm has been observed in high-energy-density rare-earth magnets with 100 to 500 kA/m value of H_a (Buschow, 1986).

Unusually, a similarly large value of $(JH)_{max}$, i.e. as much as 10 to 20 kJ/m^3 at room temperature, is being observed in a new series of bulk amorphous alloys of R-M-A (A = Al, Ga, or B) or derivatives (Inoue and Gook Zhang, 1995; Inoue et al., 1996; Inoue et al., 1998; Inoue, 2000; Zhang and Inoue, 2000; Zhang et al., 2001). An amorphous material which retains its amorphous structure in an extended dimension, i.e. of thickness as small as a few millimetres or further larger, is referred as a bulk amorphous material. Such specific materials have a large glass forming ability (GFA). This opens a new scope of developing (i) newer hard magnetic materials of bulk amorphous solids and their applications, and (ii) basic understanding and (iii) formalism of magnetism in disordered solids. Being characteristically isotropic of nature, the amorphous magnets offer new applications of electronic and optical devices with isotropic magnetic properties (Kumar et al., 2002; Inove, 2000). This is reviewed and presented here with successive development in technically useful bulk amorphous alloys and hybrid nanocomposites with selected examples. The different alloys used in this work include the R-M-A series and derivatives of nanocomposites. Pertinent metallurgical methods of the alloying of rapid quenching, elemental milling, arc melting, copper mould casting, and water quenching are explored in order to synthesize the samples of amorphous structure in variety of shapes of thin ribbons, rods, discs, ingots, or powders.

The novel experimental results of the bulk amorphous alloy phase formations in selected compositions are analyzed and compared in terms of X-ray diffraction (XRD) and thermal analysis of samples prepared with the various methods under different conditions. Microstructure and magnetic properties are studied in view to understand and model the formalism of the unusual magnetic properties in bulk amorphous structure. The hard magnetic properties often no longer persist in thin alloy ribbons or in powdering or recrystallizing the bulk alloy structure in a refined microstructure in a hybrid nanocomposite structure. The results of the magnetic properties are compared with the reported values in other hard magnetic materials.

EXPERIMENTAL PROCEDURE

Synthesis of Thin Ribbons and Bulk Alloy

The amorphous alloys used in this work are prepared by five independent methods of (i) rapid quenching of alloy melt, (ii) mechanical milling from the elemental powders, (iii) arc melting, (iv) copper mould casting, and (v) water quenching. Highly pure R, M, Al and other metals of 99.9 % purity have been used. In method (i), alloy ingots prepared by the arc melting process (iii) in a high purity argon atmosphere have been used. In the arc melting process, the metal components were mixed together in requisite amounts and then melted and alloyed in a homogeneous distribution of the components in form of ingots of the alloy. The ingots were cut and crushed in 1 to 3 mm small pieces and then were remelted three to four times in a similar manner to ensure a chemical homogeneity. The ingots were used to prepare the bulk amorphous alloys by copper mould casting. Pieces of ingots were filled in a quartz tube and melted in it by induction melting in an argon atmosphere. As used earlier (Ram and Johari, 1990; Ram, 1990; Ram and Fecht, 2000), homogenized melt is injected through a water-cooled copper mold in 3 mm diameter and the cylindrical rods, ~ 3 mm diameter and L = 30 mm length, are obtained in the final alloy. Upon contact with the copper mould, the melt rapidly cools by a fast transfer of heat through it and results in an amorphous alloy.

In order to compare and analyze the results of bulk amorphous alloy formation in the copper mould cast rods of the alloy, we prepared the alloy as extremely thin ribbons (δ = 50 to 100 μm thickness). As we described earlier (Ram and Johari, 1990; Ram, 1990; Ram and Fecht, 2000), milling of a requisite mixture of the metals (powders) in a closed reactor (bowl) with hard steel balls in method (ii) gives a similar but shear induced amorphous alloy. Amorphous ribbons obtain a similar finely divided loose alloy powder by crushing them in 1 to 2 mm pieces and then ball milling in a similar fashion. Both the types of the alloy powders were used to fabricate a bulk amorphous alloy in practical applications by a hot compaction.

It is feasible that a high viscous flow and atomic diffusivity of a supercooled liquid, as in the case in the thin amorphous ribbons, drives a bulk amorphous alloying during hot compacting a finely divided loose alloy powder (amorphous). This concept is exploited here to fabricate a large-scale bulk amorphous alloy independent of GFA, which often limits glass forming in casting method of a glass melt in a thickness above the critical value. According to it, an alloy powder was hot pressed, under an applied pressure of 350 kPa in vacuum (10^{-4} mbar pressure), at 400 to 600 °C temperatures (i.e., close to the glass transition temperature T_g) in form of a pellet in 5 or 10 mm diameter and 3 to 5 mm thickness. In order to achieve a maximum densification, the sample was held under the pressure at the elevated temperature for 120 s and then cooled to room temperature at 10 °C/s (powder compact I).

In another experiment, the sample compressed at the given temperature for 120 s was heated to a bit higher temperature as 630 to 650 °C under the pressure at a heating rate of 0.33 °C/s and held for 600 s at this temperature before cooling to room temperature at the same rate of 1 to 10 °C/s (powder compact II). A highly compact dense pellet appears of alloy in either case. In some cases, sample-II has a partial recrystallization, which often supports the hard magnetic properties in a hybrid nanocomposite structure. Other experimental details of the syntheses of the different multicomponent alloys by using the other metallurgical or chemical methods are the same as reported elsewhere (Inoue et al., 1997; Inoue et al., 1998; Inoue, 2000; Zhang and Inoue, 2000; Zhang et al., 2001; Ram and Johari, 1990; Ram, 1990; Ram and Fecht, 2000; Frankwicz et al., 1996).

Measurements and Analysis

The alloys formed in different form are analyzed by XRD and thermal analysis. XRD was obtained on a Philips PW 1050 diffractometer using CoKα radiation. DSC thermogram was obtained by heating the sample at 0.33 °C/s in Perkin-Elmer DSC-7 calorimeter under a flowing gas of pure argon. The thermal stability of the amorphous structure is determined by the analysis of the structural parameters of $\Delta T = T_m - T_x$, which is found to be as small as 90 K, and T_m/T_x ratio, which is desirably extremely high as 0.9. Here, T_m and T_x are the onset of the melting and recrystallization temperatures of the alloy. A JEOL model-840 scanning electron microscope (SEM) was used, in conjunction with an energy dispersive X-ray spectrum analyzer, to study the microstructure with in-situ analysis of chemical composition in selected regions of sample. The magnetic properties, i.e., saturation magnetization J_s, remanence J_r, Curie temperature T_C, and H_c are studied with the help of a vibrating sample magnetometer as reported earlier (Ram and Joubert, 1992; Ram, 1994, 1997, 2001; Ram et al., 1995; Ram et al., 1997; Kumar et al., 2002; Kumar et al., 2003).

RESULTS AND DISCUSSION

Development in Bulk Amorphous Alloys

The new development in bulk glass forming metallic alloys within the last few years has led to interesting advances in the science and technology of liquid metals, alloys and derived nanocomposite alloys. The first liquid metal alloy, as vitrified by rapidly cooling ($10^5 - 10^6$ K/s) from molten state to the glass transition T_g, was reported as early as in 1960 in the Au-Si system by Klement et al. (1960). The work of Turnbull and his group in the early 1960s was another critical contribution in this series. They developed rapidly quenched Au-Si, Pd-Si and Pd-Cu-Si glasses (Chen and Turnbull, 1968, 1969, 1974). The field of metallic glasses gained momentum in 1970s when Allied chemicals developed continuous casting processes for commercial manufacture of metallic glass ribbons and thin sheets (Kavesh, 1978). In this period, simple suction casting methods has been explored to form bulk amorphous Pd-Cu-Si rods in millimetre diameter at moderate 10^3 K/s cooling rate (Table 1) (Chen, 1974). This is the first example of the bulk glasses. The Pd-Ni-P bulk metallic glasses have been developed with fluxing methods (Drehman et al., 1982; Kui et al., 1984). Ingots of size up to 10 mm formed in a borosilicate glass flux at 10 K/s cooling rate. Lee et al. (1982)

Table 1: Development in nonferrous and ferrous bulk amorphous alloys

	Years
A. Nonferrous alloys (weak or nonmagnetic)	
Pd-Cu-Si	1974
Pd-Ni-P	1982
Au-Pb-Sb	1982
Mg-R-M (M : Ni, Cu or Zn)	1988
R-Al-TM (TM: VI-VIII group metal)	1989
R-Ga-TM	1989
Zr-Al-TM	1990
Zr-Ti-Al-TM	1990
Ti-Zr-TM	1993
Zr-Ti-TM-Be	1993
Zr-(Nb, Pd)-Al-TM	1995
Pd-Cu-Ni-P	1996
Pd-Ni-Fe-P	1996
Pd-Cu-B-Si	1997
Ti-Ni-Cu-Sn	1998
B. Ferrous alloys (ferromagnetic)	
Fe-(Cu, Mn, Si etc.)-B	< 1985
Fe-Nd	1989
Fe-(Al, Ga)- (P, C, B, Si, Ge)	1995
Fe-(Nb, Mo)-(Al, Ga)-(P, B, Si)	1995
R-Fe-Al and or R-Fe-Ga	1995*
Co-(Al, Ga)-(P, B, Si)	1996
Fe-(Zr, Hf, Nb)-B	1996
Co-Fe-(Zr, Hf, Nb)-B	1996
Ni-(Zr, Hf, Nb)-(Cr, Mo)-B	1996
Fe-Co-R-B	1998*
Fe-(Nb, Cr, Mo)-(P,C,B)	1999
Ni-(Nb, Cr, Mo)-(P, B, Si)-B	1999

*Hard ferromagnets depending on the composition and microstructure.

fabricated spherical Au-Pb-Sb droplets (amorphous) of millimetre diameter by drop tube experiments.

During the late 1980s, Inoue and colleagues developed fabrication of multicomponent La-Al-Cu, Mg-Y-Cu, La-Al-Ni-Cu, Zr-Al-Cu-Ni and other similar bulk amorphous alloys using metal mould casting (Inoue et al., 1990; Inoue et al., 1991; Inoue et al., 1992). These alloys are available in form of rods, bars and strips with dimension in the 1 to 10 mm range. Later, Peker and Johnson (1993) developed a number of Zr-Ti base alloys in a deliberate search for bulk glass forming materials. This led to the discovery of bulk glasses in the Zr-Ti-Ni-Cu-Be, Zr-Ti-Ni-Cu and related systems (Peker and Johnson, 1993; Johnson, 1999). A cooling rate of 1 K/s has been found sufficient enough to produce a fully amorphous structure in rods of 50 mm diameter by pouring the melt into a copper mould.

Most of the alloys mentioned in Table 1 are confined to nonferrous systems. Several ferrous alloys have been developed in early 1980s in the Fe-B, Fe-B-Si and related systems but those are limited to thin ribbons only (Inoue et al., 1998; Johnson, 1999). Moreover, they have soft magnetic properties. As mentioned above, hard magnetic amorphous alloys could be introduced as early as in 1995 in R-M-A systems, with M = Fe and A = Al or Ga (Gong et al., 1991; Inoue and Gook Zhang, 1995; Inoue et al.; Inoue et al., 1997; Inoue et al., 1998). This series has an exceptionally large GFA. Amorphous R-Fe-Al (Ga) rods of exceptionally hard magnetic properties are derived in as large diameter as 12 mm by suction casting at 10 to 0.1 K/s cooling rate. Unusually, the practically useful hard magnetic properties no longer persist after devitrifying the amorphous structure by heating the sample at elevated temperatures.

X-ray Diffractogram and Microstructure

R-rich R-M-A Type Alloys

A bulk amorphous structure easily forms in R-M-A series of alloys of hard magnetic properties over a large range of compositions, with M = Fe (or a combination with Co) and A = Al, Ga, or B, in cooling the melt in an inert atmosphere at as slow cooling rate as 10 to 0.1 K/s. No significant devitrification appears during cooling the specimen in shapes of thin rods or ingots in 3 to 5 mm diameter. An X-ray amorphous structure of alloy is easily achieved in the various kinds of the solidification techniques, e.g., the water quenching, copper mould casting, high pressure die casting, arc melting, unidirectional arc melting, or suction casting (Inoue and Gook Zhang, 1995; Inoue et al., 1997; Zhang et al., 2001). The mechanical attrition of an elemental mixture in a controlled atmosphere gives an altogether different way of vitrifying a bulk amorphous alloy without involving the melting. A small addition of a volatile additive of a hydrocarbon liquid during the attrition helps significantly the vitrification in a refined amorphous structure.

For example, Fig. 1 compares XRD patterns in $Nd_{57}Fe_{20}Co_5Al_{10}B_8$ alloys prepared by different methods of (a) the elemental milling, (b) the milling of the prealloy ingots, (c) the arc melting in alloy ingots, (d) the melt spinning in thin ribbons, and (e) the copper mould casting in thin rods or cylinders. As summarized in Table 2, three broad diffraction halos q_1, q_2 and q_3, characteristic of an amorphous structure (Ram and Johari, 1990; Ram, 1990), appear in the samples in (a) to (d). Only one major diffraction halo q_1 = 14.9 nm^{-1} remains in sample (e) of the cast rods or cylinders (3 mm diameter) in part of the surface devitrification into nanocrystallites due to α-Nd and $NdFe_2$. A minor variation in positions in three halos indicates minor changes in local amorphous structures in the alloys derived by the different methods. This accounts for differences in magnetic and other properties in these samples. The local structure seems to be highly sensitive to the method of the alloying and other experimental conditions.

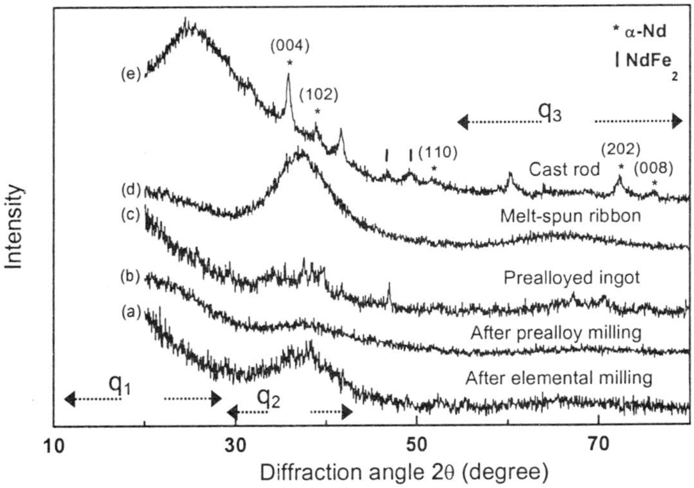

Fig. 1: XRD patterns in $Nd_{57}Fe_{20}Co_5Al_{10}B_8$ alloys prepared by (a) elemental milling, (b) prealloy milling, (c) arc melting, (d) melt spinning, and (e) copper mould casting.

In general, the samples obtained through the mechanical attrition or the melt spinning have smaller q_i values [1] (larger interatomic distance $d = \lambda/2\sin\theta \cong 2\pi\, q^{-1}$ as per the Bragg's law of $2d \sin\theta = \lambda$) if compared with those in the samples derived by the other methods. This ascribes a relatively large amount of excess energy $\Delta\varepsilon$ (or enthalpy ΔH) being stored over the equilibrium value ε_0 in forms of the structural imperfections and defects in mixing and alloying of the components in the two methods. According to the first law of thermodynamics,

$$\Delta H = \Delta\varepsilon + P\Delta V, \tag{1}$$

the excess $\Delta\varepsilon$ or ΔH results in an excess volume ΔV of the final alloy. This excess ΔV releases in a broad exothermic structural relaxation signal if heating the sample in a thermal analyzer as studied in detail with thermodynamic models and simulations by Ram and Johari (1990) in series of shear amorphized and supercooled alloys.

SEM micrograph of mechanically alloyed material in general indicates a layered structure of alloy formed by a peculiar reaction within the components and a cold welding of resulting structure in layer-by-layer structure of final alloy. As shown in Fig. 2, a similar complex microstructure of cold welding results in milling even a pure metal component such as iron powder (Ram and Fecht, 2000).

Resulting layer of alloy in continuous collisions gets as thinned as possible of mono-atomic level. As per usual microstructure in attrition of pure metal components, it assumes a definite thickness (t), which appears to be smaller in comparison to the critical diameter (D) in a stable crystallite in the material, i.e. $t < D$. An instantaneous cold-welding of a freshly created nascent layer

Table 2:. Diffraction halos in $Nd_{57}Fe_{20}Co_5Al_{10}B_8$ alloys prepared by different methods

Sample	*Wave vectors (nm^{-1})*		
	q_1	q_2	q_3
Milled elemental powder	12.8	22.9	38.2
Milled prealloy	13.4	23.2	38.9
As-prepared ingot*	12.8	22.2	39.3
Malt-spun ribbon	12.8	22.2	37.7
Copper mould cast rods**	12.9	–	–

* The ingots prepared by the arc meting have a significant fraction of devitrified surfaces.

** The cast rods have some α-Nd and $NdFe_2$ nanocrystallites precipitated during the casting.

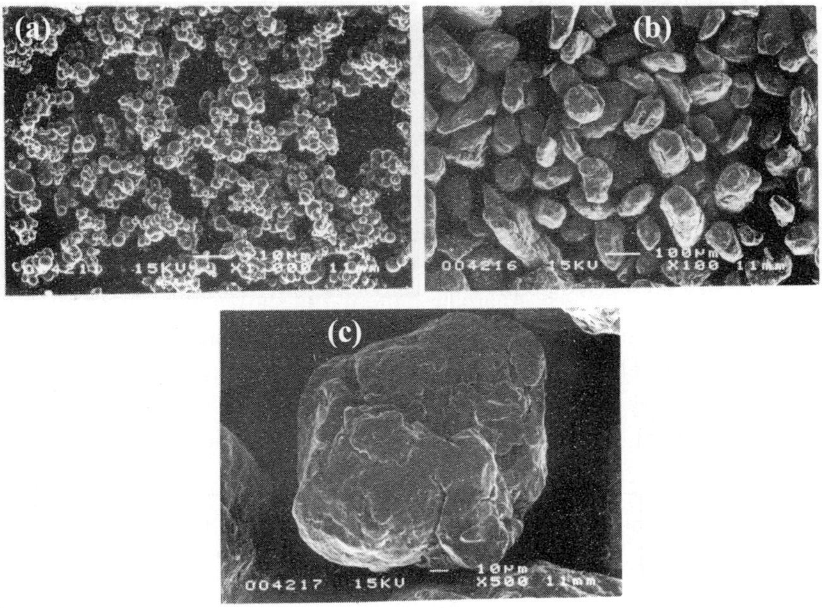

Fig. 2: SEM in Fe particles after milling (a) 0 h, (b) 72 h, and (c) 136 h in a closed container under a high purity argon gas. Sample (a) has been obtained by a thermochemical reduction of a Fe salt by calcium metal granules at 900 °C under a high purity argon gas.

structure over a stable layer supports its originally thin-layered structure. As a consequence, such thinned layers do not recrystallize so easily in course of extended milling and the sample achieves an amorphous structure of individual layers, which arrange ultimately one over others in minimizing the total energy. We analyzed TEM micrographs for selected samples with the corresponding electron diffractograms and in-situ studies of the compositional maps. A single amorphous structure of average alloy composition lies throughout the specimen, with a characteristic broad diffraction ring $q_2 \sim 22.7\ nm^{-1}$, in favourable agreement with the XRD pattern.

A peculiar microstructure of separated domains or clusters lies in acicular or spherical shapes in the alloys obtained by the arc melting or the copper mould casting. They have distinct structures with well-defined boundaries in liquid-liquid phase separated regions. Some small crystallites are also visible in the liquid-liquid phase separated regions. It appears that the alloy components are highly soluble one into others and thus easily form a stable homogeneous melt of a single phase. The melt thus easily retains a liquid structure even upon a slow cooling as ~1 °C/s.

M-RICH M-R-A TYPE ALLOYS

A bulk amorphous structure with hard magnetic properties lies in M-rich R-M-A alloys (M = Fe with Co and A = Al, Ga, or B) in cooling the melt at moderate rate as in the R-rich M-R-A alloys. Figure 3 shows XRD patterns of (a) melt-spun ribbons, (b) cast rods or cylinders, and (c) powder compacts in an alloy of $Fe_{67.0}Co_{9.5}Nd_{3.0}Dy_{0.5}B_{20}$. The XRD pattern of ribbons (a), $\delta = 50\ \mu m$, consists of two halos at wavevectors $q_1 = 14.2$ and $q_3 = 33.0\ nm^{-1}$ characteristic of an amorphous structure. The two diffraction halos describe two prominent pair distribution functions of atoms in an amorphous structure of alloy. The Halo q_3 is prominent with ~ 90 % of the integrated intensity in the total pattern. A similar pattern (not shown here) of a prominent q_3 halo lies in a powder of alloy derived by mechanical attrition of the ribbons (a) or an elemental mixture. This is common in Fe-rich amorphous alloys produced by rapid solidification or powder metallurgy (Kumar et al., 2002; Kumar et al., 2003). A modified diffractogram lies of two halos q_1 =

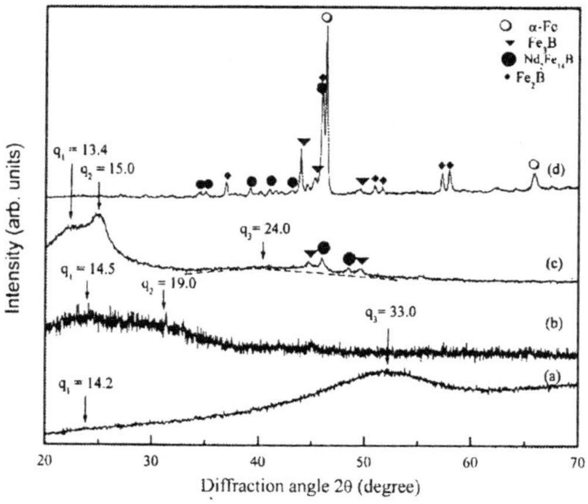

Fig. 3: XRD patterns in $Fe_{67.0}Co_{9.5}Nd_{3.0}Dy_{0.5}B_{20}$ alloy of (a) thin ribbons, (b) as cast rods, (c) consolidated powder, or (d) recrystallized rods (b) in 600 s of annealing at 630 °C in N_2 gas.

14.5 and q_2 = 19.0 nm^{-1} in the cast rods (3.0 mm diameter) in pattern (b) while of three halos q_1 = 13.4, q_2=15.0 and q_3 = 24.0 nm^{-1} in the alloy powder compacts of pellets (5 to 10 mm diameter and 3 to 5 mm thickness) of XRD pattern (c). Halo q_1 or q_2, instead of halo q_3 in ribbons, becomes to be prominent in these two samples.

The powder compacts I and II either derived from the ribbons or a milled elemental mixture have the same XRD pattern. Halo q_2 = 15.0 remains to be the most intense in either sample. Halo q_3 follows in weak intensity at average ~24.0 nm^{-1} position in superposition with extremely weak peaks in intensity in incipient nucleation and growth of $R_2M_{14}B$ and Fe_3B during the hot compaction process at 530 °C temperature. Its modified position in comparison to halo q_3 = 33.0 nm^{-1} in the ribbons infers a different amorphous structure on part of the precipitation of the alloy as $R_2M_{14}B$ and Fe_3B nanocrystallites (D ~30 nm size according to the widths $\Delta 2\theta_{1/2}$ in the peaks as per the Debye Scherrer relation (Klug and Alexander, 1974). According to value of halo q_3, this involves an average 10 % larger value of interatomic distance $r_3 = 2\pi/q_3$ of virgin ribbons.

The as-cast rods of the alloy are recrystallized readily on heating at the same temperature of 530 °C for 120 s (and then cooling to room temperature at 10 °C/s) as used in the powder compact I. XRD pattern (d) in recrystallized alloy rods is given in Fig. 3. As indexed in Table 3, the recrystallized precipitates are reflected in sharp peaks in nanocrystallites of α-Fe, Fe_3B, Fe_2B and $R_2M_{14}B$ as analyzed in comparison with their standard diffractograms (Ram et al., 1995; Ram, 1997; Ram et al., 1997; McClume, 1979). According to relative intensities in diffractogram peaks, the prominent phase lies of α-Fe with $R_2Fe_{14}B$ as the second prominent phase. The sharp peaks are superimposed over a continuous scattering background in part of the amorphous alloy, which still retains in a significant volume fraction. The amorphous part of alloy keeps α-Fe and $R_2Fe_{14}B$ in a dispersed structure and supports their stability in such small particles, and in turn controls the magnetic and other useful properties in a hybrid nanocomposite structure (Ram et al., 1995; Ram, 2000, 2001).

Assuming all of the R content in the virgin alloy to be precipitating as $R_2M_{14}B$, it results in an average B-rich composition of $M_{74}B_{26}$, which is deduced as follows

$$\underset{\text{(amorphous)}}{M_{76.5}R_{3.5}B_{20}} \longrightarrow \underset{\text{(crystallites)}}{\frac{3.5}{2}(R_2M_{14}B)} + \underset{\substack{\downarrow \\ M_{74}B_{26}\ \text{(amorphous)}}}{M_{52.0}B_{18.25}} \tag{2}$$

In fact, the amount of $R_2M_{14}B$ precipitated in this example of annealed rods is not such large. According to relative intensity in the most intense peak of the diffractogram (Fig. 3d) at 0.2394 nm in (321) reflection, it hardly imparts 6 % of the total sample. Now, if we assume that the sample

Table 3: Interplanar spacings (d_{hkl}) and relative intensities in XRD peaks in devitrified phases in $Fe_{67}Co_{9.5}Nd_{3.0}Dy_{0.5}B_{20}$ cast rods after annealing in N_2 gas

d_{hkl} (nm)	I_p	h	k	l	Phase
0.2007	100	1	1	0	α-Fe
0.2025	64	2	1	2	Fe_2B
		0	0	6	$Nd_2Fe_{14}B$
0.2119	29	3	2	1	Fe_3B
0.1613	16	3	1	0	Fe_2B
0.1631	15	2	0	4	Fe_2B
0.1432	14	2	0	0	α-Fe
0.2058	12	1	1	2	Fe_3B
0.2549	9	2	0	0	Fe_2B
0.1829	8	1	1	4	Fe_2B
0.1803	7	2	2	0	Fe_2B
0.2394	6	3	2	1	$Nd_2Fe_{14}B$
0.2164	5	2	2	4	$Nd_2Fe_{14}B$
0.2281	5	1	1	5	$Nd_2Fe_{14}B$
0.1878	5	4	1	1	Fe_3B
0.2707	4	3	1	1	$Nd_2Fe_{14}B$
0.2759	4	2	2	2	$Nd_2Fe_{14}B$

has a complete $R_2M_{14}B$ precipitation in the total R content in the virgin alloy, it has a composition of $M_{76}R_3B_{21}$, which is not so different from the starting composition. The sample has been annealed in N_2 gas at 630 °C for 600 s.

It is interesting to learn that despite the close similarity in the chemical compositions in the amorphous alloy phases before and after part of the precipitation there is a marked difference in r_3 values. The result infers that, at early stage of $R_2M_{14}B$ nucleation and growth, a high-energy metastable microstructure designs in a high V (or Gibbs free energy G) state in the amorphous counterpart of the alloy. In fact, small crystallites, which are dispersed through and embedded in the resulting amorphous alloy, raise the total energy of the system by adding their surface energies. A concomitant V expansion occurs in the amorphous alloy counterpart, owing to its relatively mobile structure at the elevated temperature, by optimizing the total V in the two components. This is according to the first and second laws of thermodynamics of conservation of the total $H = \varepsilon + PV$ with as small value of $G = H - TS$ as possible by a redistribution of the microstructure with a maximum value of entropy in the equilibrium state.

The marked differences in halos of XRD patterns of the ribbons, rods, powder, or the powder compacts of $Fe_{67.0}Co_{9.5}Nd_{3.0}Dy_{0.5}B_{20}$ alloy suggests two different types of the liquid phases of the alloy characterized by two different sets of the pair distributions of atoms primarily in halos (i) q_1 and q_2, or (ii) q_1 and q_3. Subject to crushing and milling of the amorphous ribbons, primarily the pair distribution function involving halo q_3 undergoes a redistribution of atoms by part of precipitation of $R_2M_{14}B$ of small crystallites. The latter amorphous phase, characterized primarily by the halos q_1 and q_2, maintains its amorphicity with a high GFA than the virginal alloy phase. It seems that the amorphous phase of halos q_1 and q_3 has a low GFA and is a low temperature phase and thus separates or recrystallizes easily at early stage of temperatures during the powder compaction process. In $Fe_{67.0}Co_{9.5}Nd_{3.0}Dy_{0.5}B_{20}$ composition, which is simply also expressed as $M_{76.5}R_{3.5}B_{20}$, the two liquid phases, in general, thus described as $M_{100-y}B_y$ ($y \geq 20$) and R-M-B. In the melt-spun ribbons, the two liquid phases remain to be dissolved in a common amorphous $R_zM_{80-z}B_{20}$ (z = 3.5) solution. According to the XRD patterns (Fig. 3), the amorphous $M_{100-y}B_y$ phase is prominent in the cast rods by the copper mould casting as well as in the elemental milled powder of the alloy.

In order to analyze $Fe_{67.0}Co_{9.5}Nd_{3.0}Dy_{0.5}B_{20}$ alloying process in this multicomponent system in forms of ribbons, rods, and powder compacts, we studied their microstructures at different magnifications. SEM micrograph (a) or (b) in Fig. 4 in the as-cast rods (after itching in nital) shows distribution of two different phases in whitish and dark contrasts. The whitish regions, which represent part of the recrystallized alloy, lie in the shapes of wires, thin platelets, or spheroids. A better analysis of morphologies in these crystalline phases is not feasible as they are embedding in part of the amorphous phase (in shape of thin platelets). Using a higher magnification does not help to resolve the features. The spheroids are due to α-Fe (which occurs in clusters of small crystallites of average 0.5 μm diameter) while thin wires or thin platelets (0.2 to 0.5 μm diameter) are mostly due to $R_2M_{14}B$ as could be analyzed with chemical analysis. The amorphous platelets span over a larger 2 to 10 μm scale. The results infer that a homogeneous dispersion of ingredients maintains by suppressing the miscibility gap so that only a confined liquid-liquid decomposition occurs over such small scale. Otherwise, if it once initiates, often propagates to a submillimetre or a still larger scale (Kumar et al., 2002).

An entirely different microstructure lies in parallel strips of 0.4 to 2.0 μm diameter in the melt-spun $Fe_{67.0}Co_{9.5}Nd_{3.0}Dy_{0.5}B_{20}$ ribbons (Fig. 4c). It is argued that the strips are formed along the surface of the ribbons as driven by the flow of heat along the surface during cooling the liquid in forms of ribbons in the melt spinning process. They impart a maximum volume fraction of the sample in the prominent M-rich liquid phase with a maximum intensity in the XRD halo q_3 (Fig. 3a). Fine crystallites in SEM micrograph (Fig. 5) in the alloy powder compacts occur in platelets ($R_2M_{14}B$) or near spherical shapes (α-Fe) embedded in part of the amorphous alloy in thin layers. Part of R-rich R-M-B components, which have a high GFA value, retains in a

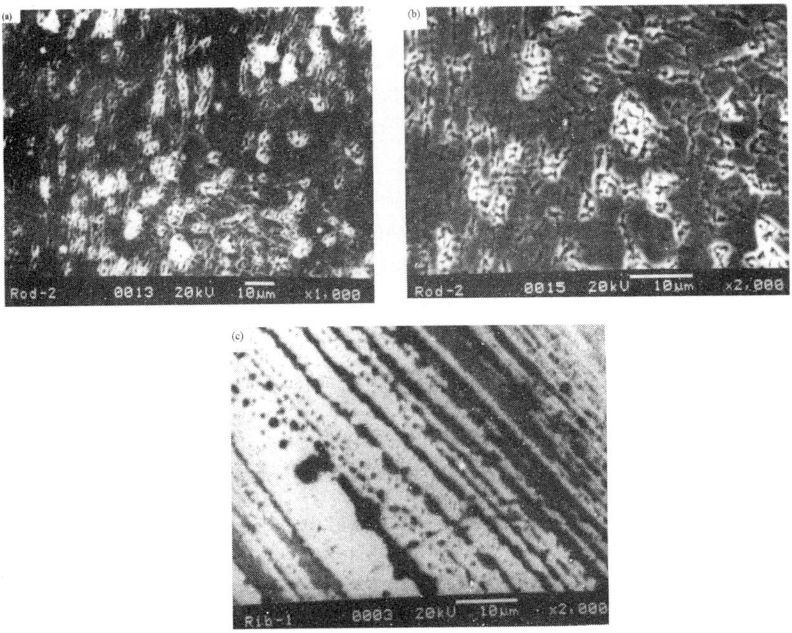

Fig. 4 SEM micrographs in as-cast rods at (a) low and (b) high magnifications and in (c) thin melt-spun ribbons for $Fe_{67.0}Co_{9.5}Nd_{3.0}Dy_{0.5}B_{20}$ alloy.

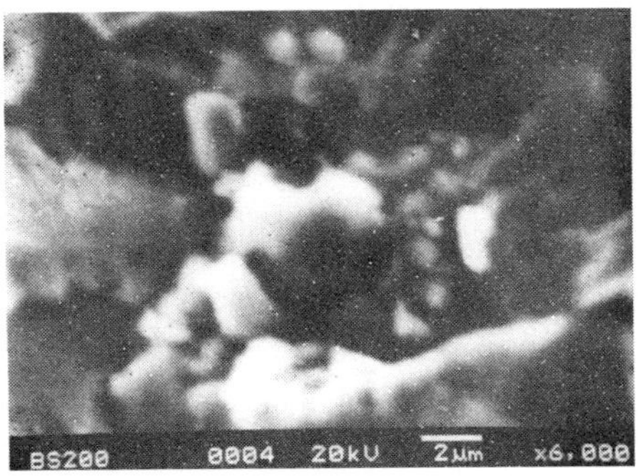

Fig. 5: SEM micrograph in a bulk consolidated $Fe_{67.0}Co_{9.5}Nd_{3.0}Dy_{0.5}B_{20}$ powder.

separate amorphous phase with manifested intensities in the q_1 and q_2 halos. M and M-rich components are recrystallized as fine particles leaving behind only a small amount of amorphous counterpart alloy with a weak diffraction halo q_3.

Phase Diagram and Glass Formability in R-M-A Alloys

An amorphous phase in R-M-A system easily forms over a wide range of compositions by application of the mechanical attrition, copper mould casting, or the arc melting process. This is studied further by thermal analysis of the final alloy. For example, a DSC thermogram for a typical $R_{90\text{-}x}Fe_xAl_{10}$, x = 30, alloy obtained in form of small ingots of 3 to 5 mm diameter by the arc melting has a broad exothermic peak over 500 to 750 K (with a change of enthalpy ΔH_1 =12 J/g) followed by a strong exothermic peak at T_p = 795 K (with enthalpy ΔH_2 =35 J/g) with T_x = 763 K and T_m= 910 K. The present T_x and T_m values are in a fairly good agreement with the values reported by other methods. The first signal is characteristic of structural relaxation with an excess energy of ΔH_1 =12 J/g above the equilibrium state ε_0 in the alloy. A further larger value of ΔH_1 =15 J/g (with T_p = 800 K, T_x = 760 K and T_m= 915 K) lies in sheared alloy by the mechanical attrition process. This excess enthalpy is stored in form of structural imperfection and defects with an enhanced volume of alloy of deformed structure in high-energy thin layers.

The T_x and T_m values for selected R-Fe-Al alloys, which show hard magnetic properties, are given in Table 4. The temperature interval, ΔT_m = T_m – T_x, of the supercooled liquid, defined by the difference between T_m and T_x, is as small as 90 K and the reduced crystallization temperature Φ = T_x/T_m is as high as 0.9. A small ΔT_m value and a large Φ value seem to be responsible for occurring of a large GFA value in this peculiar series of hard ferromagnetic alloys of the multicomponent system in a bulk amorphous structure.

Table 4: Thermal stability data in hard ferromagnetic bulk amorphous R-Fe-Al alloys

Composition	*Thermal stability**			
	$T_x(K)$	$T_m(K)$	$\Delta T_m = T_m\text{-}T_x$	$\Phi = T_x/T_m$
$Nd_{70}Fe_{20}Al_{10}$	750	866	116	0.87
$Nd_{60}Fe_{30}Al_{10}$	763	910	147	0.84
$Pr_{60}Fe_{30}Al_{10}$	765	855	90	0.90
$Er_{60}Fe_{30}Al_{10}$	820	950	130	0.86

*The data are deduced from DSC thermograms measured at 0.33 K/s heating.

A large GFA in this system, leading to the formation of bulk amorphous alloy, originates from the high thermal stability of the liquid (or solid solution) against crystallization at T_x or above. It can be described by the temperature interval, ΔT_x = T_x – T_g, of the supercooled liquid region before the primary recrystallization or the liquid-liquid decomposition in a refined microstructure. Inoue et al. (1998 and 2000) have studied the compositional dependence of ΔT_x for a series of R-Fe-Al amorphous alloys. As can be derived from the Hume-Rothery's rules of alloying (Haasen, 1986), a large GFA in this series satisfies four empirical rules for alloying as follows: (i) alloy systems consisting of multicomponents, (ii) significantly different atomic size ratios above 12% among the main constituent elements, (iii) a large negative heat of mixing among the constituent elements, and (iv) a large macroscopic interaction among the components to arrange in a polymeric structure. The fourth rule, which is proposed here, governs with the existence of a relatively large value of temperature coefficient of viscosity in this series. Instantaneous value of viscosity plays a crucial role in undercooling the liquid through a continuous growth of it at moderate rate as a function of cooling so that it slowly achieves an immobile glassy state (supercooled liquid) as T_g approaches.

The factor (iv) explains the existence of a large GFA in alloying a multicomponent system in combination with other parameters in rules (ii) and (iii). Otherwise, a typical phase diagram, as shown in Fig. 6 for a usual multicomponent alloy series of soft magnetic materials, demonstrates formation of an amorphous glass only in a very narrow region. This is reproduced after the work of Inoue et al. (1998). According to it, the largest value of ΔT_x = 49 K is obtained at an alloy composition of $Fe_{74}Al_4Ga_2P_{12}B_4Si_4$. A Fe-rich $Fe_{80}P_{12}B_4Si_4$ composition has a smaller ΔT_x = 36 K. Simultaneous dissolution of Ga and Al with larger atomic sizes and large negative heat of mixing against the other constituent elements is, therefore, very effective for extension of the ΔT_x region before the recrystallization process sets in as the T_x approaches.

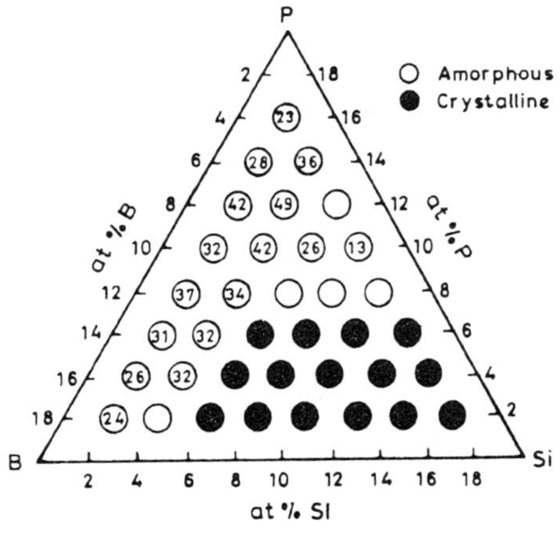

Fig. 6: Phase diagram in an amorphous phase formation (soft ferromagnetic) in $Fe_{74}Al_4Ga_2$ (P, B, $Si)_{20}$ series. The numbers represent the $\Delta T_x = T_x - T_g$ values.

The R-Fe-Al alloys achieved with the extended ΔT_x = 49 K are found to be truly amorphous in nature and involve a large value of enthalpy of relaxation ΔH_{rel} in the T_g region in accord with the above results. A large value of ΔH_{rel} (e.g., 15 J/g observed in a typical $Nd_{60}Fe_{30}Al_{10}$ alloy) masks the relatively weak endothermic step of the T_g. It is, therefore, not visible in these examples. The excess ΔH_{rel} releases with a prominent exothermic structural relaxation signal on heating the amorphous alloy through the T_g region. In general, the ΔH_{rel} signal is found to be more prominent in the shear vitrified alloys. All these alloys exist in nonequilibrium metastable states far above their equilibrium energy states. The excess ΔH_{rel} energy, therefore, releases as an exothermic relaxation signal while heating the sample through the T_g. In an isothermal heating at a temperature $T \leq T_g$, it results in a monotonically decreasing exothermic signal as a function of time. Such thermal signals have been studied and discussed with thermodynamic modeling and simulations in several examples in this series (Ram and Johari, 1990; Ram, 1990).

The hard ferromagnetic bulk amorphous alloys of the R-M-A series, with A = Al, B, or Ga, present an exceptionally extended phase diagram in formation of a fully amorphous phase over a wide compositional range. For example, a typical phase diagram is given in Fig. 7 for the R-Fe-Al series. A thin amorphous ribbon easily forms over 0 to 90 at.% Fe and 0 to 93 at.% Al by rapid quenching as first reported by Inoue et al. (1997). Also a bulk amorphous alloy forms over a pretty wide range of 20 to 40 at.% Fe and 10 to 30 at.% Al by copper mould casting and other methods (Inoue et al., 1996). The as-cast ingots of the R-Fe-Al alloys, 4 to 5 mm diameter, by the arc melting, as used in this work, thus have a smooth surface and a metallic luster at 20 to 40 at.% Fe, indicating formation of an amorphous phase of the alloy. A systematic change in the atomic size of Nd>Fe>Al and large negative heats of mixing for Nd-Al, Al-Fe, and Nd-Fe atomic pairs seem to support a favourably large GFA value in this peculiar series of alloys as per the above four empirical rules of the alloying in a multicomponent system.

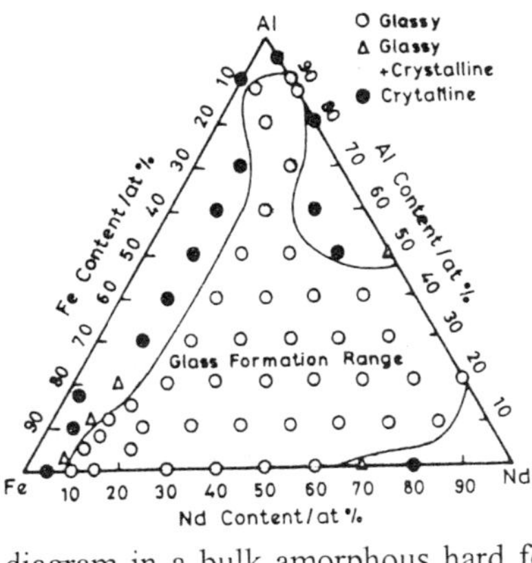

Fig. 7: Phase diagram in a bulk amorphous hard ferromagnetic phase formation in the R-Fe-Al series of multicomponent alloys.

Thermal Analysis in $Fe_{67.0}Co_{9.5}Nd_{3.0}Dy_{0.5}B_{20}$ Alloys

The thin melt-spun $Fe_{67.0}Co_{9.5}Nd_{3.0}Dy_{0.5}B_{20}$ ribbons have a prominent exothermic peak of DSC thermogram (Fig. 8a) at temperature T_{P1} = 597 °C with the onset temperature T_x of 590 °C. A weak exothermic signal accompanies the heat out-put over lower temperatures in the characteristic structural relaxation of the sample by releasing part of the structural energy or enthalpy ΔH in this region. It masks the due endothermic step of the characteristic T_g of amorphous structure of the specimen. A value of $T_g = 555$ °C, as marked by the arrow in the curve with dash, is resolved after releasing the ΔH by heating the sample at 555 °C for 20 min in the calorimeter as described earlier with series of amorphous alloys (Ram and Johari, 1990; Ram, 1990). The heating at this temperature does not influence recrystallization of the alloy, with the change in the enthalpy $\Delta H_{P1} = 2.6$ J/g in the peak T_{P1}. A value of $\Delta T_g = T_x - T_g \cong 35$ °C is obtained with $T_x = 590$ °C and $T_g = 555$ °C. Zhang and Inoue reported a similar $\Delta T_g = 40$ °C value in amorphous melt spun $Fe_{70-x}Co_{9.5}Nd_xDy_{0.5}B_{20}$ ribbons of $x = 3.5$. This is the maximum value of ΔT_g that occurs in this series. Values of $T_x = 572$ °C and $T_g = 525$ °C had been reported (Zhang et al., 2001) without a correction for the exothermic step of the signal in the structural relaxation.

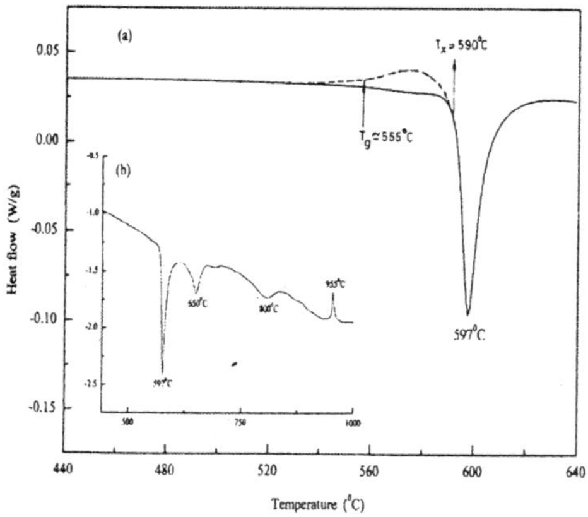

Fig. 8: (a) DSC thermogram in $Fe_{67.0}Co_{9.5}Nd_{3.0}Dy_{0.5}B_{20}$ ribbons recorded at 20 °C/min heating rate. The curve with the dash, after 20 min of annealing the sample at 550 °C in the calorimeter, shows $T_g \sim 555$ °C. Part of the thermogram over extended temperatures is given in the inset (b).

The occurrence of the T_g with such a large value of the supercooled liquid region prior to the primary recrystallization in T_{P1} peak indicates a high GFA of the alloy. Part of thermogram (b) shown in the inset in Fig. 8 indicates presence of two other exothermic peaks of T_{P2} and T_{P3} at subsequently higher temperatures of 650 and 800 °C in part of a total of three-step recrystallization of the alloy. The endothermic peak at 955 °C refers to its average melting point T_m, with a value of the enthalpy of the fusion $\Delta H_f = 3.0$ J/g. It is used to determine the value of the reduced glass transition temperature $T_{rg} = T_g/T_m$, i.e. 0.58, with $T_g = 555$ °C. This is a bit larger value in comparison to the values of 0.56 to 0.57 reported in $Fe_{70-x}Co_{9.5}Nd_xDy_{0.5}B_{20}$ alloys, at $x = 3 - 4$, by Zhang and Inoue (2000), or a value of 0.59 to 0.62 in $Nd_{55}Cu_{15}Ni_{10}Co_5Al_{15}$ alloy in forms of ribbons, rods, or cylinders (Inoue et al., 1997). Note that an amorphous supercooled liquid state of an alloy forms if the melt cools in the liquid state over T_m to T_g sufficiently fast by freezing in an immobile state at T_g. Its thermal stability against recrystallization is determined by the ΔT_g value that is the extent up to which it can be heated through T_g by retaining the amorphous structure. The T_{rg} and ΔT_g values, which determine GFA in general, are sufficiently large in this example, supporting the formation of a thermally very stable amorphous alloy structure.

A modified thermogram of a broad exothermic envelope (Fig. 9) lies in the as-cast rods of the $Fe_{67.0}Co_{9.5}Nd_{3.0}$ -$Dy_{0.5}B_{20}$ alloy on heating over 100 to 1000 °C temperature in a highly pure argon gas atmosphere. The envelope, as marked over the dashed line of AB, involves a value of the change of the enthalpy ΔH = 66 J/g, which is considerably larger over a total of 8.6 J/g value in the three exothermic peaks in the amorphous ribbons (Fig. 8). It supports the conjecture that the dispersion of small crystallites in part of the amorphous alloy designs a high-energy metastable microstructure of enthalpy H_m above the average value H_a in the amorphous state, i.e., $H_m > H_a$. Furthermore, the envelope consists of at least four overlapping exothermic signals of average positions at 265, 395, 540 and 800 °C. The 800 °C signal also lies in the ribbons. Despite a large deviation from the equilibrium thermodynamic state of the ribbons, average T_m value, which lies at 955 °C, with ΔH_f = 3.5 J/g, is not changed much. The derived powder compacts from the ribbons (after crushing, milling and hot compacting) have a very similar broad envelope of thermogram, with T_m = 955 °C (ΔH_f = 3.0 J/g).

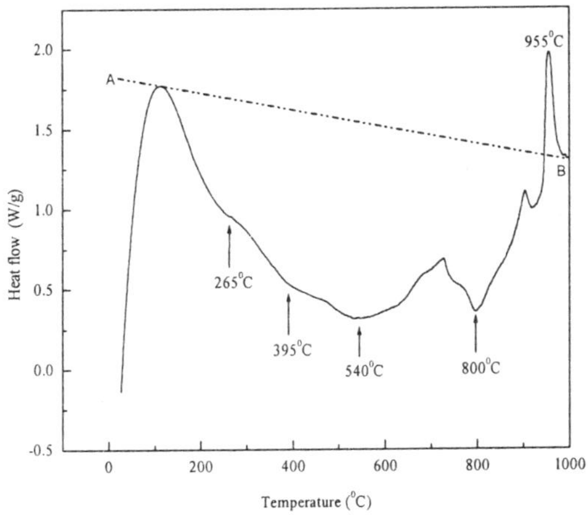

Fig. 9: DSC thermogram in $Fe_{67.0}Co_{9.5}Nd_{3.0}Dy_{0.5}B_{20}$ cast rods recorded at 20 °C/min heating rate.

The values of ΔH in recrystallization of the alloy in the three or four exothermic peaks and of other thermodynamic parameters obtained from DSC thermograms are given in Table 5. The reported data in the literature

Table 5: Thermal parameters in $Fe_{67.0}Co_{9.5}Nd_{3.0}Dy_{0.5}B_{20}$ alloy in forms of thin ribbons, rods, and powder compacts

Sample	*Temperature (°C)*						*ΔGGH (J/g)*			
	T_g	T_{P1}	T_{P2}	T_{P3}	T_{P4}	T_m	ΔGGH_{P1}	ΔGGH_{P2}	ΔGGH_{P3}	ΔGGH_{P4}
$Fe_{67}Co_{9.5}Nd_{3.0}Dy_{0.5}B_{20}$ Ribbons I	555	580	650	800	–	955	–2.6	–0.4	–1.08	–
$Fe_{67}Co_{9.5}Nd_{3.0}Dy_{0.5}B_{20}$ ribbons II	525	572	–	–	–	–	–	–	–	–
As-cast rods	–	264	393	542	800	955	–	–	3.09	5.50
Powder compacts I	–	–	–	–	–	955	–	–	–	–
Powder compacts II						955				

The data for ribbons II are reported from Zhang et al. (2000). The powder compacts I and II have a broad exothermic signal ranging from 100 to 900 °C.

(Zhang and Inoue, 2000) on a similar $Fe_{70-x}Co_{9.5}Nd_xDy_{0.5}B_{20}$ alloy, with $x = 3.5$, in the form of the ribbons are included in this table for a comparison. No endothermic step of the T_g could be resolved in the case of the cast rods or the powder compacts. A predominant thermal signal in the structural relaxation followed by a local grain growth in small precipitates masks the weak T_g signal.

Magnetic Properties

M-rich M-R-A Type Alloys

The bulk amorphous $R_{90-x}Fe_xAl_{10}$ alloys have optimal values of remanence J_r, saturation magnetization J_s, energy-product $(JH)_{max}$, and Curie temperature T_C at a specific value of $x \sim 30$ irrespective to the R-element. As given in Table 6, the alloy with R = Nd has the highest value of $(JH)_{max}$ = 19 kJ/m^3 with the highest value of T_C = 610 K. T_C is a strong function of Fe-Fe, R-Fe and R-R exchange interactions in magnetic R and Fe species in this series. As a result, it sensitively varies with a function of R as well as with a function of x in a given R-series. Pr($4f^2$), which has a smaller magnetic moment than Nd ($4f^3$), accounts for a smaller value of T_C (515 K in comparison to 610 K at $x \sim 30$) in the $Pr_{90-x}Fe_xAl_{10}$ alloy with respect to that in the $Nd_{90-x}Fe_xAl_{10}$ alloy (Table 6). Moreover, in thermomagnetograms, an increase in the value of x from (a) $x = 20$ to at (b) $x = 30$ in the $Nd_{90-x}Fe_xAl_{10}$ series leads to an increase in the T_C-value by 60 K with the final value of 610 K. According to it, the Fe-Fe exchange interactions have a more pronounced effect on T_C than the R-Fe or R-R exchange interactions. Very similar magnetic properties lie in $R_{90-x}Fe_xGa_{10}$ or $R_{90-x}Fe_xB_{10}$ alloys.

Table 6: Magnetic properties in bulk amorphous $R_{90-x}Fe_xAl_{10}$ alloys

Composition	*Magnetic properties*			
	$J_r(T)$	$H_c(kA/m)$	$(JH)_{max}$ (kJ/m^3)	T_C (K)
$Nd_{70}Fe_{20}Al_{10}$	0.08	239	13.0	550
$Nd_{60}Fe_{30}Al_{10}$	0.12	277	19.0	610
$Pr_{60}Fe_{30}Al_{10}$	0.09	300	13.0	515
$Er_{60}Fe_{30}Al_{10}$	0.10	280	12.0	590

An especial advantage with the Pr-based alloys (in particular with a partial Al→ Ga substitution) is that they are characteristically more oxidation-corrosion resistant than the Nd-based alloys in ambient atmosphere (Ram, 1997). They, therefore, present their substantially stable magnetic properties in ambient atmosphere. This is highly necessary for practical applications of these alloys in magnet technology and related devices and components. In this case, a partial substitution of Nd by Pr in the $(Nd_{1-y}Pr_y)_{90-x}Fe_xAl_{10}$ series thus could be a better alternative to have a practically useful bulk amorphous alloy with stable magnetic properties. A partial replacement of Fe by Co is preferred in order to improve the T_C value. The optimal values lie in the bulk amorphous alloy at 30 at % Fe content.

$Fe_{67.0}Co_{9.5}Nd_{3.0}Dy_{0.5}B_{20}$ Alloys

Figure 10 compares hysteresis loops for $Fe_{67.0}Co_{9.5}Nd_{3.0}Dy_{0.5}B_{20}$ alloy in forms of (a) thin ribbons, (b) rods, and (c) a consolidated powder of pellet. The as-cast rods behave to be extremely soft magnetic in nature with a very small value of H_c = 7.1 kA/m. There is no significant H_c value in the ribbons. At room temperature, the alloy in the form of thin ribbons behaves as superparamagnetic in nature. A value of H_c = 188.5 kA/m develops only after compacting and annealing a refined powder after the ribbons in the form of a rod, disc or cylinder at 630 °C for 600 s under a high purity N_2 gas. As given in Table 7, the remanence J_r is increased to be as large as 0.89 T with a similarly improved value of the saturation magnetization J_s = 1.51 T. The results suggest a substantial increase in the value of the magnetic anisotropy after consolidating the powder as a compact mass of bulk amorphous alloy.

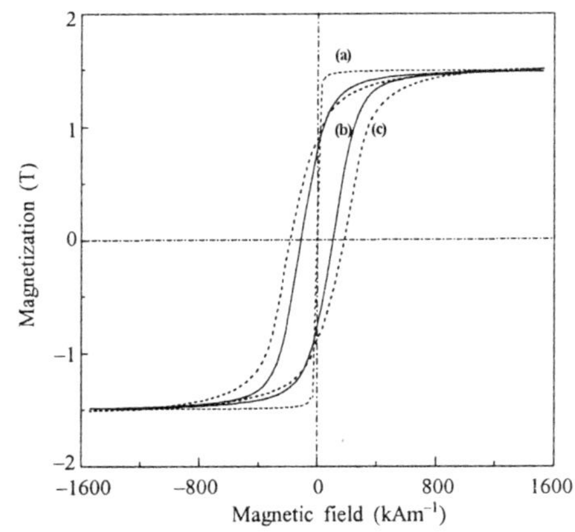

Fig. 10: Hysteresis loops of $Fe_{67.0}Co_{9.5}Nd_{3.0}Dy_{0.5}B_{20}$ alloy in forms of (a) melt-spun ribbons, (b) cast rods, and (c) a bulk consolidated powder (powder compacts I).

Table 7: Correlation in structure and magnetic properties in $Fe_{67.0}Co_{9.5}Nd_{3.0}Dy_{0.5}B_{20}$ alloy in forms of thin rods, ribbons, and powder compacts

Alloy form	*Nature*	*H_c (kAm^{-1})*	*J_r (T)*	*J_s (T)*
Ribbons	Amorphous	~ 0	~0	1.49
Cast rods	Amorphous	7.1	0.10	1.45
Powder compacts I Powder compacts II`	Semi amorphous	188.5	0.89	1.51

According to the Stoner-Wohlfarth (1948), an effectively large value of local magnetocrystalline anisotropy H_a governs H_c in a composite microstructure of ideal single domain particles of surface anisotropy H_s or shape anisotropy H_b as follows:

$$H_c = \Phi\{H_a - H_{sb}\}, \text{ with } H_{sb} = H_s + H_b \quad (3)$$

Here, Φ is a geometrical parameter, which has a value of 0.48 for thin platelet particles (Ram, 1989, 2000). An average value of $H_{sb} \sim H_b$, with $H_s \sim 0$, describes its contribution in H_c in hard magnetic particles in which it is negligibly small as compared to the H_a value. $H_b = 0$ for a spherical and $4\pi J_s$ for a thin platelet shape of particles. An intermediate value lies in other shapes. In soft magnetic particles, in which H_{sb} or $H_b \geq H_a$, a modified relation applies as (Ram, 2000),

$$H_c = \phi H_a \left[1 - \left(\frac{H_{sb}}{H_a}\right) A\right] \quad (4)$$

The function $f(H_{sb}) = (H_{sb}/H_a)A$ in the second term defines an effective structural fractional value of H_{sb} that imparts the total H_c. The coefficient A has a positive value between 0 and 1 by defining a positive $f(H_{sb}) \leq 1$ value. $A = 1$ in hard magnetic particles of $H_b/H_a < 1$. Using $H_a = 588.9$ kA/m, as in hcp Co (Chaik and Tebble, 1969), it gives a value of $H_c = 282.7$ kA/m for $H_{sb} \sim 0$, with the same value of $\Phi = 0.48$ as used in the platelet shape of particles. In fact, the sample in question has a nonzero value H_{sb}, with some magnetostatic interactions between particles, which lowers the final H_c value depending on the microstructure as observed in this example. A value of $H_c = 188.5$ kA/m, observed in the consolidated bulk amorphous alloy powder, gives a significant value of $H_{sb} = 196.2$ kA/m.

In isotropic magnets, the value for J_r is restricted to only half of the J_s at high field (Schrefl et al., 1994). A manifested value to 59 % of J_s value arises in this specific example in exchange coupling interaction (Ram et al., 1995; Zhang and Inoue, 2000) between the soft (α-Fe or Fe_3B nanocrystallites, and the amorphous alloy) and hard ($R_2M_{14}B$ nanocrystallites) magnetic phases. A similar large value of H_c develops on annealing the ribbons at 500 to 700 °C in a controlled atmosphere. The value is found to be increasing monotonically as a function of increasing the annealing temperature in this range. An optimal $H_c = 256$ kA/m value is obtained in 420 s of annealing at 700 °C in vacuum or N_2 gas. This is consistent with a maximum of $H_c = 263$ kA/m value reported in recrystallized $Fe_{70-x}Co_{9.5}Nd_xDy_{0.5}B_{20}$, $x = 3 - 4$, ribbons in 420 s of annealing at 660 °C (Zhang and Inoue, 2000). A total of 420 s of annealing time used in this example has been found to be sufficient to induce a maximum value of H_c with a maximum value of J_s or J_r. Other details of the magnetic properties achieved in the present work in the various samples are given in Table 7.

A maximum value of H_c of 256 kA/m found in this series of alloys is similar as in the usual hard magnetic materials of hexagonal ferrites (Ram, 1989; Buschow, 1986). As can be seen from the data in Table 8, an order of larger H_c value often lies in $R_2Fe_{14}B$, RM_5, or R_2M_{17} intermetallics (Schnitzke et al., 1990; Ram and Joubert, 1992; Ram, 1994, 1997; Ram et al., 1995; Ram et al., 1997). The $Fe_{70-x}Co_{9.5}Nd_xDy_{0.5}B_{20}$ alloy series of a nanocomposite structure, as in this work, offers a further scope of optimizing the final H_c value by optimizing the composition and/or microstructure with a proper combination of thermal annealing in a controlled atmosphere. As in other examples of ferromagnetic alloys (Schnitzke et al., 1990; Ram and Joubert, 1992; Ram, 1994, 1997; Ram et al., 1995; Ram et al., 1997), an addition of nitrogen or other interstitial gases may be helpful in optimizing desirably hard or soft magnetic properties in a refined semi-amorphous or nanocrystalline composite structure.

Table 8: A comparison of magnetic properties in important hard magnetic materials

*Material**	$(JH)_{max}$ (kJm^{-3})	J_s (Am^2kg^{-1})	J_r (T)	H_c $(kAcm^{-1})$	T_C (K)	H_a $(kAcm^{-1})$
Hexagonal ferrites	28.5	71.5	0.39	2.75	725	13.1
Alnico	83.5	–	1.04	1.24	–	–
$SmCo_5$	300	105	0.40	22.8 (51.4)	1020	310
Sm_2Co_{17}	–	–	0.68	0.32(0.15)	1050	–
$Nd_2Fe_{14}B$	350	150	1.35	15-25 (15.0)	590	90.0
$Sm_{20}Fe_{70}Ti_{10}$	–	–	0.30	51.6(51.4)	–	–
$Sm_2Fe_{17}C_2$	–	–	0.61	18.5(15.0)	673	–
$Sm_2Fe_{17}N_{2.5}$	–	–	0.78	24.0(20.0)	743	–
$Nd_{60}Fe_{30}Al_{10}$	19.0	~1.5	0.12	2.77	610	–

All the materials are crystalline except $Nd_{60}Fe_{30}Al_{10}$ which is a bulk amorphous alloy. The data are reported from the literature mentioned in this paper.

Large Shape Formability in Bulk Amorphous Alloys

The present bulk amorphous alloys of multicomponent R-M-A systems are highly ductile and easy to fabricate in practically important shapes of spheres, balls, thin sheets or plates, rings, wires, rods, cylinders, coils, or any other designs for use in small components and devices of permanent magnets and their applications in electronic and other devices. Liquid-liquid phase separation or local recrystallization in extremely small particles or nanoclusters promotes toughness in a hybrid nanocomposite structure (Kumar et al., 2002; Frankwicz et al., 1996). We are fabricating some of the above bulk amorphous R-M-Al alloys of $SmCo_5$, $Nd_2Fe_{14}B$ and other high-energy-density magnetic materials in different shapes.

Applications of Magnetism in Bulk Amorphous Alloys

The magnetic bulk amorphous alloys offer multifold applications in permanent magnets, motors, generators, charge accelerator, magnetic resonance imaging, defense, space applications, domestic uses (microphone, computer etc.), microwave devices, magnetotherapy, biomaterials, magnetic levitation, robotics, magnetic valves, rely and switches, and ferrofluids. Some of them show giant magnetoresistance (GMR) or giant magnetostriction (GMS) for applications in spintronic devices (such as magnetic random access memories), read heads, gas sensors/surface catalysts, low power magnetic devices, low power field sensors and memory elements, high resistance devices (varistors), optoelectronics, low energy detectors, and tunnel junction. As compared to well-known $R_2Fe_{14}B$ intermetallics and other high-energy magnetic materials, the bulk amorphous R-M-A alloys exhibit a relatively large value of electrical resistivity, which make them particularly suitable for applications in high frequency communication devices at low power loss.

Another high-tech discipline of applications of bulk amorphous alloys is in hydrogen storage, gas sensor, and related usages in nonconventional sources of production and application of energy. A hybrid nanocomposite structure is supposed to meet an improved capacity of charging with H_2 gas at room temperature or bit higher. Efforts are being made in fabricating of mesoporous or microporous alloys for such specific applications, mesoporous membranes, hot filters, surface catalysts, solid fuels, solid electrolytes, and biomaterials.

CONCLUSIONS

Bulk amorphous and/or nanocomposite alloys of multicomponent systems of R-M-A, with R as a rare-earth element, M = Fe or part of Co, and A = Al, B, or Ga offer a new class of high-energy-density magnetic materials for permanent magnets and other applications. The as-received bulk amorphous R-M-A alloys present as large energy-density as $(JH)_{max} \sim 19$ kJ/m^3, i.e. comparable to the value in the conventional hard ceramic magnets. Unusually, the R-Fe-Al, R-Fe-Ga, or R-Fe-Ga types of alloys if recrystallized behave to be extremely soft magnetic in nature and extremely hard in mechanical strength. Several unusual magnetic transitions appear at low temperatures in virgin amorphous as well as in recrystallized alloys with small crystallites of nanometer size.

Acknowledgements: A financial support by a research grant from the DRDO, Government of India, is gratefully acknowledged.

REFERENCES

Buschow, K.H.J. (1986). *Meter. Report*, **1:** 1.

Cahn, R.W. (1993). *In*: rapidly solidified alloys. Ed. Libermann, H.H. (Marcel Decker, New York), 1.

Chaik, D.J. and Tebble, R.S. (1969). Magnetic Materials. John Wiley and Sons, London.

Chen, H.S. (1974). *Acta Met.*, **22:** 716.

Chen, H.S. and Turnbull, D. (1968). *J. Chem. Phys.*, **48:** 2560.

Chen, H.S. and Turnbull, D. (1969). *Acta Met.*, **17:** 1021

Chen, H.S. and Turnbull, D. (1974). *Acta Met.*, 22: 897.

Croat, J.J., Herbst, J.F., Lee, R.W. and Pinkerton, F.E. (1984). *J. Appl. Phys.*, **55:** 2078.

Drehman, A.L., Greer, A.L. and Turnbull, D. (1982). *Appl. Phys. Lett.*, **41:** 716.

Frankwicz, P.S., Ram, S. and Fecht, H.J. (1996). *Mater. Lett.*, **28:** 77.

Frankwicz, P.S., Ram, S. and Fecht, H.J. (1996). *Appl. Phys. Lett.*, **68:** 2825.

Frankwicz, P.S., Ram, S. and Fecht, H.J. (1996). *J. Non-Cryst. Solids*, **205-207:** 522.

Gong, W., Li, H., Zhao, Z. and Chen, J. (1991). *J. Appl. Phys.*, **69:** 5119.

Haasen, P. (1986). Physical Metallurgy. Cambridge Univ. Press, Cambridge. 125.

Inoue, A. (2000). *Acta. Mater.*, **48:** 277.

Inoue, A. and Gook Zhang, J.S. (1995). Mater. Trans., JIM, **36:** 1180.

Inoue, A. and Gook Zhang, J.S. (1995). Mater. Trans., JIM, **36:** 1282.

Inoue, A., Kato, A., Zhang, T., Kim, S.G. and Masumoto, T. (1991). *Mater. Trans.*, JIM, **32:** 609.

Inoue, A., Nakamura, T., Nishiyama, N. and Masumoto, T. (1992). *Mater. Trans.*, JIM, **33:** 937.

Inoue, A., Takeuchi, A. and Zhang, T. (1998). *Metall. Mater. Trans.*, **29A:** 177.

Inoue, A., Zhang, T. and Masumoto, T. (1990). *Mater. Trans., JIM,* **31:** 177.

Inoue, A., Zhang, T. and Masumoto, T. (1990). *Mater. Trans., JIM,* **31:** 425.

Inoue, A., Zhang, T., Takeuchi, A. and Zhang, W. (1996). *Mater. Trans.*, JIM, **37:** 636.

Inoue, A., Zhang, T., Takeuchi, A. and Zhang, W. (1997). *Mater. Trans.*, JIM, **37:** 1731.

Johnson, W.L. (1999). *MRS Bull.*, **24:** 42.

Kavesh, S. (1978). In metallic glasses, Chapter 2. Ed. Gillman, J.J. and Leamy, H.J. (ASM International Metal Park, OH), 36.

Klement, W., Willens, R.H. and Duwez, P. (1960). *Nature*, **187:** 869.

Klug, M.P. and Alexander, L.E. (1974). X-ray Diffraction Procedure for Polycrystalline and Amorphous Materials. Wiley, New York, 634.

Kui, H.W., Greer, A.L. and Turnbull, D. (1984). *Appl. Phys. Lett.*, **45:** 615.

Kumar, G., Eckert, J., Roth, S., Loser, W., Ram, S. and Schultz, L. (2002). *J. Appl. Phys.*, **91:** 3764.

Kumar, G., Eckert, J., Roth, S., Loser, W., Schultz, L. and Ram, S. (2003). *Acta Mater.*, **51:** 229.

Kumar, G., Eckert, J., Schultz, L. and Ram, S. (2002). *Mater. Lett.*, **53:** 305.

Lee, M.C., Kendall, J.M. and Johnson, W.L. (1982). *Appl. Phys. Lett.*, **40:** 382.

McClume, W.F. (1979). Powder Diffraction File JCPDS (Joint Committee on Powder Diffraction Standards), International Centre for Diffraction Data, Swarthmore, PA, (a) 03-1050, aFe , (b) 39-1314, 39-1315 Fe_3B, (c) 39-1314 Fe_2B and (d) 77-1404 $Nd_2Fe_{14}B$.

Peker, A. and Johnson, W. L. (1993). *Appl. Phys. Lett.*, **63:** 2342.

Ram, S. (1989). *J. Magn. and Magn. Mater.*, **82:** 129.

Ram, S. (1990). *Phys. Rev.*, **B42**: 9582.

Ram, S. (1994). *Phys. Rev.*, **B49:** 9632.

Ram, S. (1997). *J. Mater. Sci.*, **32:** 4133.

Ram, S. (2000). *J. Mater. Sci.*, **35:** 3561.

Ram, S. (2001). *Acta. Mater.*, **49:** 2297.

Ram, S. (2001). *Ind. J. Phys.*, **84A:** 379.

Ram, S. and Fecht, H.J. (2000). *Mater. Trans.*, JIM, **41:** 554.

Ram, S. and Johari, G.P. (1990). *Phil. Mag.*, **B61:** 299.

Ram, S. and Joubert, J.C. (1992). *J. Appl. Phys.*, **72:** 1164.

Ram, S., Claude E. and Joubert, J.C. (1995). IEEE. Trans. Magn., **31:** 2200.

Ram, S., Febri, M., Fecht, H.J. and Joubert, J.C. (1995). *Nanostruct. Mater.*, **6:** 473.

Ram, S., Fecht, H.J., Haldar, S., Ramachandrarao, P. and Banerjee, H.D. (1997). *Phys. Rev.*, **B56:** 726.

Sagawa, M., Fujimura, S., Togawa, M. and Matsuura, Y. (1984). *J. Appl. Phys.*, **55:** 2083.

Schnitzke, K., Schultz, L., Wecker, J. and Katter, M. (1990). *Appl. Phys. Lett.*, **56:** 587.

Schrefl, T., Filder, J. and Kronmüller, H. (1994). *Phys. Rev.*, **B49:** 6100.

Stoner, E. and Wohlfarth, E.P. (1948). *Philos. Trans. R. Soc. London* A, **240:** 599.

Strnat, K., Hoffer, G., Olson, J., Ostertag, W. and Becker, J.J. (1967). *J. Appl. Phys.*, **38:** 1001.

Zhang, W. and Inoue, A. (2000). *J. Appl. Phys.*, **87:** 6122.

Zhang, W., Matsushita, M. and Inoue, A. (2001). *J. Appl. Phys.*, **89:** 492.

Microstructure Parameters from X-ray Line Profile Analysis

T. Ungár

Department of General Physics
Eötvös University, Budapest, Hungary

INTRODUCTION

Ever since transmission electron microscopy (TEM) has been discovered and used for looking into the microstructure of materials, X-ray diffraction peak profile analysis (XDPPA) has been a complementary, at the best an alternative tool for describing the microstructure of crystalline materials. The reason for this is, that there are still fundamental questions open as to the interpretation of the parameters provided by the X-ray method and as to the correspondence between the parameters given by X-rays and other more direct methods, especially TEM and scanning electron microscopy (SEM). However, recent developments in diffraction techniques, in both synchrotron and home laboratories, enable the measurement of diffraction profiles with high precision, where background to peak ratios are typically between 10^{-3} and 10^{-5}. Precise peak shape and breadth measurements enable the critical comparison of refined models with experimental data and the conclusion about detailed characterization of the microstructure in terms of crystallite size, size distribution and lattice defects. These two factors, improving experimental facilities and increasingly sophisticated physical modeling, are the driving force for developing the X-ray diffraction (XRD) method in order to become competent in parallel with TEM and SEM. In the present account the main emphasis will be given to the state-of-the-art physical interpretation of parameters provided by the X-ray method and their correlation with parameters obtained by TEM and SEM.

In the case of ultrafine-grained materials the physical properties are strongly related to the microstructure (Chokshi et al., 1989; Weertman, 1993; Siegel, 1994; Gubicza et al., 1999; Gubicza and Wéber, 1999; Zehetbauer et al., 2003; Arato et al., 1998; Szépvölgyi et al., 2001; Qian et al., 2001; Pantea et al., 2002; Herzer, 1995; Hernando et al., 1995; Kaptás et al., 2002;

Hughes and Hansen, 2001; Hansen, 2001). The mechanical properties are enhanced. However, the ductility is often strongly reduced (Chokshi et al., 1989; Weertman, 1993; Siegel, 1994; Gubicza et al., 1999; Gubicza and Wéber, 1999; Zehetbauer et al., 2003). Compaction by sintering without binding material is a promising procedure (Arato et al., 1998; Szépvölgyi et al., 2001; Qian et al., 2001; Pantea et al., 2002) and favourable soft magnetic properties can be achieved, especially in ultra fine grained iron or iron base alloys (Herzer, 1995; Hernando et al., 1995; Kaptás et al., 2002). The characterization of the microstructure by TEM is complicated because of the very small size of crystallites and the extremely high dislocation density (Weertman, 1993; Siegel, 1994). At the same time the microstructure of ultrafine-grained materials (especially which are prepared by severe plastic deformation) is built up hierarchically by the following sequence of units (in the order of decreasing size): (i) grains, (ii) subgrains, (iii) cell blocks and (iv) dislocation cells (Hughes and Hansen, 2001; Hansen, 2001). This complexity of the microstructure often makes a deviation between the crystallite size and grain size determined by X-ray and TEM, respectively. Consequently, for obtaining the correct picture of the microstructure in ultrafine-grained materials, the simultaneous application of XRD and TEM is useful.

XRD peaks broaden when the crystal lattice becomes imperfect. From the materials science point of view the microstructure means the extent and the quality of lattice imperfectness. In XRD, imperfectness is by far not straightforward. A modern 10 inch silicon wafer is imperfect if it contains more than one or two dislocations. XRD on such a crystal fails completely to tell anything about its microstructure. This is one of the reasons why simple X-ray topography is not used anymore for characterizing silicon wafers. The diffraction peaks are perfect. Darwin curves described in detail by the theory of dynamic scattering. Dynamic scattering for X-rays applies until the defect free crystals are considerably larger than extinction lengths, typically from several to 100 μm, depending on the wavelength and the structure factor. These distances correspond to dislocation densities between about 1×10^8 and 2.5×10^{11} m^{-2}. The dynamic contrasts of dislocations are as wide as the extinction lengths; therefore, when the dislocation densities exceed these values the dynamic scattering conditions break down. In well annealed copper single crystals the dislocation density cannot be lowered below 1×10^{12} m^{-2}. Thus, dynamic scattering conditions with X-rays in copper can practically not be realized. Since, in most of the structural materials neither the extinction lengths are considerably shorter than several microns nor the dislocation densities can be lowered below about 1×10^{12} m^{-2}, it is safe to say that XRD in materials can be described by the theory of kinematical scattering. Here we note that for slow neutrons in copper the extinction lengths is considerably shorter than for X-rays; therefore, in the case of neutron diffraction, dynamic scattering conditions in copper can easily be realized.

Within the frame of the kinematical scattering theory crystal imperfectness and the ideal specimen for powder diffraction are in close connection. The ideal powder diffraction specimen is an abstraction for the purpose of structure determination. The goal is to have as narrow diffraction peaks as possible. This is achieved by strain free crystals or crystallites with average sizes around 5 to 10 μm. Bulk metals or alloys automatically fulfill the conditions for kinematical scattering; however, the crystallites or, more appropriately, the grains are almost never strain free. Among the obvious similarities of a powder specimen prepared for the purpose of structure determination and a bulk metal or alloy sample, one of the important differences are texture. This effect can be eliminated in the case of a powder sample, but, remains an intrinsic property in the case of a bulk specimen.

According to the theory of kinematical scattering, XRD peaks broaden either when crystal or crystallite size becomes smaller than about a micron or if lattice defects are present in large enough abundance. In terms of dislocations this means a dislocation density larger than about 5×10^{12} m^{-2}, as discussed before. Kinematical scattering theory offers a rather pragmatic scheme for the correlation between diffraction and lattice imperfectness: the size broadening is diffraction order independent; however, strain broadening caused by lattice defects increases with diffraction order. This thumb rule, as any simple rule, has many exceptions depending on the specific crystallite size, especially crystallite shape, and on the specific type of strain caused by different types of lattice defects.

STRAIN, LATTICE DEFECTS AND DISLOCATIONS

The most common sources of strain can be given as : (*a*) dislocations, (*b*) stacking faults, (*c*) twinning, (*d*) microstresses, (*e*) long-range internal stresses, (*f*) grain boundaries, (*g*) sub-boundaries, (*h*) internal stresses (*i*) coherency strains, (*j*) chemical heterogeneities, (*k*) point defects or (*l*) precipitates and inclusions. Though dislocations are listed alone as one of the sources of strain, they are the major component in (*b*) to (*g*) and can play an important role in (*h*) and (*i*).

The ideal powder diffraction pattern consists of narrow, symmetrical, delta-function like peaks, positioned according to a well-defined unit cell. The aberrations from the ideal powder pattern can be conceived as: (i) peak shift, (ii) peak broadening, (iii) peak asymmetries, (iv) anisotropic peak broadening and (v) peak shape. The major correlation between peak aberrations, i.e. broadening, shifts or asymmetries, and the different elements of microstructure are summarized in Table 1. For the sake of comprehension, crystallite smallness is also listed in the table. Table 1 shows that (i) peak shift is related to the different types of internal stresses and planar faults, especially stacking faults or twin boundaries, (ii) peak broadening indicates crystallite smallness and microstresses, but stress gradients and/or chemical

Table 1: The most typical correlations between diffraction peak aberrations, i.e. broadening, shifts or asymmetries, and the different elements of microstructure

Peak Aberrations → *Sources of Strain ↓*	*Peak Shift*	*Peak Broadening*	*Peak Asymmetry*	*Anisotropic Peak Broadening*	*Peak Shape*
Dislocations		+	+	+	+
Stacking Faults	+	+	+	+	+
Twinning	+	+	+	+	+
Microstresses		+			
Long-Range Internal Stresses	+		+		
Grain Boundaries	+	+			
Sub-Boundaries	+	+			
Internal Stresses	+				
Coherency Strains	+	+	+		
Chemical Heterogeneities	+	+	+		
Point Defects					+
Precipitates and Inclusions			+		+
Crystallite Smallness		+		+	+

heterogeneities can also cause peak broadening, (iii) peak asymmetries can be caused by long-range internal stresses, planar faults or chemical heterogeneities, (iv) anisotropic peak broadening can result from anisotropic crystallite shape or anisotropic strain, and (v) Peak shape refers to the way the intensity decays into the background, mainly below the half maximum.

The correlation between peak shape and specific lattice defects is discussed in terms of the hierarchy of lattice defects based on their strain fields. Krivoglaz (1996) sorted lattice defects into three different classes: point-, linear- and planar defects. (1) The strain fields of point defects decay as $1/r^2$, where r is the distance from the defect, (2) The strain fields of linear defects decay as $1/r$, and (3) the strain field of planar defects is homogeneous, or space independent. The three different types of spatial dependencies are of short- and long-range order and homogeneous, respectively. Due to the reciprocity between crystal and reciprocal space, the scattering related to point defects is extended far from the fundamental Bragg reflections. The scattered intensity decays asymptotically as ΔK^{-2}, where $\Delta K=(2\cos\theta/\lambda)\Delta\theta$, λ is the wave length of X-rays, θ is the scattering angle and $\Delta\theta$ is measured from the centre of gravity of the diffraction peak. This is often called Huang scattering (Trinkaus, 1972). The strain field of linear defects is of long-range character. The scattered intensity decays asymptotically as ΔK^{-3} (Krivoglaz, 1996; Wilkens, 1970; Groma, 1998) and is clustered around the fundamental Bragg reflections. This is the diffraction effect called diffraction peak broadening, and the related topic is called peak profile analysis.

The strain fields of planar defects, especially stacking faults and twin boundaries, are space independent or homogeneous. Primarily they cause lattice parameter changes or shifts of Bragg reflections. As a secondary effect they also cause peak broadening and asymmetry (Warren, 1959; Treacy et al., 1991). As discussed before and shown in Table 1, lattice defects are of a great variety and it is often not straightforward to classify them within the three classes of the hierarchy of Krivoglaz (1996). Precipitates or inclusions, for instance, would classify as point defect; however, the particles can be separated from the matrix by misfit dislocations producing long-range strain field, thus mixing (1) and (2). Stacking faults produce homogeneous strain fields causing peak shifts. However, they are bounded by partial dislocations and can be confined to finite volumes, therefore mixing (1), (2) and (3). This means that stacking faults cause peak shift, peak broadening, peak asymmetries and diffuse scattering, as indicated in Table 1. In nanostructured materials, capillary pressures by curved grain boundaries can produce peak shifts, whereas high dislocation densities together with particle smallness cause complex peak broadening. Concentration gradients can make peaks asymmetric and broad without specific lattice defects. Within all these complex phenomena dislocations play a central role: either as the dominant lattice defects or as the major components in complex lattice defects.

Dislocations are usually present either as dislocation dipoles or in special configurations where screening plays an important role. The screening effect can be handled in general as the dipole character. As a result of screening, depending on its strength, the strain field will decay between $1/r$ and $1/r^2$. In the first case there is no screening at all, whereas in the second case full screening is present. Screening is also directly related to the statistical distribution or the spatial correlation of dislocations (Krivoglaz, 1996; Wilkens, 1970; Groma, 1998). In the first or second case, dislocations are distributed chaotically or with strong correlation, respectively (Groma, 1998). Wilkens (1970) used the term "restrictedly random" for the case between the two extremes and, ingeniously, worked out the diffraction profile shape for such a dislocation arrangement. The idea can definitely be considered ingenious, since in the overwhelming case of real dislocation arrangements the conditions of "restrictedly random" distribution are satisfied. The asymptotic behavior of peak profile tails is ΔK^{-3} or ΔK^{-2} for completely chaotic or completely correlated distributions, respectively. In latter case the dislocation dipoles can be called: infinitesimal dislocation dipoles. In terms of diffraction peak shapes: the more effective the screening, the scattering due to dislocations will occur more farther apart from the fundamental Bragg position. In other words, the diffraction profiles will have longer tails and their breadths, or their FWHM, will be relatively smaller. Quite opposite, when the dislocations are distributed chaotically, and have no or little dipole character, the scattering will occur close to the fundamental Bragg positions and the peaks will have short tails (the profiles will have "bell" shape). The effectiveness of screening, or the dipole character—the two have the same physical meaning—reflect the dislocation-dislocation interaction, a property not really visible in TEM micrographs. Obviously, the number or density of dislocations and the dislocation-dislocation interaction are two independent physical quantities which, inevitably, need two parameters for a complete characterization of the dislocation structure. Wilkens (1970) suggested to use the dislocation density and the effective outer cut off radius, ρ and R_e, respectively.

Kroener (1981) suggested to use a dislocation-density tensor. Since ρ and R_e can be directly related to diffraction profiles, for the interpretation of diffraction experiments, these are used as introduced by Wilkens (1970). Though R_e has a length dimension it should not be interpreted as a physically existing length in the crystal. Especially, the experimental value of R_e must not be correlated directly with the coherent domain size. In order to avoid confusion with other length scales, it is better to use the dislocation arrangement parameter $M=R_e\sqrt{\rho}$, instead of R_e, as introduced by Wilkens (1970). The value of M is smaller or larger than unity as (i) the profile tails are longer or shorter, (ii) as the average exponent in the ΔK dependence of the asymptotic behaviour of the tails of profiles is closer to -3 or -2, (iii) as the screening of strain fields of dislocations is stronger or weaker and (iv) as the dipole character of dislocations is stronger or weaker, respectively. All formulations in (i) to (iv) have the same physical meaning.

SIZE

The model for size broadening is based on the simple concept that those and only those planes participate in a diffraction experiment whose normal vectors are parallel to the diffraction vector and that the extension of these planes is immaterial for the scattering mechanism (Bertaut, 1950). According to this concept all fractions of all crystals or crystallites, which have the same depth parallel to the diffraction vector can be thought contained in a single "column" with one particular "length". Proceeding this mechanism through the whole sample, all material will be sorted into the columns with different lengths. Each column length will have a weight proportional to the volume fraction of the material sorted into a particular column. The diffraction peak corresponding to one particular column is the squared Fourier transform of the form function pertaining to that column length. The final size-broadened peak is the weighted sum of the individual peaks corresponding to the individual columns. The direct meaning of a size-broadened peak is thus related to the weighted column-length distribution in the specimen. The weighted column-length distribution in the specimen depends on the size, the shape and the size distribution of either the crystals or the crystallites or the coherently scattering domains (CSD). In order to obtain the mean size of crystallites or CSDs, assumptions have to be made about their shape and their size-distribution.

BASIC PRINCIPLES OF PEAK PROFILE ANALYSIS

Assuming that size (S) and lattice distortion (D) are the two sources of XRD peak broadening, the profile of a Bragg reflection can be given by the convolution of the size and the distortion profiles (Warren, 1959):

$$I^F = I^S * I^D \tag{1}$$

The Fourier transform of this expression is the Warren-Averbach equation (Warren, 1959):

$$\ln A(L) \cong \ln A_L^S - 2\pi^2 L^2 g^2 \langle\varepsilon_{g,L}^2\rangle, \tag{2}$$

where $A(L)$ are the absolute values of the Fourier coefficients of the physical profiles, A_L^S are the size Fourier coefficients, g is the absolute value of the diffraction vector and $\langle\varepsilon_{g,L}^2\rangle$ is the mean square strain. L is the Fourier length defined as $L = na_3$, where $a_3 = \lambda/2(\sin\theta_2 - \sin\theta_1)$, n are integers starting from zero, λ is the wavelength of X-rays and $(\theta_2 - \theta_1)$ is the angular range of the measured diffraction profile.

When strain is caused by dislocations or dislocation type defects, $\langle\varepsilon_{g,L}^2\rangle$ can be expressed as (Wilkens, 1970):

$$\langle\varepsilon_{g,L}^2\rangle = (b/2\pi)^2 \pi\rho C f(\eta) , \tag{3}$$

where $\eta = L/R_e$, b is the Burgers vector of dislocations and C is the contrast or orientation factor of dislocations. Wilkens (1970) determined the $f(\eta)$ function for parallel straight screw dislocations in the entire L range from zero to infinity, see e.g. in Eqs. A.6 to A.8 in (Wilkens, 1970). It can be shown that the Wilkens function, $f(\eta)$, has a more general validity for edge and curved dislocations and, it can be extended in a simple way for infinitesimal dislocation dipoles when M goes to zero (Groma and Csikor, 2003).

Strain Anisotropy

Strain anisotropy means that neither the breadths nor the Fourier coefficients of the diffraction profiles are monotonous functions of the diffraction angle or g (Caglioti et al., 1958). It has turned out to be a useful feature of diffraction peak broadening. The orientation or *hkl* dependence of the contrast of dislocations is well known in TEM. It is expressed explicitly in Eq. (3) by the contrast factor C. If dislocations are the major source of strain in a crystal the values of C depend on the relative orientation of the Burgers-, the line and the diffraction vectors, $\boldsymbol{b}$, $\boldsymbol{l}$ and $\boldsymbol{g}$, respectively. If either all possible slip systems in a crystal are more-or-less equally populated or, if a polycrystal is more-or-less texture free, the contrast factors corresponding to one particular *hkl* can be averaged over the permutations of these *hkl*s. It can be shown that, in a dislocated crystal, the average contrast factors, $\overline{C}$, are linear functions of the fourth order invariants of *hkl* (Hernando et al., 1995). For cubic crystals $\overline{C}$ is (Ungár and Tich, 1999):

$$\overline{C} = \overline{C}_{\mathrm{h00}}(1 - qH^2), \tag{4}$$

where $\overline{C}_{\mathrm{h00}}$ is the average dislocation contrast factor of the *h00* type reflections and $H^2 = (h^2k^2 + h^2l^2 + k^2l^2)/(h^2 + k^2 + l^2)^2$. The values of $\overline{C}_{\mathrm{h00}}$ and the q parameter can be obtained numerically for different dislocation types as functions of the elastic properties of a crystal (Ungár et al., 1999). For hexagonal crystals $\overline{C}$ is (Ungár and Tichy, 1999; Ribárik et al., 2001):

$$\overline{C}_{\mathrm{hk.l}} = \alpha + \frac{\left[\beta(h^2 + k^2 + i^2) + \gamma l^2\right] l^2}{\left[h^2 + k^2 + i^2 + \frac{3}{2}\left(\frac{a}{c}\right)^2 l^2\right]^2}, \tag{5}$$

where α, β and γ are constants depending on the elastic constants of the crystal and the type of dislocations (e.g. basal, prismatic, pyramidal and/or any sub-slip-system) and on c/a, the ratio of the two lattice constants in the hexagonal crystal. The value of α is the average contrast factor corresponding to the *hk.0* reflection: $\overline{C}_{\mathrm{hk.0}} = \alpha$. $\overline{C}_{\mathrm{hk.l}}$ in eq. (5) can be rationalized for numerical fitting purposes (Ribárik et al., 2001):

$$\overline{C}_{\mathrm{hk.l}} = \overline{C}_{\mathrm{hk.0}}\{1 + [a_1(h^2 + k^2 + i^2) + a_2l^2]l^2(h^2 + k^2 + i^2 + \lambda^{-2}l^2)^{-2}\} \tag{6}$$

where a_1 and a_2 are numerical constants and $\lambda^2 = (2/3)(c/a)^2$. For the purpose of physical interpretations Eq. (6) has been further rationalized (Dragomir and Ungár, 2002):

$$\overline{C}_{\mathrm{hk.l}} = \overline{C}_{\mathrm{hk.0}}\,[1 + q_1x + q_2\,x^2\,], \tag{7}$$

where $x = (2/3)(l/ga)^2$, $q_1 = a_1$ and $q_2 = -a_1\lambda^{-2} + a_2$. Eq. (7) means that the average contrast factors corresponding to a specific sub-slip-system (these are the 11 most likely slip systems in a hexagonal crystal according to Jones and Hutchinson, 1981) and materials constants (lattice parameters and elastic constants) have to follow a parabola as a function of x. The $\overline{C}_{\mathrm{hk.0}}$ and the q_1 and q_2 parameters have been numerically evaluated and compiled for a large number of hexagonal crystals (Dragomir and Ungár, 2002).

Breadth Methods

The two simplest measures of peak broadening are the full width at half maximum (FWHM) and the integral breadths, the two are denoted here as: ΔK_{FWHM} and ΔK_{β}, where $\Delta K=(2\cos\theta/\lambda)\Delta\theta$, θ is the Bragg angle and $\Delta\theta$ is the FWHM or the integral breadth in degrees, respectively. Since size is independent and strain is increasing with diffraction order, in order to separate size and strain effects, Williamson and Hall (1953) have suggested to plot the breadth of profiles versus sin θ. The extrapolated intercept of such a plot at sin θ = 0 should give the inverse of an "apparent" size and the slope should be proportional to strain. A typical Williamson-Hall plot of a copper specimen deformed by equal channel angular pressing (ECAP) in a single pass is shown in Fig. 1. The apparent straying of the data points is far beyond experimental error, this indicates a strong strain anisotropy. This effect makes a simple interpretation of the Williamson-Hall plot impossible. However, it is noted that the qualitative importance of strain anisotropy can be very well assessed on the basis of the classical Williamson-Hall plot.

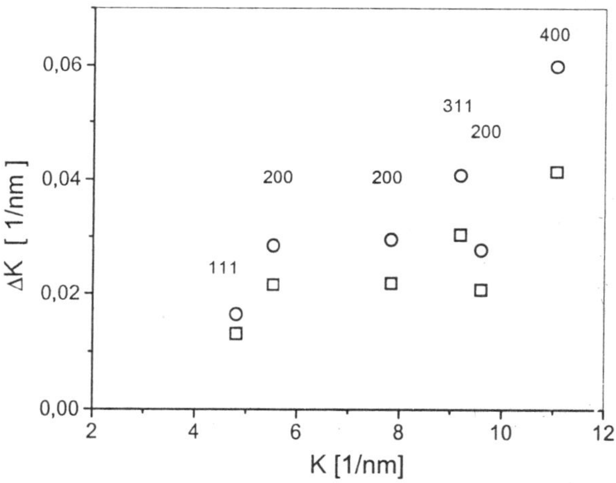

Fig. 1: The Williamson-Hall plot (Williamson and Hall, 1953) of the integral breadths and the FWHM of a copper specimen deformed by equal channel angular pressing (ECAP) in a single pass. The apparent straying of the datapoints is far beyond experimental error indicating a strong strain anisotropy.

Strain anisotropy in the classical Williamson-Hall plot can be rationalized by the contrast factors for dislocated crystals and the result is called the *modified* Williamson-Hall plot (Ungár and Borbély, 1996; Ungár et al., 2001). Before presenting this equation we show by simple dimensionality analysis that the order dependent strain part of peak breadth can only have even-power terms of K. Any expression of strain broadening, e.g. the Fourier coefficients, contain the following expression: $\exp[i\boldsymbol{u}(\boldsymbol{r}-\boldsymbol{g}n)-i\boldsymbol{u}(\boldsymbol{r}+\boldsymbol{g}n)]$, where

i is the imaginary unit, $\boldsymbol{u}$ is the displacement vector at $\boldsymbol{r}\pm\boldsymbol{g}n$, $\boldsymbol{r}$ is the position vector of the "source" of distortions in the crystal, n is a normalized length and $\boldsymbol{g}$ is the diffraction vector. (For the sake of brevity summations or integrations over $\boldsymbol{r}$ are not indicated.) Since this expression must be invariant to the sign of $\boldsymbol{g}$, only even powers of $\boldsymbol{g}$ or K can have non-vanishing contributions to the order dependent strain part of peak breadth. In the case of the integral breadths the *modified* Williamson-Hall plot is (cf. Ungár et al., 2001):

$$\Delta K_\beta = 1/d + \alpha K^2 \overline{C} + O(K^2 \overline{C})^2, \tag{8}$$

where d is an apparent crystallite size, $K = (2\sin\theta/\lambda)\Delta\theta$, α is a constant and O stands for higher order terms in $K^2 \overline{C}$, not interpreted here. For spherical crystallites the volume average diameter $<x>_{vol}$ is directly related to d: $<x>_{vol} = (4/3)d$ (Hinds, 1982). The same data as in Fig. 1 are plotted according to Eq. (8) in Fig. 2. It can be seen that the breadth values follow smooth curves for both, the integral breadths and the FWHM.

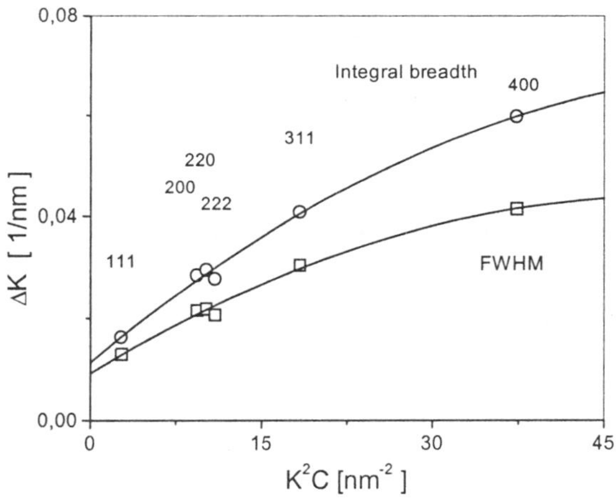

Fig. 2: The *modified* Williamson-Hall plot (Ungár and Borbély, 1996) of the same data as in Fig. 1 according to Eq. (8). The breadth values follow smooth curves for both, the integral breadths and the FWHM. In the present case the average contrast factors, C, correspond to half edge, half screw dislocations, (cf. Ungár et al., 1999).

Whole Profile Fitting Method

The idea of whole profile fitting goes back to Rietveld who suggested to produce the measured diffraction pattern by taking into account all possible effects which contribute to the intensity distribution, including all instrumental effects (Rietveld, 1967, 1969). The purpose of this method was to refine the structure parameters from powder diffraction patterns. Well defined analytical profile functions are used in the Rietveld method, such as the Voigt or pseudo-Voigt functions. However, profile shapes corresponding to dislocated crystals cannot be described by simple analytical functions (Krivoglaz, 1996; Wilkens, 1970; Groma, 1998; Groma and Csikor, 2003). Therefore a procedure has been worked out, in which the diffraction pattern is simulated by using *ab-initio* physically well established functions for both, the size- and the strain-profile (Ribárik et al., 2001, Ungár et al., 2001; Ribárik et al., 2003). The size profile, I^S, is given by assuming either spherical or ellipsoidal crystallite shape and log-normal size distribution. In the case when spherical crystallite shape can be assumed, the function is defined by the two parameters of the size distribution function: the median, m, and the variance, σ. The strain function, I^D, is given by using the $f(\eta)$ Wilkens function, as defined in Eq. (3). The I^D function is defined by two parameters: the dislocation density, ρ and the dislocation arrangement parameter, M (for latter see above). This means, that in the simplest case, together with the q parameter in the contrast factors for cubic crystals, the numerical procedure is operating with five independent parameters: m, σ, ρ, M and q. Here we note that, due to the nature of the microstructure, this is the minimum number of parameters, which have to be used. The instrumental profile, I^i, has to be measured. The numerical procedure (called Convolution Multiple Whole Profile, CMWP, fitting procedure) produces the convolution of the three functions, I^S, I^D and I^i (Ribárik et al., 2003):

$$I^{SIM} = I^S * I^D * I^i, \tag{9}$$

The background, I^{BG}, is determined before the simulated I^{SIM} function is fitted to the measured diffraction pattern (Ribárik et al., 2003). A typical diffraction pattern and the fitted I^{SIM} function are shown in Fig. 3a for an electrodeposited, ultrafine-grained nickel specimen. For the purpose of better visualization the intensity is given in logarithmic scale. A good correlation between the measured, open circles, and the simulated, solid line, data can be observed. The microstructure parameters provided by the fitted pattern are: (i) the dislocation density, $\rho = 71 \times 10^{14}\ m^{-2}$, the dislocation arrangement parameter, $M = 3$, the q parameter in the contrast factor [see Eq. (4)], $q = 2$, the median and the variance of the crystallite size distribution function, $m = 10$ nm, and $\sigma = 0.68$. In a texture free loose powder of Ni the intensity ratios of the first seven Bragg peaks would be slightly different than in Fig. 3a. This effect is due to a weak texture, as can be seen in the TEM micrograph in Fig. 3c. A weak texture has only small effects on the dislocation structure; therefore, it does not influence its determination.

THE CORRELATION OF SIZE-DISTRIBUTION DETERMINED BY THE X-RAY METHOD AND TEM IN ULTRA-FINE GRAIN Ni

The size distribution function determined from the m and σ values provided by X-ray profile analysis (Ungár et al., 2001), corresponding to the same

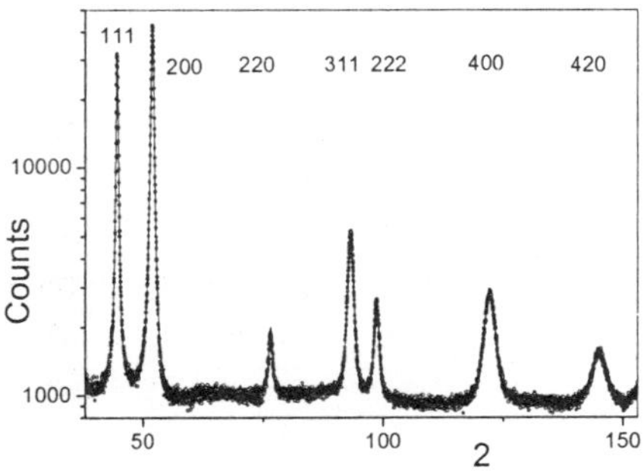

Fig. 3: (a) A typical diffraction pattern measured in a Philips X'pert diffractometer with secondary graphite monochromator (open circles) and the fitted, I^{SIM}, function [solid line, see also Eq. (9)] of an electrodeposited, ultra fine grain nickel specimen.

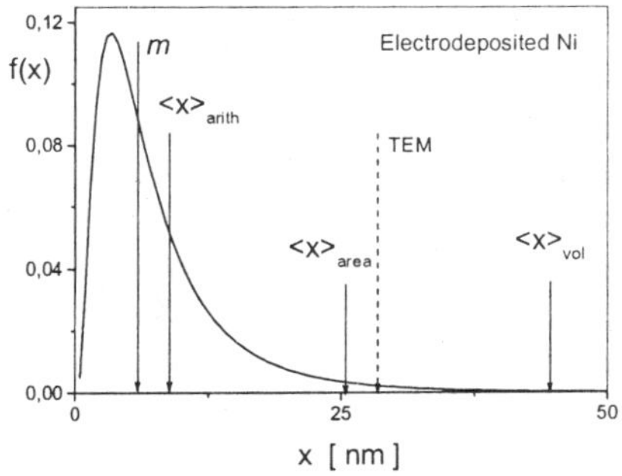

Fig. 3: (b) The size distribution function, $f(x)$, determined by the X-ray method. The mean values obtained by the different averaging procedures together with the average corresponding to the TEM micrograph in the next figure, are also indicated.

Fig. 3: (c) TEM micrograph of the same specimen (Zhilyaev et al., 2003).

electrodeposited Ni specimen as in Fig. 3a, is shown in Fig. 3b. The arithmetic, the area and the volume averaged means of the crystallite diameters, $<x>_{arith}$, $<x>_{area}$ and $<x>_{vol}$, as given elsewhere (Ungár et al., 2001; Hinds, 1982), are also indicated. The TEM micrograph of the same sample is shown in Fig. 3c (Zhilyaev et al., 2003). Since in a TEM micrograph the average size is related to area-weighting, the micrograph indicates that, in this case, the X-ray size distribution and the TEM grain size are in good correlation.

The size distribution function determined from X-ray data and the TEM micrograph of a nickel specimen deformed by high-pressure-torsion (Zhilyaev et al., 2003) are shown in Figs. 4a and 4b, respectively. This indicates that the TEM average size is considerably larger than the X-ray size values. A similar behaviour has been found earlier in a copper specimen deformed by

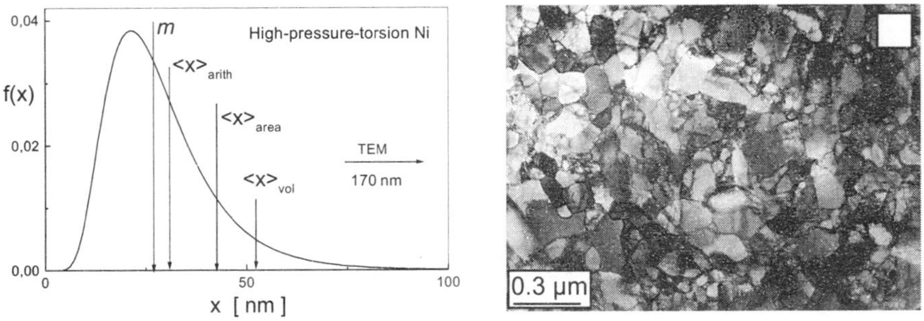

(a) (b)

Fig. 4: The size distribution function determined by the X-ray method (a) and the corresponding TEM micrograph (b) of a nickel specimen deformed by high-pressure-torsion (Zhilyaev et al., 2003).

ECAP (see Figs. 7 and 9 in Ungár et al., 2001). Both micrographs, the one here in Fig. 4b and Fig. 7 (in Ungár et al., 2001) reveal clearly, that due to tilts or twists within the specimen the contrast is changing strongly within the micrographs. In regions at good contrast conditions the fine details show small undisturbed regions, whereas in regions at poor contrast conditions, large regions reveal no boundary features. The two different appearances can be well observed in the upper central and left or the lower right parts of the micrograph in Fig. 4b.

As discussed above, the X-ray crystallite size is, in fact, the smallest undisturbed region in the crystal. The TEM evaluations, however, provide the well defined grains bounded by clearly visible boundaries. The two quantities and the corresponding size distribution functions may, however, need not be identical, as it is shown for the two different examples: (i) the electrodeposited and (ii) the high-pressure-torsion deformed Ni specimens in Figs. 3 and 4, respectively. Zehetbauer et al. (2003) have investigated recently the relation between the strength and the X-ray crystallite size (domain size) in copper specimens deformed by different methods. A good correlation has been found in the Hall-Petch relation between strength and the X-ray domain size. This indicates, to some extent, that the grain size determined from TEM micrographs may provide data in which size values corresponding to length scales of different physical meaning or relevance are mixed up. There are strong indications that the size of subgrains or dislocation cells determined from TEM micrographs have fractal character within certain length ranges (Hahner et al., 1998; Székely et al., 2000). Since a structure with fractal character has no well defined average dimension, the discrepancy between X-ray and TEM size determinations may also be due to this behaviour. A further aspect worth noting is that the X-ray size determination procedure, if carried out in a physically and mathematically correct way, in contrary to the evaluation of TEM micrographs, does not require the sometimes subjective collaboration of the experimentator.

THE STABILITY OF DIFFERENT DISLOCATION TYPES DURING ANNEALING IN FINE GRAIN Ti

A commercial purity Ti specimen was deformed by eight ECAP passes in the B mode (Zhu et al., 2003). The dislocation structure has been characterized in terms of crystallite size and size distribution, as well as dislocation densities and slip systems. The Burgers vector populations and slip system activities have been investigated on the basis of the experimentally determined dislocation contrast factors and the theoretical contrast factors defined in Eq. (7). A thin section of the specimen has been heated in a Perkin Elmer DSC2 scanning calorimeter to different temperatures and fast cooled to room temperature. The exercise was practiced up to 850 K. Between each annealing the X-ray investigations have been carried out and the experimental values of the q_1 and q_2 parameters, $q^{(m)}{}_1$ and $q^{(m)}{}_2$, were determined. The theoretical values of the q_1 and q_2 parameters corresponding to the eleven most likely slip systems according to Jones and Hutchinson (1981) and Kuzel and Klimanek (1989) are described elsewhere (Dragomir and Ungár, 2002). Based on the three different types of Burgers vectors the eleven slip systems can be classified into three groups: <*a*> type, where b_1=1/3<11$\bar{2}$0>; <*c*> type, where b_2=<0001> and <*c*+*a*> type, where b_3=1/3<11$\bar{2}$3>. The measured, $q^{(m)}{}_1$ and $q^{(m)}{}_2$ and theoretical, q_1 and q_2 parameter values can be matched by a numerical algorithm described in detail by Gubicza et al. (2003) and Máthis et al. (2003). Subsequently after the ECAP deformation the relative fractions of the <*a*>, <*c*> and <*c*+*a*> Burgers vectors are 62%, 0-4 % and 32-36 %, respectively, in good correlation with earlier X-ray (Ribárik et al., 2003) and TEM (Kuzel and Klimanek, 1989; Gubicza et al., 2003) results. The evolution of the Burgers vector population and the relative fraction of the three basic slip systems as a function of the annealing experiments is shown in Fig. 5. It can be seen that as the temperature increases the relative fraction of the <*c*+*a*>-type dislocations decreases close to between 0 and 4%, indicating that these dislocations disappear faster than the <*a*> or <*c*>-type ones. This may be related to the fact that the <*c*+*a*>-type dislocations have larger Burgers vectors than the other two types and therefore anneal out faster.

THERMAL STABILITY OF THE MICROSTRUCTURE IN UFG *Al*-Mg-Sc-Zr AND UFG-Ti

The aluminum base *Al*-5.9Mg-0.3Sc-0.18Zr alloy (at.%) was deformed by the method of HPT applying 1/2, 1, 3, 5, 10 and 15 rotations, respectively (Fátay et al., 2003). The crystallite size and the dislocation density were determined by the X-ray method. Three specimens deformed by 15 rotations were heated in a differential scanning calorimeter (DSC2, Perkin-Elmer) from 320 to 450, 520 and 800 K at a heating rate of 20 K/min. The heat flow corresponding to the specimen heated up to 800 K is shown in Fig. 6.

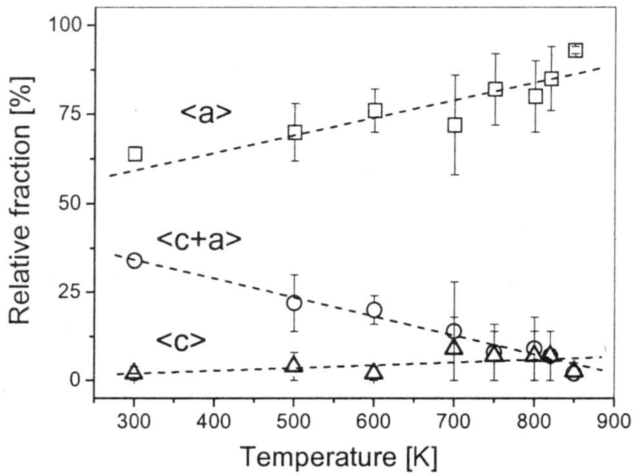

Fig. 5: Burgers vector populations and slip system activity as a function of annealing temperature in an ultra-fine-grain Ti specimen deformed by eight ECAP passes (Zhu et al., 2003).

The two specimens heated to 450 and 520 K, after reaching these temperatures were immediately cooled to room temperature by the fastest cooling rate available in the calorimeter. The volume averaged crystallite size, the dislocation density and the heat flow are shown in Fig. 6 as a function of temperature. It can be seen that at the temperature where the heat flow starts to increase (i) the major part of the dislocation density is annealed out, and (ii) however, the crystallite size is still at a low level. The crystallite size starts to increase considerably at the temperature where the heat flow has its maximum. This indicates that in the UFG alloy produced by HPT (a) the dislocation density is reduced by an order of magnitude before crystallite size starts coarsening and (b) the major part of heat flow is related to crystallite size coarsening, most probably to recrystallization, and not to dislocation annealing.

A commercial purity Ti specimen has been deformed by eight ECAP passes in the B mode (Zhu et al., 2003) (These are the same specimens in which the slip system activity has been discussed in a previous paragraph). A number of specimens were heated in a differential scanning calorimeter (DSC2, Perkin-Elmer) from 320 to 1000 K at a heating rate of 20 K/min. The heat flow corresponding to the specimen heated up to 1000 K is shown qualitatively in Fig. 7. The specimens heated up to different temperatures between 320 and 900 K, after reaching the selected temperature were immediately cooled to room temperature by the fastest rate available in the calorimeter. This indicates the same qualitative behaviour of the crystallite size, the dislocation density and the heat flow in Ti as in the case of the *Al*-Mg-Sc-Zr alloy discussed above. (a) The dislocation density is reduced by an order of magnitude before the crystallite size starts coarsening and (b) the

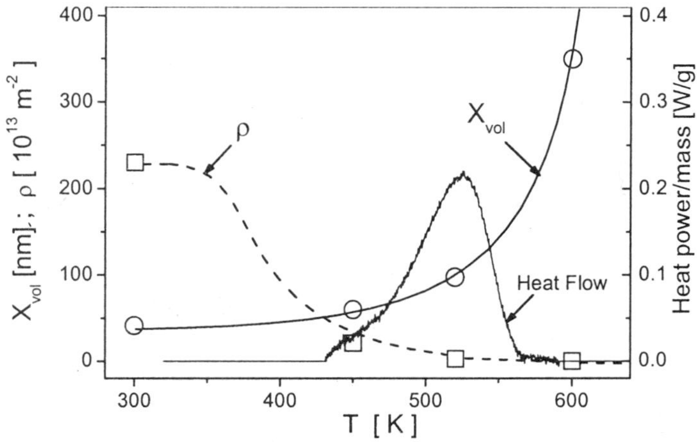

Fig. 6: The dislocation density, the area averaged mean crystallite size and the heat flow as a function of temperature in an *Al*-Mg-Sc-Zr alloy deformed by 15 rotations of high-pressure-torsion (Fátay et al., 2003).

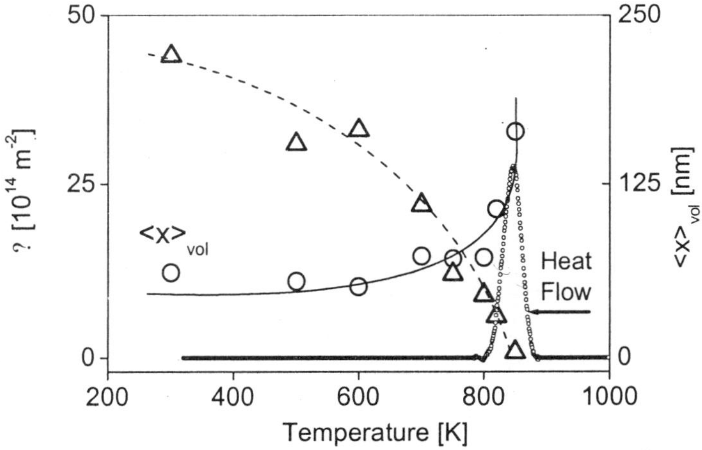

Fig. 7: The dislocation density, the volume averaged mean crystallite size and the heat flow as a function of temperature in an ultra-fine-grain commercial purity Ti specimen deformed by eight ECAP passes (Zhu et al., 2003).

major part of heat flow is related to crystallite size coarsening or recrystallization, and not to dislocation annealing.

CONCLUSIONS

Due to high resolution and high intensity diffractometers and progress in numerical facilities XRD peak profile analysis has become a competent supporting tool in parallel with TEM and SEM for the characterization of microstructure in crystalline materials, especially in ultra fine grain metallic materials. When crystallite size is small and the lattice is distorted by dislocations the X-ray method can describe the microstructure by five or six physically necessitated parameters in cubic or hexagonal crystals, respectively. The crystallite size or size distribution is given by the median and the variance of the size distribution function. These two parameters enable to determine all conceivable particle diameters weighted either by number or by area or volume. The area weighted mean crystallite size can directly be compared with size values obtained by TEM micrographs. The correlation between the X-ray and TEM determined size values depends on how well the TEM micrographs can be evaluated.

The X-ray crystallite size is always the average of the smallest undistorted volumes in the crystal. The dislocation structure is given by two parameters, the density, ρ, and the arrangement parameter, *M*, of dislocations. Latter is the measure of the average screening, or dipole character, prevailing in the dislocation structure. *M* is small or large if screening is strong or weak, or if the dipole character is strong or weak. The characterization of the dislocation structure by two parameters is in line with the tensor type of description of the dislocation density, e.g. given by Kroener. In addition to peak broadening, diffraction patterns reveal the effect of strain anisotropy. The evaluation can be done in terms of dislocation character, i.e. in terms of edge or screw character, especially in the case of cubic crystals. In the case of hexagonal crystals the relative fraction of the three fundamental dislocation types, <*a*>, <*c*> or <*c*+*a*> is possible. In a highly deformed commercial purity Ti specimen, it has been shown that deformation produces mainly <*a*> and <*c*+*a*> type dislocations. Annealing experiments of the same specimen have revealed, however, that the relative fraction of <*c*+*a*> type dislocations decreases, whereas that of <*c*> type dislocations increases with annealing temperature. This indicates that the <*a*> and <*c*> type dislocations are more stable versus temperature than the <*c*+*a*> type ones. Thermal stability experiments on ultra fine grain Al base alloys and the same Ti specimen have shown that the annealing of dislocations occurs at lower temperatures than where grain growth becomes substantial. At the same time, the major part of heat flow in a calorimeter experiment goes concomitant with grain growth.

Acknowledgement: This work was supported by the Hungarian Science Research Fund, OTKA, Grant Nos. T031786, T034666 and T046990.

REFERENCES

Arato, P., Bartha, L, Porat, R., Berger, S. and Rosen, A. (1998). *Nanostructured Mater.*, **10:** 245.

Bertaut, E.F. (1950). *Acta Cryst.*, **3:** 14.

Caglioti, G., Paoletti, A. and Ricci, F.P. (1958). *Nucl. Instrum.*, **3:** 223.

Chokshi, A.H., Rosen, A., Karch, J. and Gleiter, H. (1989). *Scripta Metall.*, **23:** 1679.

Dragomir, I.C. and Ungár, T. (2002). *J.Appl.Cryst.*, **35:** 556.

Fátay, D., Bastarash, E., Nyilas, K., Dobatkin, S., Gubicza, J. and Ungár, T. (2003). *Z.Metallkde.* **7:** 842.

Groma, I. (1998). *Phys. Rev.* B, **57:** 7535.

Groma, I. and Csikor, F. (2003). *J. Mech. Behav. Mater.*, submitted.

Gubicza, J. and Wéber, F. (1999). *Mat. Sci. Eng.* A, **263:** 101.

Gubicza, J., Arató, P., Wéber, F. and Juhász, A. (1999). *Mat. Sci. Eng.* A, **259:** 65.

Gubicza, J., Nam, N.H., Balogh, L., Hellmig, R.J., Stolyarov, V.V., Estrin, Y. and Ungár, T. (2003) *J.Met.Alloys*, In the press.

Hahner, P., Bay, K. and Zaiser, M. (1998). *Phys. Rev. Lett.*, **81:** 2470.

Hansen, N. (2001). *J. Mater. Sci. Technol.*, **17:** 409.

Hernando, A., Navarro, I. and Gorría, P. (1995). *Phys. Rev.* B, **51:** 3281.

Herzer, G. (1995). *Scripta Metall.*, **33:** 1741.

Hinds, W.C. (1982). Aerosol Technology: Properties, Behaviour and Measurement of Airbone Particles. Wiley, New York.

Hughes, D.A. and Hansen, N. (2001). *Phys. Rev. Lett.*, **87:** 135503.

Jones. I.P. and Hutchinson, W.B. (1981). *Acta Metall.*, **29:** 951.

Kaptás, D., Kiss, L.F., Balogh, J., Gubicza, J., Kemény, T. and Vincze, I. (2002). *Hyperfine Interactions*, **141/142:** 175.

Krivoglaz, M.A. (1996). Theory of X-ray and Thermal Neutron Scattering by Real Crystals. Springer-Verlag, Berlin.

Kroener, E. (1981). Physics of Defects, eds. Balian, R., Klemean, M. and Poirier, J.P., Amsterdam: North Holland, 215.

Kuzel, R. Jr. and Klimanek, P. (1989). *J. Appl. Cryst.*, **22:** 299.

Máthis, K., Nyilas, K., Axt, A., Dragomir, I.C., Ungár, T. and Lukáè, P. (2003). Unpublished work.

Pantea, C., Qian, J. and Zerda, T.W. (2002). *J. Appl. Phys.*, **91:** 1957.

Qian, J., Pantea, C., Voronin, G.A. and Zerda, T.W. (2001). *J. Appl. Phys.*, **90:** 1632.

Ribárik, G., Ungár, T. and Gubicza, J. (2001). *J. Appl. Cryst.*, **34:** 669.

Ribárik, G., Ungár, T. and Gubicza, J. (2003). *Mat. Sci. Eng.* A, submitted.

Rietveld, H.M. (1967). *Acta Cryst.*, **22:** 151.

Rietveld, H. M. (1969), *J. Appl. Cryst.*, **2:** 65.

Siegel, R.W. (1994). *J. Phys. Chem. Solids*, **55:** 1097.

Székel, F., Groma, I. and Lendvai, J. (2000). *Phys. Rev.* B, **62:** 3093.

Szépvölgyi, J., Mohai, I. and Gubicza, J. (2001). *J. Mater. Chem.*, **11:** 859.

Treacy, M.M.J., Newsam, J.M. and Deem, M.W. (1991). *Proc. Roy. Soc. London* A, 433.

Trinkaus, H. (1972). *Phys. Stat. Sol.* (b), **51:** 307.

Ungár, T. and Borbély, A. (1996). *Appl. Phys. Lett.*, **69:** 3173.

Ungár, T. and Tichy, G. (1999). *Phys.Stat.Sol.*(a). **171:** 425.

Ungár, T., Dragomir, I., Révész, Á. and Borbély, A. (1999). *J. Appl. Cryst.*, **32:** 992.

Ungár, T., Gubicza, J., Ribárik, G. and Borbély, A. (2001). *J. Appl. Cryst.*, **34:** 298.

Warren, B.E. (1959). *Progr. Metal Phys*, **8:** 147.

Weertman, J.R. (1993). *Mater. Sci. Eng.* A, **166:** 161.

Wilkens, M. (1970). *Phys. Stat. Sol.* (a), **2:** 359.

Wilkens, M. (1970). Fundamental Aspects of Dislocation Theory, eds. Simmons, J.A., de Wit, R. andBullough, R., Vol. II. Nat. Bur. Stand. (US) Spec. Publ. No. 317, Washington DC, USA; 1195.

Williamson, G. K. and Hall, W. H. (1953). *Acta Metall.*, **1:** 22.

Zehetbauer, M., Stuewe, H. P., Vorhaber, A., Schafler, E. and Kohout, J. (2003). *Adv. Eng. Mater.*, **5:** 330.

Zhilyaev, A.P., Gubicza, J., Nurislamova, G., Révész, Á., Suriñach, S., Baró, M.D. and Ungár, T. (2003). *Phys. Stat. Sol.* (a), **171:** 425.

Zhilyaev, A.P., Gubicza, J., Suriñach, S., Baró, M.D. and Langdon, T.G. (2003). *Mat. Sci. Forum*, **426:** 4507.

Zhu, Y.T., Huang, J.Y., Gubicza, J., Ungár, T., Wang, Y.M., Ma, E. and Valiev, R.Z. (2003). *J. Mater.Res.*, **18:** 1908.

Diffraction Effects on Antiphase Boundaries in Ordered Structures: An Overview

P. Ghosal

Defence Metallurgical Research Laboratory
Kanchanbagh, Hyderabad 500 078, India

INTRODUCTION

An antiphase domain is a region of an ordered crystal where the pattern of occupation of sublattice sites by various atomic species is the same. Waves diffracted from different domains usually are out of phase, hence the name antiphase domains. A region of contact between adjacent domains is a surface called on antkphase boundary (APB). These boundaries have a tendency to lie along certain crystallographic planes so as to minimize total domain boundary energy. Preference of one type of boundary over another type influences effective domain size normal to boundary planes. Antiphase domain boundaries can result not only from the growing together of different nuclei of order, but also from the passage of dislocations through an ordered crystal.

A perfect dislocation having a unit Burgers vector in the disordered alloy is only a partial dislocation in the ordered alloy, since the unit cell for the ordered state is larger than that for the disordered one. A movement of such a dislocation across a slip plane in a superlattic, therefore, leaves a plane behind it that is an antiphase boundary. In order to reduce the extra energy involved in producing this boundary, dislocations in a superlattic tend to travel in pairs, with only a narrow ribbon of APB. The passage of the second dislocation on the pair restores the ordered structure. A broadening of line profiles in powder X-ray diffraction (XRD) patterns and spots in selected area electron diffraction patterns results owing to the occurrence of such APBs if the antiphase domain size is sufficiently small. APBs can be characterized by their plane and shift vector which relates similarly occupied sites in adjacent domains across a boundary.

One of the methods for characterizing APBs is their direct observation in a transmission electron microscope. Information only from a localized area is found at a time by this technique and thus scanning of different regions and/or samples is required to gather information for the alloy as a whole. Another method of characterizing the APBs is through diffraction studies, where the information obtainable is less localized. Although an indirect technique, the X-ray method possesses the chief merits of simplicity in experiment, rapidity and reproducibility besides being capable of dealing with bulk solids at high levels of dislocation density. For utilizing diffraction methods for characterization of APBs, one needs to know the diffraction effects arising from the presence of different types of APBs in a given ordered structure. Before proceeding to discuss the diffraction effects, we will discuss the ordered structures based on the simple metallic structures, namely fcc (A1), bcc (A2) and cph (A3).

Detailed reviews of work in this area have appeared from time to time (Wilson, 1943; Wilson and Zsoldos, 1966; Warren, 1963, 1969; Mikkola and Cohen, 1966; Rothman and Cohen, 1971; Anantharaman et al., 1972; Rothman et al., 1970; Cohen and Bever, 1960; Yamaguchi and Umakoshi, 1990). Diffraction effects of faults and antiphase boundaries in fcc and bcc based structures have been investigated extensively. However, similar work on cph based structures is more recent and will thus receive more attention.

SUPERLATTICE STRUCTURES AND THEIR PLANAR FAULTS

Superstructures Based on the fcc Lattice

$L1_2$ Structure

The atomic arrangement in the $L1_2$ structure with an A_3B stoichiometry is shown in Fig. 1(a). The unit cell shown in Fig. 1 has B atoms at 000 positions and A atoms at 1/2 1/2 0 positions and consists of four interpenetrating simple cubic sublattices, each occupied by atoms of only one kind.

Typical examples are Cu_3Au, Pt_3Sn, Zr_3Al, Ni_3Fe etc. Slip in the intermetallic $L1_2$ superstructure compounds occurs on {111} and {001} planes. Three possible planar faults on {111} have been considered; antiphase boundary (APB), superlattice intrinsic stacking fault (SISF) and complex stacking fault (CSF). The shift vectors of APB, SISF and CSF are ½ $[\bar{1}10]$, 1/3 $[\bar{1}2\bar{1}]$ and 1/6 $[\bar{2}11]$ respectively. In order to investigate the stability of these planar faults, surface energy calculation along with crystal symmetry considerations are taken into account (Yamaguchi et al., 1981). Details are given in Table 1.

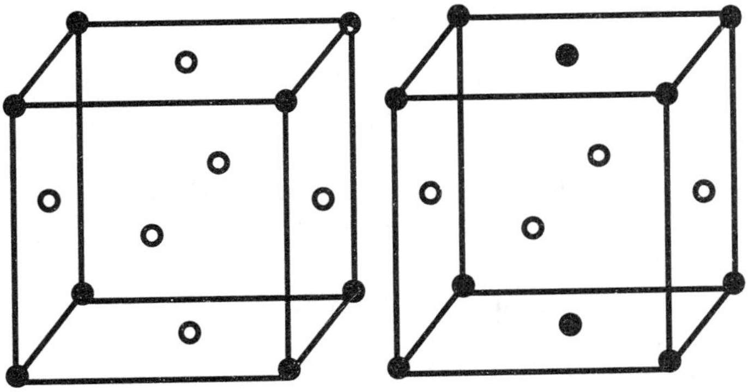

(a) $L1_2$ (A_3 B) Superstructure (b) $L1_0$ (A B) Superstructure

Fig. 1: (a) $L1_2$ (A_3 B) Superstructure and (b) $L1_0$ (A B) Superstructure where O represents A type atoms and ● represents B type atoms.

Table 1: Stability and fault vectors of possible planar faults in the $L1_2$ structure

Fault	*Fault plane*	*Fault vector*	*Stability*
APB	(111)	$1/2\ [\bar{1}10] + f'$; f' □$[\bar{1}\bar{1}2]$	Unstable
CSF	(111)	$1/6\ [\bar{1}10] + f'$; f' □$[\bar{1}\bar{1}2]$	Unstable
CISF	(111)	$1/3\ [\bar{1}2\bar{1}]$	Stable
APB	(001)	$1/2\ [\bar{1}10]$	Stable

The magnitude of f′ may vary from material to material, but its direction is shown in the table.

$L1_0$ Structure

The unit cell of $L1_0$ structure is shown in Fig. 1(b) having AB stoichiometry. Typical examples of this type of structure are TiAl, HgPt, AuCu etc. As in the case of $L1_2$ structure, three possible planar faults exist on {111} and {100} planes, i.e. APB, CSF and SISF. Stability is discussed by Yamaguchi and Umakoshi (1990) for the above faults which are summarized in the Table 2.

Table 2: Stability and fault vectors of possible planar faults in $L1_0$ structure

Fault	*Fault plane*	*Fault vector*	*Stability*
APB	(111)	$1/2[\bar{1}01]$	Unstable
CSF	(111)	$1/6[\bar{2}11]$	Unstable
CISF	(111)	$1/3[\bar{1}\bar{1}2]$	Stable
APB	(001)	1/2[011]	Stable

Superstructures Based on the bcc Lattice

B2 Structure

This type of superstructure is very frequently encountered in alloy systems due to the ordering of the beta-brass type structure. The unit cell can be represented in terms of two interpenetrating simple cubic sublattices as shown in Fig. 2(a). For AB stoichiometry, A type atoms and B type atoms segregate to cube corners and centres, respectively. Some typical examples are β′-CuZn, β′-AuCd, β-NiZn etc. No SISF are believed to exist in these structures. Only one type of APB characterized by a shift of 1/2<111> can exist on its possible slip planes of {110} and {112}. On the basis of crystal symmetry and restoring force considerations, Yamaguchi et al. (1981) have shown that 1/2<111> APB is always stable as shown in Table 3.

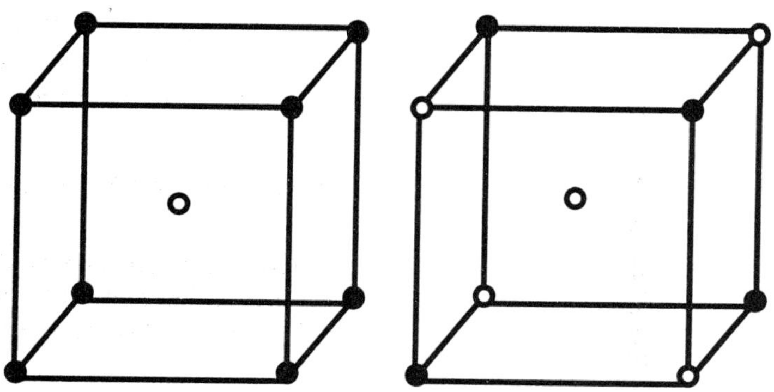

(a) B2 (AB) Superstructure (b) $D0_3$ (A_3B) Superstructure

Fig. 2: (a) B2 (AB) Superstructure and (b) $D0_3$ (A_3B) Superstructure where O represents A type atoms and ● represents B type atoms.

Table 3: Stability and fault vectors of possible planar faults in B2 structure

Fault	*Fault plane*	*Fault vector*	*Stability*
APB	{110}	1/2<111>	Stable
APB	{110}	<111>(=<001>)	Stable
APB	{112}	1/2<111>	Stable
APB	{112}	<111>	Stable

$D0_3$ Structure

The unit cell of the $D0_3$ superstructure, which is also based on the bcc lattice and has A_3B stoichiometry, is shown in Fig. 2(b). The typical examples of this type of superstructure are Fe_3Si, Fe_3Al, Mg_3La etc. The characteristic feature of the $D0_3$ structure is that each minority atom is surrounded by the maximum number of unlike nearest and second-nearest neighbour atoms.

Two types of APBs can be formed on its possible slip planes of {110} and {112} by shifts of 1/2<111> and <111>. The <111> type shift vector is crystallographically equivalent to that of the type <001>. On the basis of restoring force and crystal symmetry considerations, the 1/2<111> APB is either unstable or stable with displacement vector 1/2<111>+f′. The magnitude of f′ may vary from material to material, but its direction must be along <001> for the APB on {110} and <111> for that on {112}. The <111> APB, which is equivalent to the <001> APB, is always stable as shown in Table 4.

Table 4: Stability and fault vectors of possible planar faults in $D0_3$ structure

Fault	*Fault plane*	*Fault vector*	*Stability*
APB	{110}	1/2<111>	Unstable/Stable
APB	{110}	<111>(=<001>)	Stable
APB	{112}	1/2<111>	Unstable/Stable
APB	{112}	<111>	Stable

Superstructures Based on the cph Lattice

$D0_{19}$ and B_{19} Structures

A structure closely related to the better known $L1_2$ type and also having A_3B stoichiometry is the $D0_{19}$ superlattice with the unit cell shown in Fig. 3. However, unlike the $L1_2$ type, each of the four sublattices in the $D0_{19}$ type is close packed hexagonal with the *a* spacing corresponding to twice the spacing of the disordered structure and the *c* spacing unchanged. Of the four sublattice sites, one is occupied by B atoms, resulting in the general formula A_3B. Alloys exhibiting this type of ordering are Al_3Th, Cd_3Mg, Mg_3Cd, Pt_3U, Ti_3Al, Ti_3Sn etc. When two of the sublattices are occupied by B atoms and the remaining two by A atoms, we get the B19 superstructure with AB stoichiometry. There is also a loss of symmetry leading to an orthorhombic cell which can be formed by a small distortion of the hexagonal cell. Typical example of this structure is MgCd.

There are basically two types of APBs in the $D0_{19}$ structure. When a crystal orders, it starts ordering at a large number of positions throughout the crystal. When these ordered domains grow and meet each other, the boundary created between the domains are thermal or climb APBs. APBs can also arise when an ordered material is deformed. Deformation will take place on a particular slip plane by slip through the Burgers vector of the gliding dislocation. If this vector is half the lattice translation vector of the ordered cell, then an APB will be produced on the plane which is designated mechanical or shear APB.

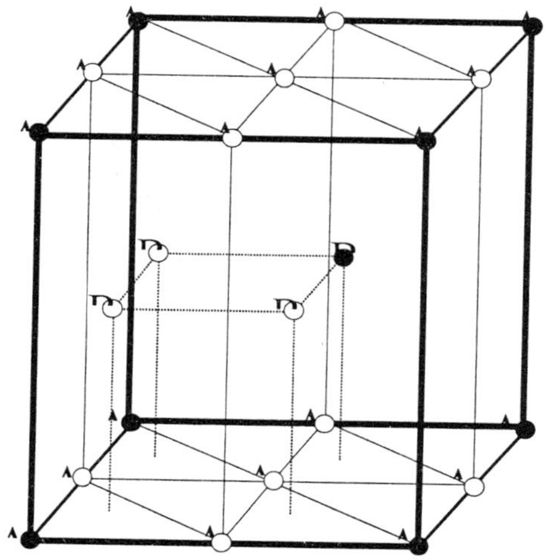

Fig. 3: $D0_{19}$ (A_3B) ordered cph structure where O represents A type atoms and ● represents B type atoms. Layer type with sublattice sites (superscript) are also indicated.

We now consider the various types of boundary and their stability in the $D0_{19}$ superstructure. We begin by considering APBs parallel to seven types of planes which contain atomic sites and have the largest interplanar spacing (do). Such planes are more clearly packed and formation of APBs parallel to these planes may be energetically favourable (Ghosal et al., 1993). Two types of antiphase vectors, namely (1/6) $<11\bar{2}0>$ and (1/2) $<10\bar{1}0>$ are considered. Bounding partial dislocations for both these cases have been shown to be stable for most situations by Umakoshi and Yamaguchi (1981). We will discuss this aspect for various planes in the subsequent discussion. The possible APBs have been found by taking suitable combinations of an APB plane and a bounding partial dislocation. Table 5 lists the 15 different types of APBs which can occur. The boundaries are designated shear or climb type respectively depending on the bounding partial dislocations lying in or out of the plane of the boundary (Anantharaman et al., 1972).

The stability of possible planar faults in $D0_{19}$ ordered structures has been investigated (Umakoshi and Yamaguchi, 1981) using crystallographic symmetry considerations and by calculating the energy of faults by counting the number of atomic bonds across the fault plane, and up to eight nearest neighbour interactions are taken into account. Restoring force, which is the force by which the crystal below the fault acts on the crystal above, is also considered to find the stability. In general, the planar fault is formed by cutting the crystal along a crystallographic plane and displacing one half of the crystal with respect to the other by a shift vector f. Vitek (1968 and 1974) has suggested that in order to investigate the stability of such a planar fault, the γ–surface in the vicinity of the vector f has to be calculated to find whether a minimum corresponding to the vector f exists in the surface.

Table 5: Types of antiphase boundaries in DO_{19} structure

Plane of the boundary	d_0/a_0	*Antiphase vector*	*Type*
(0002)	0.4082	$(1/6)[11\bar{2}0]$	Shear
$(20\bar{2}0)$	0.4330	$(1/6)[1\bar{2}10]$	Shear
		$(1/6)[11\bar{2}0]$	Climb
$(20\bar{2}1)$	0.3825	$(1/6)[1\bar{2}10]$	Shear
		$(1/6)[\bar{2}110]$	Climb
$(20\bar{2}2)$	0.2970	$(1/6)[1\bar{2}10]$	Shear
		$(1/6)[\bar{2}110]$	Climb
$(2\bar{4}20)$	0.2500	$(1/2)[10\bar{1}0]$	Shear
		$(1/6)[\bar{2}110]$	Climb
		$(1/6)[1\bar{2}10]$	Climb
$(2\bar{4}22)$	0.2132	$(1/6)[\bar{2}110]$	Climb
		$(1/6)[1\bar{2}10]$	Climb
$(2\bar{4}24)$	0.1581	$(1/2)[10\bar{1}0]$	Shear
		$(1/6)[\bar{2}110]$	Climb
		$(1/6)[1\bar{2}10]$	Climb

In ordered $D0_{19}$ structure, it has been reported that slip can occur on various planes such as (0001), $\{10\bar{1}0\}$, $\{11\bar{2}0\}$, $\{10\bar{1}1\}$, $\{10\bar{1}2\}$, $\{11\bar{2}2\}$ and $\{11\bar{2}4\}$. Umakoshi and Yamaguchi (1981) have considered the possible planar faults on these planes. On basal plane, apart from APB of fault vector $(1/6)[1\bar{2}10]$ they also have considered the complex stacking fault (CSF) with fault vector $(1/6)[01\bar{1}0]$ and superlattice intrinsic stacking faults (SISF) with fault vector $(1/3)[\bar{1}100]$. On prismatic planes $(10\bar{1}0)$, there are eight layers when viewed normal to the plane. If the two adjacent $(10\bar{1}0)$ layers occur in a unit of narrow-spacing are regarded as a single corrugated layer, then four corrugated $(10\bar{1}0)$ layers occur in a unit cell. When the crystal is cut between any two adjacent corrugated layers, APB-I can form while APB-II is formed otherwise. Their work is summarized in Table 6.

PROFILE PARAMETERS

In applications of line profile analysis, it is usual to work with certain parameters, which can be found from the line profile. To typify the profile the following parameters are usually measured: (i) integrated intensity, (ii) integral breadth and (iii) measure of asymmetry.

Integrated Intensity

If all the diffracted beams sent out by crystal as it rotates through the Bragg angle are received in a counter, the total energy of the diffracted beam can be measured and is known as the integrated intensity. An analytical expression

Table 6: Summary of the stability of APBs and other faults in $D0_{19}$ structure

Fault Plane A3 basis	*Fault Vector (R) A3 basis*	*Fault Plane $D0_{19}$ basis*	*Fault Vector $D0_{19}$ basis*	*Type of Fault*	*Remarks/ Conclusions*
(0001)	$(1/3)[\bar{1}2\bar{1}0]$ $(1/3)[01\bar{1}0]$ $(2/3)[\bar{1}100]$	(0001)	$(1/6)[\bar{1}2\bar{1}0]$ $(1/6)[01\bar{1}0]$ $(1/3)[\bar{1}100]$	APB CSF SISF	Stable Stable Always stable
$(10\bar{1}0)$	$(1/3)[\bar{1}2\bar{1}0]$	$(\bar{2}0\bar{2}0)$	$(1/6)[\bar{1}2\bar{1}0]$	APB	Always stable
$(\bar{1}2\bar{1}0)$	$[\bar{1}010]$	$(\bar{2}4\bar{2}0)$	$(1/2)[\bar{1}010]$	APB	Expected to be stable; unlikely to exist because of large magnitude of **R**
$(\bar{1}0\bar{1}1)$	$(1/3)[\bar{1}2\bar{1}0]$ $(1/3)[\bar{2}113]$	$(20\bar{2}1)$	$(1/6)[\bar{1}2\bar{1}0]$ $(1/6)[\bar{2}116]$	APB APB	May be unstable Can exist in hcp; in $D0_{19}$ structure γ-surface for these planes needs to be calculated
$(10\bar{1}2)$	$(1/3)[\bar{1}2\bar{1}0]$ $(1/3)[\bar{4}223]$	$(10\bar{1}1)$	$(1/6)[\bar{1}2\bar{1}0]$ $(1/3)[\bar{2}113]$	APB No fault	May be unstable Lattice translation vector; does not produce any fault
$(11\bar{2}2)$	$[\bar{1}100]$ $(1/3)[\bar{1}\bar{1}23]$	$(11\bar{2}2)$	$(1/2)[\bar{1}100]$ $(1/6)[11\bar{2}6]$	APB APB	Can exist in hcp; in $D0_{19}$ structure γ-surface for these planes needs to be calculated
$(11\bar{2}4)$	$[\bar{1}100]$ $(1/3)[\bar{4}223]$	$(11\bar{2}2)$	$(1/2)[\bar{1}100]$ $(1/3)[\bar{2}113]$	APB No fault	May be unstable Lattice translation vector; does not produce any fault

for the integrated intensity, T, of a reflection can be obtained by integrating the intensity over the full period of the reflection, that is

$$T \equiv \int I(h)\, dh$$
Over one period

And is also equal to the area under the intensity $I(h)$ vs h curve.

Integral Breadth

Various definitions for the breadth of a distribution are in use. A definition of breadth best adapted to calculation is the integral breadth. The integral breadth of a reflection in reciprocal space is defined by

$$\beta = \frac{\int I(h)\, dh}{I_{max}} = \frac{\text{Integrated Intensity}}{\text{Peak Intensity}}$$

Measure of Peak Asymmetry

The peak asymmetry can be measured for each peak by finding out the difference between the position of mode and mean of the distribution curve. The mode is the value of h for which the intensity $I(h)$ is maximum

$$\text{Mode} = (h)_{max} = \text{Value of } h \text{ at } I(h)_{max}$$

The mean or the centroid of the intensity distribution can be expressed as

$$< h > = \frac{\int hI(h)\, dh}{\int I(h)\, dh}$$

Thus, a measure of the peak asymmetry can be defined as the shift of centroid from the mode of any reflection,

$$\Delta h = <h> - (h)_{max}$$

The profile maximum position is the position in h at which $I(h)$ is maximum.

DIFFRACTION THEORY

The first mathematical formulation of the theory of diffraction from antiphase boundaries (APBs) was carried out by Wilson (1943) (see also Wilson and Zsoldos, 1966) in ordered fcc structure of the $L1_2$ type. Rothman et al. (1970) worked out the theory of XRD from stacking faults and APBs in the B2 type bcc structure. We have recently studied (Ghosal, 1996) the diffraction effects in $D0_{19}$ type ordered cph structure. The formulation of the diffraction theory will differ from structure to structure although the basic principle is the same. We give the formulation for ordered cph ($D0_{19}$ type) structure as an illustration and then mention results for the more important structures.

Diffracted Intensity

In terms of the conventional hexagonal basis vectors a_1, a_2, a_3 (a_1 and a_2 being double the lattice vectors of the disordered lattice and a_3 being equal to that for the disordered structure), the position vector R_m of a unit cell at the $(m_1 m_2)$ position in layer m_3 of a possibly faulted $D0_{19}$ crystal is given by

$$R_m = m_1 a_1 + m_2 a_2 + \tfrac{1}{2}\, m_3 a_3 + Q_{m3}$$

where Q_{m3} is the displacement vector in the plane of the layer. This vector takes different values in the presence or absence of an APB or CSF parallel to the basal plane. Following Warren (1969) and using his notation the diffracted intensity from such a crystal is given by

$$I = I_e \sum_{m=-\infty}^{\infty} F_m \exp\left[\frac{2\pi i}{\lambda}(s - s_0)\cdot R_m\right]$$

$$\times \sum_{m^1=-\infty}^{\infty} F^*_{m'} \exp\left[-\frac{2\pi i}{\lambda}(s - s_0)\cdot R_{m'}\right]$$

or
$$I = I_e\, |F|^2 \sum_{m=-\infty}^{\infty} \sum_{m^1=-\infty}^{\infty} \exp\left[-\frac{2\pi i}{\lambda}(s - s_0)\cdot(R_m - R_m)\right]$$

where s and s_0 are unit vectors in the directions of the diffracted and incident beams. F is the structure factor of a two dimensional unit cell in the basal plane of the $D0_{19}$ structure.

Substituting the values of R_m and R_m', performing the summations over m_1, m_1', m_2, m_2' and simplifying, we get (Warren, 1969)

$$I = \psi^2 \sum_{m3}\sum_{m'3} \exp\left\{\left[\frac{2\pi i(s - s_0)}{\lambda}\right]\cdot\left[\frac{1}{2}(m_3 - m_3')a_3 + (Q_{m3} - Q_{m'3})\right]\right\}$$

We have used the abbreviation

$$\psi^2 = I_e |F|^2 \frac{\sin^2 \frac{\pi}{\lambda}\ (s - s_0).N_1 a_1\ \sin^2 \frac{\pi}{\lambda}\ (s - s_0).N_2 a_2}{\sin^2 \frac{\pi}{\lambda}\ (s - s_0).a_1\ \sin^2 \frac{\pi}{\lambda}\ (s - s_0).a_2}$$

where N_1 and N_2 are the number of cells along the directions a_1 and a_2.

Introducing $m = m_3 - m_3'$,

$$Q_m = Q_{m3} - Q_{m'3}, \text{ and}$$

expressing diffraction vector in terms of the vectors $b_1 b_2 b_3$ reciprocal to $h_1 h_2 h_3$, we can write

$$\frac{s - s_0}{\lambda} = h_1 b_1 + h_2 b_2 + h_3 b_3\ s - s_0$$

Making the above substitutions and simplifying, we get,

$$I(h_3) = \psi^2 \sum_{m_3} \sum_{m_3'} \exp\left[2\pi i\left(h_1 b_1 + h_2 b_2 + h_3 b_3\right).Q_m\right]$$

$$\exp\left[\pi i\left(h_1 b_1 + h_2 b_2 + h_3 b_3\right).ma_3\right]$$

We note that ψ^2 and therefore $I(h_3)$ has nonvanishing values only for integer values $h_3 k$ of h_1, h_2. Let the phase difference between rays diffracted from mth layer and origin layer be denoted by ϕ_m. then

$$\phi_m = 2\pi(h_1 b_1 + h_2 b_2 + h_3 b_3).Q_m$$

Now, replacing the double sum with a single sum over m and inserting ϕ_m in the intensity equating, we obtain

$$I(h_3) = \psi^2 \sum_m N_m \left\langle \exp[i\phi_m] \right\rangle \exp(\pi imh_3)$$

where N_m is the number of pairs of layers at an interlayer separation and carets represent an average. Assuming the effects of planar defects under consideration as dominant; variation of N_m with m can be neglected. Thus we can write

$$I(h_3) = N\psi^2 \sum_m N_m \left\langle \exp[i\phi_m] \right\rangle \exp(\pi imh_3)$$

$$= N\psi^2 \sum_m J_m \exp(\pi imh_3)$$

It is evident that the problem of determining the diffracted intensity reduces to finding the values of

$$J_m = \langle \exp\ [i\phi_m] \rangle$$

To evaluate $\langle \exp\ [i\phi_m] \rangle$, we use the following technique

$$\left\langle \exp[i\phi_m] \right\rangle = \left\langle \exp[i\phi_{m-1}] \exp[i(\phi_m - \phi_{m-1})] \right\rangle$$

$$= \left\langle \exp[i\phi_{m-1}] \right\rangle \left\langle \exp[i(\phi_m - \phi_{m-1})] \right\rangle$$

The last result follows from the fact that for independent variables, the average of products is equal to the product of averages. Denoting $(\phi_m - \phi_{m-1})$ by $\delta\phi_m - \phi_{m-1}$ and the corresponding probability of the occurrence of such a phase difference by $P(m, m-1)$, we have $\left\langle \exp[i\phi_m] \right\rangle = \sum \left\langle \exp[i\phi_{m-1}] \right\rangle$ $P(m, m-1-\exp[i\delta\phi_{m,m-1}]$. This leads to a recurrence relation which can be solved by using initial conditions of the problem. As an example, let us consider the simple case of APBs on basal plane of ordered cph structured.

APBs on Basal Planes of $D0_{19}$ Structure

APBs with $(1/6)[11\bar{2}0]$ or $(1/2)[\bar{1}100]$ shift vectors cause a change in occupation of B atoms from site IV to I and vice versa (Fig. 3). Since these two shift vectors lead to identical changes in site occupation by B type atoms, they will lead to indistinguishable effects. We shall specifically treat APBs with (1/6)<11$\bar{2}$0> shift vectors with the understanding that the result for APBs with (1/2)<$\bar{1}$100> type shift vectors will be identical. We can construct probability trees which give values of $P(m, m-1)$ as well as $\delta\phi_{m, m01}$. Consider an A type layer with B atoms on site IV to be present on the $(m-1)$ layer. In the absence of an APB, the next layer will be B type with B atoms again on site IV. The displacement vector for this case is (1/3)[$\bar{1}$010] ≡ (1/3)[$\bar{2}$ $\bar{1}$0] and the corresponding phase difference is thus (2π/3)(–2h–k) ≡ (2π/3)(h–k) since h is an integer. Next consider an APB with shift vector (1/6)[11$\bar{2}$0] ≡ (1/2)[110]; the next layer is again of B type, however, with B atoms occupying type I site. The displacement vector for this case can be obtained by adding the displacement vector in the absence of an APB and the shift vector of the APB. Thus the displacement vector for this APB is (1/3)[$\bar{1}$010] + (1/6)[11$\bar{2}$0] = (1/6)[$\bar{1}$100] ≡ (1/6)[$\bar{1}$10]. The corresponding phase difference is thus (2π/6)(–h+k). Further, we denote the probability of occurrence of this PAB by α. APBs with shift vectors (1/6)[$\bar{2}$110] and (1/6)[1$\bar{2}$10] can also occur with probability α. The corresponding phase difference can be found as above. Finally, the probability for the absence of any APB is clearly (1–3α). All these possibilities are shown in the probability tree in Fig. 4. The same figure also incorporates a probability tree for the case when a B type layer occurs in m–2 position. This part of the tree has been constructed using arguments similar to those given above. Similar trees can be drawn considering B type layer with B atoms on site IV to be present on the (m–1) layer by cyclic permutation. Using these trees, as well as similar ones which are not shown, one can construct a slightly different type of tree which shows all possible ways of reaching a particular type of layer. From these probability trees, we can write down the following set of difference equations terminating with an A layer (superscript 0) and a B layer (superscript 1) respectively. By substituting and simplifying we can write for the above type of APB on this plane for the said structure.

$$J^1_{\mathrm{m}} = (1-3\alpha)\exp\left[\frac{2\pi i}{3}(h-k)\right] j^0_{\mathrm{m}-1} + \alpha\exp\left[\frac{2\pi i}{3}(-h+k)\right] j^0_{\mathrm{m}-1}$$

$$+\ \alpha\exp\left[\frac{2\pi i}{3}(2h+k)\right] j^0_{\mathrm{m}-1} + \alpha\exp\left[\frac{2\pi i}{3}(-h-2k)\right] j^0_{\mathrm{m}-1}$$

On further simplification, we get

$$J^1_{\mathrm{m}} = [(1-3\alpha)^2 + \alpha^2\,\{1+(-1)^{\mathrm{h}} + (-1)^{\mathrm{k}}\}^2 + 2\alpha(1-3\alpha)\{1+(-1)^{\mathrm{h}} + (-1)^{\mathrm{k}}\}\ \cos\pi(h-k)\ J^1_{\mathrm{m}-2}$$

Thus the general recurrence relation can be written as $J_{\mathrm{m}} = -b\,J_{\mathrm{m}-1} - c\,J_{\mathrm{m}-2}$ which is more commonly known as a characteristic equation. In the above equation b and c are functions of $P\,(m, m-1)$ and thus of α.

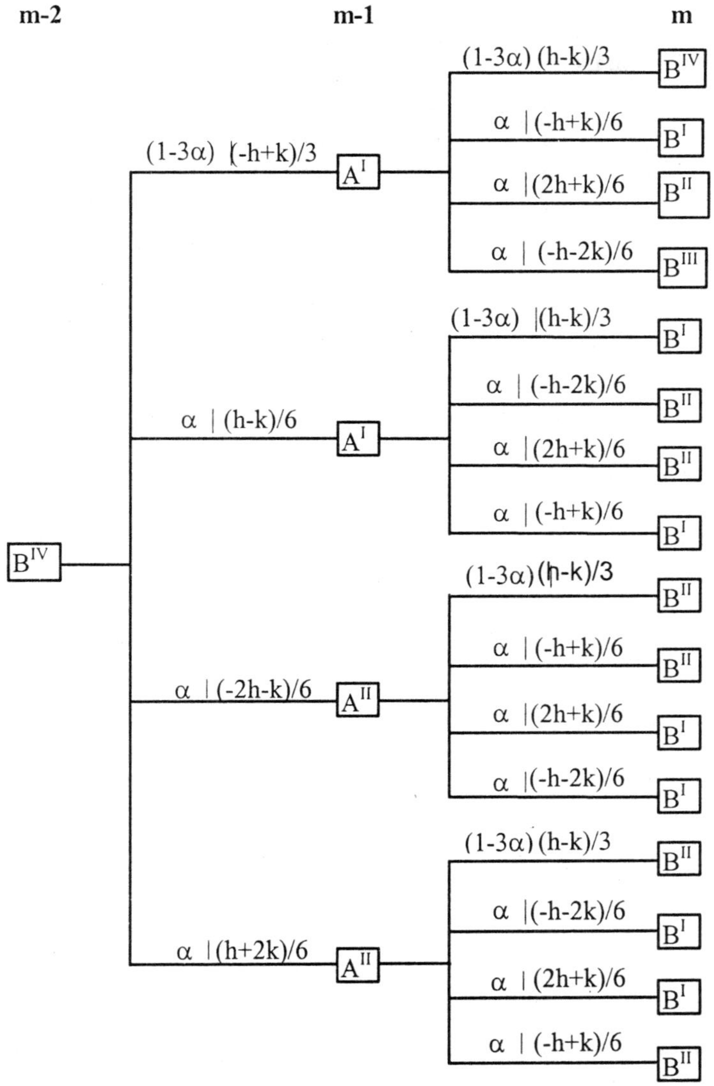

Fig. 4: Probability tree with m, $(m-1)$ and $(m-2)$ layers showing the probability of occurrence of APB and the corresponding phase difference separated by a vertical line.

Let, $J_m = a\,\rho^m$ be a solution of the characteristic equation, then $\rho^2 + b\rho + c = 0$ which has roots $\rho_1 = -b + \sqrt{(b^2 - 4c)}$ and $\rho_2 = -b - \sqrt{(b^2 - 4c)}$ and in general, J_m can be expressed as

$$J_m = a_1\rho_1{}^m + a_2\rho^m{}_2$$

To find a_1 and a_2, we use boundary conditions. For $m = 0$ and $m = 1$, we have

$$J_0 = a_1 + a_2$$

$$J_1 = a_1\rho_1 + a_2\rho_2$$

or,
$$a_1 = (J_1 - J_0\rho_2)/(\rho_1-\rho_2)$$
$$a_2 = (J_1 - J_0\rho_1)/(\rho_2 - \rho_1)$$

Substituting the values of a_1, a_2 in the expression for J_m we obtain,

$$J_m = \frac{J_1 - J_0\rho_2}{\rho_1 - \rho_2}\rho_1^m + \frac{J_1 - J_0\rho_1}{\rho_2 - \rho_1}\rho_2^m$$

Substituting these in the intensity expression and simplifying we have,

$$I(h_3)/N\psi^2 = \frac{J_1 - J_0\rho_2}{\rho_1 - \rho_2}\sum_m \rho_1^m \exp(\pi imh_3) + \frac{J_1 - J_0\rho_1}{\rho_2 - \rho_1}\rho_2^m \sum_m \rho_2^m \exp(\pi imh_3)$$

The above summations can be usually performed and give rise to maxima in the intensity for certain values of h_3. A typical intensity plot for α = 0.05, 0.1, 0.15 is shown in Fig. 5. Characteristics parameters of such a distribution govern its area, width and asymmetry. These can be found analytically as shown later.

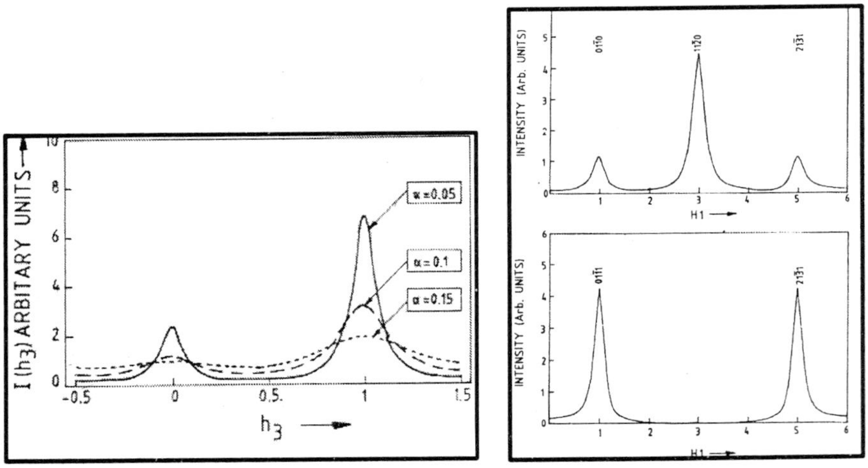

(a) (b)

Fig. 5: (a) The intensity $I(h_3)$ plotted as a function of h_3 for 1/6 <1120> type of APBs on basal plane for $h–k = 3N\pm1$ along a row of superlattice reflections using α = 0.05 (continuous line), α = 0.1 (dotted line) and α = 0.15 (dashed line). (b) Asymmetry observed in cph ordered structure with different APBs.

DIFFRACTION EFFECTS

We shall now discuss the diffraction effects of APBs in some common ordered structures. We begin with the $Cu_3Au/L1_2$ ordered fcc structure. In the $L1_2$ superstructure five types of boundaries have been considered. These are (111) 1/2 $[1\bar{1}0]$ shear, (0010) 1/2 $[1\bar{1}0]$ shear, (001) 1/2 $[10\bar{1}]$ climb, (110) 1/2 $[1\bar{1}0]$ shear and (110) 1/2 $[10\bar{1}]$ climb type boundaries. A summary of the results for the effective domain size in $L1_2$ structure due to various APBs is given in Table 7 (Wilson, 1943; Wilson and Zsoldos, 1966; Anantharaman et al., 1972). APBs generally broaden superlattice reflections. Warren (1969) has further concluded that they have no effects on the integrated intensities. It is also apparent that they do not give rise to any asymmetry. Diffraction effects of various APB's in a B2 structure (such as AgMg) have been derived by Rothman et al. (1970). These are summarized in Table 8.

Table 7: Diffraction effects of antiphase boundaries in $L1_2$ structures

Boundary Plane	*Boundary* Shift vector	*Type of boundary*	*Components unbroadened*	D_{eff}
(111)	$(1/2)[1\bar{1}0]$	Shear	(h-k) even	$\left(\sum_b h_0(u+b)/\lvert h+k+l\rvert\right)(a/\alpha)$
(001)	$(1/2)[1\bar{1}0]$	Shear	(h-k) even	$\left(\sum_b h_0(u+b)/\lvert l\rvert\right)(a/\alpha)$
(001)	$(1/2)[10\bar{1}]$	Climb	(h-l) even	$\left(\sum_b h_0(u+b)/\lvert l\rvert\right)(a/\alpha)$
(110)	$(1/2)[1\bar{1}0]$	Shear	(h-k) even	$\left(\sum_b h_0(u+b)/\lvert h+k\rvert\right)(a/\alpha)$
(110)	$(1/2)[10\bar{1}]$	Climb	(h-l) even	$\left(\sum_b h_0(u+b)/\lvert h+k\rvert\right)(a/\alpha)$

$h_0 = (h^2 + k^2 + l^2)^{1/2}$; a_0 = lattice parameter
u and b: the number of unaffected (u) and broadened (b) components of a powder reflection with multiplicity (u + b)
a = probability of the fault.

Table 8: Diffraction effects of antiphase boundaries in B2 structures

Boundary Plane	*Boundary* Shift vector	*Type of boundary*	*Components unbroadened*	D_{eff}
(110)	$(1/2)[1\bar{1}1]$	Shear	(h–k+1) even	$\left(\sum_b h_0(u+b)/2\lvert h+k\rvert\right)(a_0/\alpha)$
(211)	$(1/2)[\bar{1}11]$	Shear	(–h+k+l) even	$\left(\sum_b h_0(u+b)/2\lvert 2h+k+1\rvert\right)(a_0/\alpha)$
(321)	$(1/2)[\bar{1}11]$	Shear	(–h+k+l) even	$\left(\sum_b h_0(u+b)/2\lvert 3h+2k+1\rvert\right)(a_0/\alpha)$

$h_0 = (h^2 + k^2 + l^2)^{1/2}$; a_0 = lattice parameter
u and b: the number of unaffected (u) and broadened (b) components of a powder reflection with multiplicity (u + b)
a = probability of the fault.

Recent investigations of the diffraction effects of various APBs and CSFs in $D0_{19}$ structure (Mg_3Cd, Ti_3Al type) are shown in Table 9. The calculation of the effective antiphase domain size in powder patterns is discussed by Ghosal et al. (1993), for the most probable type of APBs in $D0_{19}$ type of superstructure. The effective domain size can be written as,

$$\frac{1}{\bar{D}} = \frac{2\alpha_n}{d_0} \cdot \frac{\sum_b \cos\phi}{(u+b)} = \left[\frac{2a}{d_0} \quad \frac{\sum_b \cos\phi}{(u+b)} \right] \frac{\alpha_n}{a}$$

where ϕ is interplanar angle between the diffracting and APB planes and is given by,

$$\cos\phi = \left[\frac{2}{3a^2} (hh_0 + kk_0 + ii_0) + \frac{ll_0}{c^2} \right] dd_0$$

Further, α_n is the probability of occurrence of APBs, d_0 the interplanar spacing of the APB planes with indices h_0, k_0, i_0, l_0 $(u + b)$ is the multiplicity of the superlarttice reflection hkil in a powder pattern and u and bear the number of unbroadened and broadened components. The interplanar spacing of the corresponding planes is d.

When several types of boundaries are present which are stable (Umakoshi and Yamaguchi, 1981) and observed (Court, 1990; Banerjee, 1994), the effective domain size can be found by an additive procedure and is given by

$$\frac{1}{D_{\text{eff}}} = \left[\frac{4d}{c(u+b)} \cdot \sum_b \left| 1 \quad \frac{a}{c} \right| \right] \left[\frac{\alpha^{0002}_{\text{APB,Shear}}}{a} + \frac{\alpha^{0002}_{\text{CSF,Shear}}}{a} + \frac{\alpha^{0002}_{\text{APB,Climb}}}{a} \right]$$

$$+ \left[\frac{8d}{3a(u+b)} \cdot \sum_b |2h+k| \right]$$

$$\left[\frac{\alpha^{20\bar{2}0}_{\text{APB,Shear}}}{a} + \frac{\alpha^{20\bar{2}0}_{APB,Climb}}{a} \right] + \left[\frac{4d}{a(u+b)} \cdot \sum_b \left| 1 \frac{a2}{c2} - 2k \right| \right] \frac{\alpha^{2\bar{4}22}_{\text{APB,Shear}}}{a}$$

Utilizing experimental measurements of integral breadths to evaluate D_{eff} for at least six independent reflections, like $10\bar{1}0$, $11\bar{2}0$, $10\bar{1}1$, $10\bar{1}2$, $12\bar{3}0$, $11\bar{2}2$ we can solve six linear simultaneous equations of the above form to yield values of the fault probabilities. This leads to a characterization of the type and frequency of occurrence of each APB and CSF in a chosen sample. Alternately, measurements of changes in integrated intensity and peak asymmetry (see Table 9) can also be used instead of some of the integral breadth measurements for the evaluation.

Table 9 : Diffraction effects of APBs and CSFs in DO_{19} structures

Plane	Shift vector	Fault	Diffraction Effects			
			Fundamental Reflections		*Superlattice Reflections*	
			$h - k = 3N$	$h - k = 3N \pm 1$	$h - k = 3N$	$h - k = 3N \pm 1$
(0001)	1/6 $<11\bar{2}0>$	APB	No diffraction effects		(i) Integrated intensity unaffected (ii) Integral breadth proportional to 4α for all reflections (iii) No peak asymmetry	
	1/6$<\bar{1}100>$	Shear CSF	(i) Integrated intensity unaffected (ii) No peak broadening (iii) No peak asymmetry	(i) Integrated intensity affected; fractional change (for α << 1) is -9α/2 for l even and +3α/2 for l odd. (ii) Integral breadth is proportional to 9α/2 (for α << 1) (iii) No peak asymmetry	(i) Integrated intensity unaffected (ii) Integral breadth (for α << 1) is proportional to 4α (iii) No peak asymmetry	(i) Integrated intensity affected; fractional change (for α << 1) is +3α/2 for l even and -α/2 for l odd. (ii) Integral breadth (for α << 1) is proportional to 5α/2. (iii) No peak asymmetry
	1/6$<\bar{1}10\bar{3}>$	Climb CSF	(i) Integrated intensity unaffected (ii) No peak broadening (iii) No peak asymmetry	(i) Integrated intensity affected; no effects to first order. (ii) Integral breadth (α << 1) is proportional to 9α/2 for l even and 3α/2 for l odd (iii) No peak asymmetry	(i) Integrated intensity unaffected (ii) Integral breadth (for α << 1) is proportional to 4α (iii) No peak asymmetry	(i) Integrated intensity affected; No effects to first order. (ii) Integral -breadth (for α<< 1) is proportional to 5α/2 for l even and 7α/2 for l odd. (iii) No peak asymmetry
			h - k = 3N	h - k = 3N ± 1	h - k = 3N	h - k = 3N ± 1

(Contd.)

Table 9 (Cont.)

$(\bar{1}010)$	$1/6[1\bar{2}10]$	Shear APB	No diffraction effects	(i) Integrated intensity affected for k odd; No effects to first order ($\alpha_1, \alpha_2 << 1$). (ii) Integral breadth for reflections with k odd is proportional to $2(\alpha_1 + \alpha_2)$. (iii) No peak asymmetry (iii) Peak asymmetry occurs for k odd; shift in centroid proportional to $(\alpha_1 \sim \alpha_2)$ to first order.
	$1/6[11\bar{2}0]$	Climb APB	(i) integrated intensity affected; no effect to first order (ii) No peak broadening (iii) No peak asymmetry	(i) Integrated intensity affected; no effects to first order. (ii) Integral breadth for h or k odd is proportional to $2(\alpha_1 + \alpha_2)$. For h and k both odd, the peaks are unbroadened. (iii) No peak asymmetry
$(\bar{2}4\bar{2}2)$	$1/6[1\bar{2}16]$	Shear APB	No diffraction effects	(i) Integrated intensity unaffected (ii) Integral breadth for k odd is proportional to α. For k even the peaks are unbroadened. (iii) No peak asymmetry

α is the probability of occurrence of the relevant APB/CSF.

DISCUSSION

Diffraction effects due to antiphase boundaries in ordered structures mainly influence the intensity distribution by broadening it but not the integrated intensity. The above conclusion is valid for ordered fcc and ordered bcc based structures. It has been found that for ordered cph structures, such a generalization is not valid for large fault probabilities. The integrated intensity of superlattic reflection and also that of fundamental reflections is changed due to some types of boundaries. Peak broadening and in some cases peak asymmetry is also observed.

The diffraction calculations discussed are based on the assumption that APBs and CSFs extend completely through a coherently diffracting domain. Contrary to this assumption, the APBs or CSFs will usually be bounded by partial dislocations, which has two consequences. The first is that the probability α of the occurrence of an APB has to be redefined. α may be considered to be equal to the fraction of faulted area i.e., $\alpha = \rho_d w d_0$, where ρ_d is the dislocation density, w is the width of the APB or CSF and d_0 is the interplanar spacing of the planes on which the faults occur. The second consequence is the occurrence of strain broadening of the reflections due to the strain distribution produced by the partials bounding the APBs. Using observed data on peak broadening and peak asymmetry for a suitable set of reflections should in principle enable an evaluation of the various fault probabilities and the strain ε. The latter is correlated with the dislocation density ρ_d. Adler et al., 1970 have carried out for such an analysis for fcc alloys. Then the width w can also be evaluated from $w = \alpha/\rho_d\ d_0 = K\ \alpha/\varepsilon d_0$, where K is a proportionality constant. Since w is inversely proportional to the fault or antiphase boundary energy, we can then evaluate this fundamental property of a material. As diffraction methods are indirect in nature, it may be useful to anchor one value of a set of evaluations for the APB energy by electron microscopy which is a direct method. Alloy design for ordered alloys is crucially dependent on variation of APB energy with composition. The diffraction approach outlined and reviewed here provides a suitable methodology for such determinations.

Acknowledgement: The author would like to thank Professor Shrikant Lele, BHU for basic training and superb intellectual guidance. He enlightened us by his idealist insights for various aspects of life. I express my sincere and warmest gratitude to him.

REFERENCES

Adler, R.P.I., Otte, H.M. and Wagner, C.N. (1970). *J. Met. Trans.*, **1:** 2375.

Anantharaman, T.R., Rama Rao, P. and Lele, S. (1972). *In:* Recent Developments in Metallurgical Science and Technology. Indian Inst. Metals, Calcutta, 407.

Banerjee, D. (1994). *In:* Intermetallic compounds, eds. Westbrook J.H. and Fleischer R.L. John Wiley, New York, USA, 91.

Cohen, J.B. and Bever, M.B. (1960). *Trans TMS-AIME*, **218:** 155.

Court, S.A., Lőfvander, J.P.A., Loretto, M.H. and Fraser, H.L. (1990). *Phil Mag* A, **61:** 109.

Court, S.A., Lőfvander, J.P.A., Loretto, M.H. and Fraser, H.L. (1989). *Phil Mag* A, **59:** 379.

Ghosal, P. (1996). Diffraction from $D0_{19}$ Type Ordered hcp Intermetallics Containing Antiphase Boundaries. Ph.D Thesis, Banaras Hindu University, India.

Ghosal, P., Ramachandra, C. and Lele, S. (1993). *Scripta Metall. Mater.*, **28:** 477.

Mikkola, D.E. and Cohen, J.B. (1966). *In:* Local Atomic Arrangements as Studies by X-rays eds. Cohen, J.B. and Hilliard, J.E., Gordon and Breach New York, USA, 289.

Mikkola, D.E. and Cohen, J.B. (1966). *Acta Metall.*, **14:** 105.

Rothman, R.L. and Cohen, J.B. (1971). *J Appl. Phys.*, **42:** 971.

Rothman, R.L., Marion, R. and Cohen, J.B. (1970). *In:* Ordered Alloys: Structural Applicationand Physical Metallurgy. eds. Kear, B.H., Sims, C.T.,Stoloff, N.S. and Westbrook, J.H. Claitors, LA, USA, 149.

Umakoshi, Y. and Yamaguchi, M. (1981). *Phys. Stat. Sol.*, **68:** 457.

Vitek, V. (1968). *Phil. Mag.*, **18:** 773.

Vitek, V. (1974). Crystal Lattice Defects. **5:** 1.

Warren B.E. (1969). X-ray Diffraction. Addison-Wesley, Rading, USA.

Warren, B.E. (1963). *Progr. Met. Phys.*, **8:** 147.

Wilson, A.J.C. (1943). *Proc. Roy. Soc.* A, **181:** 360.

Wilson, A.J.C. and Zsoldos, L. (1966). *Proc. Roy. Soc.* A, **290:** 508.

Yamaguchi, M, Vitek, V. and Pope, D.P. (1981). *Phil. Mag.* A, **43:** 1027.

Yamaguchi, M. and Umakoshi, Y. (1990). *Prog. Mater. Sci.*, **34:** 1.

Yamaguchi, M., Pope, D.P., Vitek, V. and Umakoshi, Y. (1981). *Phil. Mag.* A, **43:** 1265.

X-ray Diffraction Line Profile Analysis on the Microstructure of Cold-Worked Ni_3Al: Effect of Boron Addition

G. Bhikshmaiah, S.V. Suryanarayana[1] and A.K. Singh[2]

Department of Physics, Nizam College (Autonomous)
Osmania University, Hyderabad – 500007, India
[1]Department of Physics, Osmania University
Hyderabad – 500007, India
[2]Defence Metallurgical Research Laboratory
P.O. Kanchanbagh, Hyderabad – 500058, India

INTRODUCTION

The nickel aluminide (Ni_3Al) crystallizes in the $L1_2$ structure and is ordered up to its melting point (1663 K). Ni_3Al or γ' phase is also the major strengthening constituent in commercial nickel base super alloys that are used in marine, aircraft and industrial gas turbines, space vehicles, rocket engines, nuclear reactors, steam power plants and other high temperature applications (Pope and Ezz, 1984). The lower density and anomalous temperature dependence of the yield strength compared to the nickel base super alloys make Ni_3Al a potential alloy for high temperature applications. The alloys exhibit oxidation resistance in air at elevated temperatures forming adherent aluminium oxide films (Liu and Stiegler, 1984). However, due to low ductility in polycrystalline form it is not used as an engineering material (Baker et al., 1984). The addition of small amounts of boron was found to be more effective in improving ductility (Taub et al., 1984; Huang et al., 1984) and strength (Weihs et al., 1987; Schulson et al., 1985) of Ni_3Al.

Mohan Rao (1989) studied the effect of addition of boron on the physical properties of Ni_3Al, namely, lattice parameter, microhardness, thermal expansion, Debye temperature and order parameter. The present experimental study deals with the X-ray diffraction line profile analysis (XRDLPA) of a cold-worked Ni_3Al and six alloys of Ni_3Al containing varying amounts of boron to study the effect of boron on the microstructure in the deformed state of Ni_3Al alloy. The defect parameters have been characterized in terms of the several microstructural parameters using standard XRDLPA techniques (Bhikshmaiah and Suryanarayana, 1987; Venkat Reddy and Suryanarayana, 1986; Bhikshmaiah et al., 1988).

EXPERIMENTAL PROCEDURE

The alloys used in the present study were prepared by non-consumable arc melting in the presence of argon atmosphere using high purity (99.99%) metals Ni, Al and NiB metal alloy. The melting was repeated five times to ensure chemical homogeneity. All the ingots were sealed in a quartz tube under 1×10^{-5} torr vacuum and homogenized for 60 hrs at 1000 °C and then air cooled to room temperature. The weight loss was found to be negligible.

The alloy composition was determined by instrumental and wet chemical methods. The aluminium content was determined by atomic absorption spectroscopy while boron by D.C plasma spectroscopy. The chemical compositions of the alloys are given in Table 1.

Table 1: The Chemical compositions of Nickel aluminides

	Chemical Composition in (at.%)		
Alloy	*Ni*	*Al*	*B*
NA-1 (Ni_3Al)	76.1	23.9	-
NA-2 (Ni_3Al)	75.0	25.0	-
NAB-1	76.65	23.15	0.20
NAB-2	75.88	23.70	0.42
NAB-3	75.63	23.62	0.75
NAB-4	75.26	23.77	0.97
NAB-5	75.38	23.41	1.21
NAB-6*	75.67	22.60	1.73

*Two-phase alloy

X-ray diffraction (XRD) data were collected on all the specimens with a scanning speed of 1° per minute on a PW 1730 Philips X-ray diffractometer using CuKα radiation. The X-ray data for line profile analysis were recorded using divergence and scattering slits of 1° each and receiving slit of 0.2 mm. X-ray data on cold-worked filings (–270 mesh) for the (111), (200), (220), (311), (222) and (400) profiles were collected at room temperature using point counting technique at intervals of 0.1° in 2θ for general background and 0.01° in 2θ near peak maximum at 1000 °C for 24 hrs and cooled to room temperature at the rate of 25°C per hour. X-ray data on these materials were taken under identical conditions. The X-ray data of cold worked and annealed samples was plotted on a graph paper.

RESULTS

XRD analysis shows that all the samples were found to be single phase with $L1_2$ structure except NAB-6 (Table 1), which exhibits additional NiAl phase. Standard methods (Warren, 1961; Alder and Wagner, 1962) were used to determine peak maxima and the centre of gravity of the line profiles.

Following Warren (1969) and Wagner (1966) the defect states were characterized by evaluating different defect parameters, namely, intrinsic (α'), extrinsic (α'') and twin or growth (β) fault probabilities, change in lattice parameter ($\Delta a/a_0$), dislocation densities (ρ), domain sizes in the fault plane (D_{SF}) along [100] and [111] and minimum value of particle size (T_{min}) from peak shift, peak asymmetry and peak broadening analysis of the recorded profiles for all the alloy compositions investigated. These analyses have been done following the same procedure adopted earlier (Pope and Ezz, 1984) and results are shown in Tables 2 to 4.

Table 2: Values of deformation fault probabilities (α), asymmetry parameter ($4.5\ \alpha'' + \beta$), compound [$1.5\ (\alpha' + \alpha'') + \beta$], intrinsic ($\alpha'$), extrinsic ($\alpha''$) and twin ($\beta$) fault probabilities in deformed Ni_3Al aluminides

Alloy	$\alpha = (\alpha' - \alpha'') \times 10^3$ From peak Shift	$\alpha = (\alpha' - \alpha'') \times 10^3$ From lattice parameter	$4.5\alpha''+\beta \times 10^3$	$1.5(\alpha'+\alpha'') + \beta \times 10^3$	$\alpha' \times 10^3$	$\alpha'' \times 10^3$	$\beta \times 10^3$
NA-1	39	46	32	72	52	13	–25
NA-2	38	41	13	79	32	-6	40
NAB-1	31	45	6	77	15	-16	79
NAB-2	45	48	63	13	39	-6	37
NAB-3	46	50	12	76	49	3	–2
NAB-4	51	61	16	65	68	18	–63
NAB-5	69	51	31	73	112	42	–159
NAB-6*	50	53	–	–	–	–	–
Error Limits	±1 to ±4	±1 to ±4	±1 to ±3	±3 to ±8	±2 to ±8	±2 to ±3	±3 to ±5

* Two-phase alloy

DISCUSSION

Lattice Parameters

When a third element is added to Ni_3Al, if it goes into solution, it can occupy either Ni or Al lattice position or an interstitial, depending on the atomic size of the additive. Boron being a very small atom goes into the interstitial position of the lattice. Thereby, one can expect an increase in the lattice parameter of the matrix phase. As expected, the lattice parameter of the Ni_3Al phase increases with the addition of boron except for that of 0.20 at.% B alloy (Fig. 1). The value of the lattice parameter of aluminides

Table 3: Lattice parameters of Ni_3Al aluminides at room temperature

Alloy	*Lattice parameter (nm)*		
	annealed (a_{ann})	*coldworked* (a_{cw})	$(\Delta a/a_{ann}) \times 10^2$
NA-1	0.34748	0.35797	−1.3710
NA-2	0.35766	0.35468	0.3076
NAB-1	0.35730	0.35654	2.1271
NAB-2	0.35750	0.35653	−0.0834
NAB-3	0.35781	0.35770	0.3074
NAB-4	0.35806	0.35800	0.1676
NAB-5	0.35807	0.35794	0.3631
NAB-6*	0.35807	0.35740	1.8711
Error limit	±0.00004	± 0.00005	

*Two-phase alloy

Table 4: Values of particle sizes (D_e), rms microstrains ($<\varepsilon^2_{L=5nm}>^{1/2}$) and dislocation densities (ρ) in deformed Ni_3Al aluminides

Alloy	$D_e(nm)$		$<\varepsilon^2_{L=5nm}>^{1/2} \times 10^3$		$D_{SF}(nm)$			$\rho \times 10^{-15}(metre/metre^3)$		
	[111]	[100]	[111]	[100]	[111]	[100]	T_{min}	[111]	[100]	average
NA-1	15.5	5.6	1.47	1.78	11.48	4.97	2.36	9.45	4.09	6.77
NA-2	8.0	4.0	1.94	0.95	10.48	4.54	1.81	7.06	5.46	6.26
NAB-1	10.0	4.5	1.73	1.31	10.71	4.64	1.98	8.67	5.78	7.22
NAB-2	7.9	6.8	2.21	1.64	63.95	27.69	3.85	7.17	6.51	6.84
NAB-3	8.1	4.1	1.48	2.70	10.87	4.71	1.86	19.57	8.11	13.84
NAB-4	6.8	4.0	2.05	1.87	12.72	5.51	1.91	13.88	6.46	10.17
NAB-5	8.2	4.2	1.92	2.26	11.27	4.88	1.92	15.98	7.42	11.70
NAB-6	7.2	–	–	–	–	–	–	–	–	–
Error Limits	±0.2 to ±0.5		±0.03 to ±0.08							

* Two-phase alloy

containing 0.20 and 1.76 at.% B deviate more from linearity. The deviation might be due to the lower concentration of aluminium in these two alloys (Table 1). It can be seen from Fig. 1 that the lattice parameter of the alloy containing 1.76 at.% B is same as 1.22 at.% B, confirming the solubility limit is at least 1.22 at.% B. The solubility limit of boron in Ni_3Al, obtained in the present study, lies between the values reported by those of Baker et al. (1988) (1.12 at.%) and Huang et al. (1984) (1.50 at.%). The lattice parameter data reported in literature (Huang et al., 1984; Baker et al., 1988; Kim et al., 1985) for boron added Ni_3Al alloys are also shown in Fig. 1. It can be seen that the lattice parameter values of the boron added Ni_3Al alloys vary with heat treatment.

The data points of $(a_{hkl})_{cw}$ corresponding to the different hkl planes, showing a systematic scatter is consistent with Paterson's (1905) theory around the reference line (Fig. 2). This line is considered to be fault free

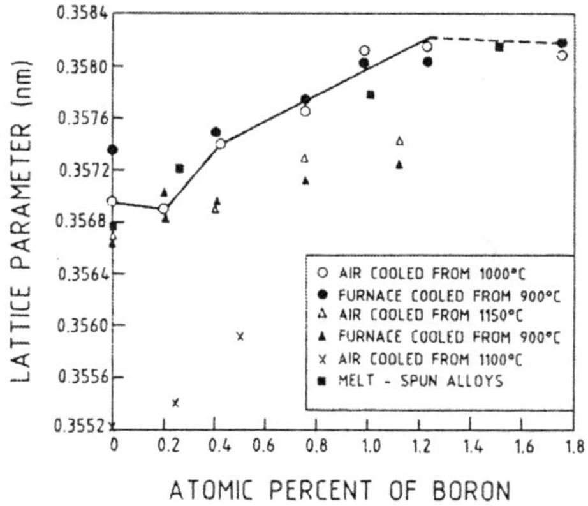

Fig. 1: Variation of room temperature lattice parameter with atomic percent of boron in Ni_3Al alloys.

(Present study: o - air cooled from 1000 °C, ● - furnace cooled form 1150 °C; Δ - air cooled from 1150 °C, ▲ - furnace cooled from 900 °C (Baker et al., 1988); x - air cooled from 1100 °C (Kim et al., 1985) ■ - melt-spun alloys (Huang et al., 1984).

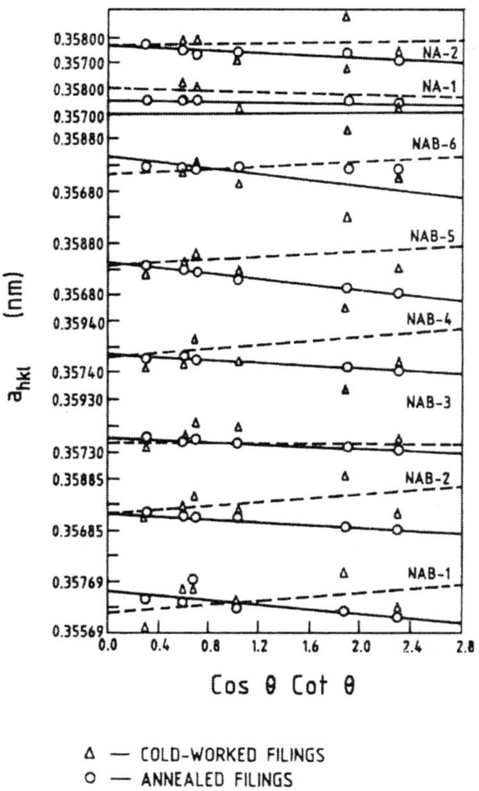

Fig. 2: Plots of lattice parameters (a_{hkl}) vs extrapolation function (cosθ cotθ) for boron added nickel aluminides (Δ - cold-worked filings, o - annealed filings).

according to the faulting coefficients and this is a clear indication of the presence of stacking faults alone, since the reference line is drawn with the consideration of $\alpha = 0$. As the extrapolated lattice parameters of (a_{cw}) with those "Standard Values (a_{ann}), it may be considered that only stacking faults produced during filing contribute principally to deformation. Further, data suggests that cold working causes contraction of lattice in these alloys.

Deformation Fault Probability

The deformation fault probability α in these alloys determined both from peak shift and lattice parameter analysis (Table 2) are comparable. A general agreement can be seen in the present work in respect of observation that deformation fault probability increases with increasing solute content. It can also be seen from Table 2 and Fig. 3 that deformation fault probability increases with the addition of boron content. However, a decrease is observed in case of NAB-1 and NAB-6. Similar trends have also been observed by Mohan Rao (1989) in the microhardness data (Fig. 3). This is because the value of α also depends upon the aluminium content present in the alloy. Further, it is well known that the value of α decreases at the phase boundary (Sen Gupta and Goswami, 1967) as seen in case of NAB-6 alloy. The deformation faults produced in NAB alloys is very high compared to those produced in Ag and Cu base alloys (Bhikshmaiah et al., 1988; Wert et al., 1983).

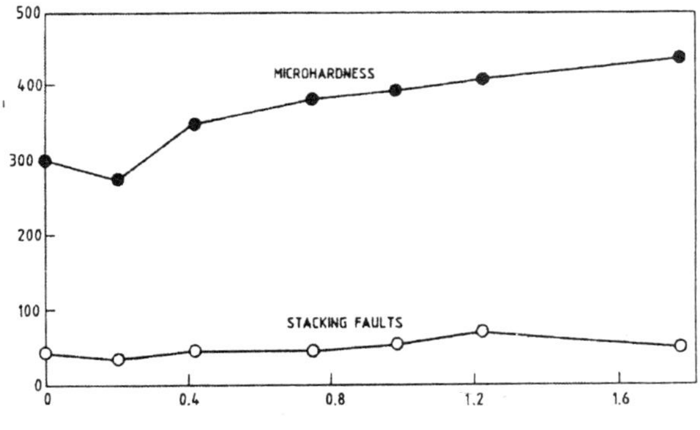

Fig. 3: Variation of stacking fault probability (α) and microhardness (kg/mm^2) with atomic percent of boron in Ni_3Al alloys.

The value of stacking fault probabilities $\alpha = (\alpha' - \alpha'')$ obtained for Ni_3Al alloys (Table 2) are comparable with the value 42×10^{-3} reported by Arnorth et al. (1977).

Microstrains

The rms microstrains along [111] and [100] for NAB alloys are shown in Fig. 4 as a function of L (the distance in the crystal normal to the reflecting

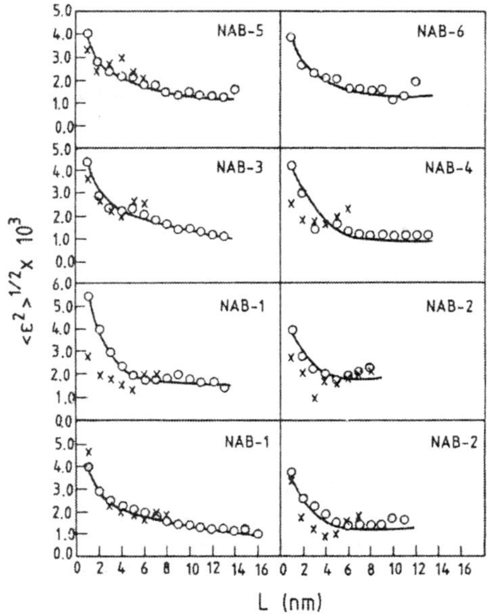

Fig. 4: Plots of the rms strain $<\varepsilon^2_L>^{1/2}$ vs L(nm) for nickel aluminides. (o - [111], x - [100]).

plane (hkl)) (Wagner, 1966). This shows that the rms strain decreases asymptomatically with increase in L and microstrains along [100] and [111] are of comparable magnitudes. However, the value of microstrain along [111] is greater for low concentrations of boron while smaller for high concentrations of boron than those along [100].

Domain Sizes

The values of the domain sizes along [111] and [100] are in general found to decrease with addition of boron (Table 4) as observed in Ag and Cu base alloy systems (Bhikshmaiah et al., 1988; Wert et al., 1983). It is also seen that the values of D_e are found to be anisotropic along [111] and [100]. The minimum value of the fault width T_{min} of these alloys shows no systematic variation with boron content.

Stacking Fault Concentrations and Dislocation Density

From Table 2, it can be seen that the values of compound fault probabilities $1.5\ (\alpha' + \alpha'') + \beta$ exhibit no systematic variation with boron concentration while asymmetry parameter $4.5\ \alpha'' + \beta$ shows no systematic variation in these alloys.

The values of intrinsic (α'), extrinsic (α'') and twin (β) fault probabilities in these alloys were evaluated from the values of $\alpha' - \alpha''$, $4.5\ \alpha'' + \beta$ and $1.5\ (\alpha' + \alpha'') + \beta$ and are given in Table 2. It is seen that the intrinsic faults are more than the extrinsic and twin faults. The value of α' in general is found to increase with boron concentration. The value of á' for Ni_3Al alloys are 0.052 and 0.032, respectively for NA-1 and NA-2 (Table 2) while Arnorth et al. (1977) have reported a value of 0.049 for Ni_3Al. The negative values for extrinsic or twin faults are physically unrealistic and show the absence of such faults.

From Table 4, it can be seen that the dislocation density is anisotropic along [111] and [100] ($\rho_{111} > \rho_{100}$) and both the values show an increasing trend with solute content. It can be seen that the rate of increase of ρ per atomic percent of solute atoms is higher for these alloys than those widely studied Ag and Cu base alloys (Bhikshmaiah and Suryanarayana, 1987; Venkat Reddy and Suryanarayana, 1986; Bhikshmaiah et al., 1988; Wert, 1983).

Solubility Limit of Boron in Ni_3Al Alloy

In Fig. 5 the variation of average effective particle size and strain along [111] and [100] is shown as a function of atomic percent of boron in Ni_3Al alloy while Fig. 3 shows the variation of deformation faults and microhardness of these alloys with boron concentration.

The data on the lattice parameters (Fig. 1), deformation faults (Fig. 3), average strain and average effective particle size (Fig. 5) of Ni_3Al alloys show a decrease or an increase up to 1.22 atomic percent of boron showing a discontinuity at this atomic percent of boron. This suggests that at least 1.22 at.% of boron goes into Ni_3Al lattice in $L1_2$ phase. However, the data on microhardness (Fig. 3) do not show how such discontinuity at 1.22 at.% boron may be due to its bulk property. The deviation from linearity in the data on microstrain and particle size (Fig. 5) at 0.20 at.% of Boron may be due to the lower concentration of aluminium in this alloy as mentioned above.

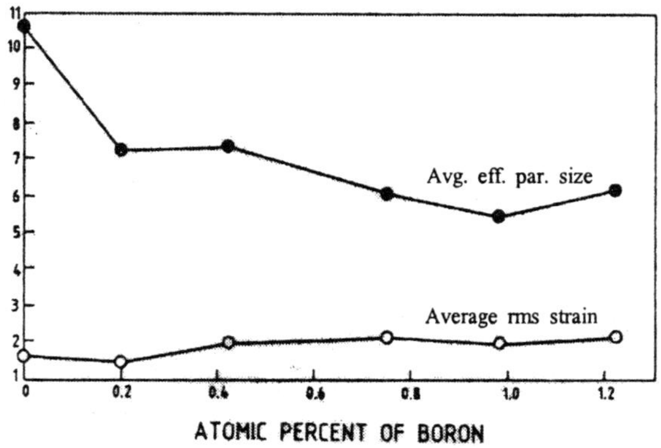

Fig. 5: Variation of average rms microstrain ($<\varepsilon^2_L>^{1/2}$) and effective particle size(D_e) with atomic percent of boron in Ni_3Al alloys.

CONCLUSIONS

(i) The lattice parameter of Ni_3Al alloy increases with addition of boron.
(ii) It is found that at least 1.22 at.% of boron is soluble in Ni_3Al.
(iii) The lattice parameter is found to decrease with cold working.
(iv) The value of deformation fault probability á of these alloys is found to be more sensitive to boron content.
(v) The value of compound fault probability is found to be independent of boron content,
(vi) The particle size along [111] and [100] are anisotropic i.e., D_e (111)>D_e (100) and both are found to decrease with boron content.
(vii) Intrinsic faults are more than those of extrinsic and twin faults and show an increase with boron content.
(viii) Microstrains are found to be insensitive to boron concentration in these alloys.
(ix) The rate of increase of dislocation density with boron concentration is more than those of widely studied fcc alloys viz. Ag and Cu.

REFERENCES

Alder, R.P.I. and Wagner, C.N.J. (1962). *J. Appl. Phys.*, **33:** 3451.

Arnorth, F.W., Calrk, J.P. and Mohanty, G.P. (1977). *J. Appl. Phys.*, **48:** 1771.

Baker, I., Hung, B. and Schulson, E.M. (1988). *Acta Metall.*, **36:** 493.

Baker, I., Viens, D.V. and Schulson, E.M. (1984). *J.Mater. Sci.*, **19:** 1799.

Bhikshmaiah, G. and Suryanarayana, S.V. (1987). *J Less Common Metals*, **120:** 29.

Bhikshmaiah, G., Suryanarayana, S.V., Nagender Naidu, S.V., Rajashekharan, T. and Gopalan, R. (1988). *Wear*, **125:** 241.

Huang, S.C., Taub, A.I. and Chang, K.M. (1984). *Acta Metall.*, **32:** 1703.

Kim, Y.G., Yoon, G.W. and Stoloff, N.S. (1985). *J.Mater. Sci. Lett.*, **4:** 1407.

Liu, C.T. and Stiegler, J.O. (1984). *Science*, **226:** 636.

Mohan Rao, P.V. (1989). Ph.D.Thesis, Osmania University, Hyderabed, (India).

Paterson, M.S. (1905). *J. Appl. Phys.*, **23:** 805.

Pope, D.P. and Ezz, S.S. (1984). *Intl. Metals Reviews*, **29:** 136.

Schulson, E.M., Weihs, T.P., Viens, D.V. and Baker, I.L. (1985). *Acta Metall.*, **33:** 1587.

Sen Gupta, S.P. and Goswami, K.N. (1967). *Brit. J. Appl. Phys.*, **18:** 193.

Taub, A.I., Huang, S.C., and Chang, K.M. (1984). *Metall. Trans.*, **A15:** 399.

Venkat Reddy, S. and Suryanarayana, S.V. (1986). *Bull. Mater. Sci.*, **8:** 61.

Wagner, C.N.J. (1966). Local Atomic Arrangements studied by X-ray Diffraction. J.B.Cohen and J.E. Hillard (eds.) Gordon and Breach, New York, NY, **36:** 7.

Warren, B.E. (1961). *J. Appl. Phys.*, **32:** 2428.

Warren, B.E. (1969). X-ray Diffraction: Addision-Wesley, Massachusetts, MA.

Weihs, T.P., Zinoviev, V., Viens, D.V. and Schulson, E.M. (1987). *Acta Metall.*, **35:** 1109.

Wert, J.J., Singerman, S.A., Caldwell, S.G. and Chaudhary, D.K. (1983). *Wear*, **92:** 213.

Structural Characterization of Catalytic Materials by Powder X-ray Diffraction

Veda Ramaswamy

Catalysis Division, National Chemical Laboratory, Pune

INTRODUCTION

Powder diffraction patterns contain poorer data as compared to single crystal data mainly due to the overlap of equivalent and overlapping reflections. The advent of computer technology and software development has made the structural solution by powder diffraction with their inherent disadvantages, possible today. As data collection is one of the most ioportant steps, factors to be considered are geometry of the diffractometer, radiation source, the wavelength, appropriate sample preparation, slit sizes and counting times. The author has made an attempt to carry out Rietveld refinement for a wide variety of different structures, which are used as catalysts. In Rietveld method (Rietveld, 1969) each data point (2θ step) is an observation. During the refinement procedure, in addition to the scale parameter, profile parameters viz., zero shift, asymmetry, peak width, shape function, background, unit cell parameters and structural parameters viz., cation positions, site occupancy factor and isotropic temperature (thermal parameter) are refined in a least square procedure until the calculated profile based on the structural model, matches the observed pattern the best. The Rietveld method employs the entire powder diffraction pattern thereby overcoming the problem of overlap and allowing the maximum amount of information that can be extracted from the pattern.

As a catalyst, while sulfated zirconia is used as a super-acid (Arata, 1990), zirconia doped with other metal oxides such as CaO, MgO, Y_2O_3, La_2O_3, Ce_2O_3, CuO, Cr_2O_3, Fe_2O_3, MnO_2 and NiO could exploit the redox properties of these multivalent metals that could be well dispersed in the zirconia lattice (Zhou et al., 1997; Ray et al., 2001; Etsell and Flengas, 1970; Keshavaraja and Ramaswamy, 1994; Hoch and Mathur, 1962). The main aim of using these metal oxides as dopents is to obtain zirconia in the cubic phase (at low temperatures) with a high surface area. These stabilized or partially stabilized zirconia phases have been exploited for a number of catalytic reactions, as oxygen pumps and as gas sensors where these properties of zirconia are structure dependent. An attempt is made to show the additional structural information that can be obtained by the Rietveld refinement of a powder pattern as compared to the data from a normal powder pattern of copper stabilized zirconia catalysts.

Perovskite type of oxides such as $La_{1-x}A_xBO_3$ with A = Ca, Sr, Ba and Ce and B = Co, Mn, Ni and Fe have been used as catalysts for the combustion of methane. An increase in activity is seen with the increase in ionic conductivity with the introduction of oxygen and due to Ce^{III}/Ce^{IV} redox couple (Arai et al., 1986). In the present work, a comparison of the results obtained by Rietveld refinement of the powder XRD profiles of $LaCo_{1-x}Cu_xO_3$ where x = 0, 5, 10, 20 mole % scanned in both reflection and geometry are discussed.

Zeolites and molecular sieves are usually synthesized as very small crystals of 1 μm or less in size, which makes structural characterization very difficult. Zeolites are microporous, crystalline aluminosilicates with a relatively rigid anionic aluminosilicate framework with well defined channels and cavities. The catalytic properties of zeolites and molecular sieves are very much dependent on both their chemical and physical features. Isomorphous substitution of a T (tetrahedrally coordinated) element in a zeolite or a molecular sieve (silcalite, alumino silicates/phosphates) material leads to the modification of its acidic or redox properties and hence to tailoring of its catalytic and shape selective properties (Millini et al., 1992; Rakshe et al., 1997; Rakshe et al., 1996). The structural characterization of these materials using powder diffraction can unequivocally prove the isomorphous substitution of metal in the framework. For illustration, a typical sample of titanium silicalite, a zeolite of the pentasil family having MFI structure (a 10 membered medium pore zeolite consisting of a three dimensional system of channels), containing ~ 2.22 % Ti substituting Si, which is an important, efficient and selective catalyst for oxidation reactions with H_2O_2, is considered and the Rietveld refinement studies are reported.

SAPO-40 (silicon aluminophosphate) has the AFR topology, designated by the structure commission of the International Zeolite Association (Meierand and Olson, 1992) consisting of a two dimensional system of channels, with free diameters of 6.7 and 3.8 Å, running parallel to the b- and c-axes, respectively (Dumont et al., 1993). The connectivity of tetrahedrally coordinated cations in the framework structure of SAPO-40 without reference to atom type is described in literature with structure refinement carried out (McCusker and Baerlocher, 1996; Meden et al., 1997). An aluminophosphate (AlPO-40) with the above AFR topology is taken for illustration in this report. To investigate the reason for the stability of the AlPO-40 framework

(AFR topology) and to understand the difference in the structural features observed in the NMR studies (Lourenco, 1997) of this sample with that of SAPO-40, a structural analysis of the as-synthesized AlPO-40 was undertaken and reported (Veda Ramaswamy et al., 1999). TPA (tetra propyl ammonium hydroxide) and TMA (tetra methyl ammonium hydroxide) are used as the structure-directing agent in the synthesis of AlPO-40. Some of the structural features derived are discussed in this report.

EXPERIMENTAL PROCEDURE

$Zr_{1-x}Cu_xO_2$–Fluorite Structure

The details of the preparation of Cu-Zr samples with copper content varying from 1-30 mol% using coprecipitation technique are given elsewhere (Mahesh Bhagwat et al., 2003). The polycrystalline zirconia samples were scanned on a Philips Powder X-ray diffractometer, Model PW 1729, using monochromatic CuKα radiation (λ=1.5406 Å). Silicon was used as an external calibrant. More details are available in literature (Mahesh Bhagwat et al., 2003). The Rietveld Refinement studies of the powder X-ray diffraction (XRD) profiles were performed using the DBWS Rietveld refinement programme with a Cerius-2 interface.

La $Co_{1-x}Cu_xO_3$–Perovskite Structure

$LaCo_{1-x}Cu_xO_3$ samples with copper content varying from 1-30 mol% (x = 0, 0.05, 0.1 and 0.2, 0.3, 0.5) were prepared using citrate complexation method reported earlier (Veda Ramaswamy et al., 2004). The powder patterns were taken from RIGAKU D-MAX III VC (Japan) Instrument with Ni filtered CuKα radiation and a graphite crystal monochromator. The sample was rotated throughout the scan for better counting statistics and to minimize preferred orientation effect if present. The *d* values were indexed to a primitive rhombhohedral cell (non primitive hexagonal cell). The programme TREOR was used to index the powder pattern. Selected typical samples were also scanned on STOE/STADI-P with transmission geometry. Rietveld refinement of their powder XRD profiles scanned was carried out using the programme FULLPROF. Since transmission geometry is recommended for samples containing lighter elements to avoid sample transparency problems arising in the reflection mode, the samples were scanned on STOE/STADI-P with transmission geometry.

Titanium Silicalite

The titanium silicalite (TS-1) samples investigated in this report have been synthesized using the method described in reference (Thangaraj et al., 1991) with some modifications. Powder XRD analysis was carried out using RIGAKU X-ray diffractometer (D-MAX III V) with Ni filtered CuKα radiation and a graphite crystal monochromator. The sample was rotated throughout the scan for better counting statistics and to minimize preferred orientation effect if present. The data were collected in the 2θ range 5-50° with a step size of 0.02° 2θ and counting time of 10s at each step. The *d* values were indexed to a primitive orthorhombic cell. The Rietveld refinement studies of the powder XRD profiles in the 2θ range 15-50° were performed using the DBWS Rietveld refinement programme with a Cerius-2 interface. Space group *Pnma* was used for the orthorhombic phase of TS-1 sample. The starting model and the coordinates were taken from literature (Marra et al., 2000). The structural model is more complex than the previous two examples with the number of parameters to be refined becoming very large due to the large number of cations in one unit cell. The restraints on (Si/Ti)-O and O-O bond distances were applied. Only two independent isotropic displacement parameters were imposed for all T sites (one for both Si and Ti) and the other for all oxygen atoms.

Aluminophosphate (AlPO-40) with AFR Structure

A high quality synchrotron diffraction data were collected on the sample (as synthesized) at room temperature in an oscillating 1 mm capillary on the Beamline at Grenoble, France. The duration of data collection was 22 hrs. The 2θ range was 1.5-60.16° with a step size of 0.007° 2θ. Indexing of the XRD peaks by the synchrotron data collection was done using the programme TREOR. Rietveld refinement was carried out using the XRS-82 (Baerlocher et al., 1993) package. The greater details of both data collection and refinement are given elsewhere (Veda Ramaswamy et al., 1999).

RESULTS AND DISCUSSION

$Zr_{1-x}Cu_xO_2$—Fluorite Structure

Powder XRD patterns of samples $Zr_{0.98}Cu_{0.02}O_2$, $Zr_{0.95}Cu_{0.05}O_2$, $Zr_{0.9}Cu0_{0.1}O_2$ and $Zr_{0.8}Cu_{x0.2}O_2$ show cubic zirconia as the major phase with small concentrations of monoclinic phase. Distinguishing between the cubic and the tetragonal phase by the splitting of the 040 reflection into 400 and 004 reflections, respectively is difficult because of the broad nature of the overlapping peaks. Rietveld refinement of powder XRD data is reported to distinguish between the cubic phase and the metastable tetragonal phase of Y stabilized zirconia (Lamas and Walsue De Reca, 1999). Hence, the Rietveld refinement technique was used to analyze the XRD patterns. Figure 1 (a-d) shows the Rietveld refinement plots of samples $Zr_{0.98}Cu_{0.02}O_2$, $Zr_{0.95}Cu_{0.05}O_2$, $Zr_{0.9}Cu0_{0.1}O_2$ and $Zr_{0.8}Cu_{x0.2}O_2$, respectively.

The Rietveld refinement of these samples improved only when all the three phases, cubic (Space Group - Fm3m), tetragonal phase (Space Group - $P4_2/nmc$) and the monoclinic (Space Group - $P2_1/c$) phases were used as

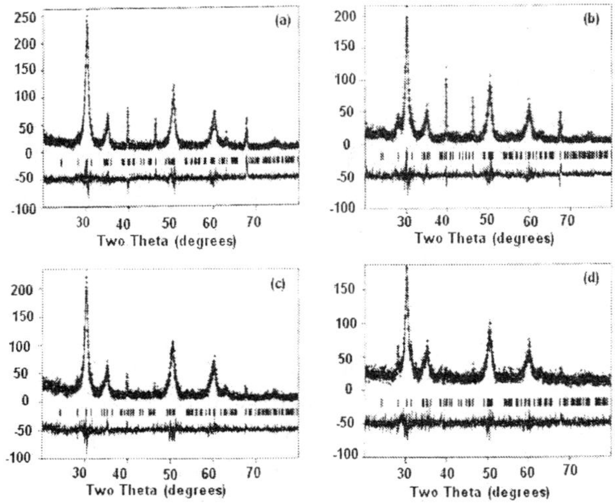

Fig. 1: Rietveld refinement plots of $Zr_{0.98}Cu_{0.02}O_2$, $Zr_{0.95}Cu_{0.05}O_2$, $Zr_{0.9}CuO_{0.1}O_2$ and $Zr_{0.8}Cu_{0.2}O_2$ samples scanned at 303 K (Mahesh Bhagwat et al., 2003).

the starting models. The background was refined using a polynomial with six refinable coefficients. The peak shape was corrected using a Pseudo–Voigt function. The peak width (FWHM) was determined using the equation, FWHM=$\sqrt{(u + v \tan \theta + w \tan^2\theta)}$ and then the crystallite size was determined using the Scherrer equation (Klug and Alexander, 1974). The Rietveld refinements results are listed in Table 1. The Rietveld refinement results infer that zirconia is stabilized partially into tetragonal phase and partially into the cubic phase. The percentage of the cubic phase seems to increase at the expense of the tetragonal phase as the copper concentration increases indicating that higher concentrations of copper is required for stabilizing zirconia in the cubic phase. A decrease in the unit cell volume of both the cubic phase and the tetragonal phase with increase in copper concentration indicates substitution of copper in the zirconia lattice. The presence of cubic phase is reported in the copper stabilized zirconia prepared by sol-gel procedure where the overlapping peaks of cubic and tetragonal zirconia could not be distinguished without refinement of the XRD profiles by Rietveld procedure and the unit cell volume data revealed that only one half of the input copper probably goes into the lattice and the remainder stays as extra lattice copper (Dongare et al., 2001). This conclusion was drawn on the basis of comparison of lattice parameters calculated using the ionic radii of copper and zirconium with the experimental value.

The present data from Rietveld refinement infers that the extra lattice copper could be in the tetragonal phase, which could not be detected by the authors in the absence of refinement studies. The crystallite size of the cubic phase decreases continuously from 26 nm to about 8 nm with the increasing copper content from 2 to 20 mole%, respectively. Thus, the Rietveld

Table 1: Rietveld Refinement data of the powder XRD patterns of the copper substituted zirconia samples

	$Zr_{0.98}Cu_{0.02}O_2$	$Zr_{0.95}Cu_{0.05}O_2$	$Zr_{0.9}Cu_{0.1}O_2$	$Zr_{0.8}Cu_{x0.2}O_2$
Cubic phase				
Composition, %	23.9	38.9	46.1	56.7
Crystallite size, Å	27	15	11	8
Unit cell Volume. $Å^3$	133.46	133.02	132.7	132.36
Tetragonal phase				
Composition, %	76.5	61.3	54.0	43.24
Crystallite size, Å	12	14	11	14
R factors				
R_P	19.5	18.19	16.50	16.0
R_{WP}	27.13	25.20	23.08	21.10
R_{EXP}	23.88	22.21	21.32	19.69
G.O.F, S	1.14	1.13	1.08	1.07
D-WD	1.75	1.74	1.82	1.86

refinement studies reveal that zirconia seems to stabilize in the cubic phase (in larger concentration as compared to tetragonal phase) at smaller crystallite size (~8 nm) and higher copper loading (~20 mol%). Recently, Wang and Caruso (2002) have reported the formation of only tetragonal zirconia by doping copper using the same co-precipitation technique, whose crystallite size was ~14-16 nm. Our results indicate stabilization of zirconia into both cubic and tetragonal phases with smaller crystallite size at similar copper loading. The value for GOF (goodness of fit given by R_{WP}/R_{EXP}) justifies a very good refinement. All the three phases were observed in all the samples, with monoclinic phase in small concentration (around 2%) and the concentration of tetragonal and cubic phases varying with the copper content.

$LaCo_{1-x}Cu_xO_3$—Perovskite Structure

Structural characterization of these samples is done by Rietveld refinement of the powder XRD profiles scanned on two different diffractometers having reflection and transmission geometry. Because the disadvantages of the reflection geometry are the preferred orientation effects associated with the flat plate specimens, which can lead to incorrect intensities and show strong *hkl* dependence. While using the reflection geometry it is assumed that the sample is infinitely thick i.e. the X-ray beam is completely absorbed by the sample. However, this might not be the case with samples containing low absorbing lighter elements and this geometry suits the samples containing heavier elements. The phase analysis of $LaCo_{1-x}Cu_xO_3$ samples with Cu content in the range 1-20 mol% exhibit the presence of perovskite phase with primitive rhombhohedral cell (non-primitive hexagonal cell). The XRD pattern of samples with 30 and 50 mol% copper reveal the presence of copper oxide phase. Although the Rietveld refinement of samples $LaCo_{1-x}Cu_xO_3$ (x = 0, 0.05, 0.1 and 0.2) were done, only two typical samples are taken for comparison of data by two different geometry. The increase in unit cell volume from undoped $LaCoO_3$ to doped (up to 20 mol%) sample is of the order of 2.2-4.8 Å^3 which compares well with the literature (Porta et al., 1999) and confirms the substitution of cobalt by copper in the lattice. The comparison of the results obtained by the Rietveld refinement of the powder XRD profiles of two typical samples viz., $LaCo_{1-x}Cu_xO_3$ where x = 0.05 and 0.1 (pervoskite structures) scanned in both reflection and transmission geometry are given in Table 2.

Table 2: Crystallographic data on $LaCo_{1-x}Cu_xO_3$ samples scanned on both reflection and transmission geometry

Unit cell	*Reflection Geometry*		*Transmission Geometry*	
	X=0.05	*X=0.1*	*X=0.05*	*X=0.1*
Chemical formula (input)	$LaCu_{0.05}Co_{0.95}O_3$	$LaCu_{0.1}Co_{0.9}O_3$	$LaCu_{0.05}Co_{0.95}O_3$	$LaCu_{0.1}Co_{0.9}O_3$
a **(Å)**	5.429	5.437	5.420	5.417
c **(Å)**	13.098	13.110	13.088	13.072
V (Å^3)	334.43	335.65	333.02	333.23
Refinement				
Number of observations	3500	3500	3500	3500
No. of contri-buting reflections	25	25	25	25
No. of profile/ structural parameters	8/6	8/6	8/6	8/6
X coordinate of O_2	0.4589	0.4466	0.446(0.0061)	0.442(0.0068)
S.O.F (La, Co, Cu, O)			6, 5.693, 0.307, 18	6, 5.34, 0.66, 18
R_{exp}	6.83	7.91	4.9	5.57
R_{wp}	13.5	10.7	6.01	6.64
χ^2	1.95	1.83	1.23	1.19
R_{Bragg} (I Phase)*	2.1	2.2	3.26	4.39
R_{Bragg} (II Phase)	17.0	13.0		

* I Phase = $LaCu_xCo_{1-x}O_3$
II Phase = Co_3O_4

The Rietveld refinements of the XRD profiles scanned on Rigaku D-MAX III VC with reflection geometry was started with the model having space group $R\bar{3}m$ with corresponding coordinates for the four atoms in the sample (La, Co, Cu and O). The refinement was not converging and satisfactory and a change in the space group to $R\bar{3}c$ was tried with constraints on the occupancy of copper and cobalt to be equal to one. Since only oxygen atoms (18e site) can change their position, the x-coordinate of oxygen was refined and found to vary in the range 0.4446 to 0.4589. The refinement improved after introducing another phase (impurity phase likely to be Co_3O_4) (Fig. 2). The Rietveld refinement results in GOF (χ^2) varying in the range 1.83 and 1.95 for both the samples mentioned above in reflection geometry.

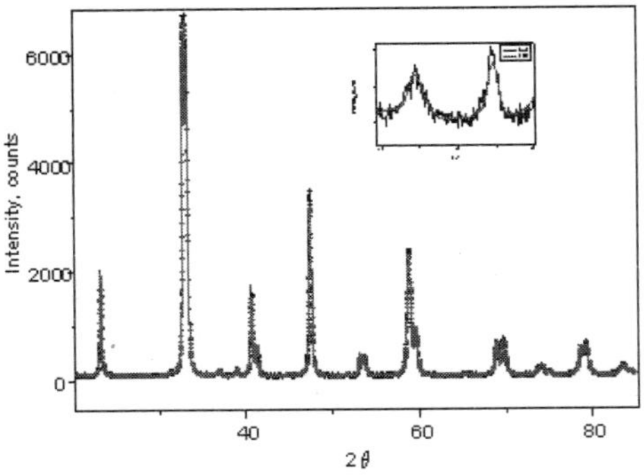

Fig. 2: Rietveld refinement profile (reflection geometry) of a typical sample La $Cu_{0.1}Co_{0.9}O_3$ showing the Co_3 O_4 phase in the inset.

The Rietveld refinement of the XRD profiles scanned on STOE/ STADI-P with transmission geometry gave precise fitting for each sample with a reliability factor (R_{wp}) less than 4.9–6.69%. The x-coordinate of oxygen was refined and found to vary between 0.442 and 0.446 with a greater error value for copper doped samples. Co_3O_4 phase was introduced in the refinement. It was not possible to identify any such phase because of the not so highly crystalline sample (the temperature of calcination of the sample was 600 °C) with the amorphous part contributing to higher signal to noise ratio (S/N). The GOF varied from 1.19 to 1.23 for all the four samples (Table 2, Fig. 3). The results infer that for complete structural data on a sample sometimes an integrated approach in data collection is required to infer the impurity present and to get good counting statistics for better refinement results. The chemical formula derived from the refined data is $LaCu_{0.052}Co_{0.948}O_{3-\delta}$ and $LaCu_{0.11}Co_{0.89}O_{3-\delta}$ for $LaCo_{1-x}Cu_xO_3$ samples with $x = 0.05$ and 0.1 respectively. In conclusion, $LaCo_{1-x}Cu_xO_3$ ($x = 0.05$ and 0.1) is perovskite phases with minor Co_3O_4 phase identified by the reflection geometry diffractometer and could not be seen in transmission geometry. Cobalt is in 3^+ oxidation state and absence of Co^{4+} and presence of

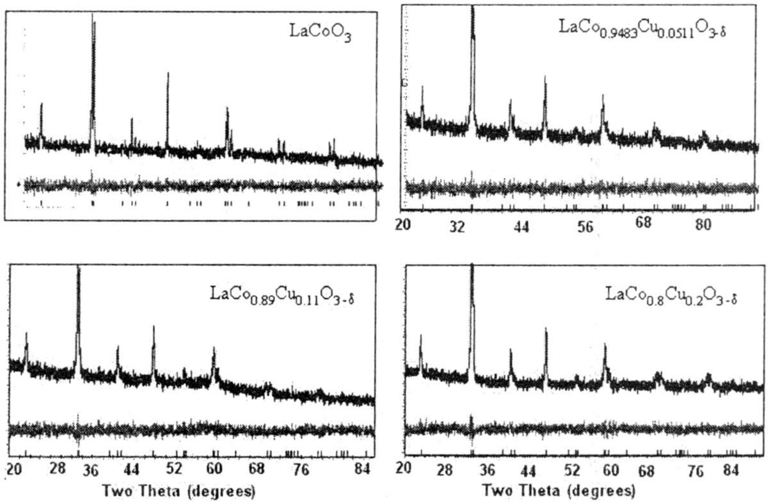

Fig. 3: Rietveld refinement profile (transmission geometry) of $LaCu_xCo_{1-x}O_3$ (x = 0, 0.05, 0.1 and 0.2).

diamagnetic Co^{III} is indicated by XPS data. In the absence of Co^{4+} species in the samples, the amount of Co present in $LaCu_xCo_{1-x}O_3$ could be low spin Co^{III} ($t^6_{2g}\ e^0_g$ electron configuration) species. The interatomic distances obtained from the refinement of the profiles gave three different La-O distances 2.396, 2.698 and 3.027Å for the doped samples and equal Co-O (1.931-1.932 Å) distances for all samples. The three different La-O distances are likely due to the tilting of adjacent CoO_6 octahedra along the cubic axis (Thornton et al., 1986; Mineshige et al., 1996) and different from those reported using neutron diffraction data ($d_{La\text{-}O}$=2.451, 2.702 and 2.992 Å). The lower bond distance of Co-O as compared to the expected on the basis of ionic radii (Co^{iii}-O=1.945 Å and Co^3-O=2.01 Å) could be due to the covalent contribution to the bond.

Titanium Silicalite

Titanium silicalite having MFI structure (TS-1) has shown remarkably high efficiency and selectivity in a number of catalytic oxidation reactions. These properties are attributed to the presence of isolated TiO_4 tetrahedral species in the shape selective silicalite framework with MFI structure (Fig. 4a). Although Ti^{4+} has a strong tendency to coordinate octahedrally with oxygen, the flexibility of the pentasil zeolite framework (Fig. 4a) has been responsible for the isomorphous substitution of tetrahedral Si^{4+} (or T atoms) positions by cations with higher ρ values (according to Pauling criterion theory for stable tetrahedral environment in the range $0.225<\rho<0.414$ ($\rho = r_c/r_o$, r_c and r_o being the ionic radii of the cation and O^{2-}). MFI-type materials show two stable structures: a low temperature phase having the monoclinic symmetry and the high temperature phase having the orthorhombic symmetry. The

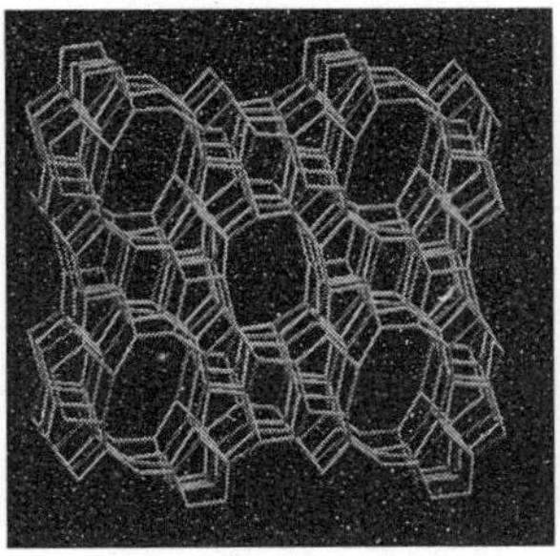

Fig. 4(a): Pentasil Zeolite (MFI) Framework viewed along (010) showing the 10 membered channels.

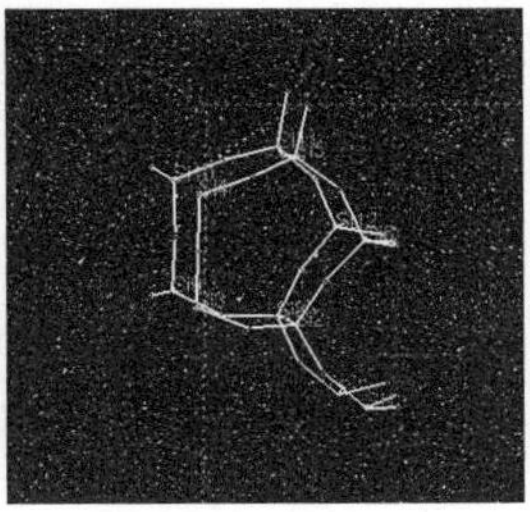

Fig. 4(b): One asymmetric unit (ASU) having 12 T atoms and 24 O_2 atoms.

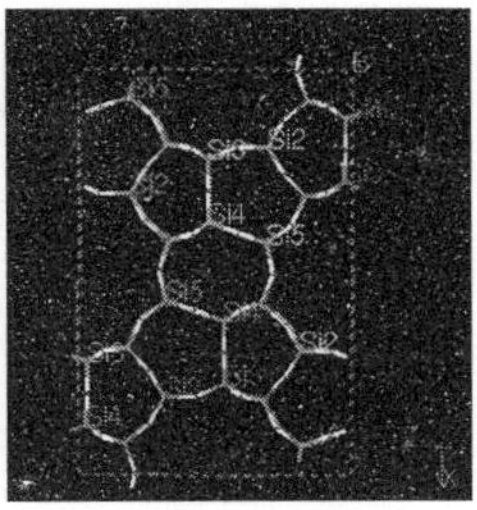

Fig. 4(c): Structure of pentasil (TS-1) showing the atomic labeling scheme (1 unit cell) 96 T atoms and 192 O_2 atoms.

powder XRD pattern of the present example TS-1 taken for study is indexed on a orthorhombic symmetry with Pnma space group. The absence of the splitting of some reflections in the XRD pattern (2θ =24.4° and 29.3°) did not support the presence of monoclinic symmetry. Figures 4b and 4c show the atomic labeling scheme of one asymmetric unit and one unit cell having 12 T atoms, 24 O_2 atoms and 96 T atoms, 192 O_2 atoms, respectively. Table 3 and Fig. 5 give the Rietveld refinement data and plot, respectively, obtained for a typical titanium silicate with MFI structure. The crystallinity of the sample was very good and there was no amorphous titania phase present. The increase in unit cell volume as compared to silicalite-1 (V = 5319Å^3) indicates the expansion of the unit cell due to the isomorphous substitution of Si^{4+} by Ti^{4+}. This substitution of tetrahedral Ti in the framework is responsible for the enhanced catalytic activity in the oxidation of phenol (Veda Ramaswamy et al., 2004). The refinement improves accuracy in the determination of the unit cell dimensions as compared to the values from

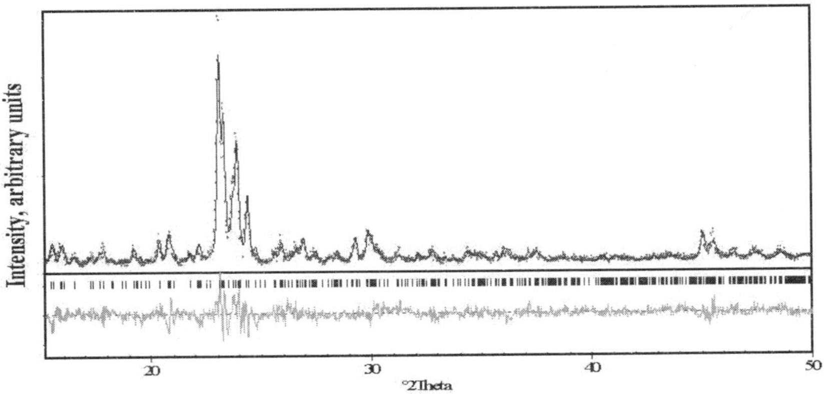

Fig. 5: Observed (....), calculated (----) and difference (bottom) profiles for the Rietveld refinement of Titanium silicalite-1.

Table 3: Crystallographic data on Titanium silicalite (TS-1) samples

Physical data	
Crystallinity-Counts for 100% peak	39555
Si/Ti (XRF)	44.7
Ti, wt% (EDAX)	2.17
Unit cell	
Space Group	Pnma(62)
a (Å)	20.102
b (Å)	19.818
c (Å)	13.405
V^* $(Å)^3$	5341.0
Rietveld refinement data	
Software used	X'PERT PLUS
Profile fit	Pseudo-Voigt
Number of observations	2248
Number of contributing reflections	508
Number of parameters refined (profile and structural)	124
$R_F = \Sigma\mid F(obs)\text{-}F(calc) / \Sigma F(obs)$	6.98
$R_{EXP} = [(N\text{-}P_1\text{-}P_2)/\Sigma wy^2(obs)]^{1/2}$	3.71
$R_{WP} = \{\Sigma w[y(obs)\text{-}y(calc)]^2/\Sigma wy^2(obs)\}^{1/2}$	8.91
Goodness of Fit $\chi^2 = R_{wp} / R_{exp} = 1$	2.4

*Unit cell volume, V of Silicalite-1(silicon analogue of MFI structure) is 5319Å^3.

least square fit. The quality of refinement is good indicated by the low values of the R_{wp} and R_F parameters and the goodness of fit has a value of 2.4 for 124 free parameters. Compared to the previous two examples of oxide samples, the number of parameters refined in this example is very large.

Aluminophosphate (AlPO-40) with AFR Structure

Software programme TREOR was used to index the XRD which gave a unit cell parameters of a = 21.7425, b = 14.2270, c = 13.7269 with a figure of merit = 90 and M_{20}=37 with all the lines indexed. The space group identification was done on the basis of systematic absences and a non-standard setting of $Pc2_1n$ was taken for the Rietveld refinement.

Rietveld refinement using the XRS-82 (Baerlocher, 1993) package was carried out with the space group $Pc2_1n$ and the framework atoms were optimised using DLS (Baerlocher et al., 1997). The lattice parameters were found to change as a function of time in the X-ray beam during the refinement of the profile parameters. Both a and b increases and c decreases as a function of time in the X-ray beam (Veda Ramaswamy et al., 1999). This indicates that perhaps the 12-rings flatten sufficient during the measurement to increase the *a* and *b* values (saddle shaped 12 rings are in the *ab* plane; Fig. 6) and decrease the *c* value. The calculated pattern was scaled to fit the high angle data and the difference Fourier map showed the location of two crystallographically equivalent TPA (tetra propyl ammonium hydroxide) ions in the 12 ring channel. There were no TMAOH ions in the channel although TMAOH is also used in addition to TPAOH in the synthesis. One N, 12 C and 28 H atoms corresponding to the organic template TPAOH in the pore, were then added to the model with geometric restraints on the bond distances and angles. The difference Fourier map at this stage showed a high electron density atom X in the framework connecting Al (2) and Al (3) atoms. This atom position was used in the structure as an oxygen atom (O33) to approximate it and the refinement of the structure was continued with the initial co-ordinates for this atom from the Fourier map.

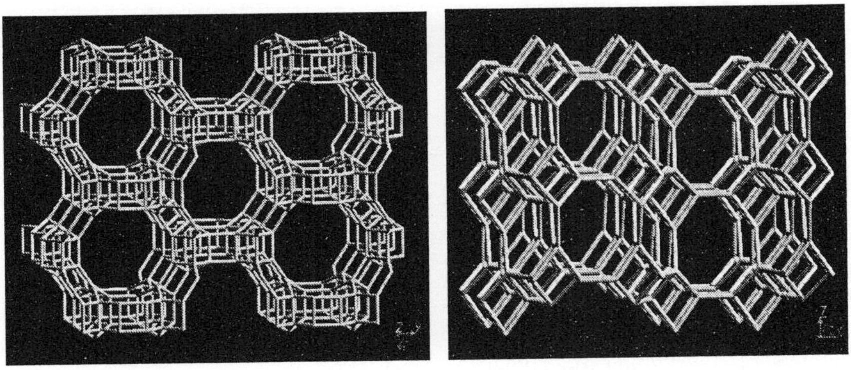

Fig. 6: AFR (AlPO-40) framework viewed in 001 and 010 direction.

To keep the number of parameters to a minimum, only four temperature factors (one each for the tetrahedral framework atoms, for the framework oxygen atoms and for the non-framework N and C atoms) were refined. Subsequent refinement of profile and the structure with fixed cell values, converged with the error indices $R_{wp} = 0.137$, $R_{exp} = 0.121$ and $R_F = 0.053$ (Table 4). The framework was refined using restraints on the bond lengths and angles of the framework atoms (22). The relative weight of these geometric restraints with reference to the X-ray data was generally reduced as the refinement continued. During the course of the refinement the weight of the geometric constraints on both framework and non-framework atoms was steadily decreased until the common weighting factor was equal to 1.0. The common weighting factor was further reduced to 0.1 and it was found that the bond distances and bond angles were not okay as compared to the expected values. Hence it was discarded and the structure was found to be stable only at a common weighting factor of 1.0 with very good agreement of the bond angles and bond distances. The bond distances of Al (2) -O (33) and Al (3)-O (33) were 1.95Å and 2.00 Å, respectively. All positional parameter shifts in the final refinement were less than their corresponding esd's. Neutral atomic scattering factors were used for all atoms. No anomalous dispersion or absorption correction was applied. The crystallographic details are given in Table 4.

Table 4: Crystallographic data for AlPO-40

Data Collection	
Synchrotron facility/ Beamline	SNBL at ESRF/ BL 1
Monochromator/ Analyzer crystal	Ge 111/ Ge 220
Wavelength (Å) and 2θ Range (°2θ)	0.99978 and 1.5-60.15
Step size (°2θ) and Time per step (s)	0.007 and 2-14
FWHM (°2θ)	0.025-0.0336
Unit Cell [a]	
Space group	$Pc2_1n$
a (Å)	21.751
b (Å)	13.730
c (Å)	14.218
Refinement	
Software used	XRS - 82
Standard Peak (hkl, °2θ)	200, 5.27
Peak Range (number of FWHM)	26
Number of observations	8225
Number of contributing reflections	2298
Number of geometric restraints	274
Framework	192
TPA (including H's)	82
Number of structural parameters (including scale)	237
Framework	146
TPAOH (non-H) and H	42 and 48
Number of profile parameters	7
$R_{exp} = [(N-P_1-P_2)/\Sigma wy^2(obs)]^{1/2}$	0.121
$R_{wp} = \{\Sigma[(y(obs)-y(calc)]^2/\Sigma wy^2(obs)\}^{1/2}$	0.137
$R_F = \Sigma\vert F(obs)-F(calc) / \Sigma F(obs)$	0.053

[a]Since the unit cell parameters changed during data collection, only nominal values are given refined.

More details on the refined final atomic co-ordinates, bond lengths are given elsewhere (Veda Ramaswamy et al., 1999). The observed and calculated powder patterns are shown in Fig. 7.

The AFR structure with 185 non–hydrogen positional parameters is the largest refined powder data so far. The topology of AFR structure is confirmed

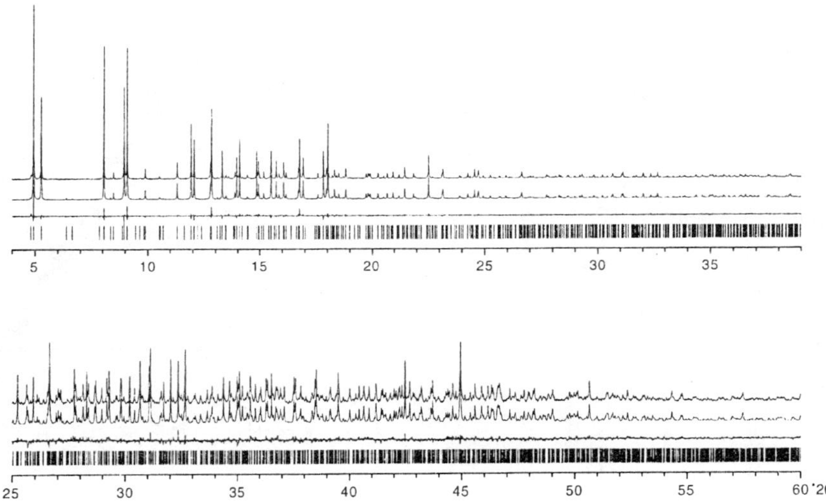

Fig. 7: Observed (top), calculated (middle) and difference (bottom) profiles for the Rietveld refinement of AlPO-40. To show more details at higher angles, the intensity scale has been increased by a factor of 10 (From reference Veda Ramaswamy et al., 1999).

for AlPO-40 also with a complete Rietveld refinement of the structure of AlPO-40. Two each of the four equivalent TPA ions lie on the pseudo two-fold axis running down the centre of the 12-ring channel and are similarly oriented with reference to the framework's topology. The two TPA^+ ions have two propyl groups pointing towards the pockets in the 12-ring channel and two groups pointing towards the 8-ring openings (Figs. 6 and 8). The difference Fourier map showed a negative charge probably OH^- ions in the framework as the refinement of the structure proceeded smoothly after the addition of the atom between two framework Al atoms. This indicates that the template TPA^+ is occluded in the pore volume of AlPO-40 ion counteracting the negative charge of the framework. The lower symmetry of the structure of AlPO-40 as compared to its silicon analogue (SAPO-40) is hence due to the defect in the framework created by the negative charge. There are four OH^- counterions to four TPA^+ ions in one unit cell supporting the number of intracrystalline water molecules per unit cell by TG data. The peak at 18 ppm in the ^{27}Al NMR spectrum of AlPO-40, which disappears on calcination, is indeed due to the difference in the co-ordination of Al(2) and Al(3) aluminium atoms in the framework modified by the presence of OH^- species [O(33)].

CONCLUSIONS

The Rietveld refinement of different kind of catalyst samples viz copper stabilized zirconia having fluorite structure, copper doped lanthanum cobaltate having perovskite structure, titanium substituted silicalite having MFI structure

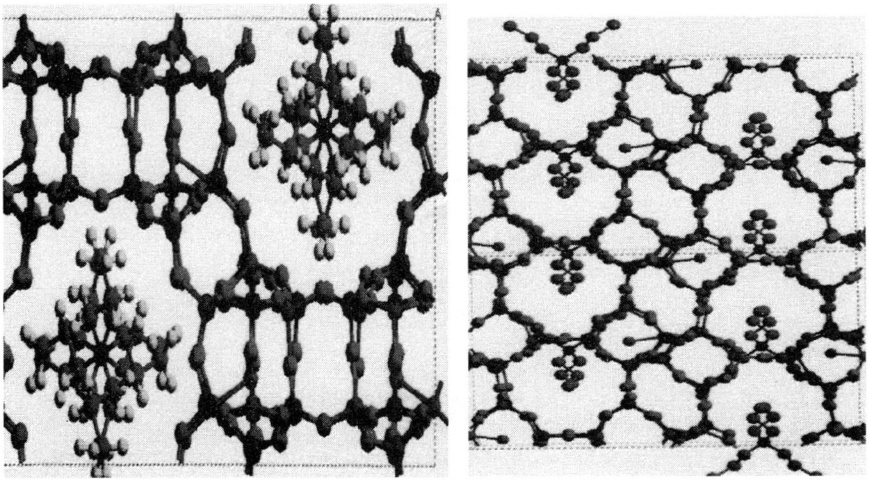

Fig. 8: AFR framework showing the TPA in the 12-ring pore opening viewed along different directions.

and aluminophosphate AlPO-40 having AFR structure has unequivocally added more value to the characterization of the material. In the case of copper stabilized zirconia, the Rietveld refinement only could infer that zirconia is stabilized partially into tetragonal phase and partially into the cubic phase, adding more value to the results obtained from normal analysis of the powder patterns. Thus, zirconia seems to stabilize in the cubic phase at smaller crystallite size and higher copper loading and these samples are likely to yield better performance in the combustion of methane. In lanthanum cobaltate samples with copper loading, the refinement results infer that for complete structural data on a sample sometimes an integrated approach using different geometry during the data collection to get complete structural data such as identification of additional phases and to get good counting statistics for better refinement results. Rietveld refinement of the powder pattern of a typical TS-1 molecular sieve sample indicates the isomorphous substitution of tetrahedral Si by Ti in the framework by the expansion in the unit cell volume and a weak evidence of the possible Si (T) sites of substitution by Ti can also be determined. In the case of aluminophosphate AlPO-40 molecular sieve the refinement and then the difference Fourier map showed a negative charge probably OH^- ions in the framework linked to two aluminium atoms [Al(2) and Al(3)] in the framework which explains the disappearance of the peak at –18 ppm in the ^{27}Al NMR spectrum of AlPO-40 after calcinations. The reason for the change in the unit cell parameters as a function of time in the synchrotron X-ray beam during the refinement is unclear.

Acknowledgements: Author is grateful to Dr. Lynne McCusker, Lab fur Kristallographie, ETH, Zurich, Switzerland and her group for the support and guidance in AlPO-40 work, Dr. A.V. Ramaswamy, Department of Chemistry, University of Pune, Pune for discussions and encouragement, Pilot plant, catalysis division, N.C.L. for TS-1 sample, Dr. T.N. Guru Rao for XRD patterns of $LaCoO_3$ samples and Mr. M.A. Bhagwat, Catalysis Division for experimental help.

REFERENCES

Arai, H., Yamada, T., Euguchi, K. and Seiyamma, T. (1986). *Appl. Catal.*, **26:** 265.

Arata, K. (1990). *Adv. Catal.*, **37:** 165.

Baerlocher, Ch. (1993). *In:* The Rietveld method. Ed. Young, R.A., Oxford Univ. Press, Oxford, IUCr, 186.

Baerlocher, Ch., Hepp, A. and Meier, W.M. (1977). Distance Lesat Squares refinement Program DLS-76, Institute fur Kristallographie und Petrographie, ETH, Zurich.

Dongare, M.K., Veda Ramaswamy, Gopinath, C.S., Ramaswamy, A.V., Scheurell, S., Brueckner, M. and Kemnitz, E. (2001). *J. Catal.*, **199:** 209.

Dumont, N., Gabelica, Z., Deroune, E.G. and McCusker, L.B. (1993). *Micro.Mater.*, **1:** 149.

Etsell, T. H. and Flengas, S.N. (1970). *Chem Rev.*, **70:** 1970.

Hoch, M. and Mathur, M. (1962). *J. Am. Ceram. Soc.*, **45:** 373.

Keshavaraja, A. and Ramaswamy, A. V. (1994). *J. Mater. Res.*, **9:** 837.

Klug, H. P. and Alexander, L. E. (1974). X-ray diffraction procedures : For polycrystalline and amorphous materials: John Wiley and Sons, New York, 618.

Lamas, D.G. and Walsue De Reca, N. E., (1999). *Mater. Lett.*, **41:** 204.

Lourenco, J.P., Ribeiro, M.F., Ramoa Ribeiro, F., Rocha, J., Onida, B., Garrone, E. and Gabelica, Z. (1997). *Zeolites*, **18:** 398.

Mahesh Bhagwat, Ramaswamy, A.V., Tyagi, A.K. and Veda Ramaswamy (2003). *Mater.Res.Bull.*, **38:** 1713.

Marra, G.L., Artioli, G., Fitch, A.N., Milanesio, M. and Lamberti, C. (2000). *Micro. Meso. Mater.*, **40:** 85.

McCusker, L.B. and Baerlocher, Ch. (1996). *Micro. Mater.*, **6:** 51.

Meden, A., Baerlocher, Ch. and McCusker, L.B. (1997). *Micro. Mater.*, **11:** 247.

Meierand, W.M. and Olson, D.H. (1992). Atlas of Zeolite structure Types. Butterworth-Heinemann, London, 30.

Millini, R., Previde Massara, E., Perego, G. and Bellusi, G. (1992). *J. Catal.*, **137:** 690.

Mineshige, A., Inaba, M., Yao, T., Ogumi, Z., Kikuchi, K. and Kawase, M. (1996). *J. Solid. State. Chem.*, **121:** 423.

Porta, P., DeRossi, S., Faticanti, M., Minelli, G., Pettiti, I., Lisi, L. and Tourco, M. (1999). *J. Solid. State. Chem.*, **146:** 291.

Rakshe, B., Veda Ramaswamy and Ramaswamy, A.V. (1996). *J. Catal.*, **163:** 501.

Rakshe, B., Veda Ramaswamy, Vetrivel, R., Hegde, S.G. and Ramaswamy, A.V. (1997). *Catal. Lett.*, **45:** 41.

Ray, J. C., Pramanik, P. and Ram, S. (2001). *Mater. Lett.* **48:** 281.

Rietveld, H.M. (1969). *J. Appl. Crystallography*, **2:** 65.

Thangaraj, A., Kumar, R., Mirajikar, S.P. and Ratnasamy, P. (1991). *J. Catal.*, **130:** 1.

Thornton, G., Tofield, B.C. and Hewat, A.W. (1986). *J. Solid, State. Chem.*, **61:** 301.

Veda Ramaswamy, Awati, P. and Tyagi, A.K. (2004). *J. alloys and Compds.*, **364:** 180.

Veda Ramaswamy, McCusker, L.B. and Baerlöcher, C. (1999). *Micro. Mater.*, **13:** 1.

Wang, Y. and Caruso, R. C. (2002). *J. Mater. Chem.*, **12:** 1442.

Zhou, R. X., Jiang, X. Y., Mao, J. X. and Zheng, X. M. (1997). *Appl. Catal. A*, **162:** 213.

Crystal Structure Refinement of $Ni_xMg_{1-x}Fe_2O_4$

R. Nithya, V.K. Mittal[1] and Santanu Bera[1]

Materials Science Division, IGCAR, Kalpakkam-603102, India
[1]Water and Steam Chemistry Laboratory, BARC Facilities Kalpakkam-603102, India

INTRODUCTION

Oxides with a general formula, AB_2O_4 (A: divalent and B: trivalent metal cations) having a cubic spinel structure exhibit a large variety of physical and chemical properties. The oxygen ions form a face centered cubic lattice and cations depending on their relative size and site preference energies occupy interstitial positions between anions. Among 96 interstitials between anions, there are 64 tetrahedral sites and 32 octahedral sites. There are eight formula units per unit cell i.e., each of which consist of 32 anions and 24 cations (Kurt et al., 1999). Of the 64 tetrahedral positions available only eight are occupied by cations and the remaining 16 cations occupy half of the 32 octahedral interstices. There are two extreme distributions possible: normal and inverse. In a normal spinel structure, divalent cations occupy tetrahedral sites and trivalent cations occupy only octahedral sites. Conversely, in an inverse spinel, divalent cations occupy only octahedral sites and trivalent cations are distributed equally among tetra and octa sites. The cation distribution can be determined by an inversion parameter, δ which specifies the fraction of A sites occupied by trivalent cations. Thus, for a normal spinel, $\delta = 0$ and for an inverse spinel $\delta=1$. The degree of inversion varies in a very complex way. Some of the factors that influence cation inversion are (i) synthesis procedure, (ii) thermal history like heating and cooling rates, (iii) cation radii etc., (O'Neill et al., 1992; Seshan et al., 1981). Spinel ferrites AFe_2O_4 represent an important class of magnetic materials used in many technological applications. The interesting physical, chemical, electrical and magnetic properties of these compounds arise from their ability to distribute the cations among tetra and octahedral sites (Blasse, 1964). $MgFe_2O_4$ is a typical spinel in which the cation distribution in the lattice is very much sensitive to heat treatment due to high diffusibility of Mg^{2+} ions (Seshan et al., 1981; Sawatzky et al., 1968; DeGrave et al., 1979). In contrast, $NiFe_2O_4$ exhibits fully inverse spinel structure and does not show any temperature dependence because Ni^{2+} has a strong octahedral site preference (Sawatzky et al., 1968). A solid solution of these two, however, is expected to show changes in cation distribution and hence, properties as a function of nickel content, x in $Ni_xMg_{1-x}Fe_2O_4$. Due to strong octahedral site preference of Ni^{2+} ions, the competition to occupy the tetrahedral sites would be between the Mg^{2+} and Fe^{3+} ions. In view of these considerations, it would be interesting to study their structural properties.

EXPERIMENTAL PROCEDURE

Polycrystalline samples of $Ni_xMg_{1-x}Fe_2O_4$ (x = 0, 0.1, 0.2, 0.3, 0.5, 0.7, 0.8 and 1) were prepared by the conventional solid state reaction in air using $MgCO_3$, $NiCO_3$, $2Ni(OH)_2.4H_2O$ and α-Fe_2O_3. These oxides, taken in appropriate ratios, were thoroughly ground in acetone medium for homogenization. The mixture was pressed into pellets and the pellets were sintered at 1200 °C for 30 hrs with one intermediate grinding.

X-ray diffraction (XRD) patterns were obtained with a STOE diffractometer in Bragg-Brentano geometry using CuKα radiation in reflection mode. Data for Rietveld refinement were collected in the 2θ range 10° to 100° with a step interval of 0.05° and a count time of 10 sec. per step. Rietveld refinements of the XRD patterns were performed by a multi-purpose programme package, General Structure Analysis System, (GSAS) (Larson and Von Dreele, 2000; Toby, 2001). This programme can be used for both neutron and XRD data obtained from both angle dispersive and energy dispersive modes.

RIETVELD ANALYSIS

Rietveld analysis is a computer based method to calculate diffraction pattern by least squares fitting from a structural model given and compare with the experimental whole powder diffraction pattern obtained from X-ray/Neutron source. Rietveld (1969) first developed this method to study crystal structures from neutron powder diffraction data. Later this method was adopted to powder XRD data by Malmros and Thomas (1977) and Khattak and Cox (1977). Since then, many versions of computer programmes were developed and some of them are DBWS, GSAS, RIETAN etc. downloadable from the internet. Rietveld analysis proceeds via a number of steps. Initially, background of the powder diffraction pattern should be fitted. There are eight functions viz., Chebyschev polynomial, Cosine Fourier series, linear interpolation etc., available in GSAS to fit background intensity. Some of the functions are useful for both X-ray and Neutron powder data. Many of these functions are quite flat but there are some functions which represent well diffused background from an amorphous phase present in samples or background from a quartz sample holder. Instrument parameters like diffractometer zero should be refined followed by refinement of profiles. Angle-dispersive powder XRD profiles obtained with powder diffractometer have a complex shape due to instrument geometry, micro-structural defects such as micro strain and small crystallite size. Hence the profile function used to model the diffraction profile is very important in obtaining the best fit to the observed diffraction data during refinement. In general, there are six analytical functions available for quantifying XRD profile peaks in Rietveld refinement. For the angle-dispersive X-ray data, pseudo-Voigt and Pearson VII functions represented the experimental diffraction pattern well (Young and Wiles, 1982).

In Ferrites, unit cell origin can be assigned at a vacant site or an occupied lattice site. In the present study, origin was chosen at an octahedral vacancy with the point symmetry $\overline{3}$ m. Crystal structure refinements were performed on the diffraction data using GSAS package assuming space group, Fd3m with cations at 8a, 16d and oxygen at 32e positions. The background was refined with a linear interpolation function with 12 coefficients. The diffraction profiles were quantified by a Pseudo-Voigt function corrected for asymmetry which gives a better fit to asymmetric profiles and shows less correlation with lattice parameters. Apart from structural parameters, there are other parameters like instrument constants, absorption parameter, asymmetry parameters, preferred orientations, background which are to be refined. The computations are done in two steps consisting of data-preparation step and then a least squares fit. The input file to the refinement programme contains the following information:

1. Powder diffraction data.
2. Structural model of the compound (from previous studies and chemistry of the compound). The structure model inputs are approximate unit cell parameters, the space group and a set of fractional atomic coordinates, site occupancies and thermal parameters.

Then powder diffraction pattern is calculated using the structure model, X-ray optics of the instrument with the profile function chosen. Following the above steps, the structural model is optimized by least squares refinement of the various parameters until suitable agreement between the experimental and calculated data is obtained. Refinement is achieved by minimizing the sum of squares of the weighted differences between the observed and calculated intensities for every point of the profile under the Bragg reflections by adjusting structural parameters (i.e., lattice parameters, fractional atom coordinates, occupancy parameters and thermal parameters) and profile parameters (U, V, W, asymmetry, sample shift, transparency coefficient, Lorentzian parameters corresponding to strain and size broadening etc.) to get a best fit of the entire calculated diffraction pattern to the whole experimental diffraction pattern. The quality of the fit is indicated by a set of residual parameters corresponding to a global minimum and they are:

Goodness-of-fit, $$\chi^2 = \frac{M_{\mathrm{p}}}{(N_{\mathrm{obs.}} - N_{\mathrm{var.}})}$$

R-structure factor, $$R(F^2) = \frac{\sum(|F_{\mathrm{o}}| - [SF_{\mathrm{c}}])}{\sum[F_{\mathrm{o}}]}$$

R-pattern, $$R_{\mathrm{p}} = \frac{\sum(I_0 - I_{\mathrm{c}})}{\sum I_{\mathrm{o}}}$$

R-weighted pattern, $$R_{\mathrm{wp}} = \frac{\sqrt{M_{\mathrm{p}}}}{\sum w I_{\mathrm{o}}^2}$$

Durbin-Watson parameter, $$D_{\mathrm{dw}} = \frac{\sum_{i=2}^{N_{obs.}} (\Delta_{\mathrm{i}}/\sigma_{\mathrm{i}} - \Delta_{\mathrm{i-1}}/\sigma_{\mathrm{i-1}})^2}{\sum_{i=1}^{N_{obs.}} (\Delta_{\mathrm{i}}/\sigma_{\mathrm{i}})^2}$$

where M_{p} is the minimization function given by $M_{\mathrm{p}} = \Sigma w(I_{\mathrm{o}} - I_{\mathrm{c}})^2$, $N_{\mathrm{obs.}}$ and $N_{\mathrm{var.}}$ are no. of observations and no. of variables considered for the refinement, I_{o} and I_{c} are observed and calculated intensities, F_{o} and F_{c} are observed and calculated structure factors, S is a scale factor and w is the weight fraction $(=1/y_{\mathrm{i}})$. A statistical measure of the serial correlation in the powder diffraction pattern differences ($\Delta = I_{\mathrm{o}} - I_{\mathrm{c}}$) is given by Durbin-Watson parameter given above. The value at the start of the refinement is usually close to zero because of strong correlations and it increases to some maximum value as the refinement progresses. The value of D_{dw} will be close to two if there are no serial correlations present in the differences. $\Delta_{\mathrm{i}}/\sigma_{\mathrm{i}}$ is the weighted residual of the least squares at the ith data.

RESULTS AND DISCUSSION

X-ray Diffraction

XRD pattern for $MgFe_2O_4$ is shown in Fig. 1. All the peaks could be indexed to a cubic spinel structure with space group, Fd3m (No. 227 in International Tables for Crystallography). No impurity phases were found within the detectable limits of powder XRD technique. Figure 2 illustrates diffraction patterns of all the samples under study. The diffraction patterns of all samples showed peaks characteristic of single phase cubic spinel structure. In this compound, it is well known that the diffraction peaks from

(111), (222) and (220), (440) planes are due to the atoms present at octahedral and tetrahedral sites, respectively (Sepelak et al., 1998). Redistribution of cations between tetra- and octahedral sites is evident from the increase in the intensities of (111) and (222) peaks and decrease in the intensity of (220) and (440) peaks with increase in nickel content in the samples. In order to determine the cation distribution, integrated intensities of the observed diffraction patterns were calculated from pattern fitting module available with WinXPow software. Thus, ratios of the intensities of the reflections considered to be sensitive to the cation distribution were determined. The ratio of integrated intensity, I_{220}/I_{222} against x is plotted in Fig. 3. The decrease in the ratio indicates the removal of Mg ions from tetrahedral sites increasing the degree of inversion.

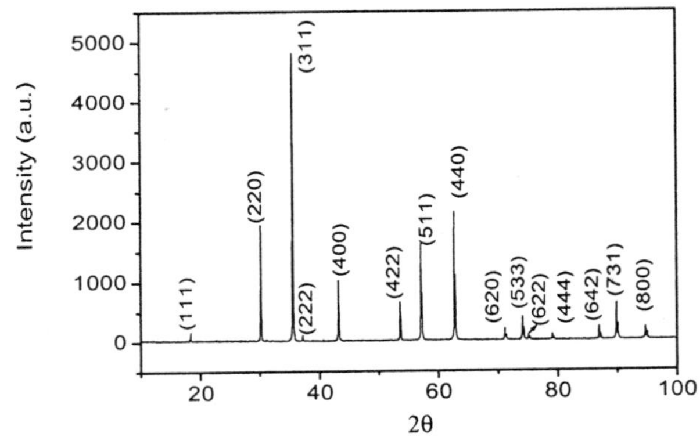

Fig. 1: Indexed XRD pattern of $MgFe_2O_4$.

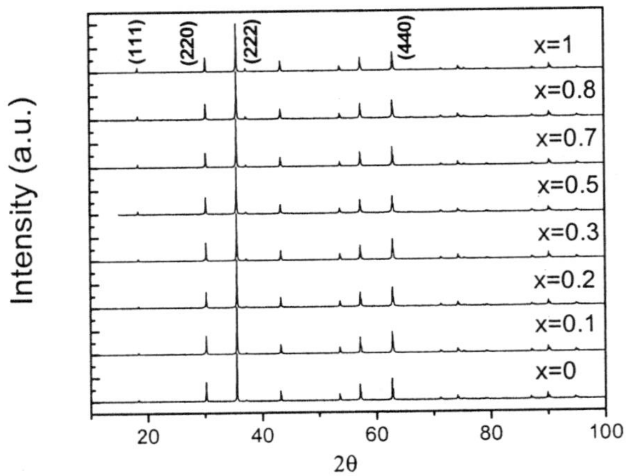

Fig. 2: Powder XRD patterns of $Ni_xMg_{1-x}Fe_2O_4$.

Structure Refinement

Structure refinement of powder XRD patterns of $Ni_xMg_{1-x}Fe_2O_4$ were done with space group Fd3m. At the start of the refinement, both Mg and Fe ions

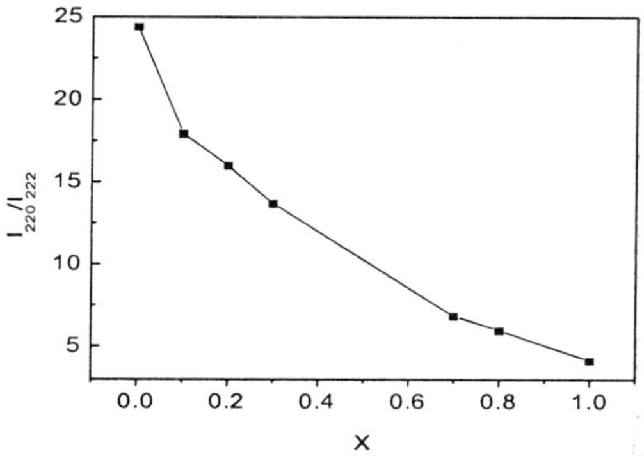

Fig. 3: Intensity ratio of (220) and (222) diffraction planes as a function of Ni concentration in $Ni_xMg_{1-x}Fe_2O_4$.

were randomly placed in both tetrahedral and octahedral sites and oxygen in a special position, 32e. In the case of nickel substituted samples, nickel ions were placed at octahedral positions due to its strong site preference to octahedral positions. Because Fe, Mg and Ni occupy the same sites, their atom parameters must be constrained to shift together during least squares refinement. Hence, the occupancies of Fe, Mg and Ni in both sites were constrained independently so as to preserve their stoichiometric value. Thermal parameters were considered to be isotropic for all the samples. While refining isothermal parameters, it was assumed that all cations have same U_{iso}. This assumption yielded low residual values compared to each cation or cations in tetrahedral/octahedral to have individual values. A good agreement between the experimental and the calculated XRD patterns for all the samples was observed with low residuals. Thus from refinement analysis, cation distribution, lattice, isothermal parameters and oxygen parameter as a function of nickel composition have been determined. Refined lattice parameters and their density from the final refinement cycle are plotted against nickel content in Fig. 4. The relationship between the lattice parameter and the concentration of nickel obeys Vegard's law. The results provide evidence for a contracting unit cell as the smaller Ni ion is substituted on the Mg site. From the crystal structure refinements, it is seen that the solid solution of nickel ferrite and magnesium ferrite compounds studied in the present work are single phase with a cubic structure. In $MgFe_2O_4$, Mg^{2+} ions are distributed both in octahedral and tetrahedral sites and the inverse parameter equal to 0.89. As nickel is substituted for magnesium, due to the strong preference it replaces Mg at octahedral site initially and for higher concentrations of Ni it pushes iron present in the octahedral site to tetrahedral sites. Thus the inverse parameter for $NiFe_2O_4$ is 1.0.

From the refined occupancy factors, the refined chemical composition is $MgFe_2O_{3.99}$ which is close to its nominal composition of $MgFe_2O_4$. The result of the refinements of cation occupancy at different sites establishes

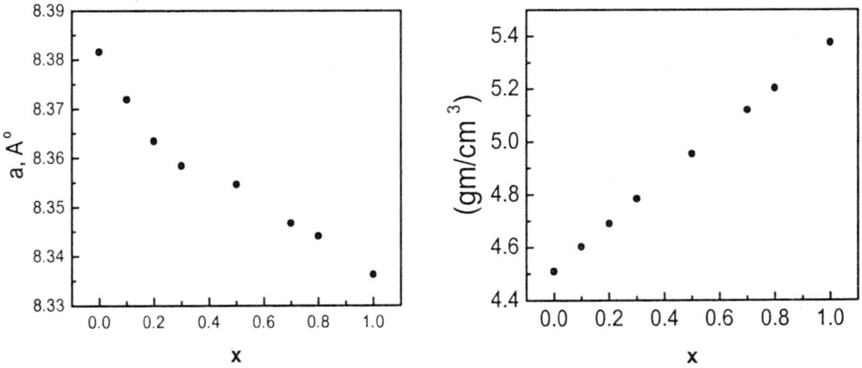

(a) (b)

Fig. 4: Variation of (a) lattice parameter and (b) X-ray density as a function of x in $Ni_xMg_{1-x}Fe_2O_4$.

the following cation distribution in this compound $(Mg_{0.11}Fe_{0.89})[Mg_{0.445}Fe_{0.555}]O_{3.9983}$ where parenthesis and square brackets represent tetrahedral and octahedral sites, respectively. Some of the representative graphs of the observed powder pattern (symbols) along with the calculated intensities (solid curve) are displayed in Figs. 5, 6 and 7. The difference curve between calculated and observed pattern is also included in all the figures at the bottom of the graph.

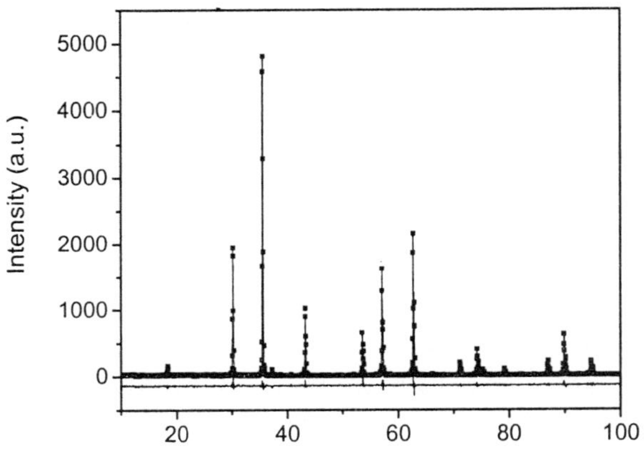

Fig. 5: Observed (symbols) and calculated XRD profile of $MgFe_2O_4$ along with a difference plot.

The structural parameters i.e., positional coordinates, occupancies, thermal displacements corresponding to $MgFe_2O_4$ are given in Table 1 and for one of the nickel substituted samples, $Ni_{0.2}Mg_{0.8}Fe_2O_4$ the parameters are given in Table 2. Table 3 presents the residuals got for the final refinement for all the samples. As can be seen from this table, the goodness of fit is low for all the samples. Cation distribution for the solid solutions of the present

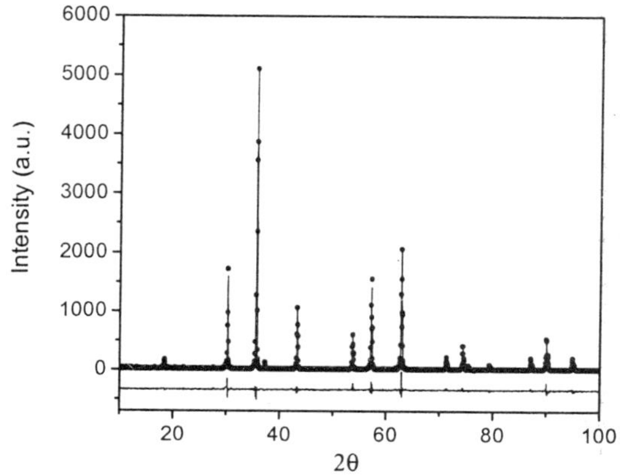

Fig. 6: Observed (symbols) and calculated XRD profile of $Ni_{0.2}Mg_{0.8}Fe_2O_4$ along with a difference plot.

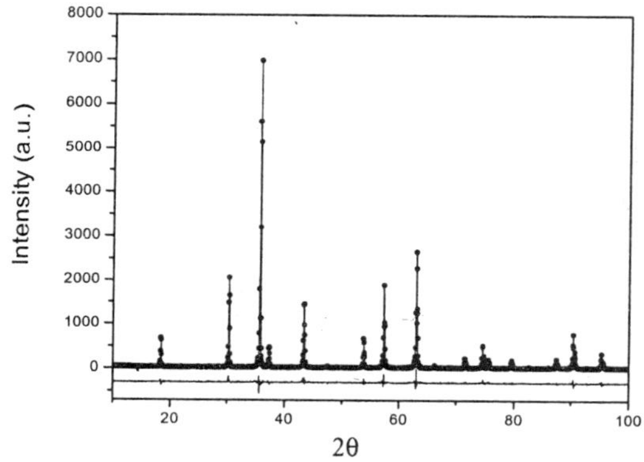

Fig. 7: Refined powder XRD pattern of $NiFe_2O_4$.

system are given in Table 4. The integrated intensity calculations of individual X-ray peaks and whole powder pattern refinement of the profile confirm that the divalent nickel ion has an octahedral preference. The positional coordinate for oxygen, u was 0.256523 for $x = 0$. It decreased to 0.253155 for $x = 0.7$ and then increased to 0.254346 for $x = 1.0$.

Table 1: Crystallographic data for $MgFe_2O_4$: x, y, z-positional coordinates, U_{iso}-isotropic thermal parameter, a = 8.381564(13) Å

Atom type	$x = y = z$	*Multi.*	*Occupancy*	*Uiso*
Mg^{2+}	0.5	16	0.445(8)	0.0002(11)
Fe^{3+}	0.5	16	0.555	0.0002
Mg^{2+}	0.125	8	0.11	0.0002
Fe^{3+}	0.125	8	0.89	0.0002
O^{2-}	0.256523(330)	32	0.9983	0.00061

Table 2: Rietveld refinement results of $Ni_{0.2}Mg_{0.8}Fe_2O_4$: x, y, z-positional coordinates, Uiso-isotropic thermal parameter, a = 8.3634 Å

Atom type	*x = y = z*	*Multi.*	*Occupancy*	*Uiso*
Mg^{2+}	0.5	16	0.3541	0.00186
Ni^{2+}	0.5	16	0.0950	0.00186
Fe^{3+}	0.5	16	0.5509	0.00186
Mg^{2+}	0.125	8	0.1018	0.00186
Fe^{3+}	0.125	8	0.8982	0.00186
O^{2-}	0.253837	32	1.0	0.00329

Table 3: Agreement factors from the final Rietveld refinement for all the samples

x	χ^2	$R(F^2)$	R_p	R_{wp}	D_{wd}
0	1.29	3.90	10.3	15.9	1.819
0.1	1.30	4.8	11.2	16.75	1.776
0.2	1.52	5.37	12.0	16.84	1.705
0.3	1.60	6.91	12.3	16.67	1.495
0.7	1.52	5.33	10.8	15.41	1.449
0.8	1.54	5.43	10.5	15.14	1.562
1.0	1.45	3.27	9.62	14.33	1.726

Table 4: Cation Distributions in $Ni_xMg_{1-x}Fe_2O_4$ obtained from Rietveld analysis

X	*Cation Distribution*
0.0	$(Mg_{0.11}Fe_{0.89})[Mg_{0.445}Fe_{0.555}]_2O_{3.9932}$
0.1	$(Mg_{0.11}Fe_{0.89})[Mg_{0.3957}Ni_{0.0493}Fe_{0.555}]_2O_{3.9932}$
0.2	$(Mg_{0.1018}Fe_{0.8982})[Mg_{0.3541}Ni_{0.095}Fe_{0.5509}]_2O_{4.0}$
0.3	$(Mg_{0.1061}Fe_{0.8939})[Mg_{0.297}Ni_{0.15}Fe_{0.553}]_2O_{4.0}$
0.7	$(Mg_{0.1}Fe_{0.9})[Mg_{0.1}Ni_{0.35}Fe_{0.55}]_2O_{4.0}$
0.8	$(Mg_{0.0638}Fe_{0.9362})[Mg_{0.0681}Ni_{0.4}Fe_{0.5319}]_2O_{4.0}$
1.0	$(Fe_{1.0})[Ni_{0.5}Fe_{0.5}]_2O_{4.0}$

CONCLUSIONS

Investigations on the cation distributions of polycrystalline solid solutions of $NiFe_2O_4$ and $MgFe_2O_4$ have been carried out. A series of compounds corresponding to $Ni_xMg_{1-x}Fe_2O_4$, x = 0 to 1.0 were prepared by high temperature solid state reaction in air. Room temperature powder XRD technique revealed the single phase nature of the compounds. Nickel substitution into $MgFe_2O_4$ brings about changes in the cation distribution among tetrahedral and octahedral sites that are reflected in the diffraction intensities of some of the peaks. There are considerable changes in the intensities of reflections with increasing Ni content, indicating variations in the distribution of three cations among tetrahedral and octahedral positions in the lattice. Rietveld refinements of diffraction patterns were performed using GSAS package. By varying a number of parameters during refinement, close agreement to the observed diffraction patterns were obtained with low reliability factors. The inversion parameter changed from 0.89 for $MgFe_2O_4$ to 1.0 for $NiFe_2O_4$.

REFERENCES

Blasse, G. (1964). Philips Res. Rep. Suppl. **3:** 91.

De Grave, E., Govaert, A. and Chambaere, D. (1979). *Physica B*, **96:** 103.

Khattak, C.P. and Cox, D.E. (1977). *J. Appl. Crystallogr.*, **10:** 405.

Kurt E. Sickafus, John, M. Wills and Norman, W. Grimes (1999), *J. Am. Ceram. Soc.*, **82:** 3279.

Larson, A.C. and Von Dreele, R.B. (2000). "General Structure Analysis System (GSAS)", Los Alamos National Laboratory Report : LAUR 86-748.

Malmros, G. and Thomas, J.O. (1977). *J. Appl. Crystallogr.*, **10:** 7.

O'Neill, H.S.T.C., Annersten, H. and Virgo, D. (1992). *Am. Mineralogist*, **77:** 725.

Rietveld, H.M. (1969). *J. Appl. Cryst.*, **2:** 65.

Sawatzky, G.A., Vander Woude, F. and Morrish, A.H. (1968). *J. App. Phys.*, **39:** 204.

Sepelak, V., Buchal, A., Tkacova, K. and Becker, K.D. (1998). *Materials Science Forum*, **862:** 278-281.

Seshan, K., Shashimohan, A.L., Chakrabarthy, D.K. and Biswas, A.B. (1981). *Phys. Stat. Sol.* (a), **68:** 97.

Toby, B.H. (2001). EXPGUI, a graphical user interface for GSAS. *J. Appl. Cryst.*, **34:** 210.

Young, R.A. and Wiles, D.B. (1982). *J. Appl. Cryst.*, **15:** 430.

XRD Rietveld Analysis Studies of the Annealing Behaviour of the Superconducting Ferromagnetic Compound $GdSr_2RuCu_2O_8$

J. Janaki, T. Geetha Kumary, M.C. Valsakumar, Y. Hariharan and V.S. Sastry

Materials Science Division, IGCAR
Kalpakkam-603 102, India

INTRODUCTION

The hybrid ruthenocuprate $GdSr_2RuCu_2O_8$ with the triple perovskite structure exhibits a variety of interesting physical properties including high temperature superconductivity and weak ferromagnetism (Tallon et al., 1999; Bernhard et al., 1999). The combination of superconductivity and ferromagnetism in the same compound is generally uncommon and many interesting consequences can arise due to such a coexistence like formation of a spontaneous vortex phase etc. (Blount and Varma, 1979). $GdSr_2RuCu_2O_8$ is isostructural with the well known high T_c superconductor $YBa_2Cu_3O_{7-\delta}$ with Y and Ba being completely replaced by Gd and Sr, respectively, and the CuO chain being replaced by RuO_2 planes (Fig. 1). The current broad understanding of the system is that superconductivity originates in the CuO_2 planes and magnetic ordering in the RuO_2 planes. The crystal structure of $GdSr_2RuCu_2O_8$ is tetragonal (Kuz'micheva et al., 2001; Chmaissen et al., 2000) (space group P4/mmm, $a = 3.84$Å and $c = 11.57$Å). Electron diffraction studies (McLaughlin et al., 1999) indicate evidence for a $\sqrt{2a}$ X $\sqrt{2a\,xc}$ superstructure related to rotation of RuO_6 octahedra and a domain structure originating from the fact that $c \approx 3a$. The changes in these and other microstructural features as a function of annealing including distribution of grain sizes and compositional mapping have not been adequately investigated in the literature.

Despite the chemical complexity of the compound $GdSr_2RuCu_2O_8$ it has been prepared in single phase form by solid state reaction. Although the formation and study of physical properties of the system $LnSr_2RuCu_2O_8$ and a related system $Ln_{2-x}Ce_xSr_2RuCu_2O_{10}$ (Ln = Sm, Gd, Eu) were reported as early as 1995 (Bauernfeind et al., 1995), the existence of superconductivity was not established in $LnSr_2RuCu_2O_8$ until much later. This is because of the inherent granularity and disorder which may be present in the samples. This may lead to ill defined superconducting characteristics. A range of values, 15-50K, has been reported for T_c in the literature. Reports also exist of non-occurrence of superconductivity as well as occurrence of zero resistivity with absence of diamagnetic signal in this compound. In this context it is desirable to study the annealing effects on T_c and ΔT_c (superconducting transition width) and their dependence on the structural and microstructural state of the sample. The outcome of such a study is expected to provide clues for synthesis of improved materials with superior superconducting properties. Strong annealing effects on T_c and ΔT_c have already been indicated in literature (Lorentz et al., 2001, Geetha Kumary et al., 2003). Further studies on the annealing behaviour of $GdSr_2RuCu_2O_8$ have been carried out by us, at present using X-ray diffraction (XRD). Rietveld analysis of the as-prepared and annealed samples, and the results correlate well with T_c measurements carried out earlier (Geetha Kumary et al., 2003).

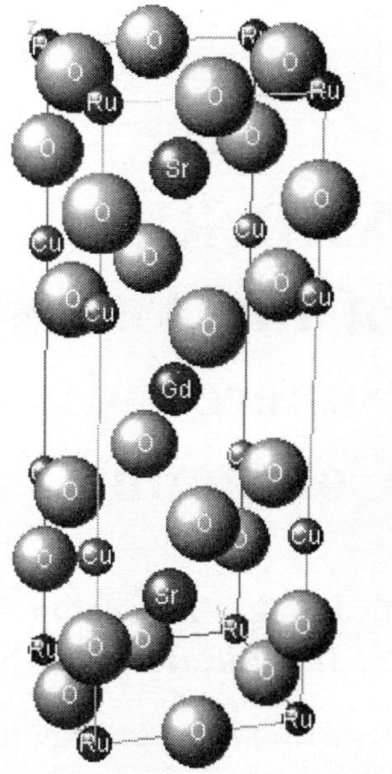

Fig. 1: Idealized structure of $GdSr_2RuCu_2O_8$.

EXPERIMENTAL PROCEDURE

Sample Preparation

The compounds have been prepared starting from high purity Gd_2O_3, $SrCO_3$, CuO and Ru powder by solid state reaction. The mixture of starting materials in stoichiometric quantities was ground and heated at 600 °C for 48 hrs to avoid RuO_2 volatility and then at 950 °C for 24 hrs in air. It was then compressed into pellets and then heated at 1000 °C in air for three days. This constituted the as-prepared sample. For optimizing the superconducting parameters annealing was found necessary. The annealing was carried out at 1060 °C in oxygen atmosphere for different time durations including 15 hrs, three days and one week followed by slow cooling up to 400 °C. The compounds have been characterized by XRD analysis.

Characterization

High resolution powder XRD data has been obtained using a STOE diffractometer with a scintillation counter detector and CuKα radiation. The runs were taken in the step scanning mode with a step of 0.02° (2θ) and counting time of 15 secs and divergence and receiving slits of 0.20 mm. A zero background specimen holder comprising a Si (911) single crystal wafer specially developed in our laboratory (Rao et al., 1996) has been used for loading the sample for XRD. An SEM-EDS study of the microstructure and chemical composition has been carried out using Phillips XL30 ESEM. The physical properties including the electrical resistance and magnetic susceptibility of all the samples have been measured using a homemade dipstick in the temperature range 4.2-300 K. A calibrated Si diode thermometer has been used for temperature measurement. A.C. susceptibility studies have been carried out by the Mutual Inductance method and resistivity measurements using the four probe method.

Rietveld Analysis

The Rietveld refinement has been carried out using GSAS-EXPGUI software (Larson and Von Dreele, 2000; Toby, 2001). The background function used was the Chebyschev polynomial of the first kind as given in the GSAS Rietveld refinement programme. The refinement was carried out in the sequence: background, scale factor, lattice parameters, profile parameters, site occupancies and thermal parameters. The profile function used for fitting was the Type 4 function among the four types of constant wavelength profile functions. This is basically a convolution of the pseudo-Voigt function and an asymmetry function due to axial divergence. It also incorporates P. Stephen's microstrain broadening treatment (Stephens, 1999). The coefficients from this profile function X, X_e and S_{hkl} can be interpreted to give both strain and particle size information. The volume-weighted average particle size is given by $p = 18000\, K\lambda/\Pi X$ where K is the Scherrer constant depending upon the crystalline shape and the units of p are in Å. The anisotropy particle size components are given by p (parallel) = 18000 $K\lambda/\Pi(X + Xe)$ and p (perpendicular) = 18000 $K\lambda/\Pi X$.

RESULTS AND DISCUSSION

Preliminary Profile Analysis studies indicated that the peaks fit well to a Lorentzian profile. Hence the condition GU=GV=GW=GP=0 was incorporated in the pseudo-Voigt function of Type 4. The Rietveld refinement plots for the as-prepared sample of $GdSr_2RuCu_2O_8$ and the samples annealed at 1060 °C in O_2 for three days and one week are shown in Figs. 2, 3 and 4, respectively. The Rietveld refinement plot for NBS, silicon is shown in Fig. 5. Because the NBS Si sample has very large and strain-free crystallites,

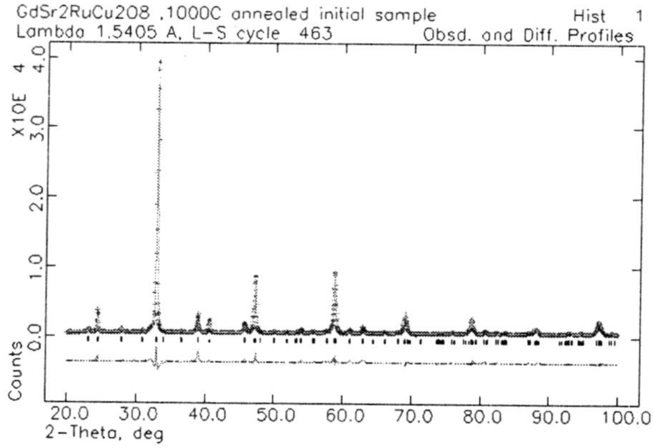

Fig. 2: Rietveld refinement plot for as prepared $GdSr_2RuCu_2O_8$.

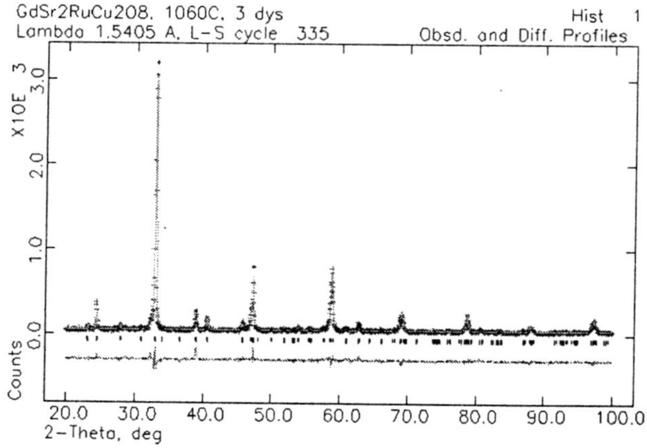

Fig. 3: Rietveld refinement plot for $GdSr_2RuCu_2O_8$ annealed in O_2 at 1060 °C for three days.

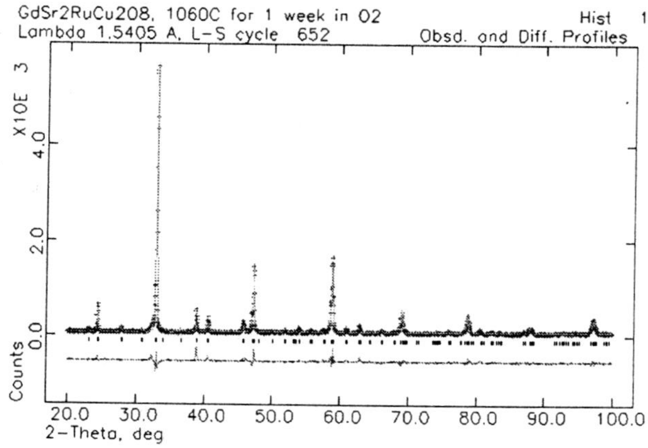

Fig. 4: Rietveld refinement plot for $GdSr_2RuCu_2O_8$ annealed in O_2 at 1060 °C for one week.

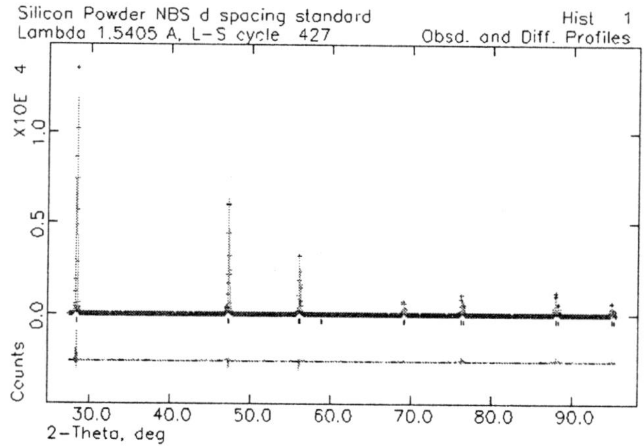

Fig. 5: Rietveld refinement plot for a standard sample of NBS Silicon powder.

the broadening observed for this sample represents the instrumental contribution to the peak broadening. The residuals obtained on Rietveld refinement using GSAS were around 10% for R_{wp} and R_p for all the refinements (Table 1). The lattice parameters and site occupancies obtained on Rietveld refinement are listed in Table 2. There are no significant changes in lattice parameters on annealing. Rietveld refinement of site occupancies (Table 2) indicates that the sample becomes closer towards ideal stoichiometry on annealing for three days. Further annealing leads to deficiencies at the Gd and Sr sites which may arise due to segregation of very low concentrations of Gd and Sr rich impurity phases at the grain boundaries. The crystallite size has been determined using Rietveld profile parameters and correcting for instrumental broadening. They are presented in Table 3. The data indicates that there is an overall increase in size of the coherently diffracting domain (grain) from 1425Å to 2985Å on annealing at 1060 °C for one week, compared to the as-prepared sample. Anisotropy is also observed in the grains. As the broadening of the X-ray lines is dominated by the smaller particles the above grain sizes obtained from XRD represent the lower limit of grain size in the sample. SEM studies on the same sample (not shown) indicate that larger grains of micron sizes are present apart from smaller grains. The wide distribution of grain size from micron to submicron range is characteristic of some ceramics with fractal nature.

Resistivity and a.c. susceptibility studies of the annealing behaviour of $GdSr_2RuCu_2O_8$ carried out by us indicate that the superconducting transition becomes sharper and T_c increases upon increasing the annealing time (Geetha Kumary, 2003). The susceptibility studies as a function of annealing on the bulk sample are presented in Fig. 6. Comparison of these results with the results on microstructure and composition using Rietveld refinement shows

Table 1: Collection of intensity data and Rietveld R factors for $GdSr_2RuCu_2O_8$

Characteristics	*As-Prep. Sample*	*1060 °C, O_2 three days*	*1060 °C, O_2 one week*
2θ angular range	20-100	20-100	20-100
Steps (°2θ)	0.02	0.02	0.02
Exposure (s)	15	15	15
X-ray reflection no.	160	160	160
WR_P (Fitted)	11.8	15.2	14.8
R_P (Fitted)	7.9	11.4	11.0
WR_P (-Bknd)	13.49	14.35	15.03
R_P (-Bknd)	9.25	11.06	11.28
Chi^2	1.29	1.95	2.69
R (F^2)	13.67	16.67	15.11

Table 2: Refined structural parameters for As-prepared and annealed $GdSr_2RuCu_2O_8$

Structural Parameter		*As-prepared sample*	*1060 °C, O_2, three days*	*1060 °C, O_2, one week*
Lattice Constant [a,c (Å)]		3.838, 11.570	3,838, 11.576	3.837, 11.570
Site Occupancies, Uiso(Å^2)				
Atom	Site			
Gd	1d	0.923, 0.008	1.0, 0.009	0.837, 0.012
Sr	2h	0.978, 0.010	1.0, 0.011	0.882, 0.016
Ru	4l	1.000, 0.007	1.0, 0.008	1.00, 0.002
Cu	2g	0.999, 0.011	1.0, 0.013	0.975, 0.002
O	4i	1.000, 0.012	1.0, 0.013	1.00, 0.002
O	8r	1.000, 0.440	1.0, 0.534	1.00, 0.393
O	4n	1.000, 0.012	1.0, 0.014	1.00, 0.009

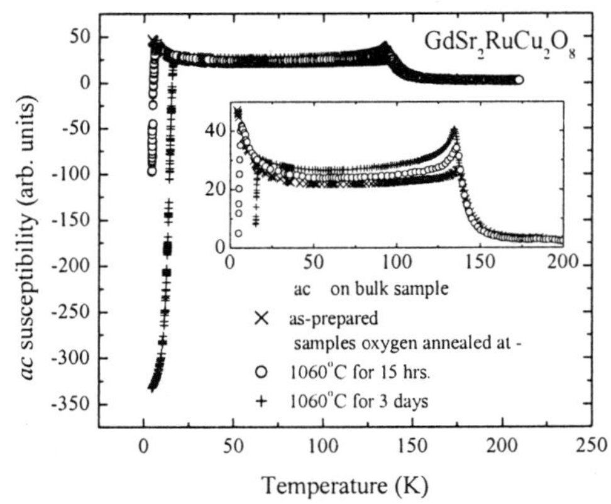

Fig. 6: a.c. susceptibility of a bulk sample of $GdSr_2RuCu_2O_8$ as a function of annealing.

Table 3: Refined profile and strain parameters and crystallite sizes

Profile/Strain Parameter	*As-Prepared Sample*	*1060°C, O_2 three days*	*1060°C, O_2 one week*
Profile Parameters	U=V=W=P=O X=0.117E+02 Ptec=0.492E 01	U=V=W=P=O X=0.117E+02 Ptec=0.158E+01	U=V=W=P=O X=0.846E+01 Ptec=0.144E+01
Strain Parameters			
S400	0.0528	0.02636	0.0222
S004	0.00184	0.00495	0.00203
S220	-0.0735	-0.0698	0.00926
S202	0.01337	0.0472	0.00843
Crystallite Size			
p (Å)	1425	1425	2985
Anisotropy			
p // (Å)	848	1136	2007
p ⊥(Å)	1425	1425	2985

Note: The profile and strain parameters correspond to profile No. 4 in the Rietveld refinement programme GSAS.

that improved superconducting properties correlates with an increase in grain size, decrease of lattice strain and shift of stoichiometry towards the ideal one. A more detailed microscopic picture can be obtained from detailed electron microscopy studies which are in progress.

REFERENCES

Bauernfeind, L., Widder, W. and Brown, H.F. (1995). *Physica C*, **254:** 151.

Bernhard, C., Tallon, J.L., Niedermayer, Ch., Blasius, Th., Golink, A., Brucher, E., Kremer, R.K., Noakes, D.R., Stronach, C.E. and Ansaldo, E.J. (1999). *Phys. Rev. B*, **59:** 14099.

Blount, E.I. and Varma, C.M. (1979). *Phys. Rev. Lett.*, **42:** 1079.

Chmaissen, O., Jorgensen, J.D., Shaked, H., Dollar, P. and Tallon, J.L. (2000). *Phys. Rev.B*, **61:** 6401.

Geetha Kumary, T., Janaki, J., Sastry, V.S., Hariharan, Y., Radhakrishnan, T.S. and Valsakumar, M.C. (2003). *Solid State Physics*, India.

Kuz'micheva, G.M., Luparev, V.V., Khlybov, E.P., Kostyleva, I.E., Andreenko, A.S. and Gavrilov, K.N. (2001). *Physica C*, **350:** 105.

Larson, A.C. and Von Dreele, R.B. (2000). GSAS: General Structure Analysis System, Los Alamos National Laboratory, Report No. LAUR 86-748.

Lorentz, B., Meng, R.L., Cmaidalka, J., Wang, Y.S., Lenzi, J., Xue, Y.Y.and Chu, C.W. (2001). *Physica C*, **363:** 251.

McLaughlin, A.C., Zhou, W., Attfield, J.P., Fitch, A.N. and Tallon, J.L. (1999). *Phys Rev.B*, **60:** 7512.

Rao, G.V.N., Sankara Sastry, V., Gopalakrishna Murthy, H.S., Seshagiri, V. and Radhakrishnan, T.S. (1996). *Powder Diffraction*, **11**: 200.

Stephens, P.W. (1999). *Journal of Applied Crystallography*, **32**: 281.

Tallon, J., Bernhard, C., Bowden, M., Gilberd, P., Stoto, T. and Pringle, D. (1999). IEEE *Trans. Appl.Supercond.*, **9**: 1696.

Toby, B.H., (2001). EXPGUI : A graphical user interface for GSAS. *Journal of Applied Crystallography*, **34**: 210.

Rietveld Refinement of α_2 Phase in Binary Ti_3Al Base Alloys

P. Ghosal, S. Banumathy and A.K. Singh

Defence Metallurgical Research Laboratory
Kanchanbagh P.O., Hyderabad-500058, India

INTRODUCTION

Intermetallics are one of the few class of materials along with ceramics and composites that have the potential to be used in demanding industrial applications including many critical components in aircraft and aerospace industries for service at elevated temperatures and in hostile environments (Westbrook and Fleischer, 2002). Ordered intermetallics belong to a class of materials that are blessed with unusual combination of lightness and high temperature strength. On the other hand, they lack room temperature ductility and toughness. Further, among the well-studied intermetallic compounds, Ti_3Al is suitable for next generation high temperature materials along with TiAl. Efforts initiated in 1960 (McAndrew and Simcoe, 1961; Winter, 1968) and intensified in the 1980 (Blackburn and Smith, 1982) to develop high temperature engineering alloys based on the Ti_3Al have been largely successful (Banerjee et al., 1993) in late 1990.

Ti_3Al is an ordered α phase, designated as α_2 phase, and a long range ordered close packed hexagonal (cph) phase with $D0_{19}$ structure and $P6_3/mmc$ symmetry. The composition range and critical temperature for order-disorder transition is well defined (Murray, 1987; Kattner et al., 1992). It has been assumed in most of the studies that atoms in the α_2 phase are located at regular cph positions cs given by Pietrokowsky for Ti_3Sn phase (Pietrokowsky, 1952). Gehlen (1970) has reported that the position of titanium atoms in a single crystal of the α_2 phase are slightly displaced from the sites they would occupy in regular cph lattice. This has resulted in very large changes in the intensities of some of superlattice peaks. Apart from this, the study of stability of planar faults and other defects do not consider the shift in Ti atoms, which seems to have an effect on energy that in turn affect the stability of the faults. Thus, the aim of the present investigation is to study the crystallographic structure of the α_2 phase and related atomic shifts of titanium atoms as function of aluminium concentration using Rietveld refinement (Young, 2000). These results will be useful to study the deformation behavior of $D0_{19}$ structures in general and the α_2 phase in particular.

CRYSTALLOGRAPHY OF α_2 PHASE

The basal plane projection of α_2 phase and corresponding symmetry elements are given in Fig. 1. Structural details such as lattice, number of atoms per unit cell, space group, lattice parameters, occupancies, equivalent position of atoms and their respective site symmetries of the α_2 phase are given in Table 1. The unit cell for this structure can be described in terms of four interpenetrating cph sub lattices. The 'a' spacing of the ordered cell is twice that of the cell for corresponding disordered structure, while the 'c' spacing changes marginally.

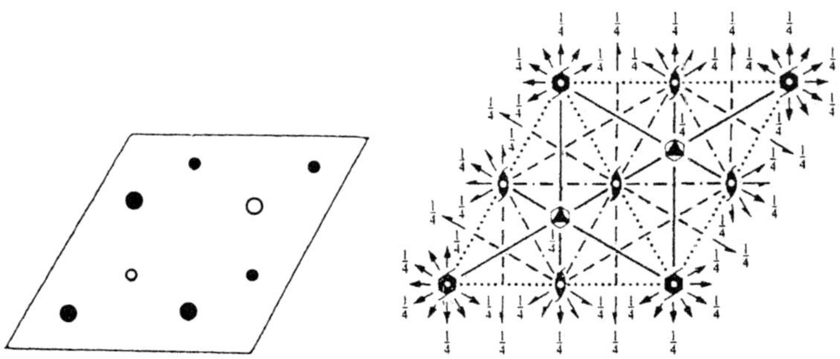

(a) (b)

Fig. 1: (a) Basal plane projection of Ti_3Al structure. Filled and open circles represent Ti and Al atoms, respectively. Large and small circles correspond to sites in alternate layers.
(b) Symmetry elements corresponding to Ti_3Al structure ($D0_{19}$ type with $P6_3/mmc$ symmetry).

EXPERIMENTAL PROCEDURE

Buttons of three binary alloys with nominal compositions Ti-20Al, Ti-25Al and Ti-30Al (all in at.%) were prepared by non-consumable vacuum arc melting technique. The melting was repeated six to seven times to ensure chemical homogeneity. The analyzed chemistry of all alloys is given in Table 2. The small pieces of the first three alloys were heat treated at 1000 °C for 24 hrs (homogenization treatment) and then water quenched (WQ). The powders of all heat treated alloys were made by filing and then subjected to stress relieving treatment at 425 °C for 100 hrs.

The X-ray diffraction (XRD) profiles of annealed powder samples were recorded using CuKα radiation in a Philips PW3020 diffractometer equipped

Table 1: Structural details of Ti_3Al (α_2) Phase

Ti_3Al : Hexagonal

$P6_3/mmc$ or D^4_{64} ; 8 atoms per unit cell hp8

a = 5.7975 (Å) c = 4.6413 (Å)

Atomic Positions:

Atoms	*Wycknoff Notation*	*Symmetry*	*x*	*y*	*z*
Al	2(c)	6m2	1/3	2/3	1/4
Ti	6(h)	mm2	x	2x	1/4
Regular hcp atomic positions			1/6	2/6	1/4

Table 2: Chemical analysis of the alloys

Alloy	*Elements (atom %)*		
	Ti	*Al*	*Al*
Ti-20Al	81.3	18.7	18.7
Ti-25Al	74.5	25.5	25.5
Ti-30Al	71.1	28.9	28.9

with a graphite monochromator operated at 40 kV and 30 mA in a step size of 0.015° (2θ) and counting time of four seconds/step. The XRD data for all the alloys were subjected to Rietveld refinement programme using Philips Xpert plus. The specimens for microscopy were prepared from bulk samples following standard metallographic technique used for titanium and its alloys. The microstructure of heat treated alloys were examined by optical microscope.

RESULTS AND DISCUSSION

The experimental alloys were examined by optical microscope for microstructural characterisation. A typical representative of these microstructures for the alloys Ti-25Al is shown in Fig. 2. The XRD patterns

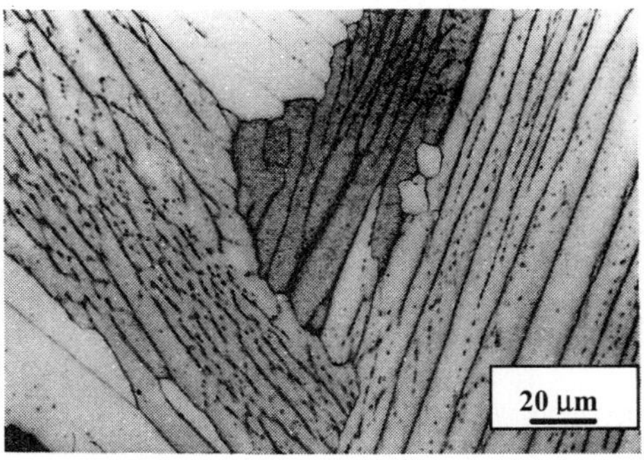

Fig. 2: Optical microstructures of the alloy Ti-25Al.

of the binary alloys exhibit single α_2 phase only. The XRD patterns of all the alloys have been subjected to Rietveld refinement. The initial models and the results of refinement are given in the Table 3.

Table 3: Initial models and results of Rietveld refinement for experimental alloys

	Ti-20Al	*Ti-25Al*	*Ti-30Al*
Initial Mode	2(c): 0.252Ti+ 0.748Al 6(h):1.0Ti	2(c) :1.0Al 6(h): 0.993Ti+ 0.007Al	2(c) :1.0Al 6(h): 0.948Ti+ 0.052Al
Lattice Constants (Å)	a = 5.8193(2) c = 4.6639(2)	a = 5.7975(6) c = 4.6413(4)	a = 5.7802(7) c = 4.633(5)
Output chemistry	$Al_{1.50}Ti_{6.50}$ or Ti-18.8 Al	$Al_{2.04}Ti_{5.96}$ or Ti-25.5 Al	$Al_{2.31}Ti_{5.69}$ or Ti-28.9 Al
R (expected)/%	14.169	13.480	14.120
R (profile)/%	16.396	18.535	17.195
R (weighted)/%	21.549	24.582	21.073
Goodness of Fit (GOF)	2.313	3.325	2.227
6(h) Wyckoff position	0.807(1)	0.7913(6)	0.797(1)
Ti-Ti bond length (Å)	2.903	2.932	2.897
Ti-Al bond length (Å)	2.873	2.873	2.857
Volume of tetrahedron ($Å^3$)	2.84	2.88	2.81

The alloy Ti-20Al lies in two-phase $\alpha + \alpha_2$ field (Murray, 1987; Kattner et al., 1992) although the volume fraction of α phase is very small (around 5%). The volume fraction of α phase has been ignored in the initial model of this alloy (Table 3). The other two alloys Ti-25Al and Ti-30Al lie in single α_2 phase field (Murray, 1987; Kattner et al., 1992). Although, the three experimental alloys exhibit a systematic change in Al concentration but they are three different types of alloys in terms of their respective site occupancy. As mentioned in Table 1, the stoichiometric Ti_3Al phase has two Wyckoff positions i.e. 2(c) and 6(h), fully occupied by aluminium and titanium, respectively. Any deviation from stoichiometric composition results in two situations: (1) When Al<25 at.%, 6(h) sites are occupied by titanium and 2(c) by aluminium and excess titanium, and (2) when Al >25 at.%, 6(h) sites are occupied by titanium and excess aluminium and 2(c) by aluminium only. Based on these arguments, the initial models for these alloys have been selected and given in Table 3. The input for site occupancy of the experimental alloys has been calculated based on analyzed alloy chemistry (Table 2).

As reported by Murray (1987), the lattice constants 'a' and 'c' of the Ti-Al alloys over a large range of composition (up to 40 at.% of aluminium), which covers both the range of disordered (α) and ordered (α_2) phases. This shows a continuous reduction of 'a' and 'c' parameters with increase in aluminium contents. The lattice parameters of the binary α_2 phase of all experimental alloys lie in same range and also exhibit similar trend as a function of aluminium as those reported in literature (Murray, 1987).

After Rietveld refinement the 2(c) positions remain same as given in Table 1, while the 6(h) positions are shifted from regular hexagonal positions. Similar displacements in 6(h) positions have been reported in a single crystal of the α_2 phase (Gehlen, 1970). A careful observation of the α_2 structure and corresponding symmetry arrangements indicates that 2(c) positions lie at the intersection of three mirrors, which are defined an unique point in the space. As a result, no atomic shift of 2(c) positions is possible without breaking of the $P6_3/mmc$ symmetry.

In one unit cell, the 6(h) positions form two tetrahedrons with 2(c) positions orientated in opposite directions along the 'c' axis (Fig. 1). In each tetrahedron, three of 6(h) positions lie on three different mirrors. These three atoms can move along their respective mirrors without breaking any symmetry. Therefore, atomic shift corresponding to 6(h) positions in the α_2 phase is possible. In fact the atomic shift is related to atomic potential, which in turn is governed by its surrounding atoms. Depending upon the site occupancies of both the sites i.e. 2(c) and 6(h), potential varies and result in atomic shift in 6(h) positions. As a consequence, bond length in the respective neighbours changes.

One of the 6(h) site of the regular cph position is 0.833. The corresponding 6(h) positions obtained in all experimental alloys are less than 0.833, which in turn changes the Ti-Ti and Ti-Al first neighbor within and out of plane bond lengths, respectively. It is to be noted that 2(c) and 6(h) atomic sites are predominantly occupied by aluminium and titanium atoms, respectively. Although these bond lengths show a systematic trend in binary alloys but it would not be correct to correlate the same as a function of Al content. This is due to the fact that the three binary alloys exhibit entirely different pattern of site occupancy and hence these alloys and related potential should be treated differently.

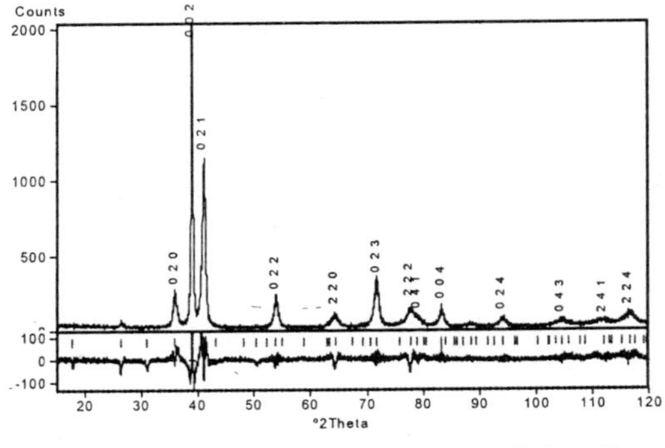

Fig. 3: Observed and difference XRD patterns of the alloy Ti-25Al powder after homogenization and stress relieving treatments.

The volumes of tetrahedron of experimental alloys have been calculated using aluminium in the apex position (Table 3). This shows that the volume of tetrahedron is maximum for the alloy Ti-25Al and its size decreases from the deviation of stoichiometric composition in binary alloys. It is to be noted that the overall volume of the unit cell reduces with increase in aluminium content. The sizes of tetrahedron and octahedron holes are expected to influence interstitial diffusion and solubility of interstitial in α_2 phase (Gehlen, 1970). In fact it has been reported that the presence of Ti_3Al phase in Ti-Al alloys affects the solubility of H_2 (Portisch and Margolin, Final Report).

Any small shift of titanium atoms affects the intensities of super lattice reflections. Gehlen (1970) has also observed the change in the intensities of superlattice reflections by static displacements of titanium atoms. The intensity of $11\bar{2}0$ reflections increases with increase in aluminium content. This indicates that the atomic shift has to be considered during the measurement of long-range order parameter. The magnitude of atomic shift is a function of extent of ordering present in the material as well as temperature. The measurement of the amount of order in partially ordered materials would depend on site occupancy and spatial position of the lattice sites.

The presence of small shift may also affect the physical properties of the alloys (Collings, 1984; Collings and Hoe, 1969; Collings and Hoe, 1970). The sudden drop in the magnetic susceptibility in the Ti_3Al composition range might be explained by the presence of cluster of atoms (Collings and Hoe, 1970). The tighter bonding of titanium atoms (smaller the volume of tetrahedron) in tetrahedron may results in broadening of titanium band. This in turn introduces a large change in the density of states corresponding to the drop in magnetic susceptibility. Apart from this, determination of the small shift associated with 6(h) positions along with the site occupancy will be helpful to calculate accurately the energy of planar faults in these alloys.

CONCLUSIONS

The shift in titanium atoms corresponding to 6(h) Wyckoff positions has been calculated in different alloys containing α_2 phase. This is related to the presence of symmetry elements in Ti_3Al phase, which allows the displacement along the mirror. The shift may be attributed to different site occupancies resulting in different atomic potentials. These atomic coordinates are useful to calculate the energy of planar faults in these alloys.

Acknowledgements: The authors acknowledge the financial support of the Defence Research Development Organization, Government of India. Thanks are also due to the members of the Titanium Alloy Group and Structure and Failure Analysis Group for their help in carrying out the experimental work.

REFERENCES

Banerjee, D., Gogia, A.K., Nandy, T.K., Muraleedharan, K. and Mishra, R.S. (1993). *In:* Structural Intermetallics, R. Darolia, J.J. Lewandowski, C.T. Liu, P. L.Martin, D.B. Miracle, M.V. Nathal (Eds). The Mineral, Metals and Materials Society, Warrendale, pp.19-33.

Blackburn, M.J. and Smith, M.P. (1982). AFML-TR-82-4086, USA.

Collings, E.W. (1984). The Physical Metallurgy of Titanium alloys, ASM, Metals Park, OH.

Collings, E.W. and Hoe, J.C. (1969). *Phys. Letters*, **29A:** 306-307.

Collings, E.W. and Hoe, J.C. (1970). Science, *In:* Technology and Application of Titanium, Jaffee, R.L. and Promisel, N.E. (Eds). Pergamon Press, New York, 1970, pp. 331-347.

Gehlen, P.C. (1970). *In:* Science, Technology and Application of Titanium, Jaffee, R.L. and Promisel, N.E. (Eds). Pergamon Press, New York, pp. 349-357.

Kattner, U.R., Lin, J.C. and Chang, Y.A. (1992). *Metall. Trans.*, **23A**: 2081.

McAndrew, J.P. and Simcoe, C.R. (1961). ASD TR-61-466, USA.

Murray, J.L. (1987). Phase diagram of binary titanium alloys, ASM International, Metals Park, Ohio.

Pietrokowsky, P. (1952). *J. Metals*, **194:** 211.

Portisch, M. and Margolin, H. Final Report on Contract No. DA-ARO(D)-31-124-G519, Army Reseach Office, Durham.

Young, R.A. (2000). The Rietveld method, ICUR, Oxford Science Pub.

Westbrook, J.H. and Fleischer, R.L. (2002). (Eds). Intermetallic Compounds: Principles and Practice. John Willey & Sons Ltd., West Sussex, England.

Winter, H. (1968). US Patent No. 3411 901.

Cristobalite Formation of an Indian Kaolinite Investigated by Rietveld Technique

Akshoy Kumar Chakraborty

Central Glass and Ceramic Research Institute
Kolkata–700 032, India

INTRODUCTION

Mullite is the ultimate crystalline product of thermal decomposition of kaolinite. The kaolinite to mullite reaction series was extensively investigated using various physico chemical techniques and was reviewed (McConville et al., 1998; Lee, et al., 1999; McConville, 2001). The phase transformation of kaolinite is presented chronologically in the flow chart (Fig.1).

KAOLINITE $\xrightarrow{\text{1st Stage}}$ METAKAOLINITE + H_2O

(i) SILICA PHASE
+
$\xrightarrow{\text{2nd Stage}}$ (ii) Al–Si SPINEL PHASE (Cubic) $\xrightarrow{\text{3rd Stage}}$ MULLITE + β-CRISTOBALITE (Orthorhombic)
+
(iii) ALUMINA RICH SILICATE PHASE (Amorphous MULLITE)

Fig. 1: Three Stages of Decomposition of Kaolinite.

The literature shows the following three major issues:

1. The crystal structure and chemistry of the spinel phase formed at 980 °C exotherm of kaolinite.
2. Nature and the composition of mullite phase
   (a) Mullite formation process
   (b) Nature of mullite changes with temperature.
3. High temperature DTA characteristics of kaolinite,
   (a) Exotherm relevant to Al-Si spinel formation
   (b) Exotherms related to mullitization of kaolinite.

These problems have been discussed elaborately by Chakraborty and Ghosh (1991) and Chakraborty (1992) in the context of the exhibition of the latest DTA thermogram of kaolinite as shown in (Fig. 2). The 980 °C reaction of kaolinite is emerged as follows.

$$\text{Metakaolinite} \xrightarrow[\text{Exotherm}]{\text{1st}} \begin{array}{l} \text{i) Al-Si spinel phase, 20–30\%} \\ \text{ii) Mullite phase, 5\%} \\ \text{iii) Alumino silicate phase (A), 30–40\%} \\ \text{iv) } SiO_2 \text{ (A) phase, 36\%} \end{array} \qquad (1)$$

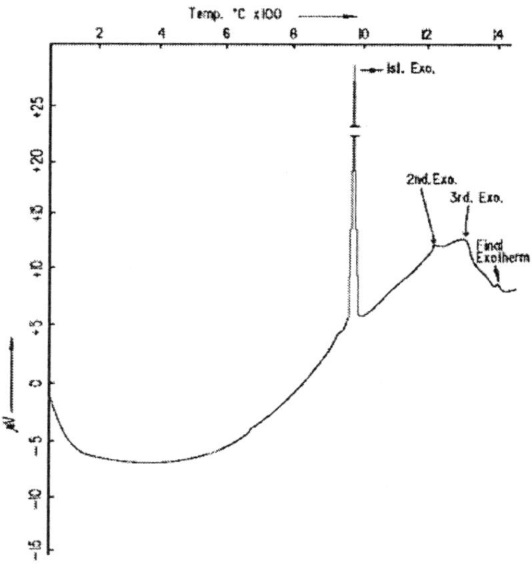

Fig. 2: DTA curve of well crystallized kaoline.

The aim of present investigation is concerned with the crystallization behaviour of amorphous silica phase, which is generated at the occurrence of 1st exotherm as shown in Eq. (1). Various investigators have calculated change in crystal size and lattice strain of different compounds during grinding by X-ray line broadening analysis utilizing Willamson and Hall's equation (Williamson and Hall, 1953). In this study, we have observed the change in lattice constant, crystal size during the cristobalite formation from kaolinite. The changes in crystal size are calculated by using Quasar software programme of Philips.

EXPERIMENTAL PROCEDURE

Heat Treatment

The koalinite clay was collected from Madhya Pradesh (M.P.), India and it was elutriated with adequate amount of pyrophosphate solution, centrifuged, washed and finally dried. Washed kaolinite was heated in an electric muffle furnace at a heating rate 100 °C/hr to different temperatures with one hr soaking time and cooled normally. After cooling, the heated clay samples were ground to below 10 μm and finally stored for X-ray analysis.

XRD Analysis

All samples were analyzed in X Pert PRO X-ray Generator of Philips, attached with Hybrid Monochromator, Programmable anti-scatter (PASS) and receiving slits (PRS) and coupled with a computer. The tube was operated at 40 kV and 20 mA, with CuKα radiation. The X-ray diffraction (XRD) pattern is recorded in a step size of 0.02° (2θ) and counting eight seconds/ step, within the 2θ angle 5 to 75°. The lattice parameter, crystal size etc. are calculated step by step after lots of refinement in the raw scan data. The software namely, Profit, Dicvolt, Quasor based on various mathematical and statistical methods are used in this connection. Quantitative analysis and size calculations are done from refined data with the help of Rietveld refinement procedure, where preferred orientation, atomic displacement factor etc. are taken into consideration.

RESULTS

Phase Developments of Kaolinite

The comparison of standard data of well crystallized kaolinite, chemical and mineralogical compositions of washed M.P. Kaolinite are shown in Fig. 3 and Table 1. Formation of high temperature phases were monitored by XRD pattern and is shown in Fig. 4. At room temperature, XRD pattern of the kaolinite shows a medium crystallized variety with trace amount of free quartz and anatase. The XRD pattern is amorphous at 800 °C with 3-4 nm phase band recognized as metakaolinite, with trace quartz. The XRD pattern at 1050 °C also exhibits amorphous phase along with decomposition product

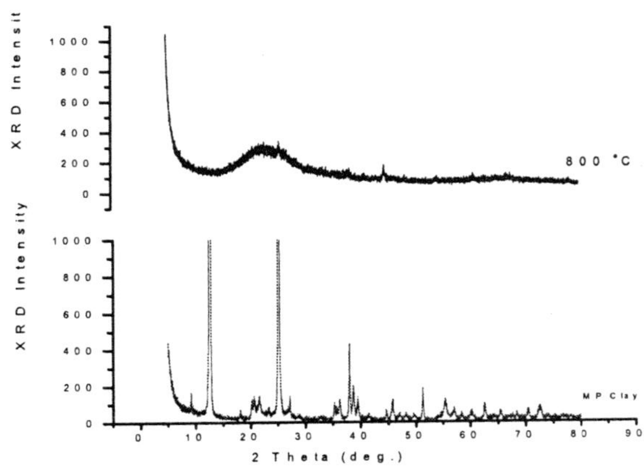

Fig. 3: XRD patterns of M.P. Kaolinite clay. (a) Washed variety; (b) Raw clay heated to 800 °C shows quartz and anatase as impurity phases.

Table 1: Chemical analysis and Mineral composition of M.P. Kaolinite

Chemical constituents			*Mineral constituents*	
Raw Kaolinite	*wt. %*		*Raw Kaolinite*	*Washed Kaolinite*
SiO_2	56.92	Kaolinite	Kaolinite	Kaolinite
Al_2O_3	30.41	Illite/Mica	Illite	Illite
Fe_2O_3	0.87	Montmorilonite	Nil	
TiO_2	1.34	Chlorite	Nil	
MgO	trace	Quartz	Quartz	Quartz (trace)
CaO	0.28	Rutile/Anatase	Rutile/Anatase	Anatase (trace)
K_2O	0.27	Feldspar	Nil	
Na_2O	0.21	Fe_2O_3	Hematite	Nil
L.O.I	9.70			

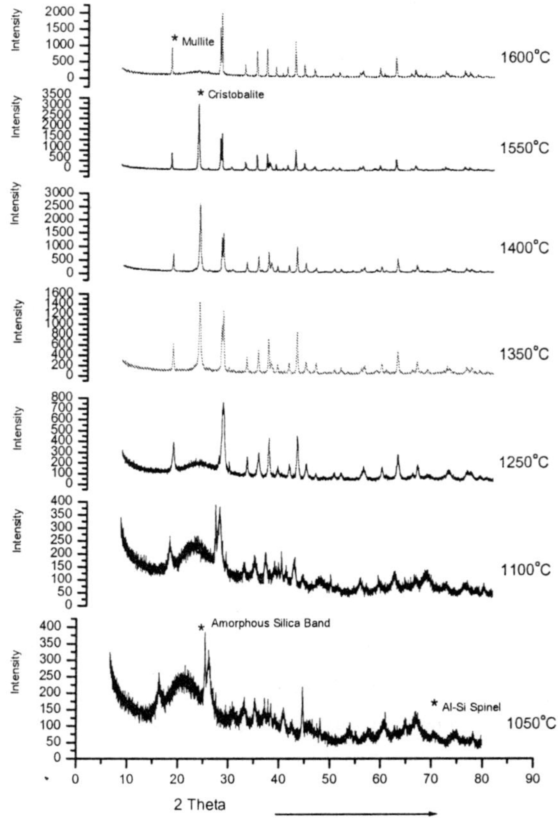

Fig. 4: XRD patterns of decomposition phases of M.P.clay; Mixtures of Al-Si spinel, mullite and amorphous silica band formed at 1050 °C; Mixtures of mullite and amorphous silica band formed at 1100–1250 °C; Mixtures of mullite and cristobalite noted at 1350–1550 °C; Cristobalite disappears and mullite exist at ~ 1600 °C.

of metakaolinite. Weakly crystalline spinel phase comprising three broad XRD peaks at 0.238 nm, 0.197 nm and 0.137 nm, respectively are seen. Very poorly crystalline mullite is observed by showing Bragg reflection peaks at 0.537 nm, 0.340 nm etc.

The characteristic peaks of mullite, observed at 1100 °C are very broad indicating poorly crystalline mullite. The broadness of mullite peaks decreases and the amorphous band of metakaolinite diminishes at 1250 °C. Due to secondary mullite formation the intensities of mullite peak increase much and all peaks (specially 0.41 nm and 0.252 nm) become sharper at 1350 °C. Intensity of mullite peaks increases in the temperature range 1450-1550 °C and peaks become sharper. Cristobalite formation also occurs to a great extent in the temperature range 1450–1550 °C. Peaks due to cristobalite almost vanish at 1600 °C. A band is again reappeared in the XRD pattern due to glassy phase formation.

The effect of progressive heat treatment of M.P. clay as shown in Fig. 4 describes how kaolinite decomposes to metakaolinite in 1st stage then to Al–Si spinel phase in 2nd stage and finally to mullite and cristobalite during firing process. The quantity of phase developed is proportional to the peak height of the related phase peak at that temperature. From nature of the phase band, one can get an approximate idea about the start of crystallization temperature, its range of existence and the temperature of its disappearance.

Relative Changes in Surface Area

Surface area of kaolinite heat treated to different temperatures was measured in micrometrics and is shown in Fig. 5. There is slight increase in surface area followed by a rapid decrease at higher temperatures.

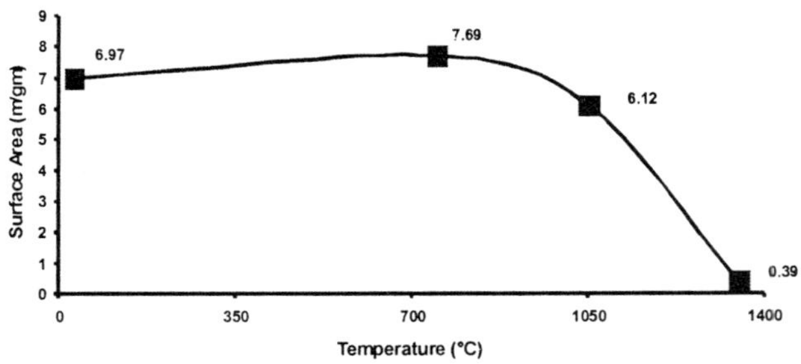

Fig. 5: Change in relative surface area of kaolinite during heat treatment.

Cristobalite Formation

Semi Quantitative Analysis of Cristobalite

Figure 6 shows the growth curve of cristobalite both in static and dynamic heating schedules. It starts on heating M.P. clay at ~ 1250°C in static condition in comparison to its formation at higher temperature in case of dynamic heating condition. It increases slowly with rise of heat treatment temperature up to 1300 °C. It further increases to a considerable amount on heating at 1450 °C. This increment continues to a smaller extent up to 1550 °C and then attains more or less a steady level and then decline at 1600 °C. Heating effect of kaolinite also shows a relative change in structural parameter of cristobalite. For example, in addition to change in relative development and transition to glassy phase, the peak position, integral breadth and FWHM of say 0.405 nm peak of cristobalite are changed with increase in heating temperature (Fig. 7).

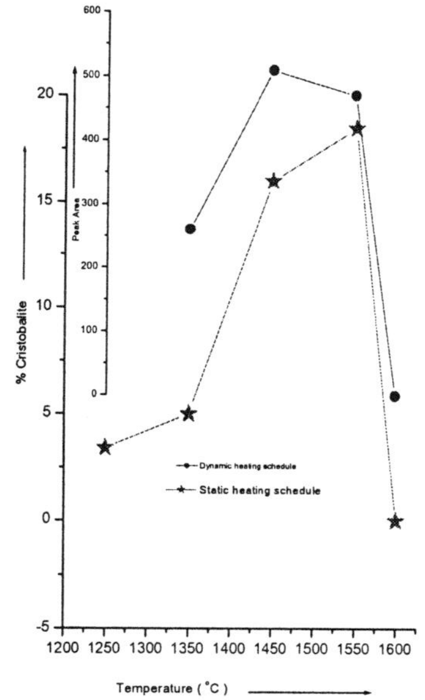

Fig. 6: Growth curve of cristobalite at two different heating schedules.

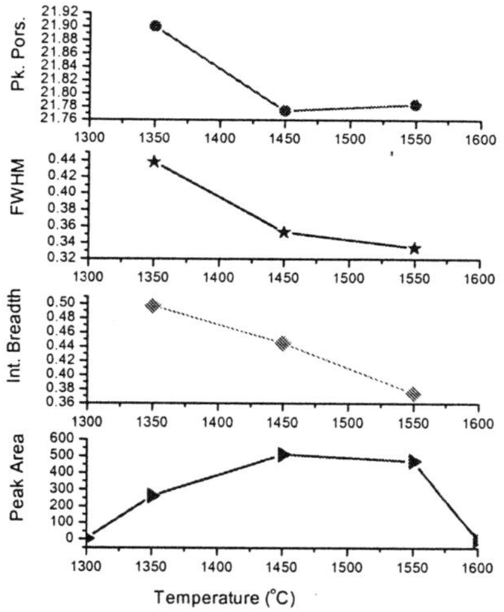

Fig. 7: Relative change in 0.404 nm peak parameters of cristobalite.

Relative Changes of Crystallite Size of Cristobalite

The effect of heat treatment on the changes of the size value of cristobalite is shown in Fig. 8. The crystallite size of cristobalite increases with increase in annealing temperature.

Lattice Parameter of Cristobalite

Changes in lattice parameter of cristobalite are shown in Fig. 9a and b. It shows that the values of lattice constants ***a*** and ***c*** of the cristobalite changes during heating process. This change is obvious since the values of peak position, FWHM, and integral breadth are found to decrease with increase of calcination temperatures as shown in Fig. 7. Crystallite size increases in the reverse direction. Therefore, all these parameters explain the phenomena related to formation behaviour of cristobalite. The nature of this shifting explains a similar reason and is relevant to increase in crystallinity of

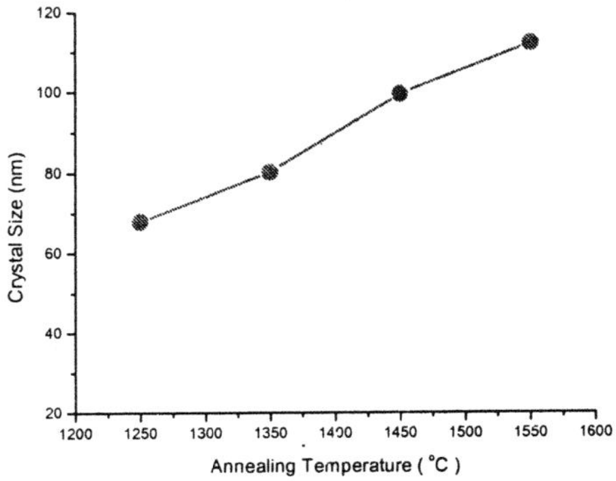

Fig. 8: Changes in crystallite size of cristobalite during heating process.

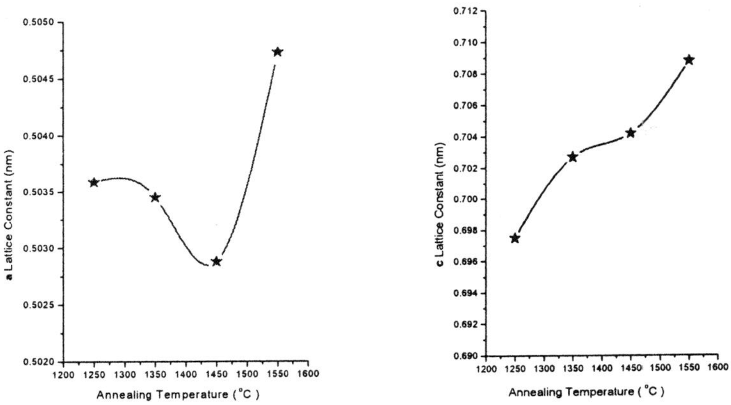

(a) (b)

Fig. 9: (a) (b) Changes in lattice parameters of cristobalite at different temperatures.

cristobalite during heating process. All those data converge sharply at ~ 1450 °C and thereafter these parameters change slowly with further rise of temperature.

DISCUSSIONS

In this study we note both the changes in lattice constants and crystallite size of cristobalite during its formation and growth process out of amorphous silica released in the decomposition of kaolinite at ~ 1000 °C. The results obtained in the present study can be discussed in following three parts.

Growth Curve

Figure 6 shows that upon heating crystallization of amorphous silica to crystobalite takes place slowly at the beginning of 1250 °C, increases rapidly at ~1400 °C, attains a steady level and finally disappears due to glassy phase formation at high temperature where surface area measurement shows a low value due to sintering (Fig. 5). Tomura et al. (1986) previously calculated percent cristobalite formation from its peak intensity during heating a spherical kaolinite. The present data is found comparable with those of Tomura et al. (1986) (Table 2).

Table 2: Formation of β-Cristobalite during Heating Process

Temp. Calcination (°C)	*wt. % of β-Cristobalite Obtained by Different Ways*	
	(1)	*(2)*
	By Tomura et al. (1986) by Internal Standard method	*By present study, using Rietveld refining technique*
1250	0	3.4
1350	16	5.0
1450	22	16.0
1550	0	18.5
1600	0	0

In the DTA thermogram of kaolinite (Fig. 2), Chakraborty (2003) has shown a broad exotherm at ~ 1400 °C. The temperature of formation of cristobalite as shown in the present study corroborates with the exhibition of the above mentioned DTA exotherm. However, crystallization temperature of amorphous silica to β-cristobalite is found to be at ~ 1060 °C when silica gel is analyzed by DTA (Fig. 10).

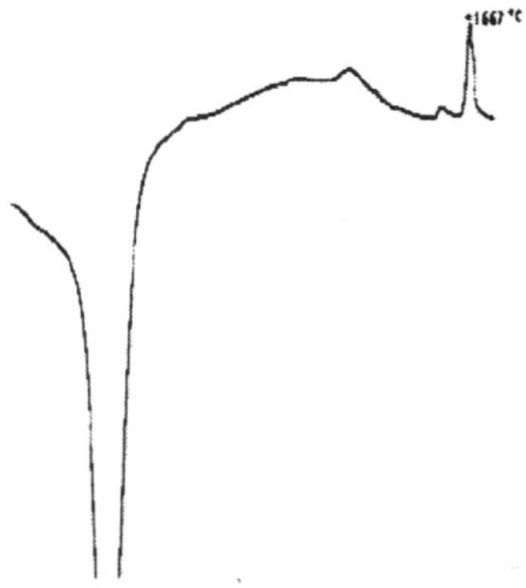

Fig. 10: DTA pattern of silica gel.

Constitution of Silica Phase

By chemical alkali leaching study, Chakraborty and Ghosh (1978) and Chakraborty (1992) have estimated the amount of silica exolved out of the decomposition of kaolinite to be ~ 37 wt.% and proved the Al-Si spinel hypothesis of Brindley and Nakahira (1959). They modified the composition of Al-Si spinel ($3Al_2O_3.2SiO_2$) as analogous to the composition of 3:2 mullite and designated it as cubic form of mullite – a polymorph of high temperature orthorhombic form. The exolution data of silica was later verified by RED analysis by Leonard et al. (1979). Using similar alkali leaching technique, Moya et al. (1985) have verified the exolved silica data. By EDS study of the leached residue, Srikrishna et al. (1990) have shown that the ratio of Al/Si peak confirmed the composition evaluated by Chakraborty (1992). Chakraborty and Ghosh (1978) have shown that the amorphous silica which is generated out of the exolution of metakaolinite structure contains 5-6 wt % Al_2O_3. Subsequently, Okada et al. (1986) and Sonuparlak et al. (1987) have conjectured that the residual phase consisting of major amorphous phase is nearly of the composition of $Al_2O_3 6SiO_2$ instead of free silica. Therefore, the composition of silica phase is to be called as high siliceous rich alumina phase instead of free silica.

Prediction from Lattice Constant Data

Both *a* and *c* axes data of cristobalite increase only slightly with increase in firing temperature. The crystallite size value of it, however, increases with temperature in a regular manner. On the other hand, Verduch (1959) has heat treated silicic acid for his kinetic studies and observed that the cristobalite attains maximum crystallinity during heating silicic acid at ~ 1425 °C and its interplaner spacing attained to 0.4042 nm (Table 3).

Table 3: Comparative *a* (nm) Value of β-Cristobalite Obtained on Heating Two Different Sources of Silica

Parent SiO_2 Source	(1) *a* (nm) Value by Verduch (1959)	(2) *a* (nm) Value by Tomura et al. (1986)	(3) *a* (nm) Value by Present Study	(4) Difference of *a*(nm)Value of β-Cristobalite between two Sources
1. Silica Gel	0.4042	–	0.4038	0.5028-0.4042 = 0.1026
2. Kaolinite Clay	–	0.5022-0.5018	0.5028	–

Lattice constant value *a* of the cristobalite obtained by Rietveld analysis in the present study is 0.503 to 0.504 nm and the same evaluated by Tomura et al. (1986) are 0.5022 to 0.5018 nm, respectively. These data are noted to be in the higher side than that of 0.404 nm of cristobalite synthesized out of silicic acid by Verduch (1959) and in the present study. These results pose two important questions: (1) why crystallization temperatures of amorphous silica from the two different sources of amorphous silica are different and (2) why the lattice constant of cristobalite obtained from kaolinite is more? Florke (1955) and Yamuchi and Kato (1943) have shown that some foreign cations play a great role in build up of cristobalite lattice. Tamela (1950) has reviewed the earlier literature and shown the effect of heating a series of silica alumina gels to 1160 °C by examination of the X-ray analysis. In the presence of alumina, a weak pattern of cristobalite is obtained. With 2wt.% alumina, pattern of cristobalite is greatly strengthened and the full cristobalite pattern is seen. With increasing alumina, the pattern of cristobalite is lost altogether.

As the amorphous silica which is exolved out of the decomposition of kaolinite at the 1st exotherm contains some quantity of alumina, it is conjectured that the increase of *a* and *c* values may be due to crystallization of this high silica rich alumina phase to form cristobalite during the heating process. The following transformation might have taken place at the last exotherm.

$$\underset{\text{(Amorphous)}}{\text{High silica rich alumina phase}} \xrightarrow[\sim 1400\ ^\circ\text{C}]{\text{Exotherm}} \beta\text{-cristobalite}$$

In other words, high formation temperature of cristobalite (of the order of 1400 °C) and showing large increase in *a* and *c* values indicate that the silica (A) which exolves out of decomposition of kaolinite at ~980 °C contains alumina as other constituent as noted in Chakraborty's alkali leaching study of heated kaolinite (Chakraborty, 1992). The presence of a few percentage of Al_2O_3 in siliceous phase has resulted in to enhancement of its crystallization temperature and increments of its lattice constant values. The final transformation of kaolinite is extended and is shown as flow chart in Fig. 11.

CONCLUSIONS

The following conclusions have been drawn out of the results of comparative crystallization studies of silica gel with kaolinite with help of DTA, XRD and Rietveld analysis.

1. The outstanding problems namely characterization and chemistry of 980 °C spinel phase, formation of mullite out of spinel and binary aluminosilicate phases by two consecutive thermal changes during the decomposition of kaolinite are reviewed.
2. Silica phase which exolves out of the decomposition of metakaolinite contains some alumina in the network as per the previous alkali leaching study is concerned. In consequence of it, the following observations are made.
   (i) Cristobalite formation is delayed and occurs significantly not below 1250 °C in static heating and 1300 °C in case of dynamic heating conditions.
   (ii) Lattice parameters of Cristobalite, calculated by Rietveld technique show higher values in comparison to crystallization of pure silica system. Direction of changes of other structural parameters indicates that crystallinity of cristobalite increases on progressive heating of kaolinite.

It is concluded that the so-called free amorphous silica liberated out of structural decomposition of kaolinite is to be designated as high silica rich alumina amorphous phase instead of pure silica phase. The whole thermal reaction processes of kaolinite are shown.

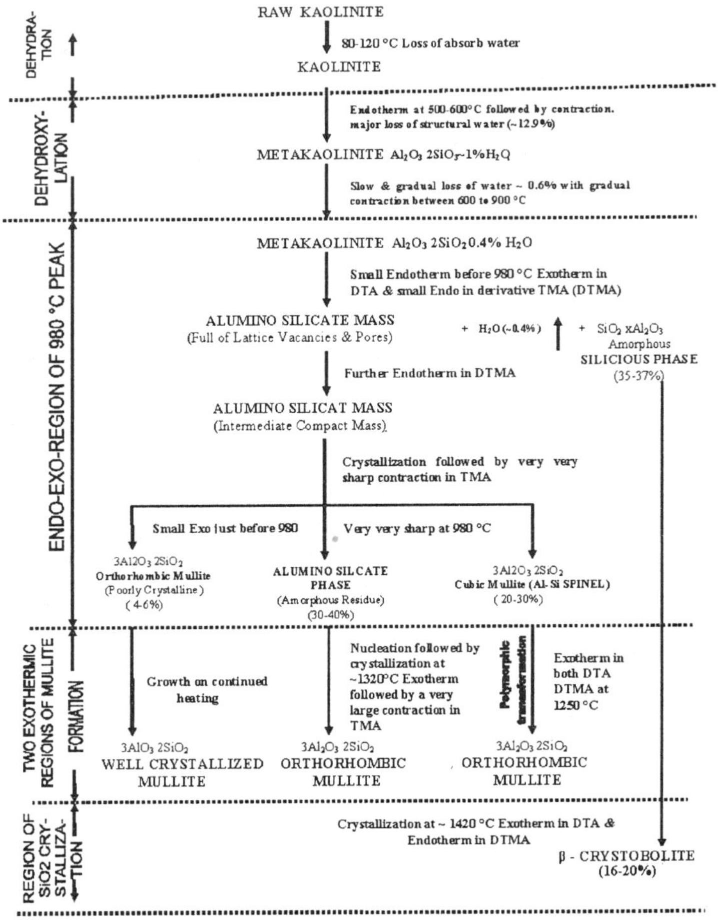

Fig. 11: The flow chart showing final transformation of kaolinite.

Acknowledgement: We are thankful to our Director, Dr. H.S.Maiti for his kind permission to publish this paper.

REFERENCES

Brindley, G.W. and Nakahira, M. (1959). *J.Am.Ceram.Soc.*, **42:** 319.

Chakraborty, A.K. (1992). *J.Mater.Sci*, **27:** 2075.

Chakraborty, A.K. (2003). *J.Therm.Analysis and Calorimetry*,**71:** 799.

Chakraborty, A.K. and Ghosh, D.K. (1978). *J. Am. Ceram. Soc.* **61:** 170.

Chakraborty, A.K. and Ghosh, D.K. (1991). *J.Amer.Ceram.Soc*, **74:** 1401.

Chakraborty, A.K. (1992). *J.Amer.Ceram.Soc*, **75:** 2013.

Chakraborty, A.K. (2003). *Clay Science*, **12:** 63.

Florke, O.W. (1955). *Ber. Deut.Keram.Ges.*, **32:** 369.

Lee, S., Kim, Y.J. and Moon, H. (1999). *J.Amer.Ceram.Soc*, **82:** 2841.

Leonard, A.J., Genet, M.J., Lemaitre, J., Bulens, M. and Delmon, B. (1979). *J.Amer. Ceram.Soc*, **62:** 529.

McConville, C., Lee, W.E. and Sharp, J.H. (1998). *Br.Ceram. Proc.* 75.

McConville, C.J. (2001). *Ceram. Eng. Sci. Proc*, **22:** 149.

Moya, J.S., Serna, C.J. and E.Iglesias, J. (1985). *J.Mater.Sci*, **20:** 32.

Okada, K., Otsuka, N. and Ossaka, J. (1986). *J.Amer.Ceram.Soc*, **69C:** 251.

Sopnuparlak, B., Sarikaya, M. and Aksay, I.A. (1987). *J.Amer.Ceram.Soc*, **70:** 837.

Srikrishna, K., Thomas, G., Martinez, R., Correl, M.P., De.Aza, S. and Moya, J.S. (1990). *J.Mater. Sci*, **25:** 607.

Tamele, M. W. (1950). *Disk. Farad. Soc.*, **8:** 270.

Tomura, S., Shibasaki, Y., Mizuta, H. and Maeda, T. (1986). *Yogo Kyokai Shi*, **98:** 917.

Verduch, A.G. (1959). *J.Amer.Ceram. Soc*, **38:** 703.

Williamson, G.K. and Hall, W. H. (1953). *Acta Met*, **1:** 22.

Yamuchi, T. and Kato, S.J. (1943). *Japan Ceram.Asso.*, **51:** 319.

A New Dimension to X-ray Powder Diffraction with in-situ High Temperature Measurement

B.S. Acharya, L.D. Pradhan and B.B. Nayak

Regional Research Laboratory, Bhubaneswar-751013, India

INTRODUCTION

It is well known that the structural state of a material (volume, symmetry deviation, lattice constants, elastic constants etc.) is directly linked to the thermodynamics (free energy, entropy and heat capacity) of the crystal and both can be described as a function of temperature and pressure. Polymorphic phase transitions, in which low temperature form belongs to a symmetry group that is a subgroup of the high temperature form, can be studied with in-situ high temperature or high-pressure X-ray diffraction (XRD). These studies have been facilitated with the advent of position sensitive detectors as well as high-energy synchrotron sources. However, conventional XRD at elevated temperature along with differential thermal analysis or thermo gravimetric analysis remains unsurpassed when transformation of phase studies are concerned. The paper presents here the results of our XRD measurements on some of the important industrial carbides (iron carbide, tungsten carbide), hydrotungstite, copper chloride, ammonium molybdate tetrahydrate and goethite carried out at high temperatures under in-situ condition and discusses various aspects of structural properties in the light of XRD findings.

EXPERIMENTAL PROCEDURE

All measurements were carried out on powder samples placed on the Pt-Rh strip of the heater of HTK-2 diffraction camera attached to Philips X-ray diffractometer. For this purpose PW1830 generator, Cu tube and PW1710 electronic system control and detection were used. In some cases post diffracted graphite monochromator was used for $K\alpha_2$ removal whereas in others nickel filter was placed in front of the target, which cuts off β radiation of copper. Both scintillation and proportional counters were used depending upon the situation. The samples were heated at a rate of 20 °C/min when the selected temperature was reached; the material was held at that temperature for five minutes before the scanning was performed. In some cases selected few reflections were chosen to see the decay and growth of new phase formation.

RESULTS AND DISCUSSIONS

Tungsten oxide (WO_3) shows polymorphism at room temperature and present as stable monoclinic (m-WO_3) form (Christian, 1975). It transforms into several metastable form like hexagonal and cubic structure. Intercalation of H^+ ion into WO_3 lattice leads to several new bronze structure H_xWO_3 with x varying from 0.1 to 0.5. Widespread interest for determining the phase transformation and reaction kinetics taking place in amorphous systems and in dynamic conditions have resulted in several publications in this area (Cazzanelli et al., 1999; Purans et al., 1997; De Haver et al., 1984; Baucke, 1987). Phase transformation kinetics in amorphous to crystalline form in several hydride systems have also been studied (Baucke, 1987). Hydrogen tungsten bronzes have received considerable attention for their exotic behaviour as electrochromic material. Hydrogen tungsten oxide or H_2WO_4 powder has been taken up for high temperature in-situ study to see the phase transformation taking place in this system and thus to throw light on the reaction kinetics involved in the process of such transformation. Commercially available analytical reagent grade H_2WO_4 powder was taken up for in-situ high temperature XRD studies in this investigation.

XRD patterns of H_2WO_4 taken at four different temperatures are given in Fig. 1. The diffraction patterns change as the temperature is raised and finally at 1000 °C, WO_3 phase is found to be formed. The 'd' values obtained from these patterns when matched with the standard values of PDF cards show the sequence of phase transformation taking place in this system to be,

$$H_2WO_4 \xrightarrow{(210\ ^\circ C)} H_xWO_3 \xrightarrow{(237\ ^\circ C)} WO_3(\text{cubic}) \xrightarrow{(600\ ^\circ C)} (\text{monoc})WO_3$$

Figure 2a depicts the decay of peak intensity at 150 °C for <020> reflection of H_2WO_4 phase. This decay curve has been fitted to the exponential decay function and the same has also been given in the Fig. 2. The area under this peak with respect to temperature has been found to follow two stages of decay; first stage up to 150 °C where there is no decrease in peak area and a second stage where the fall is very sharp (Fig. 3). The <100> reflection of H_xWO_3 phase was found to have two stage growth like <020> reflection of H_2WO_4 phase (Fig. 3b). Both these data were fitted with several equations and were found to have best fit for reverse sigmoidal equation of the form

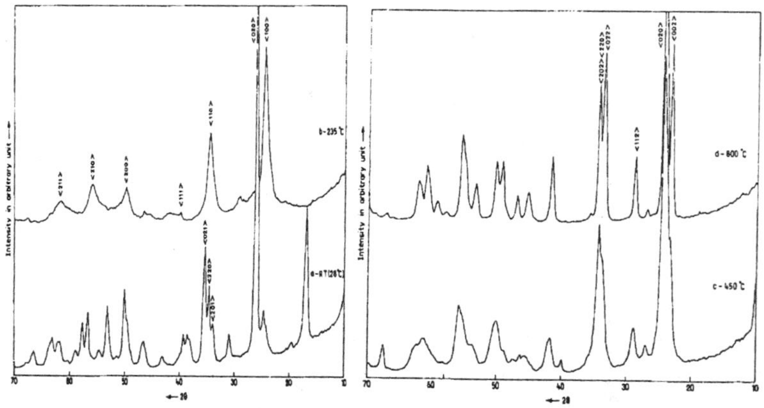

Fig. 1: In-situ XRD patterns of hydrotungstite at different temperatures.

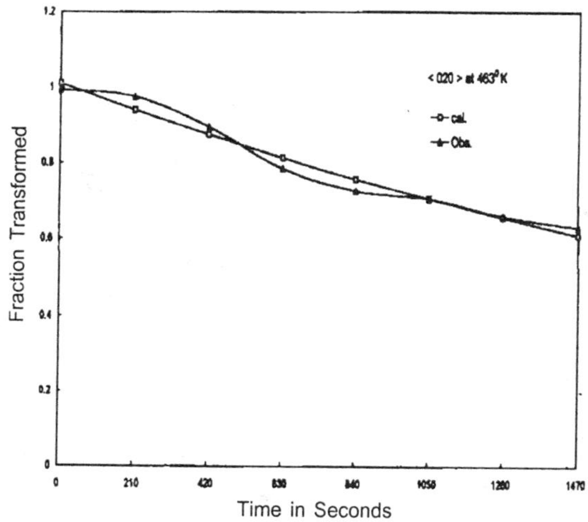

Fig. 2: Experimental decay of <020> reflection at 463 °K for H_2WO_4.

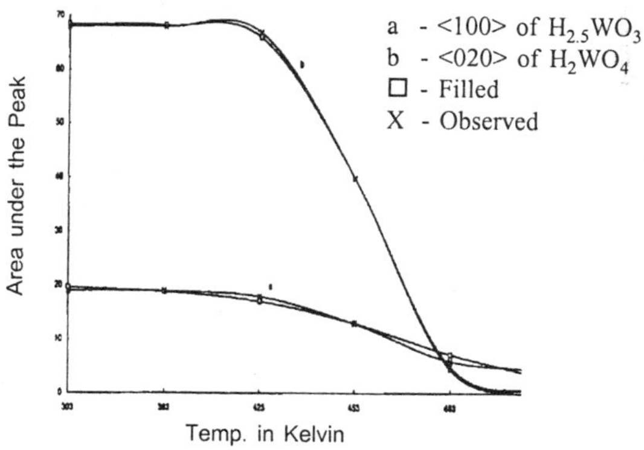

Fig. 3: Decay curves for <020> reflection of H_2WO_4 and <100> reflection of H_xWO_3 phase with temperatures.

$$y = a_0\ [1-1/1 + \exp\ \{-(x-a_1)/a_2\}]$$

where y = area under the peak, a_0 = amplitude, a_1 = centre point and a_2 = width.

The growth of <100> reflection for H_xWO_3 phase of 210 °C with time has been shown in Fig. 4. The growth equation has been fitted to Mehl-Avarami, and Reckitt–Austin empirical equations and also to the equation proposed by us. All these fitted curves have been given in Fig. 4 for comparison. It can be seen that the equation proposed by us fits the best to the observed data.

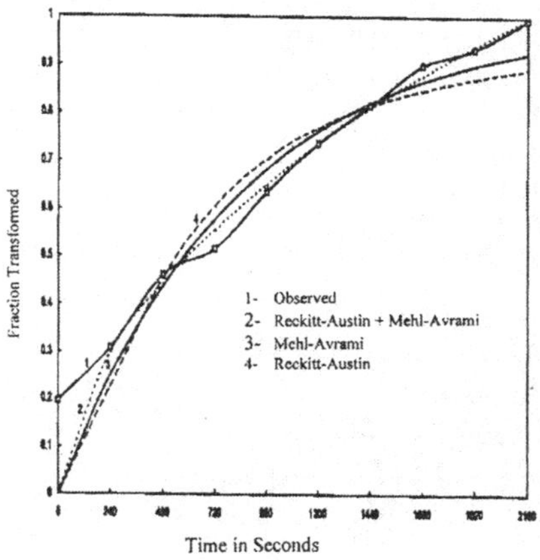

Fig. 4: Growth of <100> reflection of H_xWO_3 phase at 483 °K fitted to different empirical equations.

Hydrotungstite, H_2WO_4, crystallizes in monoclinic form having a, b and c as 7.50, 6.93 and 3.70 Å, respectively and hydrogen tungsten bronze, $H_{0.5}WO_3$, crystallizes with cubic structure having lattice constant 3.75 Å (International diffraction data (ICDD) USA, 2003 and Wykoff, 1962). When hydrotungstite is heated, the complex crystal structure breaks and the atoms reorganize themselves to give many intermediate metastable phases finally leading to cubic structure. In the process, water is liberated and when heated to 600 °C and above, WO_3 is formed. H_xWO_3 shows metallic conductivity whereas WO_3 is an insulator. Thus, there is a metal to insulator transition temperature, which brings out the phase change in this system. The XRD patterns at room temperature (≈ 28 °C) and at 110 °C remain same (curve not shown), whereas the pattern taken at 235 °C and 450 °C are quite different. These two patterns identify themselves with the patterns of H_xWO_3 and WO_3 (PDF card No.18-1420 and 6-707). Heat treatment of the sample produces considerable changes in the crystal structure. At an optimum temperature of 210 °C, the hydrotungstite changes its structure from monoclinic to cubic i.e. from H_2WO_4 to $H_{0.5}WO_3$ and when heated at 800 °C, WO_3 (monoclinic) is formed again.

Two well isolated reflections <020> of H_2WO_4 and <100> of H_xWO_3 were chosen for isothermal and isochronal studies. The decay of <020> reflection intensity at 463 °K has been found to follow exponential decay, signifying the kinetics to be monomolecular (Fig. 2). The growth of <100> reflection intensity of H_xWO_3 phase has been fitted with Johnson Mehl-Avrami, Austin-Reckett, and the equation proposed by the authors. The best fit has been obtained for our proposed equation. The plot of fraction transformed with (time)$^{1/2}$ (not given) indicated the reaction to be diffusion controlled. Further, this diffusion controlled mechanism takes place through two stages also.

Linear and parabolic rate equations have been found to describe many solid state and metallic reactions (Christian, 1975) and especially it has been shown that if the reaction is controlled through nucleation at the interface it will follow a linear growth (Chung et al., 1993). However, in both the cases the driving forces are extremely complex. In the present investigation, it has been observed that a plot of fraction transformed with (time) $^{1/2}$ gives a straight line (not shown), which substantiates the growth mechanism to be diffusion controlled, but when Austin-Reckett equation is fitted, it does not conform to our experimental results. Thus, we are led to fit the profile with the equation: Y (fraction transformed) = 1–exp (k't)" + (k't)" which is nothing but the sum of Johnson-Mehl-Avrami and Reckett-Austin equation. From this we believe that in this system both nucleation and diffusion mechanism compete with each other and in the initial stage when nucleation of the H_xWO_3 phase starts the reaction kinetics is found to be linear and with the transformation taking place this reaction becomes non-linear. It has been argued that in rare-earth hydride system, the diffusion of hydrogen is complete within a temperature range of 250 °C (Johnson, 1988). Thus, our observation of 210 °C as the complete transformation temperature for H_2WO_4 to become H_xWO_3 is justified. The reaction kinetics found from the fit is 1.5, which speaks about the complexity of the transformation process taking place in this system.

Iron Carbide

Iron carbide is an important material, which is prepared industrially from blue dust of iron ore in fluidized bed reactor or by smelt reduction. Iron carbide so produced is hard, non friable and resistant to oxidation up to 300 °C and has potential to replace direct reduction of iron ore, scrap or molten iron to produce steel at a low cost. Iron carbide studied in this investigation has been prepared through a novel route like plasma smelting. Iron ore slime concentrate containing about 93% Fe_2O_3 was taken with 10 wt.% extra carbon addition over stoichiometry (for carbothermic reduction of Fe_2O_3) and smelted in a 35 kW dc extended arc plasma produced between two graphite electrodes. Methane and hydrogen in the volume ratio 1:2.5 (in lpm) were used in the form of a mixture to produce plasma in the inter electrode space in the beginning then the plasma swelled into the entire volume of the reactor space. Samples were prepared at 700 gm scale and the iron carbide product was produced in the form of lumps showing hardness in the range 570-660 $VHN_{0.1}$. Chemical analysis of the important constituent elements in the smelted product is found to be C 6.71–6.89 wt.%, S <0.01 wt.%, P 0.11–0.12 wt.%. This carbide has been subjected to in-situ high temperature study to see its resistance to oxidation and conversion to any iron oxide form.

Figure 5 shows the XRD patterns taken at different temperatures. Room temperature investigation of several samples produced through plasma route revealed the material to contain several iron carbide phase like Fe_2C, Fe_3C, FeC, FeO and SiO_2:Fe_2O_3 etc. The sample chosen for this study contains Fe_2C, Fe_3C only. In-situ XRD patterns taken at 300 °C do not reveal any change of phases for this sample. However, there is a reduction in intensities of the reflections observed at room temperature. In-situ measurement at 500 °C brings out new reflections and reduction of the most intense line. The peaks broaden signifying decrease in crystallite size. New phase formation starts and as the temperature is raised to 900 °C distinct iron oxide (Fe_2O_3) reflections were observed. New reflections appearing for 500 and 700 °C in-situ XRD patterns have been identified to be hematite and magnetite. In case of a sample studied at 900 °C, three phases—wustite, magnetite and hematite—are observed.

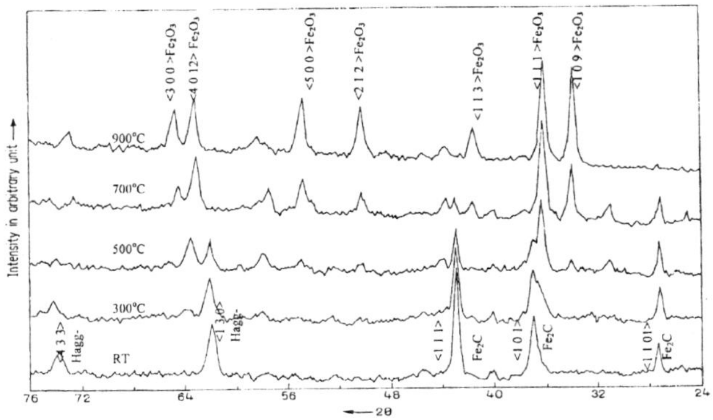

Fig. 5: In-situ XRD patterns of iron carbide at different temperatures prepared through plasma route.

In a recent study of low carbon steel and extra low carbon steel by Caudron and Buscail (2000), three iron oxide layers were observed: FeO at the greatest depth, Fe_3O_4 at the lowest depth and Fe_2O_3 near the oxide gas interface. Similar studies on three faces (111), (110) and (100) of magnetite single crystal reveals formation of α-Fe_2O_3 (hematite) phase on all the samples when annealed at 300 °C in air (Zhou et al., 2004). Maghemite

(ν-Fe_2O_3) is metastable and is a modification of α-Fe_2O_3 at low temperature. It transforms to α-Fe_2O_3 or hematite at 300 °C (Schimanke and Martin, 2000). Crystallite size has been found to play a major role for stabilization of this maghemite phase (Schimanke and Martin, 2000). Our results on in-situ transformation of iron carbide phase to different iron oxide phase support this view. Accordingly the reaction can proceed as under

$$Fe_2C + Fe_3C \xrightarrow{300\ ^\circ C} \text{No change} \xrightarrow{500\ ^\circ C} \text{Maghemite} \xrightarrow{700\ ^\circ C}$$

$$\text{Maghemite + Hematite + Magnetite} \xrightarrow{900\ ^\circ C} \text{Wustite + Hematite + Magnetite}$$

Oxidation of magnetite has been studied extensively and has been found to be size dependent. Diffusion of iron cation was observed to be important in the transformation process (Gallagher et al., 1979). Magnetite particles with size greater than 300 nm were oxidized to α-Fe_2O_3 below 180 °C (Gallagher, 1968) whereas particles with size less than 300 nm were oxidized to maghemite.

Fused Tungsten Carbide: (WC + W_2C)

Fused tungsten carbide was prepared by melting (W+WC) pellets (mixed in the wt.% ratio 55.5:44.5) in a 100 kW dc extended arc plasma reactor using argon as the plasma generating gas. The micro hardness of the sample showed value in the range 2000-2100 $VHN_{0.1}$. The total carbon in the fused tungsten carbide sample was found to be around 3.1 wt.% whereas in the grain, carbon was found to be 1.91 wt.%.

Tungsten carbide exists in three forms: WC, αWC_{1-x} and W_2C. It has been observed from the phase diagram of tungsten-carbon system that below 2530 °C αWC_{1-x} decomposes to WC + W_2C, which is called fused tungsten carbide (FTC) (Massalski, 1990). Hexagonal W_2C crystallizes in three forms denoted by β, β' and β'' with PbO_2, Fe_2N and CdI_2 structure. WC_{1-x} crystallizes in NaCl structure and is called γ form and WC crystallizes in hexagonal structure denoted as 'α' form. Technically important α-WC is the only phase stable at room temperature. Good wear resistance of WC makes it suitable to be coated on engine components and parts. However, oxidation of the carbide limits use of this coating to temperature below 480 °C. Hence, it will be worthwhile to discuss the phase changes during in-situ high temperature XRD study. It is reported that β-W_2C and β-WC_{1-x} are unstable at low temperature ($\approx$ 430 °C) and the transformation takes place around 470 °C (Kong et al., 1983). Further, W_2C decays to α-WC when annealed in argon atmosphere at 1000 °C. Accordingly, few reflections of WC and W_2C were recorded at different temperature for different timings. The behaviour of these reflections are given in Figs. 6, 7 and 8 at 450 °C, the intensities of reflections <002> and <004> decrease linearly with time, but

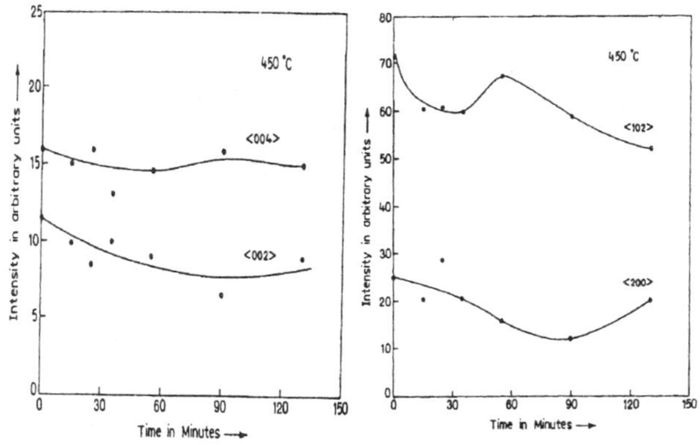

Fig. 6: Variation of intensities of selected reflections of tungsten carbide measured at 450 °C.

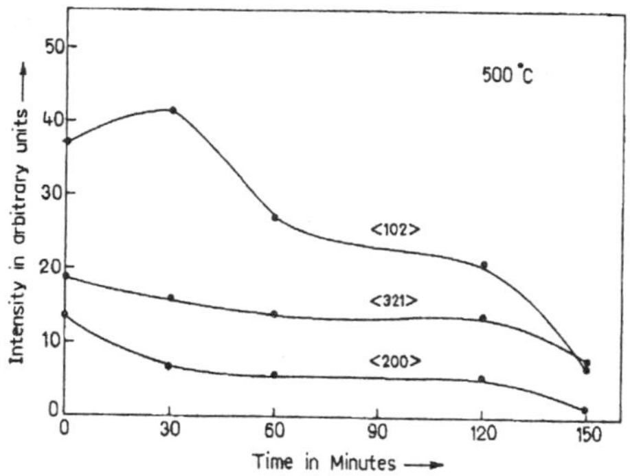

Fig. 7: Variation of intensities of selected reflections of tungsten carbide measured at 500 °C.

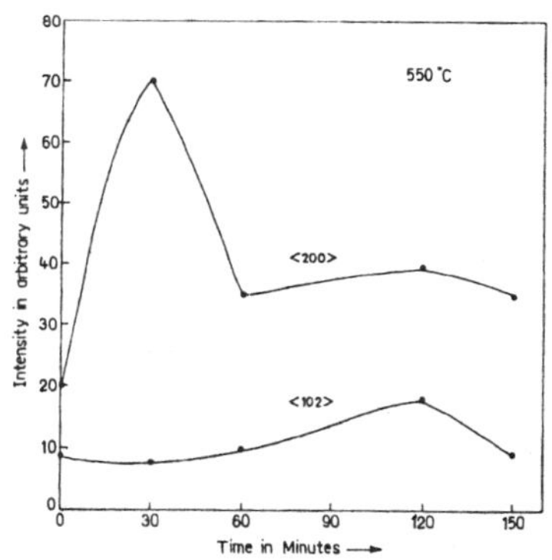

Fig. 8: Variation of intensities of selected reflections of tungsten carbide measured at 550 °C.

the process is slow. The behaviour of <102> and <200> reflections are quite different. Though the intensity for <200> reflection decreases with time, after 90 minutes, it increases, thereby signifying the growth of a new phase along this direction. However, there is a mixed behaviour for <102> at 450 °C. At 500 °C, the intensities of three reflections <102>, <321> and <200> decrease with time. The behaviour of <102> is not linear. Similarly in-situ study of fused tungsten carbide (FTC) at 550 °C gives total different picture for both <102> and <200> reflections. <200> intensity rises sharply for 30 minutes in-situ study, whereas subsequent annealing decreases the intensity. The XRD patterns recorded at different temperatures have also been given in Fig. 9. The phases identified at different temperatures are given in Table 1.

Even though earlier studies (Kong et al., 1983) indicate formation of WO_3 from WC at comparatively low temperature around 430 °C, we have not observed such transformation even up to temperature of 700 °C. In a study (Lofaz and Kagonovskii Yu, 1995) on WC-Co oxidation, it has been

Table 1: Phases identified for FTC at different temperatures

Sample	*Temp. in °C*	*Phase identified*
FTC	26	W_2C Major WC Minor
FTC	500	W_2C Major WC Minor
FTC	700	W_3C Major WC Minor WO_3 Minor

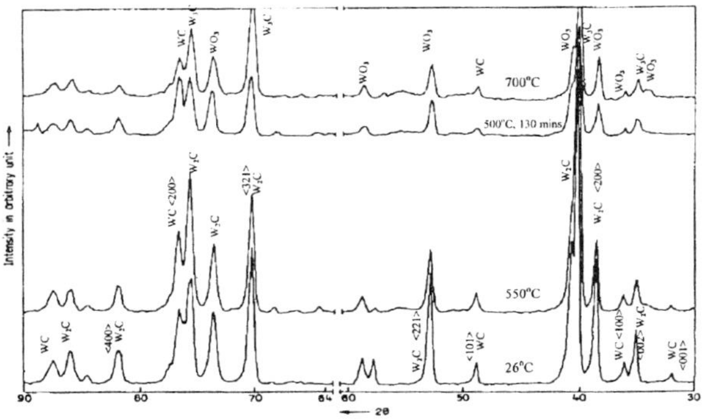

Fig. 9: In-situ XRD patterns of FTC prepared by plasma route.

reported that partial brittle layer of WO_3 phase is formed within a temperature range of 650 to 800 °C. Addition of Co decreases the oxide layer and the process of oxidation takes place through extensive swelling of porous oxide layer at the interface. The fused tungsten carbide of this present investigation was analysed for its carbon content and it was found to have 1.91 wt.% of carbon in the grain and 3.1 wt.% total content in sample. Thus the sample was thought to be easily oxidisable. However, on the contrary, it was observed to be hard and resistant to oxidation, a good sign for tungsten carbide to be used at high temperature. The variation of intensities can be explained on the basis of grain growth and variation of strain near grain boundaries. Further, the overlapping reflections of different tungsten carbide phases may vary according to the change in their ratio. However, it is difficult at this stage to pin point the definite mechanism involved.

Ammonium Molybdate Tetrahydrate

Ammonium salts have been studied by differential thermal analysis method for their phase transformation (Erdy et al., 1964). Three ranges of temperature have been identified for phase transformation. Thus it was thought appropriate to study the XRD of this molybdate at different temperatures. In-situ XRD patterns of the sample recorded at different temperatures are given in Fig. 10. There is a distinct change in the intensities of low angle reflections when the diffractogram of 210 °C is compared with that of room temperature. The <040> reflection has decreased and the triplet structure around 12° 2θ corresponding to <100> <110> and <121> has reduced to doublet structure. The DTA and TG curve recorded for this sample show four sharp transitions and reduction in weight. These temperatures were found to be 125, 228, 310 and 790 °C (Fig. 11). Accordingly, we have recorded the patterns near about these temperatures. In the initial stage due to loss of water, the structure breaks to $(NH_4)_6Mo_7O_{24}$ and this leads to endothermic peak at 125 °C of DTA curve. In XRD pattern the intensities of all the peaks decrease only (not shown). As the heating continues, the second stage of decomposition starts around 180 °C and is completed around 280 °C. The XRD pattern recorded at 210 °C reveals new feature and broadening of some reflections takes place throughout the range of 2θ values. Again at 310 °C, the doublet structure of the reflections appearing in the low angle range 12°–16° 2θ disappear completely and new reflections appear. The weight of the sample

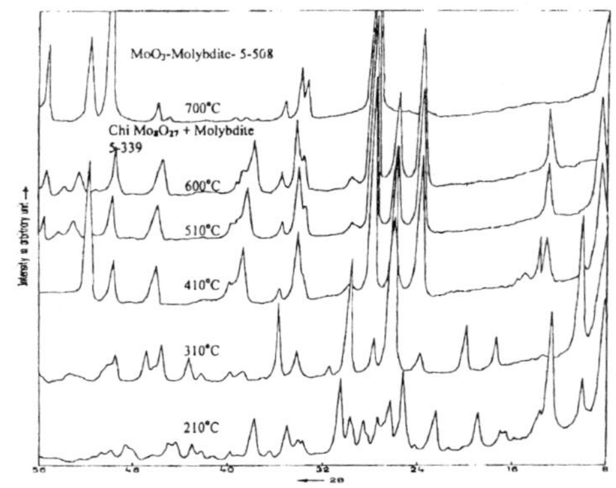

Fig. 10: (a) In-situ high temperature XRD patterns of Molybdenum oxide from ammonium molybdate $(NH_4)_6Mo_7O_{24}{\cdot}4H_2O)$.

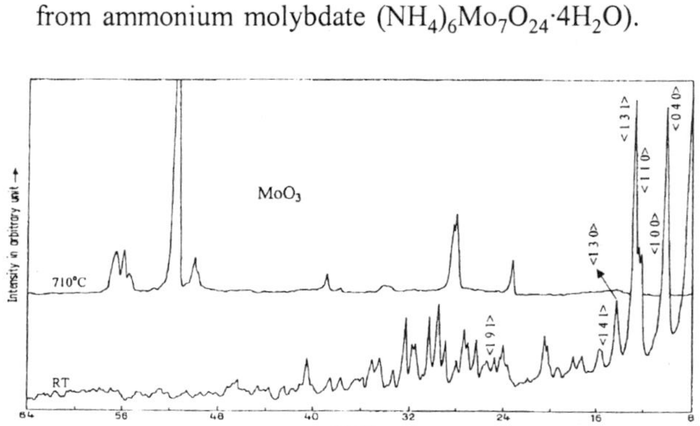

Fig. 10: (b) In-situ XRD patterns of ammonium molybdate tetrahydrate at room temperature (28°) and at 700 °C for one hr.

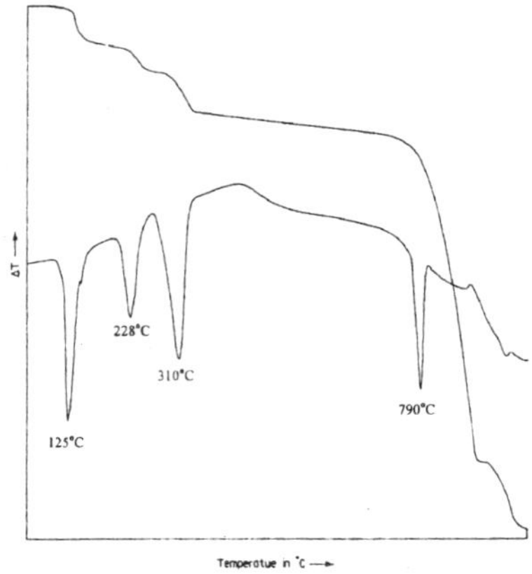

Fig. 11: TG/DTA curves for ammonium molybdate tetrahydrate (heating rate 20 °C /minute).

decreases more rapidly and the bound water molecule in $(NH_4)_6\ Mo_7O_{24}$ are lost. The reaction can be written as

$$(NH_4)_6\ Mo_7O_{24} \xrightarrow{230\ ^\circ C} 6NH_3.7MoO_3 + 3H_2O$$

In the third stage, NH_3 molecules are removed. Thus the patterns recorded at 310 and 410 °C are totally different. In the high angle range well developed sharp reflections appear. The process indicates the formation of MoO_3 compound. At 600 °C, XRD pattern shows presence of two MoO_3 phase such as Chi Mo_8O_{23} (PDF Card No. 5-339) and molybdite (MoO_3) (PDF Card No. 5.508). When the sample is heated to 700 °C only MoO_3 phase is observed. Sharp, well isolated peaks appear and the number of reflections decrease. When the sample heated at 700°C for 2 hrs cooled down to room temperature and again the XRD pattern recorded at 710 °C the structure of the pattern differs, although same MoO_3 reflections appear. In the TG curve, a fourth stage has been observed and the endothermic peak appears at 790 °C (Fig. 11). Earlier DTA study (Erdy et al., 1964) on ammonium salts has not gone beyond 600 °C and in the present study we have observed two phases of MoO_3. β-MoO_3 crystallises in monoclinic form and changes to α-MoO_3 above 350 °C. α-MoO_3 is hexagonal and changes to MoO_3 orthorhombic form around 450 to 475 °C. So it is likely that heating beyond 600 °C, the structure breaks and there is a tendency to become amorphous, as the melting point of the compound is 795 °C. Thus the whole process of transformation can be written as

$$(NH_4)_6\ Mo_7O_{24},\ 4H_2O \xrightarrow{125\ ^\circ C} (NH_4)_6\ Mo_7O_{24} \xrightarrow{210\ ^\circ C} 6NH_37MoO_3 + 3H_2O$$

$$\xrightarrow{310\ ^\circ C} 7MoO_3 \xrightarrow{410\ ^\circ C} ChiMo_8O_{23} + Molybdite(MoO_3) \xrightarrow{700\ ^\circ C} MoO_3$$

The decay of intensity of reflections within a range of 9 to 30° (2θ) has been given in Figs. 12 and 13. One can observe the fall in intensities of all the reflections at 130 °C with the increase in annealing time. Similarly the growth of two reflections at 210 °C has been shown in Fig.13b. This growth is linear whereas the growth of two reflections at 210 °C becomes exponential with temperature (Fig. 12).

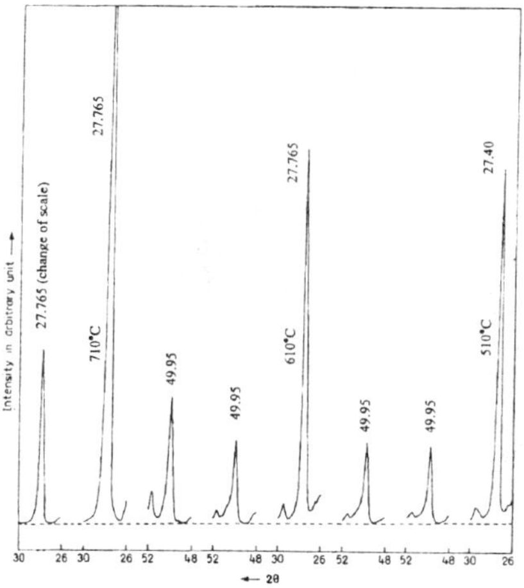

Fig. 12: Growth of two reflections of ammonium molybdate with temperature.

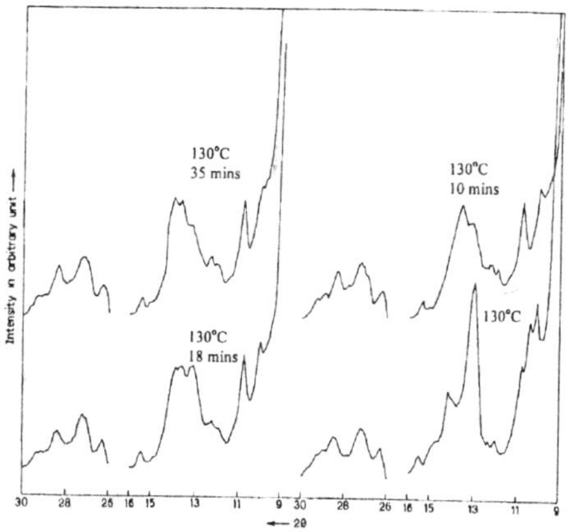

Fig. 13: (a) Decay of two reflections of ammonium molybdate at 130 °C.

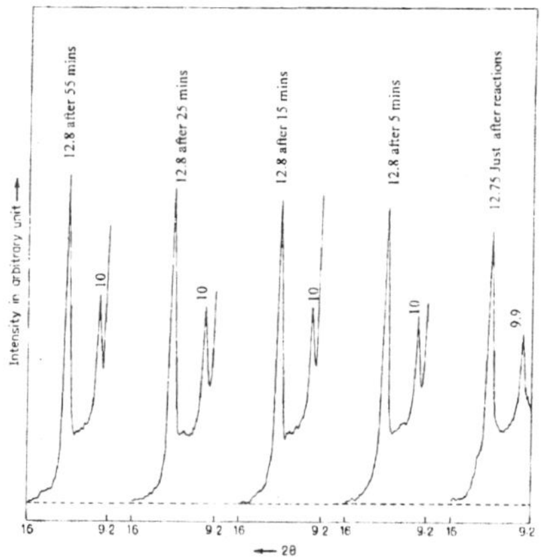

Fig. 13: (b) Growth of two reflections of ammonium molybdate at 210 °C for different timings.

Copper Chloride

Normal copper chloride available as laboratory reagent grade contains water molecules and has formula $CuCl_2.2H_2O$. This compound loses its water molecules when dried at 100 °C for one hr. This dried chloride sample was used in this investigation to look for its sequence of phase transformation to CuO. This reagent grade chemical after drying at 100 °C for 2 hrs was exposed to X-ray for recording its diffraction pattern at room temperature. The powder diffraction data base searched for this compound could not match any $CuCl_2$ with water of crystallization but matched with PDF Card No. 34-0775, $CuCl_2$ for which no structure or planes have been identified. Further, the diffraction pattern recorded by us revealed the presence of minor amount of $CuO.3H_2O$. In-situ XRD of the sample at 300 °C revealed the formation of $CuCl_2$ with broad and intense peaks, indicating smaller crystallite size. However, presence of CuO phase was not ruled out completely. XRD pattern at 400 °C shows new reflections due to presence of CuCl phase which is reported to be stable at 407 °C and possesses hexagonal structure. Heating the sample at 450 °C does not bring about radical changes in the diffraction pattern except that relative intensities have been changed for various reflections. When the sample was heated to 500 °C and 550 °C monophasic copper oxide (PDF Card No. 41-254) has been found to be formed (Figs. 14a and b). So the process of phase transformation can be enumerated as:

$$CuCl_2 2H_2O \xrightarrow{100\ ^\circ C} CuCl_2 + CuO,\ H_2O \xrightarrow{300\ ^\circ C} CuCl_2 + CuO \xrightarrow{400\ ^\circ C}$$

$$CuCl\ (Hexag) \xrightarrow{450\ ^\circ C} CuCl \xrightarrow{500\ ^\circ C,\ 550\ ^\circ C} CuO$$

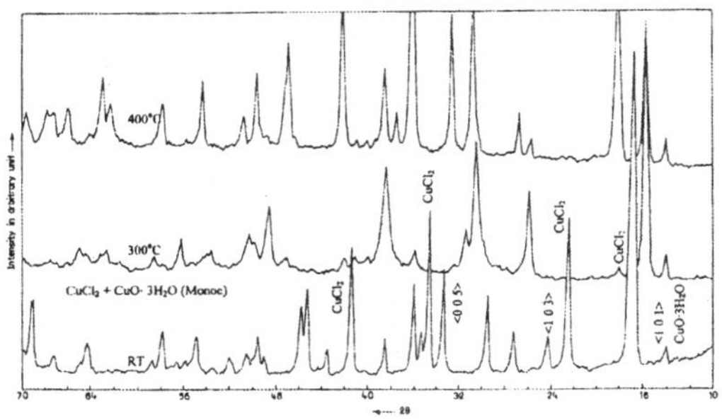

Fig. 14: (a) In-situ XRD patterns of copper chloride at different temperatures (RT to 400 °C).

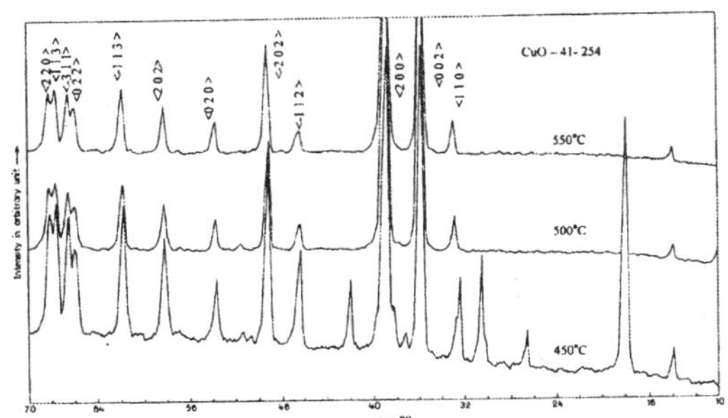

Fig. 14: (b) In-situ XRD patterns of copper chloride at different temperatures (400-550 °C).

Goethite

In nature iron is not only associated with hematite, maghemite or magnetite but is found to be associated with goethite also. Nickeliferrous laterite contains nickel associated with goethite and in many cases goethite is amorphous in nature. In one of our earlier study on over burden chromite ores from Sukinda Mines of Orissa, it was observed that the phase transformation of FeO(OH) to α-Fe_2O_3 or hematite takes place around 280 °C (Zhou et al., 2004). Goethite prepared through chemical route has been found to loose its crystallinity within few hrs of preparation. Thus an effort has been made to synthesize the compound through chemical route (chloride route) and the goethite so produced was studied for in-situ high temperature XRD to see the phase transformation taking place in this system. The XRD patterns recorded at different temperatures have been shown in Fig.15. The low intensity in the diffraction pattern of goethite at room temperature can be due to its poor crystallinity. Heating the sample at 200 °C does not alter the features of the pattern, except that intensities of all the peaks increase marginally. Pattern taken at 300 °C speaks about the changes taking place in the system. The low angle reflection representing <110> for goethite disappears completely and new reflections appear in the pattern. However, all the peaks are broadened which can be attributed to smaller crystallite sizes. Heating the sample at temperatures like 400, 500, 600 and 700 °C does not alter the main features of the diffraction pattern. The intensities of all the reflections increase and the peaks become sharp. The crystallinity of the sample increases. Unlike iron carbide high temperature study where transformation to hematite takes place through maghemite, here in this case of goethite no such maghemite phase was observed.

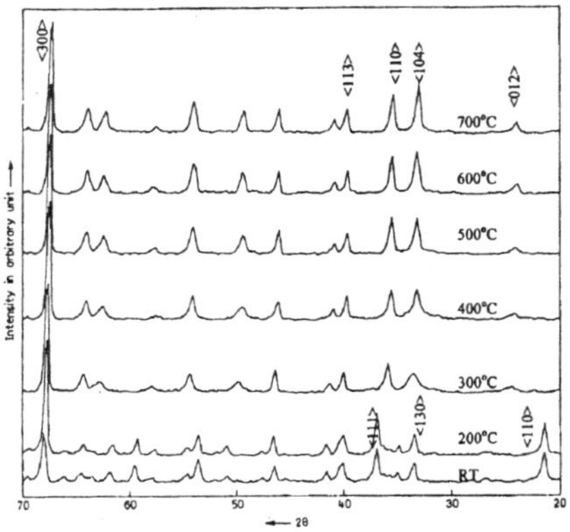

Fig. 15: In-situ high temperature XRD patterns of goethite.

The intensity variation for three reflections of goethite such as <111>, <130> and <110> were followed for several minutes after the requisite temperature (i.e 250 °C) is reached. This has been shown in Fig. 16. The nature of decay of such curves for <110> and <111> were found to follow power law type decay. The intensity of <130> does not change with time. This indicates that decay kinetics of different reflections differ. Such type of preferential oxidation mechanism has been observed for maghemite to hematite in the nanocrystalline powders of iron oxide (Schimanke and Martin, 2000) and also for two different surfaces of iron oxide single crystal (Lauer et al., 2003). In a recent study (Jiang et al., 2000) on well characterized needle shaped goethite (FeOOH) acicular shaped magnetite has been found to be formed when the sample was heated to 200 °C and then cooled to ambient temperature (cooling rate 200 °C/hr). The DSC heat flow curve showed first exothermic reaction around 225 °C where conversion of magnetite to maghemite takes place and the second exothermic peak at 590 °C depicting the conversion of maghemite to hematite. Further with the rise in formation temperature of acicular magnetite, the DSC peak temperature shifts to high temperature side linearly and there is an increase in enthalpy for conversion of magnetite to maghemite. Thus, depending upon the crystallinity and purity of the sample the conversion from magnetite → maghemite → hematite takes place. From TEM and XRD studies, Jiang et al. (2000) have shown that non-uniform X-ray line broadening takes place in goethite derived hematites due to plate like shape of hematite crystals.

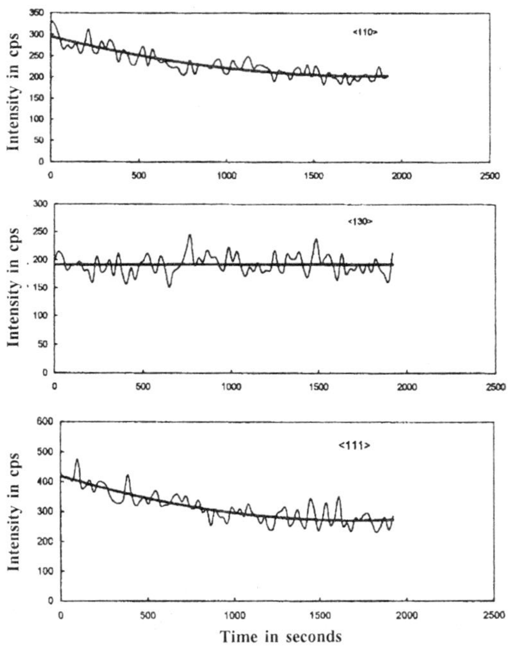

Fig. 16: Intensity variation for three different reflections of goethite measured at 250 °C.

Our study indicates non-uniform X-ray line broadening for hematite phase. The crystallite size measured for our goethite was found to be 30 to 35 nm. Nanocrystallite Fe_2O_3 produced from reverse micelle method showed the presence of maghemite when the crystallites are below 30 nm while crystallite size above this size gives hematite. Thus our observation of absence of magnetite phase in the conversion process of synthesized goethite to hematite is justified.

Acknowledgements: The authors are thankful to the Director, RRL (Bhubaneswar) for permitting to present the paper in this conference.

REFERENCES

Christian, J.W. (1975). The theory of transformations in metals and Alloys. Pergamon Press, Oxford.

Baucke, F.G.K. (1987). *Solar Energy Mat.*, **16:** 67-69.

Caudron, E. and Buscail, H. (2000). *Materials Chem. & Phys*, **64(1):** 29-36.

Cazzanelli, E., Vinegoni, C., Mariatto, G., Kuzmin, A. and Purans, J. (1999). *Solid State Ionics*, **123:** 67-74.

Chung, D.D.L., Dettaven, P.W., Arnold, H. and Ghosh, D. (Eds.) (1993). X-Ray diffraction at elevated temperatures. VCH Publishers, New York.

De Haver, P.W., Walker, G.A. and O'Neil, N.A. (1984). *Adv. in X-ray Analysis*, **27:** 89-393.

Erdy L., Gal, S. and Liptay, G. (1964). *Talanta* **11:** 913-940.

Gallagher, K.J., Feit Knecht, W. and Mannweiler, U. (1968). *Nature*, **217:** 1118.

Gallagher, P.K., Gyorgy, E.M. and Bair, H.E. (1979). *J. Chem. Phys.* **71:** 830.

Herman, H. and Cochrane, R.W. (Ed.) (1988). *Materials Science and Engg.*, **97:** 650.

Jiang, J.Z., Strahl, K., Nielsen, K. and Da Costa, G.M. (2000). *J. Phys (Condens Matter)*, **12:** 4893-98.

Johnson, W.L. (1988). *Mater. Sc. and Engg.* **97:** 1-6.

Kong, P.F., Suzuki, M., Young Rao Pfender, E. (1983). *Plasma Chem. and Plasma Process*, **3(1):** 115-133.

Lauer Jr, H.V., Ming, D.W. and Golden, D.C. (2003). *Lunar and Planetary Sc.*, **XXXIV:** 1341-1342.

Lofaz, F. and Kagonovskii Yu, S. (1995). *J. Material Sc.* **30:** 1811-1817.

Massalski, T.B. (Ed) (1990). Binary Alloy Phase Diagrams, Vol I: Alloys, Vol.2. Phasediagrams: ASM Ints. Materials Park, Ohio.

Our report on extraction of nickel from nickel bearing Chromite overburden of TISCO Mines. Vol. 3, 1993. Regional Research Laboratory, Bhubaneswar.

Powder Diffraction Data File No. 18-1420, 6-707 (2003). Prepared by International diffraction data (ICDD), USA.

Purans, J., Kuzmin, A. and Guercy, C. (1997). *Proc. SPIE*, **2968:** 174-179.

Schimanke G. and Martin M. (2000). *Solid. St. Ionics*, **136-137:** 1235-1240.

Tu, K.N. and Berry, B.S. (1972). *J. Appl. Phys.*, **43:** 3283-3290.

Wykoff, R.W.G. (1962). Crystal structures – 2nd Edn. Wiley, New York, **2:** 82.

Zhou, Y., Jin Xuesong, Mukovskii Y.M. and Shvets I.V. (2004). *J. Phys.* (Condens. Matter), **16:** 1-12.

Ziolo, R.F., Giannelis, E.P., Weinstein, B.A., O' Horo, M.P., Ganguly, B.N., Mehrotra, V., Russel, M.W. and Huffman, D.R. (1992). *Science*, **257:** 219.

Thermal Expansion Studies on Selected Polycrystalline Materials by High Temperature X-ray Powder Diffractometry

K.V. Govindan Kutty and R. Asuvathaman

Materials Chemistry Division, Indira Gandhi Centre for Atomic Research
Kalpakkam 603 102, India

INTRODUCTION

The coefficient of thermal expansion, or thermal expansivity, is an important thermophysical property of materials. It is defined as the fractional change in length per unit rise in temperature. It governs the suitability of materials for many applications in research and industry. The theory and measurement methods of thermal expansion of solids have been discussed in detail in the literature (Ho, 1998; Krishnan et al., 1979). High temperature X-ray powder diffraction is a convenient method for studying the thermal expansion behaviour of materials (Ho, 1998; Krishnan et al., 1979; Kirby, 1992). It is a micro-method of measurement, as opposed to macro-methods such as dilatometry on bulk solids. The expansion of the unit cell dimensions is measured here. The occurrence of thermally generated vacancies in the crystal at elevated temperatures does not, therefore, affect the expansion measurements. Only a small amount of material, often less than 50 mg, is enough for carrying out the experiment; or the specimen can be a small strip in the case of metal and alloy samples. The presence of small amounts of extraneous phases generally does not interfere with the measurements on the desired major phase. Additionally, thermal expansion measurements can be carried out along the different crystallographic axes in non-cubic systems. This enables an assessment of the thermal expansion anisotropy of the material; such anisotropy, in extreme cases, can lead to microcracking due to internal thermal stresses in the material, thus impairing its mechanical strength. The critical grain size, before micro-cracking takes place in a material, has an inverse relationship with its thermal expansion anisotropy (Yamai and Ota, 1993; Senbhagaraman and Umarji, 1990). The salient features of the high temperature X-ray powder diffraction (HTXRD) technique and how it was employed in our thermal expansion studies on various materials, are considered in the following sections.

EXPERIMENTAL PROCEDURE

High Temperature X-ray Powder Diffractometry

The description of a standard HT X-ray powder diffractometer and its calibration experiments are given elsewhere (Rajagopalan et al., 1988). Figure 1 shows the schematic diagram of the system. It shows the HT chamber attached to the goniometer of a vertical diffractometer. The HT attachment consists of a cylindrical, double-walled brass chamber provided with a gas-tight lid. A metallic strip held tight between two electrodes passing through the lid is situated at the centre of the chamber, and serves as the sample carrier. The powder sample is applied as a thin layer on the flat strip which is then electrically heated. The temperature is measured by means of a thermocouple spot-welded at the rear of the sample-carrying strip. The output of the thermocouple is also used to control the heating and cooling at programmed rates by means of a temperature controller. A beryllium window provided on the body of the cylindrical chamber allows the passage of the incident and diffracted X-ray beams. Through the annular space between the walls of the chamber, cooling water is circulated when the system is at high temperature. The chamber can be maintained at a vacuum or an inert atmosphere during experiments.

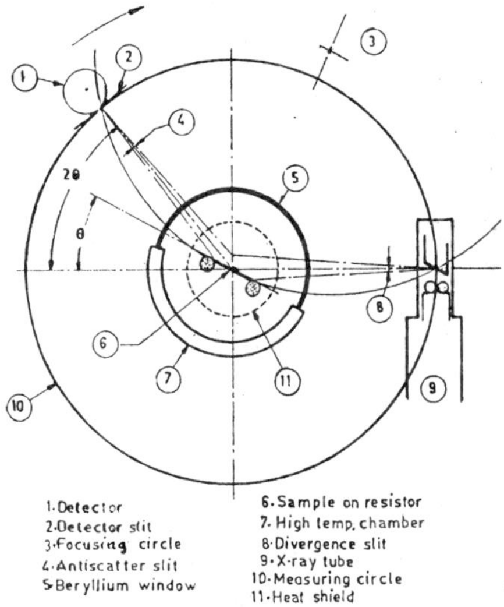

Fig. 1: Schematic diagram of a High Temperature X-ray Powder Diffractometer.

In order to carry out thermal expansion measurements, the lattice parameters at different temperatures are calculated by determining the exact positions of high-angle reflections by step-scanning. The linear thermal expansion between the temperatures T_2 and a reference temperature T_1 (commonly the room temperature or 273 K) is then calculated as:

$$\text{Thermal expansion (\%)} = \frac{100 \times [L_2 - L_1]}{L_1}$$

where L_2 is the lattice parameter at T_2 and L_1 that at T_1. The corresponding average linear thermal expansion coefficient (α) can be obtained by the expression:

$$\alpha = \frac{[L_2 - L_1]}{L_1(T_2 - T_1)}.$$

An instantaneous thermal expansion coefficient at a temperature T can also be defined as follows:

$$\text{Instantaneous thermal expansion coefficient at } T = \left[\frac{1}{L}\right]\left[\frac{dL}{dT}\right]$$

A Seifert HDK2 high temperature chamber coupled to a Siemens D500 X-ray powder diffractometer (Rajagopalan et al., 1988) was employed for most of our measurements. An Anton Parr HTK16 chamber mounted on a Philips X'Pert Pro MPD diffractometer was also made use of in some studies.

Materials

A family of rare earth oxide pyrochlores covering the light lanthanides, a series of uranium – bearing solid solutions of gadolinium pyrozirconate, uranium – rare earth ternary oxides, sodium zirconium phosphate (NZP) – type compounds, and a couple of metal and alloy systems are covered in this study. The ceramic substances were synthesized by different methods such as solid state reactions (for oxide pyrochlores and NZP compounds) and solution combustion or citrate polymer decomposition (for oxide pyrochlores and the uranium-containing compounds). The products were characterized by powder X-ray diffraction (XRD).

RESULTS AND DISCUSSION

Rare Earth Oxide Pyrochlores

These compounds have many technological applications (Subramanian et al., 1983). They have the cubic structure (space group: Fd3m) with a unit cell content of eight formula units (lattice parameter ~ 10 Å). Some of them are candidate crystalline matrices for radioactive waste immobilization (e.g., $Gd_2Ti_2O_7$ and $Gd_2Zr_2O_7$), while some are good oxide ion conductors (e.g., $Gd_2Zr_2O_7$) and neutron absorbers (e.g., $Eu_2Hf_2O_7$ and $Gd_2Hf_2O_7$). A study was undertaken (Govindan Kutty, 1994) to measure the thermal expansion behaviour of the system of compounds $R_2M_2O_7$ (R = La, Nd, Sm, Eu, Gd; M = Zr, Hf). Figure 2 shows the variation of the lattice parameter of $Gd_2Zr_2O_7$ with temperature. The lattice parameter increases from the room temperature value of about 10.53Å to about 10.65 Å at 1373 K; this gives a mean expansion coefficient of 11.04 x 10^{-6} K^{-1} in this temperature range. The

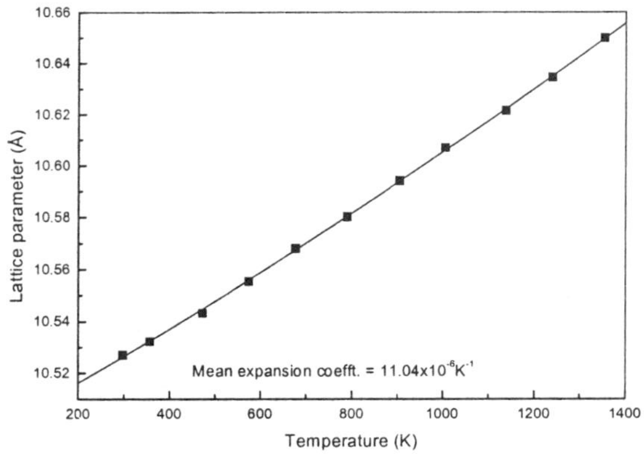

Fig. 2: Variation of the lattice parameter of $Gd_2Zr_2O_7$ with temperature.

thermal expansion behaviour of the family $R_2Zr_2O_7$ is given in Fig. 3. The magnitude of thermal expansion with reference to the room temperature can be seen to be about 1 % for all of them. However, at temperatures above about 800 K, the expansion at a given temperature tends to increase from the La compound to the Gd compound. This is especially clear when the expansion coefficients are plotted versus temperature (Govindan Kutty et al., 1994). Figure 4 displays this in the case of the rare earth pyrozirconates; the same behaviour is exhibited by the corresponding pyrohafnates also. This expansion behaviour can be explained in terms of the ionic bonding model of the oxide pyrochlores (Pannetier, 1973). The rare earth pyrozirconate structure is built up of edge-sharing RO_8 distorted cubes and ZrO_6 distorted octahedra, the latter being the major contributor to the Madelung energy. As the trivalent rare earth ionic radius decreases from La to Gd, and the value approaches that of tetravalent Zr, the ZrO_6 octahedra get progressively deformed. The structure thus gets weakened, with a consequent increase in the expansivity.

$Gd_2Zr_2O_7$ is a good crystalline matrix to host actinide elements in radioactive waste immobilization (Wang et al., 1999). In this context, we have studied the solubility of uranium in this matrix by synthesizing a series of uranium-containing $Gd_2Zr_2O_7$ compositions, and found that large concentrations of uranium can indeed be substituted at the Gd and Zr positions of the crystal (Govindan Kutty et al., 2003). We have also studied the

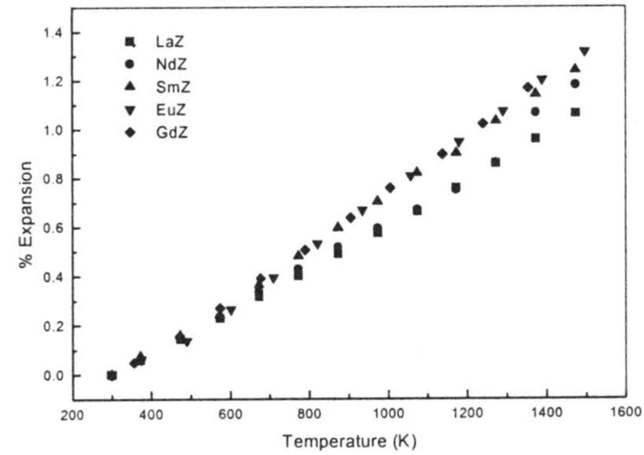

Fig. 3: Thermal expansion behaviour of $R_2Zr_2O_7$ (RZ) with temperature.

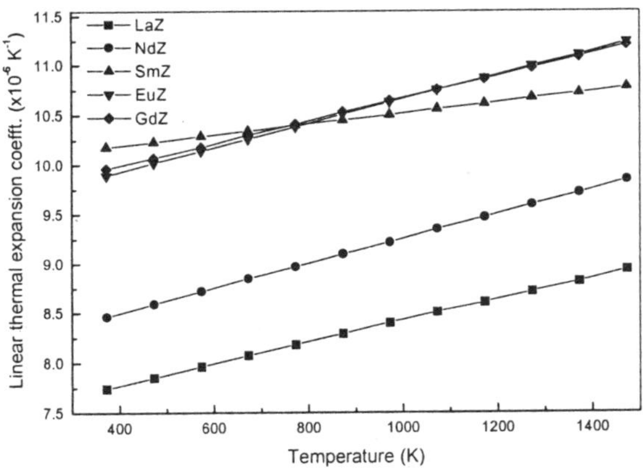

Fig. 4: Variation of the thermal expansion coefficient of the $R_2Zr_2O_7$ (RZ) with temperature.

thermal expansion characteristics of these compounds since their long term performance as a radioactive waste form will be decided by this property. Figure 5 delineates the results; practically there is no difference in the thermal expansivity of the compounds. The thermal expansion is about 1 % at about 1300 K (Govindan Kutty et al., 2003). Though there are small differences in the expansion values of the compounds at high temperatures, there is no trend in the variation of expansion with respect to the uranium content in the solid solutions. The observed differences could have arisen due to small differences in sample characteristics such as the oxygen stoichiometry; the compounds were synthesized in air, so at least a small fraction of uranium could have got oxidized to the hexavalent state.

NZP Compounds

Sodium zirconium phosphate ($NaZr_2(PO_4)_3$, NZP) is the typical member of a family of isostructural compounds crystallizing with a rhombohedral-

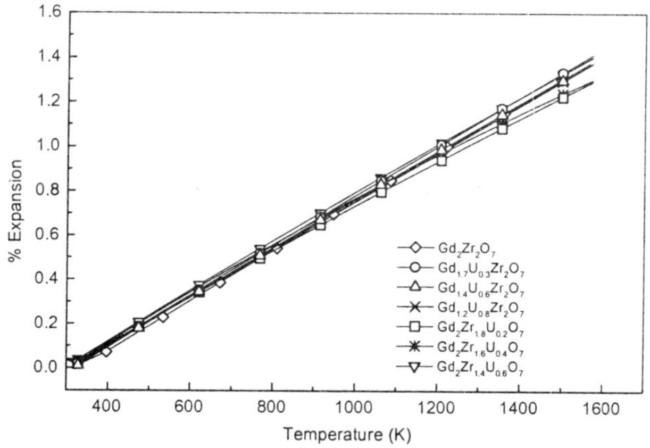

Fig. 5: Thermal expansion of $Gd_2Zr_2O_7$ containing uranium.

hexagonal structure (space group $R\overline{3}c$), built up of a flexible framework of corner-sharing ZrO_6 octahedra and PO_4 tetrahedra (Oota and Yamai, 1986; Hong, 1976). The three-dimensionally linked interstitial space so created is partially occupied by sodium. The compounds are characterized by low and controllable thermal expansivity and sodium ion conductivity. The phase is amenable to a wide variety of chemical substitutions, and is thus a potential crystalline matrix for nuclear waste immobilization. Under thermal stresses, the crystal expands anisotropically; it expands along the hexagonal c-axis and contracts along the a-axis. The average expansivity, as per the formula given below, becomes very low because one of the terms is negative.

$$\text{Average linear expansivity } (\alpha) = -(2/3)\ \alpha_a + (1/3)\ \alpha_c$$

This kind of axial expansion is brought about by bond angle distortions in the crystalline framework; the ZrO_6 and PO_4 polyhedra undergo coupled rotation under the stress created by the large thermal expansion of the Na–O bond (Lenain et al., 1987). The bonds within the polyhedra exhibit negligible expansion (Hazen et al., 1987). Figure 6 depicts the axial expansion behaviour of $Na_2Hf_2(PO_4)_3$ measured in our laboratory (Govindan Kutty et al., 1994). Between the room temperature and 1273 K, the expansion coefficient along the crystallographic a-axis is $- 6.5 \times 10^{-6}\ K^{-1}$ and that along the c-axis is $+ 18.3 \times 10^{-6}\ K^{-1}$; the average expansivity of a polycrystalline sample thus works out to be $+ 1.8 \times 10^{-6}\ K^{-1}$. This is the typical expansion behaviour of an NZP compound. It is controlled by the concentration, size, and the crystallographic site occupancy of the interstitial ion (Oota and Yamai, 1986). However, in some compounds, the framework ions play a dominant role, an example of which is $NbSn(PO_4)_3$. The d electron subshell of the tetravalent tin ion is fully filled, and it becomes easy to bend and distort the Sn–O–P bonds, so much so that the room temperature space group of ths compound is $R\overline{3}$ (against $R\overline{3}c$ of a typical NZP

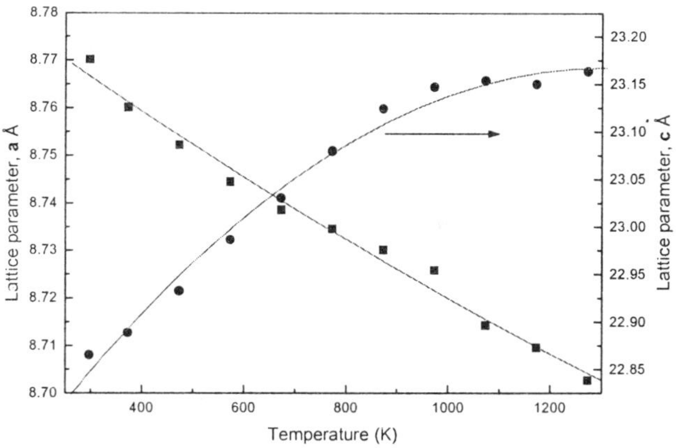

Fig. 6: Axial expansion behaviour of $NaHf_2(PO_4)_3$.

compound); this structure is a little more close-packed than the NZP structure. Figure 7 displays the axial expansion of this compound (Govindan Kutty et al., 1994). Expansion along the a-axis is the predominant thermal effect at temperatures up to 650 K; under this stress, a contraction takes place along the c-axis. At about this temperature, thermal agitations open up the structure and the space group transfoms to $R\bar{3}c$, typical of NZP; thereafter, the crystal expands along the hexagonal c-axis and contracts along the a-axis.

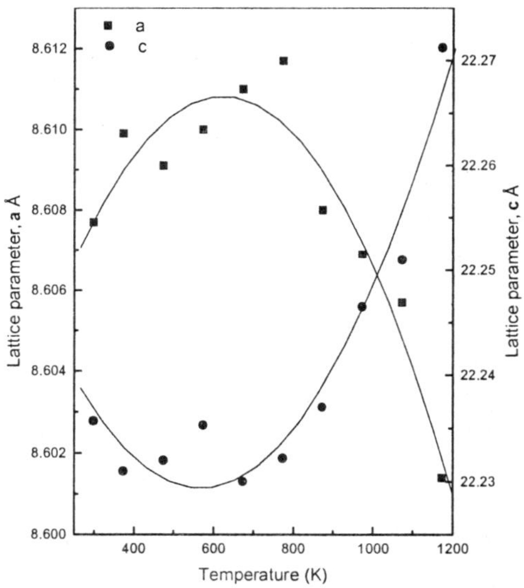

Fig. 7: Axial expansion behaviour of $NbSn(PO_4)_3$.

In NZP compounds, the room temperature space group can be $R\bar{3}$ when the size of the interstitial ion is large. This is exemplified by the series of compounds $ACrSn(PO_4)_3$ [A = Ca, Sr, Ba], whose axial expansion is shown in Fig. 8. The compound $CaCrSn(PO_4)_3$ has the space group $R\bar{3}c$ at room

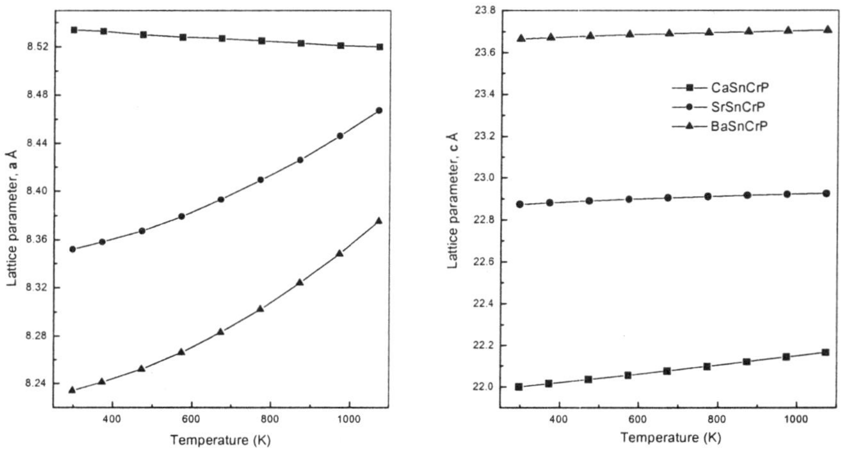

Fig. 8: Axial expansion behaviour of $ACrSn(PO_4)_3$ [A = Ca, Sr, Ba].

temperature, the ionic radius of divalent Ca being very close to that of Na; the compound, therefore, shows thermal expansion along the c-axis and contraction along the a-axis. In the Sr and Ba analogues, the divalent interstitial ions are larger, and the strain is accommodated by the coupled twist of the framework polyhedra; the space group thus turns $R\overline{3}$ in the case of these compounds, and remains so to high temperatures. This is reflected in their axial expansion, where both the lattice parameters register an increase with temperature, the increase in the lattice parameter *a* being more than that in *c*.

This kind of controllability of axial thermal expansion by altering the framework and interstitial composition of the NZP system also enables us to control the axial thermal expansion anisotropy (maximum difference in the axial expansivities) of the materials. Tables 1 and 2 illustrate this with two series of compounds. In Table 1, it is seen that both $CaCrSn(PO_4)_3$ and $CaFeSn(PO_4)_3$ have low average expansivities (α_{av}) between room temperature and 1073 K, but they also have high expansion anisotropy (Buvaneswari et al., 2004). Bulk expansivities measured by dilatometry (α_{dilato}) are also shown for comparison. While dilatometry reveals the low thermal expansivity of these two compounds, the aspect of anisotropy is brought out only by HTXRD. Table 2 shows that $Ca_{1/2}Zr_2(PO_4)_3$ and $Ca_{1/2}Hf_2(PO_4)_3$ have low average thermal expansivities between room temperature and 1273 K (Govindan Kutty et al., 1998), but they are also characterized by high expansion anisotropy; whereas $Sr_{1/2}Zr_2(PO_4)_3$ and $Sr_{1/2}Hf_2(PO_4)_3$ are characterized by low expansivity and low anisotropy.

Uranium—Rare Earth Ternary Oxides

Drastic expansion anisotropy involving thermal expansion along one axis and contraction along another, occurs due to bond angle changes under

Table 1: Axial and bulk thermal expanson coefficients of $AMSn(PO_4)_3$ [A = Ca, Sr, Ba; M = Cr, Fe] in the temperature range 298–1073 K

Compound	*Thermal expansion coefft.* $\times 10^{-6}K^{-1}$		$\alpha_{av} \times 10^{-6}K^{-1}$	$\|\alpha_a - \alpha_c\|$ $\times 10^{-6}K^{-1}$ *Anisotropy*	α_{dilato} $\times 10^{-6}K^{-1}$ *Dilatometry*
	α_a	α_c			
CaFeSnP	-3.37	6.57	–0.06	9.94	2.48
SrFeSnP	17.67	0.56	11.97	17.11	9.32
BaFeSnP	23.57	2.39	16.55	21.18	10.84
CaCrSnP	-2.12	9.74	1.83	11.86	1.99
SrCrSnP	17.77	2.88	12.81	14.89	11.59
BaCrSnP	22.10	2.24	15.48	19.86	10.48

Table 2: Axial expansion behaviour of $A_{1/2}M_2(PO_4)_3$ [A = Ca, Sr; M = Ti, Zr, Hf, Sn] in the temperature range 298–1273 K

Compound	$\alpha_a \times$ $10^{-6}K^{-1}$	$\alpha_c \times$ $10^{-6}K^{-1}$	$\alpha_{vol} \times$ $10^{-6}K^{-1}$	$\alpha_{av} = \alpha_{vol}/3$ $\times 10^{-6}K^{-1}$	$\|\alpha_a - \alpha_c\|$ $\times 10^{-6}K^{-1}$ *Anisotropy*
CaZrP	-2.57	7.74	2.55	0.85	10.31
SrZrP	2.24	2.28	6.65	2.22	0.04
CaHfP	-3.29	7.52	0.92	0.31	10.81
SrHfP	1.66	1.67	5.01	1.67	0.01
CaTiP	6.99	1.31	15.41	5.14	5.68
SrTiP	9.88	-0.14	19.73	6.58	9.74
CaSnP	8.95	2.05	20.05	6.68	6.90
SrSnP	8.07	0.50	16.88	5.63	7.57

thermal stresses, as already seen in the case of NZP compounds. The changes in bond angle can be arrested, the effect due to bond length changes will be the main contributor to expansion. An example is provided by the system of compounds UR_6O_{12} (R = La, Nd, Sm). Figure 9 shows the axial expansion of these compounds (Hrudananda Jena et al., 2000) whose crystal structure is rhombohedral–hexagonal. The crystals expand along both the hexagonal axes almost equally, hence their axial expansion anisotropy is small. The crystal structure is built up of UO_6 and RO_6 octahedra, with considerable edge-sharing between pairs of the latter (Hinatsu et al., 1988). The edge-sharing imparts rigidity to the structure and precludes polyhedral rotations.

Metal and Alloy Samples

Metals and alloy samples in the form of thin flat strips can be directly connected to the electrodes of the HTXRD chamber and heated. We have measured the thermal expansion of the alloy permandur (49Fe – 49Co – 2V) by this method. The sample exhibits the XRD pattern of α-iron (BCC).

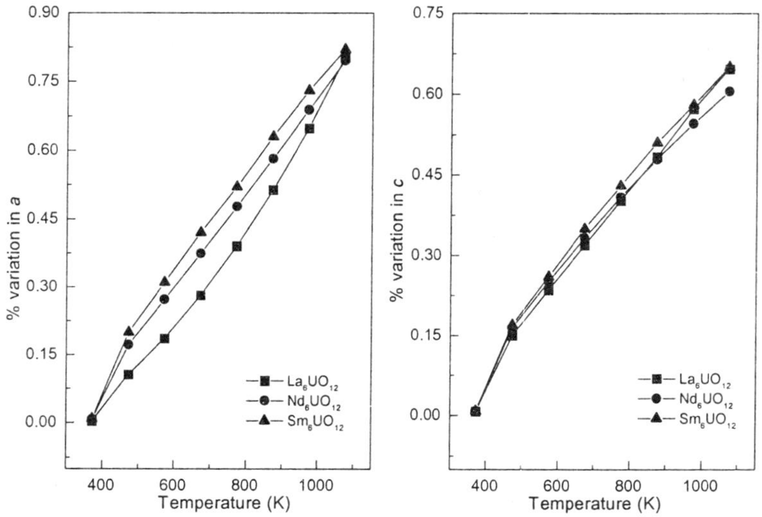

Fig. 9: Axial expansion behaviour of UR_6O_{12}.

Table 3 lists the measured lattice parameters of the material at different temperatures, and Fig. 10 shows the thermal expansion behaviour. At 873 K, the thermal expansion with respect to the room temperature is about 0.6 %; this amounts to a linear thermal expansion coefficient of 10.6 x 10^{-6} K^{-1} in this temperature interval (Govindan Kutty et al., unpublished results).

Table 3: Lattice thermal expansion of permandur

Temperature (K)	*Lattice Parameter (Å) ± 0.0005 Å*
298	2.8593
373	2.8609
473	2.8639
573	2.8669
673	2.8704
773	2.8739
873	2.8769

We have also studied the lattice expansion of nickel (Mohammad Yousuf et al., 1986) about its ferromagnetic-to-paramagnetic transition at ~ 630 K. The λ-type anomaly marking this transition could be seen clearly in the variation of the linear thermal expansion coefficient with temperature (Fig. 11).

Limitations and Errors in Measurement

The main experimental difficulties in HTXRD are sample displacement from the focussing circle, accuracy of the measured sample temperature owing to heat losses, and temperature gradients through the sample

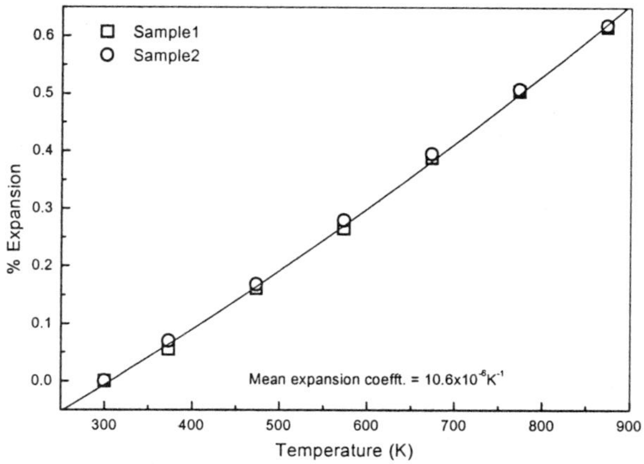

Fig. 10: Thermal expansion behaviour of permandur alloy.

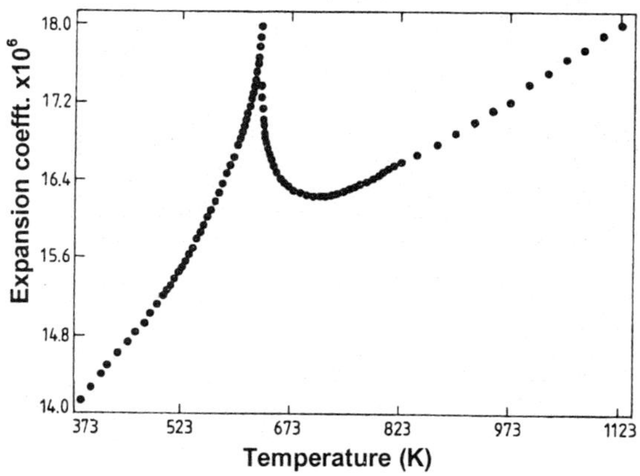

Fig. 11: Thermal expansion of nickel about the Curie magnetic transition.

(Rajagopalan et al., 1988; Chung et al., 1993). Due to the thermal expansion of the thin heater strip (which also carries the sample), sample displacement from the X-ray para-focussing geometry can occur. This would alter the reflection angles and lead to erroneous lattice parameter values. This problem can be avoided by first going to a higher temperature and then coming down to the desired temperature. In some equipment, we will be able to rotate the electrodes slightly and make the strip taut after reaching the required temperature. Temperature gradients across and through the sample can be minimized by employing a very thin sample layer. To avoid heat losses from the heater strip, some HTXRD chambers will be equipped with a surroundings heater around the sample. This is very effective and, for most of our measurements, we used such a system. When the surroundings heater is not provided, the heat losses can be significant. In such cases, a temperature calibration can be made by measuring the thermal expansion of standard substances like MgO with the equipment.

Figure 12 shows such a calibration curve made by us. It can be seen that above about 670 K, the actual temperature lags considerably behind the set temperature; at 1673 K, the difference is about 100 K. Figure 13 shows our thermal expansion study of ThO_2; it illustrates how temperature correction in this manner can mitigate the problem of measuring the exact sample temperature.

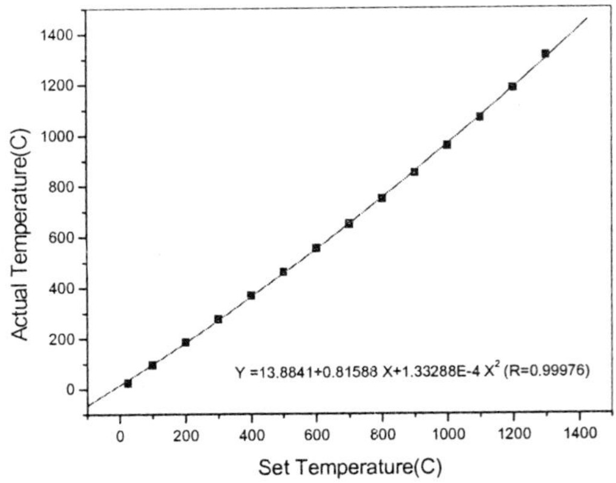

Fig. 12: Temperature calibration by measuring thermal expansion of MgO.

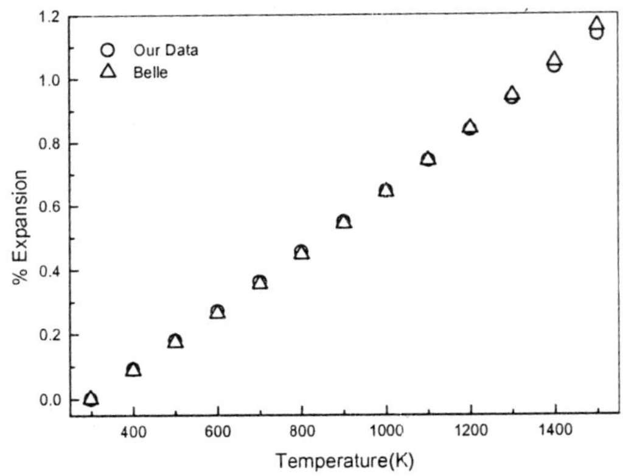

Fig. 13: Linear thermal expansion of ThO_2 (after temperature correction).

The errors in thermal expansion measurement mainly depend on the precision of the lattice parameter values and the magnitude of the thermal expansion coefficient (Ho, 1998). Since the change in the unit cell dimension is a difference between two nearly equal quantities, it can be known accurately only if the individual lattice dimensions are measured with high precision. Depending on the expansion coefficient of the material, the change in length also varies. It has been shown that the expected error in the measured thermal expansion coefficients of various ranges by assuming a typical precision of 0.0001 Å in lattice parameter measurement; for a material of initial lattice parameter 5 Å and having a thermal expansion coefficient of 10 x 10^{-6} K^{-1}, is 4 % if the temperature difference is 100 K (Ho, 1998). The error is a few % for high expansivities, but it is higher for low expansivity systems. In general, the method of high temperature XRD is sufficiently accurate for meaningful thermal expansion measurements on materials in research and industry.

Acknowledgement: We thank our colleagues Shri Hrudananda Jena, Shri R. Raja Madhavan and Dr S. Rajagopalan for their significant contributions to the work discussed here. We also thank Dr G. Periaswami, Head, Materials Chemistry Division, IGCAR, Kalpakkam, for constant encouragement.

REFERENCES

Belle, J. (1984). *In:* Thorium Dioxide: Properties and Nuclear Applications, Belle, J. and Berman, R.M. (eds). Govt. Printing Office, Washington, D.C., 167.

Buvaneswari, G., Govindan Kutty, K.V. and Varadaraju, U.V. (2004). *Mater. Res. Bull.*, **39:** 475.

Chung, D.D.L., DeHaven, P.W., Arnold, H. and Debashis Ghosh (1993). X-ray Diffraction at Elevated Temperatures : A method for in-situ process analysis. VCH Publishers, New York.

Govindan Kutty, K.V., Asuvathraman, R. and Sridharan, R. (1998). *J. Mater. Sci.*, **33:** 4007.

Govindan Kutty, K.V., Asuvathraman, R., Mathews, C.K. and Varadaraju, U.V. (1994). *Mater. Res. Bull.*, **29:** 1009.

Govindan Kutty, K.V., Asuvathraman, R., Raja Madhavan, R. and Hrudananda Jena (2003). XI IUPAC Conf. on High Temperature Materials Chemistry, The University of Tokyo, Japan.

Govindan Kutty, K.V., Rajagopalan, S. and Mathews, C.K. Unpublished results.

Govindan Kutty, K.V., Rajagopalan, S., Mathews, C.K. and Varadaraju, U.V. (1994). *Mater. Res. Bull.*, **29:** 759.

Hazen, R.M., Finger, L.W., Agrawal, D.K., McKinstry, H.A. and Perrotta, A.J. (1987). *J. Mater. Res.*, **2:** 329.

Hinatsu, Y., Masaki, N. and Fujino, T. (1988). *J. Solid State Chem.* **73:** 567.

Ho, C.Y. (ed.) (1998). Thermal Expansion of Solids, ASM International, Materials Park.

Hong, H. Y-P. (1976). *Mater. Res. Bull.*, **11:** 173.

Hrudananda Jena, Asuvathraman, R. and Govindan Kutty, K.V. (2000). *J. Nucl. Mater.*, **280:** 312.

Kirby, R.K. (1992). *In:* Compendium of Thermophysical Property Measurement Methods, Recommended Measurement Techniques and Practices. Maglic, K.D., Cezairliyan, A. and Peletsky, V.E. (eds). Plenum Press, New York, **2:** 549.

Krishnan, R.S., Srinivasan, R. and Devanarayanan, S. (1979). Thermal Expansion of Crystals, Pergamon Press, Oxford.

Lenain, G.E., McKinstry, H.A., Alamo, J. and Agrawal, D.K. (1987). *J. Mater. Sci.*, **22:** 17.

Mohammad Yousuf, Sahoo, P.Ch., Jajoo, H.K., Rajagopalan, S. and Govinda Rajan, K. (1986). *J. Phys. F : Met. Phys.*, **16:** 373.
Oota, T. and Yamai, I. (1986). *J. Am. Ceram. Soc.*, **69:** 1.
Pannetier, J. (1973). *J. Phys. Chem. Solids*, **34:** 583.
Rajagopalan, S., Kutty, K.V.G., Jajoo, H.K., Ananthakrishnan, S.K.and Asuvathraman, R. (1988). IGCAR Report No. 96.
Senbhagaraman, S. and Umarji, A.M. (1990). *J. Solid State Chem.*, **85:** 169.
Subramanian, M.A., Aravamudan, G. and Subba Rao, G.V. (1983). *Prog. Solid State Chem.*, **15:** 55.
Wang, S.X., Begg, B.D., Wang, L.M., Ewing, R.C., Weber, W.J. and Govindan Kutty, K.V. (1999). *J. Mater. Res.*, **14:** 4470.
Yamai, I. and Ota, T. (1993). *J. Am. Ceram. Soc.*. **76:** 487.

Investigation of High Alumina Cement Forming Phases by High Temperature XRD

Kaushlesh Kumar, J. Prasad, P. Nandi and B.D. Chattraj

Research and Development Centre for Iron and Steel
Steel Authority of India Limited, Ranchi–834002, India

INTRODUCTION

Refractory cements, presently, are obtained by fusing or sintering a mixture of aluminous and calcareous materials in suitable proportions and grinding the resultant product to a fine powder. Refractory cement comprises calcium aluminate (CA) and calcium dialuminate (CA_2) phases. Other phases formed from interaction of CaO and Al_2O_3 do not have any cementing properties. Therefore, for obtaining good cement it is imperative that these phases should be sufficiently present. Quantity of individual phases are dependenv on stoichiometry and firing temperature. Quality of cements thus depends on relative quantity of calcium based phases like CA, CA_2, $C_{12}A_7$, C_5A_3, C_2S etc.

When Dolomite is taken as source of CaO instead of limestone, the sintered product should produce both calcium aluminate cement and spinel. From CaO-MgO-Al_2O_3 system, it is found that an addition of alumina in dolomite results in a slow drop in refractoriness from 2400 °C to 1700 °C. A similar addition of alumina to lime yields melting points as low as 1500 °C (Chesters, 1973).

Braniski et al. (1973) has described a refractory cement composition consisting of a hydraulic component in the form of 75 to 40% by weight of calcium monoaluminate (CaO.Al_2O_3) or calcium dialuminate or a combination of both, and 20 to 55% by weight of magnesium oxide alumina spinel (MgO-Al_2O_3), the later being a nonhydraulic constituent, which does not destroy the binder characteristics of the composition and in fact constitute a super refractory. Preferably the cement consists of 75 to 50% by weight of calcium aluminate and 25 to 45% by weight of spinel (MgO.Al_2O_3).

It is reported by William Vance et al. (1994) that sintering of fine alumina in castable is enhanced when spinel is added along with calcium aluminate cement. The dense matrix so formed by better sintering resist considerably slag attack. Hot strength of such castable is also reported to be much high.

High alumina cement containing spinel may be useful for formulation of castable. This type of cement may, therefore, serve better in operating conditions of ladle and other refractory lining. With these purposes it was decided to develop a cement containing spinel phase. High temperature XRD of dolomite–alumina system were carried out for the formation of high alumina cement and spinel.

EXPERIMENTAL PROCEDURE

Raw dolomite and calcined alumina were taken for making batches. Initially, different batches were made containing 10 to 90 wt.% dolomite in dolo-alumina system. Chemical analysis of raw dolomite, calcined alumina and various formulations are given in Table 1. Samples were made in the form of small discs of diameter 29 mm and around 5 mm thick under a specific pressure of 1.5 t/cm^2. Powder mixes were granulated using polyvinyl chloride as binder for better flow and compaction.

After pressing samples were dried at 110 °C for 24 hours in a hot air oven and then tested for green density by taking weights and geometrical dimensions. Samples were sintered at 1400 °C for 180 minutes and then analysed for phases present using Bruker D8 Advance X-ray Diffraction (XRD) Unit operated at 40 mA and 40 V with CuKα radiation. Visual examination of sintered samples was also done before testing for phases.

Based on the phases observed from initial studies, further formulation was made to produce more quantity of cement phases consisting of 30-60 wt.% dolomite. Compositions of these batches are given in Table 2. Samples were prepared and sintered at 1350, 1400 and 1500 °C for different soaking times ranging from 10–240 minutes. Some samples were also sintered at 1600 and 1650 °C for identification of phases. Sintered samples were tested for bulk density and identification of phases.

The extent of formation of cement phases during heating were studied with composition of 40–50 wt.% dolomite samples and tested using high temperature XRD at different temperatures from 1000 to 1550 °C.

RESULTS AND DISCUSSION

Samples containing 10 to 90 wt.% of raw dolomite were sintered at 1400 °C for 180 minutes. On physical observation, it has been observed that the samples containing 40 to 60 wt.% dolomite are well sintered while samples containing 70 to 90 wt.% dolomite completely melted away. Samples containing 10 to 30 wt.% dolomite are not well sintered. These compositions require higher sintering temperatures.

Table 1: Initial formulation of batches

S. No.	*Identification*	*Dolomite (wt.%)*	*Calcined Alumina (wt.%)*
1.	DA9	10	90
2.	DA4	20	80
3.	D3A7	30	70
4.	D2A3	40	60
5.	DA	50	50
6.	D3A2	60	40
7.	D7A3	70	30
8.	D4A	80	20
9.	D9A	90	10

Chemical analysis

	Dolomite	*Calcined Alumina*
CaO	55.58 %	–
MgO	38.38 %	–
SiO_2	02.76 %	0.06%
Al_2O_3	02.56 %	99.2%
Fe_2O_3	00.52 %	0.03%
TiO_2	–	0.010%
Na_2O	–	0.03%

Table 2: Formulation of Cement Forming Batches

S. No.	*Identification*	*Dolomite (wt.%)*	*Calcined Alumina (wt.%)*
1.	I	30	70
2.	II	35	65
3.	III	40	60
4.	IV	45	55
5.	V	50	50
6.	VI	60	40

Various phases present in all the fired samples was identified by XRD using CuKα radiation. It is observed that the samples containing 10 to 30 wt.% dolomite do not contain any cementing phase. Compositions containing 70 to 90 wt.% dolomite, though contains some cementing phases but these samples also contain large amounts of free MgO. Hence samples containing 10-30 and 70-90 wt.% dolomite are not considered for further investigation. Compositions containing 30 to 60 wt.% have been selected for further investigation.

In the second phase, some more compositions were made containing 35 and 45 wt.% dolomite, besides compositions containing 30, 40, 50 and 60 wt.% dolomite.

X-ray Diffraction Studies

XRD patterns of different samples are taken. As inferred from the diffraction patterns, the amounts of various phases present in different samples, sintered at 1450 and 1500 °C for 2 hrs, are shown in Figs. 1 and 2. From Fig. 1, it is observed that the compositions IV and V contains sufficient amount of

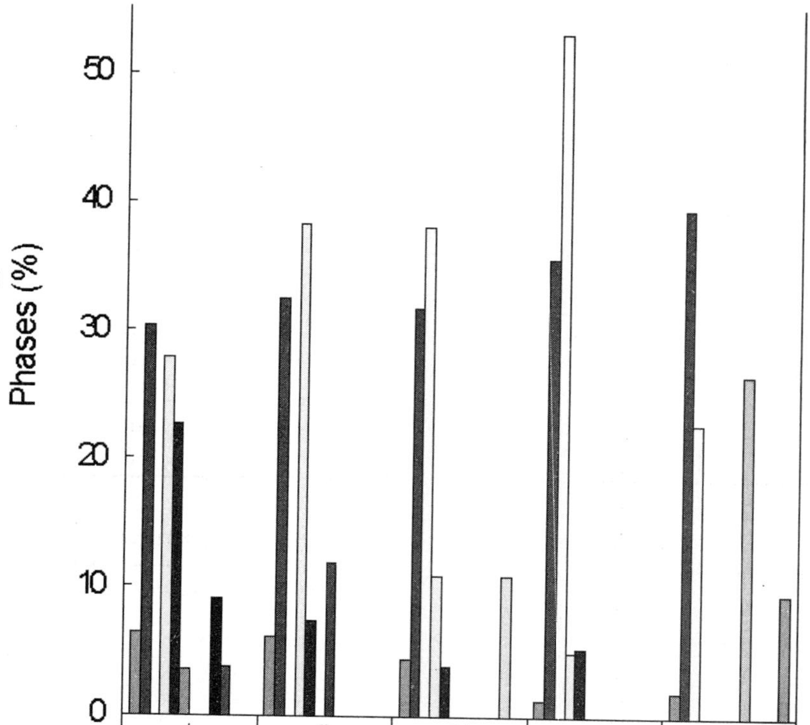

	I	II	IV	V	VI
Al_2O_3	6.4	6.1	4.4	1.3	1.9
MA	30.3	32.5	31.7	35.6	39.4
CA	0	0	38.1	53.1	22.7
CA2	27.9	38.3	10.9	4.9	0
CA6	22.7	7.4	3.8	5.2	0
C3A	3.5	0	0	0	0
C5A3	0	11.9	0	0	0
C12A7	0	0	0	0	26.5
CMA	9.1	0	0	0	0
C9A3	3.7	0	0	0	0
C3A5	0	0	10.9	0	0
CXA11	0	0	0	0	9.5

Composition code

Fig. 1: Amount of phases present in different samples, sintered at 1450 °C/2 h.

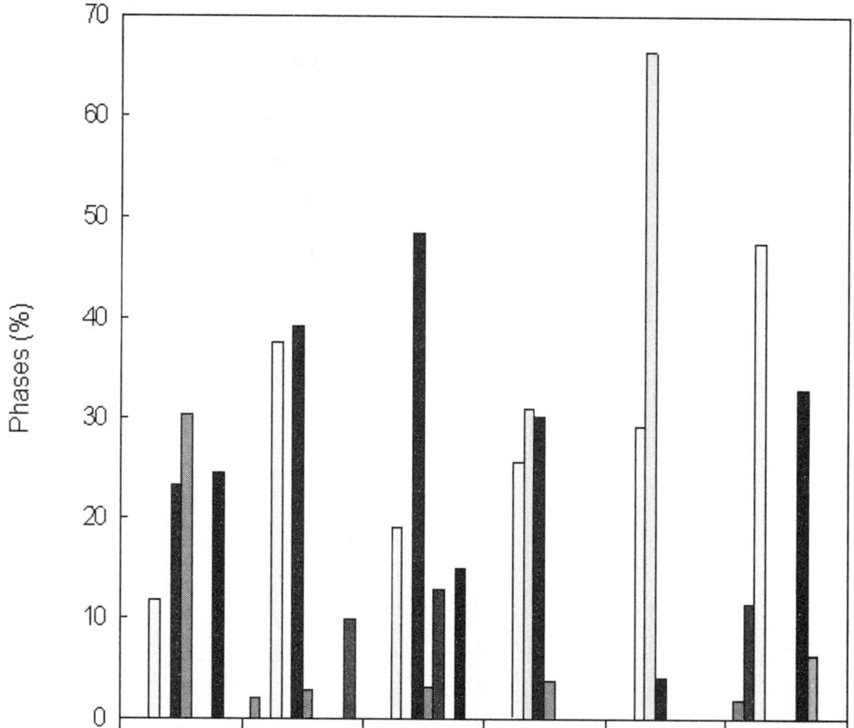

	I	II	III	IV	V	VI
Al2O3	0	2.08	0	0	0	1.9
MgO	0	0	0	0	0	11.4
MA	11.8	37.5	19.1	25.7	29.3	47.5
CA	0	0	0	30.9	66.5	0
CA2	23.4	39.06	48.3	30.2	4.2	0
CA6	30.3	2.86	3.2	3.8	0	0
C3A	0	0	12.9	0	0	32.9
C12A7	0	0	0	0	0	6.4
CMA	24.6	0	15	0	0	0
C3A2	0	9.9	0	0	0	0

Composition code

Fig. 2: Amount of phases present in different compositions, sintered at 1500 °C/2 h.

cementing phases as well as Mg-Al spinel. Sample No. VI, though it contains cementing phases and spinel, but it also contain free MgO. So it cannot be used as a refractory cement. In composition III, the amount of the main cementing phase, i.e. CA, is nil. Compositions I and II do not contain cementing constituents and hence unsuitable as cement.

High Temperature XRD

Samples with 40 and 45 wt.% of dolomite were studied with high temperature XRD at temperatures ranging from 1000 to 1550 °C and XRD pattern were taken at different temperature by using XRD high temperature unit. Phases were identified at interval of temperature which are shown in Figs. 3 to 5. It was observed that by increasing temperature above 1500 °C the cementing phase CA were decreasing and CA2 phases were increasing which are shown in Fig. 6. Some other non-cementing phase were also observed. So it is better to sinter the sample at 1500 °C.

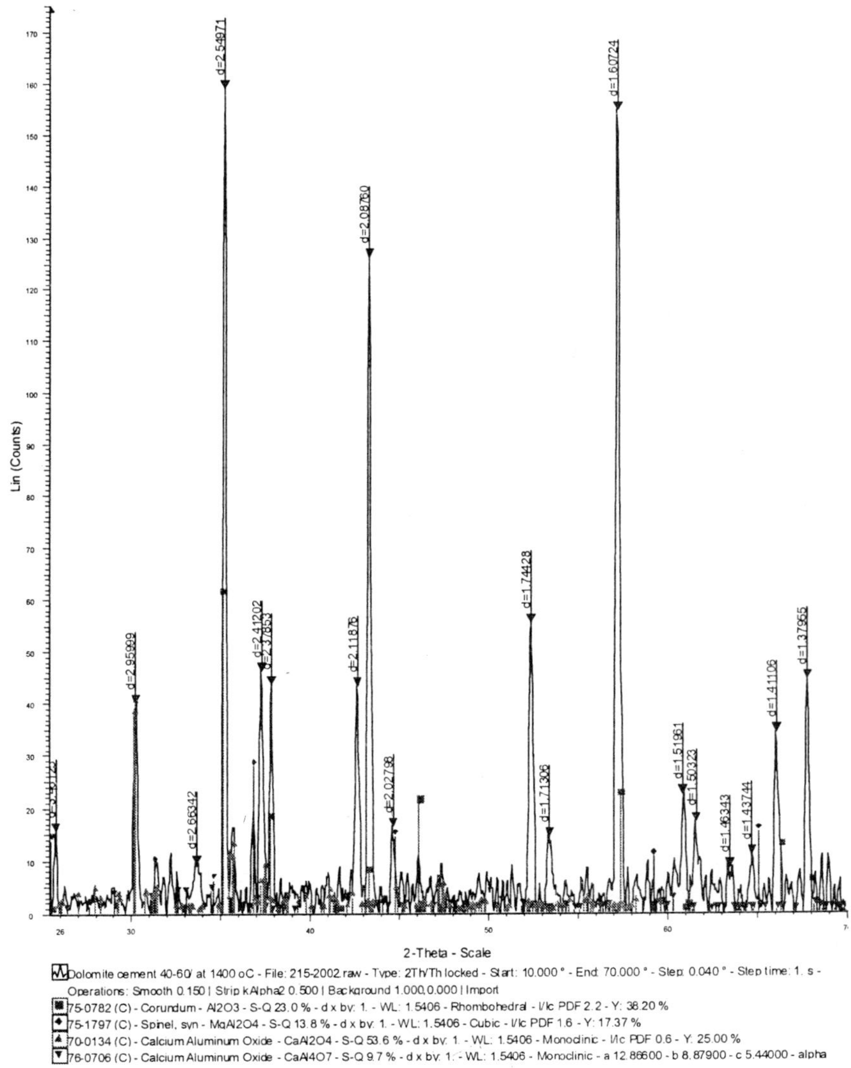

Fig. 3: XRD pattern of 40-60 composition at 1400 °C.

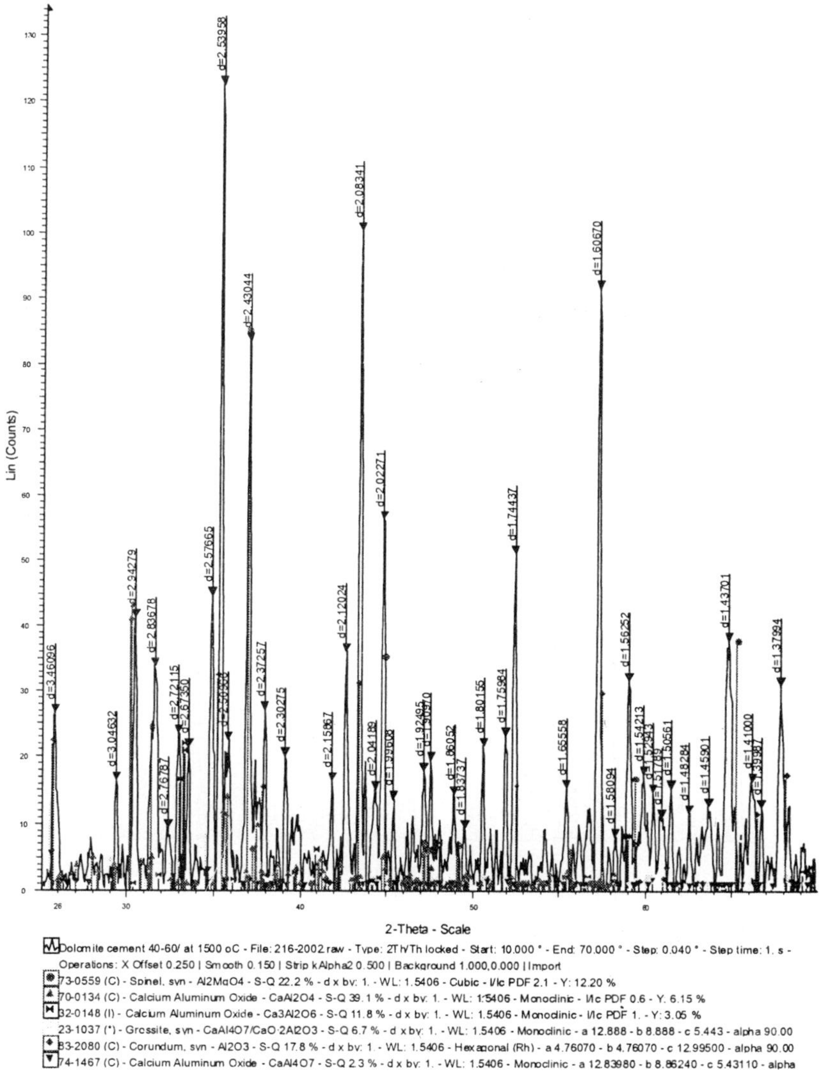

Fig. 4: XRD pattern of 40-60 composition at 1500 °C.

It was evident that when sintering temperature is increased to more than 1500 °C, CA phase formation reduced drastically with marginal increase of CA2 phases. Spinel phase, however, increased substantially with higher temperature. Therefore for obtaining good cement the temperature of firing should be restricted to 1500 °C.

CONCLUSIONS

(i) Dolomite and technical alumina when admixed in proper ratio produced spinel phases embedded in cement after firing/sintering at around 1500 °C.

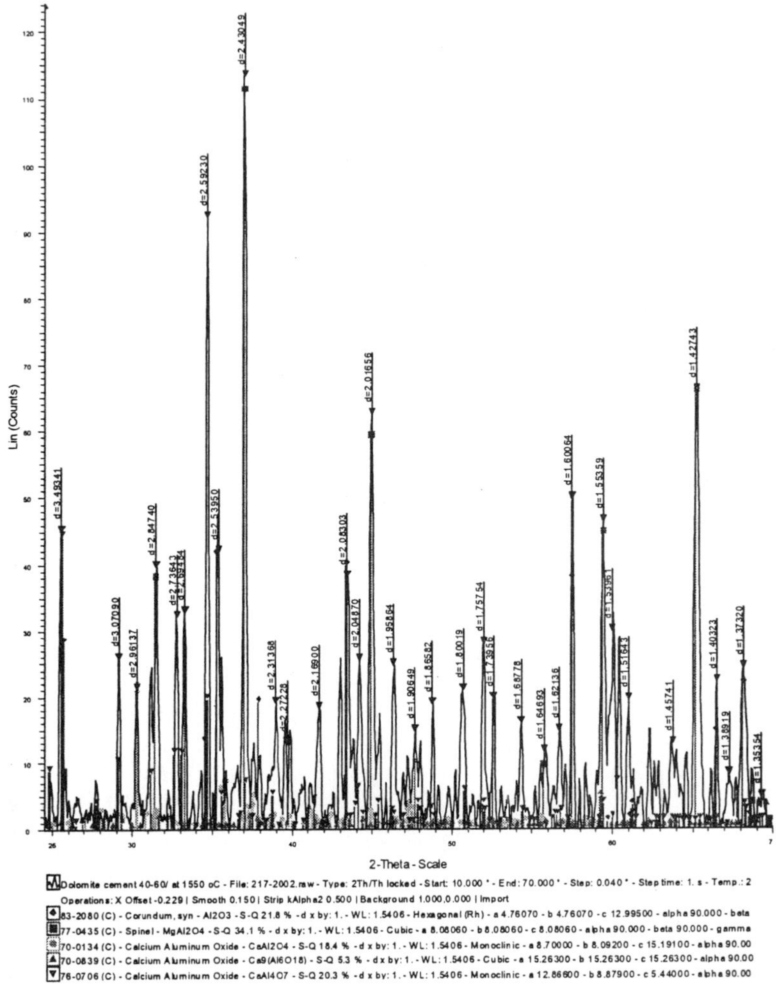

Fig. 5: XRD pattern of 40-60 composition at 1550 °C.

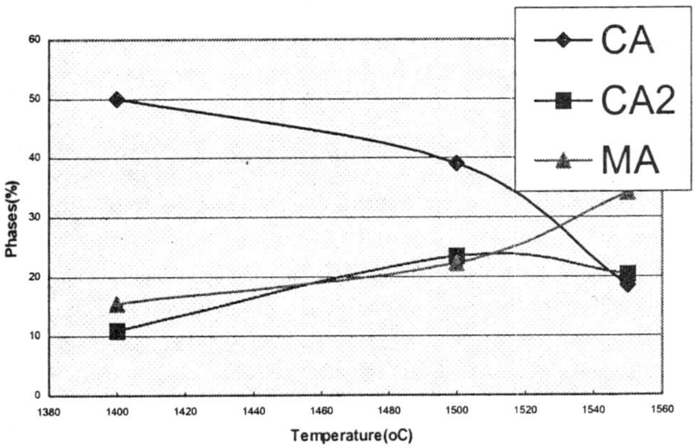

Fig. 6: Effect of temperature on phases

(ii) A mixture of 45-50 wt.% of dolomite and 50-55 wt.% of calcined alumina show better cementing properties.
(iii) Firing temperature for this type of cement should be within 1500 °C to generate sufficient quantity of cement phase.

RECOMMENDATIONS

(i) The work has established that refractory cement containing spinel could be made by sintering good quality dolomite with technical alumina at a temperature of around 1500 °C .
(ii) Spinel produced by this route is only 35% and cement phase is about 60%. Since requirement of spinel is much more compared to cement in spinel containing castable, additional amount of spinel may be required for its commercial utilization.
(iii) Advantages of using this cement need to be found out by further investigation in form of manufacturing castable incorporating this cement followed by using this castable for lining in ladle, tundish etc.
(iv) Effect of additives on properties of these cement required to be investigated.

Acknowledgements: The authors sincerely acknowledge with thanks the help rendered by S/Shri P.K. Giri, P.N. Bhatia and S.K. Dikshit in carrying out various tests and operations in this laboratory. They are also thankful to all other members of RLG for their help and suggestions from time to time during different stages of this project. The authors express their sincere gratitude and thanks to Sri S. Chakraborty, HOA (Steel) for his keen interest in taking up this project. They are also thankful to the management of RDCIS for allowing to pursue this project and bringing out this report.

REFERENCES

Braniski, A.I., Ionescu, T.D. and Deíca, N.D. (1973). U.S. Patent 3,748,158, Assigned to Institutul de Ceretari Metallurgice, Rumania.

Chesters, J. H. (1973). Refractories, Production and Properties. Iron and Steel Institute, London, 183.

William Vance, M., Gangolf, Kriechbaum, W., Ray, Henrichsen, A., George MacZura, Kenneth, Moody, J. and Stefan Munding (1994). Influence of spinel additives on High-alumina/Spinel castables. *American Ceramic Society Bulletin*, **73(11):** 70.

Applications of X-ray Diffraction in Aluminium Industry

S.K. Anantha Krishnan

Senior Manager (Analytical Research), R&D Centre
Indian Aluminium Co. Limited
Belgaum – 590 010, Karnataka, India

INTRODUCTION

The role of X-Ray Diffractometry (XRD) in the production control and quality assurance activities in any process industries has assumed significant importance, due to its better instrument performance and automation coupled with data analysis. The modern X-Ray Diffractometers are very user friendly, and do not warrant for high technical skill at operators level, hence have become an essential industrial tool for production control and quality assurance activities.

The application of XRD in Bayer and Smelting processes is known for more than two decades and indeed dedicated XRD is emerging with position sensitive detector and geometry to enhance speed of analysis. In Indian Aluminum Company Limited (Indal), Belgaum, the application of XRD started in the year 1986. Indal has upgraded its XRD, to meet demands of its internal and external customers.

This paper explains the applications, in sequence, by XRD from mining plans, process control, product certification up to dispatch of right product to the right customer according to the specifications. All these activities form production control and quality assurance using XRD. Also an approach for alpha alumina analysis in alumina and figures of merit are summarized. New concepts in computing bath ratio for smelter control and preferred orientation studies in annealed aluminium alloys are discussed.

QUALITY ASSURANCE ACTIVITIES

Quality assurance activities start sequentially from mining, process/production control, up to dispatch.

Mining Plan

Mining plan starts with mapping the area and estimate the abundance of the deposit with respect to alumina (Al_2O_3) content and aluminous minerals content. At a well defined interval, bauxite samples are drilled, collected area wise and depth wise; and number of such samples might be from a few hundred to thousands of samples depending on the mining area. A fast and accurate analysis of these samples is essential for mining purpose. A conventional mineralogical analysis would take anywhere between six months to one year. The current demand of analysis is within a few weeks. This can be easily met by XRD due to its speed and automation coupled with accuracy and precision. With the development of position sensitive detectors and Rietveld software programmes, one can think of on-line mineralogical analysis. Rietveld refinement technique has been used to simulate powder diffraction pattern providing all necessary structural parameters of the individual phase. The calculated and observed patterns and selected parameters (both profile and structural) are refined at each point on the XRD pattern by full-matrix least squares refinement to get the final structure. A typical mineralogical phase analysis of bauxite, based on this method, is presented in Table 1 (Taylor, 1991).

Table 1: Typical Mineralogical Analysis of Bauxite

Mineral	*Composition %*
Gibbsite	30.2
Boehmite	31.4
Hematite	27.2
Anatase	5.7
Rutile	1.4
Quartz	1.7
Maghemite	2.3

Process Control

One of the process controls in any Bayer plant for metal grade alumina is degree of calcination of alumina tri-hydrate ($Al_2O_3.3H_2O$). This is reflected in phase transformation of tri-hydrate to corundum (Al_2O_3), known as alpha. The alpha control during calcination is achieved by XRD measurement. Faster estimation of alpha content helps plant to have accurate control over temperature of calcination. XRD is calibrated using standards. Area under the peak at 3.48 Å in terms of counts is correlated with assigned concentration values for the purpose of calibration. Instrument conditions employed for measurement are presented in Table 2 and a typical calibration graph is given in Fig. 1.

Table 2: Instrument conditions used for recording XRD pattern

X-Ray Tube Target Voltage Current Spinner	Co K alpha 40 kV 30 mA ON				
Peak Positions and Measurement Conditions.	"d" Value Å	Start Angle (2ϑ)	End Angle (2ϑ)	Step (2ϑ)	Time at background (sec)
	3.48	29.0	31.0	0.08	2

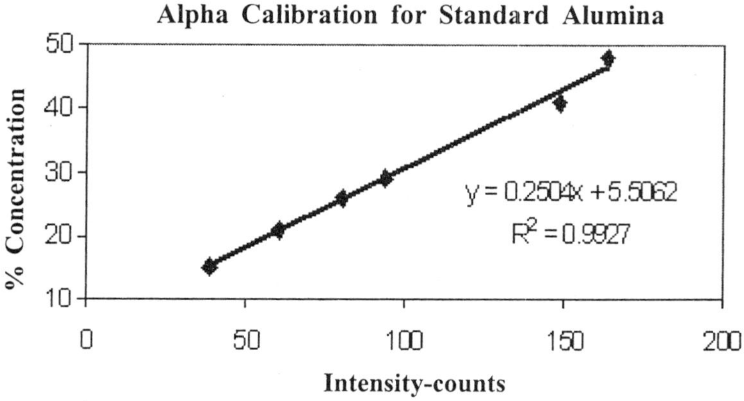

Fig. 1: Calibration graph.

Crystallinity of alpha phase is calculated using full width at half maxima (FWHM) measured for the peak at d = 3.48 Å. To correct for instrument width a high pure silicon was used. The % crystallinity was calculated using the following relation.

$$\% \text{ Crystallinity} = \{\text{FWHM of Si } (2\vartheta) \ / \ \text{FWHM of Sample } (2\vartheta)\}$$

For smelting operations, excess AlF_3 in the electrolytic bath and bath ratio NaF/AlF_3 are to be controlled. Modern concepts make use of multi-linear regression approach for computing excess AlF_3 and bath ratio. For this purpose both phase and elemental compositions are measured in bath sample. Typically, phase content of chiolite, fluorite and elemental Ca expressed as CaF_2 are computed. The multi-linear regression approach is as follows:

$$\text{Excess } AlF_3 = A1\ (CaF_2) + A2\ (I_{Chiolite}) + A3\ (I_{Fluorite})$$

Suitable electrolytic bath standards are selected and XRD peak area under chiolite and fluorite phases is measured. With assigned excess AlF_3 and CaF_2 values multi-linear regression analyses are performed. An example of multi-linear regression data is presented in Table 3.

Table 3: Multi-linear Regression Data

Dependent Variable		*Independent Variables*					
Ex.AlF_3	% CaF_2	$I_{chiolite}$	$I_{Fluorite}$	*Cal.Ex.* AlF_3	*Difference*		
12.2	6.1	2.6514	0.1461	12.3	0.1		
12.2	6.1	2.6271	0.1518	12.2	0.0		
12.2	6.1	2.6589	0.2002	12.1	–0.1	Summary output	
10.8	5.7	2.342	0.1555	10.8	0.0		
10.8	5.7	2.341	0.1665	10.7	–0.1	*Regression Statistics*	
10.8	5.7	2.3543	0.1364	10.9	0.1	R Square	0.9997
6.2	7.2	0.954	0.1389	6.2	0.0	Observations	18
6.2	7.2	0.9584	0.1449	6.2	0.0		
6.2	7.2	0.9727	0.1295	6.3	0.1		
3.1	8.6	0.4109	0.4920	3.1	0.0		
3.1	8.6	0.4046	0.4853	3.1	0.0		
3.1	8.6	0.4078	0.4999	3.0	–0.1	*Coefficients*	
1.2	9.0	0.2424	0.7570	1.3	0.1	Intercept	-0.8681
1.2	9.0	0.2219	0.7604	1.2	0.0	X Variable 1	0.5521
1.2	9.0	0.2383	0.7651	1.2	0.0	X Variable 2	3.9688
4.9	6.4	0.773	0.1719	4.9	0.0	X Variable 3	-4.9831
4.9	6.4	0.7815	0.1782	4.9	0.0		
4.9	6.4	0.7811	0.1785	4.9	0.0		
				Mean	0.0		
				1 Sigma	0.1		

Bath ratio is calculated from the following relation.

$$\text{Bath Ratio} = \frac{(100\text{–Free Alumina–\% } CaF_2) - Ex.AlF_3}{(100\text{–Free Alumina–\% } CaF_2) + Ex.AlF_3}$$

where free alumina is un-dissolved alumina.

Metallurgical Applications

Metallurgical applications using powder X-ray diffractometer include identification of inter-metallic compounds, solid solution formations, preferred orientations and strain measurements. Also it is well known that orientation and texture change with respect to temperature variations during heat treatment. Preferred orientation was observed in rolled aluminium sheets during annealing. Aluminium sheet specimens annealed at 220, 320 and 400 °C have shown orientation effect for (200) reflection predominantly.

This has been shown in Fig. 2, XRD diffraction pattern. An attempt is under progress to make use of preferred orientation effect to predict recrystallization behaviour of the alloys.

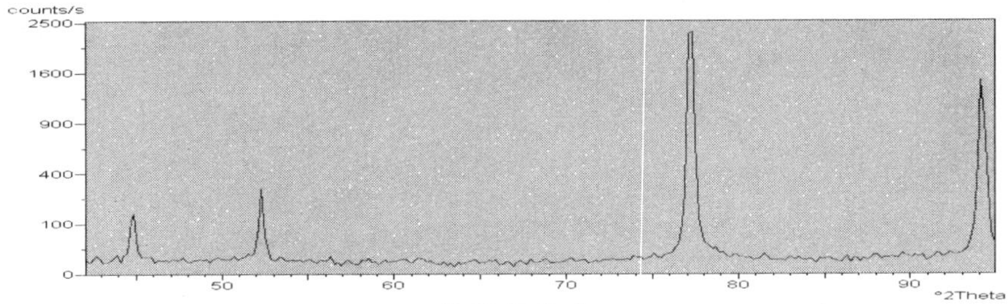

Cold Rolled

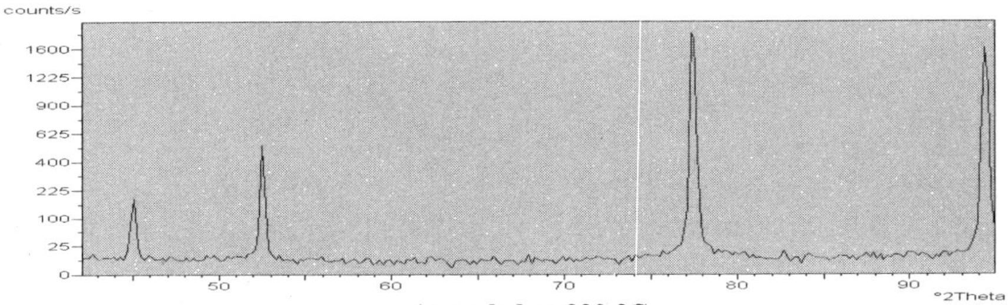

Annealed at 220 °C

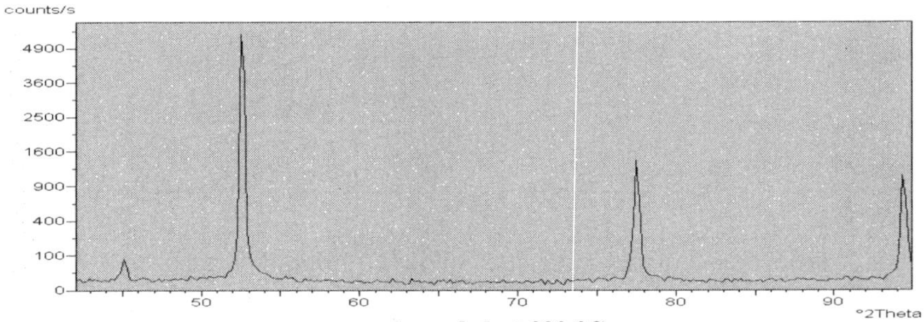

Annealed at 320 °C

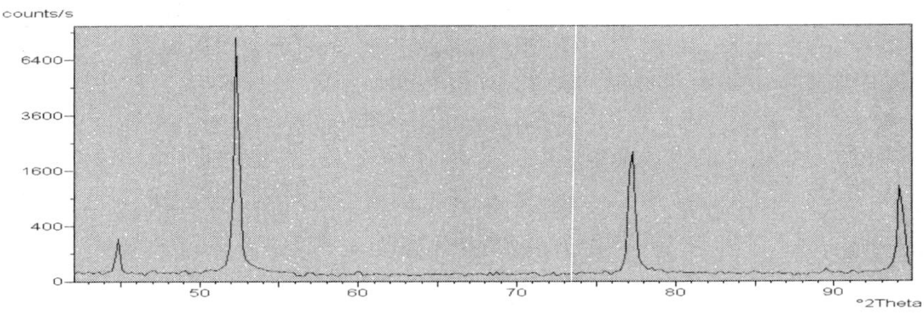

Annealed at 400 °C 1

Fig. 2: XRD Response to Textural Changes of Annealed Specimens (Al-Alloy).

Dedicated XRD Equipments

Recently, dedicated combined XRD and XRF equipments for aluminium industries are becoming popular. The author had an opportunity to calibrate one such equipment namely TAXA (Total Aluminium X-ray Analyzer) sold by Thermo ARL 9800 XP. This instrument performs all the needed analysis for smelter operations. Concept of monochromatic X-rays filtered from polychromatic rays and coherently scattered X-rays from the specimen (XRD Reflection) diffracted by single crystal have made the X-ray geometry very compact and promotes faster analysis

However, selection of desired wavelength from polychromatic radiation is not possible, which restricts the measurement to very narrow angles and desired peaks of XRD cannot be seen. Also peak profile analysis, which is dependent on sample characteristics, cannot be carried out since each reflected peak is the rocking curve of the diffracting crystal. Since these instruments are dedicated to perform only identified tasks, one should be judicious in selection and needs must be thoroughly discussed with suppliers.

CONCLUSIONS

XRD applications to aluminium industry are expanding with new instrument and software development. Modern mathematical model coupled with software reduce time of analysis and provides automation thereby increasing productivity. There exists a potential to use preferred orientation effects to study recrystallization temperature of aluminium alloys.

REFERENCE

Taylor, J.C. (1991). Computer Programs for Standardless Quantitative Analysis of Minerals Using the Full Powder Diffraction Profile. *Powder Diffraction* **6**, No. 1.

Application of X-ray Technique to Surface Modification: Case Studies

Manish Roy

Defence Metallurgical Research Laboratory
Kanchanbagh, Hyderabad-500 058, India

INTRODUCTION

Demanding and stringent engineering requirement in recent days resulted in variety of surface modification processes whereby the surface properties of a component is altered significantly without making any change of the bulk properties (Lipsitt, 1985; Pons et al., 1986; Islam and Hindam, 1993). In order to explore the suitability of modified surface for a particular application, it is necessary to evaluate the microstructural features and mechanical properties of the concern surface. One of the most powerful tool for characterization of both the microstructural features and mechanical properties of a material or a surface is X-ray diffraction (XRD) techniques. The objective of the present work is to describe some of the case studies related to microstructural characterization of modified surface layer using XRD techniques. The first section of the paper is concerned with the type of phase transformation takes place during thermal spraying of Ni_3Al, WC-Co and Cr_3C_2-25(Ni20Cr) coating. The presence and distribution of various phases on laser surface nitrided layer is discussed in second section. In the third section, grain size measurement of thermal sprayed, nanocrystalline Cr_3C_2-25(Ni20Cr) coating is presented.

EXPERIMENTAL DETAILS

Air plasma spraying was carried out using a METCO type 7MB plasma spraying system, which consisted of a water-cooled copper anode and a thoriated tungsten cathode. The coating was deposited on sand blasted samples. The gun was operated at around 30–40 KW. For this power range, temperature in excess of 10,000K was available near the nozzle tip of the plasma gun. The gas velocity near the nozzle tip was around 1000–3000 m/s.

Detonation sprayed coating was obtained with the help of DENIPER-3 D-Gun. The gun was designed and manufactured by Institute of Material Science Problem, Ukrain. Sand blasted samples were coated with this technique. Detonation was obtained using a combustible gas mixture of O_2 and C_2H_2. A gas temperature of around 3500K was obtained by detonation of the oxyfuel mixture. The coating powder attained a velocity of approximately 800-1000 m/s as reported by the supplier.

Nanocrystalline coating was deposited employing TAFA HVOF system JP-5000 gun. The system was equipped with a gun, a control console, a powder feeder and an X-Y manipulator. The gun has a nozzle and a combustion chamber at right angle to the nozzle. Oxygen and liquid fuel kerosene were combusted under high pressure in the chamber and the combustion product was allowed to exit into the atmosphere. A pilot flame was used to ignite the main flame. The powder was fed radially into the hot stream of gases. Argon was used as carrier gas to transport the powder from the feeder to the gun. The powders were heated in the gas stream and got deposited on the substrate at very high speed.

Commercially pure titanium was laser gas nitrided using a 5KW continuous wave CO_2 laser (Spetra Physics, 975) coupled with a 5 axes CNC work station and a transmittive optics of Zn-Se with Fransel No. 5,8. The alloying was performed by melting the substrate in the nitrogen atmosphere. The nitrogen gas was supplied through two nozzles. One nozzle was coaxial with the laser beam and the other was parallel to the substrate to be alloyed.

RESULTS AND DISCUSSION

Transformation in Thermal Spraying

The Ni_3Al is an important layer used as a bond coat for oxide coatings such as Al_2O_3 and Cr_2O_3. The XRD patterns of powder, plasma sprayed coating and detonation sprayed coating are shown in Fig.1. It is clear from Fig.1 that the diffraction pattern of the powder contains super lattice reflections i.e. reflections from the planes having mixed hkl indices indicating that the powder have an ordered Ni_3Al structure. In contrast, the super lattice reflections are not present in diffraction pattern obtained from plasma sprayed or detonation sprayed coatings. This feature is indicative of the fact that the spraying has transformed ordered structure to disordered. These oxide (Al_2O_3, Cr_2O_3) coatings are potential layers for elevated temperature tribological application (Hirose et al., 1993). One reason for using Ni_3Al layer as bond coat is its ordered structure. The ordered structure is able to retain the strength at elevated temperature (Huntz and Puig, 1989). Consequently, coating with disordered structure will exhibit poor performance at elevated temperature application. Hence it is imperative to carry out the optimization of spraying process, which will result in formation of coating with ordered structure.

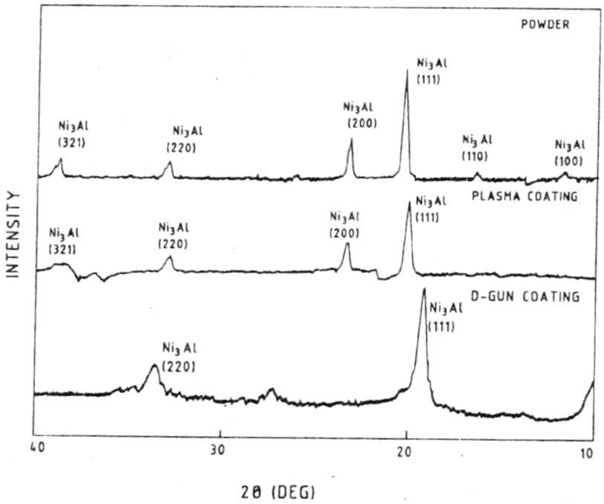

Fig. 1: The XRD patterns obtained from Ni_3Al powder, plasma sprayed coating and detonation sprayed coating.

The NiCr-25(Cr_3C_2) is a very promising cermet variety coating for tribological application (Li, 1980). This coating is expected to find application in metal forming tools, brake discs of high speed train, fuel rod mandrel of nuclear power plant, exhaust fan blades of gas cleaning plant of heavy industries etc. This coating is essentially replacement of most potent tribological layer, WC-Co coating at elevated temperature where WC-Co coating starts decarburising. The diffraction pattern obtained from the powder, plasma sprayed coating and detonation sprayed coating are presented in Fig.2. As can be seen from Fig.2, the powder is completely crystalline. However, on spraying by both the method, it loses crystalline structure. The coated layer is predominantly amorphous. The amorphous layer is sometimes beneficial for tribological application due to reduction of friction coefficient (Nertz et al., 1991). Since the amorphous layer is unstable at high temperature, its application at elevated temperatüre is limited. It will be demonstrated subsequently that with proper selection of spraying powder and spraying process, it is possible to avoid the amorphous layer and get completely crystalline coating.

The WC-Co coating has demonstrated extremely well resistance to several types of tribological degradation (Taylor et al., 1983; Manish Roy et al., 1999). The XRD pattern from such coatings obtained by plasma spraying and detonation spraying is illustrated in Fig.3 along with diffraction pattern of the powder. There is clear indication of partial decarborisation of the powder on spraying. Such decarborisation is harmful for tribological application (He et al., 2000). Hence spraying condition needs to be optimised to ensure minimum decarborisation.

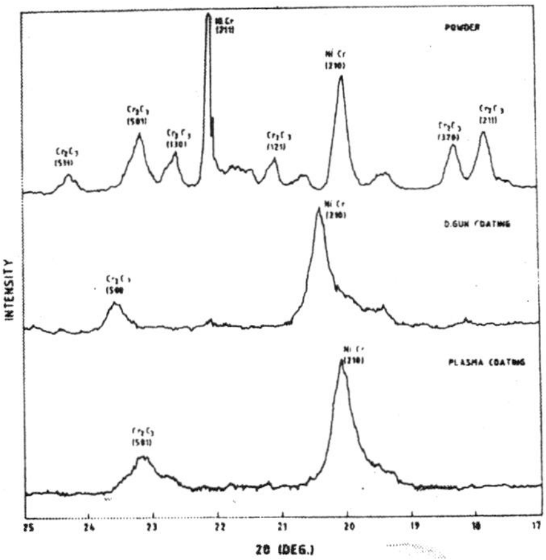

Fig. 2: The XRD patterns obtained from Cr_3C_2-25(Ni20Cr) powder, plasma sprayed coating and detonation sprayed coating.

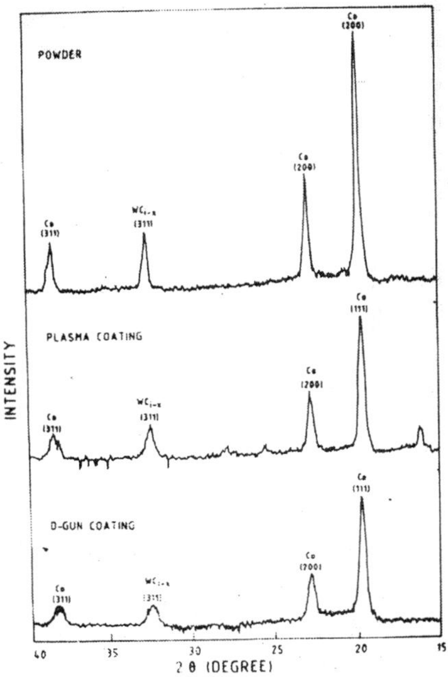

Fig. 3: The XRD patterns obtained from WC-Co powder, plasma sprayed coating and detonation sprayed coating.

Phase Transformation and Distribution during Laser Nitriding of Ti

Commercial pure titanium was nitrided using CO_2 laser. The XRD study from the alloyed surface reveals the presence of two phases viz. TiN and α-Ti. The distribution of phases along the cross section of the layer has been studied by recording the XRD patterns after removing the material from the alloyed surface in the steps of 150 μm. Results of this study on a substrate that was alloyed at 2.1kW laser power and at a scan speed of 0.0043m/s are given in Fig.4. It can be seen that as one moves away from the surface, peaks of TiN diminish and those of α-titanium become more prominent. Thus, it is clear that the surface contains primarily hard bcc TiN. The volume fraction of TiN phase reduces moving away from the surface. At a depth of 0.45 mm below the surface, the intensity of TiN phase starts decreasing and hexagonal Ti(α) phase starts appearing. Beyond the depth of 0.45 mm below the surface, there is hardly any TiN phase and Ti(α) is the dominant phase.

Using direct comparison method, an attempt has been made to estimate the volume fraction of TiN and Ti(α) phases across the nitrided layer quantitatively. According to this method, the ratio of the intensities from a particular plane of TiN and Ti(α) phases is related to their concentration as

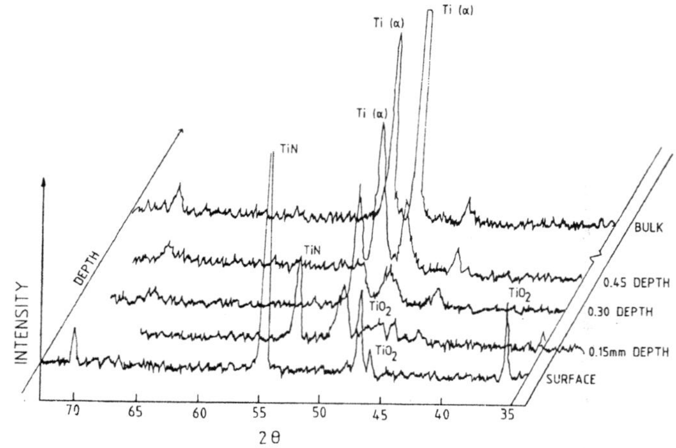

Fig. 4: The XRD pattern obtained from and across the nitrided layer.

$$\frac{I_{\mathrm{TIN}}}{I_{\mathrm{TI}(\alpha)}} = \frac{R_{\mathrm{TiN}} C_{\mathrm{TiN}}}{R_{\mathrm{Ti}(\alpha)} C_{\mathrm{Ti}(\alpha)}} \tag{1}$$

$$C_{\mathrm{TiN}} + C_{\mathrm{Ti}(\alpha)} = 1 \tag{2}$$

where $$R = \left(\frac{1}{v^2}\right)\left[|F^2|\, p\left(\frac{1+\cos^2 2\theta}{\sin 2\theta\ \cos\theta}\right)\right]\left(e^{-2M}\right) \tag{3}$$

where v is volume of unit cell (m^3), F is structure factor, p is multiplicity factor, θ is Bragg angle and e^{-2M} is temperature factor. The appropriate values of these factors and further details of phase analysis is given elsewhere (Manish Roy). On the basis of this analysis, the variation of volume fraction of these two phases across the nitrided layers are given in Fig.5. The decrese in volume fraction of TiN phase as one goes beneath the surface is rather steep. In contrast, the increase in volume fraction of Ti(α) phase across the nitrided layer is quite smooth. Such behaviour is reflected in Fig.6 wherein the variation of microhardness with the depth beneath the surface is shown (Manish Roy and Bharti, 2000). In this figure a steep decrease in microhardness is evident. The microhardness is maximum near the surface where predominantly TiN phase is present (Fig.6). At the end of the nitrided layer, hardness equals the hardness of the substrate as Ti(α) is the predominant phase.

Grain Size Measurement of Nanocrystalline Cr_3C_2-25(Ni20Cr) Coating

Nanocrystalline Cr_3C_2-25(Ni20Cr) coating was deposited on 253 MA alloy using JP 5000 spraying unit (Manish Roy et al., 2003). A representative TEM micrograph of the coating is given in Fig.7. From the micrograph, it

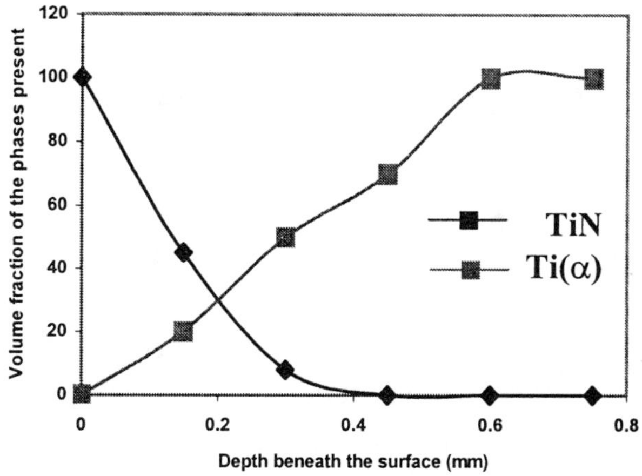

Fig. 5: The distribution of TiN and Ti(α) phases in the nitrided layer.

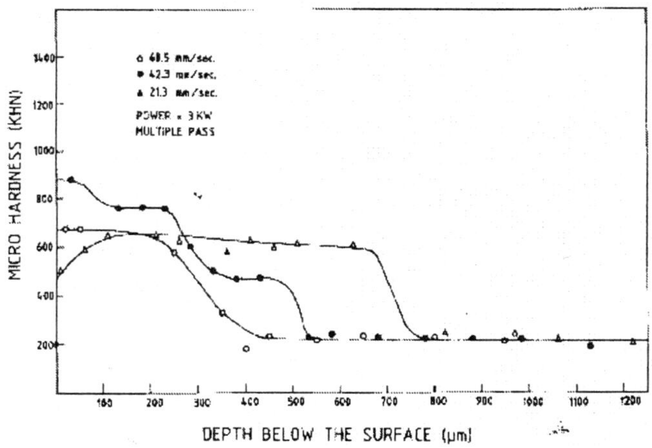

Fig. 6: The microhardness profile across the nitrided layer.

is clear that the grain size is in the range of 80 to 100nm. The electron diffraction pattern also confirms crystallinity and the nanometer size range of the coating. XRD patterns were obtained from the powders and the coated surfaces. The grain sizes of the coated samples or the powder were measured using the same diffractometer with Scherrer grain size measurement technique. XRD of the standard silica specimen was used to measure instrumental broadening and Ziegler's (Ekstron et al., 1985) method was used to remove instrumental broadening to get true crystal broadening of the specimen. Using the Scherrer relationship for size effect and Cauchy's correction for stress effect, the following relationship is obtained:

$$\beta \cos \theta = \frac{K\lambda}{D} + \frac{4\Delta d}{d \sin \theta} \tag{4}$$

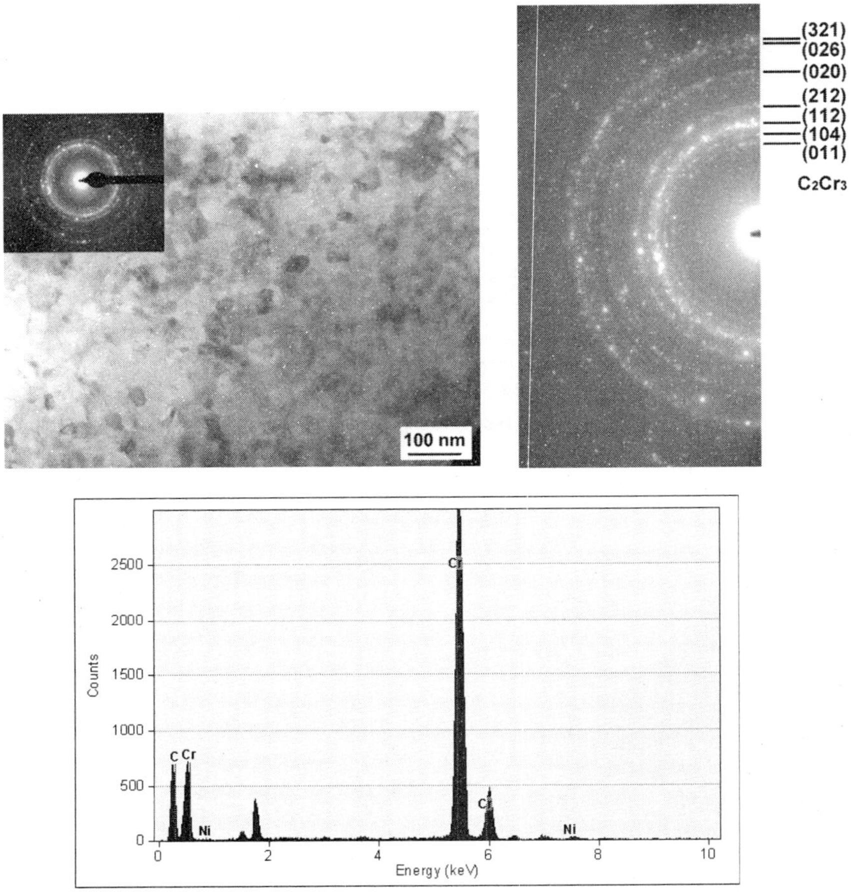

Fig. 7: The TEM images of the Cr_3C_2-25(Ni20Cr) coating along with electron diffraction pattern and EDAX spectrum.

where β is the corrected true full width at half maximum (FWHM) of the diffraction peak. θ is the diffraction angle, *K* is the shape constant (close to unity), λ is the wave length of the radiation, *D* is the size of the crystallite, Δd is the lattice strain and *d* is the lattice dimension. The parameter β cos θ was plotted against sin θ and a straight line was obtained. The intercept as per equation (1) gives the crystal size and the slope reveals the stress.

The XRD pattern of the powder and coating with nanocrystalline grains is shown in Fig.8. Most important point to be noticed from the XRD pattern is that, there is no change of the phases present in the powder. This is quite in contrast to what is commonly observed for thermal sprayed WC-Co coatings described previously. Further, the amorphous phase was not formed on spraying. Thus, with selection of proper powder and the spraying condition amorphisation can be avoided. One possible reason for these phenomena is that the residence time of the particles in the flame is much less for the

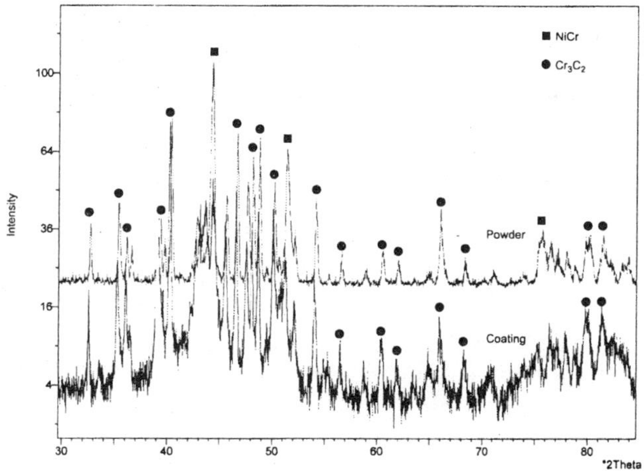

Fig. 8: The XRD pattern obtained from Cr_3C_2-25(Ni20Cr) nanocrystalline powder, HVOF sprayed nanocrystalline coating.

spraying system with liquid fuel. X-ray study also reveals that the grain sizes of both the phases in the powder are in the range of 100nm. Main peak of both the phases are analysed to estimate the grain size. For NiCr phase NiCr (111) peak and for Cr_3C_2 phase Cr_3C_2 (203) peaks have been considered for the calculation. However, it should be stated that this estimation is purely qualitative and consistent with TEM micrograph.

CONCLUSIONS

1. XRD technique can be used meaningfully and to beneficial purpose to understand the phase transformation during thermal spraying.
2. XRD technique is a useful method to study the distribution of phases and their effect on the micro hardness of the nitrided layer.
3. This technique estimates the grain size of nanocrystalline Cr_3C_2-25(Ni20Cr) coating quite accurately.

Acknowledgement: The author is grateful to Defence Research Development Organisation for financial support and Dr. A.K. Singh for technical discussion.

REFERENCES

Ekstron, T., Chatfield, C., Wruss, W. and Schreiber, M.M. (1985). *J. Mater. Sci.*, **20:** 1266.

He, J., Ice, M., Dallek, S. and Lavernia, E.J. (2000). *Metal. Mater. Trans.*, **31A:** 541.

Hirose, A., Ueda, T. and Kobayashi, K.F. (1993). *Mater. Sc. and Engg.*, **A 60:** 143.

Huntz, A.M., Puig, T., Confignal, L., Carpentier, F. and Condat, M. (1989). *Mater. Sci. Engg.*, **A121:** 555.

Islam, M.U. and Hindam, H. (1993). *Materials and Manufacturing Processes*, **5(1):** 93.

Li, C.C. (1980). *Thin Solid Films*, **73:** 59.

Lipsitt, H.A. (1985). High Temperature Ordered Intermetallic Alloy Koch, C.C., Liu, C.T. and Staloff, N.S. (eds). MRS, Pittsburg, PA, 351.

Manish Roy and A. Bharti (2000). Proc. of 3rd Workshop on Application of Laser in Mechanical Industry, Jadavpur, Calcutta, India.

Manish Roy, DMRL Technical Report, under preparation.

Manish Roy, Nerkhadeand, B.E. and Paul, S.N. (1999). Proc. of the 2nd Int. Conf. on Industrial Tribology, Ed. Haraprasad. Hyderabad, India, 205.

Manish Roy, Pauschitz, A. and Franek, F. (2003). Tribological Research and Design for Engineering System, Tribology Series, 41, Dowson, D. (ed). Elsevier Science, BV Amsterdam, ISBN 0444512438, 141.

Nertz, J.E., Kushnor Jr., B.A. and Rotolico, A.J. (1991). Proc. of the 4th National Thermal Spraying Conf. Pittsburgh, PA, USA, 115.

Pons, M., Galerie, A. and Cailet (1986). *Journal of Material Science*, **21:** 2697.

Taylor, A., Overs, M.P., Quets, J.M. and Tucker Jr., R.C. (1983). *Thin Solid Films*, **107:** 427.

Industrial Application of X-ray Diffraction in Determining Residual Stress

B. Ravi Kumar

Materials Science and Technology Division
National Metallurgical Laboratory
Jamshedpur – 831007, India

INTRODUCTION

In industries fast response is necessary to counter problems such as characterisation of new or improved materials, monitoring process or understanding the cause of lack of performance of products. Most methods can not deliver the results of analysis rapidly to solve urgent problems. Successful characterisation requires full understanding and analysis of the problems and conveying the appropriate solution to those who posed the initial problem or question. The form and size in which samples occur in the industry can be a challenge. In addition, specimen preparation, such as cutting, dicing or grinding may influence certain chemical or physical transformation. Therefore, the specimen preparation techniques must also be tailored to avoid with other analysis. Further the knowledge of sample history is essential to understand the reasons behind the lack of performance of the product. A large number of industrial problems directly or indirectly are associated with residual stresses. Determination of residual stress by X-ray diffraction (XRD) is not a new technique. However, engineers of design and manufacturing communities are still unaware of it. This technique has many advantages. In particular, the ability to measure stresses in small areas in specific direction is of great advantage. During the development of new manufacturing process residual stress analysis can often be very useful. The nature and magnitude of residual stresses present in the finished component to a significant extent can be tailored by controlling manufacturing process parameters. For example additional treatments, such as shot peening or heat treatment commonly used in the automobile industry significantly alters the surface residual stress pattern of the component. Residual stress measurement can verify the effectiveness of the process by measuring the stress gradient in the finished product in producing the expected results. In the present paper, the use of XRD technique is discussed with some suitable examples related to problems associated with residual stresses in automobile industries.

CASE STUDY I

Evaluation of Residual Stress in Gray Cast Iron Cylinder Block

Engine cylinder blocks of gray cast iron, shown in Fig. 1, have been found to develop cracks during manufacturing process at various stages of production. Cracks were mostly found to occur around Rib 2 to Rib 4 region (Fig. 1). These defects were considered due to residual casting stresses generated in the castings during manufacturing process. To study the residual stress pattern in the castings, it was planned to investigate casting at various stages of manufacturing and compare with an imported casting of similar type. Following four gray cast iron engine cylinder blocks have been selected for the stress measurement:

1. Imported casting in stress relieved condition
2. Indigenous casting in stress relieved condition
3. Indigenous casting in stress relieved and machined condition
4. Indigenous casting in stress relieved condition with a crack

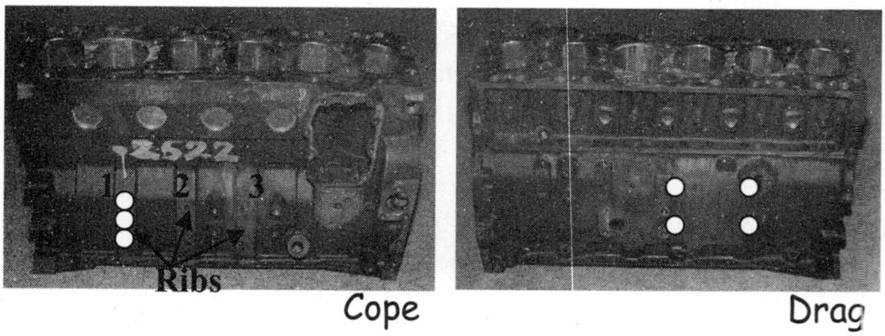

Fig. 1: Dots indicate the measurement locations on ribs 1, 2, 3 from top on (a) cope side and (b) drag side.

Experimental Studies

As the cracks were mostly found to occur around Rib 2 to Rib 4 region, it was decided to assess the stress profile at Rib2, Rib3 and Rib4 on the cope side and at four sites on the drag side by XRD technique (Society of Automotive Engineerings, Inc. USA, 1971; Noyan and Cohen, 1987). A portable X-ray Stress Analyser AST X2001 has been used for the measurements. Due to geometrical hindrances this technique was restricted to outside surface only. All the stress measurement locations have been cleaned thoroughly to degrease before the measurements. CrKα radiation and $\sin^2\psi$ method has been chosen for the XRD studies. Additionally, microstructural as well as hardness studies have been also carried out to correlate it with the stress values.

Observations

Results of XRD are represented in Table 1 and graphically in Fig. 2. Microstructures of the three castings are shown in the Fig. 3 along with hardness values. XRD studies on the outer surface of the castings in general indicated compressive stress in all the castings. On comparing the results from the *Imported casting* and *its equivalent indigenous casting* (stress relieved casting), it was noted that latter showed a better uniformity in stress values within the Ribs 2, 3 and 4. On the other hand, imported casting showed uniformity in stress values within the ribs only. Both the castings, in general, showed uniform stress distribution on cope and drag side but one-to-one correspondence did not exist.

Table 1: Residual stress values [MPa] measured on cylinder engine block

Measurement site	*Imported casting*	*Stress Relieved casting*	*Stress Relieved and Machined casting*	*Stress Relieved casting with Crack*
Rib2				
0	–	–	–	
1	–73	–86	–85	–62
2	–63	–87	–119	–61
3	–23	–84	–100	–35
Rib3				
0	–	–	–	–
1	–108	–82	–79	–33
2	–93	–90	–127	–84
3	–81	–85	–105	–80
Rib4				
0	–	–	–	–
1	–141	–74	–81	–43
2	–141	–64	–88	–78
3	–146	–100	–95	–84
Drag				
1	–77	–83	–116	–68
2	–60	–75	–73	–43
3	–96	–72	–108	–57
4	–60	–89	–126	–67
Front	–54	–45	–26	–141
Rear	–62	–73	–26	–123

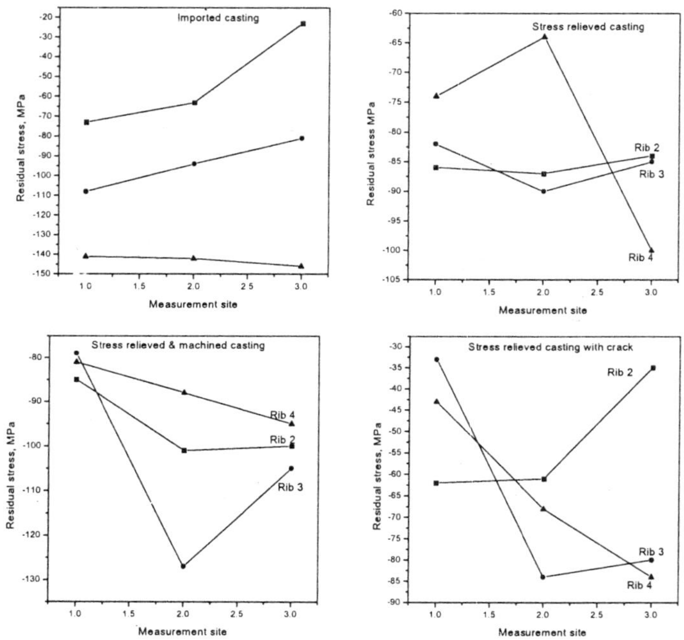

Fig. 2: Residual stress profile of the cylinder block at various stages of manufacturing processes along the ribs.

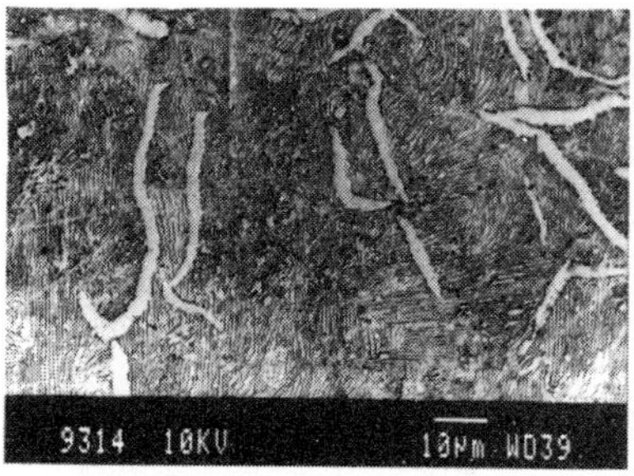

(a) Imported Casting 224 HB

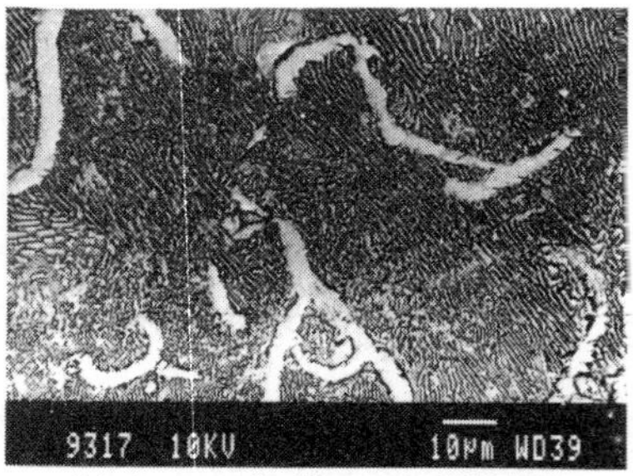

(b) As-cast Indigenous 170 HB

(c) Indigenous Casting 156 HB

Fig. 3: Microstructure and hardness of three different cylinder blocks showing variation in graphite morphology.

Casting in stress relieved and machined condition did not exhibit any one-to-one correspondence with either of the above two castings. Moreover, the scatter in data in this case was higher on both cope and drag side. Front and rear ends showed almost similar stress values indicating uniform effect of machining. *Stress relieved casting with cracks* showed better match with indigenous stress relieved casting compared to machined one. However, front and rear ends showed highest value compared to all other castings. Residual stress pattern did not show any alarming variation, which could lead to cracking of the castings. However, microstructure and hardness studies showed difference between imported and indigenous castings. From the microstructures, the difference in graphite morphology can be very clearly noted.

Remarks

- Residual stress measurement did not give any conclusive pattern, which could indicate or relate it to the development of crack in the castings during its manufacture. Moreover, the stresses are compressive in nature, which is considered beneficial under most circumstances.
- The variations in the microstructure at the critical locations from one casting to another should be looked into in more details.

CASE STUDY II

Residual Stress Gradient Analysis of Automobile Crankshaft

Crankshafts are highly stressed component of the internal combustion engine. Production of crankshafts requires careful selection of material, forging practice, forging heat treatment and surface heat treatment of the machined component. All the above factors largely contribute towards the improvement of the fatigue properties of the component and also help in near elimination of the straightening after hardening and tempering. During manufacturing induction hardening process of the crankshaft, distortion in the shape and microcracks near the fillet region have been found to occur. These manufacturing defects were attributed to the thermal residual stresses generated in crankshaft during induction hardening stage of the machined crankshaft by the manufacturer.

To understand the nature of residual stress, an investigation is undertaken to determine the residual stress of the crankshafts manufactured through different routes. Details of the manufacturing routes of the four crankshafts considered for analysis are mentioned below:

1. Imported in machined condition
2. Forged and controlled, cooled and machined (CC)
3. Forged and hardened and tempered and machined (HT)
4. Forged and normalised, hardened and tempered and machined (NHT)

Experimental Studies

Residual stresses have been measured near the fillet regions of the main bearing and crankpins (Fig. 3). All the stress measurement locations have been cleaned thoroughly to degrease before the measurements. CrKα radiation and $\sin^2\psi$ method has been chosen for the XRD studies. After the stress measurement on the as-received surface condition, top surface layers have been removed incrementally by electrochemical polishing for stress gradient analysis. An electrolytic bath of acetic acid of 10% perchloric acid has been used for the layer removal. The thickness of the layer removed is determined by using a dial gauge (2.5 μm accuracy) set up.

Results and Observations

Residual stresses as measured by XRD at main bearing 4 and crankpin 2 are shown in the Tables 2 and 3. Tensile stresses of same magnitude have been found in all the three indigenous crankshafts in the as-received surface conditions on the main bearing 4. Whereas, the crankpin 2 shows much higher value nearly double to that of the main bearing. This was due to the larger designed stock removal at the crankpin by machining. Further, stresses were similar near both the fillet region. The imported crankshaft, however, showed much lower values of tensile stress as compared to other crankshaft. Measurement after layer removal of nearly 200 μm from the main bearing 4, produced very low tensile stresses. Machining operation, which is the case here, generally produces stresses of shallow layer. Stress measurements after stress relieving operation at 600 °C for 2 hrs and 30 min and furnace cooled yielded nearly negligible stress values. The variation of stress values at various surfaced conditions is shown in Fig. 4.

Table 2: Residual stress values [MPa] measured on Main bearing 4 of Crankshaft

Crankshaft number	*Residual Stress near the fillet*			
	On as-received surface		*After 100 μm layer removal*	*After 200 μm layer removal*
	Near shank end	*Away from shank end*	*Near shank end*	*Near shank end*
Imported	+180	+155	+47	+27
CC	+463	+479	+55	+59
NHT	+330	+483	+204	+44
HT	+462	+435	+254	+109

The low inherent/bulk residual stress as found by the sub surface stress measurements, which is produced by heat treatment/straightening process may not lead to the distortion in the shape of the crankshaft during induction

Table 3: Residual stress values (MPa) measured on Crank Pin 2 of Crankshaft

Crankshaft number	*Residual Stress near fillet*		
	On as-received surface		*After 100 μm layer removal*
	Near shank end	*Away from shank end*	*Near shank end*
Imported	+148	+225	+20
CC	+752	+751	+161
	+619	+675	+90
HT	+801	+787	NHT

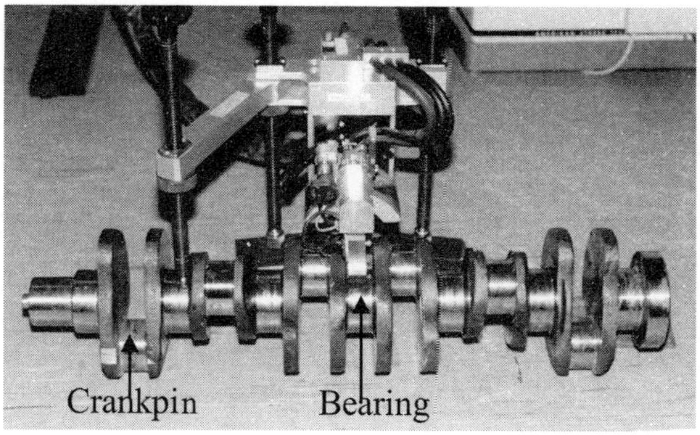

Fig. 4: Crankshaft with X-ray diffractometer goniometer sitting on its top. Arrows indicate the crankpin and bearing.

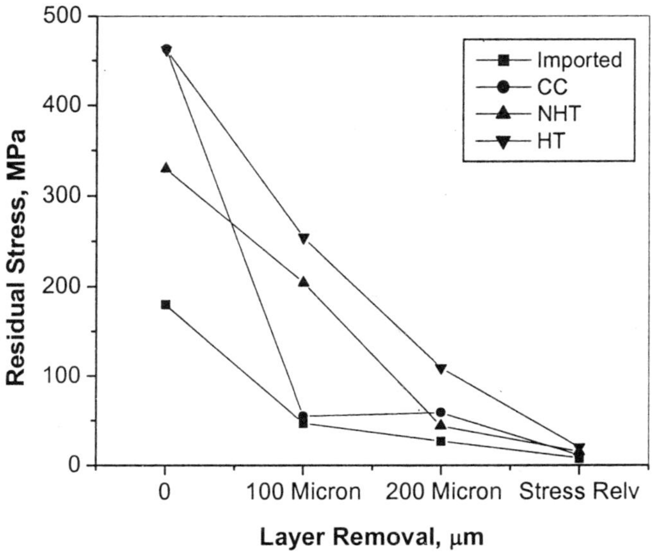

Fig. 5: Residual stress variation in the main bearing of the crackshaft at different depths.

hardening after machining. Distortion, if any due to induction hardening, may arise due to the presence of high bulk stresses rather than shallow highly stress surface layer generated by machining.

Microcracks observed near the fillet region may be due to improper induction hardening process. This is generally encountered due to inconsistent and non-uniform hardness depth pattern by induction hardening. Further, the non-uniformity in the microstructure may also be responsible for these cracks.

CONCLUSIONS

The two case studies discussed in the present paper shows that XRD technique could be used in the industry to sort out shop floor problem. This technique can be used for trouble shooting as well as a tool for quality control.

REFERENCES

Noyan, I.C. and Cohen J.B. (1987). Residual Stress Measurement by X-ray Diffraction and Interpretation. Springer Verlag, N.Y. Inc., USA.

Residual Stress Measurement by X-ray Diffraction (1971). Report: SAE J84a, Society of Automotive Engineerings, Inc., USA.

X-ray Diffraction and Transmission Electron Microscopic Investigations of Amorphous to Icosahedral Phase Transformation in Al-Cu-(V, Ti) Alloys

R.S. Tiwari, D.K. Misra and O.N. Srivastava

Department of Physics, Banaras Hindu University
Varanasi 221005, India

INTRODUCTION

Since the discovery of the icosahedral quasicrystalline phase (I-phase) in rapidly quenched $Al_{86}Mn_{14}$ alloy (Shechtman et al., 1984), there have been several studies on the transformation of alloys from amorphous phase to the icosahedral phase (Chen and Sapepen, 1982; Shen et al., 1986; Köster et al., 1999; He et al., 1988). Large number of alloys exhibiting glass to quasicrystalline transformation has been broadly classified into Pd-based (Poon et al., 1985), Al-based (Inoue et al., 1988), Ti-based (Molokanov and Chebotnikov, 1990), Zr-based (Koster et al., 1996) and Hf-based (Li et al., 2000) amorphous alloys.

Koster and co-workers (1996) were the first to report the formation of an i-phase upon crystallization in $Zr_{63}Cu_{25}Al_{12}$, $Zr_{65}Cu_{17.5}Ni_{10}Al_{7.5}$ and $Zr_{69.5}Cu_{12}Ni_{11}Al_{7.5}$ alloys. Xing et al. (1999) have reported the stabilization of the i-phase by Ti in Zr based alloys during crystallization of the amorphous phase. Inoue et al. (1999) have reported that a number of additional elements such as Pd, Ag, Au and Pt stabilize the i-phase in alloys based on Zr-Cu-Ni-Al. Three criteria have been proposed by Saida et al. (2000) for the formation of the i-phase in Zr-based alloys namely (i) a multicomponent alloy system, (ii) significant atomic size mismatch (>10%) and (iii) atomic pairs with negative heat of mixing with at least one atomic pair having nearly zero or positive heat of mixing. Murty et al. (2000) have found i-phase formation in $Zr_{65}Cu_{15}Ni_{10}Pd_{10}$, $Zr_{65}Cu_{25}Pd_{10}$, $Zr_{70}Cu_{10}Ni_{13}Pd_7$ and $Zr_{70}Cu_{23}Pd_7$ during primary crystallization of metallic glass. This study indicates that Ni and Al are not essential for i-phase formation in Pd containing Zr-based alloys. It has been pointed out by Murty et al. (2000) that an atomic pair with near-zero or positive enthalpy of mixing is not necessary for i-phase formation since all three binary phases in Zr-Cu-Pd system have negative enthalpies. Yet another example of nanoquasicrystalline systems are Zr-Pt (Saida et al., 2000) and Zr-Ni-Mn (Li et al., 2000). Li et al. (2000) have reported that Mn is essential for i-phase formation in Zr-Ni-Mn alloy.

Beside Zr-based amorphous alloys new systems containing Ti as basic element have also been reported in which amorphous to nanoquasicrystalline transformation has been studied (Murty et al., 2001). It has been reported that nanocrystalline i-phase forms directly from the melt spinning at low wheel velocity of 20 m/s. Nanoquasicrystallization of amorphous phase has been observed on subsequent annealing after melt-spinning at higher wheel velocities of 30 and 40 m/s. The i-phase is stable upto 873 K in Ti-rich alloy, while it transforms to crystalline phase in Zr-rich alloy below this temperature. Murty et al. (2001) have reported that the i-phase formed both directly from the melt and from the amorphous phase is much finer in Zr-rich alloy compared to Ti-rich alloy.

Li et al. (2000) have recently reported precipitation of an icosahedral quasicrystalline phase in $Hf_{59}Ni_8Cu_{20}Al_{10}Ti_3$ metallic glass. This alloy has been designed by taking into consideration the fact that atomic radii and electronic configuration of Zr and Hf are similar. Therefore, just like Zr-based amorphous alloy, Hf based amorphous alloy is feasible to form and its subsequent annealing is expected to give rise to precipitation of quasicrystals. The crystallization process of $Hf_{59}Ni_8Cu_{20}Al_{10}Ti_3$ metallic glass has been investigated by differential scanning calorimetry (DSC), X-ray diffraction (XRD) and transmission electron microscopy (TEM). It has been reported that two exothermic peaks correspond to two reactions namely (i) amorphous$\rightarrow$ icosahedral phase + fcc phase and (ii) icosahedral phase + fcc phase$\rightarrow Hf_2Cu+Al_{16}Hf_6Ni_7+Hf_3Al_2$.

Quite recently Saida and Inoue (2000) have reported the nano-icosahedral phase formation in the devitrification stage of Zr- and Hf-based glassy alloys. The primary phases observed are metastable fcc Zr_2Ni and fcc Hf_2Ni phases in the $Zr_{65}Al_{7.5}Ni_{10}Cu_{17.5}$ and $Hf_{65}Al_{7.5}Ni_{10}Cu_{17.5}$ glassy alloys, respectively. They have reported that by substitution of 5 at.% Pd for Cu, the primary crystalline phase changes to an icosahedral quasicrystalline phase in both alloys. The important formation factor of the icosahedral phase during devitrification has also been attributed to enhancement of nucleation frequency. It has been suggested that the structure of these metastable phases are correlated with the local icosahedral order of the glass.

Amongst several amorphous alloys, Al-based systems have attracted considerable attention as studies in simple systems are expected to throw light on the mechanism of these transformations (Upadhya and Tiwari, 1996; Popescu et al., 1995; Zhang et al., 2001). The investigation on Al-based amorphous alloys has also become important because a homogeneous dispersion of nanocrystalline particles embedded in an Al-based amorphous matrix has been found to induce a significant increase in tensile strength as compared to that for the corresponding amorphous phase (Inoue et al., 1992; Watanabe et al., 1993; Inoue et al., 1995).

Amongst several Al-based alloys, Al-Cu-V system has been of special interest because this belongs to Al-Cu-3dTM (TM = transition metals) series of alloys which have been systematically studied and show chemical dependence of stability of amorphous, icosahedral and decagonal phases (Watanabe et al., 1993). For example as the atomic number of transition metal increases in these alloys, the as-quenched alloy exhibits amorphous phase (Al-Cu-V, Al-Cu-Ti), metastable (Al-Cu-Mn, Al-Cu-Cr) and stable (Al-Cu-Fe, Al-Cu-Ru, Al-Cu-Os) icosahedral phases and stable decagonal phase (Al-Cu-Co, Al-Cu-Rh) (Tsai et al., 1994; Tsai, 1996; Tsai et al., 1988; Tsai et al., 1987; Tsai et al., 1989; He et al., 1988; He et al., 1990).

The Al-Cu-V system which was shown by Tsai et al. (1994) to form an amorphous phase is very interesting from view point of studies pertaining to the amorphous to quasicrystalline transformation (Tsai et al., 1989; Dubois et al., 1989; Knapp and Follstaedt, 1985; Lilienfeld et al., 1985; Levine et al., 1989). The electron energy loss spectroscopy study of $Al_{75}Cu_{15}V_{10}$ suggests that amorphous and icosahedral phases in this system have similar short-range ordering (Levine et al., 1989). In recent studies a close structural interrelation between amorphous and quasicrystalline structures has been shown (Tsai et al. 1994). Dubois et al. (1989) have shown that an icosahedral phase grows within Al-Cu-V amorphous matrix in competition with conventional·crystalline compounds. Tsai et al. (1989) have investigated the transformation kinetics of the amorphous $Al_{75}Cu_{15}V_{10}$ alloy employing DSC and TEM and have mentioned that the transformation occurs by homogeneous nucleation and diffusionless growth. The nature of transformation reaction has been found to be polymorphic. Holzer and Kelton (1991) have analyzed the kinetics of transformation in this system with the help of a kinetic model assuming a constant nucleation rate and growth on the quenched in nuclei. The low interfacial energy ($\sigma = 0.002$ J/m^2) between the icosahedral phase and the glass has been attributed to the similarity between the short range order of amorphous and icosahedral phases. The $Al_{75}Cu_{15}V_{10}$ alloy has been investigated in regard to the structures of amorphous and icoashedral phases (Tsai et al., 1988). The high resolution image of the icosahedral phase in $Al_{75}Cu_{15}V_{10}$ shows an atomic arrangement similar to that described by the icosahedral glass (I-G) model. The highly resolved lattice image presents a two-dimensional tilling pattern comprised mainly of pentagons joined together side by side randomly and filled with unique tear like and crack like defects exhibited in I-G model. Tsai et al. (1994) have also pointed out that despite the high degree of structural defects of the I.G. phase, the characteristic physical properties of quasiperiodic lattices such as high electrical resistivity, high hardness and large elastic modulus strongly persists in the I.G. phase. Thus investigation of amorphous to icosahedral transformation in $Al_{75}Cu_{15}V_{10}$ alloy provides an impetus to substitute Ti in place of V in Al-Cu-V alloy to find out its effect on the stabilization of amorphous phase and the transformation into icosahedral/crystalline phases. This is because of the fact that V and Ti are neighbouring transition metals in periodic table and are similar in their atomic sizes and atomic number. Furthermore, Murty et al. (1992) have reported the glass forming composition ranging from 40 to 90 at.% Ti in Al-Ti system. Tsai (1996) has given a table for the formation tendency of stable phases in Al-Cu-TM system. He has shown that both Al-Cu-V and Al-Cu-Ti alloys exhibit the formation of amorphous phase. Thus, it appears that substitution of Ti in place of V in Al-Cu-V alloy may stabilize amorphous phase, which after subsequent annealing may form icosahedral phase.

Thus, keeping above facts in view, we have partially substituted Ti in place of V in order to stabilize the amorphous phase. Partially substituted $Al_{75-x}Cu_{15}V_5Ti_5$ (with x = 5 and 10) and fully substituted $Al_{45+x}Cu_{45}Ti_{10-x}$ (with x = 0, 2, 4, 5, 6 and 8) as well as $Al_{50-x}Cu_{45+x}Ti_5$ (with x = 3 and 5) alloys were synthesized and characterized in the present investigation. A composition namely $Al_{70}Cu_{20}Ti_{10}$ which is almost similar to amorphous $Al_{75}Cu_{15}V_{10}$ alloy has been also prepared to stabilize the amorphous phase in Al-Cu-Ti alloy. Following the Hume-Rothery rule (e/a ratio), we have deduced two specific compositions e.g. $Al_{70}Cu_{20}Ti_5V_5$ and $Al_{65}Cu_{25}Ti_5V_5$ for which electron concentration per atom e/a ratios were close to that for $Al_{75}Cu_{15}V_{10}$ alloy which is known to form amorphous phase after rapid solidification. It is interesting to note that when V is partially substituted by Ti in the present investigation amorphous phase gets stabilized for lower concentration of Al. However, amorphous to icosahedral phase transformation is observed only in case of $Al_{50}Cu_{45}Ti_5$ alloy. The present work is also important due to the fact that out of various Al-Cu-3d-TM alloys, information regarding the stabilization of quasicrystalline phase in Al-Cu-(Ti-V) alloy system is still missing. The aim of the present investigation is to report the results concerned with the formation of amorphous phase and transformation from amorphous to quasicrystalline/crystalline phases in Al-Cu-(V,Ti) alloys. Growth of icosahedral phase in α-Al matrix has been investigated during isothermal annealing at different temperatures. After different periods of time at different annealing temperatures, a curious microstructure in the form of banded structures surrounding the icosahedral phase have been observed. The relative stability of icosahedral phase in terms of size effect in Al-Cu-TM series of alloys has also been discussed.

EXPERIMENTAL PROCEDURE

Al-Cu-(V, Ti) and Al-Cu-Ti alloys with compositions $Al_{75-x}Cu_{15+x}V_5Ti_5$ (with x = 5 and 10), $Al_{70}Cu_{20}Ti_{10}$, $Al_{45+x}Cu_{45}Ti_{10-x}$ (with x = 0, 2, 4, 5, 6 and 8), and $Al_{50-x}Cu_{45+x}Ti_5$ (with x = 3 and 5), were synthesized in the present work. The master alloy ingots were prepared by R.F. induction melting, a mixture of pure Al (99.99% purity), Cu (99.99% purity), V (99.99% purity), Ti (99.99% purity), in a purified argon atmosphere. The master alloy ingots prepared in this way have been repeatedly melted (2-3 times) to ensure homogeneity. The homogenized pre alloyed ingots have been subsequently crushed and placed in a quartz tube of ~1 cm internal diameter with a circular orifice of ~ 1 mm diameter and then melt-spun using an over pressure (≈ 50 atmosphere) of argon on to a copper wheel (dia ~13 cm), rotating at a speed of about 40 m/sec. The ribbons so produced were ~2 mm in width and about ~ 40 μm in thickness. During melt-spinning, the entire apparatus was enclosed in a steel enclosure where argon gas was continuously flown so as to prevent oxidation of the ribbons after ejection from the nozzle. The density measurements were performed by Archimedean method. A single pan precision balance with sensitivity of 0.001 mg was employed. Sample used weighed about 1 gm.

The gross structural characterization was done at room temperature employing X-ray diffractometer (Philips PW-1710) with graphite monochromator and CuKα radiation in the range of 20° to 80°. Specimens for TEM were electrolytically thinned using an electrolyte of 67% methanol and 33% nitric acid at –20°C and examined in a Philips CM-12 electron microscope. The crystallization behaviour of these amorphous alloys has been studied using DSC 2910 (TA instruments Inc. USA) system in modulated DSC mode.

RESULTS AND DISCUSSION

Figure 1(a) shows XRD pattern corresponding to $Al_{70}Cu_{20}Ti_5V_5$ (e/a = 2.21) alloy. All the peaks of XRD corresponding to this alloy have been indexed in terms of Al_3Ti type tetragonal phase with lattice parameter a = 3.84Å, c = 8.61Å. One can notice from this diffraction pattern that the $Al_{70}Cu_{20}Ti_5V_5$ alloy neither exhibit amorphous nor icosahedral phase. However XRD profiles [Fig. 1(b)] corresponding to $Al_{65}Cu_{25}Ti_5V_5$ for which e/a =2.11 reveals a broad diffuse peak which is indicative of amorphous nature of the alloy. Thus the partial Ti substitution in this alloy stabilizes the amorphous phase. In order to explore the transformation behaviour of $Al_{65}Cu_{25}Ti_5V_5$ alloy, DSC was carried out. Figure 2 shows the DSC curve with the heating rate of 5°C/min. The continuous heating trace does not show the characteristic patterns of glass transition. There is only one peak at 460 °C, which clearly indicates single stage crystallization of this alloy. It may be noticed that the crystallization temperature has considerably been enhanced in $Al_{65}Cu_{25}V_5Ti_5$ alloy (T_{onset}=430 °C) in comparison to $Al_{75}Cu_{15}V_{10}$ alloy (T_{onset}=280 °C) (Misra et al., 2003). In order to explore the transformation of amorphous to quasicrystalline phase in $Al_{65}Cu_{25}V_5Ti_5$ alloy, the sample was isothermally annealed at temperature around the exothermic peak in DSC graph (Fig. 2). Figure 3 represents the XRD pattern after annealing the alloy at 400 °C for 2 hrs. All the XRD peaks have been indexed in terms of Al_3Ti type tetragonal phase. Thus there is no trace of icosahedral phase in the transformed alloy. For an unambiguous identification of phases in as-cast and annealed samples of $Al_{65}Cu_{25}Ti_5V_5$ alloy, samples were investigated by TEM techniques with particular emphasis on the microstructure. Figure 4(a) shows the representative microstructure of as-quenched $Al_{65}Cu_{25}Ti_5V_5$ alloy. The absence of any contrast in this microstructure clearly indicates the amorphous nature of the sample. The corresponding selected area diffraction (SAD) pattern shown in Fig. 4(b) exhibits broad diffuse halos confirming the presence of amorphous phase. Figure 5(a) depicts representative bright field electron micrograph corresponding to $Al_{65}Cu_{25}Ti_5V_5$ alloy annealed at 400 °C for 2 hrs. The SAD patterns corresponding to annealed $Al_{65}Cu_{25}Ti_5V_5$ alloy are shown in Figs. 5(b-c). These patterns have been indexed in terms of Al_3Ti type tetragonal phase with lattice parameter a = 3.84Å, c = 8.61 Å. The zone axes of patterns [shown in Figs. 5(b-c)] are [001] and [011], respectively. On the basis of above observation it may be inferred that partial substitution of Ti in place of V in Al-Cu-V alloy namely $Al_{65}Cu_{25}Ti_5V_5$ alloy stabilizes the amorphous phase but transformation of amorphous phase does not lead to the formation of icosahedral phase. The transformation of $Al_{65}Cu_{25}Ti_5V_5$

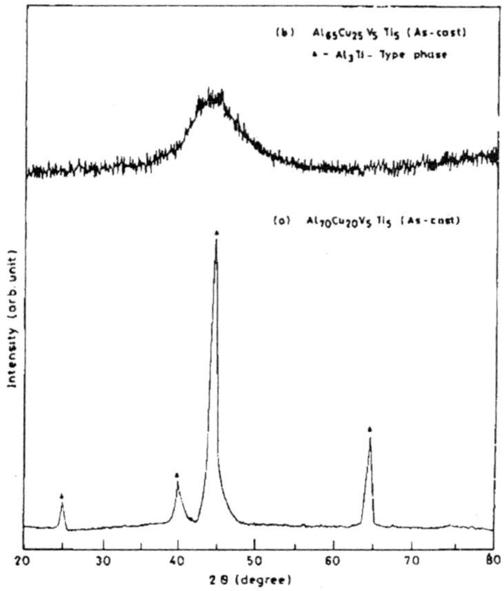

Fig. 1: The powder XRD profile corresponding to partially Ti substituted (a) $Al_{70}Cu_{20}Ti_5V_5$ alloy and (b) $Al_{65}Cu_{25}Ti_5V_5$.

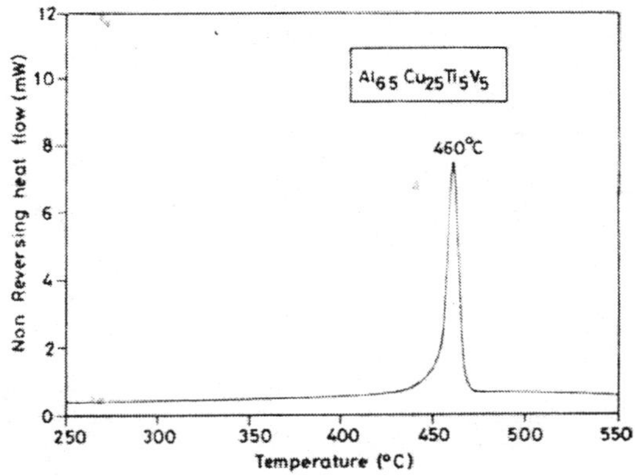

Fig. 2: DSC curve taken corresponding to $Al_{65}Cu_{25}V_5Ti_5$ alloy with the heating rate of 5 °C/min.

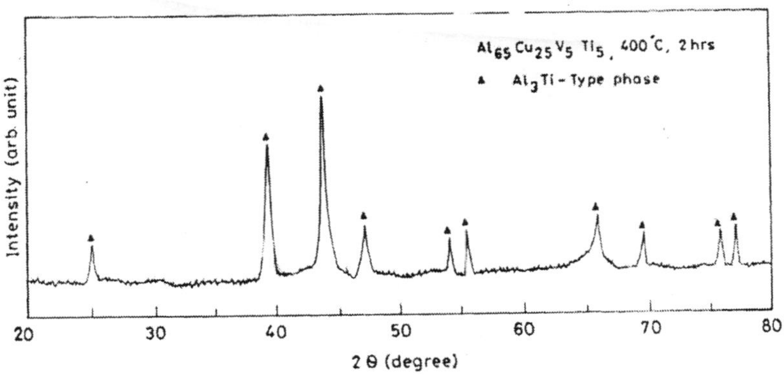

Fig. 3: XRD patterns of annealed $Al_{65}Cu_{25}V_5Ti_5$ alloy at 400 °C for two hours showing the presence of Al_3Ti type tetragonal phase.

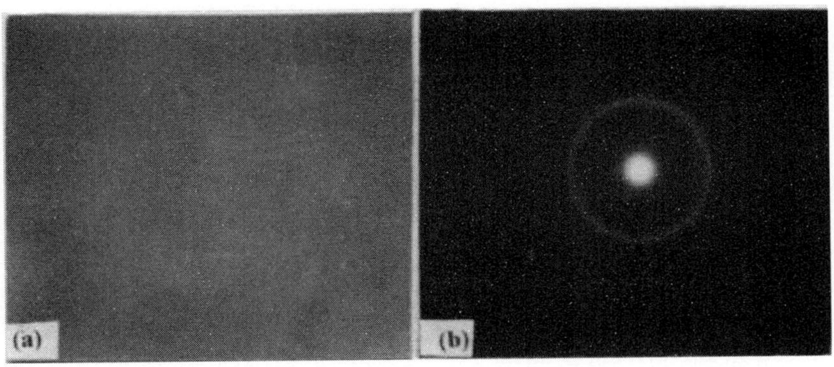

Fig. 4: (a) Depicts representative microstructure of as-quenched $Al_{65}Cu_{25}V_5Ti_5$ amorphous alloy showing contrast free feature of amorphous phase.
(b) SAD pattern exhibiting broad diffuse halos confirming the presence of amorphous phase.

Fig. 5: (a) Depicts representative bright field electron micrograph corresponding to $Al_{65}Cu_{25}V_5Ti_5$ alloy annealed at 400 °C for 2 hrs. (b and c) SAD pattern corresponding to annealed $Al_{65}Cu_{25}V_5Ti_5$ alloy at 400 °C for 2 hrs showing Al_3Ti type tetragonal phase with zone axes [001], $[0\bar{1}1]$ respectively.

alloy to crystalline phase rather than to icosahedral phase indicates that possibly the chemical and topological short range ordering in this amorphous alloy may be different from $Al_{75}Cu_{15}V_{10}$ amorphous alloy.

In order to explore the stabilization of amorphous phase and its subsequent transformation to icosahedral phase, a systematic investigation of fully Ti substituted alloys with a wide composition range was also carried out. Based on the concept of Hume-Rothery rule (Fermi-surface-Brillouin-zone interaction) which is broadly known to predict the specific composition at which electron phases are most likely to stabilize, we have estimated the specific composition at which the amorphous and icosahedral phases get stabilized in the present system. For example in Al-TM based system, e/a varies from 1.5 to 2.5 for icosahedral phase. Thus, we have deduced several specific compositions e.g. $Al_{45+x}Cu_{45}Ti_{10-x}$ (with x = 0, 2, 4, 5 and 6) at which amorphous phase is expected most likely to occur. For the range of e/a values 2.14 to 2.20, the occurrence of amorphous phase has actually been found. A list of Al-Cu-Ti alloys with the specific compositions along with the corresponding average *e/a* concentration, peak position and peak width have been given in Table 1.

Table 1: Compositions of Al-Cu-Ti alloy along with the corresponding *e/a* values, peak position and peak width

Chemical Compositions	*e/a*	*Peak position (nm^{-1})*	*Peak width (FWHM) (Rad)*
$Al_{45}Cu_{50}Ti_5$	2.05	29.3	0.036
$Al_{47}Cu_{48}Ti_5$	2.09	29.6	0.018
$Al_{53}Cu_{45}Ti_2$	2.12	29.94	0.014
$Al_{51}Cu_{45}Ti_4$	2.14	29.90	0.06
$Al_{50}Cu_{45}Ti_5$	2.15	30.07	0.11
$Al_{49}Cu_{45}Ti_6$	2.16	30.05	0.12
$Al_{47}Cu_{45}Ti_8$	2.18	29.9	0.13
$Al_{45}Cu_{45}Ti_{10}$	2.20	29.7	0.14

Since in the present alloy transition metal Ti is involved, a charge transfer to their *d* bands is allowed for the compensation of the unpaired spins of *d*-electrons. We have, therefore, used elemental valence assigned by Raynor (1949). It may be pointed out that the range of determined *e/a* values are close to that of $Al_{75}Cu_{15}V_{10}$ which is known to exhibit amorphous phase.

Figures 6(a-e) show XRD traces obtained from the as melt-spun $Al_{45+x}Cu_{45}Ti_{10-x}$ (with x = 0, 2, 4, 5 and 6). All the traces show the broad amorphous peak around an angle 43.5°. The peak position and peak width corresponding to different compositions are given in Table 1. The subtle differences in the broad peaks (e.g. width of the peak and peak position) may be noticed. These differences may be attributed to the different type of short range ordering in these alloys. It will be interesting to mention here that during present investigation we have also chosen Cu rich Al-Cu-TM system e.g. $Al_{47}Cu_{48}Ti_5$ and $Al_{45}Cu_{50}Ti_5$ to stabilize amorphous phase in these compositions. The average *e/a* (2.09 and 2.05, respectively) was kept very close to that for $Al_{75}Cu_{15}V_{10}$ alloy. However, XRD patterns corresponding to these alloys [Figs. 7 (a-b)] do not show amorphous phase. All the peaks of XRD patterns have been indexed in terms of Al_3Ti type tetragonal phase.

Figure 8 shows XRD pattern corresponding to $Al_{70}Cu_{20}Ti_{10}$ alloy which reveals only Al_3Ti type tetragonal phase. No trace of amorphous or icosahedral phases have been noticed. It should be mentioned here that the *e/a* value for this particular alloy composition is 2.70 which is much higher than the value for $Al_{75}Cu_{15}V_{10}$.

During present investigation attempt has been made to examine Fermi surface-Brillouin zone interaction through the criterion $K_p = 2K_F$ for the stabilization of amorphous and icosahedral phases. For this purpose three

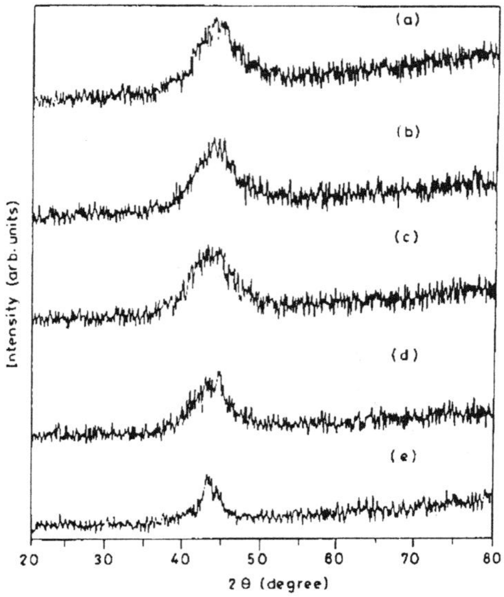

Fig. 6: XRD pattern of as-quenched – (a) $Al_{45}Cu_{45}Ti_{10}$ (b) $Al_{47}Cu_{45}Ti_8$ (c) $Al_{49}Cu_{45}Ti_6$ (d) $Al_{50}Cu_{45}Ti_5$ (e) $Al_{51}Cu_{45}Ti_4$ alloys showing broad amorphous peak.

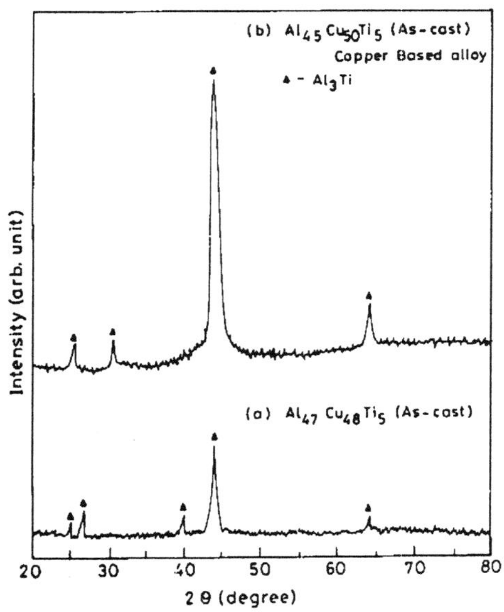

Fig. 7: (a-b) The powder XRD pattern corresponding to Cu rich completely Ti substituted $Al_{50-x}Cu_{45+x}Ti_5$ (x = 3 and 5) alloy showing crystalline phase.

alloys namely $Al_{65}Cu_{25}V_5Ti_5$, $Al_{50}Cu_{45}Ti_5$ and $Al_{45}Cu_{45}Ti_{10}$ have been chosen. It can be seen here that the amorphous phase formation for these compositions can be understood in terms of FS-BZ interaction. By using

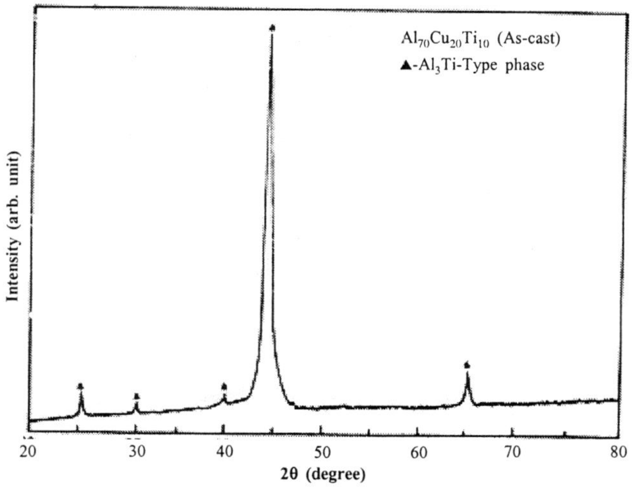

Fig. 8: The powder XRD pattern corresponding to rapidly quenched $Al_{70}Cu_{20}Ti_{10}$ alloy showing single Al_3Ti type phase.

valences for Al, Cu, V and Ti as proposed earlier (Raynor, 1949) for crystalline Al-TM alloys, one calculates the electron concentration *e/a* to be 2.11, 2.15 and 2.20, respectively. According to free electron model, the Fermi vector is given by

$$K_f = \left(\frac{3\pi^2 \, e/a}{V} \right)^{1/3} \tag{1}$$

where V is the average atomic volume. The values of K_f calculated for the three amorphous alloys are given in Table 2. The density of the alloys has been determined by Archimedean method described in section of experimental procedure. The reciprocal vectors K_P corresponding to the maximum peak intensity for amorphous phase in these alloys as determined from XRD are 29.39, 30.7 and 29.7 nm^{-1} respectively. The parameters used to calculate $K_p/2K_f$ for amorphous phase is given in Table 2. It is interesting to note that for these alloys, density is decreasing from 3.96 to 3.56 gm/cm^3 while *e/a* is increasing from 2.11 to 2.20.

Table 2: Density ρ, electron concentration *e/a*, Fermi wave-vector K_f and reciprocal vector K_p for amorphous phase

Alloy Phase	$Al_{65}Cu_{25}V_5Ti_5$ *(amorphous)*	$Al_{50}Cu_{45}Ti_5$ *(amorphous)*	$Al_{45}Cu_{45}Ti_{10}$ *(amorphous)*
ρ (gm/cm^3)	3.96	3.84	3.56
e/a	2.11	2.15	2.20
$2K_f$ (nm^{-1})	31.44	29.74	29.05
K_p (nm^{-1})	29.39	30.70	29.70
$K_p/2K_f$	0.93	1.03	1.02

It can easily be discerned that for these three alloys the ratio $K_p/2K_f$ remains nearly equal to unity and consequently amorphous phase gets stabilized. For $Al_{45+x}Cu_{45}Ti_{10-x}$ (with x = 0, 2, 4, 5 and 6) alloys, the variation of K_P with e/a is shown in Fig 9. It is evident from this graph that K_p is decreasing with increasing *e/a* ratio which in turn is related to K_F through relation (1). Therefore only in the limited composition range where the variation of ρ, K_P and *e/a* ratio satisfies the condition $K_P \approx 2K_F$, the amorphous phase is getting stabilized.

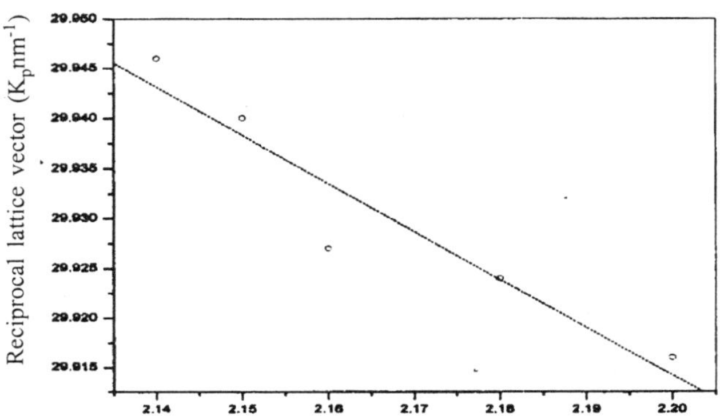

Fig. 9: Variation of K_p with *e/a* for $Al_{45+x}Cu_{45}Ti_{10-x}$ (x = 0, 2, 4, 5 and 6).

Figure 10 shows the DSC curve taken corresponding to the alloy $Al_{50}Cu_{45}Ti_5$ with the heating rate of 5 °C/min. The DSC curve corresponding to this alloy contains a wide exothermic peak extending from 288 to 550 °C. This wide peak consists of two overlapping maxima at least. The peak temperature of two peaks are 393 °C and 440 °C respectively. Figures 11(a-b) show XRD traces of $Al_{50}Cu_{45}Ti_5$ alloy obtained from isothermal annealing of the sample at 280 °C for 5 hrs and 400 °C for 8 hrs, respectively. The XRD patterns corresponding to $Al_{50}Cu_{45}Ti_5$ amorphous alloy annealed at 280 °C shows only peaks which can be identified as a mixture of crystalline α-Al and icosahedral phases. The identification of icosahedral phase was made on the basis of six independent miller indices proposed by Elser (1985). The $Al_{50}Cu_{45}Ti_5$ alloy when annealed at 400 °C i.e. near the strong exothermic DSC peak (~393 °C) for 8 hrs, shows only diffraction peaks which have been identified as the Al_3Ti-type tetragonal phase with lattice parameter a = 3.84Å, c = 8.61Å (Fig. 11b). The above observation clearly indicates that in the first stage, amorphous alloy $Al_{50}Cu_{45}Ti_5$ starts transforming to icosahedral and α-Al phases around 280 °C. At 400 °C, α-Al and icosahedral phases transform to Al_3Ti type tetragonal phase.

Figure 12(a) depicts a representative microstructure of $Al_{50}Cu_{45}Ti_5$ alloy after isothermal annealing at 280 °C for 5 hrs. The microstructure shows that

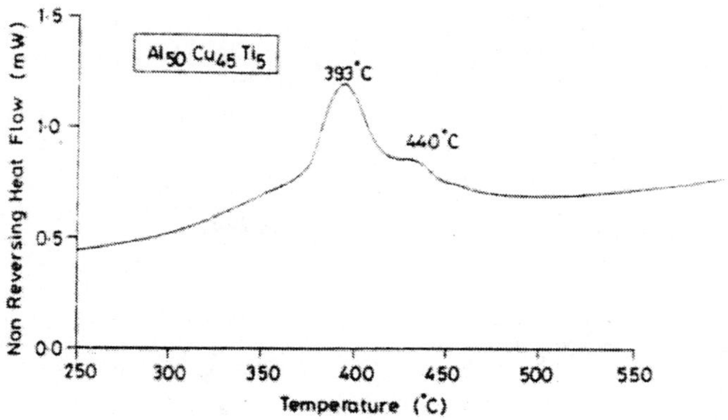

Fig. 10: DSC curve of $Al_{50}Cu_{45}Ti_5$ amorphous alloy at scanning rate 5 °C/min.

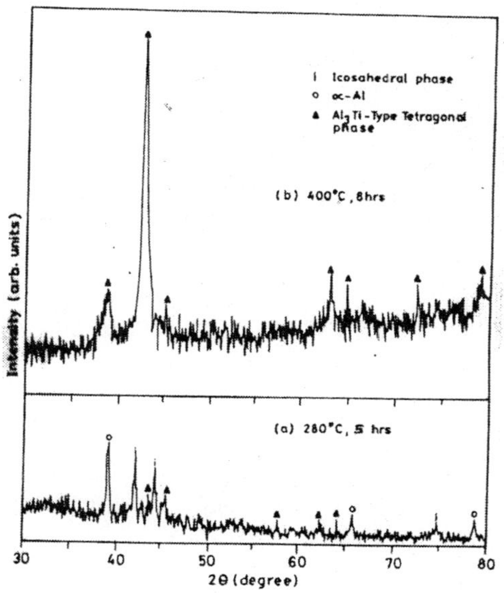

Fig. 11: XRD patterns of $Al_{50}Cu_{45}Ti_5$ alloy annealed in vacuum (10^{-6} torr) at (a) 280 °C for 5 hrs showing α-Al and icosahedral phases. (b) 400 °C for 8 hrs showing the presence of Al_3Ti type tetragonal phase.

during transformation of amorphous phase coarser intermetallic phase and α-Al phase form simultaneously. The SAD patterns taken from coarser particles in the microstructure are shown in Figs. 12b (i-iii). These SAD patterns correspond to five fold, three fold and two fold of icosahedral phase. Figure 13(a) represents bright field electron microstructure obtained from the sample of $Al_{50}Cu_{45}Ti_5$ alloy annealed at 310 °C for 8 hrs. The microstructure consists of domains or colonies of bands. The bands are arranged in the form of concentric rings, centering around the coarse icosahedral particle. Observation at high magnification [Fig. 13(b)] indicates

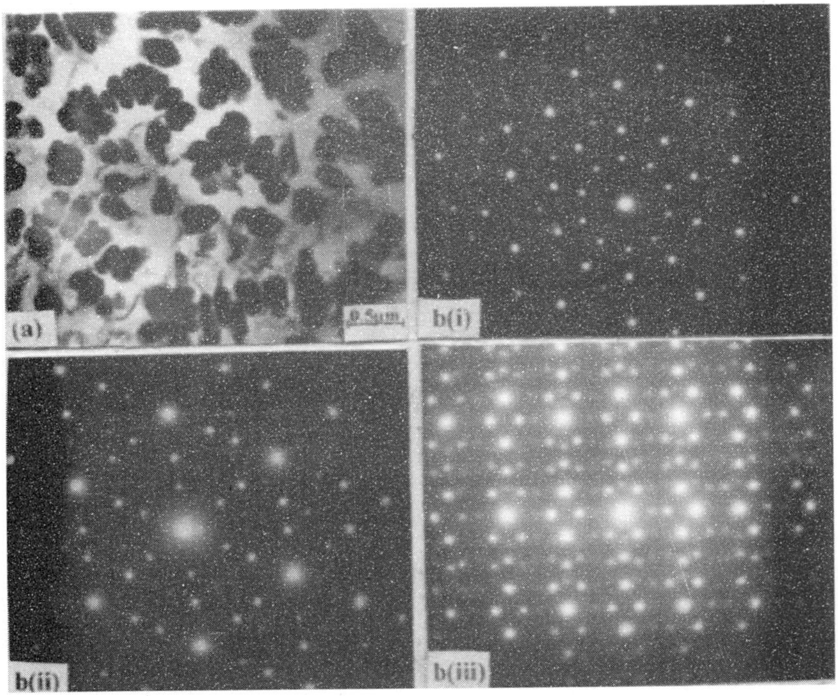

Fig. 12: (a) Electron micrograph showing coarser icosahedral grain. (b) SAD patterns exhibiting (i) five fold, (ii) three fold and (iii) two fold axes parallel to electron beam corresponding to coarser icosahedral grains.

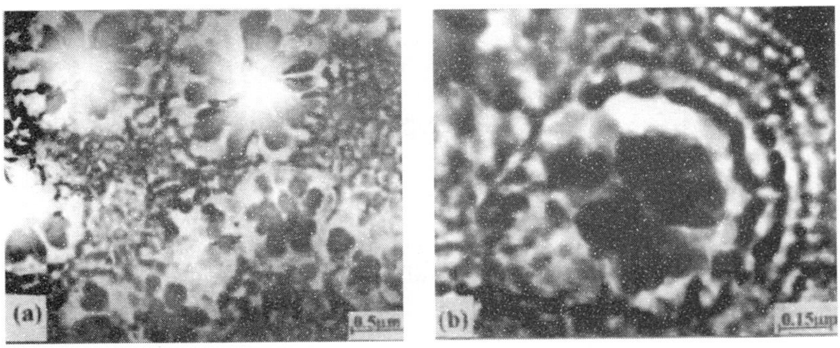

Fig. 13: (a) Electron micrograph corresponding to the alloy $Al_{50}Cu_{45}Ti_5$ annealed at 310 °C for 8 hrs showing banded microstructure surrounding the coarse icosahedral grain. (b) Magnified electron micrograph indicating that band consists of large particle of 70 nm size embedded in the matrix.

that the bands consist of large particle of 70 nm size embedded in the matrix. The SAD patterns from these regions show single crystal pattern of icosahedral and/or α-Al phase. However, we have not noticed a well defined orientational relationship between icosahedral phase and α-Al matrix. In order to monitor the growth behaviour of icosahedral phase in α-Al matrix, isothermal annealing of amorphous $Al_{50}Cu_{45}Ti_5$ alloy has been carried out at 280, 310, 320, 330 and 350 °C for different interval of time.

Table 3: The temperature and time of isothermal annealing of $Al_{50}Cu_{45}Ti_5$ amorphous alloy

Temperatures	*Annealing time (in hours)*					
280 °C	5	20	40	56	73	100
310 °C	½	1	3	6	8	
320 °C	½	1	3			
330 °C	½	1	3			
350 °C	½	1	3			

These temperatures were selected because at lower temperatures the reaction proceeds exceedingly slow and at higher temperatures too rapid. Figures 14(a-c) show bright field electron micrograph of annealed $Al_{50}Cu_{45}Ti_5$ alloy at 280 °C for 5, 20 and 40 hrs, respectively. One can notice that the size of icosahedral grain increases with time. Figures. 15 (a-d) depict electron micrographs obtained from the sample of $Al_{50}Cu_{45}Ti_5$ alloy annealed at 310 °C for ½, 1, 3 and 6 hrs, respectively. The microstructures reveal that the size of icosahedral grain increases with time. Microstructural analysis regarding variation in grain size with time at higher temperatures reveals similar trend as in the case of sample annealed at temperatures 280 and 310 °C. It may be mentioned that the size of icosahderal particle $d(t)$ was taken as the largest distance between the tips of dendrites. Figure 16 shows a graph representing the variation of grain size of icosahedral particle at 280 °C with different intervals of time. From this graph, it is evident that the icosahedral particle in α-Al matrix follows parabolic nature of its growth. Thus the growth of icosahedral particle in α-Al matrix satisfies a relation (Freed and Vander Sande, 1980)

$$d(t) = \alpha\sqrt{Dt}$$

where d = diameter of the particle, D = diffusion coefficient and t = time of growth. The dimensionless parameter α is given by

$$\alpha = \sqrt{2}\left[2\frac{C - C_\beta}{C_{\beta\alpha} - C_{\alpha\beta}}\right]^{1/2}$$

where C = concentration of solute in matrix, $C_{\alpha\beta}$ = concentration at interface in matrix and $C_{\beta\alpha}$ = concentration in particle.

Figure 17 depicts the variation of $d^2(t)$ with time at above mentioned temperatures showing linear variation of $d^2(t)$ with time. With the help of Fig. 17, we have calculated the slope. The Arrhenius plot for $\ln(\alpha^2 D)$ is shown in Fig. 18. From the slope of Fig. 18 we obtain the estimate of the

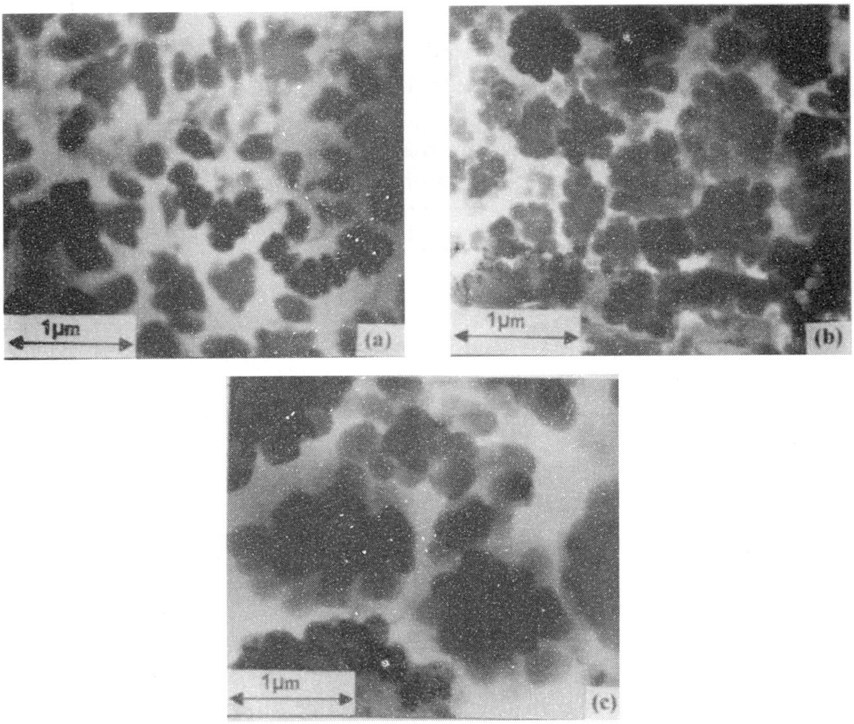

Fig. 14: (a-c) Bright field electron micrograph of annealed $Al_{50}Cu_{45}Ti_5$ alloy at 280 °C for 5, 20 and 40 hrs, respectively showing increase in grain size with time.

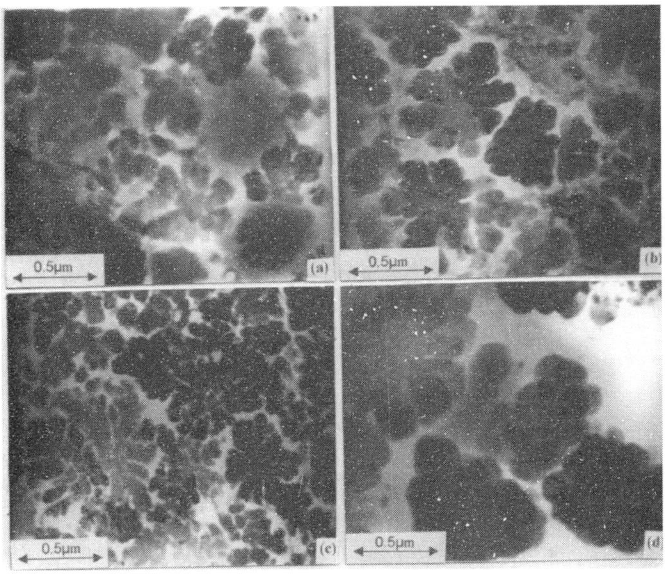

Fig. 15: (a-d) Electron micrographs obtained from the sample of $Al_{50}Cu_{45}Ti_5$ alloy annealed at 310 °C for ½, 1, 3 and 6 hrs, respectively indicating gradual increase in grain size with time.

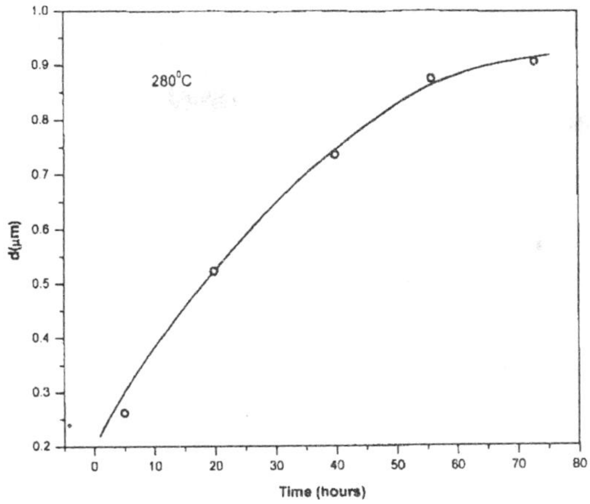

Fig. 16: Variation of grain size of icosahedral particle at 280 °C with different intervals of time indicating parabolic nature of its growth.

activation energy ≈110±20 kJ/mol. Since the atomic compositions of icosahedral phase and α-Al is not exactly known at present, therefore, it is difficult to associate this activation energy with the diffusion of either of Al, Cu or Ti. However, it is interesting to mention that the value of activation energy for diffusion of Cu in Al has been reported to be 121 kJ/mol (Minamino et al., 1985), which is closer to the above mentioned value. Also in comparison to other alloys of Al-Cu-TM the Cu content of $Al_{50}Cu_{45}Ti_5$ is much higher. Thus it may be possible that during growth of icosahedral particle in α-Al, diffusion of Cu from matrix to the interface may be governing this process.

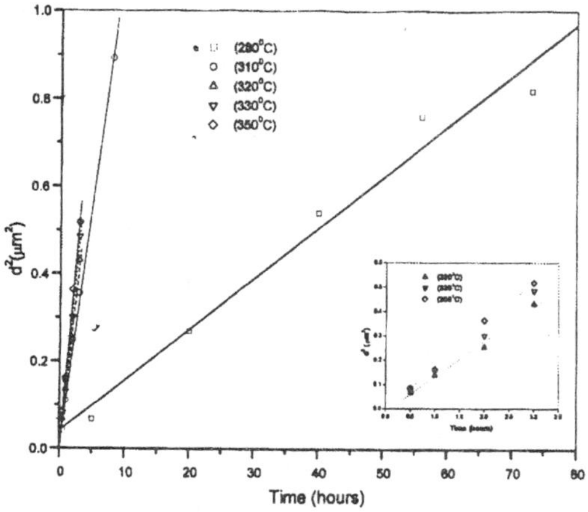

Fig. 17: Variation of maximum grain size $d^2(t)$ vs. time at different temperatures showing linear variation of $d^2(t)$ with time.

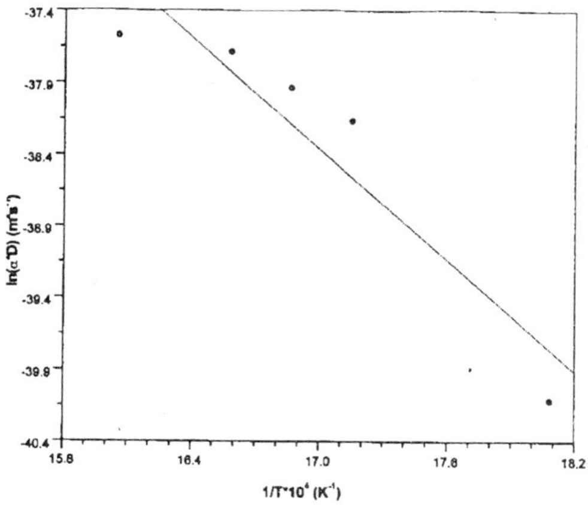

Fig. 18: Arrhenius plot for $\alpha^2 D(T)$ which was obtained from Fig. 17. From the slope of this figure, the activation energy E_g follows and comes out to be ≈110 ± 20 kJ/mole.

For comparison of diffusivity with other systems, if we take the value of α in the range of 2 to 4 as estimated by Köster et al. (1980) for Fe-B alloys, an approximate estimate of diffusion coefficient comes out to be in the range of 2.0×10^{-19} to 3.0×10^{-17} m^2/sec for temperature range 280 to 350 °C. These values are somewhat higher than the value of diffusion coefficients given for Fe-B alloys.

The banded microstructures shown in Fig. 13 have also been observed at all the above mentioned temperatures. Figures 19(a-c) show bright field electron micrograph showing banded structures obtained from the sample annealed at 280, 310 and 320 °C, respectively. This type of banded microstructure evolving during solid state transformation is different from the banded microstructure reported by Chattopadhyay and Mukhopadhyay (1990) in rapidly solidified $Al_{90}Mn_{10}$ quasicrystalline alloy. They have reported the formation of banded structure with bands perpendicular to the growth direction in the rapidly solidified Al-10 at.% Mn alloy and explained the formation of banded microstructure on the basis of interplay of kinetics of solute transport and the nature of free energy curves. In the present case banded structures are appearing at later stage of transformation during the growth of icosahedral grain in α-Al matrix.

A plausible explanation may be given with the help of a model shown in Fig. 20. In the first stage of crystallization amorphous phase is transforming to icosahedral phase and α-Al phase. As we have explained earlier, growth of the icosahedral phase proceeds by the diffusion of solutes from α-Al to icosahedral phase. The dendritic growth of icosahedral grain is radial in nature. The nearly spherical annular region becomes rich in solutes as shown by dense dotted region in Fig. 20, while the region in the vicinity of the icosahedral grain becomes depleted in solute. In the second stage solute rich

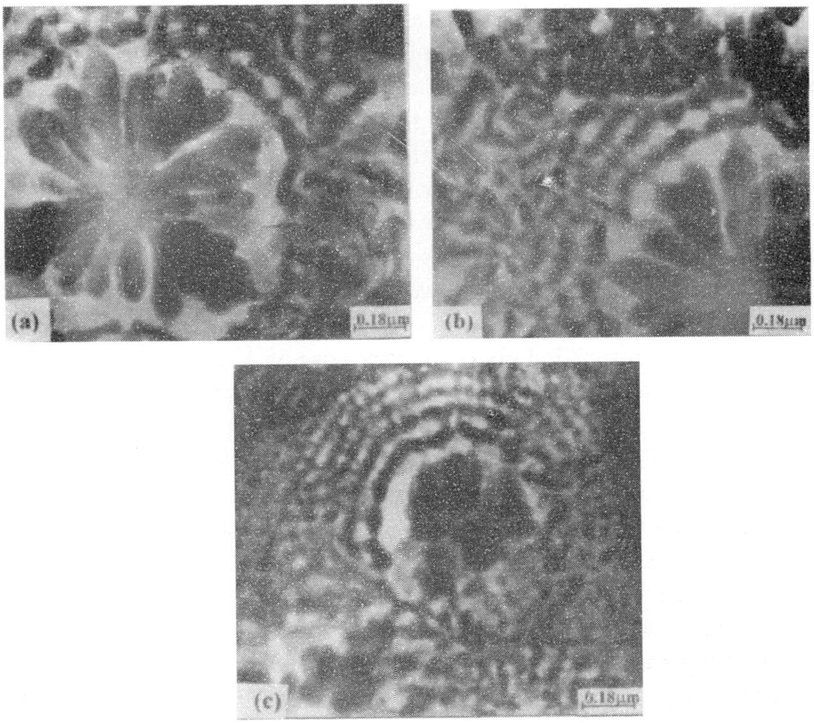

Fig. 19: (a-c) Bright field electron micrograph showing banded structures obtained from the sample annealed at 280, 310 and 320 °C, respectively.

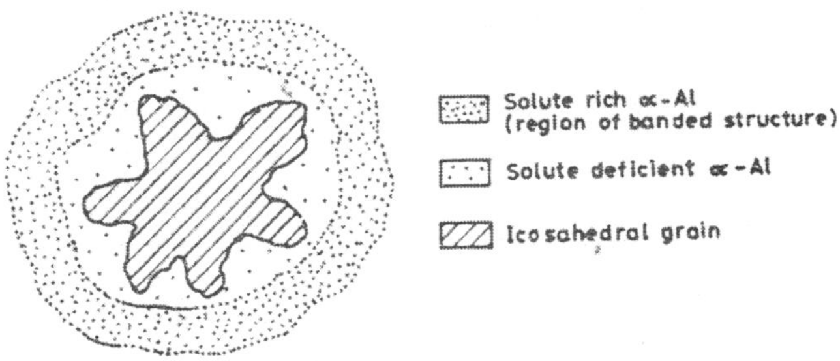

Fig. 20: Schematic diagram showing growth of icosahedral (I) grain in α-Al matrix. The solute rich α′-Al region where cellular precipitation of I and α-Al takes place, has also been indicated.

region which may be denoted as α′-Al would be transforming by cellular precipitation (Manna et al., 2001). This may be expressed as:

$$\alpha' - \text{Al} \rightarrow \alpha - \text{Al} + \text{I (Icosahedral phase)}$$

The circular appearance of banded structure is resulting due to the fact that around the spherullitic icosahedral grain (which becomes dendritic at later stage) the solute rich region (α′-Al) becomes in the form of thick spherical shell. The cellular precipitation of α′-Al →α-Al+I phase is taking place in this region only. When solute rich two spherical shells intersect each other, a complex region of banded structure would also appear. Such regions are marked by arrow in Fig. 13. It is not possible to quantify the band separation at different temperatures due to wide variation in its value. However, in Fig. 19 it appears that band separation becomes coarser at lower annealing temperature. It is worthwhile to mention that as the Ti concentration in Al-Cu-Ti system is increased e.g. in $Al_{45}Cu_{45}Ti_{10}$ alloy, it still exhibits amorphous phase. However, the amorphous phase in this alloy is more stable and it transforms directly to Al_3Ti type phase without transforming to icosahedral phase as in the case of $Al_{50}Cu_{45}Ti_5$ alloy. Figure 21a shows microstructure corresponding to $Al_{45}Cu_{45}Ti_{10}$ alloy annealed at 400 °C for 8 hrs. The nanosized crystals of Al_3Ti type phase can easily be noticed. The SAD pattern corresponding to this region is shown in Fig. 21b. The indexing of powder lines indicates that transformed phase is Al_3Ti type tetragonal phase. It can be seen here that the amorphous phase formed for the compositions $Al_{50}Cu_{45}Ti_5$, $Al_{45}Cu_{45}Ti_{10}$ and $Al_{65}Cu_{25}V_5Ti_5$ alloys is behaving as an electron phase as shown in Table 2. It can easily be discerned that for these two alloys the ratio $k_P/2k_F$ remains nearly equal to unity and consequently amorphous phase gets stabilized. However it is quite intriguing that unlike $Al_{45}Cu_{45}Ti_{10}$ amorphous alloy which directly transforms to Al_3Ti-type phase, $Al_{50}Cu_{45}Ti_5$ amorphous alloy initially transforms to icosahedral phase. At this stage it is not clear why these two alloys which have nearly same *e*/*a* behave differently in regard to amorphous to icosahedral phase transformation. A plausible explanation may be given in terms of hindrance of nucleation of icosahedral phase in $Al_{45}Cu_{45}Ti_{10}$ alloy. Even in $Al_{50}Cu_{45}Ti_5$ alloy, large size of icosahedral grains (Fig. 12a) indicates that nucleation rate is low thus allowing the growth of icosahedral grains before impingment. However, with slight variation in stoichiometry (e.g. in $Al_{45}Cu_{45}Ti_{10}$), the nucleation rate of icosahedral phase might be considerably reduced. This would result in the transformation of amorphous phase in $Al_{45}Cu_{45}Ti_{10}$ alloy to Al_3Ti-type phase. Further, investigations would be required to get insight of the factors which govern the nucleation of icosahedral phase in these amorphous alloys. It should be mentioned that as in the case of Al-Cu-V alloy, Al-Cu-Ti alloy also exhibits amorphous to icosahedral phase transformation. In comparison to Al-Cu-V, however, the morphology of icosahedral phase in the present case is more dendritic in nature.

Present investigation of synthesis and transformation of amorphous phase in Al-Cu-Ti alloy also provides an opportunity to compare the relative stability of amorphous and icosahedral phases in Al-Cu-TM alloy system. In this relation it should be mentioned that the stability of icosahedral phase has been attributed to various factors like average *e*/*a*, size of the atomic species and icosahedral cluster packing scheme (Guo and Tsai, 2002; Pauling and Kamb, 1986). It has been recently pointed out that in case of Cd-Ca and

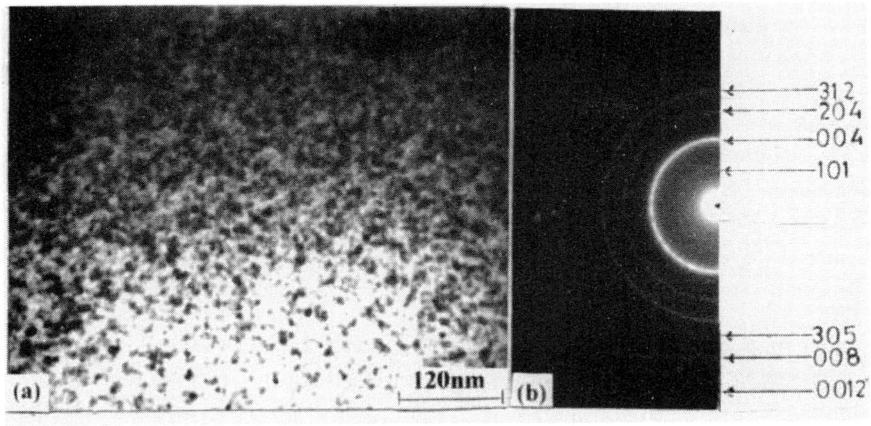

Fig. 21: (a) Nano-sized electron micrograph corresponding to Al_3Ti -type tetragonal phase. (b) Powder ring corresponding to nano-sized grain which has been indexed with Al_3Ti-type tetragonal phase.

Cd-Yb systems, the atomic diameter ratio $d_{Cd}/d_{Ca\text{-}Yb} \approx 0.8$ and average *e/a* of 2.0 favours the formation of stable icosahedral phase (Guo and Tsai, 2002). In the present work nominal composition of the alloys $Al_{65}Cu_{20}Fe_{15}$, $Al_{68.5}Cu_{19}Mn_{12.5}$, $Al_{68}Cu_{17}Cr_{15}$, $Al_{75}Cu_{15}V_{10}$ and $Al_{50}Cu_{45}Ti_5$ have been considered for comparing the relative stability of amorphous and icosahedral phases. We have categorised the atoms of ternary alloy Al-Cu-TM on the basis of their atomic sizes into two categories namely Cu (d_{Cu}~2.56Å) and Al-TM ($d_{A\text{-}Ti}$=2.88Å) to compare the stability of icosahedral phase. Figure 22 shows the variation of quasilattice constant a_R with the atomic diameter ratio $d_{Cu}/d_{Al\text{-}TM}$. The value of quasilattice constant a_R has been taken from reference for Al-Cu-TM systems shown in Table 4.

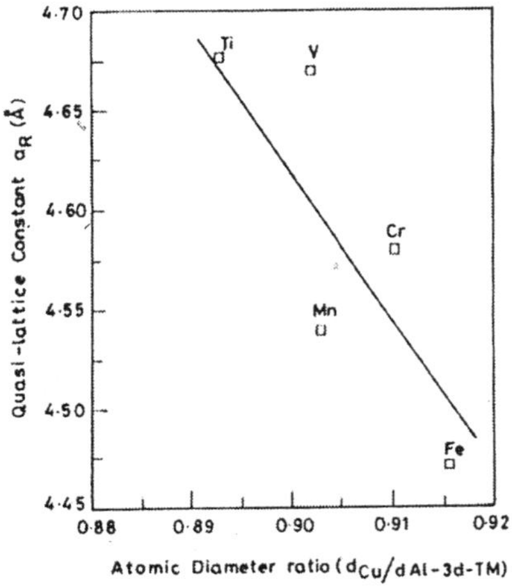

Fig. 22: Relation between quasilattice parameter (a_R) vs atomic diameter ratio $d_{Cu}/d_{Al\text{-}TM}$ for different transition metals in Al-Cu-TM series of alloys showing linearity relation.

Table 4: The alloy composition, quasilattice parameter a_R and corresponding references

Alloy	a_R (Å)	*References*
$Al_{65}Cu_{20}Fe_{15}$	4.47	Popescu et al. (1997)
$Al_{68.5}Cu_{19}Mn_{12.5}$	4.54	Popescu et al. (1997)
$Al_{68}Cu_{17}Cr_{15}$	4.58	Popescu et al. (1997)
$Al_{75}Cu_{15}V_{10}$	4.67	Holzer and Kelton (1991)
$Al_{50}Cu_{45}Ti_{5}$	4.65	Present Work

It may be seen that as the atomic diameter ratio increases from 0.89 to 0.91, the quasilattice constant a_R decreases linearly. In other words as the size of transition metal increases in the order of Fe→Mn→Cr→V→Ti there is more deviation in the ratio of sizes of the two type of atomic species in comparison to ideal value for icosahedral packing i.e. 0.95 (Pauling and Kamb, 1986) and consequently stability of icosahedral phase would also decrease. From the graph (Fig. 22) it is clear that the atomic diameter ratio d_{Cu}/d_{Al-TM} in Al-Cu-Fe icosahedral alloy is around 0.91 which is maximum as compared to that for other transition metals in Al-Cu-TM series of alloys and very close to that for ideal icosahedral packing value (0.95), thereby indicating the most stable system for icosahedral phase in Al-Cu-TM series of alloys. Thus it appears that besides Fermisurface-Brillouin zone interaction leading to lowering of electronic energy, size effect also determines the stability of icosahedral phase.

CONCLUSION

1. Partial substitution of Ti in place of V in $Al_{65}Cu_{25}V_5Ti_5$ alloy has been found to stabilize the amorphous phase. However, this alloy does not exhibit amorphous to icosahedral transformation.
2. The alloy compositions in Al-Cu-Ti system at which the amorphous phase does really occur have been shown to be $Al_{45+x}Cu_{45}Ti_{10-x}$ with $x = 0, 2, 4, 5$ and 6. These compositions have been deduced based on the Hume-Rothery *e/a* criterion.
3. It has been found that amorphous to icosahedral transformation proceeds in two stages in case of $Al_{50}Cu_{45}Ti_5$ alloy. In the first stage of crystallization amorphous phase transforms to icosahedral phase and α-Al phase.
4. The icosahedral phase in the present investigation has been found to grow with time with parabolic mode of growth. Activation energy for the diffusion has been found to be ≈ 110±20 kJ/mol.
5. In the second stage of transformation banded structure surrounding the icosahedral grains has been found. It has been pointed out that these bands are formed in the solute rich region by cellular precipitation.

6. It is also concluded that besides *e*/*a*, the atom size is another dominant factor for the formation and stability of icosahedral phase in Al-Cu-TM series of alloys. In the increasing order of atomic number and decreasing order of quasilattice parameter $a_{R,}$ the stability of Al-Cu-TM series of icosahedral phase increases.

Acknowledgement: The authors are grateful to the Ministry of Non-Conventional Energy Sources and Department of Science and Technology for providing the financial support. Authors are also thankful to Dr. A.M. Awasthi, Inter University Consortium, Indore for useful discussions and providing DSC facilities for the present work. One of the authors (D.K. Misra) gratefully acknowledges the financial support provided by the Council of Scientific and Industrial Research (CSIR).

REFERENCES

Chattopadhyay, K. and Mukhopadhyay, N.K. (1990). *J. of Crystal Growth*, **106:** 387.

Chen, L.C. and Sapepen, F. (1982). *Nature*, **336:** 366.

Dubois, J.M., de Boissieu, M. and Pianelli, A. (1989). *Scripta Metallurgica*, **23:** 1069.

Elser, V. (1985). *Phys. Rev.* B, **328:** 4892.

Freed, R.L. and Vander Sande, J.B. (1980). *Acta Metallurgica*, **28:** 103.

Guo, J.Q. and Tsai, A.P. (2002). *Philosophical Magazine Letters*, **82(6):** 349.

He, L.X., Wu, Y.K. and Kuo, K.H. (1988). *J. Mater. Sci. Letters*, 7: 1284.

He, L.X., Wu, Y.K., Meng, X.M. and Kuo, K.H. (1990). *Phil. Mag. Letters*, **61:** 15.

He, Y., Poon, S.J. and Shiflet, G.J. (1988). *Science*, **241:** 1640.

Holzer, J.C. and Kelton, K.F. (1991). *Acta. Metall.*, **39:** 1833.

Inoue, A., Bizen, Y. and Masumoto, T. (1988). *Metall Trasn.* A., **19:** 383.

Inoue, A., Kimura, H.M., Sasamori, K. and Masumoto, T. (1995). *Mater. Trans.* JIM, **36:** 6.

Inoue, A., Watanabe, M., Kimura, H.M., Takhashi, F., Magata, A. and Masumoto, T. (1992). *Mater. Trans. JIM*, **33:** 723.

Inoue, A., Zhang, T., Saida, J., Matsushita, M., Chen, M.W., Sakurai, T. (1999). *Mater. Trans. JIM*, **40:** 1181.

Inoue, A., Zhang, T., Saida, J., Matsushita, M., Chen, M.W., Sakurai, T. (1999). *Mater. Trans. JIM*, **40:** 1137.

Knapp, J.A. and Follstaedt. D.M. (1985). *Phys. Rev. Lett*, **55:** 1591.

Koster, U., Meinhardt, J., Roos, R. and Liebertz, H. (1996). *Appl. Phys. Lett.*, **69:** 179.

Köster, U., Meinherdt, J., Roos, R. and Liebertz, H. (1999). *Appl. Phys. Lett.*, **74:** 2110.

Koster, U., Herald, U. Hillenbrand, H.G. and Devis, J. (1980). *J. Materials Science*, **15:** 2125.

Koster, U., Meinhardt, J., Roos, R. and Liebertz, H. (1996). *Appl. Phys. Lett.*, **69:** 179.

Levine, L.E., Gibbons, P.C. and Kelton, K.F. (1989). *Phys. Rev. B.*, **40:** 9338.
Li, C., Saida, J., Matsusheta, M. and Inoue, A. (2000). *Phil. Mag. Letts.* **80(9):** 621.
Li, C., Saida, J., Matsusheta, M. and Inoue, A. (2000). *Phil. Mag. Letts.*, **80(9):** 621.
Li, C., Saida, J. and Inoue, A. (2000). *Scripta Metall.*, **42:** 1077.
Lilienfeld, D.A., Nastasi, M., Johanson, H. Ast, D.G and Mayer, J.W. (1985). *Phys. Rev. Letter*, **55:** 1587.
Manna, I., Pabi, S.K. and Guest, W. (2001). *International Materials Reviews*, **46(2):** 53.
Minamino, Y., Yamane, T. and Takahashi, T. (1985). *J. Mater. Sci. Letts.* **46:** 797.
Misra, D.K., Tiwari, R.S. and Srivastava, O.N. (2003). Phys. Stat. Sol (a), 17.
Molokanov, V.V. and Chebotnikov, V.N. (1990). *J. Non-crystalline Solids*, **117-118:** 789.
Murthy, B.S., Naik, M.D., Mohan Rao, M. and Ranganathan, S. (1992). *Materials Forum*, **16:** 19-26.
Murty, B.S., Kim, W.T., Kim, D.H. and Hono, K. (2001). *Materials Transactions*, **42(2):** 372-375.
Murty, B.S., Ping, D.H. and Hono, K. (2000). *Appl. Phys. Letts.*, **77(8):** 1102.
Pauling, L. and Kamb, B. (1986). *Proc. Nat. Acad. Sci., USA*, **83:** 3569.
Poon, S.J., Drehman, A.J. and Lauless, K.R. (1985). *Phys. Rev.Lett.*, **55:** 2324.
Popescu, R., Jianu, A., Manciu, M., Nicula, R. and Manaila, R. (1995). *J. of Alloys Comp.*, **221:** 240C.
Popescu, R., Macovei, D., Manciu, M., Zavaliche, F., Fratiloiu, D., Jianu, A., Devenyi, A., Manila, R., Xie, Y., Hu, T., Orton, B.R., Cernik, R.J. and Tang, G.C. (1997). *J. Phys : Condens. Matter*, **9:** 7523.
Raynor, G.V. (1949). *Progr. Metal Phys*, **1**.
Saida, J. and Inoue, A. (2000). *J. Non-Cryst. Solids*, **317:** 97-105.
Saida, J., Matsushita, M., Li, C. and Inoue, A. (2000). *Appl. Phys. Lett.*, **76:** 3558.
Shechtman, D., Blech, I., Gratias, D., Cahn, J.W. (1984). *Phys. Rev. Lett.*, **53:** 1951.
Shen, Y., Poon, S.J. and Shiflet, G.J. (1986). *Phys. Rev. B.*, **34:** 3516.
Tsai, A.P. (1996). New Horizons in Quasicrystals (Research and Applications, Eds. Goldman, I.A., Danial, J., Sordelet, A., Patrica, A. Thiel and Dubois, J.M. *World Scientific*, **17:** 253.
Tsai, A.P. (1996). New Horizons in Quasi-crystals (Research and Application). *World Scientific*, **17:** 253, ed. Alan Goldman, I., Danial, J., Sordelet, Patrika Thiel, A. and Dubois J.M.
Tsai, A.P., Hiraga, K., Inoue, A., Masumoto, T. and Chen, H.S. (1994). *Physical Review B*, **49(5):** 3569.
Tsai, A.P., Inoue, A. and Masumoto, T. (1987). *Jpn. J. Appl. Phys.*, **27:** L1587.
Tsai, A.P., Inoue, A. and Masumoto, T. (1988). *Jpn. J. Appl. Phys.*, **26:** L1505.
Tsai, A.P., Inoue, A. and Masumoto, T. (1989). *Mater. Trans.*, **30:** 3000.
Tsai, A.P., Inoue, A., Bizen, Y. and Masumoto, T. (1989). *Acta. Metall.* **37:** 1443.
Tsai, P., Inoue, A. and Masumoto, T. (1988). *J.Mater. Sci. Lett.*, **7:** 322.
Upadhya, R. and Tiwari, R.S. (1996). *Phys. Stat. Sol.* (a), **39:** 156.
Watanabe, M., Inoue, A., Kimura, H.M., Aiba, T. and Masumoto, T. (1993). *Mater. Trans. JIM*, **34:** 162.
Xing, L.Q., Eckert, J., Loser, W. and Schultz, L. (1999). *Appl. Phys. Lett.*, **74:** 664.
Zhang, Wu, Y., Cai, X., Zhou, G., Shi, Y., Yang, H. and Wu, S. (2001). *J. Phys : Condens. Matter.*, **13:** L647.

Indexing of Quasicrystalline Diffraction Patterns

Rajiv Kumar Mandal

Centre of Advanced Study, Department of Metallurgical Engineering
Institute of Technology, Banaras Hindu University, Varanasi-221 005, India

INTRODUCTION

Solids with long range translational order give rise to discrete diffractogram. Periodic and quasiperiodic translational orders are displayed by underlying atomic arrangements. A crystalline solid has periodic translational order in all the three directions. The incommensurate phase lacks translational periodicity in either one of the three directions or more than one direction. It gives rise to diffraction patterns that are invariant under one of the thirty two crystallographic point groups. In contrast, quasicrystals possess non-crystallographic symmetry. Icosahedral quasicrystals display three dimensional quasiperiodic translational order and their diffraction patterns are invariant under icosahedral point group. Similarly, decagonal phases have quasiperiodic translational order in two dimensional plane and their diffraction patterns are invariant under 10/mmm or $\bar{5}$ m symmetries. Please note that (m$\bar{3}\bar{5}$), (10/mmm), ($\bar{5}$ m) point group symmetries contain a rotation axis that is incompatible with periodic lattice translations. Hence, these are termed as non-crystallographic symmetries. The relevant concepts involved in the area of quasicrystals are given elsewhere (Janot, 1994). The purpose of this article is to make self contained and all the necessary details required to index powder diffraction or single crystal diffraction patterns. We shall begin this presentation by introducing terms that are important for the subsequent deliberations of this article. This will be followed by expression of reciprocal lattice vectors and one example of indexing each for the icosahedral and decagonal phases.

RUDIMENTS OF INDEXING

Any translationally ordered solid has sharp diffraction patterns. Diffraction peaks or spots in single crystal diffraction experiment can be assigned indices. Such an assignment requires proper definition of basis vectors in reciprocal space. Any reciprocal lattice vector corresponding to a diffraction spot/peak can be written as an integral linear combination of the basis vectors. Number of integrally independent basis vectors, required for a given solid, is known as the rank of solid. All three dimensionally periodic solids have rank equal to three. In general, the reciprocal lattice vector (**G**) corresponding to a 3d crystal can be written as

$$\mathbf{G} = h\mathbf{a}_1 + k\mathbf{a}_2 + l\mathbf{a}_3 \tag{1}$$

where (hkl) are indices of reflection and $\mathbf{a}_1$, $\mathbf{a}_2$, $\mathbf{a}_3$ are integrally independent basis vectors. Thus, 3d crystals possess a rank of three. The magnitude of **G**(hkl) is given by

$$|\mathbf{G}(hkl)| = [h^2a_1^2 + k^2a_2^2 + l^2a_3^2 + 2hk\mathbf{a}_1 \cdot \mathbf{a}_2 + 2hl\mathbf{a}_1 \cdot \mathbf{a}_3 + 2kl\mathbf{a}_2 \cdot \mathbf{a}_3]^{1/2} \tag{2}$$

The direct space distance corresponding to this plane

$$|d(hkl)| = \frac{2\pi}{|\mathbf{G}(hkl)|} \tag{3}$$

The left hand side of equation (2) corresponding to a given spot is known in $Å^{-1}$ experimentally. However, assignment of (hkl) to this spot requires information about magnitudes and the relative orientations of the basis vectors. In crystallographic literature, these are known as lattice parameters. They are related to the unit cell parameters a, b, c and α, β, γ corresponding to the conventional cell in three dimensional crystallography. For a cubic case,

$|a_1| = |a_2| = |a_3| = \frac{1}{a}$; $\mathbf{a}_i \cdot \mathbf{a}_j = \delta_{ij}$ and equation (2) takes the form of

$$|\mathbf{G}(hkl)| = [h^2 + k^2 + l^2]^{1/2}\left(\frac{1}{a}\right)$$

and

$$|d(hkl)| = \frac{2\pi a}{\sqrt{h^2 + k^2 + l^2}}.$$

However, as we have seen, solution of equation (2) for unknown crystal system is a non-trivial exercise and association of (hkl) indices to a diffraction spot need to be done judiciously. We refer the readers to some of the classical texts for this and we shall not discuss them further. Our primary goal in this section is to understand the modification of equation (1), when solids possess quasiperiodic translational orders. Let us define this term. We say a set of points on a line is quasiperiodically arranged if the one dimensional lattice vector corresponding to any point on the line is expressible by

$$X_n = n\,a + \frac{a}{\tau}\left\lfloor\frac{n}{\tau}\right\rfloor \tag{4}$$

where n is an integer; $\tau = \dfrac{\sqrt{5}+1}{2}$ = Golden mean; $\lfloor x \rfloor = x -$ fractional part of x and is read as floor function of x or minimum integer of x and a = length scale in Å. Please note that na and $\dfrac{a}{\tau}\left\lfloor\dfrac{n}{\tau}\right\rfloor$ are periodic independently but X_n does not generate a set of vectors that are spaced periodically. However, existence of equation (4) ensures translational order. Such an order is not periodic and is known as quasiperiodic translational order.

As mentioned above, $\left\lfloor\dfrac{n}{\tau}\right\rfloor$ changes by integer only and let us replace this term by another integer 'm'. Hence, equation (4) can be rewritten as

$$X_{n,m} = a\left[n + \frac{m}{\tau}\right] \tag{5a}$$

$$= n\mathbf{v}_1 + m\mathbf{v}_2 \tag{5b}$$

where $|V_1| = a$ and $|V_2| = a/\tau$ and $V_1 \parallel V_2$. Thus, equations (5b) and (5a) are equivalent.

In an analogous way, we may write the expression of reciprocal lattice vector (G) for one dimensional quasiperiodic case by

$$G_{p,q} = a^*[pV_1 + qV_2] \tag{6}$$

where p, q are integers; V_1, V_2 are integrally independent basis vectors with different magnitudes but are parallel to each other as in equation (5b) and a^* is a constant parameter in $Å^{-1}$. Thus, one dimensional quasiperiodic solid has rank equal to two. The ratio of magnitudes of V_1 and V_2 can be set by looking at the diffraction pattern. The first two strong spots or peaks shall permit us to choose this. For a three dimensional quasiperiodic case, we need to choose six-integrally independent basis vectors. Unlike the notion of conventional unit cell in 3d periodic crystal, we have to rely on the symmetry information available in the single crystal diffraction patterns for defining the basis vectors whose integral linear combination will map the entire reciprocal space. For example, the six basis vectors for the icosahedral case can be taken parallel to the vertex vectors of an icosahedron (Fig. 1).

The icosahedral solid has twelve 5-, twenty 3- and thirty 2-fold vectors. Having defined six vertex vectors, all other vectors can be known by their integral linear combinations. We shall explain this by expressing vertex vectors in terms of Cartesian components

$$\begin{bmatrix} q_1 \\ q_2 \\ q_3 \\ q_4 \\ q_5 \\ q_6 \end{bmatrix} = \begin{pmatrix} 1 & \tau & 0 \\ \tau & 0 & 1 \\ \tau & 0 & \bar{1} \\ 0 & 1 & \bar{\tau} \\ \bar{1} & \tau & 0 \\ 0 & 1 & \tau \end{pmatrix} \begin{pmatrix} x \\ y \\ z \end{pmatrix} \tag{7}$$

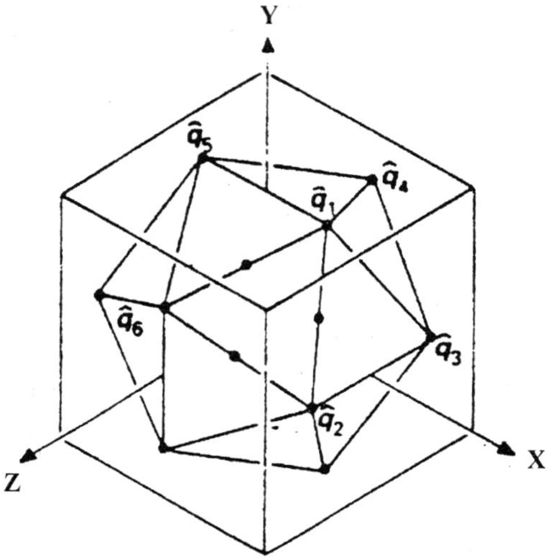

Fig. 1: Icosahedral vertex vectors q_1, q_2, q_3, q_4, q_5 and q_6 along with the Cartesian bases.

Any experimentally observed diffraction spots can be indexed by six indices (n_1 n_2 n_3 n_4 n_5 n_6) and the corresponding reciprocal lattice vector will be expressed by

$$G_{\text{exp.}} = Q_I^* \sum_{i=1}^{6} n_i\, q_i \tag{8}$$

$$= Q_I^* \,[G_x\, x + G_y\, y + G_z\, z] \tag{9}$$

with

$$G_x = (n_1 - n_5) + \tau(n_2 + n_3)$$

$$G_y = \tau(n_1 + n_5) + (n_4 + n_5)$$

$$G_z = (n_2 - n_3) + \tau(n_5 - n_4)$$

Q_I^* = Reciprocal lattice constant for the icosahedral phase.

We observe that G_x, G_y and G_z are essentially a combination of indices of the type $m + n\tau$ where m and n are integers. Owing to this, G_x, G_y and G_z generate dense set of points. However, we get discrete set of points experimentally due to the large variations in intensity corresponding to specific set of points. It has been found that those peaks for which intensity is large, the ratio of $m/n \rightarrow \tau$. The vertex vectors that are parallel to five fold axis are given by (100000) and its permutation. (211111) is a vector parallel to (100000) but has magnitude τ^3 times more than this. We may, therefore, choose any one of these vectors as the first vector along 5-fold axis. All others can be known by inflation and deflation methods. These are needed for indexing of quasicrystals and we do not encounter this while indexing diffraction peaks from 3d periodic crystals. We may note here that quasicrystals, in general, have rank > 3 unlike crystalline solids which have rank = 3.

INDEXING OF DIFFRACTION PATTERNS FROM ICOSAHEDRAL QUASICRYSTALS

We should remember that first identification of IQC has been done based on the studies of selected area electron diffraction patterns. The stereogram that has been arrived at, based on this, is shown in Fig. 2. This is like a single crystal diffraction experiment where various zones permit us to construct the entire reciprocal space. The relative orientations of five, two and three folds can also be known. Icosahedral phase in rapidly quenched Al-Mn alloy normally shows a strong peak around a reciprocal distance of ~ 2.9Å^{-1}. This corresponds to the interplanar spacing ~ 2Å in direct space. A representative X-ray diffractogram with composition $Mg_{32}(Ag,Al)_{49}$ is shown in Fig. 3. As mentioned in previous section, the diffraction peak at a reciprocal distance ~ 2.9 Å^{-1} (or interplanar spacing of 2Å) refers to a reciprocal vector whose indices are (211111) or their permutations. Substitution of above in equation (9) permits us to evaluate reciprocal lattice constant Q^*_I. Once we know this, we may compute **G** for all other indices (n_1 n_2 n_3 n_4 n_5 n_6) in Å^{-1}. We match the experimentally observed ones with our computed indices. We have indexed the X-ray diffraction (XRD) pattern of the icosahedral phase found in $Mg_{32}(Ag\text{-}Al)_{49}$ composition by adopting this procedure (Fig. 3).

For icosahedral phase, we need to identify only one parameter experimentally. In this respect, indexing of IQC is analogous to 3d cubic crystal where we need to fix up the lattice parameter 'a*' experimentally. Assignment of indices to other peaks, then follows. All the initially known IQC in various alloy systems were belonging to primitive icosahedral (PI) phase and reflections were admitting indices without any parity conditions. In contrast, many of the stable IQCs have been found to belong to face centered icosahedral (FCI) phase. This could be distinguished from PI type by noting the parity conditions being followed by indices (n_1 n_2 n_3 n_4 n_5 n_6). They refer to a set that is either all odd or all even. We may recall that such a parity condition on (hkl) indices is encountered for face centered cubic crystal in three dimensions.

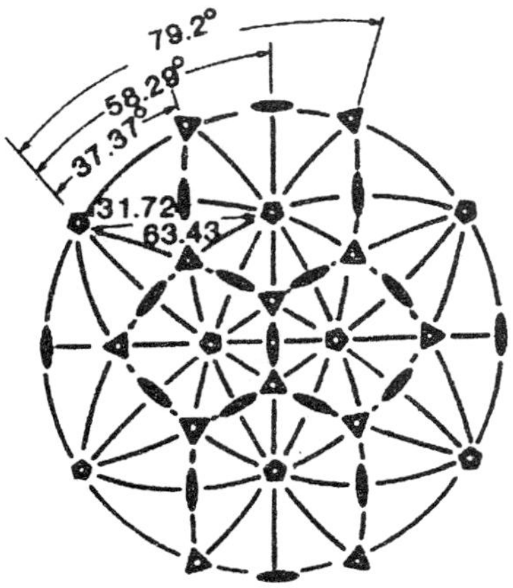

Fig. 2: Stereographic projection showing icosahedrally related 5-, 3- and 2-fold vectors.

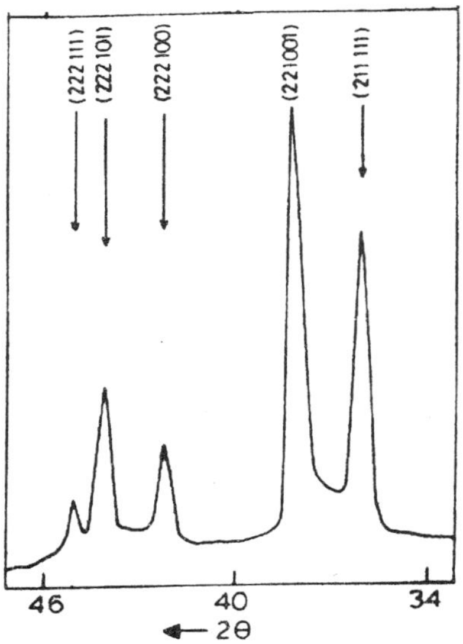

Fig. 3: Trace of XRD pattern from $Mg_{32}(Al\text{-}Ag)_{49}$ icosahedral phase.

INDEXING OF DECAGONAL PHASE

The selected area diffraction patterns from single grain of a decagonal phase give rise to a quasiperiodically arranged spots in a plane and this reciprocal section is stacked periodically along the 10-fold axis. The reciprocal space, in this respect, is quite distinct from that of icosahedral phase described above. It is, therefore, important to select new set of basis vectors for decagonal phase to conform to the above demand. Please note that the rank of decagonal phase is '5' and five basis vectors will do. However, we would like to use six index notations for this. Such a choice introduces problem of redundancy and may be surmounted by adopting procedures akin to those encountered in Miller-Bravais indexing scheme of Hexagonal crystal. In Miller-Bravais indexing scheme, we choose four index notation so that vectors related by a symmetric operation can be expressed by the permutation of indices. The choice of quadruplet of integers (hkil) is made unique by invoking h + k + i = 0. In a parallel way, for decagonal phase too, we enforce a condition on indices so that the sextuplets of indices for a given peak is unique. We shall illustrate it further now.

We refer to the model of Mandal and Lele (1989) for this purpose. The physical reciprocal space vector **G** in their model is expressed as

$$\mathbf{G} = G_x x + G_y y + G_z z \tag{10}$$

where

$$G_x = Q_D^*[(2n_1 - n_2 - n_5) + \tau(n_2 + n_5 - n_3 - n_4)] \tag{10a}$$

$$G_y = Q_D^* \sin\frac{\pi}{5}[(n_3 - n_4) + \tau(n_2 - n_5)] \tag{10b}$$

$$G_z = Q_D^* \left(\frac{1}{\sqrt{3}}\right)[n_5 + 2n_6] \tag{10c}$$

$$N_5 = n_1 + n_2 + n_3 + n_4 + n_5$$

$$Q_D^* = \text{Constant for decagonal phase}$$

Here x, y, z are Cartesian bases. The z-axis is chosen parallel to the 10-fold axis and x, y axes are chosen parallel to two 2-fold axes known as P and D directions, respectively in quasicrystalline literature (see Ranganathan et al., 1997). It is obvious from equations (10a)-(10c) that $\mathbf{G}(2222\bar{5}) = 0$. This means that addition and subtraction of multiples of this vector will lead to different indices of the same diffracted spot. The difference of sum of indices $N_6 = n_1 + n_2 + n_3 + n_4 + n_5 + n_6$ will be given by a multiple of 5. Thus, if we evolve a criterion, that all sextuplets chosen for indexing of Decagonal phase be such that $N_6 = N_6$ modulo 5. Then, $(n_1\ n_2\ n_3\ n_4\ n_5\ n_6)$ will remain unique and their summation will admit values –2, –1, 0, 1, 2 only. Please note that this feature is inbuilt in our model and it has also been shown by us that all the equivalent indices corresponding to a reciprocal vector give rise to same diffracted intensity. However, the purpose of this paper is different and we shall not discuss them further.

INDEXING OF ELECTRON DIFFRACTION PATTERNS OF DECAGONAL PHASE

Figures 4(a-c) display D, P, 10-f diffraction patterns from T_4 phase respectively from our earlier work on electron microscopic studies of rapidly solidified Al-Co-Cu-Ni alloys (Prasad et al., 1997).

The tracings for these diffraction patterns are respectively shown in Figs. 5(a-c). Decagonal phases have 2, 4, 6 and 8 layer periodicities along the 10-fold axis and accordingly, they are designated as T_{2n} (for $n = 1, 2, 3, 4$). T_{2n} phase has been shown to display a periodicity of ~ $4n$ Å.

Positions of a selected set of diffracted spots are given by alpha-numeric designation. Indexing of these spots from D, P and 10-f patterns are given in Tables 1 to 3. All other spots can be found by vector summation. For the sake of comparison, Tables 4 and 5 give indices of important reflections at the same reciprocal distances for T_2, T_4, T_6 and T_8 phases, respectively. The indices of important spots in the above for these phases are different as the periodicities along 10-fold axis are different. Diffracted spots at same reciprocal distances for various DQCs have to be assigned different indices to avoid non-integral entry in the sextuplets of integers.

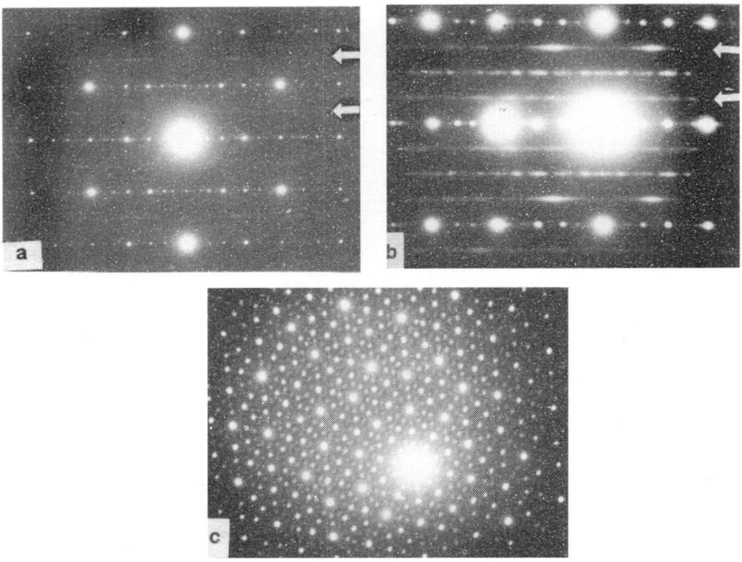

Fig. 4: Typical selected area diffraction patterns (SADP) in Decagonal phase.

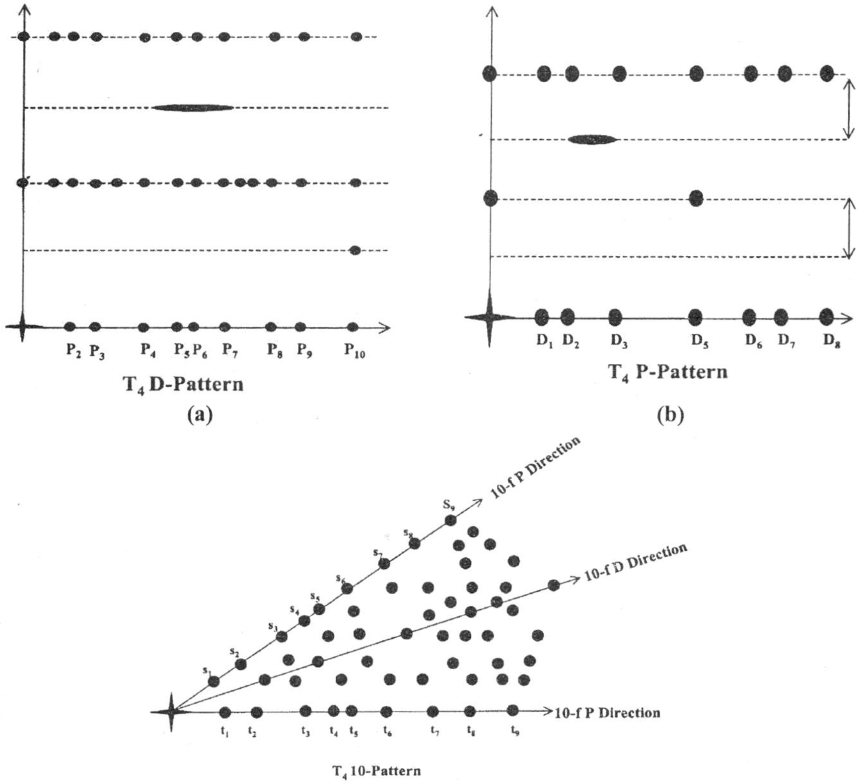

Fig. 5: Trace corresponding to Fig. 4.

Table 1: Indices of spots present in a D-pattern
(zeroeth row of T_4 phase, Figs. 4a and 4a)

Sl. No.	*Spot No.*	*Index*	N_5	N_6	N_h
1	P_1	$(6\bar{4}22\bar{4}\bar{1})$	2	1	0
2	P_2	$(\bar{4}2\bar{2}\bar{2}22)$	–4	–2	0
3	P_3	$(2\bar{2}00\bar{2}1)$	–2	–1	0
4	P_4	$(\bar{6}6\bar{2}\bar{2}6\bar{1})$	2	1	0
5	P_5	$(\bar{2}4\bar{1}03\bar{2})$	4	2	0
6	P_6	$(0200\bar{2}\bar{2})$	4	2	0
7	P_7	$(4\bar{4}00\bar{4}2)$	–4	–2	0
8	P_8	$(3\bar{1}10\bar{1}\bar{1})$	2	1	0
9	P_9	$(20000\bar{1})$	2	1	0

Table 2: Indices of spots present in a P-pattern
(zeroeth row of T_4 phase, Figs. 4b and 5b)

Sl. No.	*Spot No.*	*Index*	N_5	N_6	N_h
1	D_1	$(0\bar{6}10\,\overline{10}\,60)$	0	0	0
2	D_2	$(04\bar{6}6\bar{4}0)$	0	0	0
3	D_3	$(0\bar{2}4\bar{4}20)$	0	0	0
4	D_4	$(0\,8\,\overline{12}\,12\,\bar{8}\,0)$	0	0	0
5	D_5	$(02\bar{2}2\bar{2}0)$	0	0	0
6	D_6	$(0\bar{4}8\bar{8}40)$	0	0	0
7	D_7	$(0100\bar{1}0)$	0	0	0
8	D_8	$(002\bar{2}00)$	0	0	0

Table 3 : Indices of different spots present along two P direction (Considered as 's' and 't' directions) of a 10-fold pattern $Al_{67}Co_{15}Cu_8Ni_{10}$ (T_4 Phase, Figs. 4c and 5c)

Along t-direction					*Along s-direction*				
Sl. No.	*Spot No.*	*Index*	N_5	N_6	*Sl. No.*	*Spot No.*	*Index*	N_5	N_6
1	t_1	$(6\bar{4}22\bar{4}\bar{1})$	2	1	1	s_1	$(\bar{2}\bar{2}4\bar{6}41)$	–2	–1
2	t_2	$(\bar{4}2\bar{2}\bar{2}22)$	–4	–2	2	s_2	$(22\bar{2}4\bar{2}\bar{2})$	4	2
3	t_3	$(2\bar{2}00\bar{2}1)$	–2	–1	3	s_3	$(002\bar{2}2\bar{1})$	2	1
4	t_4	$(\bar{6}6\bar{2}\bar{2}6\bar{1})$	2	1	4	s_4	$(22\bar{6}6\bar{6}1)$	–2	–1
6	t_6	$(02002\bar{2})$	4	2	6	s_6	$(00\bar{2}0\bar{2}2)$	–4	–2
7	t_7	$(4\bar{4}00\bar{4}2)$	4	–2	7	s_7	$(004\bar{4}4\bar{2})$	4	2
8	t_8	$(\bar{4}4\bar{2}\bar{2}40)$	0	0	8	s_8	$(22\bar{4}4\bar{4}0)$	0	0
9	t_9	$(20000\bar{1})$	2	1	9	s_9	$(000\bar{2}01)$	–2	–1

Table 4: Indices of zeroeth row in P-pattern for all the decagonal phases (T_2, T_4, T_6 and T_8)

Spot designation	*Indices*			
	T_2	T_4	T_6	T_8
D_3	$0\bar{1}\bar{2}\bar{2}10$	$0\bar{2}4\bar{4}20$	$0\bar{3}6\bar{6}30$	$0\bar{4}8\bar{8}40$
D_5	$01\bar{1}1\bar{1}0$	$02\bar{2}2\bar{2}0$	$03\bar{3}3\bar{3}0$	$04\bar{4}4\bar{4}0$
D_8	$001\bar{1}00$	$002\bar{2}00$	$003\bar{3}00$	$004\bar{4}00$

Table 5: Indices of zeroeth row in D-pattern for all the decagonal phases (T_2, T_4, T_6 and T_8)

Spot designation	*Indices*			
	T_2	T_4	T_6	T_8
P_3	$2\,0\,0\,1\,1\,0\,\bar{2}$	$2\,\bar{2}\,0\,0\,\bar{2}\,1$	$4\,\bar{2}\,1\,1\,\bar{2}\,1$	$4\,\bar{4}\,0\,0\,\bar{4}\,2$
P_6	$0\,1\,0\,0\,1\,\bar{1}$	$0\,2\,0\,0\,2\,\bar{2}$	$\bar{2}\,1\,\bar{2}\,\bar{2}\,1\,2$	$\bar{2}\,2\,\bar{2}\,\bar{2}\,2\,1$
P_9	$0\,\bar{1}\,\bar{1}\,\bar{1}\,\bar{1}\,2$	$2\,0\,0\,0\,0\,\bar{1}$	$2\,\bar{1}\,\bar{1}\,\bar{1}\,\bar{1}\,1$	$4\,0\,0\,0\,0\,\bar{2}$

The N_6 and N_h values for each set of indices are also given in Tables 1 to 3. They are respectively defined as

$$N_6 = \sum_{1}^{6} n_i \tag{11a}$$

and

$$N_h = N_6 + n_6 \tag{11b}$$

N_h in the table represents the height of a layer of reciprocal lattice points in the periodic direction. N_6 denotes the unique value of the sum of indices in the present indexing scheme as mentioned in the preceding section and necessarily lies between –2 and +2. Decagonal phases having ~$4n$ Å (n = 1, 2, 3, 4) periodicity along ten-fold axis display $2n$ equally spaced rows in P and D patterns. Thus, value of N_h is fixed within a reciprocal period for a given T_{2n} phase. Further, the fact that N_6 has to lie between –2 to +2 in conjunction with the former assertion yield choices that are apriori fixed. We shall demonstrate in the next section that based on the above scheme of Mandal et al. (2004), we will be able to index the powder XRD pattern of the decagonal phase.

INDEXING OF XRD PATTERN FROM DECAGONAL PHASE

Figure 6 shows the diffraction pattern from a decagonal phase formed in $Al_{67}Co_{15}Cu_{13}Ni_5$ alloy. This has a periodicity of ~8Å along the 10-fold axis and corresponds to T_4 type decagonal phase. As mentioned earlier, while indexing icosahedral phase, there is a strong spot corresponding to the *d*-spacing ~2Å and the vector conforming to such a spot is parallel to five-fold vector of an icosahedron. For decagonal phase too, we have a strong spot corresponding to the *d*-spacing of ~2Å. In the model of Mandal and Lele (1989), they have considered their basis vectors to be parallel to the vertex vectors of an icosahedron that has been given distortion along one of the five-fold axes of the icosahedron. This distorted icosahedron has a unique 5-fold axis parallel to the axis of distortion and all other five 5-fold axes are oriented at an angle θ = 60° unlike the perfect icosahedron (Fig. 2) with θ = 63.43°. Please note that $\cos 60° = \frac{1}{2}$ and $\cos\theta = \frac{1}{\sqrt{5}}$. Latter is

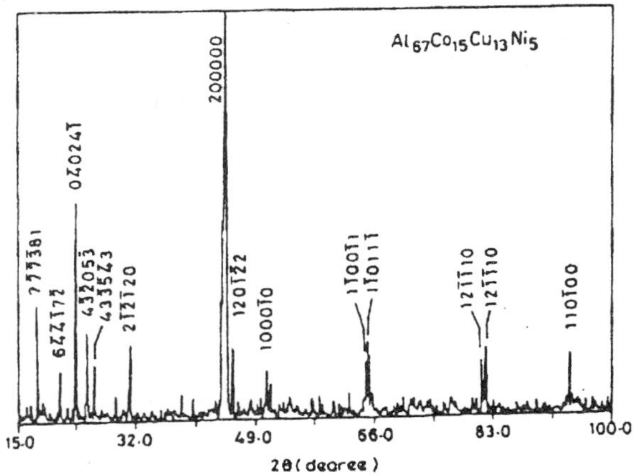

Fig. 6: XRD pattern of a decagonal phase.

irrational whereas former is a rational number. Thus, the projection of all the five vertex vectors on the unique five-fold axis in the model of Mandal and Lele (1989) divides the length by a rational fraction. This leads to periodicity along the unique five-fold axis that in the sense of diffraction may behave as ten-fold. This is owing to Friedel's law.

In view of the aforesaid discussion, we associate indices (n00000) or their permutations to strong spot corresponding to d-spacing of ~2Å for T_{2n} phase. Figure 6 is an XRD pattern from T_4 phase and (200000) has been associated with the above spot. This is a powder diffraction pattern and same spot has a multiplicity of 12. The (000002) reflection also has the same d-spacing in real space. **G** corresponding to such a spot has only z-component (cf. Eqs. 10(a)-10(c)) and we determine Q_D^*. Once we know Q_D^*, we may compute |G| for all other indices. The computed d values can then be matched with experimental ones.

DISCUSSION ON INDEXING OF APERIODIC CRYSTALS

We have presented a very simplified picture of the models for icosahedral and decagonal phases. They are sufficient for indexing but they do not hold any promise to give the readers any hint about the real space structures. The complexity of problem in commenting about the atomic arrangements of aperiodic solid arises owing to lack of periodic translational order. Suppose, we have 5000Å size specimens (single grain) of periodic and aperiodic crystals and we wish to compute the intensity in both the cases. For the former, the Fourier Transform (FT) of unit cell multiplied by the number of unit cells contained in 5000Å single grain crystal will permit us conclude about the nature of intensities for a corresponding reciprocal lattice vector **G**(hkl), where (hkl) are indices of reflection. Essentially we need Wyckoff positions of atoms in the unit cell.

But for the aperiodic case, we do not have a notion of repeat unit and atomic coordinates of all the atoms contained in 5000 Å size grain have to be known for calculating intensity. This is, where, in aperiodic literature we encounter a different domain of crystallography known as Higher Dimensional Crystallography (Janot, 1994). We shall not elaborate this point further in this presentation.

Icosahedral quasicrystals discovered so far under various experimental conditions fall in two categories mentioned earlier i.e. PI and FCI. Decagonal phases, apart from displaying variation in periodicities along periodic direction are known to possess 10_5 screw, true 10-fold and five-fold axes along the unique symmetric axis. The model of Mandal and Lele (1989) for $\theta = 60^o$ shall work for 10_5 and five-fold cases but we need to choose basis set of vectors corresponding to $\theta = 90^o$ for decagonal phases having true 10-fold axis along the symmetric direction. Apart from our work, we refer the readers to the methods due to Choy et al. (1988), Fitz Gerald et al. (1988), Singh and Ranganathan (1996a, 1996b) for knowing various technical details of indexing as well as assigning zone axes for selected area diffraction experiments.

CONCLUSIONS

We have presented in this communication all the relevant details of indexing of quasicrystals with particular reference to icosahedral and decagonal phases. We have demonstrated the method of indexing of these phases by taking examples. The essential difference between periodic and aperiodic crystals has been dealt with succinctly.

Acknowledgements: I thank Dr. N.K. Mukhopadhyay (BHU) and Dr. A.K. Pramanick (NML, Jamshedpur) for providing me with X-ray diffraction patterns (Figs. 3 and 6, respectively). I express my deep sense of gratitude to my teacher Professor Shrikant Lele for teaching me fundamentals of crystallography. I could complete this article after missing lot of deadlines given by the editor. I apologise for the delay in the submission of this manuscript. I thank Dr. A.K. Singh (DMRL) for motivating me to write this paper.

REFERENCES

Choy, T.C., FitzGerald, J.D. and Kalloniatis, A.C. (1988). *Philos. Mag.*, **B58:** 35.

Fitz Gerald, J.D., Withers, R.L., Stewart and A.C. Kalloniatis (1988). *Phil. Mag.*, **B58:** 15.

Janot, C. (1994). Quasicrystals—A Primer. Clarendon Press, Oxford.

Mandal, R.K and Lele, S. (1989). *Phys. Rev. Letts.*, **62:** 2695.

Mandal, R.K., Pramanick, A.K., Sastry, G.V.S. and Lele, S. (2004). *Mat. Res. Soc. Sym. Proc.*, **805:** LL1.6.1-11.

Prasad, S.N., Mandal, R.K., Sastry, G.V.S. and Lele, S. (1997). *Prog. Cryst. Growth and Charact.*, **34:** 1751.

Ranganathan, S., Chattopadhyay, K., Singh Alok and Kelton, F. (1997). *Prog. Mater. Sci.*, **41:** 195.

Singh, A. and Ranganathan, S. (1996a). *Phil. Mag.*, **A74:** 821.

Singh, A. and Ranganathan, S. (1996b). *Phil. Mag.*, **A74:** 841.

Milling of Oxide Blends—A Possible Clue to the Mechanism of Mechanical Alloying

G.V.S. Sastry*, S.Sudhir, Ankur Gupta and B.S. Murty[1]

Centre of Advanced Study, Department of Metallurgical Engineering
Institute of Technology, Banaras Hindu University, Varanasi-221005, India
[1]Department of Metallurgical and Materials Engineering
Indian Institute of Technology Madras, Chennai-600036, India

INTRODUCTION

Intense activity of research over twenty five years in the field of mechanical alloying and mechanical milling (MA and MM) has led to considerable understanding of many aspects of the process and at the same time left many gaps as well. Most important of these being lack of understanding of a clear mechanism by which alloying or metastable phase formation takes place under milling conditions. There appears to be different operative mechanisms for different products of MA/MM as proposed by different authors.

Originally, the technique was adopted by Benjamin et al. (1970) for oxide dispersion strengthening of Ni–base superalloys. The objective was to coat the oxide particles with a thin layer of Ni in a way similar to that by Hoyt (1930) much earlier. The objectives of mechanical milling had changed since then, with report of an amorphous phase by mechanical alloying of elemental blends of Ni and Nb corresponding to a composition of $Ni_{60}Nb_{40}$ by Koch et al. (1983). The technique of mechanical milling as it stands today may be applied to achieve:

(i) mechanical alloying and solid solubility extensions in otherwise immiscible systems,
(ii) synthesis of intermetallic compounds,
(iii) solid state amorphisation including disordering of intermetallics,
(iv) nanostructured materials, and
(v) mechano-chemical synthesis.

Extensive parametric studies governing these processes and propositions of the mechanisms that possibly explain the phenomena are undertaken by several researchers. In-depth reviews of the subject matter are published by Suryanarayana (1995), Murty and Ranganathan (1998) and lately by Suryanarayana again (2001). A reference to these reviews gives the details of the above-mentioned techniques of advanced materials processing by mechanical milling. The mechanisms of transformation offered by different researchers, as reviewed by the above authors are classified into

(i) ductile-ductile,
(ii) ductile-brittle, and
(iii) brittle-brittle components.

The objectives of the present investigation and of this communication are to bring into focus the various arguments, at times contradictory, offered by different researchers for explaining the mechanism in the above three broad classifications of components. In the ductile-ductile case, the component powders initially get plastically deformed into flakes and get cold-welded together. Upon increase of dislocation and defect density and work hardening, the material fragments into smaller particles. Alloying takes place in the final stages of milling these cold-welded and plastically deformed layers of the two ductile-ductile components of the blend to give rise to a homogenous solution when the grain size in the agglomerates (flakes) approaches nanometric scale. Thus, it is opined that (i) plastic deformability for achieving cold-welding and (ii) diffusion is necessary pre-requisites for alloying. Diffusion in this case of mechanical alloying is understood to have been mostly assisted by high defect density rather than thermal factors.

In brittle-brittle systems these arguments do not hold good due to lack of any reasonable degree of plastic deformation. On the contrary brittle-brittle component systems such as Si-Ge (Davis and Koch, 1987) and Mn-Bi (Davis et al., 1988) were found to alloy together and mixtures of brittle intermetallics such as $NiZr_2$ and $Ni_{11}Zr_9$ were found (Lee et al., 1988; Lee and Koch, 1987) to transform to amorphous alloy during milling. The diffusion of the atomic species, which is essential for either alloying or even for amorphization, is proposed to be assisted by a rise in temperature in this case due to lack of any high degree of defect density (Lee, 1990). While this postulate of local-rise-in-temperature was supported by the observation that milling of these blends at liquid nitrogen temperatures did not lead to alloying or amorphization even up to 15 hrs (Shultz, 1988). A counter observation that slight rise in milling temperature from sub-ambient to room temperature leads to the crystallization of amorphous phase obtained at lower temperatures (Miller et al., 1986) was also made. Further, a later study on the same systems (Ni-Ti and Ni-Zr) by Koch et al. (1993) showed totally opposite results that milling at liquid nitrogen temperatures took 2 h for amorphization while it took 18 h at 493K. This result was justified on the basis that alloying during mechanical milling is basically driven by nanocrystallization, which provides large fraction of nanocrystalline grain boundaries that can provide short circuit path for diffusion and that kinetics of nano-particle formation are enhanced at liquid nitrogen temperatures. Therefore, Bellon and Martin (1996) proposed an altogether different approach; that a ballistic component of diffusion coefficient adds to the contribution of thermal diffusion coefficient for material transport. However, this proposal needs verification of its applicability in the wider context.

These conflicting arguments were the motivating factor for the present investigation wherein we have selected completely brittle components such as oxides. The selection of oxides (highly stable) also leaves no scope for any stabilizing role to be played by ambient oxygen. In the case of brittle–brittle components such as intermetallic compounds oxygen was suspected to be playing a role in the stabilization of various phases including amorphous phase (Koch et al., 1983).

EXPERIMENTAL PROCEDURE

Of the many glass-forming oxide systems selected for study, the results of borosilicate glass and soda lime glass systems are presented in this communication. Requisite quantities of analytical grade oxides were weighed according to the well known glass forming composition to make batches of 100 gm each. The composition of the glasses is given in Table 1.

Table 1: Compositions of the batches

Borosilicate Glass		*Soda Lime Glass*	
Component	*Wt.%*	*Component*	*Wt.%*
Sodium oxide	3.0 %	Sodium oxide	16.0 %
Boric oxide	12.0 %	Calcium oxide	15.0 %
Alumina	5.0 %	Alumina	0.5 %
Silica	80.0 %	Silica	68.5 %

One of the batches is melted in an alumina crucible and cast into a glass button while the other batch was mixed thoroughly and milled for 20 hrs using Fritsch P5 Planetary Ball Mill at 300 rpm in Tungsten Carbide milling media. The ball to powder weight ratio was 10:1. The glass button, that was conventionally cast, was crushed to powder and the powder was then subjected to the same number of hrs of milling keeping all other conditions similar. Samples were collected for X-ray diffraction (XRD) and electron microscopy studies at 10, 15 and 20 hrs milling times.

RESULTS AND DISCUSSION

Figure 1 shows the XRD pattern of the oxide mixture of the soda lime batch (Fig. 1(a)) and that of the borosilicate batch (Fig. 1(b)) milled for 10 hrs.

The XRD pattern in Fig. 1(b) is a clear indication of the formation of amorphous phase upon 10 hrs of milling of the borosilicate mixture while in the case of soda lime batch it shows (Fig. 1(a)) the presence of numerous crystalline peaks. It is possible that the time of milling (10 hrs) was inadequate in the case of soda lime mixture for transformation to amorphous phase. One can also observe in Fig. 1(b) the superposition of some crystalline peaks over the diffuse maxima in intensity due to the amorpous phase indicating that the transformation has not gone to completion. The results of XRD needed to be verified by electron microscopy and electron diffraction and thus samples from the borosilicate batch milled for 10 hrs were collected for study. The bright field electron micrograph in Fig. 2(a) shows round particles of different sizes in the milled powder. The clustering of certain particles together might have taken place while sprinkling the sample on the carbon coated grids. No organic solvent was used in the present investigation either to sonicate the samples or to spread the particles on the grid for electron microscopy. The electron diffraction pattern obtained from the particles (Fig. 2(b)) shows clear diffuse rings confirming the formation of an amorphous phase. The diffuse rings from the supporting holy carbon grids do not interfere with this pattern and can be distinguished easily from the amorphous phase. Presence of a few crystalline particles in the cluster shown in the micrograph is indicated by the sharp spots as shown by the arrow.

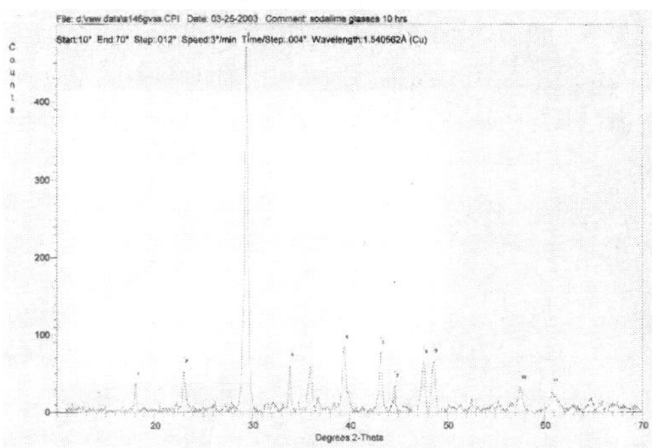

(a)

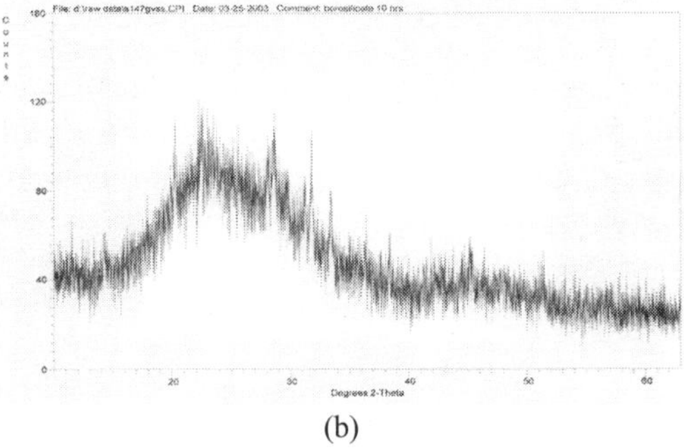

(b)

Fig. 1: XRD patterns of (a) Soda lime mixture and (b) borosilicate mixture milled for 10 hrs.

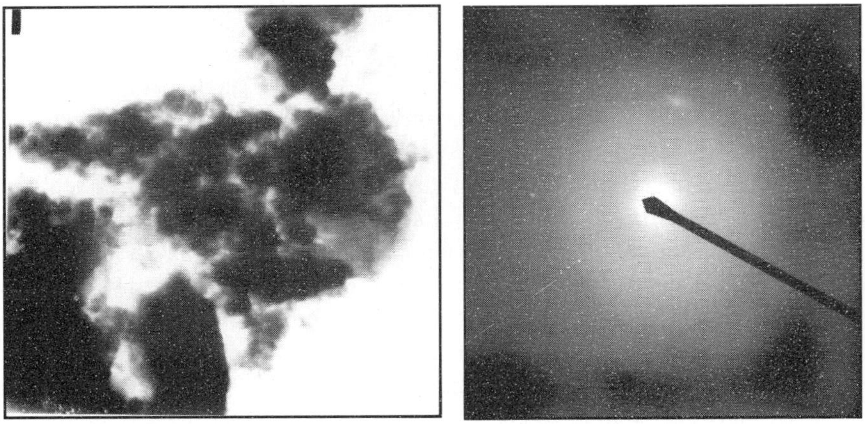

(a) (b)

Fig. 2: (a) Transmission electron micrograph of the borosilicate mix milled for 10 hrs showing featureless nano particles and (b) corresponding diffraction showing diffuse ring of the amorphous phase. Arrow indicates a few crystallite spots.

CONCLUSIONS

One can, therefore, conclude that it is possible to form amorphous phase by milling brittle oxide blends that are non-metallic. The role of significant temperature raise during milling in stabilizing an amorphous phase is vitiated in the present investigation as milling was carried out at room temperature in presence of toluene. The as-cast glass did not devitrify for the same time period of milling.

The present experiment also clearly demonstrates that mechanisms such as plastic deformation leading to cold welding of the components and generation of high defect density cannot be fundamental requisites for mechanical alloying or amorphization but at best be supplementary phenomenon. Thus, we believe that formation of an amorphous phase was only possible through anomalous diffusion processes similar to ballistic diffusion. Further study is in progress to characterize the amorphous phase and verify the effect of various other process parameters on the amorphization.

Acknowledgements: The authors would like to express their thanks to Dr. R.K. Mandal for the help rendered in carrying out X-ray diffraction. They also would like to thank Professor C. Suryanarayana, University of Central Florida, Orlando for fruitful discussions.

REFERENCES

Bellon, P. and. Martin, G. (1996). *Solid State Phys.*, **50:** 189.

Benjamin, J.S. (1970). *Metall. Trans.*, **1:** 2943.

Davis, R.M. and Koch, C.C. (1987). *Scripta Metall.*, **21:** 305.

Davis, R.M., McDermott, B. and Koch, C.C. (1988). *Metall. Trans.*, **19A:** 2867.

Hoyt, S.L.(1930). *IAME Trans.*, **89:** 9.

Koch, C.C., Cavin, O.B., Mc Kamoy, C.G. and Scarbrough, J.O. (1983). *Appl. Phys. Lett.*, **43:** 1017.

Koch, C.C., Pathak, D. and Yamada, K. *In:* Mechanical Alloying for Structural Applications. deBarbadillo, J.J., Froes, F.H. and Schwarz, R. (eds) (1993). ASM International, Materials Park, Ohio, USA, 205.

Lee, D., Cheng, J., Yuan, M., Wagner, C.N.J. and Ardell, A.J. (1988). *J. Appl. Phys.*, **64:** 4742.

Lee, P.Y. (1990). *Japan. J. Appl. Phys.*, **29:** 540.

Lee, P.Y. and Koch, C.C. (1987). *Appl. Phys. Lett.*, **50:** 1578.

Miller, P.J., Coffey, C.S. and deVost, V.F. (1986). *J. Appl. Phys.*, **59:** 913.

Murty, B.S. and Ranganathan, S. (1998). *Inter. Mater. Rev.*, **43:** 101.

Shultz, L. (1988). *Z. Phys. Chem.*, **157:** 203.

Suryanarayana, C. (1995). Bibliography on mechanical alloying and milling: Cambridge, UK: Cambridge International Science Publishing.

Suryanarayana, C. (2001). *Prog. Mater. Sci.*, **46:** 1.

X-ray Characterisation and Electrical Conductivity of Co-Cd Mixed Ferrite

G. Bhikshamaiah, S. Annapurna[1], J. Sadanandam[2] and A.K. Singh[3]

Department of Physics, Nizam College (Autonomous)
Osmania University, Hyderabad–500 001, India

[1]Department of Physics, University College for Women (Autonomous)
Osmania University, Hyderabad, India

[2]Department of Physics, College of Engineering (Autonomous)
Osmania University, Hyderabad – 500 007, India

[3]Defence Metallurgical Research Laboratory
Kanchanbagh, Hyderabad – 500258, India

INTRODUCTION

Ferrite materials are known for their technological and industrial applications such as ferromagnetic insulators. It is a metal oxide with the formula $M^{2+} Fe^{3+}O_4$, where M stands for divalent metal such as Ni, Cu, Co, Zn, Cd etc. A majority of ferrites crystallize with a cubic structure similar to that of the mineral spinal ($MgAl_2O_4$). A survey of literature shows that magnetic, electrical and dielectric properties have been widely studied on various pure and mixed ferrite materials (Ni-Zn, Co-Zn, Li-Zn etc. (Verwey et al., 1948; Jonker, 1959; Parker et al., 1968; Heikes, 1957; Satyanarayana et al., 1982). However, the mixed Co-Cd ferrites have drawn limited attention except on their magnetization behaviour (Ghani et al., 1991). Ferrites have many applications in microwave devices and therefore it is advantageous to improve their dielectric properties. The electrical conductivity has strong influence on the dielectric and magnetic behaviour of ferrites, because the low electrical conductivity of these materials results in low losses and makes them attractive for microwave applications (Manjula et al., 1986; JCPDS file no 22-1063, 1972). In addition to their use in magnetic applications, ferrites can also be used as thermistors. But control of resistivity in ferrites is a serious problem.

There are two general approaches to this: (1) controlling sintering temperature and (2) addition of minor constituents to increase or decrease the conductivity. The aim of the present work is to study the effect of addition of cadmium on electrical conductivity of $CoFe_2O_4$.

EXPERIMENTAL PROCEDURE

Six compositions of $Co_{1-x}Cd_xFe_2O_4$ with $x = 0, 0.2, 0.4, 0.6, 0.8$ and 1.0 were selected and high purity (AR Grade) powders of $CdCO_3$, Fe_2O_3 and Co_3O_4 were mixed together according to their molar ratio. The mixed powders of each composition were crushed to fine powders and pre-sintered in air at 900 °C for 12 hrs and slowly cooled to room temperature. The steps followed in the preparation route of Co-Cd ferrites are shown in Fig. 1. Circular shaped pellets of about 1.354 cm diameter and 0.30 cm thickness were prepared using these powders at a pressure of three tonnes per square centimetre. Poly Vinyl Alcohol (PVA) was used as binding material. The pellets were again sintered at 1200 °C for 6 hrs and then cooled slowly (at the rate of about 100 °C per hr) to room temperature.

MIXING OF HIGH PURITY POWDERS
$CdCO_3 + Fe_2O_3 + Co_3O_4$ IN THEIR MOLAR RATIO

↓

GRINDING INTO FINE POWDER IN AN AGATE MOTOR

↓

PRE-SINTERING AT 900 °C FOR 12 HOURS

↓

PRESSED INTO PELLETS AT A PRESSURE OF 3 tonnes/cm^2 WITH BINDER (PVA)

↓

FINAL SINTERING AT 1200 °C FOR 6 Hrs THEN SLOWLY COOLED

Fig. 1: Process route for the preparation of mixed Co-Cd ferrites.

XRD data on all compositions of Co-Cd ferrites (pellets) were collected in the Bragg angle range (2θ) 10°–80° using MoKα radiation with zirconium filter on a Phillips diffractometer with a proportional counter at room temperature. The scanning speed used was 1° per minute. A chart speed of 1 cm/minute was used. A standard silicon sample was used to calibrate the diffractometer.

Electrical conductivity measurements were carried out by means of well-known two-probe method using spring loaded copper electrodes. The electrical contacts were made using air-dried silver epoxy. The current was measured by a digital nanometer as a function of temperature ranging from 300 to 900 K for all the compositions. For the measurement of current (I), the specimens were kept inside a heating cell and sufficient time was allowed for it to attain the thermal equilibrium. All the measurements were carried out at an applied field of 4 volts/cm. From the variation of conductivity (σ) with temperature the activation energies were calculated using the following formula.

$$\sigma = \acute{\sigma}_0 \exp(-E/KT) \tag{1}$$

where E is activation energy needed to release an electron from hopping from one ion to neighbouring ion, K is Boltzmann Constant, T is absolute temperature and σ_0 is the conductivity at room temperature. A Chromel-Alumel thermocouple along with a digital multi-voltmeter was used to measure the temperature of the sample.

RESULTS AND DISCUSSION

X-ray Analysis

XRD patterns for the samples are shown in Fig. 2. These patterns show the reflection planes (220), (311), (222), (400), (422), (511), (440), (620), (533), (444), (642), (731) and (800). It was observed that the (111) plane completely disappeared in all the samples, but (444) disappeared at high cadmium concentrations in agreement with JCPDS data (1972). All the profiles were indexed to single-phase cubic spinel structure. The lattice parameter a can be calculated directly using the following relation:

$$a = \frac{\lambda}{2 \sin \theta} (h^2 + k^2 + 1^2)^{1/2} \tag{2}$$

where λ is the wave length of the radiation used (λ = 0.0710688 nm), θ is the Bragg angle and h, k,1 are the Miller indices of the diffraction planes. The measured values of lattice parameters are plotted against the extrapolation function cos θ cot θ in Fig. 3 for all compositions. The values of a_0 was obtained by least square fitting of the data to a straight line and extrapolating it to cos θ cot θ (error function for diffractometer) tending to zero. These values are given in Table 1. The lattice parameters can also be obtained precisely by well known Cohen's least square method (Cohen, 1935). The values of lattice parameter thus obtained with a suitable computer programme using all the profiles are given in Table 1. The error in the lattice parameter was evaluated following the method suggested by Jette and Foote (1935) and included in Table 1.

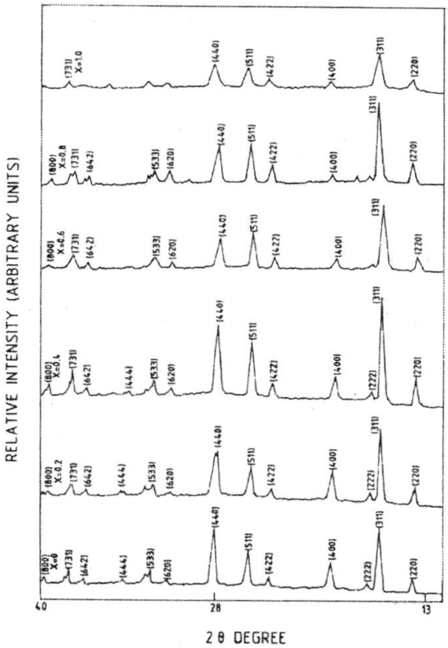

Fig. 2: XRD patterns of $Co_{1-x}Cd_xFe_2O_4$.

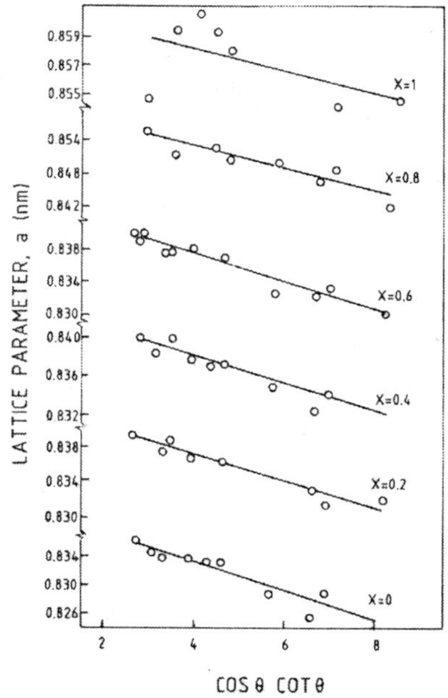

Fig. 3: Variation of lattice parameter with extrapolation function cos θ cot θ for $Co_{1-x}Cd_xFe_2O_4$.

It can be seen from the Table 1 that the data on the lattice parameters obtained by both the methods are in good agreement. The variation of lattice parameter as a function of cadmium composition is shown in Fig. 4(a). From this, it is seen that the lattice parameter increases linearly with cadmium content in the Ferrites following Vegard's law (1921). This observation may be accounted for the replacement of cobalt by cadmium ion because the ionic radius of Cd^{2+} (0.097 nm) is bigger than that of Co^{2+} 0.064 nm) (Hand Book of Chemistry and Physics, 59th Ed., 1978). The data on lattice parameters on this Ferrite system was also studied earlier by Gahani et al. (1991) which is also shown in Fig. 4(a) for comparison. It can be seen that the lattice parameter data reported by Ghani et al. (1991) is in conformity with the present results that the lattice parameter increases with cadmium composition. However, the present data on the lattice parameter is slightly lower than that of Ghani et al. (1991). This may be due to the difference in the procedure followed and the number of profiles included in the evaluation of lattice parameter. It is known that there is a correlation between ionic radius of metal ion and the lattice parameter of the Ferrite material (Vemaas and Scmidt, 1959). It has been observed that the distance between magnetic ions of the tetrahedral sites (i.e. the lattice constant) increases with the introduction of larger ions for Ni-Zn, Ni-Cd and Mn-Cd Ferrites (Pascard et al., 1977). In the present work the relation between the tetrahedral ionic radii and the lattice parameter $Co_{1-x}Cd_xFe_2O_4$ Ferrite has been studied. It is known that the Cd^{2+} ions occupy only tetrahedral (A) sites in $CdFe_2O_4$ (Devale and Kulkarni, 1978). Starting from the basic $CoFe_2O_4$ non-magnetic Cd^{2+} ions are substituted to replace Co^{2+} ions in the A-sites and Fe^{3+} ions, which in turn move into octahedral (B) sites. For a given composition (x), the mean ionic radius for a molecule on the given A-sites is obtained as (Mazen et al., 1987).

$$r_{\text{tet. ion}} = (1-x)\ r_{\text{tet.Fe3+}} + (x)\ r_{\text{tet.Me2+}} \qquad (3)$$

Table 1: Lattice parameter (a_0), tetrahedral ionic radius ($r_{tet.}$), x-ray density (d_x), conductivity (σ) and activation energies of Co-Cd mixed Ferrites

Composition (x)	*Lattice parameter a(nm)* Extrapolation method	*Lattice parameter a(nm)* Cohen's least square method	*Tetrahedral Radius (nm)*	*Density (gcm^{-3})*	*Conductivity at RT $\sigma \times 10^{11}$ (ohm^{-1} cm^{-1})*	*Activation energy (eV)*	
0.0	0.84118	0.84148	0.064	5.236	2.62	0.3175	0.3679
0.2	0.84329	0.84364	0.071	5.434	1.50	0.2593	0.3355
0.4	0.84331	0.84374	0.077	5.670	5.10	0.2728	0.2751
0.6	0.84395	0.84445	0.084	5.893	1.70	0.1541	0.5037
0.8	0.86079	0.86103	0.090	5.777	–	–	–
1.0	0.86571	0.86739	0.097	5.898	1.12	0.2166	0.2889
error limits	±0.00003 to ±0.00005	±0.00002 to ±0.00005		±0.002 to ±0.004			

where Me^{2+} denotes the substituted divalent metal. The tetrahedral ionic radius was calculated using the above formula for all the compositions are included in Table 1. A plot of lattice parameter (a_0) against tetrahedral radius (r_{tet}) of $Co_{1-x}Cd_xFe_2O_4$ is shown in Fig. 5. This shows that the lattice parameter increases with increase of tetrahedral radius in Co-Cd ferrites.

The X-ray density (d_x) was calculated using the lattice parameter (a_0) with the following relation (Mazen et al., 1987).

$$d_x = Z\ M/N\ a_0^3\ \text{gm cm}^{-3} \qquad (4)$$

where M is the molecular weight of a mixed $Co_{1-x}Cd_xFe_2O_4$ Ferrite sample. This was obtained $M=(1-x)\ M_{Co} + (x)\ M_{Cd} + 2\ M_{Fe} + 4\ M_O$, where M_{Co}, M_{Cd}, M_{Fe} and M_O are the atomic weights of cobalt, cadmium, iron and oxygen, respectively and x is the cadmium content, N is Avagadro's number, Z is the number of molecules per unit cell and is 8 in this case. The density thus calculated for all the compositions are included in Table 1, and plotted against cadmium composition (x) in Fig. 4(b). It is seen that the X-ray

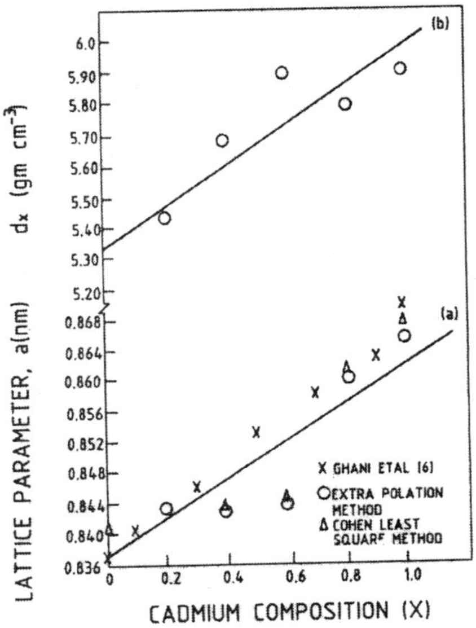

Fig. 4: Composition dependence (x) of (a) lattice Parameter a_0 (b) X-ray density for Co-Cd ferrite.

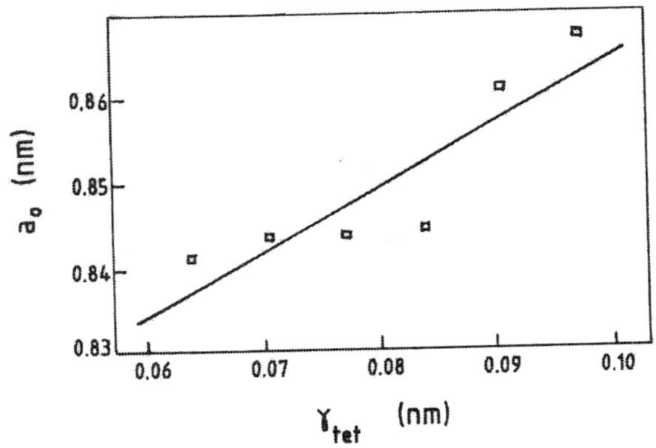

Fig. 5: Variation of lattice parameter (a_0) with tetrahedral ionic radius (r_{tet}) for Co-Cd ferrite.

density (d_x) increases linearly with increases of cadmium content, as the atomic weight of cadmium (112.4) is higher than that of cobalt (58.93).

D.C. Electrical Conductivity

The electrical conductivity (σ_0) at room temperature calculated by application of a constant potential difference (V) of 1.5 volts and also using a nano-ammeter for the measurement of the current (I) as

$$\sigma_0 = (I/V) \times (t/A) \tag{5}$$

where t is the thickness and A is the area of cross section of pellet.

The measurement of σ was extended from room temperature to an elevated temperature of about 900 K with an interval of 5°. A plot of log (σT) versus $10^3/T$ is shown in Fig. 6. It can be seen that each plot consists of two regions (except for $CdFe_2O_4$). From the plots, the activation energy (E in eV) was calculated in each region from the following relation and data is given in Table 1.

$$E = 2.303 \times K \times \text{slope} \tag{6}$$

where K is Boltzmann constant.

Cobalt has octahedral (B) while cadmium has tetrahedral (A) preferential occupation sites in Co-Cd mixed ferrites. Devale and Kulkarni (1978) investigated the d.c. conductivity of $FeCoCrO_4$ and concluded that the presence of Co^{2+} ions at the B sites lowers the conductivity. Similar trends were also observed in Co-Zn mixed ferrites in literature (Josyulu and Sobhanadri, 1980). In the present study, it has been observed that the conductivity at room temperature is minimum for $Co_{0.8}Cd_{0.2}Fe_2O_4$ than the remaining (Table 1) ferrite materials. This may be due to the presence of relatively more number of Co^{2+} ions present at the octahedral sites.

It can be seen from Fig. 6 that the electrical conductivity increases with temperature for all the composition of the Co-Cd ferrites studied. It can also be seen that each plot (except $CdFe_2O_4$) consists of two regions with different slopes. One region (ferri) was due to low conductivity containing Co^{2+} and Co^{3+} ions and other region (para) was of high conductivity containing Fe^{2+} and Fe^{3+} ions as the presence of cobalt at the tetrahedral sites favours a conduction mechanism $Co^{2+} + Fe^{3+} \rightarrow Co^{3+} + Fe^{+2}$ (Josyulu and Sobhanadri, 1980). Similar trends were also observed in several other ferrite systems namely $Co_{1-x}\ Fe_{2+x}\ O_4$ and $Co_xZn_{1-x}Fe_2O_4$ (Mousa et al., 1989).

The existence of the kinks (change in the slope) in conductivity plots or transition in the neighbourhood of the Curie Point have been explained by the theory suggested by Irkhin and Turvo (1957). It has shown theoretically that while passing through the Curie point a change in the slope of the straight line should occur (Smit and Wijn, 1959) and the magnitude of the effect depends on the exchange interaction between inner and outer electrons, which alters at the Curie point. The present results show that the magnitude of change in the slope of straight line of the conductivity plot varies with composition and it appears that maximum variation in the slope at Curie point seen in case of $Co_{0.4}\ Cd_{0.6}\ Fe_2O_4$ may be due to the relatively strong interaction between the inner and outer electrons which alters the Curie point. It should be noted that the saturation magnetization is reported (Ghani et al., 1991) to be maximum in the Co-Cd ferrites for the cadmium composition $x = 0.6$. In case of pure $CdFe_2O_4$, no transition in the conductivity plot is noticed (Fig. 6) as it is non-magnetic.

The activation energy for the Co-Cd mixed ferrites in the para magnetic region is higher than in ferri magnetic region (Table 1) and is in the range

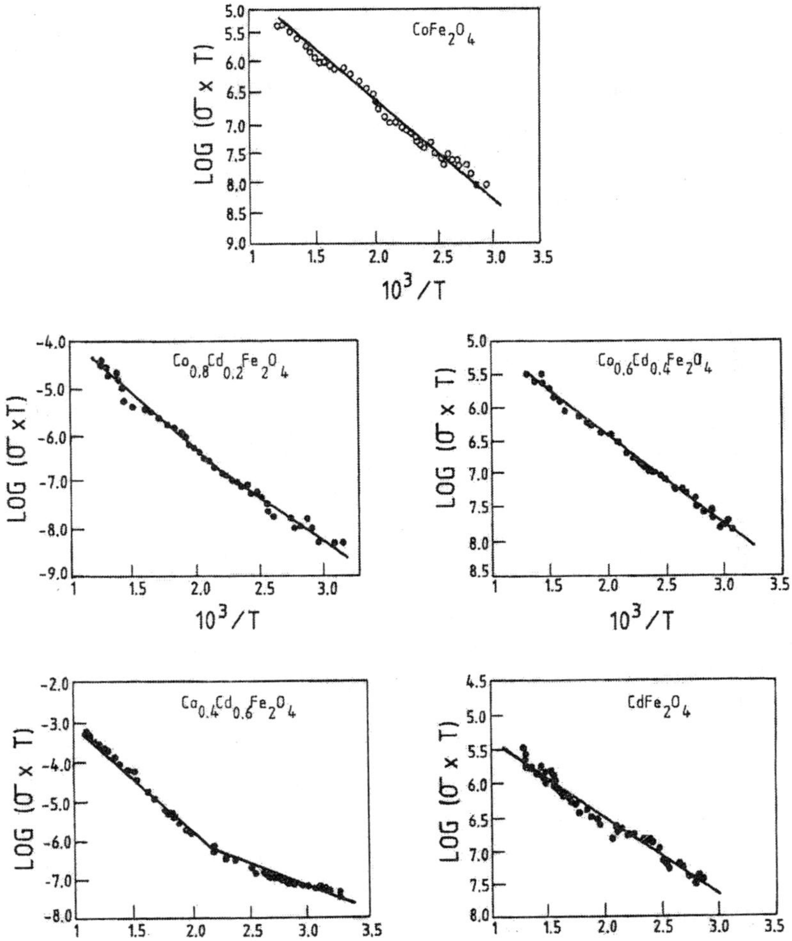

Fig. 6: Plots of log (σT) as a function of $10^3/T$ for Co-Cd ferrite.

0.1541 to 0.5037 eV, which is in conformity with the range (0.1 to 0.5 eV) suggested by Smit and Wijn (1959) for ferrite materials.

CONCLUSIONS

1. The lattice parameter increases with increase of Cd composition in Co-Cd mixed ferrites.
2. The lattice parameter is found to increase with tetrahedral molecular radius.
3. Electrical conductivity plots of Co-Cd mixed ferrites show two regions associated with the para and ferri magnetism.
4. The activation energies are in the range 0.1541- 0.5037 eV for Co-Cd mixed ferrites.

REFERENCES

Cohen, M.U. (1935). *Rev. Sci. Instru.*, **6:** 68.

Devale, A.B. and Kulkarni, D.K. (1978). *IJPAP*, **16:** 697.

Ghani, A.A., Sattar, A.A. and Pierre, J. (1991). *J.Magn. Mat.*, **97:** 141.

Hand Book of Chemistry and Physics (59th Ed.). CRC Press, USA, Ed. Robert C. Weast (1978-79).

Heikes, R.P. (1957). *J. Chem. Phys.*, **26:** 582.

Irkhin, Yu.P. and Turov, E.A. (1957). *Scot. Phys. – JETP*, **33:** 673.

JCPDS file no 22-1063 (1972).

JCPDS file no. 22-1086 (1972).

Jette, E.R. and Foote, F. (1935). *J.Chem. Phys.*, **3:** 605.

Jonker, G.H. (1959). *J.Phys. Chem. Solids*, **9:** 165.

Josyulu, O.S. and Sobhanadri, J. (1980). *Phys. Stat.Sol* (a), **59:** 323.

Manjula, R., Murthy, V.R.K. and Shobhanadri, J. (1986). *J. Appl. Phys.*, **59:** 2829.

Mazen, S.A., Abd-El-Rahiem, A.E. and Sabrah, B.A. (1987). *J. Mater. Sci.* **22:** 4177.

Mousa, M.A., Summan, A.M., Ahmed, M.A. and Badawy, A.M. (1989). *J. Mater. Sci.*, **24:** 2478.

Parker, R., Elewell, D. and Griffiths, B.A. (1968). *Brit. J.Appl. Phys.*, **17:** 1269.

Pascard, H., Globus, A. and Cabon, V. (1977). *J. Phys (Paris) Collog.*, **38:** C1 163.

Satyanarayana, R., Ramana Murthy, S. and Sheshagiri Rao, T. (1982). *J. Less Common Metals*, **86:** 115.

Smit, J. and Wijn, H.P.J. (1959). Ferrites, Phillips Tech. Library series, Cleaves–Home, London.

Vegard, L. (1921). *Z. Physics*, **5:** 17.

Vemaas, F.H.S. and Scmidt, E.R. (1959). *Beitr. Mineral, Petrogra*, **6:** 219.

Verwey, E.J.W., De Boer, F. and Van Santen, J.H. (1948). *J. Chem. Phys.*, **16:** 1091.

Study of Order-Disorder in Electrodeposited Ni-Mn Alloy by X-ray Diffraction

A. Stephen, M.V. Ananth[1] and V. Ravichandran

Materials Science Centre, Department of Nuclear Physics
University of Madras, Chennai-600025
[1]Central Electrochemical Research Institute-Chennai Unit
CSIR Complex, Taramani, Chennai

INTRODUCTION

Electrodeposition is an important cost-effective, fast route for preparing pure metals and alloys as polycrystalline films or in bulk form. Electrodeposition introduces disorder during deposition, which acts as the main source of energy. This paper reports the results of X-ray diffraction investigations into the effect of deposition current densities (CD) and bath compositions on the microstructural atomic order-disorder properties of Ni-Mn binary phase. Thermal analysis results on ordering transformations and studies on magnetic behaviour of the film are also reported.

EXPERIMENTS

The Ni-Mn alloys were electrodeposited on stainless steel substrates from simple acidic sulphate baths for a complete range of Ni/Mn salt ratios (Stephen et al., 2000). A 50 gl^{-1} ammonium sulfate solution was used as an additive to the bath. Pretreatment of the substrates was carried out in the sequence of mechanical buffing/polishing, degreasing with benzene followed by acetone and washing with detergent powder in water followed by distilled water. Bagged graphite blocks were used as anodes. The alloys were deposited at RT from the bath after adjusting its pH at 3.0. Three different current densities (CD) were employed. The deposits were mechanically removed as flakes/foils for investigations. X-ray Diffraction (XRD) measurements were carried out on a Huber diffractometer using $CuK\alpha_1$ radiation. The differential scanning calorimeter (DSC) studies were carried out on a Perkin-Elmer

DSC-7 in argon gas flux at a flow rate of 10 cc/min. The samples for Mössbauer spectroscopy were prepared by adding 5 gl^{-1} of $FeSO_4(NH_4)_2SO_4.6H_2O$ to the baths in order to substitute some of the nickel atoms with iron in the deposited Ni-Mn alloy films.

RESULTS AND DISCUSSION

The deposits of Ni-Mn were obtained for a wide range of bath compositions from Ni10Mn150 (i.e. 10 gl^{-1} of $NiSO_4.6H_2O$ and 150 gl^{-1} of $MnSO_4.H_2O$) having a Ni/Mn ratio of 0.06 to Ni150Mn25 having a ratio of 6.0. The elemental composition in each film was determined by atomic absorption spectroscopy. Figure 1 displays the at.% of Ni in the various deposits as a function of the compositional ratio in the bath. It is interesting to note that upto a Ni/Mn bath ratio of unity the Ni at.% composition shows a definite linear trend and significantly the Ni composition in these deposits occurs in a narrow band of 70– 80 at.% irrespective of the large variation of Ni in the baths. This indicates that the deposit compositions are rather driven by structural and solid solubility constraints than the source compositions, the alloying phenomenon thus being evident.

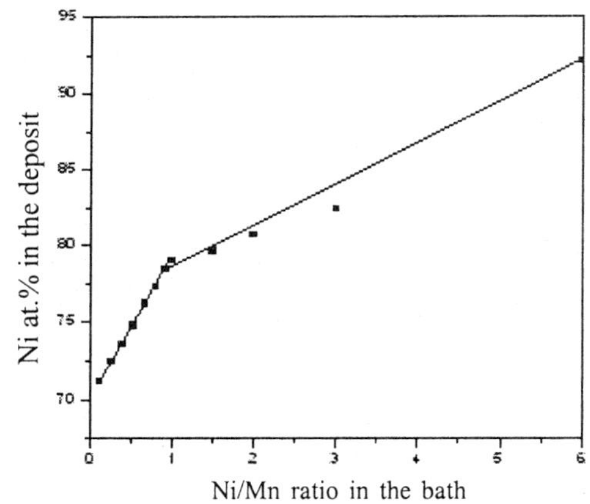

Fig. 1: Ni/Mn in the bath vs. Ni at.% in the deposit (Lines are meant only as guides to the eyes).

The alloys show an fcc structure with a = 0.35239(4) nm corresponding to the XRD pattern of the polycrystalline deposits shown in Fig. 2. These patterns for samples obtained for the deposition CDs of 30, 60 and 80 $mAcm^{-2}$ possess identical lattices but exhibit increasingly broad line profiles as CD increases, signifying grain size reduction. The average crystallite size was determined from a Fourier analysis of the broadened line profiles. The instrumental profile was obtained from a diffraction pattern of SRM standard silicon of large grain size. The size-strain analysis was carried out using WINFIT (Krumm, 1997). Figure 3 shows the results of the analysis as obtained in the standard output of the programme. The effect due to lattice

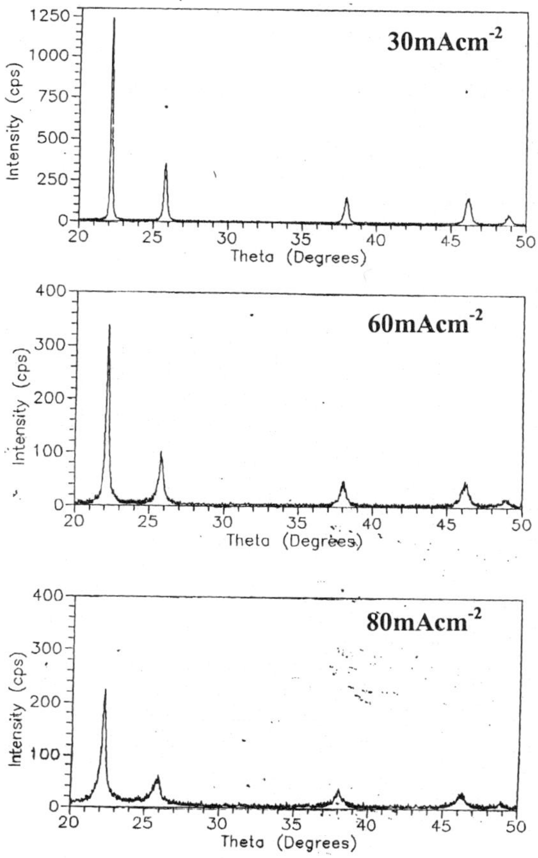

Fig. 2: XRD patterns of Ni-Mn films at different CDs showing increasing line broadening.

strain was separately determined from suitable pair reflections and was found to be very small. The average crystallite size obtained from the above results by variance analysis was 28, 14 and 8 nm for CDs 30, 60 and 80 mAcm^{-2}, respectively. This dependence of grain size on the deposition CD makes electrodeposition an useful method for producing nanocrystalline alloy films.

The existence of a narrow compositional band for the deposits irrespective of reservoir concentrations of Ni suggested that atomic ordering phenomenon in the Ni-Mn electrodeposits may show interesting behaviours. Therefore, studies on structural as well as thermodynamic behaviours of the samples were carried out at elevated temperatures. Thus, XRD under *in situ* heating and DSC were performed on the samples. The DSC scans of all the alloys done at a heating rate of 10 K/min showed two endothermic peaks one at ~570 K (hydrogen evolution) and another beyond 833 K. For alloys with Ni at.% ranging 70-85, a broad exothermic region from 623 to 833 K was also noticed. Since ordering is known to occur for a range of 60-82 at.% of Ni

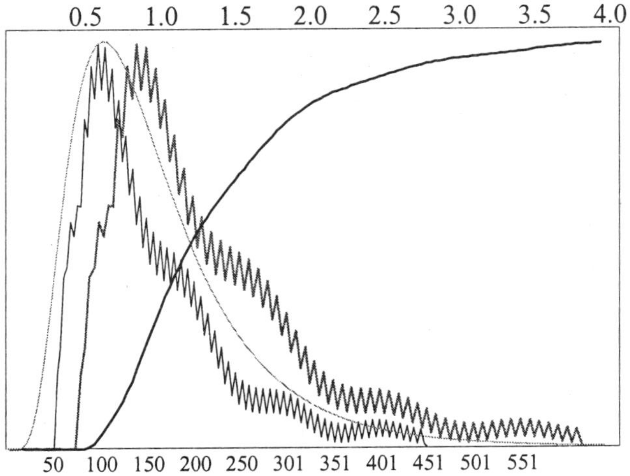

Relative and cumulative size distribution.

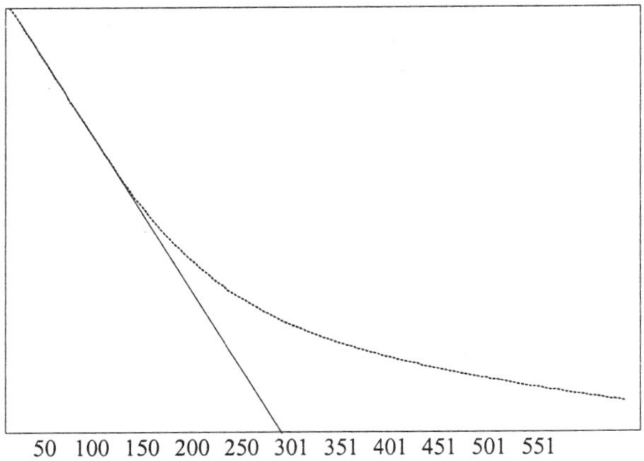

Plot of A_n versus domain size.

Fig. 3: Crystallite size distribution by Fourier analysis of line profile for 30 $mAcm^{-2}$.

(Kaya and Kussmann, 1935; Massalshi, 1992) in Ni_3Mn alloys, this exothermic region was analysed further. The DSC scan (Fig. 4) at a slow heating rate of 5 K/min for a representative Ni-Mn alloy deposit with 82.4 at.% Ni clearly shows that the broad exothermic region is composed of three individual profiles viz., Exo I, II and III. These components are respectively attributed as (i) relaxation of defects and stresses, (ii) development of a long-range order and (iii) crystallite growth from nanocrystalline to microcrystalline state (Zhou and Bakker. 1994). The Endo II transition signifies the phase transformation as a result of ordering. These transitions are observed to be irreversible.

Figure 5 shows XRD patterns taken dynamically as the samples were heated to various temperatures *in situ* in normal atmosphere. Figure 5a

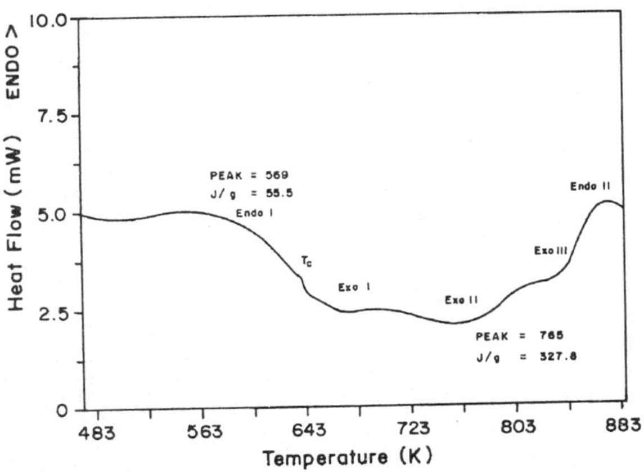

Fig. 4: DSC scan of the deposit with composition 82.4 at.% Ni.

corresponds to the deposit with 76.2 Ni at.% and Fig. 5b to that with 82.4 at.%. These two were chosen because one with 76.2 % is very close to the ordered case for Ni_3Mn alloy as per the phase diagram and the other with 82.4 % is just outside the ordered case for Ni-Mn system. The 76.2 % sample does not show any major changes in its XRD pattern on heating up to 1073K. But, a peak split with a (left) shoulder at lower θ occurs prominently for the most intense (111) peak at about 873K and grows further on heating to 1073K. It is likely that structural modifications begin to occur consequent to temperature driven large atomic displacements. It is interesting that on furnace cooling this sample to RT, the system does not

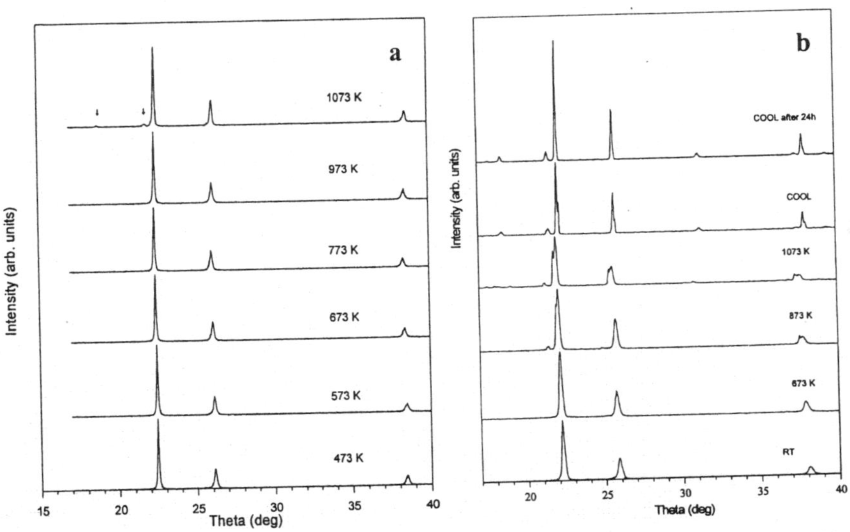

Fig. 5: The XRD patterns for the deposits with (a) 72.6 and (b) 82.4 Ni at.% taken over an in situ heating regimen and subsequent ambient cooling to RT.

return to its normal structure but some kind of phase reversal takes place leading to the switching of the shoulder to (right side) higher θ. This is perhaps due to a peculiar structural reordering of this disordered system. This behaviour in the XRD at 873K is also corroborated with the Endo peak at 873K in DSC scan (Fig. 4). However, DSC could not be done at higher temperature in our facility.

XRD of the cooled sample also shows small new peaks which correspond to a new fcc phase. This is likely to be due to oxide formation, as the heating could not be done in vacuum for XRD analysis. The *d* values and also intensities of these small peaks agree very well with the PDF entry (2000) for an oxide phase with composition $(NiO)_{0.75}$ $(MnO)_{0.25}$ having an fcc structure with a = 0.4210 nm as shown in Table 1. Other oxide phases such as NiO, $(NiO)_{0.5}(MnO)_{0.5}$, $(NiO)_{0.25}(MnO)_{0.75}$ and MnO do not match with the experimental data.

Table 1: Phase identification of the oxidization product of Ni-Mn alloy

Material	*(hkl)*	*(111)*	*(200)*	*(220)*	*(311)*	*(222)*
$(NiO)_{0.75}(MnO)_{0.25}$	d_{hkl}(nm)	0.2430	0.2105	0.1488	0.1269	0.1215
(PDF #78-0425)	I_{hkl} (rel.)	67	100	48	15	10
Our sample	d_{hkl} (nm)	0.2432	0.2102	0.1484	–	–
	I_{hkl} (rel.)	61	100	42	–	–

Figure 6 shows the Mössbauer spectra obtained at RT for the above two deposits. The data were fitted with one sextet, one doublet and a singlet for 82.4 at.% Ni, whereas the singlet is absent for 76.2 at.% Ni. The sextet corresponds to the long-range order (ferromagnetic phase) and the singlet of small intensity to the short-range order (paramagnetic phase). The hyperfine field (H_{hf}) is minimum for 76.2 at.% Ni. An asymmetry in the sextet pattern due to included hydrogen is noticed for all the deposits. The doublet arises due to the presence of the quadrupole splitting, indicating an asymmetry in the environment around the resonant Fe atoms.

CONCLUSIONS

Electrodeposited Ni-Mn alloys are formed within a narrow compositional band, irrespective of the bath composition. In the case of electrodeposited Ni-Mn, though the ordering is visible for Ni at.% range of 70-80, the alloy with 76.2 at.% Ni, is the most stable with pure ferromagnetic ordering. The alloy with 82.4 at.% Ni is disordered. This disordered phase tends to stabilize towards the ordered Ni_3Mn phase on annealing. The nature of the disorder produces peculiar thermal effects an understanding of which requires further analysis.

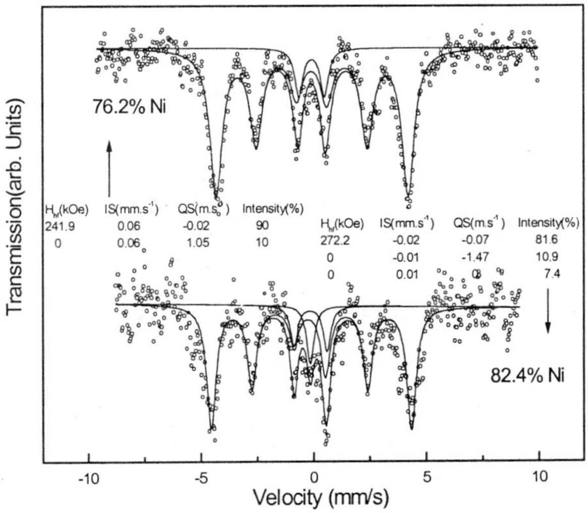

Fig. 6: Mössbauer spectra of deposits with Ni at.% 76.2 and 82.4.

REFERENCES

Kaya, S. and Kussmann, A. (1935). *Z. Physik*, **72:** 293.

Krumm, S. (1997). WINFIT version 1.2.1, Institute für Geologie, Erlangen, Germany.

Massalshi, T.B. (1992). Binary Alloy Phase Diagrams. American Society for Metal, Park, **3:** 2582.

Powder Diffraction File (PDF2) # 78-0425, (2000). JCPDS, ICDD, Pennsylvania, USA.

Stephen, M.V., Ananth, V., Ravichandran, B.R.V. and Narashiman, J. (2000). *Appl. Electrochem.*, **30:** 1313.

Zhou, G.H. and Bakker, H. (1994). *Phys. Rev.B.*, **49:** 12507.

Characterization of Phases in Diffusion Aluminide Coatings and Laser Surface Alloyed Layers by Using Conventional X-ray Diffraction Methods

D.K. Das

Defence Metallurgical Research Laboratory
Hyderabad 500058, India

INTRODUCTION

Conventional X-ray diffraction (XRD) techniques have long been used for characterization of wide varieties of structural and functional coatings and surface modified layers. Such characterization has involved identification of phases and determination of their lattice parameters etc. The basic principle that is used for phase identification in surface layers and coatings is same as that used in case of bulk materials. When X-rays irradiate the surface of a material, depending on its mass absorption coefficient, the X-rays penetrate up to a certain depth of the material. The diffracted rays emanating from the material over the above mentioned depth are used to determine the characteristics of the material such as its phase constitution (Cullity, 1978). The inherent assumption in this method of phase detection is that the material has the same phase constitution over its entire bulk so that the information obtained by XRD from a thin layer at surface is representative of the entire material. Although such an assumption remains valid for most bulk engineering materials, it may not be so for materials showing graded structures. Coatings and other surface modified layers usually constitute the graded structures. In case of thin modified layers, where sub-micron thicknesses are often involved, XRD can be used for phase identification because X-rays can easily penetrate such low film thicknesses. In case of thick surface modified layers where the layer thickness can be up to 100 μm or more, however, X-rays often fail to penetrate the entire coating thickness. If such layers have a variation of phase constitution across the thickness, phases present below penetration depth of the X-rays are unlikely to be detected. In the present paper, the author provides examples of two types of thick surface modified layers where it has been shown how conventional XRD technique can often prove inadequate for phase identification across the entire thickness of the layers.

LASER SURFACE ALLOYED LAYERS

Industrial use of high power lasers for applications such as cutting, welding, and modification of surface properties of materials is becoming increasingly common. In laser surface modification, the high energy density of the laser beam is used to alter the properties of the surface of a material up to a certain depth, typically a few hundred microns, without affecting its bulk properties (Draper, 1981; Belmondo and Castagna, 1979). Surface properties are tailored to combat surface related degradation processes such as corrosion, oxidation, wear and fatigue. Laser surface alloying (LSA) is carried out by injecting the alloying material in powder form directly into the molten pool produced on the surface by laser action so that in-situ alloying takes place. Alloying can also be achieved by applying a coating of the alloying material on the substrate and then melting the substrate up to a certain depth along with the coating so that an alloyed layer is produced on the surface. These two methods of LSA are schematically described in Fig. 1.

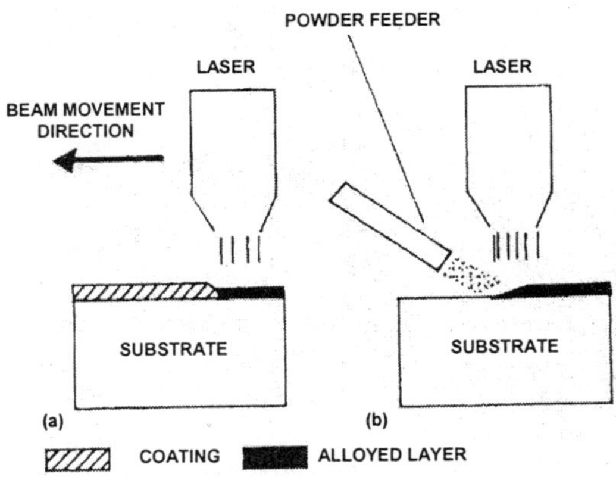

Fig. 1: Schematic representation of the two methods used for laser surface alloying of materials: (a) by depositing the alloying element in form of a coating prior to alloying and (b) adding alloying element directly into the molten pool created by laser.

Example 1: LSA of Al with Ni

In one study carried out by the present author (Das et al., 1994), a commercial pure Al was surface alloyed with Ni by using a pulsed Nd-YAG laser. A thin layer of approximately 50 μm Ni was deposited on the Al substrate which was subsequently melted using the above laser to form a surface alloyed layer. Laser tracks were overlapped by approximately 50% of their width for covering the entire surface of the substrate (Das et al., 1994). The thickness of the alloyed layer, which was determined by observing the cross-section of the alloyed layer, was measured to be ~1 mm. It has been shown in the above study that Ni distribution is quite non-uniform across the depth of the alloyed layer. This fact is also evident from the back scattered electron (BSE) image shown in Fig. 2 which clearly shows that the alloying element, i.e. Ni, remains confined to the top portion of the alloyed layer (seen as brighter region of the alloyed layer). Even in the above Ni-rich top portion of the alloyed layer, there occurs extensive non-uniformity in Ni concentration (Das et al., 1994). As a result, the top regions of the alloyed layer show different microstructures depending on the Ni concentration at that location during solidification after laser alloying. Some locations reveal Al_3Ni dendrites surrounded by the eutectic mixture of Al_3Ni and α-Al as shown in Fig. 3(a). Other locations reveal only colonies of the above eutectic structure as presented in Fig. 3(b). The lower portions of the alloyed layer where Ni concentration is quite low mostly revealed dendrites of α-Al having Al_3Ni as the interdendritic phase (see Fig. 3(c)).

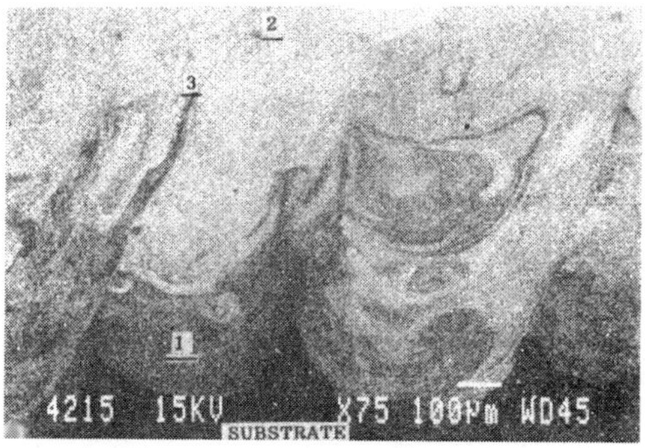

Fig. 2: Back scattered electron image of the cross-section of the laser alloyed layer obtained by alloying aluminium with Ni (Das et al., 1994).

When XRD of the above mentioned alloyed surface is carried out, X-ray peaks of both the phases, namely Al_3Ni and α-Al, are detected. In this particular case, despite the non-uniformity in Ni concentration across the depth of the alloyed layer, only the above two phases have developed although their form and distribution have been quite non-uniform across the layer. Therefore, the presence of Al_3Ni and α-Al phases in the alloyed layer as detected by XRD, which actually should represent a very small thickness near the surface of the alloyed layer, coincidentally becomes the representative of the entire alloyed layer thickness because of the reason mentioned above.

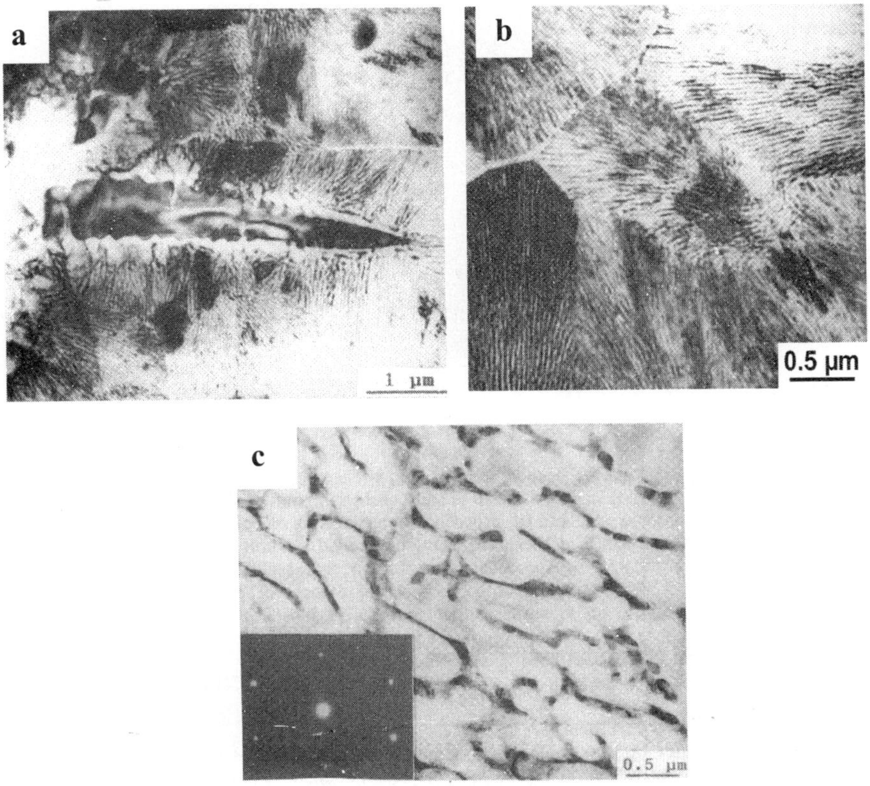

Fig. 3: Microstructure at various locations of the Ni alloyed layer on Al substrate obtained by laser surface alloying: (a) an Al_3Ni dendrite surrounded by $Al_3Ni+\alpha$-Al eutectic structure, (b) colonies of eutectic and (c) α-Al dendrites with interdendritic Al_3Ni phase (Das et al., 1994).

The present example illustrates the fact that, if the phase constitution remains the same across the thickness, XRD can be used for identification of phases even in case of thick surface modified layers.

Example 2: LSA of an Al-Si Alloy with Ni and Ti

This example demonstrates how surface alloying of LM-9 alloy with Ni under a different set of conditions has led to a completely different scenario in terms of identification phases by XRD in the alloyed layer. LM-9 is essentially an Al-12 wt.% Si alloy having very small amounts of other alloying elements such as Mg and Cu (Das et al., 1996). A Ni coating of approximately 300 μm was plasma spray deposited on LM-9 substrate which was subsequently laser alloyed using a continuous wave CO_2 laser (Das et al., 1996). The distribution of Ni across the alloyed layer was found to be extremely non-uniform. In Fig. 4, which shows the cross-sectional view of the alloyed layer, several locations where severe Ni segregation has occurred have been marked as S_1, S_2 etc. The regions showing reasonably uniform

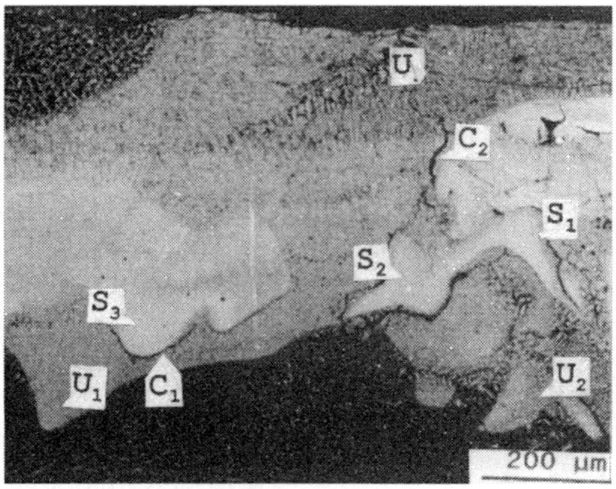

Fig. 4: Cross-sectional view of the laser alloyed layer produced by alloying LM-9 with Ni. S_1, S_2 and S_3 indicate severe Ni segregated areas. While U_1 and U_2 indicate areas where Ni distribution is reasonably uniform, C_1 and C_2 denote locations where cracks are present (Das et al., 1996).

Ni distribution have been marked as U_1, U_2 etc. Because of Ni segregation, several high Ni containing Ni-Al phases form at locations marked as S_1, S_2 etc. Figures 5(a) and (b) present the magnified images of two typical Ni segregated zones where the various phases formed have been marked as 1, 2, 3 etc. The compositions of these phases were determined by using an electron probe micro-analyser (Das et al., 1996). The XRD pattern obtained from the surface of the above Ni-alloyed layer is presented in Fig. 6. It is evident from the above pattern that only Al_3Ni and α-Al phases are detected by XRD while EPMA composition analysis of the alloyed layer clearly indicates the presence of much higher Ni containing phases such as $(Al, Si)_3Ni_5$ and (Al, Si)Ni along with $(Al, Si)_3Ni$ and α-Al. It may be mentioned here that the Al_3Ni phase that is detected by XRD is in fact $(Al, Si)_3Ni$ in which Si has replaced Al in the lattice without altering the crystal structure (Das et al., 1996). Non-detection of higher Ni-containing phases by XRD, as mentioned above, is because of the limitation of X-rays to penetrate deep enough into the alloyed layer to detect these phases. Thus, the present example of laser alloying of LM-9 alloy with Ni clearly illustrates the inadequacy of conventional XRD method to detect phases in the sub-surface region of a thick laser alloyed layer.

DIFFUSION ALUMINIDE COATINGS

The major objective in the design of advanced gas turbine engines, whether for industrial, marine or aero-engine application, is to achieve improvement in performance/efficiency and reduction in cost. With the incessant demand for increasing engine thrust and efficiency, there has been a progressive increase in engine operating temperatures (Meetham, 1986). Such increase in operating temperatures has been accompanied by a corresponding increase

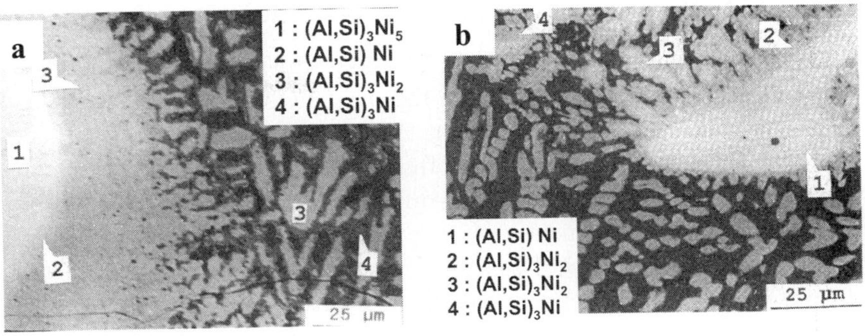

Fig. 5: Magnified view of two locations where severe Ni segregation in laser surface alloyed layer has occurred (Das et al., 1996). The various phases, as determined by EPMA composition analysis, have been marked in the figures.

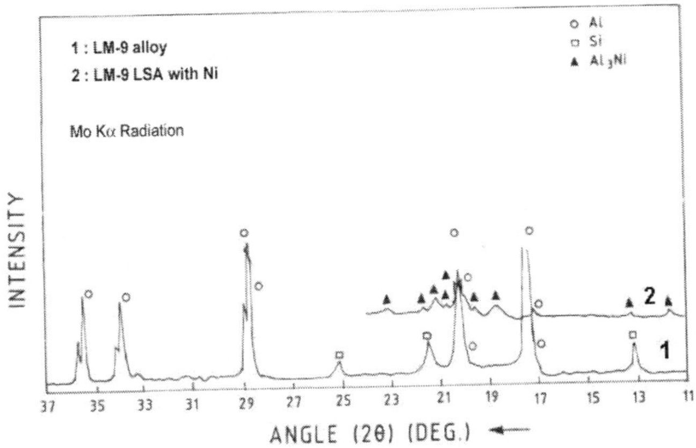

Fig. 6: XRD obtained from the surface of LM-9 alloy in as-received and laser alloyed conditions (Das et al., 1996).

in the severity of surface-related damage mechanisms such as hot corrosion and oxidation which has led to the use of protective coatings on gas turbine components. Diffusion aluminide coatings are a category of oxidation resistant coatings that are used on Ni-base superalloy turbine components. They are formed by incorporating Al onto the surface of the components by diffusion. The β-NiAl phase that constitutes these coatings acts as a reservoir of Al for the formation and sustenance of the protective alumina scale on the surface during high temperature exposure. Addition of small amounts of Pt to these coatings has been found to enhance their oxidation resistance several folds (Tatlock et al., 1987; Allam and Whittle, 1994; Sun et al., 1994; Farrell et al., 1987). The diffusion aluminide coatings containing Pt are referred to as Pt-modified aluminide coatings, or simply Pt-aluminide coatings. Currently, Pt-aluminide coatings are being widely used on turbine components of aeroengines.

Example 1: Pt-Aluminide Coating in As-Coated Condition

Usually, Pt-aluminide coatings on Ni-base superalloys are formed by depositing a layer of Pt (5-8 μm) on the substrate and then subjecting it to a pre-aluminizing diffusion treatment. During the pre-aluminizing diffusion treatment, certain amount of interdiffusion between the Pt layer and the substrate occurs. Subsequently, the diffusion treated substrate is given a pack aluminization treatment to form the final coating (Streiff and Boone, 1985; Streiff et al., 1987; Das, 2001). Depending on the extent of prior diffusion which is dependent on the temperature-time schedule of the treatment, Pt-aluminide coatings develop one of the two microstructures as presented in Figs. 7(a) and (b) (Das, 2001; Das et al., 2000). In both the microstructures, the coating consists of three layers. The outer layer is Pt-rich and consists of either $PtAl_2$ alone or a two-phase structure having $PtAl_2$ and β-NiAl, as shown in Fig. 7. The other two layers are very similar in both the coatings. The intermediate layer consists of β-NiAl phase and the inner layer is the interdiffusion layer that is typically found in all aluminide coatings.

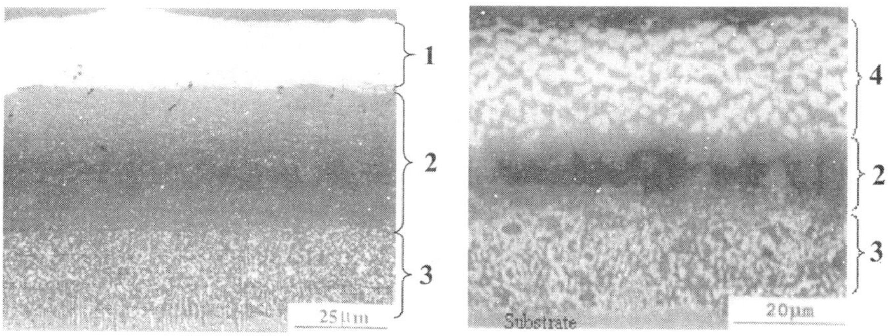

(a) (b)

Fig. 7: Typical microstructures of Pt-aluminide coating on Ni-base superalloys corresponding to (a) no/limited prior diffusion treatment and (b) extensive prior diffusion treatment (Das et al., 2000).

Figure 8 presents the XRD pattern obtained from the surface of the above two Pt-aluminide coatings. It is evident from Fig. 8 that while $PtAl_2$ is the only phase detected in case of coating corresponding to Fig. 7(a), $PtAl_2$ and β-NiAl are the major phases detected for the coating shown in Fig. 7(b). This clearly suggests that XRD has detected phases present in the outer layer of the coatings. No information on the phase constitution of the intermediate layer and the interdiffusion layer are obtained by XRD method in case of the above thick coatings. Although β-NiAl phase is detected in case of coating shown in Fig. 7(b), it is from the $PtAl_2$+β-NiAl structure of the outer layer and not from the intermediate layer. The present case of Pt-aluminide coatings is a typical example where it is clearly evident that for

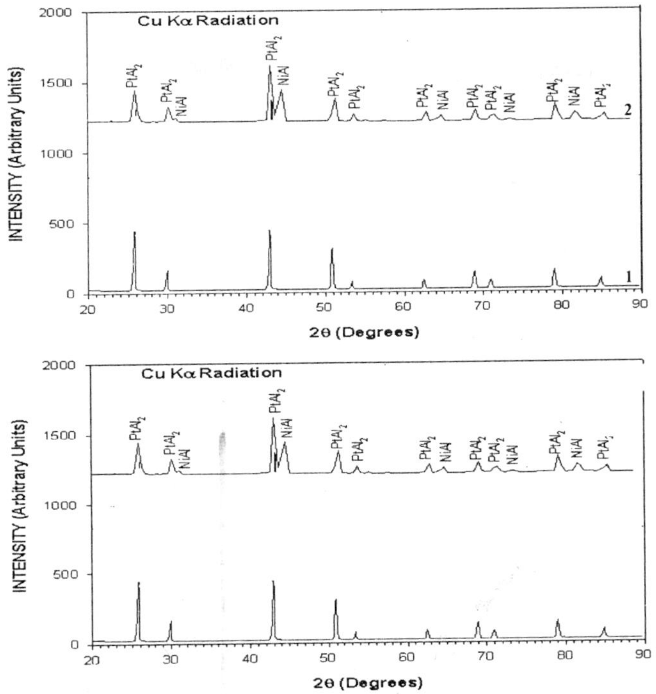

Fig. 8: XRD patterns for Pt-aluminide coatings. (a) corresponds to Fig. 7(a) and (b) to Fig. 7(b).

a multi-layered thick coating with each layer possessing different phase constitutions, it is not possible to detect the phases present in the sub-surface layers of the coating by conventional XRD method. In fact, in most cases, phases present only in the surface layer are detected. Also, in case of Pt-aluminide coating on Ti-based alloy IMI-834 (Fig. 9), phases from the top layer of the coating, namely $PtAl_2$ and Al_3Ti, are detected by XRD as indicated from Fig. 10.

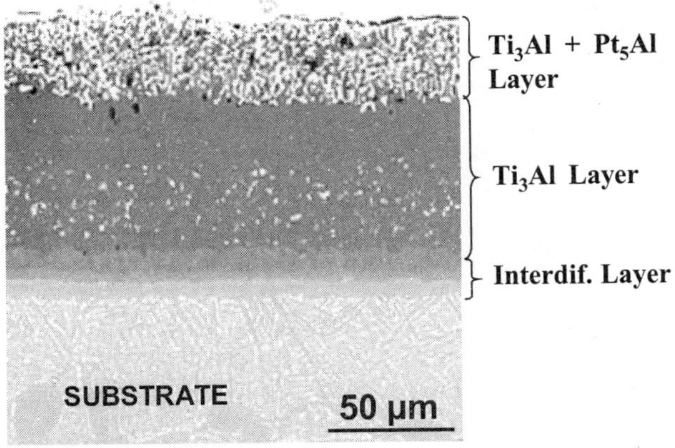

Fig. 9: Microstructure of Pt-aluminide coating on Ti-based alloy IMI-834.

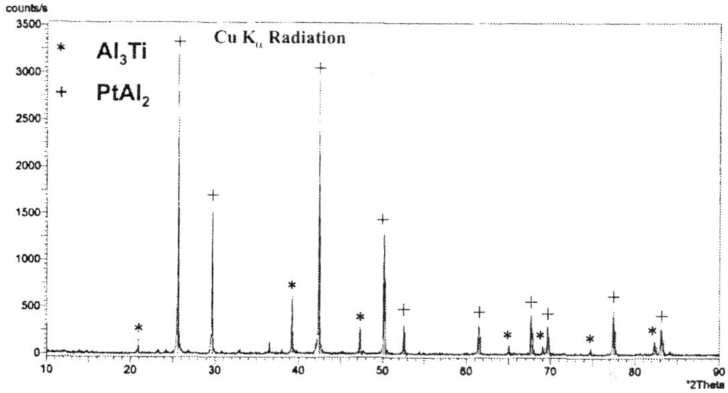

Fig. 10: XRD pattern corresponding to Pt-aluminide coating on IMI-834 alloy, as shown in Fig. 9.

Example 2: As-Oxidized Pt-Aluminide Coating

Figure 11 shows the cross-sectional microstructure of the Pt-aluminide coating (of the type shown in Fig. 7(b)) on CM-247LC super alloy, which has been oxidized at 1100 °C for 15 minutes in air (Das, 1999). A thin alumina layer that has formed due to oxidation can be seen in this figure. This layer has got detached from the surface of the sample and got embedded in the Ni layer possibly during cutting of the samples or during metallographic polishing. It may be noted here that the Ni layer was deposited on the oxidized sample prior to metallographic sample preparation for protection of the sample edges during cutting and polishing (Das, 1999). The XRD pattern taken from the surface of the oxidized coating, as presented in Fig. 12, indicates that both the coating phases $PtAl_2$ and β-NiAl are detected along with α-Al_2O_3 despite the presence of the α-Al_2O_3 layer on the surface. The detection of sub-surface phases such as $PtAl_2$ and β-NiAl is possible because

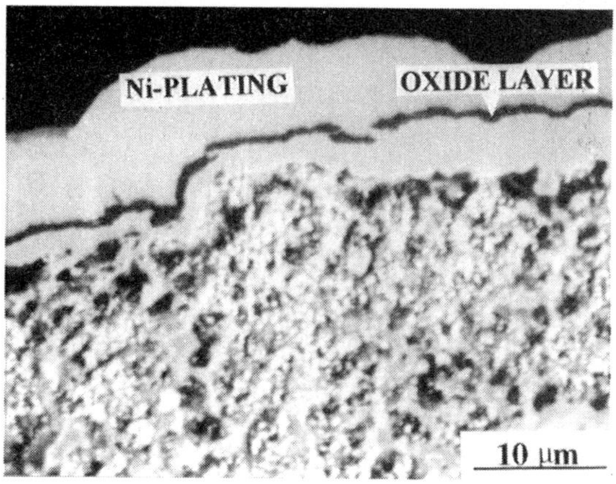

Fig. 11: Pt-aluminide coating after oxidation for 15 minutes at 1100°C in air. The thin oxide layer that has formed can be seen (Das et al., 1999).

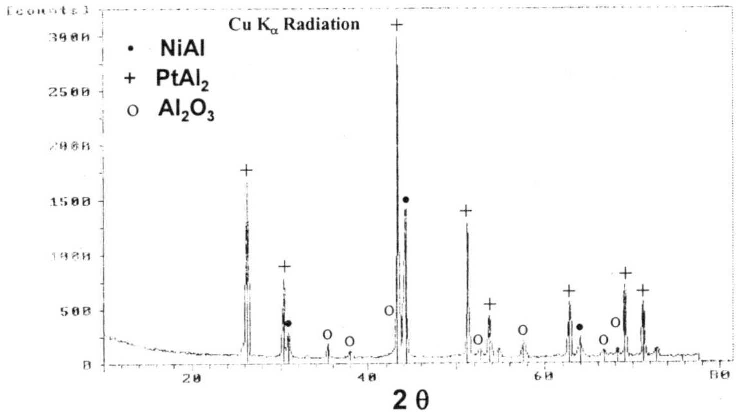

Fig. 12: XRD pattern obtained from the surface of the oxidized Pt-aluminide coating shown in Fig. 11 (Das et al., 1999).

of the low thickness of the oxide scale (~ 1 μm), which enables X-rays to penetrate beyond the oxide layer thickness. It has been found that, as the scale thickness increases, the level to which sub-surface coating phases is detected decreases (Das, 1999; 2002). This example demonstrates that, in case of a multi-layer surface structure, if the outermost layer is relatively thin to X-ray penetration, the phase constitution of the underlying layers can be detected along with that of the outer layer.

CONCLUDING REMARKS

XRD technique has been utilized extensively for identification of phases in both thick and thin coatings. It has proved to be a handy method for quick phase identification in coatings. However, because of the nature of identification that is associated, this technique can have certain limitations in detecting phases in thick multi-layer coatings and surface modified layers. In the examples given in this paper on laser surface alloying, it is amply clear that the phases detected by XRD method becomes representative of the entire alloyed layer only if the layer has the same phase constitution across its entire thickness. Otherwise, the phases that are present close to the surface are detected and those lying deeper inside the alloyed layer are missed. Similarly, in case of aluminide coatings which have multi-layer structures, only the phases present in the outermost layer are detected by XRD and those present in the sub-surface layers are missed. This is primarily because of the fact that the outer coating layer is too thick for X-rays to completely penetrate and reach the layers below. However, if the outer layer is relatively thin, as shown in the example of alumina scale present on aluminide coatings in oxidized condition, XRD can detect phases in the coating layer below the scale. Despite the above limitation, XRD has and will remain a good technique for identification phases in both thick and thin coatings.

Acknowledgements: The author acknowledges DRDO for funding this work. He is also grateful to Director, DMRL, for permitting him to publish this work.

REFERENCES

Allam, I.M. and Whittle, D.P. (1978). *J. Stringer, Oxid. Met.*, **12:** 35.

Belmondo, A. and Castagna, M. (1979). *Thin Solid Films*, **64:** 249.

Cullity, B.D. (1978). Elements of X-ray Diffraction, 2nd Ed. Addison-Wesley Publishing Inc., London.

Das, D.K. (2001). Microstructural Evolution and Oxidation Behaviour of Pt-Aluminide Coatings on Cast Ni-Base Superalloy CM-247LC. Ph.D. Thesis, Department of Metallurgical Engineering, Institute of Technology, Banaras Hindu University, Varanasi, India.

Das, D.K., Vakil Singh and Joshi, S.V. (2002). *Oxid. Met.*, **57:** 245.

Das, D.K., Manish Roy, Singh, A.K. and Sundararajan, G. (1996). *Mater. Sci. Technol*, **12:** 295.

Das, D.K., Manish Roy, Vakil Singh and Joshi, S.V. (1999). *Mater. Sci. Technol.*, **15:** 1199.

Das, D.K., Prasad, K.S. and Paradkar, A.G. (1994). *Mater. Sci. Eng.* A, **A174:** 75.

Das, D.K., Vakil Singh and Joshi, S.V. (2000). *Metall. and Mater. Trans.* A, **31A:** 2037.

Draper, C. (1981). Lasers in Metallurgy. Mukherjee, K. and Majumdar, J. (eds.) TMS-AIME, Warrendale, PA, 67.

Farrell, M.S., Boone, D.H. and Streiff, R. (1987). *Surf. Coat. Technol.*, **32:** 69.

Meetham, G.W. (1986). *In:* High Temperature Alloys for Gas Turbines and Other Applications. Betz, W., Brunetaud, R., Coutsouradis, D., Fischmeister, H., Gibbons, T.B., Kvernes, I., Lindblom, Y., Marriott, J.B. and Meadowcroft, D.B. (eds.) D. Reidel Publishing Company, Dordrecht, 1.

Streiff, R. and Boone, D.H. (1985). *In:* Coatings and Bimetallics for Aggressive Environments. Sisson Jr. R.D. (ed) ASM, Metals Park, OH, 159.

Streiff, R., Cerclier, O. and Boone, D.H. (1987). *Surf. Coat. Technol.*, **32:** 111.

Sun, J.H., Jang H.C. and Chang, E. (1994). *Surf. Coat. Technol.*, **64:** 195.

Tatlock, G.J., Hurd, T.J. and Punni, J.S. (1987). *Platinum Metals Rev.*, **31:** 26.

X-ray Diffraction Characterization of Mechanically Milled SmCoFe Nano Magnetic Powders by Rietveld Refinement

K. Suresh, A.K. Singh, R. Gopalan and V. Chandrasekaran

Defence Metallurgical Research Laboratory
Kanchanbagh, Hyderabad–500 058, India

INTRODUCTION

In recent years, formation of nanocrystalline compounds has attracted a great interest in many fields of theoretical and industrial research due to their unusual properties. High-energy ball milling is one of the conventional methods of mechanical processing for production of nanocrystalline or amorphous materials (Suryanarayana, 2001; Murthy and Ranganathan, 1998; Begin-Colin et al., 1995). The intermetallic powders processed by high-energy ball mill usually undergo several mechanochemical phase transitions and the final products are composed of different phases along with contaminations from milling media. During this process, sometimes equilibrium between breakage and re-welding of powders are reached and particle size remains almost unaltered on further milling due to agglomeration of smaller particles but the quantity of amorphous phase increases and crystallite size decreases.

Most of the X-ray diffraction (XRD) profiles of individual phase in the pattern of ball-milled powder are completely or partially overlapped. As a result, without a reliable profile fitting method, it is almost impossible to extract the information regarding angular full-width at half-maxima (FWHM), intensity maximum, integrated intensity, Gausianity, etc. of individual profile by any conventional XRD method for particle size and strain analyses by the Scherrer formula (Cullty, 1967). Furthermore, the XRD peak broadening due to small crystallite size and root mean square strain $<\varepsilon^2>^{1/2}$ can be modeled by Lorentzian and Gaussian type profile fitting functions, respectively. The pseudo-Voigt (pV) profile function is a combination of these two profile functions. A crystal structure model for polycrystalline material based on these considerations namely, the Rietveld least square fit would be the best method, for microstructure characterization as well as for quantitative estimations of constituent phases and amorphous phase (Williamson and Hall, 1953; Young, 2000). The present work is thus concerned with the characterization of mechanically milled powder of $SmCo_5$/Fe nanocomposite alloys as function of different Fe content using Rietveld's refinement procedure. The $SmCo_5$/Fe nanocomposite alloys are of interest because the presence of $SmCo_5$ (hard) and Fe (soft) magnetic phases at nano level, are predicted to enhance the magnetic properties through spring exchange coupling mechanism (Kneller and Hawig, 1995; Zeng et al., 2002; Gopalan et al., 2003; Suresh et al., 2004). Thus, a detailed analysis of particle size, lattice strain and volume fraction of amorphous phase will be useful to optimize the processing of these alloys.

EXPERIMENTAL PROCEDURE

Initially, the alloy of compositions $SmCo_5$ (1:5) was prepared from high purity (>99.5%) constituent elements by melting under argon atmosphere. The ingot was crushed and subjected to ball milling along with different Fe content (for 0, 10, 30 and 40) for 72 hrs in Toluene media to prevent oxidation. These alloys are designated as alloy 1, alloy 2, alloy 3 and alloy 4, respectively. The starting sizes of the 1:5 alloy and Fe powders are around 50 microns and 3 microns, respectively. The ball milling has been carried out in a planetary ball mill (FRITSCH pulversitte) for 72 hrs. The XRD profile of the samples were recorded in a Philips PW3020 diffractometer equipped with a graphite monochrometer operated at 40 kV and 25 mA in a step size of 0.01° (2θ) and counting time of five seconds/step. The XRD data for all alloys were subjected to Rietveld refinement programme from Philips Xpert plus. Reference FWHM have been taken from well sintered and single phase 1:5 and α-Fe powders to eliminate the line broadening due to instrument and Lorenz factor. The quantification of amorphous phase has also been carried out by mixing known amount of Si powder in the final mechanically milled powder using Rietveld refinement procedure.

RESULTS AND DISCUSSION

Rietveld refinement technique has been used widely to simulate powder diffraction pattern providing all necessary structural parameters of the individual phases. The calculated experimental pattern at each point on the XRD pattern and selected parameters (both profile and structural) are refined by the least square technique to get the final structure.

The alloy powders were milled from 4 hours to 72 hours and powders have been collected at periodic interval for XRD characterization. It has been observed that after 72 hours of milling, XRD patterns exhibit presence of sufficient amount of amorphous phase. Therefore, a detailed analysis of 72 hours milled powders of all alloys has been undertaken in the present study. The XRD pattern of alloy 1 shows the presence of 1:5 and amorphous phases (Fig. 1a) while other alloys exhibit 1:5, α-Fe and amorphous phases after milling. A typical representative of XRD pattern consisting of these three phases is shown in Fig. 1b for alloy 2. It is to be noted that the intensity of (011) peak of α-Fe phase is increasing with increase in the Fe content in the alloy. This indicates that the amorphous phase of Fe is not formed during milling. Ito et al (2001) have also studied the mechanical alloying of 1:5 and α-Fe powders with different Fe content and reported Sm-Co amorphous phase and α-Fe. Based on these observations and also since α-Fe is a ductile phase, it has been assumed that the amorphous phase coming out during milling is from the 1:5 phase.

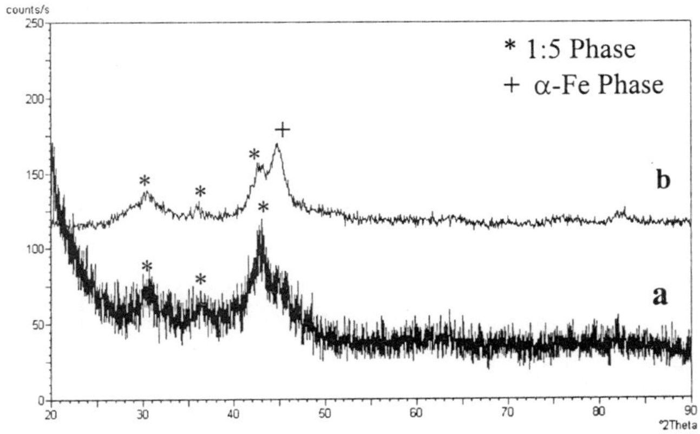

Fig. 1: The XRD patterns of (a) alloy 1 and (b) alloy 2.

Since the high intensity peaks of 1:5 and α-Fe phases in mechanically milled powder lie in 2θ range 39 to 47°, XRD patterns of alloy powders were recorded in this range in step scan mode. These XRD patterns are simulated and fitted (Fig. 2) with the low goodness of fit (GOF) value ≈1.0. It is important to mention here that (002) peak of 1:5 phase and (011) peak of α-Fe phase superimpose each other. The de-convolution of these peaks has been done after Rietveld refinement. The data have been used to calculate lattice parameters, particle size and lattice strain in both the 1:5 and α-Fe phases, respectively and are given in Table 1.

It is found that the lattice constants **a** and **c** of 1:5 phase show increasing trend with increase in the Fe content (Fig. 3). The lattice constant **a** of the α-Fe decreases with increase in Fe content in the alloy (Fig. 4). The observed variations in lattice parameters of the 1:5 and the α-Fe phases with the varying Fe content could be due to structural defects induced by milling as well as Fe-Co solid solution formation during milling. Expansion of **a** and

Table 1: Microstructural parameters of the milled alloys revealed by Rietveld refinement analysis.

Alloy	*Phase present*	*Weight frac.*	*Lattice Parameter (A°)*		*Crystal size (nm)*					*Lattice Strain%*				
			a	*c*	*[020]*	*[111]*	*[022]*	*[011]*	*[002]*	*[020]*	*[111]*	*[022]*	*[011]*	*[002]*
Alloy 1 ($SmCo_5$)	$SmCo_5$	36.4	4.979416	3.95725	10.6	10.5	10.2			1.274	1.26	1.22		
	Fe													
	Amorphous	59.6												
Alloy 2 ($SmCo_5$+10%Fe)	$SmCo_5$	25.58	4.983513	3.960245	7.1	7.1	6.9			1.748	1.713	1.622		
	Fe	8.58	2.8633					9.1	8.4				1.29	0.953
	Amorphous	62.77												
Alloy 3 ($SmCo_5$+30%Fe)	$SmCo_5$	13.5	4.994784	3.968462	9.5	9.6	9.2			1.342	1.355	1.306		
	Fe	21.7	2.859393					9.6	8.9				1.24	0.908
	Amorphous	60.94												
Alloy 4 ($SmCo_5$+40%Fe)	$SmCo_5$	17.5	4.99588	3.96926	10.9	10.8				1.135	1.12			
	Fe	20.9	2.856079					27.4	20.8				0.575	0.477
	Amorphous	58												

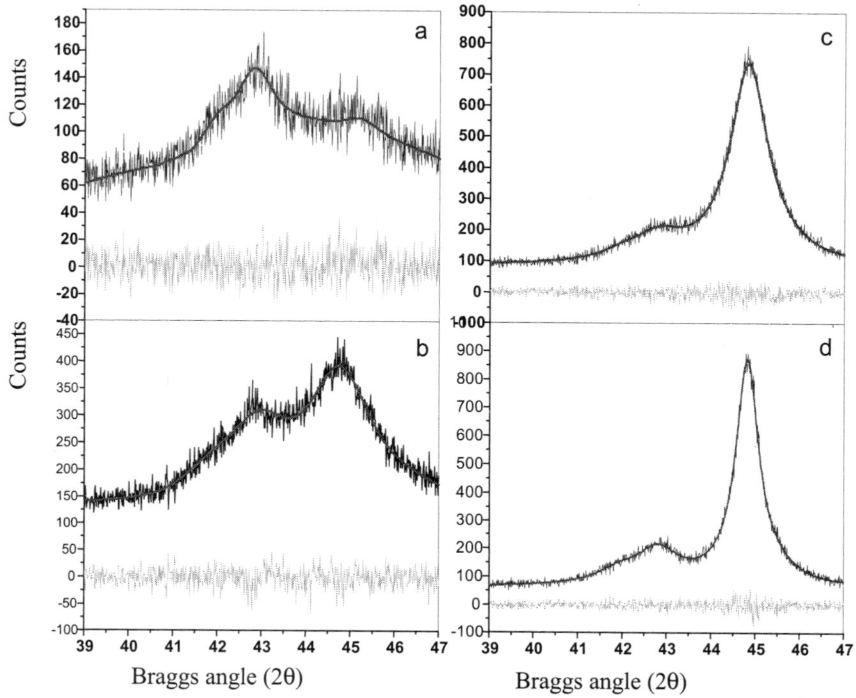

Fig. 2: Observed, calculated and difference XRD patterns of (a) alloy 1, (b) alloy 2, (c) alloy 3 and (d) alloy 4.

c parameters of pure Co have been observed during milling (Sort et al., 2003). This has been attributed to structural defects, which is likely to play some role in the expansion of the Co lattice. Apart from this, there is a possibility of forming solution of Co atoms into the α-Fe phase. It is believed that a part of Co-atoms coming out from the 1:5 phase forms solid solution with Fe. Therefore, the change in the lattice constants of both the 1:5 and α-Fe phases observed in the present study can be attributed to the structural defects and Fe-Co solid solutions

The average crystallite size and lattice strains of 1:5 phase have been calculated from {020}, {111} and {022} reflections for the first three alloys and (020) and (111) reflections for the alloy 4. The (022) peak of 1:5 phase could not be resolved in alloy 4 due to high intensity (011) Fe peak. The crystallite size and strain of 1:5 phase is nearly constant as a function Fe content (Fig. 5). The slightly low value of crystal size and strain of phase 1:5 in alloy 2 may probably be attributed to the deconvolution of 1:5 and α-Fe phases due to small difference in the intensity of both the phases. It is important to mention here that the values obtained for particle size and lattice strains for 1:5 phase are similar for all the three planes mentioned above. Since this phase is being a brittle phase, it breaks into small particles during mechanical milling without expansion/contraction in any particular

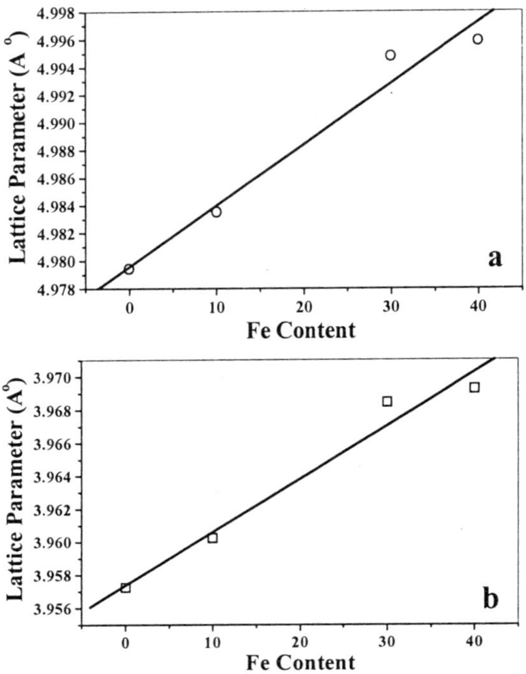

Fig. 3: The variations of lattice constants of 1:5 phase as a function of the Fe content in the alloy powders (a) **a** and (b) **c**.

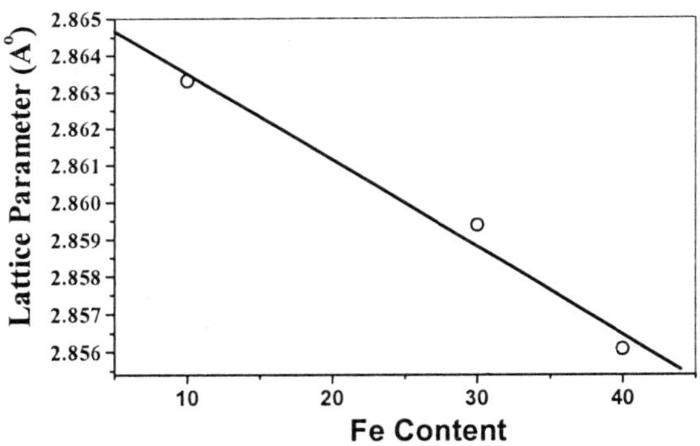

Fig. 4: The variation of lattice constant **a** of the α-Fe phase as a function of the Fe content in the alloy powders.

crystallographic direction. As a result, it exhibits similar crystallite size in different directions and hence the powder of 1:5 phase is isotropic in nature.

The particle size of the α-Fe is increasing and lattice strain is decreasing with increase in the Fe content (Fig. 5). However, the particle size and lattice strain of Fe calculated along [011] and [002] directions are different. This indicates that the milled α-Fe powders of the alloy are anisotropic in nature. The α-Fe powder being a ductile phase expands/contracts in some crystallographic directions during the initial stages of mechanical milling and then breaks into smaller particles on further milling. The different crystallite size and strain in the different directions [011] and [002] can be attributed to the development of crystallographic texture during initial stage of mechanical milling. The large value of strain along [011] direction in the α-Fe phase is in agreement with the fact that [011] is the typical preferred orientation after rolling in bcc materials (Butron-Gullien et al., 1994). However, further reduction of grains (more than 72 hrs milling of experimental alloys) will reduce anisotropy of the α-Fe powders.

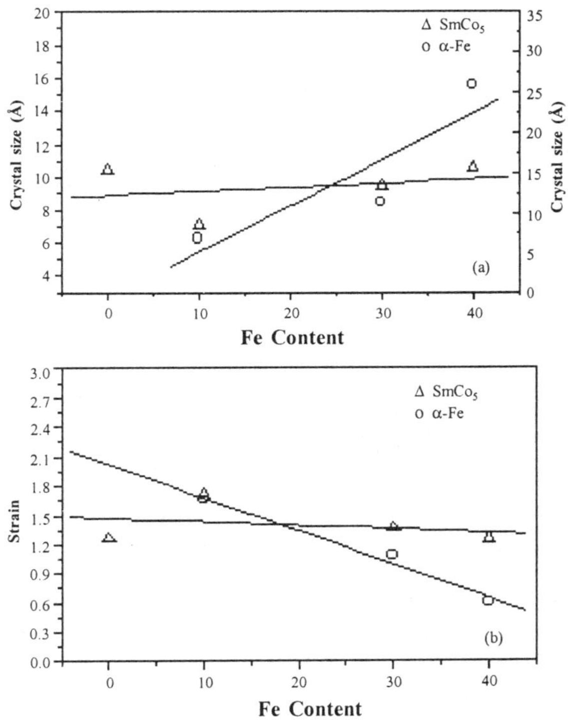

Fig. 5: The variations of (a) crystallite size and (b) strain of 1:5 and α-Fe phases as a function of Fe content in the alloy powders.

The volume fraction of the amorphous phase lies in the range of 58 to 63 % in all alloys. This shows that the volume fraction of amorphous phase appears to be independent of the Fe content. This is due to the fact that the α-Fe phase is ductile and 1:5 phase is brittle. As a result, the amorphous phase is coming out from 1:5 phase only and it remains nearly same in all alloys after 72 hrs milling irrespective of different volume fractions of α-Fe phase. It is also important to mention here that no other phase has been observed in the experimental alloys during 72 hrs mechanical milling due to milling induced phase transformation.

CONCLUSIONS

- The alloy 1 shows the presence of 1:5 and amorphous phases while the other alloys exhibit 1:5, α-Fe and amorphous phases after 72 hrs mechanical milling.
- The crystallite size and strain of 1:5 phase remains nearly constant and shows opposite trend in α-Fe phase with increasing Fe content in the alloy powders.
- The milled powder of the 1:5 phase is isotropic and the α-Fe is anisotropic in nature.

Acknowledgments: The authors are grateful to Defence Research and Development Organization, Government of India for financial support to carry out this work. The keen interest shown by the Director DMRL in this work is gratefully acknowledged. The authors also acknowledge the staff members of SFAG and MMG of the DMRL for their necessary help.

REFERENCES

Begin-Colin, S., Le Caer, G., Zandona, M., Bonzy, E. and Malaman, B. (1995). *J. Alloys Comp.*, **227:** 157.

Butron-Gullien, M.P., Jonas, J.J. and Ray, R.K. (1994). *Acta Metall.*, **42:** 3615.

Cullty, B.D. (1967). Elements of X-ray diffraction, 3rd Edition, Addison-Wesley Massachusetts, USA.

Gopalan, R., Suresh, K., Singh, A.K. and Chandrasekaran, V. (2003). *Scripta Materialia*, **48:** 1555.

Kneller, F.E. and Hawig, R. (1995). *IEEE Trans. Magn.*, **31:** 3596.

Mikio Ito, Kazuhiko Majima, Takashi Umemoto, Shigeru Katsuyama and Hiroshi Nagai (2001). *J. Alloys Comp.*, **329:** 272.

Murthy, B.S. and Ranganathan, S. (1998). *Int. Mater. Rev.*, **43:** 101.

Sort, J., Nogues, J., Surinach, S. and Boro, M.D. (2003). Phill. Mag, **83:** 439.

Suresh, K., Gopalan, R., Singh, A.K., Muraleedharan, K., Sridhara Rao, D.V. and Chandrasekaran, V. (2004). *Indian J. Phy.* A, **78A:** 115.

Suryanarayana, C. (2001). Progress in Materials Science, **46:** 1.

Williamson, G.K. and Hall, W.H. (1953). *Acta Metallurgica*, **1:** 52.

Young, R.A. Ed. (2000). The Rietveld Method, IUCR. Oxford Science Pub., New York, USA.

Zeng, H., Li, J., Liu, J.P., Wang, Z.L. and Sun, S. (2002). *Nature*, **420:** 395.

X-ray Characterization of a Biocompatibility Material from Natural Sources

S. Annapurna, T. Jagadeesh Kumar[1] and Shankar[1]

Osmania University College for Women
Department of Physics, Hyderabad-500195, India
[1]Defence Metallurgical Research Laboratory
Kanchanbagh, Hyderabad-500058, India

INTRODUCTION

Among the hard calcified tissues, biological shells play an important role in protecting the animal from environmental influences. It is well known that the formation of the shells is complex in nature and can be described in terms of two major processes: (1) cellular process of ion transport, protein synthesis and secretion and (2) a series of physicochemical process in which crystals of $CaCO_3$ are nucleated, oriented and then grow in intimate association with a secreted organic matrix. The understanding of process of crystal depositional phase of shell formation is fragmentary. However, through the analysis of fluid at the site of crystallization, the conditions governing the initiation, control of crystal growth and the formation of complex structure of the crystal units assembled in a large variety of patterns will be further defined.

Biological composites such as bones and teeth contain hydroxy apatite (HAP), a mineral component with the composition $Ca_{10}(PO_4)_6(OH)_2$. HAP is very useful biological material in dental implantation and it is a major component of bones and teeth (Ishikawa et al., 1995; Ishikawa, 1995). Synthetically prepared HAP is biocompatible and has been used successfully in dental implants for restorative dental surgery and also filling cavities. HAP has been synthesized for manufacturing various forms of implants. In the present work, HAP was synthesized by removing of CO_3 from $CaCO_3$ and adding of nitric acid (HNO_3) and hydroxy phosphate (H_3PO_4). An attempt has been made in this work to identify the structure of the HAP by X-ray diffraction (XRD) and chemical analysis data.

A brief mention of clinical application of HAP is relevant in the present context in the light of its role as a biological material of extensive contemporary interdisciplinary research. While some aspects related to dental caries have been mentioned the $F^- \rightarrow OH^-$ exchange on HAP, attempts have been made to shed light on the trends in the clinical aspects concerning the incidence of these diseases (Guobo and Mingzhu, 1989; Choung, 1973). Fluorosis is a disease caused by excessive exposure to fluorine while fluorine is one of trace element essential for life. It has been shown to be deleterious in large quantities. Fluoride when given in large doses, has been shown to reduce collagen in bones and has been found to lead to a pathological calcification. Pains in the joints, stiffness of the spine, difficulty in walking, irregular bone disposition, narrowing of the spinal canal etc. are due to excess fluoride leading to a retardation of the physical activity.

MATERIALS AND METHODS

For the present investigation minerals of calcite, marble, limestone are collected from Osmania University Geology Museum, Hyderabad and marine molluscan shells were collected from different seashores. The shell samples were cleaned to remove the foreign material and made into fine powder by grinding in agate motor.

X-ray studies were made using Siemen's X-ray diffractometer with scintillation counter and Fekα radiation. General scan of the powder sample was taken with a scanning speed of 1 degree/min for characterization. Chemical composition of Ca and P was analyzed by titration method.

In the present investigation we have considered two types of sources, which are marine shells (biological form) and rock samples (mineralogical form). In the case of marine shells we have considered the two types of samples namely conus SP and mercineria mercineria, while in case of rock samples we have considered aragonite and calcite crystals. According to previous reports the transformation temperature of aragonite to calcite is 465 °C. But in the case of marine shells the transformation temperature of aragonite to calcite is 300 °C (Annapurna and Nagander Naidu, 1989).

PREPARATION OF HAP FROM NATURAL SOURCES

Pure hydroxy apatite can be synthesized by dissolving calcium carbonate in nitric acid, which will give the product of calcium nitrate. This calcium nitrate can be precipitated with the help of phosphoric acid in the presence of ammonium hydroxide and result in pure hydroxy apatite $Ca_{10}(PO_4)_6(OH)_2$.

Example

Raw material → marine shell ($CaCO_3$)

STEP 1

$$CaCO_3 + 2HNO_3 \rightarrow Ca(NO_3)_2 + H_2O + CO_2$$

STEP 2

$$10Ca\ (NO_3)_2 + 6(H_3PO_4) + 20NH_4\ OH \rightarrow Ca_{10}\ (PO_4)_6\ (OH)_2 + 20\ NH_4NO_3 + 18\ H_2O$$

It is to be noted that HAP has also been prepared by following method (Hayek and Stadlmn, 1955).

$$10Ca(NO_3)_2 + 6(NH_4)_2HPO_4 + 8NH_4OH \rightarrow Ca_{10}\ (PO_4)_6(OH)_2 + 6H_2O + 20NH_4NO_3$$

RESULTS AND DISCUSSION

X-ray data of the HAP (natural sources) is presented in Table 1. This exhibit well defined characteristics of synthesized HAP and this is comparable to those reported in literature for both biological and mineralogical forms and agrees with the ASTM data of HAP. The X-ray intensities and *d* spacing of the reflections observed for the synthesized HAP samples are compared with those reported in the literature for the HAP structures.

Table 1: XRD Data on HAP (Natural Sources)

Reported HAP ASTM Data			*Imported HAP*		*Observed Values Biological Form*				*Observed Values Mineralogical Form*			
					Pure Ca		*Marine Shell*		*Calcite*		*Lime Stone*	
D	I/I_o	*hkl*	*d*	I/I_o	*D*	I/I_o	*d*	I/I_o	*d*	I/I_o	*d*	I/I_o
3.440	55	002	3.440	55	3.435	50	3.435	55	3.426	55	3.445	55
2.817	100	211	2.813	100	2.81	100	2.819	100	2.812	100	2.810	100
2.779	59	112			2.743	43	2.722					
2.723	81	300			2.713	81	2.722	81	2.722	81		
2.631	35	202	2.637	35			2.637	35	2.631	35		
2.263	25	130	2.276	25			2.265	25				
1.945	41	222	1.949	41	1.932	41	1.945	41	1.945	41	1.951	41
1.847	32	213	1.845	32	1.845	32	1.847	32			1.845	32
1.73	18	402	1.723	22	1.790	13	1.736	18			1.730	18

Sources of raw materials: Calcium, Marine Shell, Calcite and Lime Stone

As mentioned above, good agreement is found in the observed 'd' spacing of the sample with those in the intensity data. Particularly in mineralogical samples like marbles and Ca pure metals, the strong reflections (211) and (002) are almost same with ASTM data. When compared to the other intensities similar phenomenon was observed in biological samples. The anomalous variations in the intensity of certain reflections might be related to the species, growth characteristic of the HAP. However, the other synthesized HAP samples are in agreement with the reported data.

A small quantity of HAP powder was taken for chemical analysis and the results are presented in Table 2. It is clear that although the pH value of milk teeth, normal teeth and deceased teeth are same but the ratio of Ca/P is quite different. Tables 3 and 4 show the Ca/P ratios (observed) and also the pH reported values in the case of mineralogical and biological forms, respectively. Tables 3 and 4 show that the Ca/P ratio is same for limestone and conus sp, which is also in agreement of reported HAP data (Table 2). It is clear from Tables 2, 3 and 4 that Ca/P ratio plays a very important role in the formation of HAP.

Table 2: Comparison of the Ca/P Ratio (observed) and pH (reported) of HAP

S. No.	*Identification*	*Ca/P*	*pH*
1	Synthetic HAP Imported from Japan	1.62	7.40
2	Milk Teeth	2.46	7.00
3	Normal Teeth	2.22	7.00
4	Deceased Teeth	2.38	7.00

Table 3: Comparison of the Ca/P Ratio (observed) and pH (reported) of HAP of mineralogical form

S. No.	*Identification*	*Ca/P*	*pH*
1	Calcite	1.80	7.00
2	Marble	1.89	7.00
3	Lime Stone	1.67	7.00
4	Ca (Metal)	1.32	6.90

Table 4: Comparison of the Ca/P Ratio (observed) and pH (reported) of HAP of biological form

S. No.	*Identification*	*Ca/P*	*pH*
1	Conus sp	1.67	7.00
2	Mercineria	1.52	7.00
3	Conus sp	2.05	7.00

The gram atom ratio of Ca/P saturated solution is not consistent at the stoichiometric value of 1.67. Because natural biomaterials are established by physicochemical properties, these investigations gave a convincing explanation to account for the results on solubility of HAP.

The cost of raw material used in the present study is low and are available in nature. Apart from this the major toxic elements Ba, Pb, Fe, Sr, Al, Mg, Hg etc. are not present in synthesized HAP. Hence it can be used as a biocompatible material in medical applications. We have also tried to assess setting time for synthesized HAP. The setting times lie between 5 to 8 minutes.

For the present HAP study we have taken the samples from natural sources both mineralogical and biological forms. Finally, according to the XRD and chemical analysis data of Ca/P ratio, we synthesized a biomaterial HAP from marine shells, which is also useful for artificial bone implants even in defects and deficiencies in relatively vascular areas of madubular bone. The synthesized HAP obtained in the present study has been successfully used in dental implants and also filling cavities.

CONCLUSIONS

1. Pure HAP has been synthesized by precipitating calcium nitrate with phosphoric acid in the presence of ammonium hydroxide.
2. Reported pH values of HAP are nearly same as those observed in synthesized HAP.
3. The cost of raw materials used in the present study is low and available in nature.
4. The setting time of synthesized HAP is nearly same as that of HAP cement setting time.
5. The synthesized HAP is biocompatible and has been successfully used in dental implants and also filling cavities.

Acknowledgement: One of us (SA) is grateful to Dr. A.K. Singh, DMRL for extending laboratory facilities and inspiring discussions.

REFERENCES

Annapurna, S. and Nagander Naidu, S.V. (1989). *Ind. J. Physics*, **63A:** 142.

Chuong, (1973). *J. Dent. Res.*, **52:** 911.

Guobo, N. and Mingzhu, M. (1989). *Chin. J. Biotechnol.*, **5:** 97

Hayek. E. and Stadlmann, W. (1955). *Anguas. Chem.*, **67:** 1.

Ishikawa, K. (1995). *J. Biomed. Mater. Res.* **29:** 1537.

Ishikawa, K., Takagi, S., Chow, L.C., Ishikawa, Y. (1995). *J. Mater. Sci. Mater. Med.*, **6:** 528

Forensic Examination of Some Textile Fabrics Using XRD and XRF—A Preliminary Study

Johny T. Abraham

Central Forensic Science Laboratory
Government of India
Ramanthapur, Hyderabad 500 013, India

INTRODUCTION

There are many reports dealing with the examination of individual fibres in the forensic literature (Grieve, 2000). However, somewhat less attention has been directed towards the forensic examination of fabrics. In principle, there are two ways to tackle the examination of fabrics: either individual fibres can be plucked from the fabric and then analyzed using conventional techniques; or the entire fabric can be analyzed. The former approach suffers in that it is tedious, individual fibre techniques require a certain level of skill on behalf of the operator, and care must be taken to ensure the fibres extracted are representative of the entire fabric. This paper introduces simple techniques relating to the latter approach.

X-ray Diffractometry (XRD) and X-ray Fluorescence Spectrometry (XRFS) have long been used in the textile industry to evaluate the crystallinity of fibres and estimation of elemental composition, respectively (Sreenivasan, 1998). The methods used in industry, however, are destructive in nature. From the literature available for fibre examination, we found that not much work has been carried out in studying the scope of XRD and XRFS as a non-destructive analytical tool for the forensic examination of fibres. In this study we have attempted to characterize some commonly found textile fabrics using XRD and XRFS methods that allow the samples to be recovered intact after analysis.

MATERIALS AND METHODS

Three types of textile fabrics namely unblended cotton, unblended polyester and blended cotton-polyester (terri-cotton) fabrics that are commonly found at scenes of crime were examined. For the XRD studies, three pieces of a fabric were stacked one over the other on the X-ray sample holder of dimension of 30 mm diameter. These pieces were held together by means of adhesive tape. Care was taken to keep the edges of the cellophane tape totally away from the path of X-ray beam since characteristic peaks of cellophane tape also fall in the same Bragg angle region of fibres. For the XRFS studies, four fabric pieces of a fabric were stacked one over the other and kept inside the specimen holder of 50 mm diameter dimension. A plastic disc was kept on the top of the stacked fabric pieces to keep it wrinkle free. Prior to the analysis of fabrics, the plastic disc was first analyzed and its elemental spectrum recorded. During the analysis of fabric samples, the spectrum of plastic disc was subtracted from that of the fabric samples and resulting spectra are reported in this study.

The XRD studies were carried out using a Philips X'Pert MPD system using CuKα radiation (40 kV, 50 mA). The samples were analyzed from a Bragg angle (2θ) of 10° to 40°, while the programmable divergence slit and receiving slit were kept at 1° and 0.1 mm, respectively. The X-ray fluorescence studies were carried out using a Philips PW2400 model Wavelength Dispersive X-ray Fluorescence Spectrometer. The samples were analyzed from a Bragg angle (2θ) of 10° to 150° and the secondary X-rays were scattered using LiF200, Ge111, PX1, PX3, PX4, PX6 and PE002 crystals and detected using scintillation detector, flow detector and a duplex detector.

All the results mentioned below are reported based on the observations of minimum three repetitions under the same test parameters.

RESULTS AND DISCUSSION

X-ray Diffraction

Textile fibres are classified into two basic groups, namely natural fibres and synthetic fibres. The basic constituents of natural fibres are protein (animal fibres), cellulose (vegetable fibres) and asbestos (mineral fibres). Synthetic fibres are manufactured from either naturally occurring fibre forming polymers (for example viscose) or synthetic fibre forming polymers (for example polyester). Irrespective of the fibre type, the basic constituents have a percentage of crystallinity in their chemical composition, but are not totally crystalline. In other words, textile fibres are a mixture of crystalline and amorphous regions. The extent to which either region, i.e., crystalline and amorphous, predominates within the fibres largely determines the overall properties of the fibre (Robertson, 1992). It is this property that provides useful data while characterizing textile fibres using XRD.

Textile fabrics (woven) of unblended cotton, unblended polyester and blended cotton-polyester of two brands locally available (Bombay Dyeing, India and Binny, India) were selected for the study. Samples initially selected were white (colourless) and free of any stains/contaminants. The XRD patterns obtained for the three samples are depicted in Fig. 1. Figures 1a to 1c represent the X-ray diffraction patterns of cotton, cotton-polyester blend and polyester, respectively. The variations of diffraction patterns of fabrics, if any, with respect to brand and colour were then investigated. In addition, variation with respect to fabric samples taken from different portions of the same roll of fabric was also studied. Figures 2a and 2b represent the diffraction patterns of cotton fabric samples (colourless) belonging to Bombay Dyeing and Binny brand. Figures 2b and 2c depict the diffraction patterns of two cotton fabric samples of white and brown colour of Binny brand. Figure 3 illustrates the XRD patterns of cotton fabric samples taken from two different portions of the same roll of fabric.

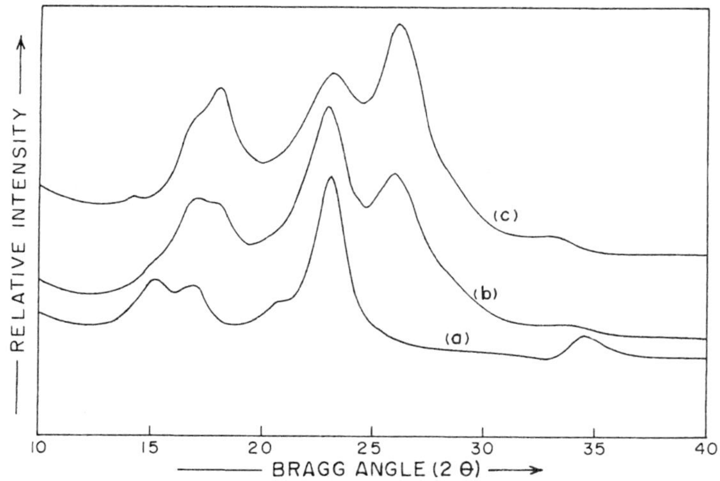

Fig. 1: XRD patterns of textile fabrics: (a) cotton, (b) cotton-polyester and (c) polyester.

When the XRD patterns (Figs. 2a and 2b) of fabric samples of two different brands are studied, it could be observed that the diffractograms obtained are characteristics of the constituent fibre, namely cotton. The variation due to the brand is reflected in the intensity and peak width of the peaks at Bragg angles of 15.5°, 17° and 23°. Similarly, when samples of cotton fabric of different colours (Figs. 2b and 2c) and from different portions of same roll (Fig. 3) were analyzed, variation due to colour and location was reflected in the intensity and peak width of the peaks obtained.

The variation in the peak width and intensity of the X-ray peaks observed with respect to brand, colour and location within a roll could arise as a result of variation in size of the crystallites and their concentration present in the

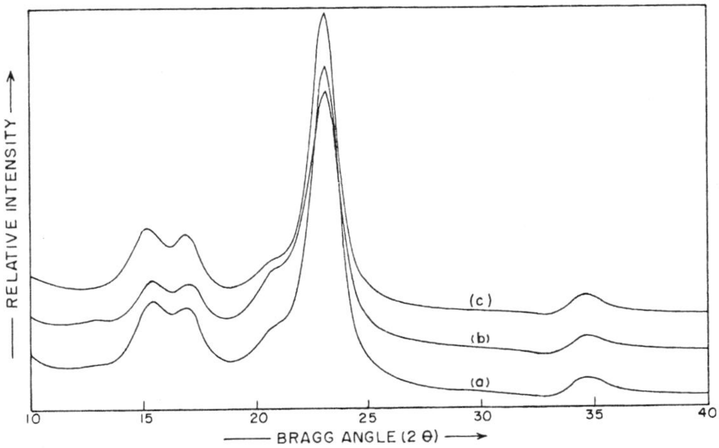

Fig. 2: XRD patterns of (a and b) white cotton fabric samples of two different brands and (c) brown cotton fabric sample.

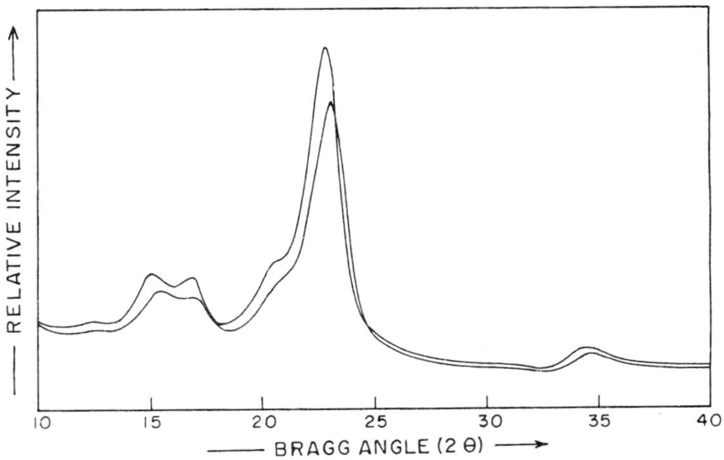

Fig. 3: XRD patterns of cotton fabric taken from two different portions of the same roll of cloth.

fibres. It is known that the width of peaks in a XRD pattern provides an indication of the average crystallite size (Cullity, 1978). Large crystallites give rise to sharp peaks while the peak width increases as crystallite size decreases. The variation in crystallite size and its concentration observed with respect to brand, colour and portion could be due to the difference in the sources (localities) from which the fibres (which went into the construction of the fabrics) had come from and also due to manufacturing process.

In blended fabrics, the constituent fibres are present in varying percentage. Hence, a study was conducted on blended fabrics so as to ascertain whether the X-ray profiles could identify the constituent fibres and at the same time used as a measure of their relative abundance. Figure 4 depicts the XRD patterns of cotton-polyester blended fabrics. Figures 4a to 4d represent profiles of unblended cotton, blended fabric having 58% cotton and 42% polyester, blended fabric having 33% cotton and 67% polyester and unblended polyester, respectively. If the profiles of the blended fabrics (Figs. 4b and 4c) are observed, it could be seen that the diffractograms are different with respect to the amount of constituent fibres. Moreover, the diffractograms of the two blended fabric samples (Figs. 4b and 4c) exhibit a tendency towards pure cotton or pure polyester X-ray characteristics depending upon the greater percentage of cotton or polyester present in the fabric.

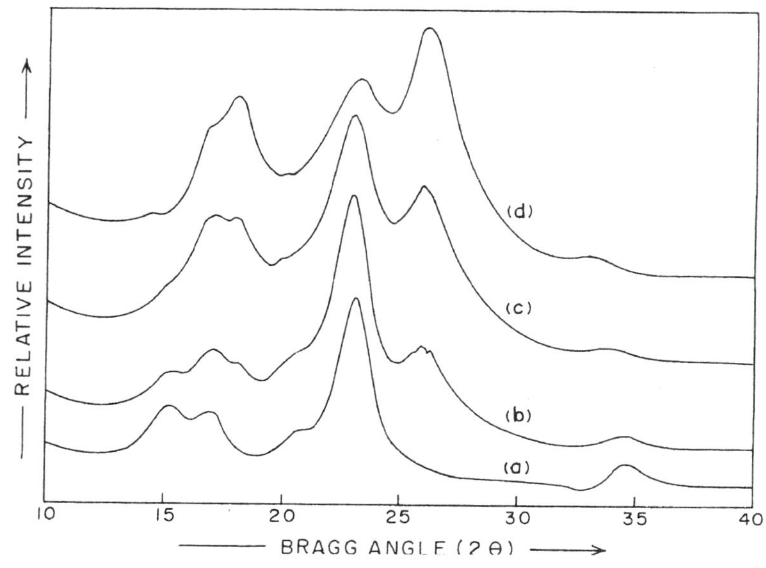

Fig. 4: XRD patterns of (a) cotton, (b) 58% cotton 42% polyester, (c) 33% cotton 67% polyester and (d) polyester.

We also analyzed knitted fabric made of wool and nylon fibres. The XRD patterns of nylon and wool fabrics along with that of cotton and polyester fabrics are depicted in Fig. 5. Figures 5a to 5d represent the profiles of unblended wool, cotton, polyester and nylon fabrics, respectively. In case of woollen fabrics, the XRD profile obtained did not have any characteristic peaks, unlike any other fabric examined in this study. This can be attributed to the amorphous structure of wool fibres.

Fabric pieces received for forensic examination might be contaminated with many and various substances. Hence it was examined whether the XRD techniques can identify the constituent fibres when the fabric is soiled. For this we selected a polyester fabric sample free of any stains/contaminants and recorded its XRD pattern (Fig. 6a). Then the surface of the fabric piece was soiled by placing about 1 mg of soil on it, then making it into a paste by mixing it with a few drops of water. The paste was then allowed to dry under normal conditions, after which the soiled fabric piece was analyzed

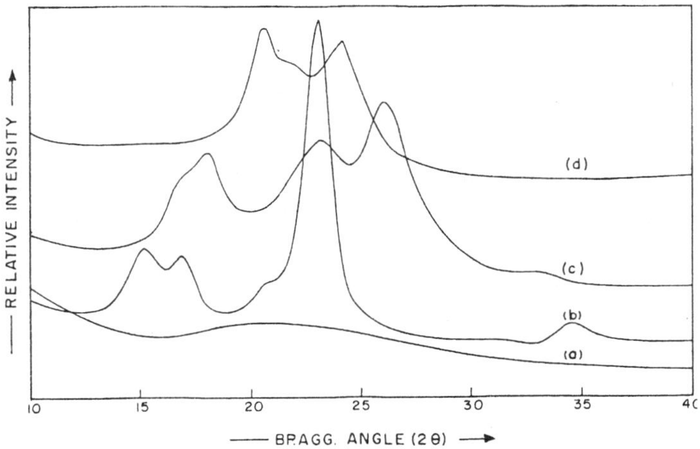

Fig. 5: XRD patterns of textile fabrics (a) wool, (b) cotton, (c) polyester and (d) nylon.

using the X-ray diffractometer. The XRD pattern obtained (Fig. 6b) was then compared with that of the unstained polyester fabric (Fig. 6a) and that of the soil used for staining (Fig. 6c). As can be seen from Fig. 6, the X-ray profile obtained (Fig. 6b) was similar to that of the fabric before soiling (Fig. 6a).

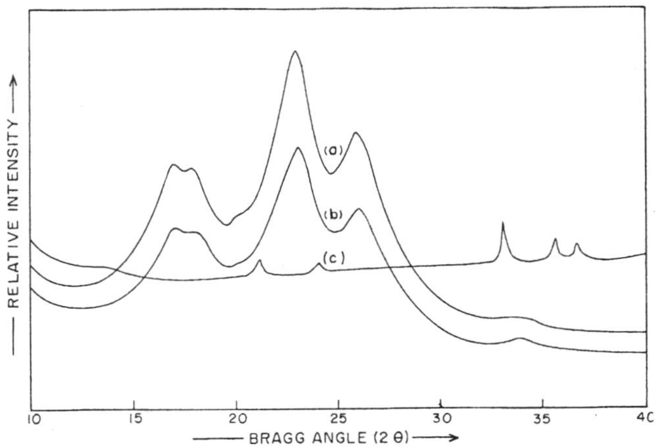

Fig. 6: XRD patterns of (a) polyester fabric, (b) soiled polyester fabric and (c) soil.

XRD is a technique applicable only to macro-scale pieces of fabric. A practical minimum sample size using our technique is a single piece of fabric of about 20 x 2 mm. The minimum length required was obtained by measuring the dimensions of the sample being exposed to the X-ray beam when the programmable divergence slit was kept at its minimum slit value.

Once the type of fibre is identified, another useful feature of the XRD in case of blended fabrics is that the same patterns could be used in estimating the relative percentage of different fibres. This can be achieved by comparing (peak area matching) the patterns of the unknown fabric with that of the certified reference standards of blended fabrics. Such certified reference standards of fabrics is supplied by the LGC Office of Reference Materials, United Kingdom (Item code: LGC 9001 and LGC 9001 in CRM Catalogue). This type of quantification is less tedious and time consuming compared to traditional gravimetric methods. Though this aspect was not thoroughly investigated in this study, we intend to investigate it further.

X-ray Fluorescence

In routine fibre examinations, techniques like Thin Layer Chromatography, UV-Visible and IR Spectrophotometers are adopted to ascertain the similarity of the questioned and control samples on the basis of colour, dyes, etc. Another characteristic property that can provide useful clue for the source correspondence of the samples is elemental profile (major and minor), for which an XRFS is a suitable analytical technique.

The elemental profile of fabrics should provide a means of comparison for forensic purposes. In the production of fabrics, fibres and yarns undergo various physical and chemical treatments in order to improve the physical properties and colour of the final product, fabric. During these processes various elements in varying proportions become part of the final fabric products. It is also well known that cotton lint that reaches a textile mill contains a high level of elements arising from foreign matter. For example, the presence of toxic inorganic elements, particularly heavy metals, in foreign matter associated with cotton has been reported (Sreenivasan, 1998). An elemental analysis of plant parts and lint carried out using XRFS (Warrier et al., 1992) indicates the presence of as many as 26 inorganic elements. The origin of these chemicals can be soil, fertilizers, fungicides and herbicides applied to cotton plant or seed.

The XRFS technique in the present study has been used as a means of comparing fabrics on the basis of their elemental profiles. Clean samples of cotton, cotton-polyester blend and polyester fabrics (Bombay Dyeing, India and Binny, India) were selected for testing. Unlike the XRD study discussed previously, the XRFS study did not involve stained fabrics, in order to avoid adding extra complexity to the spectral data of the spectrum involving the fabric and the contaminants like soil, so that the usefulness of the elemental profile of fabric samples in its source correspondence could be better understood.

Fabric samples were qualitatively analyzed using XRFS; the elemental profiles obtained with respect to 18 elements for the three types of fabric of two brands are given in Table 1. For easy identification, we have labeled the samples in the order of brand name, type of fabric and serial number of sample and details are mentioned along with the table. The presence of an element in a sample is indicated if the peak intensity (counts/sec) allows it to be distinguished from the "background" counts for the element in the plastic disc. From the results shown in Table 1, it is clear that all the six fabric samples can be discriminated on the basis of their elemental profiles. The foregoing suggests that if a piece of questioned fabric and a reference fabric share identical elemental profiles, then there is very strong evidence that the two pieces of fabric share a common source. The results given above are achieved using samples of 50 mm in diameter. However, a practical minimum quantity of fabric that can be profiled is about 6 mm in diameter.

CONCLUSIONS

This study indicates that XRD and XRFS can be powerful, non-destructive tool for fibre identification and fabric comparison. Unlike single-fibre techniques such as infrared microspectoscopy, fabric pieces can be analysed in-situ without the need for individual fibre selection and preparation.

Table 1: Qualitative analysis of fabric samples with respect to various brands and type of fibres

Elements → *Sample* ↓	*Na*	*Mg*	*Al*	*Si*	*P*	*S*	*Cl*	*K*	*Ca*	*Ti*	*Cr*	*Fe*	*Co*	*Ni*	*Cu*	*Zn*	*Sb*	*Mn*
BDC1	Y	Y	Y	Y	Y	Y	Y	Y	Y			Y			Y			
BDTC1		Y		Y	Y	Y		Y	Y	Y	Y	Y			Y			
BDP1	Y	Y	Y	Y	Y	Y		Y	Y	Y		Y			Y		Y	Y
BC1	Y	Y	Y	Y	Y	Y	Y	Y	Y		Y	Y	Y		Y			
BTC1	Y	Y		Y	Y	Y	Y	Y	Y	Y		Y		Y	Y			
BP1	Y	Y	Y	Y	Y	Y	Y	Y	Y	Y		Y	Y	Y	Y			Y

BDC1–Bombay Dyeing cotton; BDTC1–Bombay Dyeing Terri-cotton, BDP1–Bombay Dyeing polyester ; BC1–Binny cotton; BTC1–Binny terricotton; BP1–Binny polyester

Y *stands for Yes; element is present.*

The XRD patterns are unique for each type of fibre and can therefore be used as a means of identification. Although it is difficult to interpret the XRD data from first principles, searchable, digital databases can be constructed. Identification of fibres under these circumstances does not demand high expertise. Fabrics may be a source of other types of physical evidence like soil, paint, blood, etc. In case of soiled fabric, the studies reveal that the type of fibre constituent in a fabric piece could easily be identified non-destructively even if the samples are soiled. Further, if the amount of soil adhering to the sample is sufficient, X-ray profiles of the adhering soil can be compared with that of the control soil for source correspondence.

In case of XRFS, elemental profiles can be used to compare and discriminate fabrics. In the event that two fabric samples are found to have identical elemental profiles, there is a strong indication that the two pieces share a common origin. However, unlike XRD techniques, soiling on fabrics will have a detrimental effect since those of the soil will irreversibly alter the elemental characteristics of the fabric. Under these circumstances the only solution is to analyse representative samples from unsoiled areas of the fabric.

Acknowledgements: The author is grateful to the Directorate of Forensic Science, Ministry of Home Affairs, New Delhi for giving necessary sanction to carry out this work. The author wishes to thank Dr. J.S. Chauhan, Senior Scientific Assistant (Ballistics) and Shri P.N. Ramakrishnan, Senior Scientific Assistant (Physics) for their technical assistance.

REFERENCES

Cullity, B. D. (1978). Elements of X-ray Diffraction, Addison-Wesley, Massachusetts, USA.

Grieve, M. C. (2000). Back to the future – 40 years of fibre examinations in forensic science. *Science and Justice*, **40:** 93-99.

Robertson, J. (1992). Forensic examination of fibres. Ellis Horwood Limited, Sussex, England.

Sreenivasan, S. (1998). Proceedings of the National Workshop on X-ray Analysis, Nagpur, India.

Warrier, J.K.S., Chidambareswaran, P.K. and Sundaram, V. (1992). Indian Textile Journal, **5**: 54.

Microstructure and Optical Properties in Mesoporous Eu^{3+}:Al_2O_3 of a New Series of Optical Sensors

P. Mohanty and S. Ram

Material Science Centre
Indian Institute of Technology, Kharagpur-721302, India

INTRODUCTION

Amorphous Al_2O_3 or SiO_2 offers a promising host for doping of rare earth cations (R^{3+}) in novel optical materials with superior linear and nonlinear properties (Nogami and Hayakawa, 1997; Jessensky et al., 1998; Cai et al., 1998; Meltzer and Hong, 2000; Gedanken et al., 2000; Hayakawa and Nogami, 2001; Mohanty and Ram, 2002; Ishizaka et al., 2002; Nogami et al., 2002; Moadhen et al., 2002). The optical properties strongly depend on the R^{3+} concentration and their distribution and local microstructure with host. A controlled electron-phonon coupling and/or ionic interactions in R^{3+} cations monitor optimal properties in doped Al_2O_3 or SiO_2 useful for microelectronics, phosphors in coloured televisions, optical sensors, and other devices (Nogami and Hayakawa, 1997; Jessensky et al., 1998; Cai et al., 1998; Meltzer and Hong, 2000; Gedanken et al., 2000; Hayakawa and Nogami, 2001). All the examples studied so far had a nonporous structure. The interest in studies of R^{3+} optical properties is renewed with introduction of a mesoporous structure of host of Al_2O_3 or SiO_2 (Jessensky et al., 1998; Moadhen et al., 2002). Since the particles in a mesoporous solid are arranged through pores, there exist both the interface between pore wall and particles and the free surface of the particles within pores (Jessensky et al., 1998; Mohanty and Ram, 2002). In particular, depending on the experimental conditions, the pores generate a specific microstructure of R^{3+} cations in divided centres in self-confined size at a nanometric scale. Hence, it is significantly different in microstructure and in turns the optical and other properties from nonporous solids.

R^{3+} cations have $4f^n$ valence cloud shielded by the $5s^2p^6$ electrons. This renders a characteristically sharp spontaneous absorption or emission spectrum in the 4f-4f electronic transitions. A series of 4f-4f transitions, extending from far IR to near IR, visible and UV regions, occur in pure and doped $Eu^{3+}(4f^6)$ in Al_2O_3 or other hosts (Meltzer and Hong, 2000; Gedanken et al., 2000; Gschneid Jr. and Eyring, 1979; Ram et al., 1984; Ram and Sinha, 1987; Mochizuki et al., 2001). All these properties, which are derived from electronic structure, undergo a significant change with a change in electronic structure in nanoparticles (Siegel, 1991; Stern et al., 1995). In such small particles, a semiconductor or insulator, as in this example, has a quantum confinement of carriers (Meltzer and Hong, 2000; Gedanken et al., 2000). Several unusual features of hole burning and energy migration in Eu^{3+} sites are being reported in Eu^{3+} doped SiO_2 or Al_2O_3-SiO_2 glasses (Nogami and Hayakawa, 1997; Jessensky et al., 1998; Cai et al., 1998; Meltzer and Hong, 2000; Gedanken et al., 2000; Hayakawa and Nogami, 2001). A chemical cluster exists of Eu^{3+} and O^{2-} in the tetrahedral SiO_2 network. A small modification in the local structure results in a significant change in the optical properties in Al_2O_3 modified SiO_2 glasses.

Here, we report novel experimental results of enhanced photoemission in dispersed Eu_2O_3 nanoparticles in amorphous Al_2O_3 in a mesoporous structure. The results are analyzed and presented in correlation of microstructures of selected samples prepared under selected experimental conditions. The implications are useful for applications of the mesoporous R^{3+}: Al_2O_3 solids as new optical and gas sensor materials.

EXPERIMENTAL PROCEDURE

Synthesis Process

Three different Eu^{3+}:Al_2O_3 compositions were synthesized by a reactive doping of 0.5, 1.0 and 1.5 wt% Eu^{3+} in order to examine influence of the local structures and/or the Eu^{3+} carrier concentrations on the Eu^{3+} optical properties. This involves a controlled chemical reaction of dispersed Eu^{3+} cations (in a dilute solution of a highly pure (99.99 %) $EuCl_3 \cdot 6H_2O$ in distilled water in 0.04 to 0.40 M concentration) with a hydrogenated precursor of mesoporous $AlO(OH) \cdot \alpha H_2O$ powder at room temperature. It has been observed that an exothermic reaction follows by converting $EuCl_3$ to Eu_2O_3 in divided reaction centres in the pores. As described earlier (Mohanty and Ram, 2002), the local temperature evolved during the reaction (the average value could be as high as the boiling point of the solution) is sufficient enough to cause in-situ molecular decomposition of mesoporous precursor powder from $AlO(OH) \cdot \alpha H_2O$ to Al_2O_3 in divided reaction centres.

After the reaction, the obtained sample is washed in distilled water, to remove HCl and other byproducts, and dried at 450 K under a reduced pressure of 2 to 5 mbar. It is pulverized, by grinding in a mortar with a pestle, as a fine mesoporous $Eu^{3+}:Al_2O_3$ powder (porosity $\Phi \approx 50$) of a characteristic grayish white colour. A freshly prepared $AlO(OH)\cdot\alpha H_2O$ powder, $\Phi \approx 90$, by a surface hydrolysis of a pure (99.99 %) aluminium metal in a humid air (Ram and Rana, 1999), is used. The synthesis process is shown schematically in Fig. 1.

Measurements and Analysis

The mesoporous structure of $Eu^{3+}:Al_2O_3$ is confirmed with transmission electron micrograph and BET studies of average pore size and the fractional volume in pores in the sample as described earlier (Mohanty and Ram, 2003). There has been as much porosity as $\Phi \sim 51\%$ with 25 to 35 nm size of crystallites of Eu_2O_3 arranged through small pores in the mesoporous Al_2O_3 matrix (amorphous) of few nanometres of diameter. The reported content for Eu^{3+} in the final sample is analyzed by the elemental analysis (Mohanty and Ram, 2002; Mohanty and Ram, 2003).

The $Eu^{3+}:Al_2O_3$ phase analysis is carried out with X-ray diffraction (XRD). The diffractogram has been recorded with P.W. 1710 X-ray diffractometer with filtered CoKα radiation. The photoluminescence (PL) spectrum was

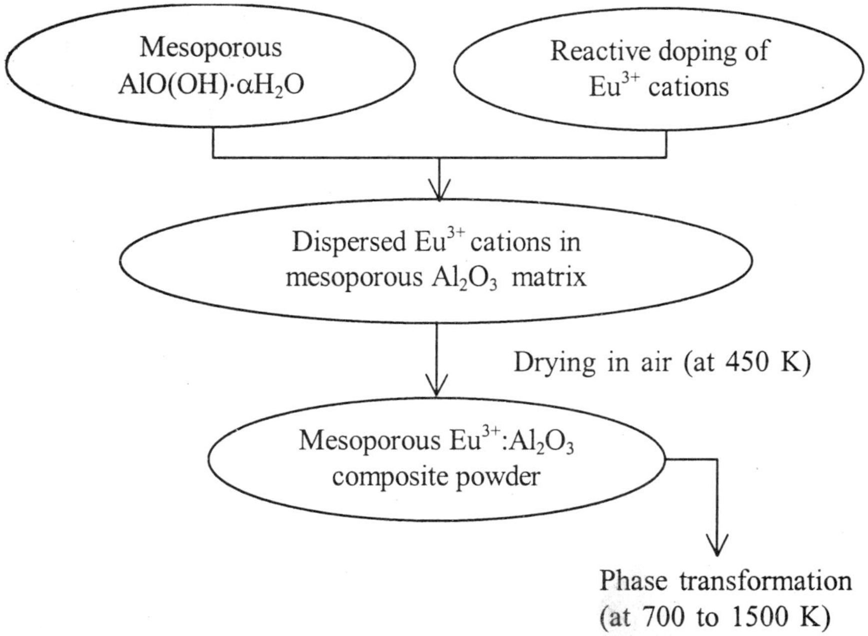

Fig. 1: A schematic diagram in forming of a mesoporous $Eu^{3+}:Al_2O_3$ composite with a reactive doping of dispersed Eu^{3+} cations through a mesoporous $AlO(OH)\cdot\alpha H_2O$ precursor.

studied by exciting the sample at λ_{exc} = 465.8 nm wavelength by an Ar^+-ion laser (Omnichrome model IEC 8251), with 90 mW power and 3.0 mm beam diameter. The data were collected through a 50 cm monochromator (SPEX model 500M) fitted with a 600-lines/mm grating and an InGaAs detector. Other experimental details are the same as reported earlier (Mohanty and Ram, 2003).

RESULTS AND DISCUSSION

Eu^{3+}: Al_2O_3 Structure

In 0.5 to 1.5 wt% Eu^{3+} doped Al_2O_3, used in this work, Eu^{3+} exists as Eu_2O_3 nanocrystals, which are dispersed through Al_2O_3 of an amorphous structure. XRD patterns of the samples have sharp peaks of Eu_2O_3 with a strong scattering background of amorphous Al_2O_3. Figure 2 presents a typical XRD patterns in a Eu^{3+}-rich sample of 1.5 wt% Eu^{3+}: Al_2O_3. Positions and relative intensities retain in sharp Eu_2O_3 peaks in a change of the composition as can be seen with a comparison of those in pattern in 1.0 wt% Eu^{3+}: Al_2O_3 sample given in the inset to Fig. 2. The background does not allow a better resolved structure in these peaks. This is a new pattern in comparison to those in bulk Eu_2O_3, which exists in five polymorphs of C, B, A, H and X at atmospheric pressure (Gschneid Jr. and Eyring, 1979; Mochizuki et al., 2001). The C phase usually exists at ambient temperature in T_h^7 cubic crystal structure. It transforms to the B phase of C_{2h}^3 monoclinic structure at 1373 K while to the A phase of D_{3d} hexagonal structure at 2313 K. The H and X phases appear at still higher temperatures at 2413 and 2553 K, respectively.

This new XRD pattern represents a different Eu_2O_3 structure in a new hexagonal crystal lattice of lattice parameters a = 0.5468 nm and c = 1.6950

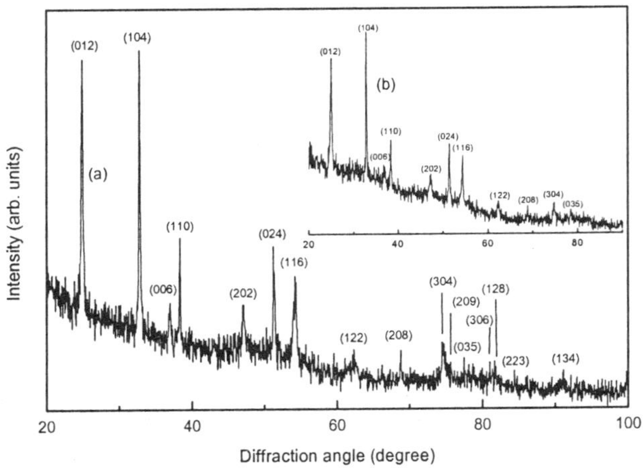

Fig. 2: X-ray patterns of (a) 1.5 wt% Eu^{3+}:Al_2O_3 and (b) 1.0 wt% Eu^{3+} : Al_2O_3 powders. The peaks marked with (hkl) are from the Eu_2O_3 nanocrystals of $R\bar{3}c$ hexagonal crystal structure.

nm and lattice volume V_0 = 1.3167 nm^3. All the peaks in Fig. 2, as marked by the (hkl) values, are indexed in terms of this new structure (Mohanty and Ram, 2002). This XRD pattern is very similar to that in the $R\overline{3}c$ hexagonal corundum structure of Cr_2O_3 (McClume, 1979). In both examples of Eu_2O_3 and Cr_2O_3, the most intense peaks are in (104) reflection, which lies at 0.3157 and 0.2666 nm, respectively, as per their a and c values [a = 0.4957 and c = 1.3584 nm in Cr_2O_3 (McClume, 1979)]. This new Eu_2O_3 polymorph should not be confused with α-Al_2O_3 of corundum structure (McClume, 1979). It has a much smaller V_0 = 0.7641 nm^3 with a = 0.4758 and c = 1.2991 nm (McClume, 1979). It has the first and second most intense peaks in (113) and (104) lattice reflections at 0.2085 and 0.2552 nm, respectively (McClume, 1979). The difference in V_0 lies in the difference in ionic radii (r_i) in the metal cations, e.g., 0.0675 nm in Al^{3+}, 0.1087 nm in Eu^{3+}, while 0.0755 nm in Cr^{3+} in the usual octahedral sites (Shannon, 1976).

TEM micrographs in Fig. 3 have fibrils of wavelike structures of (a) 1.0 wt% Eu^{3+}:Al_2O_3 and (b) 1.5 wt% Eu^{3+}:Al_2O_3 samples with curled surfaces distributed through pores (whitish contrast) in a typical mesoporous structure (Φ ~ 51 %). A similar mesoporous structure has been reported with carbon fibrils in amorphous carbon (Harris et al., 2000). The fibrils in micrograph (a) or (b) are due to self-arranged Al_2O_3 in an interconnected network, which extends 50 to 300 nm long in 5 to 10 nm diameter. The Eu_2O_3 doped in the Al_2O_3 matrix appears in near cubical shape of particles (in rather deep dark contrast) in 25 to 40 nm diameter, which compares average (a) 35 nm and (b) 38 nm crystallite size as calculated by peak-widths in the XRD peaks by the Debye Scherrer relation. A typical electron diffractogram (c), which is taken from a selected region from the sample of micrograph (b), in Fig. 3 has four concentric rings in (012), (104), (202) and (122) reflections at 0.4130, 0.3160, 0.2240 and 0.1740 nm interplanar spacings in agreement with the 0.4138, 0.3165, 0.2244 and 0.1742 nm respective values in the XRD pattern (Fig. 2a).

The corundum structure of Eu_2O_3 is not common. It is a metastable phase, which decomposes on heating at temperature as early as 700 K, resulting in an amorphous Eu^{3+}: Al_2O_3. As we described earlier (Mohanty and Ram, 2002), in this example, it formed under specific conditions by a fast nucleation and growth in divided reaction centres embedded in amorphous Al_2O_3 in pores. We believe that amorphous Al_2O_3, which embeds Eu_2O_3 in small crystallites, supports this high-energy metastable structure with a rather stable Al_2O_3 surface layer or interface. Otherwise, such large Eu^{3+} cations are hard to arrange through O^{2-} anions (of not much larger r_i ~ 0.126 nm[21]) in this specific structure. As in the original α-Al_2O_3 corundum structure (Kronberg, 1957; Bilde-Sorensen et al., 1996), the normally vacant cationic sites (octahedral), which are as large as one-third (or even more) of the total cationic sites, most likely extend the excess volume in distributing the Eu^{3+} cations in this lattice in small particles.

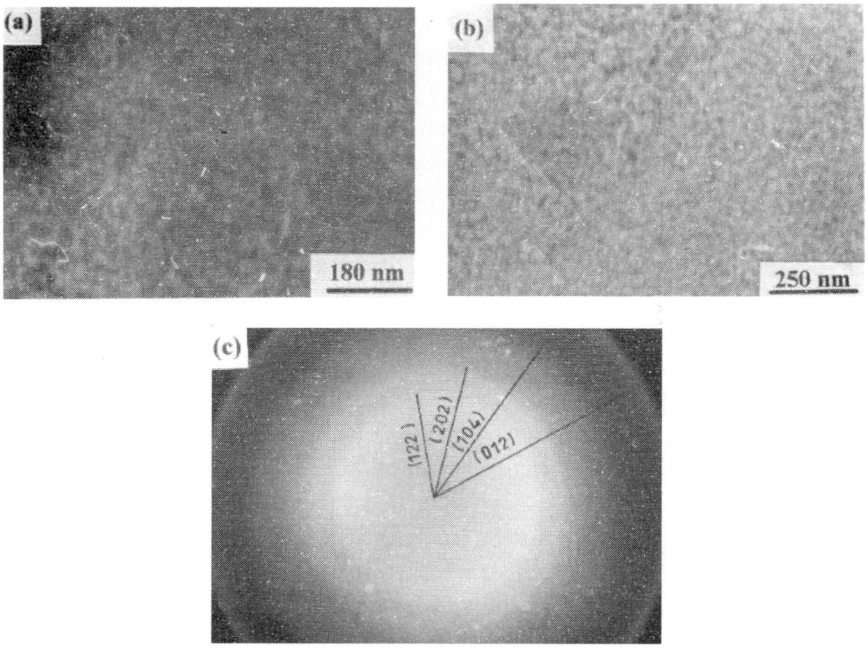

Fig. 3: TEM micrographs in (a) 1.0 wt% $Eu^{3+}:Al_2O_3$ and (b) 1.5 wt% $Eu^{3+}:Al_2O_3$ powders, and (c) the electron diffractogram from a region of sample of micrograph (b).

Photoemission in $Eu^{3+}:Al_2O_3$ Structure

Figure 4 shows photoemission spectra in three $Eu^{3+}:Al_2O_3$ specimens of (a) 0.5, (b) 1.0 and (c) 1.5 wt% Eu^{3+} with λ_{exc} = 465.8 nm. The energy levels in 4f-4f transitions in Eu^{3+} cations are well known (Gschneid Jr. and Eyring, 1979; Ram et al., 1984; Ram and Sinha, 1987). As can be seen by part of the energy level diagram given in the inset, the photoemission in this example occurs by a resonance excitation with $^5D_2$ state of Eu^{3+} cations. As shown in Fig. 5, no spectrum appears in a significant intensity in off-resonance excitation at 488.0 or 514.5 nm with the Ar^+-ion laser. Either of these lines has slightly off positions unable to resonate with the $^5D_2$ state at 465 nm or the $^5D_1$ state at 525 nm in Eu^{3+} cations. The $^5D_0$ state of Eu^{3+} that gives the predominant emission in this series lies at further larger 578 nm value to unfulfilling a resonance excitation. The results demonstrate the fact that dispersed Eu_2O_3 in nanoparticles through Al_2O_3 in pores in a mesoporous structure retains a self-confined structure of ^5D_J states about their average values. The Eu_2O_3 vibronic structure or the phonon-phonon interaction with Al_2O_3 host causes no excessive broadening in the 2J+1 spectrum. Otherwise, the ^5D_J electronic states have an extended structure, which extends the excitation over a wide range of λ_{exc} (Ram and Sinha, 1987).

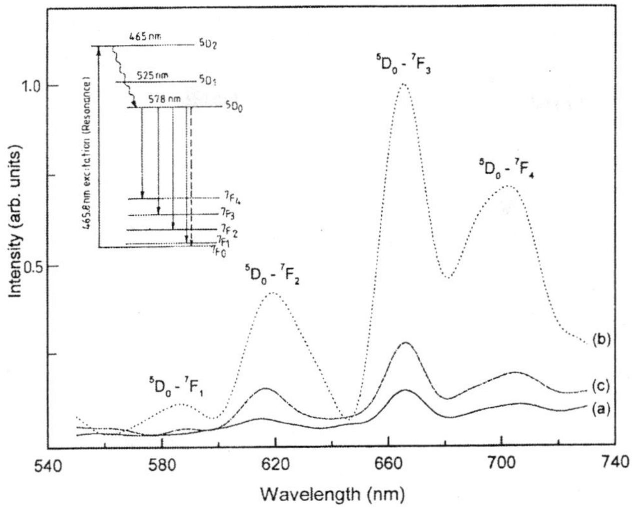

Fig. 4: Photoemission in Eu^{3+}:Al_2O_3 powders in 465.8 nm excitation by an Ar^+ - ion laser; (a) 0.5, (b) 1.0 and (c) 1.5 wt% Eu^{3+}. The inset is the schematic energy level diagram in the excitation and emission processes. No emission occurs in $^5D_0 \rightarrow {}^7F_0$ transition (the dashed line).

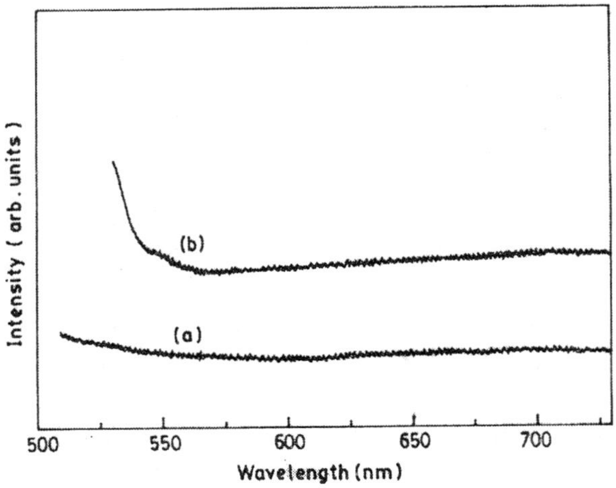

Fig. 5: PL excitations in 1.0 wt % Eu^{3+}:Al_2O_3 powder (dried at 450 K) at (a) λ_{exc} = 488.0 and (b) λ_{exc} = 514.5 nm by an Ar^+-ion laser. No PL appears in these excitations.

Furthermore, according to observed PL spectrum in Fig. 4, a manifested electron-phonon coupling in Eu_2O_3 nanoparticles renders intermixing in J-sublevels in a given ^7F_J state of the ground state multiplet to which a spontaneous emission occurs in the excitation via the $^5D_0$ excited state. As a result, in the $^5D_0 \rightarrow {}^7F_J$ transitions, no J-sublevels could be resolved in the three spectra in Fig. 4. Otherwise, a ^7F_J level splits in as many components as '2J+1' as per the Eu^{3+} site symmetry (Hayakawa and Nogami, 2001; Ram and Sinha, 1987). In Fig. 4, four distinct bandgroups occur in the (i) 560 - 600, (ii) 600 - 645, (iii) 645 – 680, and (iv) 680 -720 nm regions. Unusually, the most intense bandgroup (in terms of the peak intensity I_p) of the spectrum lies at bit longer λ in bandgroup (iii) at 666 nm irrespective to the Eu^{3+} contents in these three Eu^{3+}:Al_2O_3 samples.

In a non-mesoporous structure, the most intense band lies in $^5D_0 \rightarrow {}^7F_2$ transition at 625 nm in Eu_2O_3 thin films (Mochizuki et al., 2001) (polycrystalline in possibly T_h^7 cubic crystal structure) while at 617 nm in Eu_2O_3 doped in α-Al_2O_3 nanoparticles of 50 nm diameter (Gedanken et al., 2000). In bulk Eu_2O_3 powder (nonporous), sintered at 1273 K for 24 hrs, it shifts at 612 nm (Mochizuki et al., 2001). The results, if put together, demonstrate the fact that (i) the Eu_2O_3 size and crystal structure, (ii) the microstructure with a porous or nonporous Al_2O_3 matrix, and (iii) the Eu^{3+} concentration in a doped Al_2O_3 sample sensitively influence the $^5D_0 \rightarrow {}^7F_j$ photoemission features in 4f-4f transitions in Eu^{3+} cations.

As marked in Fig. 4, the four bandgroups observed in this example arise in the $^5D_0 \rightarrow {}^7F_1$, $^5D_0 \rightarrow {}^7F_2$, $^5D_0 \rightarrow {}^7F_3$ and $^5D_0 \rightarrow {}^7F_4$ transitions, respectively, in Eu^{3+} cations. According to it, the $^5D_0 \rightarrow {}^7F_3$ transition lies in the most intense band of the spectrum. In principle, in bulk Eu_2O_3, it is a forbidden transition by electric dipole, electric quadrupole, as well as the magnetic dipole moments (Gschneid Jr. and Eyring, 1979). As a result, it does not have a significant intensity in pure or doped Eu_2O_3 in Al_2O_3, SiO_2, or other hosts in a nonporous equilibrium structure (Gedanken et al., 2000; Hayakawa and Nogami, 2001; Ishizaka et al., 2002; Nogami et al., 2002; Moadhen et al., 2002; Gschneid Jr. and Eyring, 1979). The $^5D_0 \rightarrow {}^7F_2$ transition (electric quadrupole transition), which occurs predominantly in other examples, has a poor intensity as the third most intense bandgroup of the spectrum. The second most intense bandgroup lies in the $^5D_0 \rightarrow {}^7F_4$ transition (forced electric dipole transition). The magnetic dipole $^5D_0 \rightarrow {}^7F_1$ transition retains its features of weak intensity in this series. The $^5D_0 \rightarrow {}^7F_0$ transition is not visible in this new Eu_2O_3 polymorph of corundum structure according to its basic characteristics of an electric dipole, electric quadrupole, as well as the magnetic dipole forbidden transition (Gschneid Jr. and Eyring, 1979).

As given in Table 1, integrated intensity I_t in the four bandgroups is a nonlinear function of Eu^{3+} content in the Eu^{3+}:Al_2O_3 samples. In order to compare the results, the spectra were studied under identical conditions of other parameters. A maximum I_t value appears in the sample of 1.0 wt% Eu^{3+}. It is as large as 20 times the one in the 0.5 wt% Eu^{3+} sample or 10 times that in the 1.5 wt% Eu^{3+} sample. This presents a peculiar example of the concentration quenching of the photoemission by the ionic interaction between Eu^{3+} cations via Al_2O_3 surface layer in divided pores.

Table 1: Photoemission bands in dispersed Eu_2O_3 nanoparticles in an amorphous Al_2O_3 matrix of mesoporous structure.

Transitions	*Eu^{3+} content (wt%)*					
	0.5		*1.0*		*1.5*	
	Band position		*Band position*		*Band position*	
	(nm)	*(cm^{-1})*	*(nm)*	*(cm^{-1})*	*(nm)*	*(cm^{-1})*
$^5D_0 - {}^7F_1$	–	–	587.0	17035 (08)	588.0	17006 (01)
$^5D_0 - {}^7F_2$	615.0 624.0	16260 16025 } 02	612.0 619.0 632.0	16340 16155 } (46) 15822	616.5 624.0	16220 16025 } 10
$^5D_0 - {}^7F_3$	666.0	15015 (06)	665.5	15026 (80)	666.0	15015 (16)
$^5D_0 - {}^7F_4$	694.0 706.0	14409 14164 } (04)	692.0 702.0	14450 14245 } (100)	690.0 704.0	14492 14204 } (12)

The values of the integrated intensity in the individual bandgroups are given in the parentheses on a relative scale assuming a value of 100 units in the most intense bandgroup. Average crystallite size in the three samples varies marginally as 30, 35 and 38 nm, respectively.

In order to have a further insight in the PL properties in the two transitions in this example, we analyzed the variation of the ratio of I-values of I_3 and I_2 in the $^5D_0 \rightarrow {}^7F_3$ and $^5D_0 \rightarrow {}^7F_2$ two bandgroups, respectively, with a function of the Eu^{3+} carrier concentration (x in wt %). The plot is portrayed in Fig. 6. It represents a monotonically decreasing value of I_3/I_2 as a function of x. A value of $I_3/I_2 = 2.7$ is obtained from this plot at $x = 0.005$, and it decreased to 1.5 on increasing the x to 0.015. A plot for the total $I_t = I_2 + I_3$ value against x is given in Fig. 7. It shows a peak in I_t value (which is normalized to be a value of 100 units) at specific value of $x = 0.01$. This is a bit smaller value in comparison to a value of $x \sim 0.06$ reported in ordered Ba_2GdMeO_6 (Me = Nb or Ta) perovskites of the humite structure (Blasse, 1967). In small Eu_2O_3 particles, the Eu^{3+} cations can be treated to be rather separated in small Eu_2O_3 groups via Al_2O_3 and pores so that no concentration quenching occurs to as large Eu^{3+} content as 1.0 wt%. The Eu^{3+} cations in Eu_2O_3 nanoparticles are in indirect contact with neighbouring Al_2O_3, which promotes energy transfer from the excited Al_2O_3 in addition to the direct excitation. This mechanism is active especially in doped Al_2O_3 in a composite structure in an alloy (Blasse, 1967). Since the Eu^{3+} content is small, we can describe each group of Eu^{3+} cations as surrounded by many Al_2O_3 molecules or clusters. Moreover, according to the Meltzer and Hong (2000), an enhanced strain in small Eu_2O_3 particles causes a manifested electron-phonon coupling. It plays a crucial role in optimizing I_t at a specific Eu^{3+} content as found to be 1.0 wt% in this example.

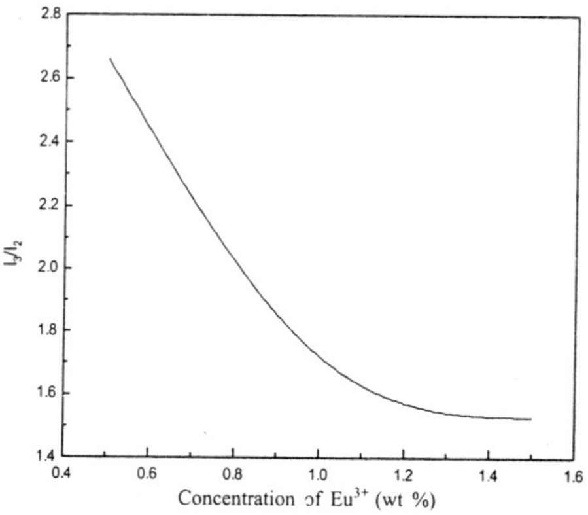

Fig. 6: Variation of the I_3/I_2 ratio in the two bandgroups of the PL spectrum with the Eu^{3+} carrier concentration in Eu^{3+}:Al_2O_3 powder.

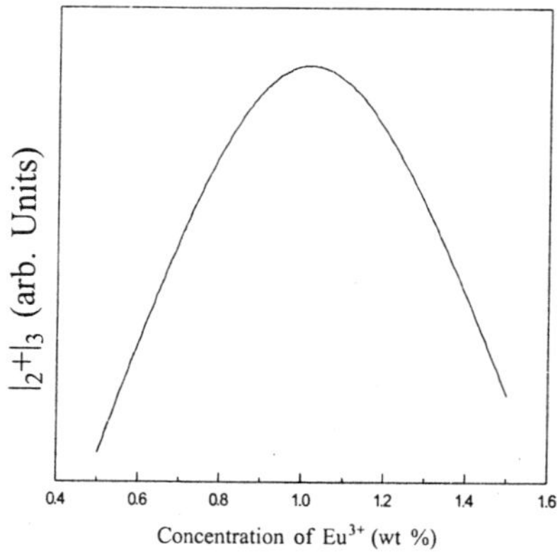

Fig. 7: Variation of the total intensity in the PL spectrum of the two bandgroups with the Eu^{3+} carrier concentration in Eu^{3+}:Al_2O_3 powder.

We studied the structures in the individual $^5D_0 \rightarrow {}^7F_1$ bandgroups in Figs. 8, 9 and 10 at an expanded scale of wavelength λ. The $^5D_0 \rightarrow {}^7F_1$ group (Fig. 8) is asymmetric in shape at shorter λ-side, indicating a multiple structure of it. Generally, $^5D_0 \rightarrow {}^7F_1$ transition splits into three components due to local field around Eu^{3+}. Their separations vary with the Eu^{3+} content and/or the microstructure so as the average position of the bandgroup varies at 587 nm in (b) 1.0 wt% Eu^{3+} sample while at 588 nm in (c) 1.5 wt% Eu^{3+} sample. It no longer appears in such distinct shape or intensity at (a) the

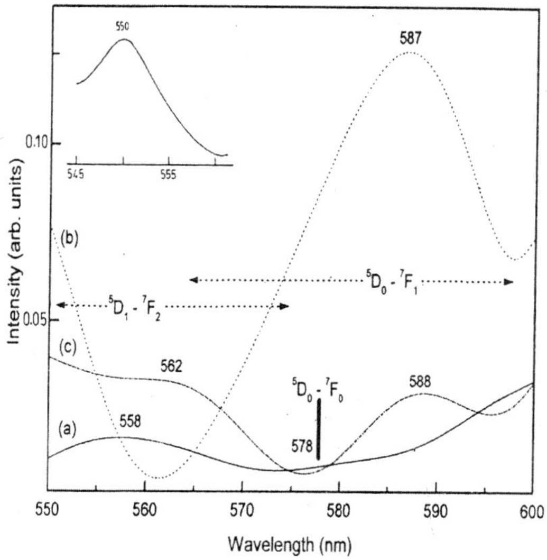

Fig. 8: $^5D_0 \rightarrow {}^7F_1$ photoemission in Eu^{3+}:Al_2O_3 powders; (a) 0.5, (b) 1.0 and (c) 1.5 wt% Eu^{3+}. Part of spectrum (b) at lower wavelengths in $^5D_1 \rightarrow {}^7F_2$ transition is given in the inset.

lower Eu^{3+} contents. An extensive intermixing in eigen functions in '2J+1' components by the enhanced electron-phonon coupling obscures their individual features in a broad asymmetric band in Eu_2O_3 nanoparticles embedded in a mesoporous Al_2O_3 structure.

The $^5D_0 \rightarrow {}^7F_2$ and $^5D_0 \rightarrow {}^7F_4$ bandgroups have similar features of asymmetric shape with at least two or three unresolved components. Average position in $^5D_0 \rightarrow {}^7F_2$ bandgroup varies (Fig. 9) from 615.0 nm in (a) 0.5 wt% Eu^{3+} sample to 619.0 nm in (b) 1.0 wt% Eu^{3+} sample or 616.5 nm in (c) 1.5 wt% Eu^{3+} sample. Two satellite components lie at 612 and 632 nm as marked therein. An opposite trend is observed in $^5D_0 \rightarrow {}^7F_4$ bandgroup (Fig. 10) whose average position varies from 706 nm in (a) 0.5 wt% Eu^{3+} sample to 702 nm in (b) 1.0 wt% Eu^{3+} sample or 704 nm in (c) 1.5 wt% Eu^{3+} sample. A satellite band lies at 694, 692, or 690 nm in the respective samples. It is interesting to note that the $^5D_0 \rightarrow {}^7F_3$ group has a sharp symmetric shape, with $\Delta\lambda$ = 20 nm half-bandwidth, in comparison to the $^5D_0 \rightarrow {}^7F_4$ ($\Delta\lambda$ = 30 nm) or $^5D_0 \rightarrow {}^7F_2$ ($\Delta\lambda$ = 25 nm) group. No significant change occurs in its average position as a function of the Eu^{3+} content used here.

The enhanced value of I_t in $^5D_0 \rightarrow {}^7F_3$ transition in this example can be explained by invoking a mixing of $^7F_3$ electronic level with the $^7F_2$ level so that it has an induced $^7F_2$ effect of a forced electric dipole allowed $^5D_0 \rightarrow {}^7F_3$ transition. As demonstrated with the energy level diagram in Fig. 11, extended vibronic levels v_2^n in F_2 raise its average energy to the F_3 value. Generally, an electronic state accompanies a vibronic spectrum of normal modes of vibration n in the structural units (EuO_6 in Eu_2O_3 in octahedral sites). The electron-phonon interaction, which is reported to be enhanced in

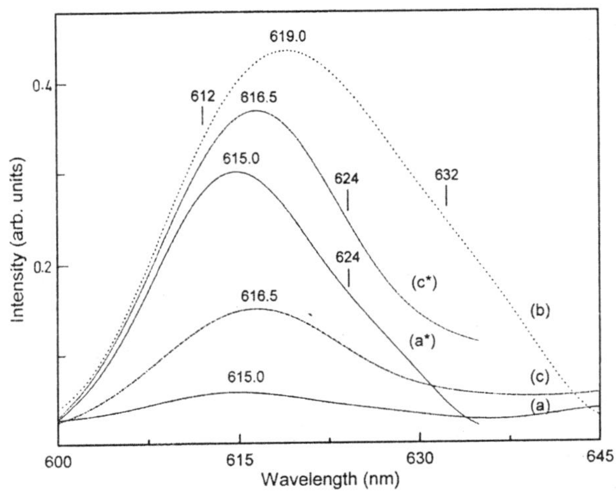

Fig. 9: $^5D_0 \rightarrow {}^7F_2$ photoemission in Eu^{3+}:Al_2O_3 powders; (a) 0.5, (b) 1.0 and (c) 1.5 wt% Eu^{3+}. Spectra (a^*) and (c^*) are 10 and 3 times expansions of those in (a) and (c), respectively, along y-axis.

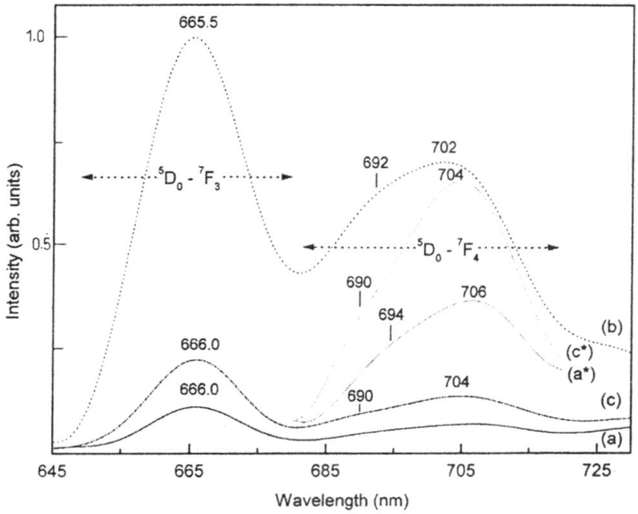

Fig. 10: $^5D_0 \rightarrow {}^7F_3$ and $^5D_0 \rightarrow {}^7F_4$ photoemissions in Eu^{3+}:Al_2O_3 powders; (a) 0.5, (b) 1.0 and (c) 1.5 wt% Eu^{3+}. Spectra (a^*) and (c^*) are 10 times expansions of those in (a) and (c), respectively, along y-axis.

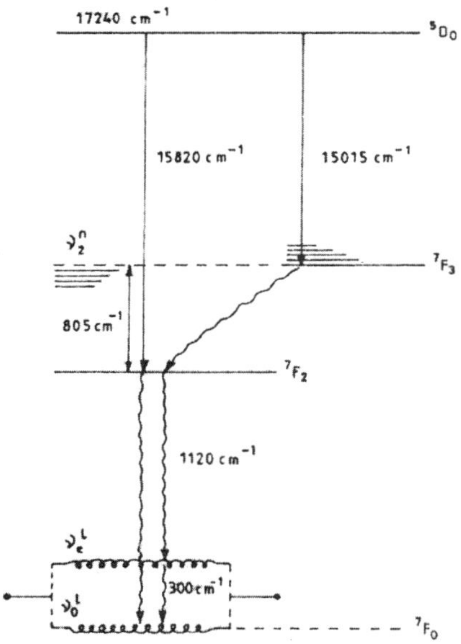

Fig. 11: A model distribution of ν_0^1, ν_e^1 and ν_2^n phonon levels over $^7F_0$ or $^7F_1$ electronic energy level in a small Eu_2O_3 particle embedded in a different matrix of amorphous Al_2O_3. Radiative emission and cross relaxation are shown by the solid and curled lines with arrows.

Eu_2O_3 nanoparticles (Meltzer and Hong, 2000), determines the structure with the localized ν_2^n vibrations and in turn operates the derived processes. It can be argued that a Eu_2O_3 nanoparticle embedded in Al_2O_3 in pore, as in this example, oscillates with a frequency ν_e^1 under the field of the neighbours. $\nu_e^{\,l}$ assumes a derived value of the lattice vibrations (the external modes of vibration l). As small value of $\nu_e^{\,l}$ as $\nu_0^{\,l} \cong 0$ lies in the low vibration states.

Under the conditions of interaction with radiation, $\nu_e^{\,l}$ governs the processes of the Eu^{3+} electronic excitation and the photoemission from the excited state(s) followed by a cross relaxation of part of the energy. In this model, the ν_2^n levels give a channel to a $^7F_3 \rightarrow {}^7F_2$ cross relaxation of the excess energy with the lattice during the emission process and thus favours the $^5D_0 \rightarrow {}^7F_3$ transition according to the experimental observation. Average values of $\nu_2^n = 805$ cm^{-1} and $\nu_e^{\,l} = 300$ cm^{-1} are considered according to the vibrational spectrum of Eu_2O_3 (Ross, 1972; Mohanty and Ram, 2002). They give an effective energy value of $^7\nu_2 = \nu_e^{\,l} + 1120 \cong 1420$ cm^{-1} for the $^7F_2$ level or $^7\nu_3 = {}^7\nu_2 + \nu_2^n \cong 2225$ cm^{-1} for the $^7F_3$ level. Favourably comparable $^7\nu_2 \cong 1420$ and $^7\nu_3 \cong 2225$ cm^{-1} values are obtained by subtracting the 15820 and 15015 cm^{-1} values in $^5D_0 \rightarrow {}^7F_2$ (at 632.0 nm) and $^5D_0 \rightarrow {}^7F_3$ (at 666.0 nm) bands in Fig. 10 from a common 17240 cm^{-1} value in $^5D_0 \rightarrow {}^7F_0$ band (at 580.0 nm).

A spectrum of the localized $\nu_e^{\,l}$ and ν_2^n phonons in Eu_2O_3 nanoparticles serve as energy carriers to carry a recurrent emission in this specific $^5D_0 \rightarrow {}^7F_3$ transition. As demonstrated in Fig. 12 with the energy levels, a recurrent $^5D_0 \rightarrow {}^7F_3$ excitation lies in several ways, resulting in a multiplied intensity of it. This proceeds via a recurrent resonance in cross relaxation processes in combination with the electric or magnetic dipole allowed transitions. In Fig. 12, in $\lambda_{exc} = 465.8$ nm (energy $h\nu_o$) excitation, the sample has a resonance excitation with its $^5D_2$ (Eu^{3+}) level. In one possible process (I), part of the $4f^6(Eu^{3+})$ electrons exited to this level decay to $^5D_1$ or $^5D_0$ level (by releasing $h\nu_1$ or $h\nu_1 + h\nu_1'$ energy in nonradiative processes), which emits a radiation of energy $h\nu_2$ or $h\nu_3$ in the transition to $^7F_0$ or $^7F_3$ level. The $^5D_0 \rightarrow {}^7F_3$ transition (marked by the solid line AB) involves a cross relaxation of the excess energy (which can be treated to be a nearly same as $h\nu_1$ or $h\nu_1'$ in combination with the $\nu_e^{\,l}$ or ν_2^n phonons) to the ground state $^7F_0$. In subsequent process, the $h\nu_1$ energy instantly recombines to $h\nu_2$ photon by performing a recurrent $^5D_2$ resonance excitation, and in turn a recurrent $^5D_0 \rightarrow {}^7F_3$ emission (the dotted AB line) occurs. The whole process is thus expressed as

$$\begin{aligned} h\nu_0 \text{ (Ar}^+\text{-laser)} &\rightarrow h\nu_1 + h\nu_1' + h\nu_3 + h\nu_1 \text{, or} \\ &\rightarrow 3h\nu_1 \text{ (relaxation)} + h\nu_3 \uparrow\text{, if } h\nu_1 \sim h\nu_1', \end{aligned} \tag{1}$$

followed by

$$\begin{aligned} h\nu_1 + h\nu_2 \rightarrow\ & h\nu_0 \\ & \downarrow \\ & 3h\nu_1 \text{ (relaxation)} + h\nu_3 \uparrow. \end{aligned} \tag{2}$$

A combination of equation (1) and (2) implies, $\nu_2 \cong \nu_3 + 2\nu_1$.

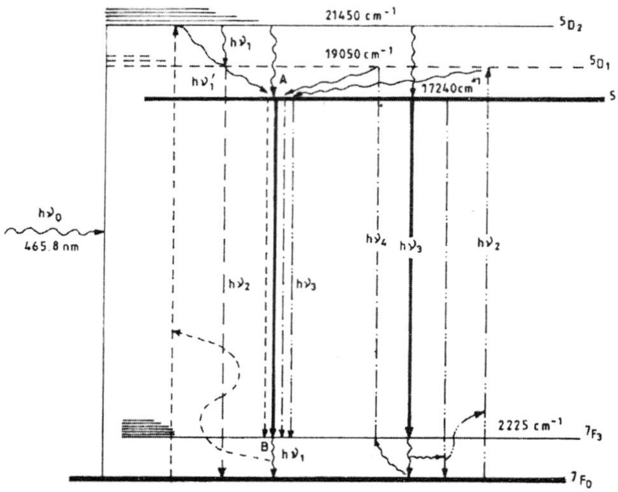

Fig. 12: A schematic energy level diagram in the resonance excitation with $^5D_2$ state and the recurrent emission processes via $^5D_1$ state in the $^5D_0 \rightarrow {}^7F_3$ transition in Eu^{3+} cations.

In another process II, the energy $h\nu_1$ instantly recombines to $h\nu_3$ photon by exciting $^5D_1$ level ($h\nu_4$), which relaxes to $^5D_0$ and emits the same $h\nu_3$ photon (dotted line AB) in $^5D_0 \rightarrow {}^7F_3$ transition,

$$\begin{array}{l} h\nu_1 + h\nu_3 \rightarrow h\nu_4 \\ \qquad\qquad\quad \downarrow \\ \qquad\qquad\quad h\nu_1 \text{ (relaxation)} + h\nu_3 \uparrow \end{array} \tag{3}$$

Similarly, it is also possible that the photon, which emits in $^5D_0 \rightarrow {}^7F_0$ transition, imparts in a re-excitation of $^5D_1$ level in combination with the $h\nu_1$ energy. After relaxing to $^5D_0$ level it adds to $^5D_0 \rightarrow {}^7F_3$ emission (dotted line AB). In general, these processes function at expense of the loss of the total energy in nonradiative processes in nonelastic interaction of the spontaneous emission of the radiation with the lattice. As a result, a manifested fraction of final intensity lies of the final radiation.

CONCLUSIONS

Eu_2O_3 nanocrystallites, 25 to 40 nm size, dispersed in an amorphous Al_2O_3 matrix of mesoporous structure have strong photoemission in $^5D_0 \rightarrow {}^7F_J$, $J = 1$ to 4, electronic transitions of the Eu^{3+} cations. Study of selected Eu^{3+}:Al_2O_3 samples, with Eu^{3+} content varied up to 1.5 wt%, in correlation with reported results in Eu_2O_3 thin films, bulk Eu_2O_3 powder and Eu_2O_3 doped in α-Al_2O_3 of nanoparticles demonstrates that the Eu^{3+} - photoemission features sensitively vary depending on (i) the Eu_2O_3 size and crystal structure, (ii) the microstructure with a porous or nonporous Al_2O_3, and (iii) the Eu^{3+} concentration. Improved quantum efficiency lies in sample of dispersed

Eu^{3+} in a mesoporous Al_2O_3 structure, with a maximum value at 1.0 wt% Eu^{3+} in an optimal electron-phonon coupling. This supports the concentration quenching of photoemission due to the ionic interactions in Eu^{3+} cations. The mesopores mediate a dispersed Eu^{3+} structure so that the photoemission is intensified further. The electron-phonon coupling, which is prevalent in this example, is a strong function of structure and morphology of Eu_2O_3 nanoparticles. In such small Eu_2O_3 particles, the morphology, which is found to be a near cubical in the electron microscopic images, controls the crystal field and exchange energy by macroscopic interaction with surroundings (Al_2O_3 and pores).

Acknowledgements: This work has been financially supported by the Council of Scientific & Industrial Research (CSIR) and the Defence Research and Development Organization (DRDO), Government of India.

REFERENCES

Bilde-Sorensen, J.B., Lawlor, B.F., Geipel, T., Pirouz, P., Heuer, A.H. and Lagerlof, K.P.D. (1996). *Acta. Metall. Mater.*, **44:** 2145.

Blasse, G. (1967). *J. Chem. Phys.*, **46:** 2583.

Cai, W., Zhang, Y., Jia, J. and Zhang, L. (1998). *Appl. Phys. Lett.*, **73:** 2709.

Gedanken, A., Reisfeld, R., Sominski, L., Zhong, Z., Koltypin, Y., Panczer, G., Gaft, M. and Minti, H. (2000). *Appl. Phys. Lett.*, **77:** 945.

Gschneid Jr., K.A. and Eyring, L.R. (1979). Handbook on the Physics and Chemistry on Rare-earths, **Vol. 3**, North-Holland, Amsterdam.

Harris, P.J.F., Burian, A. and Duber, S. (2000). *Phil. Mag. Lett.*, **80:** 381.

Hayakawa, T. and Nogami, M. (2001). *J. Appl. Phys.*, **90:** 2200.

Ishizaka, T., Nozaki, R. and Kurokawa, Y. (2002). *J. Phys. Chem. Solids*, **63:** 613.

Jessensky, O., Muller, F. and Gosele, U. (1998). *Appl. Phys. Lett.*, **72:** 1173.

Kronberg, M.L. (1957). *Acta. Metall.*, **5:** 507.

McClume, W.F. (1979). Powder Diffraction File (*Joint Committee on Powder Diffraction Standards*) (International Centre for Diffraction Data, Swarthmore, PA,), cards 6-0504 (h-Cr_2O_3), 10-173 (α-Al_2O_3).

Meltzer, R.S. and Hong, K.S. (2000). *Phys. Rev.* B, **61:** 3396.

Moadhen, A., Elhouichet, H., Oueslati, M. and Ferid, M. (2002). *J. Lumin.*, **99:** 13.

Mochizuki, S., Suzuki, Y., Nakanishi, T. and Ishi, K. (2001). *Physica* B, **308-310:** 1046.

Mohanty, P. and Ram, S. (2003). *J. Mater. Chem.*, **13:** 3021.

Mohanty, P. and Ram, S. (2002). *Phil. Mag. B*, **82:** 1129.

Mohanty, P. and Ram, S. (2002). *J. Euro. Ceram. Soc.*, **22:** 933.

Mohanty, P. and Ram, S. (2002). *Mater. Lett.*, **53:** 287.

Nogami, M. and Hayakawa, T. (1997). *Phys. Rev. B*, **56:** R14235.

Nogami, M., Enomotom T. and Hayakawa, T. (2002). *J. Lumin.*, **97:** 147.

Ram, S. and Rana, S. (1999). *Current Sci.*, **77:** 1530.
Ram, S. and Sinha, S.K. (1987). *J. Solid State Chem.*, **66:** 225.
Ram, S., Lamba, O.P. and Bist, H.D. (1984). *Pramana*, **23:** 59.
Ross, S.D. (1972). Inorganic Infrared and Raman Spectra. McGraw - Hill, England.
Shannon, R.D. (1976). *Acta. Cryst.* A, **32:** 751.
Siegel, R.W. (1991). *Annu. Rev. Mater. Sci.*, **21:** 559.
Stern, E.A., Siegel, M., Newville, R.W., Sanders, P.G. and Haskel, D. (1995). *Phys. Rev. Lett.*, **75:** 3874.

Synthesis and Microstructure of Half-metallic CrO_2-PVA Composite Films as New Granular GMR Materials

S. Biswas and S. Ram

Materials Science Centre, Indian Institute of Technology Kharagpur–721302, India

INTRODUCTION

Recently, many studies have been focused on extrinsic magnetotransport effects in ferromagnetic oxides with half metallicity, because there is an intensive demand of highly spins polarized ferromagnetic materials for spintronic devices (Hwang and Cheong, 1997; Manoharan et al., 1998; Gupta et al., 2000; Chen et al., 2003; Chen et al., 2004). CrO_2 is one of the technologically most important transition metal oxides for wide applications in magnetic and electronic devices. It behaves as a half-metallic ferromagnet with a very special property of giant magnetoresistance (GMR) useful for magnetic random access memories (MRAM) in computers, speed sensors, read sensors for magnetic disc drives, angular and displacement measuring instruments, phosphor, landmine detectors and spin-polarized scanning tunneling microscope (Van Leuken and deGroot, 1995).

Granular CrO_2 has Rutile type tetragonal structure in which Cr^{4+} cations order ferromagnetically with a Curie temperature T_C = 391 K (Schwarz, 1986). It is a nonstoichiometric compound, which is very sensitive to temperature and pressure. At about 440 °C under atmospheric pressure, it degrades as Cr_2O_3, which is an antiferromagnetic insulating phase (Chen et al., 2003; Barry et al., 2000). Its electronic structure is not resolved so far. A qualitative model for it had been proposed by Goodenough (1971), but the first self-consistent band structure calculation was not published until 1986 by Schwarz (1986). This theoretical study yields the interesting result that CrO_2 is a 'half-metallic' ferromagnet, i.e., a system which is metallic for one spin direction while at the same time insulating for the opposite spin direction. Within an ionic model, the Cr^{4+} ($3d^2$) cations in the ferromagnetic ordering yield a resultant magnetic moment of 2 μ_B per formula unit (Barry et al., 2000; Korotin et al., 1998).

In favourable consistence with the prediction of Schwarz (1986) of a ferromagnet CrO_2 band structure, experimental results of the spin-polarized photoemission (Kämper et al., 1987) and the superconducting point contact (Soulen Jr. et al., 1998) confer that CrO_2 has a high degree of spin polarization, close to 100%, mainly due to the spin-split d-band and the absence of sp conduction electrons. The nearly perfect spin polarization of CrO_2 suggests that it would be ideal for applications in ferromagnetic spin polarized tunneling (SPT) junctions, with a large value of MR. SPT, discovered by Meservey and Tedrow (1970 and 1971), laid the foundation to a new field of research of spintronics. In this regard, efforts are being made to derive the material that can act as a perfect spin polarized electron source and having a high value of MR to meet these applications.

Within the last few years, especial efforts have been made to synthesize and fabricate CrO_2 powder compacts or nanocomposite films with a polymer matrix in order to improve the room temperature MR (Chen et al., 2003; Liang et al., 2002). A thin surface coating of a polymer, Cr_2O_3, or a nonmagnetic metal on CrO_2 particles behaves as a surface barrier layer and improves the GMR, optical, and other properties (Moodera and Mathon, 1999). Such samples display a significant enhancement in the low field MR (LFMR) much larger than the intrinsic MR in a CrO_2 powder compact (Ju and Li, 2002), due to the physical mechanism of SPT through the barrier. The barrier layer plays a crucial role in the LFMR value. Here, we report synthesis and modified GMR in CrO_2–PVA (poly-vinyl alcohol) composite films, with a phenomenal study of the influence of the polymer on LFMR.

EXPERIMENTAL PROCEDURE

Synthesis of CrO_2 and CrO_2–PVA Composite Films

The CrO_2–PVA composite films of selected compositions were prepared by dispersing a finely divided loose CrO_2 powder over PVA polymer molecules in aqueous solution. It was deduced by a novel chemical method of a metal ion-polymer composite developed in our group (Biswas and Ram). The starting materials were of CrO_3, PVA, and sucrose of purity better than 99.9%. The reaction is carried out as follows:

An aqueous solution of 2.0 M concentration was obtained by dissolving CrO_3 in distilled water. Similar aqueous solutions were obtained for PVA (4.0 in 100 ml) and sucrose (40.0 g in 100 ml). A 10 to 15 hrs of magnetic stirring at 50-60 °C promotes dissolution of PVA in a clear solution. This is highly necessary in order to derive a homogeneous precursor solution by mixing with Cr^{6+} metal cations of aqueous solution. As summarized in

Fig. 1, the Cr^{6+} solution (50 ml in average 2.0 M concentration) was added dropwise to an aqueous polymer solution (200 ml) of PVA (4.0 g) and sucrose (40.0 g) at 70 to 80 °C temperatures with stirring. The amounts of the solutions used are given in Table 1. In this process, the Cr^{6+} metal cations disperse over polymer molecules in a specific configuration according to the polymer surface structure and surfactant behaviour. A highly viscous solution of a characteristic blackish colour results in a Cr^{4+} metal ion-polymer composite product of the precursor. The PVA reacts with Cr^{6+} and converts it into Cr^{4+} in the solution under the present conditions.

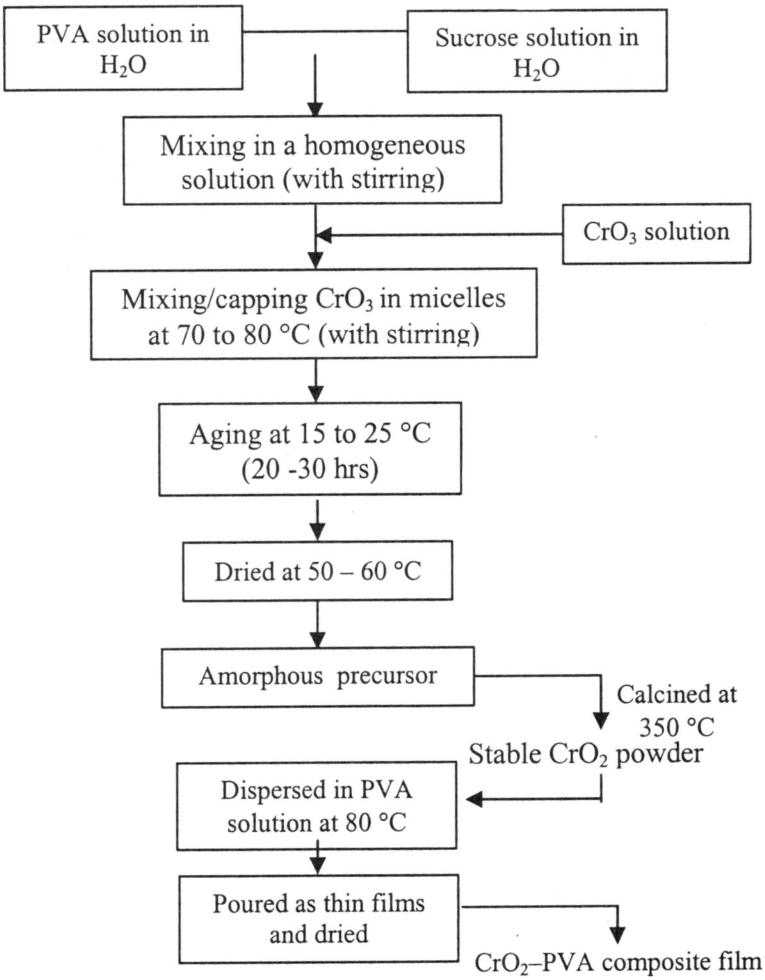

Fig. 1: Schematic diagram showing synthesis of granular CrO_2–PVA composite films.

After the reaction, the precursor solution has been cooled to room temperature. 20 to 30 hrs of aging at room temperature yields the Cr^{4+} metal ion-polymer precursor, of the blackish characteristic colour, embedding the

Table 1: Experimental conditions for forming a Cr^{4+} metal ion-polymer composite precursor

Chemical	*Solvent*	*Volume (ml)*	*Concentration*
CrO_3	Water	50.0	2.0 M
Sucrose	Water	100.0	0.4 g/ml
PVA	Water	100.0	0.04 g/ml

Cr^{4+} cations in small groups of micelles. An amorphous structure retains of sample dried at room temperature or bit higher of 50 to 60 °C in air. When heating in air, a spontaneous combustion reaction occurs (with a strong exothermic peak at 320 °C temperature in thermal analysis), forming a recrystallized sample of CrO_2 in locally directional reactions. A single-phase sample lies in 1 h of heating at 300 to 400 °C in air. A gently grinding in a mortar with a pestle pulverizes the sample in a refined powder by breaking small agglomerates. The obtained CrO_2 powder is boiled in hot water for 2 hrs to get stable CrO_2 particles with a rather stabilized surface layer of $Cr^{4+} \rightarrow Cr^{3+}$ with the impurities of carbon deduced by the decomposition of the PVA-sucrose in the precursor.

As given in Fig. 1, in order to fabricate a CrO_2–PVA composite film, the CrO_2 powder so obtained is dispersed, in a predetermined amount of 1.0, 2.0, or 3.0 wt.%, in a freshly prepared hot PVA solution at 50 to 60 °C temperature with a continuous stirring. After 1 h of stirring, the temperature has been raised to 80 °C in order to increase the viscosity of the mixture by part of evaporation of the water. A viscous mixture (of viscosity as much as near to stop the motion of the magnetic stirrer) is poured on to a glass plate in the form of a thin layer or film by sandwiching with the help of a similar glass plate. The obtained sample is dried in vacuum (at 10 to 20 mbar pressure) at room temperature. The dried films are cut into suitable sizes for the structural characterization and the electrical and other studies.

Measurements and Analysis

Thermal decomposition of the metal ion-polymer precursor and the CrO_2 nucleation and growth following the precursor decomposition were studied by DTA-TG differential thermal and thermogravimetric analysis. The DTA-TG thermograms were recorded by heating a specimen of the precursor powder or a derived CrO_2 powder at a heating rate of 10 °C/min in a thermal analyzer of Shimadzu model TA 501. Phase analysis of the samples is carried out with X-ray diffraction (XRD) using a Philips P.W. 1710 diffractometer with CrKα radiation. A SEM microscope of JEOL JSM-5800 has been used to study morphology and size in virgin CrO_2 particles or those embedded in a PVA polymer as thin composite films. Magnetic properties were measured of selected samples at room temperature with a vibrating sample magnetometer with an applied field to 1.5 kOe. A very sensitive homemade set-up has been used to measure the MR properties. A specific cryostat, with a coolant of liquid N_2, has been used to vary and control the temperature below the room temperature. Other experimental details are the same as described elsewhere (Ram, 1989; Ram and Joubert, 1991; Ram and Frankwicz, 2001).

RESULTS AND DISCUSSION

DTA-TG Analysis for Cr^{4+} Metal Ion-polymer Precursor

On heating, the Cr^{4+} metal ion-polymer precursor has a controlled reconstructive molecular decomposition over 100 to 500 °C temperature with a prominent exothermic peak in DTA at 320 °C, with a total of 30-40% loss in mass in the TG curve. In this specific example, the exothermic signal lies in preponderance of in-situ combustion of the precursor, otherwise, a pure decomposition results in an endothermic process (Mohanty and Ram, 2002; Mohanty and Ram, 2002). According to it, a pure CrO_2 ceramic sample results in 1 h of heating the precursor at temperature as low as 350 °C in air. The DTA-TG analysis, as used for the metal ion-polymer precursor, is performed to demonstrate thermal stability of so obtained CrO_2 of small particles. For example, Fig. 2 shows typical DTA-TG curves measured during heating a 3 to 5 mg of sample in air. The sample is stable as CrO_2 up to a temperature of ~ 400 °C, with a minor mass loss of 1.5 to 2.0 %, which lies in desorption of chemisorbed moisture, air and residual untreated carbon species (after combustion of the precursor) in the form of a shallow and broad desorption curve in TG.

In the TG curve (Fig. 2b), the steep falls, with a well-defined exothermic peak in the DTA at 442 °C in a release of the heat, in the phase transformation of the sample from CrO_2 to Cr_2O_3. It accompanies a chemical reduction process of Cr^{4+} cations, primarily $CrO_2 \rightarrow \frac{1}{2} Cr_2O_3 + \frac{1}{2} O_2$. Thus, releasing oxygen results inevitably in a concomitant decrease in the initial mass of the sample. The process lasts over as high temperatures as ~ 600 °C, with no significant loss in mass further in occurrence of a stable Cr_2O_3 phase over the subsequent temperatures. There is a close crystallographic correlation between the two phases of CrO_2 and Cr_2O_3 that $\{001\}_{Cr2O3} \parallel \{100\}_{CrO2}$ and $\{100\}_{Cr2O3} \parallel \{001\}_{CrO2}$ [4].

Crystal Structure and Microstructure in CrO_2 Granules

The CrO_2 crystals have the Rutile type crystal structure of the $P4_2$ mnm space group (Schwarz, 1986). The distribution of the Cr^{4+} and O^{2-} ions in the CrO_2 crystal unit cell is shown in Fig. 3. In this crystal structure, the Cr^{4+} ions form a body-centered tetragonal lattice, with an octahedral CrO_6 site of Cr^{4+} with the surrounding O^{2-} ions. The lattice is slightly distorted.

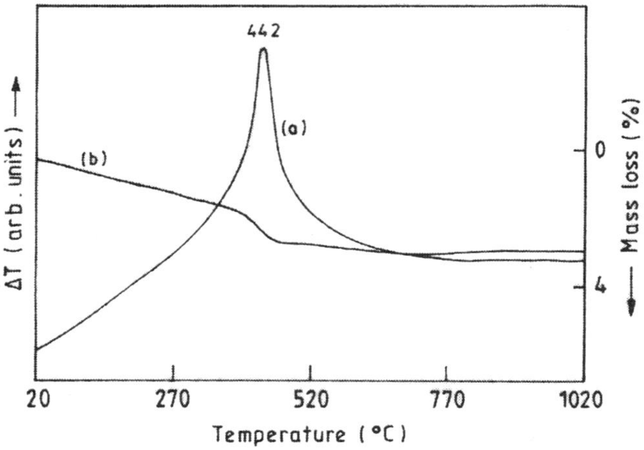

Fig. 2: (a) DTA and (b) TG curves in a CrO_2 nanopowder derived from a Cr^{4+} metal ion-polymer precursor by heating at 350 °C in air.

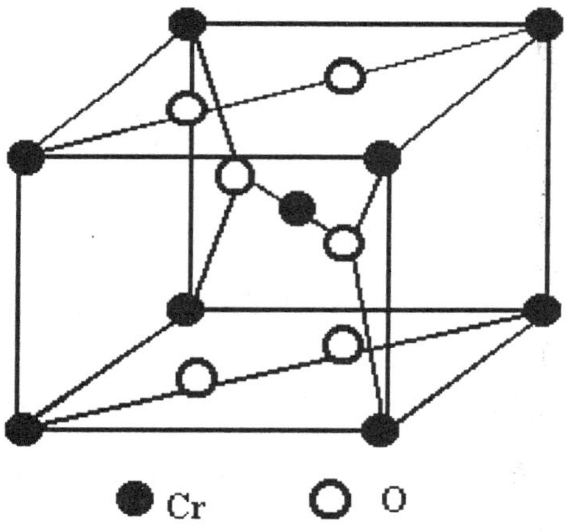

Fig. 3: Rutile crystal structure of ferromagnetic CrO_2.

The dried precursor powder of Cr^{4+} metal ion-PVA polymer composite presents an amorphous structure. A typical XRD pattern for sample derived by drying the precursor at 50–60 °C temperature in air is shown in Fig. 4. It consists of a broad halo at wave vector (defined as $q_i = 4\pi\sin\theta/\lambda$) q_1 = 6.0 nm^{-1} with a rather weak halo q_2 at 27.0 nm^{-1}, confirming an amorphous structure. Two halos arise in two distinct pair distribution functions of the Cr^{4+} and O^{2-} ions in a common amorphous structure of the sample. The results confirm a homogeneous mixing and distribution of the metal cations in the precursor solution via the polymer molecules. A considerably increased viscosity of the precursor solution by addition of sucrose, during the primary reaction, inhibits separation of the metal cations in a homogeneous distribution. Otherwise, under present experimental conditions, they result in a recrystallized compound by separating from the composite structure.

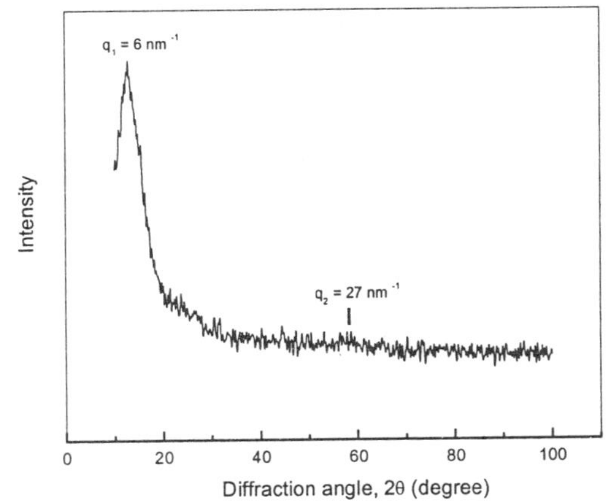

Fig. 4: XRD pattern of Cr^{4+} metal ion-polymer precursor powder.

On heating in air, the precursor decomposes, leaving behind a recrystallized CrO_2 powder of the $P4_2$ mnm tetragonal crystal lattice. A pure single-phase compound of CrO_2 forms in 1 h of heating at temperature as low as 350 °C. This powder has been used to derive CrO_2–PVA polymer films of selected chemical compositions in the two components. Figure 5a portrays XRD pattern for the composite film of 2.0 wt.% CrO_2. The XRD pattern of the powder CrO_2 is given in the inset (b). Assignments of the individual peaks in terms of (hkl) values in reflections from different lattice planes in the two samples are given in Tables 2 and 3, respectively. It is interesting to learn that the characteristic diffraction peaks are broadened along with a considerable change in the intensity distribution in the different peaks on dispersing CrO_2 in the film. Also the lattice parameters of a = 0.4198 nm and c = 0.3082 nm are modified in comparison to those of a = 0.4268 nm and c = 0.3180 nm in the virgin CrO_2 powder. The lattice parameters reveal a reduced value of the lattice volume $V_0 = a^2c \equiv 0.0543$ nm^3 relative to the 0.0579 nm^3 value in the virgin powder. Bulk CrO_2 sample has a value of V_0 = 0.0570 nm^3, with a = 0.4421 nm and c = 0.2916 nm.

Table 2: Interplaner spacing (d_{hkl}) and relative intensities (I) in XRD peaks in CrO_2 nanopowder prepared through a polymer precursor

d_{hkl} (nm)		*I*	*h*	*k*	*l*
Observed	*Calculated*				
0.3018	0.3018	100	1	1	0
0.2531	0.2549	70	1	0	1
0.2251	0.2189	30	1	1	1
0.1617	0.1636	90	2	1	1

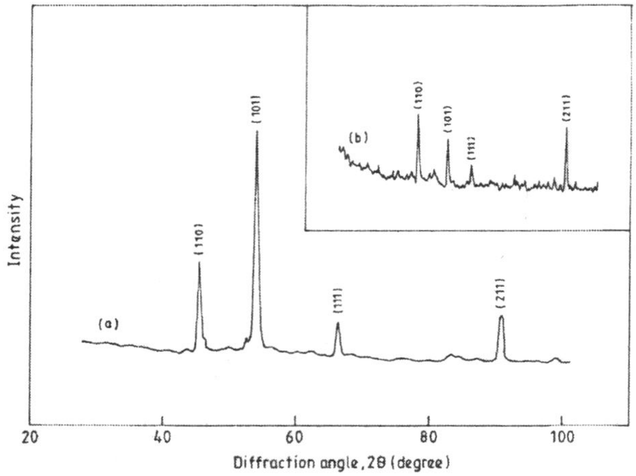

Fig. 5: XRD patterns of (a) CrO_2–PVA composite film and (b) CrO_2 powder.

Table 3: Interplaner spacing (d_{hkl}) and relative intensities (I) in XRD peaks in 2.0 wt% CrO_2–PVA nanocomposite film, with the tetragonal CrO_2 structure

d_{hkl} (nm)		*I*	*h*	*k*	*l*
Observed	*Calculated*				
0.2968	0.2968	41	1	1	0
0.2527	0.2484	100	1	0	1
0.2096	0.2137	15	1	1	1
0.1609	0.1603	27	2	1	1

The sample has been calcined at 350 °C for 1 h. The calculated d_{hkl} values refer to average lattice parameters a = 0.4268 nm and c = 0.3180 nm in the tetragonal structure.

Furthermore, a close-up of individual XRD peaks in samples (a) and (b) in Fig. 5 reveals that they are nearly symmetrical in shape. It rules out the contribution of significant defects like dislocations and stacking faults in either example (Subramanian et al., 2001). As a result, the instrumental broadening in these XRD peaks can be neglected. A comparison between the XRD patterns in the two samples shows that the (101) peak has been intensified in the CrO_2-PVA composite films. The results indicate that an orientational growth occurs in <101> domains in the CrO_2-PVA films. The PVA polymer molecules seem to govern the domain structure. This is very important in analyzing and understanding magnetic and GMR properties in such examples in thin films.

Figure 6 shows SEM micrographs for CrO_2 powder and the CrO_2-PVA composite film. Most of CrO_2 particles in micrograph (a) are in platelet morphology of 25-100 nm size. Small particles have aspect ratio $\varphi \sim 1$.

Agglomeration in small group of particles is removed on dispersing the CrO_2 powder (after boiling in water followed by sonication) in PVA polymer in composite films as demonstrated with micrograph (b) for the sample of 2.0 wt% CrO_2. The present micrograph thus consists of uniformly dispersed CrO_2 particles of 30-50 nm sizes of $\varphi \sim 1$. A Debye-Scherer analysis of peakwidths, $2\Delta\theta_{1/2}$, in the XRD peaks suggests an average crystallite size D of 45 nm in the virgin CrO_2 powder while a refined value of 35 nm in the CrO_2-PVA thin films in neglecting the contribution of the strains in the total $2\Delta\theta_{1/2}$ value. A close similarity of the D-value to the size in small particles in the two SEM micrographs indicates that they are representing single CrO_2 crystallites. The PVA polymer matrix, which exists in an amorphous state, does not contribute a fine structure at this scale.

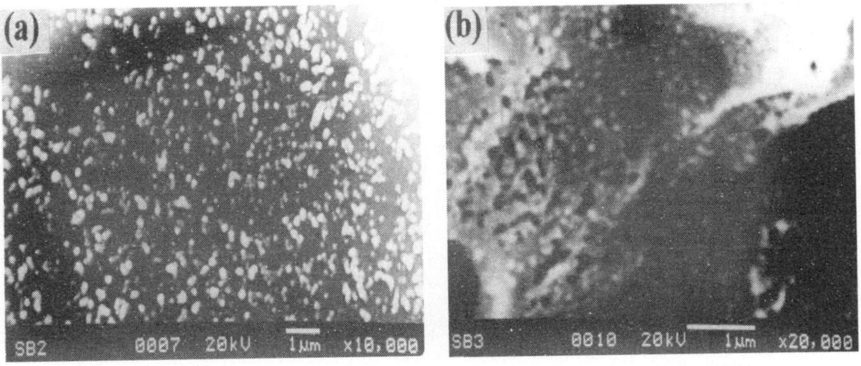

Fig. 6: SEM for (a) CrO_2 powder and (b) 2.0 wt% CrO_2-PVA (CrO_2) composite film.

Magnetic Properties in CrO_2 Granules

The magnetic or GMR studies are performed on compacted CrO_2 powder in thin pellets and that of dispersed in PVA polymer in thin CrO_2-PVA composite films. Since the magnetic behaviour of nanomaterials is largely unknown, here we check whether the measured properties of CrO_2 nanoparticles are consistent with the present theories explaining the behaviour of conventional magnetic materials (Kumar et al., 2002; Wohlfarth, 1980; Kumar et al., 2003). For instance, a characteristic feature of ferromagnets is the existence of stable magnetic domains. A magnetocrystalline anisotropy is required for such a domain structure to exist in a stable configuration. A very simple way to characterize a ferromagnetic material is by measuring its spontaneous magnetization. Due to the domain structure, a ferromagnetic sample does not have a net magnetization in absence of an external magnetic field. In an applied magnetic field H, the magnetization of the sample lies along the field direction. In a sample of small crystallites, this grows with field and tends to be saturated at a specific value of the field H equivalent to the magnetocrystalline anisotropy H_a. Thus, a hysteresis loop occurs of the magnetization process as a function of H.

A hysteresis loop measured for a sample of CrO_2 nanopowder is given in Fig. 7. It determines the magnetic parameters of the saturation magnetization M_s = 30.8 emu/g; remanence M_r = 25.2 emu/g and coercivity, H_c = 269 Oe. Bulk CrO_2 has a bit larger value of M_s = 40-45 emu/g at room temperature (Chamberland, 1977). A smaller value observed in this specific example of small particles can be described by assuming a spin canting in preponderance of the large surface or surface anisotropy. In fact, this is responsible for superior MR properties with an improved value of MR in such small particles of granular CrO_2.

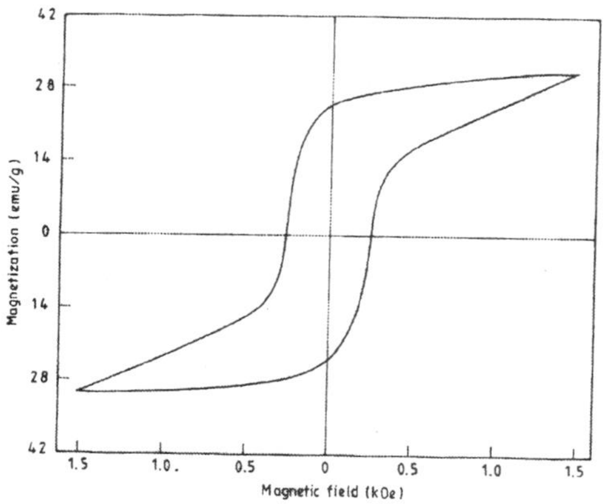

Fig. 7: The hysteresis loop for CrO_2 powder of nanogranules.

The variations of MR with a function of temperature in the CrO_2-PVA composite films of three selected compositions of (a) 1.0, (b) 2.0, and 3.0 wt.% CrO_2 are displayed in Fig. 8. The films of 1.0 wt.% CrO_2 show a maximum value of MR of 11.8% at 77 K. On raising the temperature, the value decreases exponentially to a value of as small as 1.2% at 300 K. A more or less similar temperature dependent behaviour of MR is maintained in the other two samples. This is consistent with a similar temperature dependence of MR reported in a bulk CrO_2 powder compact (Coey et al., 1998). The suppression of MR at high temperatures above 77 K is mainly due to the suppression of spin-dependent contribution of the conductance by the increase of the spin-independent hopping channel with increase in the temperâture (Shang et al., 1998).

The MR curves with applied field for the above three specimens at room temperature are displayed in Fig. 9. It is seen from Fig. 9 that the MR reaches to be as high as 6.8% at 1.1 kOe of applied field (1.0 wt.% CrO_2 film). It is also observed that the field dependence of MR is sensitive within 3.5 kOe. The nature of the polymer barrier between CrO_2 particles is most likely responsible for the enhancement of the MR in this type of composite

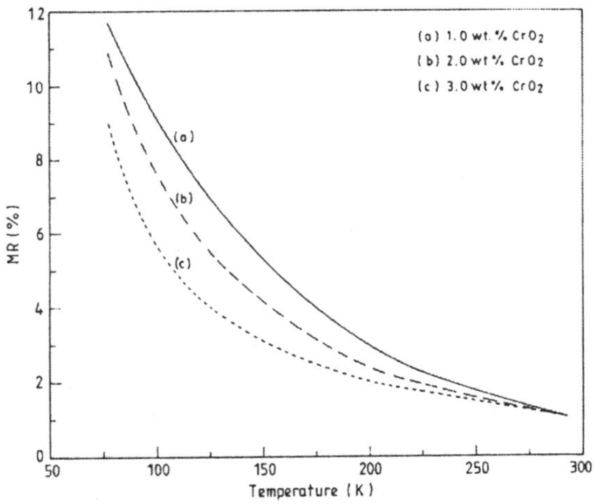

Fig. 8: Temperature dependence of MR for (a) 1.0 wt.%, (b) 2.0 wt.%, and (c) 3.0 wt.% CrO_2–PVA composite films.

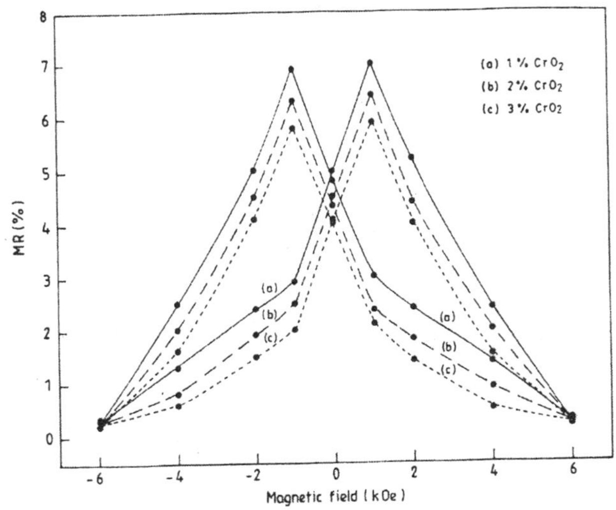

Fig. 9: Dependence of MR on applied magnetic field for the samples of (a) 1.0 wt.%, (b) 2.0wt. %, and (c) 3.0 wt.% CrO_2–PVA composite films.

films. It is found that the MR value is enhanced roughly by an order of magnitude of the value in a bulk CrO_2 powder compact. Once the CrO_2 granules are appropriately encapsulated with insulating polymer the electrical resistivity exhibits a rapid increase. Hwang et al. (1996) suggested that the low field MR in this type of coated granular materials is due to the presence of weak links between the granules. A diamagnetic surface barrier layer of polymer in individual CrO_2 granules supports spin polarized tunneling between misaligned CrO_2 grains.

Table 4: Magnetic properties in CrO_2 nanogranules embedded in PVA in thin films.

Samples	*Saturation magnetization (emu/g)*	*Remanence (emu/g)*	*Coercivity (Oe)*	*Maximum MR at room temp. (%)*
CrO_2 powder	30.8	25.2	269	–
1.0 wt.% CrO_2–PVA film	–	–	–	6.8
2.0 wt.% CrO_2–PVA film	–	–	–	6.3
3.0 wt.% CrO_2–PVA film	–	–	–	5.8

CONCLUSIONS

A novel series of GMR materials of CrO_2-PVA composite films is developed using a novel chemical method. XRD patterns reveal a nanocrystalline structure of CrO_2 with an amorphous structure of the PVA in a nanocomposite structure. At room temperature, the MR value in CrO_2-PVA composite films (1.0 wt.% CrO_2) reaches to be as large as 6.8% at 1.1 kOe of applied field. This high value of MR is due to a special microstructure of the specimen with a diamagnetic surface layer of PVA polymer, which enhances the intergranular tunneling of spin-polarized electrons. The results are analyzed in correlation with the microstructure. The implications are useful for (i) practical applications of GMR sensors at room temperature, (ii) development of GMR devices and components, and (iii) understanding and modeling of GMR properties in granular solids.

Acknowledgements: This work has been financially supported by Defence Research and Development Organization, Govt. of India.

REFERENCES

Barry, A., Coey, J.M.D. and Viret, M. (2000). *J. Phys.: Condens. Matter*, **12:** L173.

Biswas, S. and Ram, S. (patent filed).

Chamberland, B.L. (1977). *Critical Rev. Solid State Sci.*, **7:** 1.

Chen, Y.J., Zhang, X.Y., Cai, T.Y. and Li, Z.Y. (2004). *Mater. Lett.*, **58:** 262.

Chen, Y.J., Zhang, X.Y. and Li, Z.Y. (2003). *Chem. Phys. Lett.*, **375:** 213.

Coey, J.M.D., Berkowitz, A.E., Balcells, L., Putris, F.F. (1998). *Phys. Rev. Lett.*, **80:** 3815.

Goodenough, J.B. (1971). Progress in Solid State Chemistry. Reiss, H. (ed). Pergamon, Oxford. **5:** 145.

Gupta, A., Li, X.W. and Xiao, G. (2000). *J. Appl. Phys.*, **87:** 6073.

Hwang, H.Y. and Cheong, S.W. (1997). *Science*, **278:** 1607.

Hwang, H.Y., Cheong, S.W., Ong, N.P. and Batlogg, B. (1996). *Phys. Rev. Lett.*, **77:** 2041.

JCPDS X-ray powder diffraction file no. 9-332.

Ju, S. and Li, Z.Y. (2002). *Phys. Lett.*, **A 293:** 199.

Kämper, K.P., Schmitt, W., Güntherodt, G., Gambino, R.J. and Ruf, R. (1987). *Phys. Rev. Lett.*, **59:** 2788.

Korotin, M.A., Anisimov, V.I., Khomskii, D.I. and Sawatzky, G.A. (1998). *Phys. Rev. Lett.*, **80:** 4305.

Kumar, G., Eckert, J., Roth, S., Loser, W., Ram, S. and Schultz, L. (2002). *J. Appl. Phys.*, **91:** 3764.

Kumar, G., Eckert, J., Roth, S., Loser, W., Schultz, L. and Ram, S. (2003). *Acta Mater.*, **51:** 229.

Liang, J.J., Lee, S.F., Yao, Y.D., Wu, C.C., Shyu, S.G. and Yu, C. (2002). *J. Magn. and Magn. Mater.*, **239:** 213.

Manoharan, S.S., Elefant, D., Reiss, G., Goodenough, J.B. (1998). *Appl. Phys. Lett.*, **72:** 984.

Meservey, R. and Tedrow, P.M. (1971). *Phys. Rev. Lett.*, **26(4):** 192.

Meservey, R., Tedrow, P.M. and Fulde, P. (1970). *Phys. Rev. Lett.*, **25 (18):** 1270.

Mohanty, P. and Ram, S. (2002). *J. Europ. Ceram. Soc.*, **22:** 933.

Mohanty, P. and Ram, S. (2002). *Philos. Magn.*, **B 82:** 1129.

Moodera, J.S. and Mathon, G., (1999). *J. Magn. and Magn. Mater.*, **200:** 248.

Ram, S. (1989). *J. Magn. and Magn. Mater.*, **82:** 129.

Ram, S. and Frankwicz, P.S. (2001). *Phys. Status Stat. Sol.* (a), **188:** 1129.

Ram, S. and Joubert, J.C. (1991). *Phys. Rev.*, **B 44:** 6825.

Schwarz, K. (1986). *J. Phys. F: Met. Phys.*, **16:** L 211.

Shang, C.H., Nowak, J., Jansen, R., Moodera, J.S. (1998). *Phys. Rev.*, **B58:** 2917.

Soulen Jr., R.J., Byers, J.M., Osofsky, M.S., Nadgorny, B., Ambrose, T., Cheng, S.F., Broussard, P.R., Tanaka, C.T., Nowak, J., Moodera, J.S., Barry, A. and Coey, J.M.D. (1998). *Science*, **282:** 85.

Subramanian, R., Shankar, P., Kavithaa, S., Ramakrishnan, S.S., Angelo, P.C. and Venkataraman, H. (2001). *Mater. Lett.*, **48:** 342.

Van Leuken, H. and deGroot, R.A. (1995). *Phys. Rev.*, **B 51:** 7176.

Wohlfarth, E.P. (1980). *In:* Ferromagnetic Materials (2), North-Holland, New York.

Importance of Optimizing X-ray Beam Optics for High-Quality Data Acquisition Using Conventional X-ray Diffraction Techniques

G.V. Narasimha Rao

International Advance Research Centre for Powder Metallurgy and New Materials, Hyderabad – 500005, India

INTRODUCTION

X-ray diffraction (XRD) equipments have become a routine experimental tool not only in research and development establishments and well established industries around the world but also in the institutions like schools and colleges and even in the small scale industries for testing of the process materials for phase studies at various stages of manufacturing of the materials. Most of the installation engineers demonstrate the five-finger pattern of α-quartz around 67-69° 2θ range, or (101) Bragg peak of α-quartz pattern using CuKα or monochromatic $K\alpha_1$ radiation. In majority cases the quartz (101) diffraction line does not produce the $K\alpha_2$ diffraction line, or even if it is present, it just appears as a shoulder and never shows as a completely resolved diffraction peak. This is due to improper X-ray beam optics as well as from the parasitic scattering effects of various instrumental parts present in the path of the beam. The intensities of the most of the Bragg peaks in the patterns are not high enough to produce good statistical data. In a properly fine tuned setup the intensities of the peaks are found to be several times high when compared to the same peak before tuning. Over 95% of the laboratories equipped XRD systems show very poor performance. The reasons for such poor performance could vary on various reasons, from supplier to supplier, on the specifications, and among the same supplier the equipment to equipment, even though both equipments are of the same make and same model. This is mainly due to critical alignments of various parts of the equipment and the manner the equipment is handled. At this stage it is better not to discuss on the performance of the electronics, as it is considered as unique and reproducible. The most important features which affect the intensity of the pattern(s) is mainly due to the alignment of the mechanical parts. Most of the service personnel as well as routine operators take it for granted that they perform well, once it is commissioned and the service person demonstrates recording of the standard XRD graph's like quartz etc., with its five finger pattern near 67-69° 2θ region. Let us ask a question that mere reproducing a graph recorded earlier at the supplier's site is just sufficient or some thing more can be done to achieve enhanced performance of the equipment to get best results? In this paper we wish to describe various methods of fine-tuning Bruker's D8 advanced XRD system, which should be taken as an example, but similar methods could be applied to any XRD instruments manufactured by other manufacturers. The fine-tuning of the equipment is in fact, equipment dependent and not necessarily make or model number dependent. Even if same sample pattern is from two different equipments of the same manufacturer, the peak statistics need not be exact, due to several reasons and the most important of that could be beam optics.

If the XRD equipments are properly fine tuned, it is observed that the signal to noise ratio (S/N) enhances many times, resolution of data obtained increases tremendously there by clear separation of closely spaced Bragg peaks. This helps the analyst for precision measurements of parameters like peak position, FWHM, grain size calculations, identification of preferred orientation directions etc. It is practically observed that the background level drastically reduces, thereby making visible of peaks corresponding to impurity peaks of very low concentrations (some times close to 0.01%).

EXPERIMENTAL PROCEDURE

Lanthanum Hexaboride (LaB_6) sample in the form of fine powder was dusted as a very thin layer at the centre on a Si (911) Zero Background Plate (ZBP) (Narasimha Rao et al., 1996; Narasimha Rao et al., 1991) and placed at the sample holder position on the Bruker's D8 Advance system mounted in the vertical mode with θ/θ geometry, for recording the XRD patterns. There are several advantages of choosing LaB_6 as the standard material for XRD calibration purposes. Most important features of LaB_6 powder are (1) the Bragg peaks are very evenly spaced over a large range of 2θ starting from about 20° 2θ to 150° 2θ when used with CuKα radiation, (2) it produces peaks with very narrow FWHM, (3) the peaks present in the spectrum are of high intensity as the element Lanthanum with atomic No. 57 (relatively high Z material) is responsible for scattering process, (4) it is very stable compound and not affected by environmental conditions, (5) it can be obtained in relatively pure form, (6) its crystal structure is cubic and hence further calculations are very easy, (7) it does not show any preferred orientation properties in the fine powder form and (8) the powder particles do not have strains associated with it, etc. The instrumental data recording conditions for LaB_6 sample are listed in Table 1.

Table 1: XRD recording conditions of LaB_6

Property	*Associated Value*
Instrument make and Model	Bruker's D8 Advance XRD system in vertical mode
Wave length used	CuKα (1.5418 Å)
Sample area spread on the sample holder plate	15 mm diameter circular
Method of sample loading	Centre of a mirror polished Si (911) ZBP is wetted with hand moisture to stick sample particles on to the surface. Several hundreds of extremely fine and shiny particles are poured on to the surface of the plate and excess powder is tapped back to the sample to avoid heap formation on the sample plate. This process repeated several times to make thin but even layer of the sample of about 15 mm diameter area on surface.
Sample thickness used for XRD work	Very thin layer (Approximately a few grains thick)
Sample surface smoothness	May not be smooth when viewed from about 90° to the sample surface, but this roughness is only a few grains thick, hence it is negligible. No leveling or surface flattening is done which could influence the grain orientation.
Sample rotation speed	60 rpm (one revolution/sec.), to reduce preferred orientation effects.
Slits opening	Auto divergence and anti-scatter slits for constant area irradiation on the sample V12.
Foot print of the X-ray beam on the sample	Approximately 12 mm diameter, when the sample is continuously rotated
Soller slits	2°, both at incident and diffracted beam sides
Angular range of scan recorded	3° 2θ to 90° 2θ
Step size	0.02° 2θ
Step time	1 second

Earlier to recording the XRD pattern of the sample the X-ray beam optics of the Bruker AXS D8 Advance XRD system has been very carefully defined by making sure that the maximum diffracted X-ray intensity goes to the detector, without any undue reduction of intensity during the entire length of beam travelling path and also taking care that there is no parasitic scattering from any of the parts present in the path of the beam. This is done by fine tuning the equipment carefully adjusting/aligning various parts of the instrumentation such as X-ray tube, its positioning by using all the three degrees of freedom available (U-swing in the direction of the X-ray beam, tube height, and tube tilt perpendicular to the beam direction) from the tube shield holder, exactly matching the Soller slits present at the incident beam side as well as on the diffracted beam side, orientation of the slit present in the divergent slit assembly, sample table height and its tilt, orientation of the slit present in the anti-scatter slit assembly, Soller slits at the diffracted beam side, orientation and tilt of the receiving slit etc. Smallest available slits are used at every stage in such manner, thereby making sure that maximum intensity of the X-ray beam is observed at the detector assembly, right from the X-ray tube focus to the detector end. It is ensured that even if very small pressures/tilts applied on any of the above referred part(s), intensity reduces drastically when observed at the detector end by measure of count rate meter with very small counting time for integration of the intensity.

Direct Beam Profile Recording

It is often necessary to measure the direct beam profile of the equipment to define the 'zero' positions, both at the tube as well as at the receiving slit, which is coupled to the detector assembly via graphite diffracted beam monochromator. This 'zero' alignment is most important and crucial part of the exercise when the XRD equipment was procured newly or when the instrument drives mismatched. This procedure is also known as the direct beam profile checking of the equipment. In this case a glass slit assembly (as supplied by Bruker AXS) is placed at the sample holder position and turned/placed in such a manner that the narrowest X-ray beam ribbon emerging from the X-ray tube and divergent slits passes through the entire length of opening of the glass slit (50 mm long 25 microns high and 10 mm wide) present in between the two glass plates to form a narrow ribbon of X-ray beam. The emerging very narrow X-ray beam in the form of ribbon from the glass slit assembly is made to pass into the detector assembly. It is absolutely necessary to place appropriate thick metallic sheets/absorbers in front of the detector assembly to reduce the direct incident beam intensity into the detector, thereby preventing the damage of the detector. Several independent scans of tube, detector profiles are recorded, with the glass slit assembly in its place, to monitor the intensity maximum of the scan and the positions of maximum intensity of the tube as well as detector arms are stored as the 'zero' position(s) for that particular drive. This procedure is repeated by rotating the glass slit assembly by 180° in its place. The intensity variation of the emerged X-ray beam from the glass slit assembly, before and after rotation of 180° should not exceed more than 1% when the narrowest (fixed or auto divergence/anti-scatter) slits are used. This means that the photon count at the detector before and after the rotation of the glass slit should more or less remain same without any decrease in intensity. This ensures that the sample holder plate (central portion of the glass slit assembly through which the X-ray beam is passing) is exactly parallel to the ribbon joining X-ray tube focus and the receiving slit, through the glass slit. This procedure is repeated several times to ensure that the X-ray intensity recorded at the detector is same.

A typical direct beam profile pattern should appear like the one shown in Fig. 1. It is observed that there exist two humps in the direct beam profile on either side of the most intense peak corresponding to the direct beam. This is probably for the reason that a X-ray beam bounces once between the entrance point of the glass slit and the centre of the glass slit and similarly at the centre of the glass slit and the exit point of the glass slit due to the natural divergence of the X-ray beam, within the glass slit path. The un-bounced direct beam produces the central most intense peak for the direct beam. It should be noted that the humps on both sides of the direct beam must have as equal intensities as far as possible in order to achieve maximum intensity of the diffracted beam and ideal resolution. It can be noticed from Fig. 1 that the maximum intensities of the left and right side humps of the most intense direct beam (404615 counts at the top of the peak with an FWHM of 0.012°, when the X-ray tube is operated at 40 kV and 40 mA, using the narrowest available slits in the X-ray path) through the Bruker's glass slit are 7138 counts and 6608 counts with their centroid's at –0.72°2θ and +0.09°2θ, respectively. This amounts a difference of only 530 counts between the maximum intensities of the humps. If this difference is converted into percentage on dividing by the intensity of the direct beam it works out to be 0.0956 %. This indicates that the modulus of the difference of centroids of both peaks equals 0.018° 2θ. If the system is further fine-tuned carefully this difference of centroids can be brought down to 0.0° 2θ, when the intensities of both left and right humps becomes equal and producing most strongest direct beam. It means that the central point of diffracting area on the sample exactly lies on the Bragg-Brentano circle to produce maximum intensities for all the diffraction lines of the sample. Any diffraction pattern(s) recorded using such a fine tuned powder diffractometer produces, symmetrical intensities on both sides of the Bragg peaks, both at the start of the Bragg peak and at the end of the Bragg peak except in the case of sample inherent properties like stacking faults etc. In other words, the normal Bragg peaks do not produce sudden rise or fall at the foot, on both sides of the Bragg peak.

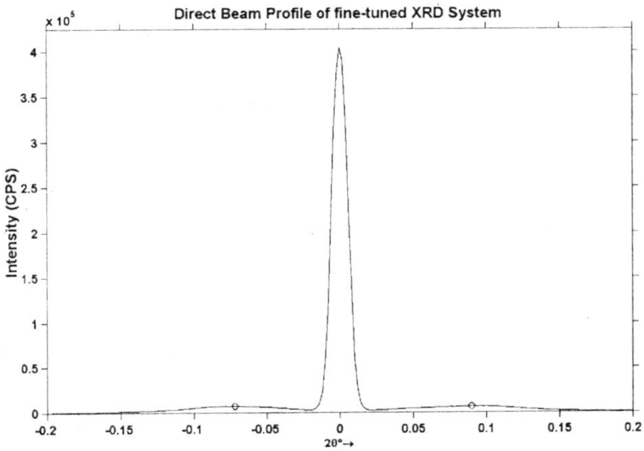

Fig. 1: Direct beam profile of the XRD equipment.

Sample Positioning Errors in Bragg-Brentano (BB) Geometry

Errors in sample positioning would result in shift in the 2θ position of the Bragg peak as per the following equation (Cullity, 1978)

$$\frac{\Delta d}{d} = -\frac{D}{R}\frac{\cos^2\theta}{\sin\theta}$$

where 'd' is the interplanar spacing corresponding to the Bragg angle θ, R is the radius of the focusing circle of the goniometer used and D is the sample displacement parallel to the reflecting-plane normal. This is normally the single largest error in most of the diffractometer measurements. Figure 2 shows the change in d spacings as a function of sample positional displacement above the focusing circle, while positioning the sample on the sample holder table. It should be ensured that the top layer of the sample under investigation, which is exposed to X-ray beam, must be flat and as smooth as possible for producing intense scattered beam. As a rule of thumb sample powders whose grain sizes are of the order of 1 to 5 micrometres produce good diffraction patterns. In the case of bulk materials like metallic samples, sintered specimens, the exposed surface if polished to very smooth or mirror finish would produce extremely sharp and most intense Bragg peaks. However, it is to be noted that the bulk materials referred above, may get preferred orientation of grains (texture) in the process of manufacture, hence the sample has to be rotated in every diffraction plane at least at a speed of 60 rpm to minimize the texture effects.

Testing the Accuracy of BB Geometry in a Given Equipment

Using such an optimized X-ray beam path, a small mirror polished (111) oriented single crystal of Silicon wafer of 0.5 mm thick and approximately 3 mm square piece, in the form of sample was placed at the sample position,

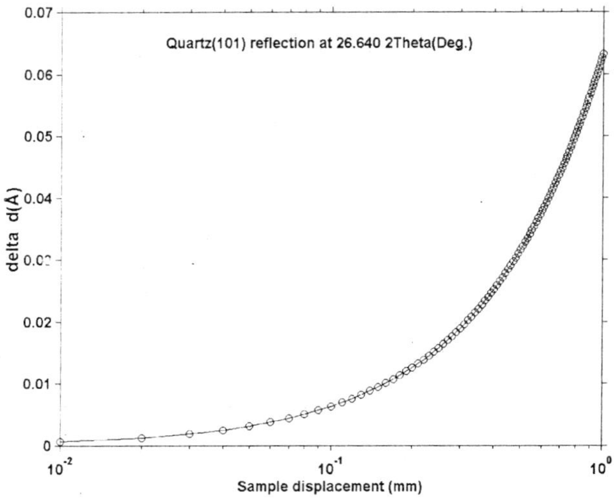

Fig. 2: Change in d-spacings resulting from sample positional displacement errors.

using a specially designed sample holder disk, wherein the sample thickness is adjusted by slowly moving on groves a rod like object which goes back and forth (according to the thickness of the sample under investigation) in a disk of 50 mm diameter. Maximum care must be taken while positioning the sample, as the top surface of the sample must fall on the focusing circle of the BB geometry. The tube and the detector arms were moved to position of 14.221° θ, in both cases such that the scattering angle is at 28.442° 2θ. The narrowest available slits (0.1°) are used at various slit positions and 0.05° slit was used at receiving slit position, before opening the X-ray window. The voltage and current settings were maintained at 40 kV and 40 mA, which are normal operating settings for recording the XRD patterns for regular samples using 2.2 kW LFF ceramic Cu target tube. At this position the X-ray beam was turned on and scattered intensity of the Si (111) Bragg peak is monitored using count rate meter. The photon count rate was continuously monitored for maximum intensity while very slowly turning the sample stage in such a manner that the maximum counts are registered in the count rate meter, of course, after positioning appropriate absorbers before the detector. Once the maximum intensity is observed for a particular orientation of the crystal, sample rotation stage is carefully adjusted by rotating crystal orientation back and forth till the maximum intensity is noted and that orientation is fixed and the X-ray tube window is closed.

At this position of the crystal, sample rotation stage was temporarily arrested by pasting tapes around the stage to avoid any possible movement of the sample spinner, which can alter the orientation of the crystal. At this time, both the tube and detector scans are independently carried out several times to optimize for maximum scattered intensity of the sample and their positions are redefined (locked). If any variations in θ or 2θ positions are observed as compared to the already locked positions at the direct beam profile, it infers that the sample stage is not exactly at the centre of the goniometer and appropriate corrections are carried out and the above procedure is repeated once again. Figure 3 shows an XRD pattern of Si (111) wafer from 28.35 to 28.6° 2θ recorded on a θ/θ vertical goniometer using BB geometry in a continuous scan mode. From this, it is very clear that FWHM for Si (111) peak is 0.01697° 2θ (corresponding to 61 arc-seconds wide). It is clear that the doublet of $CuK\alpha_{1,2}$ peaks are well separated even at as low angles as 28.44° 2θ. The recording of Si (111) peak with such narrow FWHM using conventional BB geometry with sealed X-ray tube is the lowest reported so far to the best of our knowledge (Louer, 1993). Most importantly the valley between the two peaks is very deep and the height of the valley is 96.314% when compared with the maximum height of the $K\alpha_1$ peak intensity, if taken as 100% (Klug and Alexander, 1974; Jenkins and Snyder, 1996). This way of fine-tuning the XRD equipment helps to detect weak peaks for quantitative estimation of very small quantities (some times < 0.01% concentrations) of phase(s).

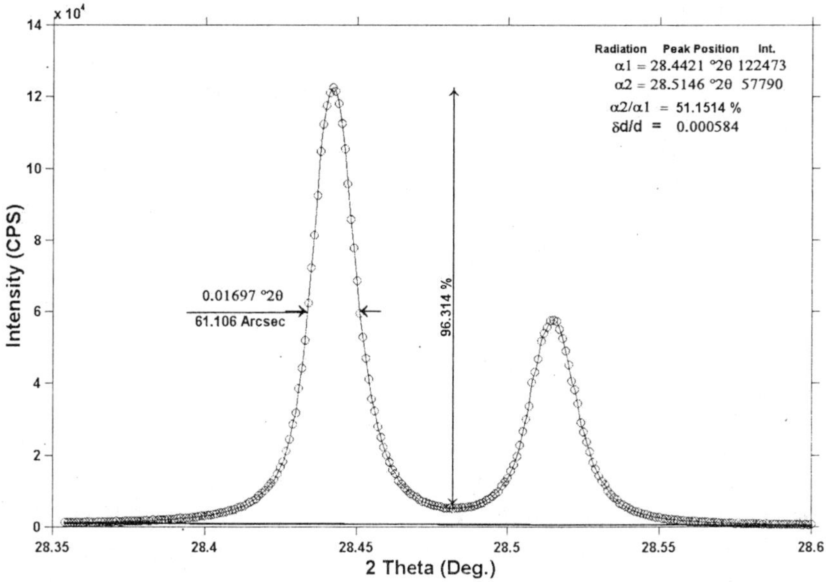

Fig. 3: XRD pattern of Si (111) reflection using CuKα radiation.

In our own work of 300-400Å coating of thin films of Tantalum Nitride on Silicon wafer (to be published), we were able to easily identify the required phases formed and able to match comfortably with patterns listed in the ICDD database. When we tried to measure for the diffraction lines required phases using grazing incidence attachment the quality of the data recorded was extremely poor as compared with the data recorded using the BB geometry. This is a clear example for measurement of very low concentration of phases in the BB geometry. Figure 4 shows XRD pattern of an uncoated and as polished 0.5 mm thick Silicon (111) crystal wafer. One can see a very small portion of CuKβ radiation, which is never observed in the normal diffraction mode. It is also to be noted that the peak marked as Si (222) is in fact Si (444) using λ/2 wavelength, as (222) is a forbidden reflection in diamond type of cubic material like silicon. The intensity of CuKβ peak, which can be seen very clearly in the XRD pattern has only around 80 counts at the top of the peak as compared to the strongest Si (111) peak, whose intensity is around 850000 counts. It means that around 0.0094 % intensity of Kβ peak can also be seen clearly using our setup.

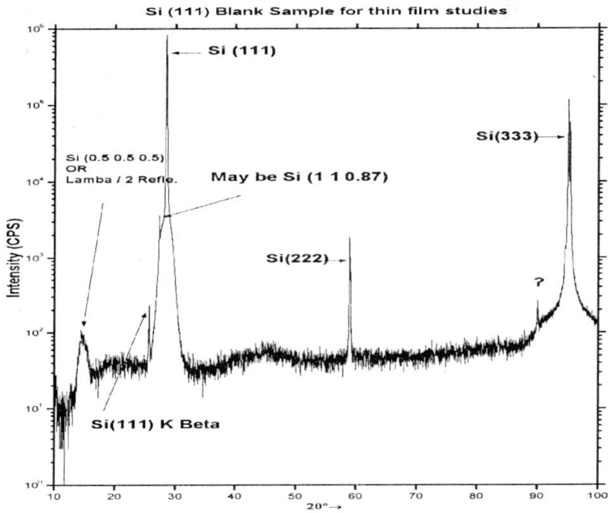

Fig. 4: XRD pattern of polished single crystal of Si (111) wafer.

Reasons for Choosing Si (111) Single Crystal Wafer for Equipment Alignment Studies

Silicon is a diamond type cubic lattice and is most often used in the semiconductor industry. Silicon with its lattice spacings of 5.430 Å belong to the space group Fd3m, produces its first (111) peak at 28.442° 2θ when $CuK\alpha_1$ radiation is used. Silicon single crystals are not much affected by environmental variations, such as humidity, changes occurring in room temperature because of the climatic conditions etc. The mirror polish given to the crystal remains untarnished for several years, if handled properly. Formation of scratches on the polished surface of wafer must be avoided as far as possible. Any dust particles/finger prints present on the surface can be easily wiped off with tissue paper wetted with normal laboratory cleaning solutions like acetone etc. Since Si (111) produces a very strong intensity, it works as an excellent X-ray beam reflector. The diffracted intensity of Si (111) peak is so high that the scattered X-ray beam can be seen easily with normal handheld fluorescent screens. Therefore, tracking the direction of diffracted X-ray beam can be done visually in minimum time to minimize the radiation exposure. This tracking of the diffracted beam of Si (111) may be necessary only in rare occasions when the tube/detector settings are grossly wrong. The procedure for fine-tuning the XRD equipment must be done only by experienced scientists or service personnel taking maximum precaution/care for possible X-radiation exposures. Once the alignment procedure is completed and all appropriate screws are tightened, the equipment alignment remains undisturbed for several years. In the unlikely event of crashing of drives or undergoing any mechanical shocks the entire procedure of alignment must be redone.

DISCUSSION

The full range of XRD data for LaB_6 recorded using BB geometry is shown in Fig. 5. The maximum intensity recorded for the strongest Bragg peak with Miller indices (110) at 30.36 is 14078 cps. It is observed that, according to Philips report, the peak count on the same Miller indices plane is around 50000 cps. It is to be noted that data collection time for 1° 2θ with Philips X'Celerator detector (area detector) is 10 times as compared to our (point detector) YAP detector, whereas the intensity count is only four times higher compared with YAP detector after fine-tuning of the equipment. An arbitrary scaling factor which is generated by evaluating maximum intensity of the peak count in the entire pattern divided by time to scan 1° 2θ by the detector shows only 103 counts/second for X'Celerator detector whereas it is 292 counts/second for fine tuned system using YAP point detector. This again clearly demonstrates the importance of instrument fine-tuning.

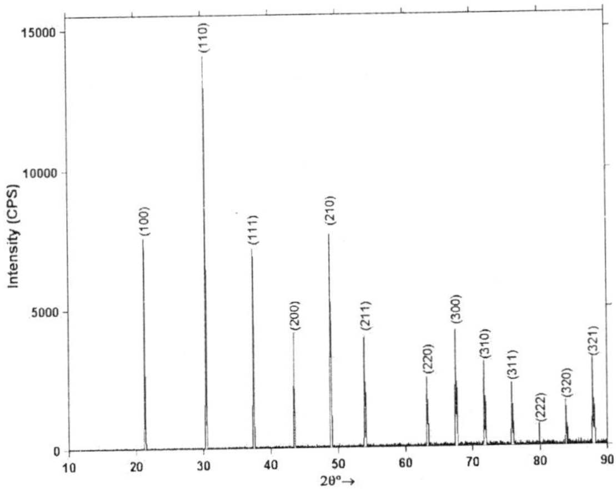

Fig. 5: Indexed XRD pattern of LaB_6 powder sample.

In its report a reputed firm also mentioned that it takes more than a month to collect the information using point detector when the data acquired with

X'Celerator took only 18 hours. Our data clearly shows that a similar maximum count rate as obtained using X'Celerator can be achieved by a fine tuned equipment like ours with about four times the present data collection time, i.e., with data collection time as 5 sec/step instead of the present scan time of 1 sec/step. We do agree that the area detector is capable of collecting the information from several scattering angles simultaneously at the same time, but more importantly equipment fine-tuning is the key factor for coherently scattering the diffracted intensity to the exact Bragg positions. Such fine tuned system produces minimum scattered intensity in the vicinity Bragg positions for the slightly disoriented grains, and exactly at Bragg position, the peak intensity rises to maximum, making peak intensity very high with narrow FWHM. If the FWHMs of most peaks are quite high then particle size calculations, using Scherrer equation ends up with large. errors, as the true widths of peaks are much smaller as compared to the observed peaks as the instrumental broadening for the peaks are much larger. In most cases the strain effects of the particles are generally ignored, considering it as a part of the instrumental width. In an ideal case, the XRD patterns of the samples must be recorded twice, once before annealing and once after annealing at appropriate temperatures for sufficiently long time to completely remove strains present in the sample. In properly fine tuned XRD equipment the observed FWHM of any Bragg peak should be very nearly equal to the FWHM of the same Bragg peak after subtraction of the instrumental broadening effect, in other words the instrumental width must be minimum.

Figure 6 shows the expanded view of (100) reflection of LaB_6 sample and it can be clearly seen the $CuK\alpha_{1,2}$ peak separations even at as low angle as 21° 2θ using sealed ceramic LFF tubes with powder samples. If the XRD equipment is not fine-tuned such separations are not possible to observe. In practice it is to be noted that generally used CuKα X-ray beam cannot penetrate more than 5 to 10 microns from the top layer of most of the samples because of the sample absorption effects. Any sample thickness greater than the maximum penetration depth of the sample is infinite thick for that sample. The powder sample thickness must be less than the maximum penetration depth in order to obtain excellent patterns. It is also important to note that if the powders used for XRD work have particle sizes greater than 10 μm, the sample thickness must be restricted to a single particle layer thereby minimizing sample absorption effects. This way of fine-tuning the equipment helps to identify very closely spaced diffraction peak(s) as the peak intensities are quite strong and well separated.

Figure 7 shows the Cagliotti function (FWHM = square root ($U^*\tan^2\theta$ + $V^*\tan\theta + W$), where U, V and W are the coefficients in the equation and θ is the semi-scattering angle, which produces the least squares fitted curve of experimentally observed FWHMs for all the peaks. In an ideal case all the experimentally observed FWHMs should lie on the fitted curve with no

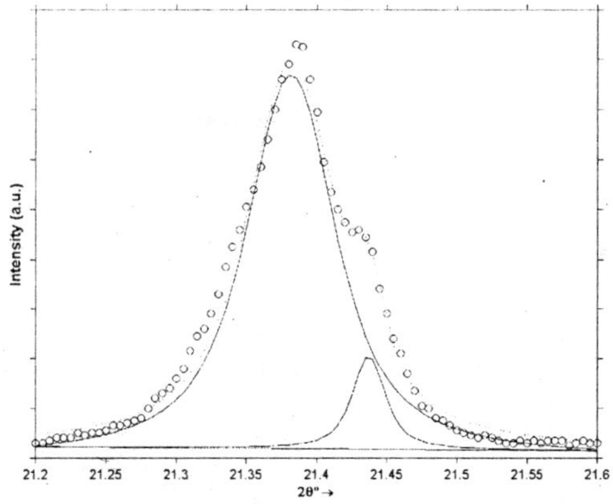

Fig. 6: Expanded view of (100) Miller plane of LaB_6 sample showing clear separation of $K\alpha_{1,2}$ peaks at as low as 21.35 °2θ.

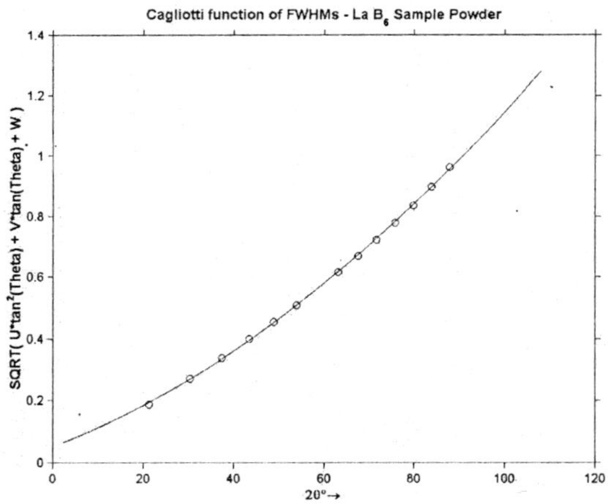

Fig. 7: Plot of Cagliotti function of FWHMs as a function of scattering angle.

scatter. Due to statistical nature of data collection some of the points may slightly deviate from the fitted equation. It is observed from Fig. 6 that our experimental points mostly fall on the fitted curve with minimum statistical errors, as compared to the similar data presented in the Philips report. In the case of Philips data, the scatter of points on the Cagliotti plot is relatively high, in spite of very high count rate of the data collection using X'Celerator. The peak widths and the errors associated with the measurement of FWHMs are also high and do not behave in a systematic order as a function of the scattering angle.

It is well known that the background intensity increases exponentially as the scattering angle decreases and background noise approaches very high value as the 2θ becomes lesser and closer to zero. This is because of air scattering and is clearly observed when the XRD measurements are carried out in open-air condition. In the vicinity of such low angle measurements most of the peaks merge with the high level of background noise level making identification of peak position and exact intensity measurements impossible. If one has to measure Bragg peaks in the materials as in large sized organic compounds, super-cells, etc., with large interplanar spacings (approximately around 40 Å and above) it is difficult or almost an impossible task. Figure 8 shows an XRD pattern of Stearic acid sample powder (ICDD pattern No. 381923) with (100) peak of 39.62 Å positioned at approximately 2.2°2θ. A fine-tuned XRD equipment minimizes this problem and enables to measure peaks even close to 0.5° 2θ.

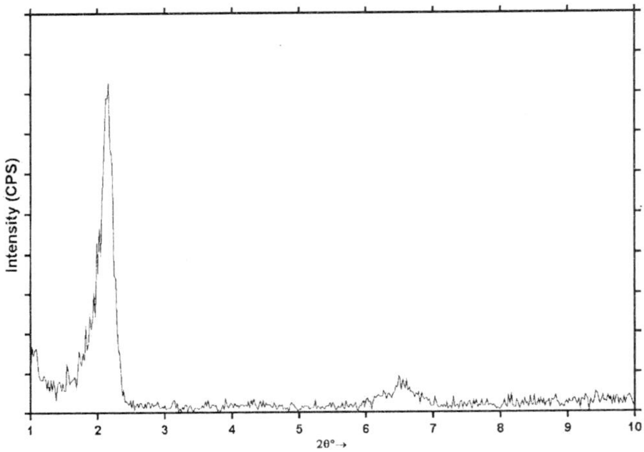

Fig. 8: XRD pattern of Stearic Acid sample powder showing (100) Bragg peak.

Figures 9 and 10 show XRD patterns of quartz sample provided by the M/s. Bruker AXS along with the equipment supply. The XRD patterns of both the patterns are recorded under identical conditions. In fact same batch file (in terms of M/s. Bruker's file extension terminology, same *.DQL), was run to maintain identical power ratings, scan speed, exposure timings, slit openings, including the same sample. It is clearly discernible from these figures that quality of data greatly enhanced in Fig. 10, which has been recorded after fine-tuning the equipment.

It can be easily noticed that the intensity of the strongest peak of the α-quartz (ICDD Pattern 461045) with (101) peak at 26.64° 2θ in Fig. 10 is much higher (2300 cps) after fine tuning compared to the same peak intensity (750 cps) in the XRD pattern before its dispatch from the suppliers of the equipment, which was crudely aligned. In Fig. 9, one can also see very broad and high intensity hump near 14° 2θ and high background level at lower 2θ. With gradual decrease in background level as 2θ decreases is definitely due to the parasitic scattering from various parts present in the beam path. One can also observe that at lower 2θ the scattered intensity

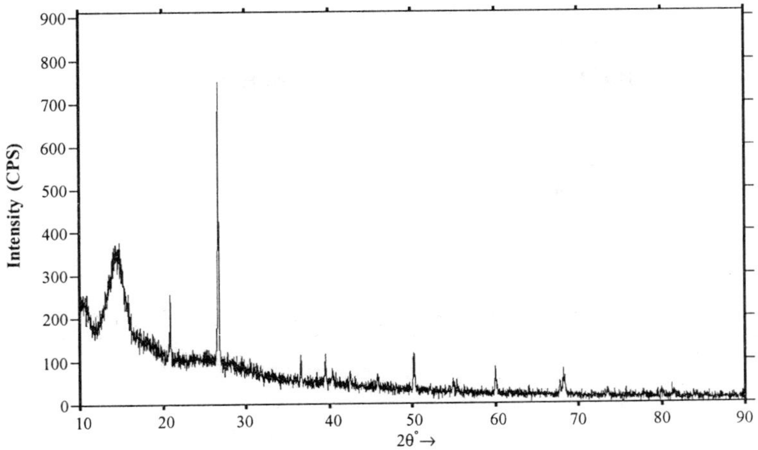

Fig. 9: XRD pattern of Quartz sample with grazing incidence attachment before equipment fine tuning (as received equipment).

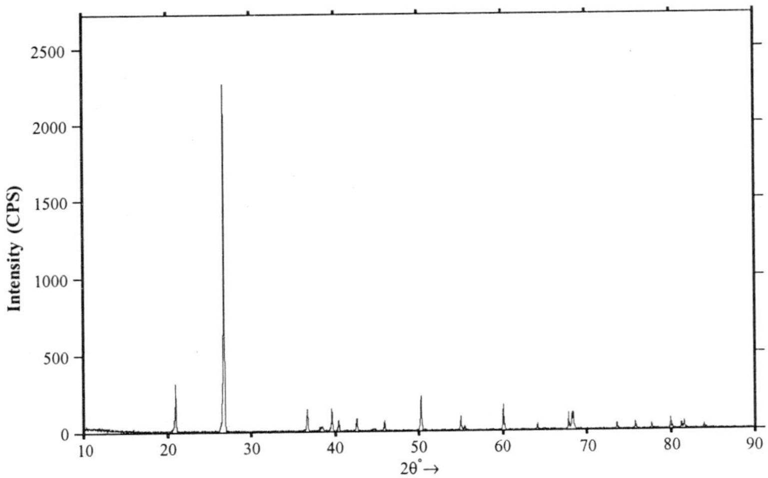

Fig. 10: XRD pattern of Quartz sample with grazing incidence attachment after fine-tuning the equipment.

increases at very high rate because of poor tuning. On the other hand, in Fig. 10, this background level is so low that it could not be observed. The peaks above 80° 2θ in Fig. 10 are much clearly defined as compared to those peaks in Fig. 9. It is also to be noted that in the Grazing Incidence (GI) attachment the X-ray beam conditions are same as in the case of normal BB geometry generally used for recording powder patterns and the diffracted beam from sample to the detector gets modified in the GI setup. That is to say, even though the incident beam is generally fixed at very small angle, and 2θ is varied, there are no collimators used between the sample and the receiving slit in the GI studies, and long Soller slits with its plates parallel to the length of the line focus of the beam are used, unlike in normal BB geometry where these plates are placed perpendicular to the length of the line focus of the beam. From our experience we strongly feel that for thin film investigations, if the researcher uses zero background plates such as Si (911) instead of Si (111) or Si (100) as substrates, the XRD pattern recorded using fine-tuned equipment produces much stronger peaks in the BB geometry itself, than that would be produced using a Grazing Incidence attachment.

CONCLUSIONS

Our studies of optimizing X-ray beam optics of XRD system indicate that fine tuning of X-ray beam optics of any diffraction equipment is most essential part in order to achieve important features like (1) very high intense peaks at correct 2θ positions, (2) the high intense Bragg peaks make the background intensity to appear negligible, because of very high signal to noise ratio, (3) the peaks produced in such a setup have the narrowest FWHMs, (4) the particle size estimations using Scherrer formula found to be in good agreement on comparison with other techniques as the instrumental width is almost negligible, (5) closely spaced peaks get well separated, making complex patterns to index with relative ease, (6) short experimental data collection times are sufficient for producing high quality XRD data of the sample, (7) the short experimental time reduces the X-ray tube operation time and enhances tube life, (8) quantitative measurement of very small intensity (≤ 0.01%) peaks corresponding to impurity/doped materials/ unreacted phases can be measured with ease, (9) very small changes occurring in the lattice constant(s) measurements as a function of change of any parameter can be easily determined, (10) because of intense diffraction peaks it is easy to locate exact peak locations, (11) materials having very large inter planar spacings which have their peaks in the vicinity of as low as 0.5° 2θ can be measured comfortably (because of very small air scattering effects and high beam collimation) and (12) produces negligible parasitic scattering from the parts of the instrument which defines the beam. It is emphasized that a carefully aligned XRD equipment produces XRD spectra almost equivalent to that of the equipments which have very expensive gadgets like beam conditioners, area detectors etc.

Acknowledgements: The author would like to thank Dr. G. Sundararajan, Director, and Dr. R. Sundaresan, Associate Director of ARCI, for their encouragement. The author also wishes to thank Mr. K. Ramesh Reddy for his assistance in the experimental work and Dr. Y. Srinivasa Rao for critical comments.

REFERENCES

Cullity, B.D. (1978). Elements of X-ray Diffraction, 3rd edition. Addison-Wesley, Massachusetts, USA.

Harold, P., Klug, and Leroy, E., Alexander (1974). X-ray diffraction procedures for polycrystalline and amorphous materials, John Wiley & Sons Inc., New York, USA.

Louer, D. (1993). *Material Science Forum*, **133-136:** 7.

Narasimha Rao, G.V., (2001). Accuracy in Powder Diffraction III, NIST, Gaithersburg MD, USA.

Narasimha Rao, G.V., Sastry, V.S., Gopala Krishna Murthy, H.S., Seshagiri, V. and Radhakrishnan, T.S. (1996). *Powder Diffraction*, **11:** 200.

Ron Jenkins and Snyder, R.L. (1996). Introduction to X-ray powder Diffractometry. Wiley-Interscience.

Powder Diffraction Geometry: Application Aspects

S. Prugovecki and A. Stefanovic[1]

PANalytical B.V., Almelo, The Netherlands
[1]PANalytical, Singapore

INTRODUCTION

X-ray diffraction (XRD) is applicable over a very wide range of material types and has applications in many subject areas (Cullity, 1978; Fewster, 1997; Guinier, 1963; Hammond, 1997; Pickworth Glusker and Trueblood, 1985; Schultz, 1989; Tanner, 1976; Taylor, 1987; Young, 2000), for example:

- structure determination of proteins, biological and new organic molecules
- phase identification in minerals
- phase, texture and stress determination in metals and alloys,
- microstructure in thin films.

X-ray experiments (both measurement and analysis) are named and classified in an apparently random manner. Sometimes named after the innovator who designed the experiment, e.g. Laue, sometimes named after the state of the material, e.g "powder diffraction", sometimes named after the X-ray scattering process e.g. "reflectometry" or sometimes named after the instrumental arrangement e.g. "Bragg-Brentano". Some traditional methods are dying out in favour of newer methods and so XRD is in a continuous state of change.

In this introduction a basic overview of the diffraction process has been presented. Different techniques focus in on different parts of the diffraction pattern and hence use different apparatus to collect the data and different methods with which to analyze the data. We will try to give an overview of different aspects of XRD geometries, namely powder XRD geometries, in respect to the wide range of applications.

HISTORICAL BACKGROUND

There are many arrangements of optical components for powder diffractometry. Figure 1 shows a typical example. Many powder diffraction experiments are based on *focussing geometry* - this is in order to capture as many crystallites as possible. The angular divergence is relatively large (say 0.1°) making this a low resolution technique, but one that is good for maximising on the available intensity in the signal. The wavelength spread includes $K\alpha_1$ and $K\alpha_2$ and some white radiation; these can be modified to optimize for $K\alpha_1$ by using a suitable monochromator. The focussing geometry means that the specimen height must be set very carefully, to minimize inaccuracies in 2θ measurements.

In order to give more systematic overview on different diffraction geometries various instrumental and application aspects will be discussed. The geometries will be considered in terms of instrument design, beam geometry and geometry of the diffraction on samples.

INSTRUMENTAL CONSIDERATIONS

"Once upon a time" XRD instruments, diffractometers, were designed with rigid system configuration. For their use quite high expertise of specialists was required. Powder XRD was mainly used for the phase analysis, mostly qualitative and in cases when some crystallographic work would be done, it would be applied on high symmetry materials, metals or metal alloys and with only few elements. Today, almost a century from the first use, powder XRD is carried out on sophisticated systems, which satisfy wide range of applications and levels of expertise. Modern XRD instruments can be characterized by:

- Multi–functionality and flexibility–different sources, X-ray optics and detectors.
- Reconfigurability–system design enables quick and simple change of system configuration best suited for different applications.
- High and low expertise users – systems can be used either by scientists or can be fully automated for industrial applications.
- Powerful software–due to the versatile software information depth was substantially changed.
- Wide area of applications–from R and D to industry.
- Crystallography: the quality of data and data collection time enable scientists to apply crystallographic analyses from inorganic materials to large molecules.

Although this overview will consider XRD methods based on diffractometers, it is worth mentioning methods which are based on various XRD cameras still in use. Some of them are still used in the same fashion as earlier and some exist more in geometrical sense of experiments rather than in their previous instrumental form. Let us briefly mention here Debye–Scherrer/Hull, Guinier and Gandolfi type of diffraction cameras. It is also worth mentioning here some other rather special techniques based on various X-ray scattering phenomena. They are addressed here only because of their application in material characterization and will not be discussed in greater details.

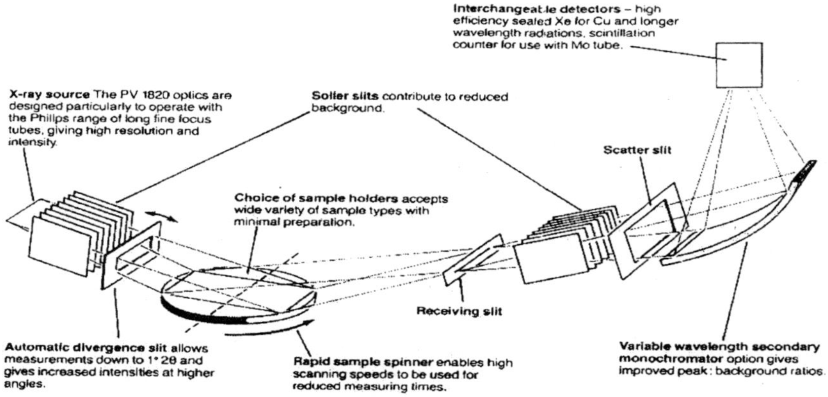

Fig. 1: Optical components for powder diffractometry.

SPECIAL X-RAY BASED TECHNIQUES

Lang and Berg Barret Topography

Topography uses collimated beam, used for single crystals set to a Bragg condition. Departures from the Bragg condition give rise to change in intensity. Contrast due to dislocations in single crystals is recorded on the photograph. Incident beam close to Bragg condition is at a shallow angle with respect to the sample. Intensities are sensitive to small tilts of crystal. The method is useful for mosaic blocks and grains of small areas of sample, but the resolution is limited by grain size of film (several microns).

Laue Method

White radiation is used, the back reflected beams are collected on photographic film. This method is used for the measurement of crystal orientation. The geometry of the back-reflection Laue method is shown in Fig. 2.

X-ray Scanning Method

X-ray scanning apparatus (Schwarzer, 1989), based on a commercial two-circle diffractometer equipped with open Eulerian cradle, is used for imaging the spatial distributions of crystal texture, residual lattice strain and measurement of pole figures on small areas.

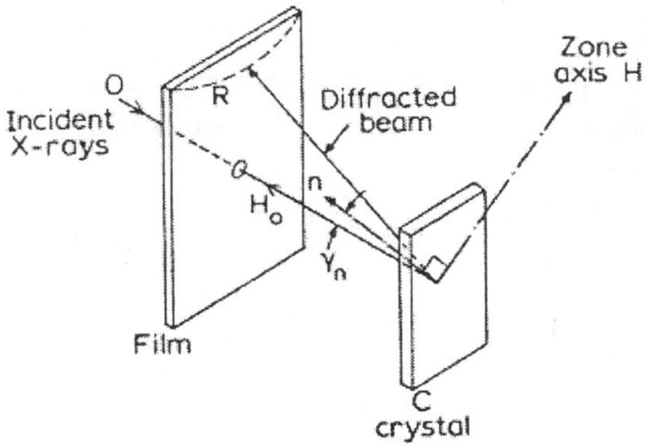

Fig. 2: The geometry of the back-reflection Laue method.

Small Angle X-ray Scattering (SAXS)

SAXS measurements typically are concerned with scattering angles < 1°. As dictated by Bragg's Law, the diffraction information about structures with large *d*-spacings lies in the region. Therefore the SAXS technique is commonly used for probing large length scale structures such as high molecular weight polymers, biological macromolecules (proteins, nucleic acids, etc.), and self-assembled superstructures (e.g. surfactant templated mesoporous materials).

SAXS measurements are technically challenging because of the small angular separation of the direct beam (which is very intense) and the scattered beam. Large specimen-to-detector distances (0.5-10 m) and high quality collimating optics are used to achieve good signal-to-noise ratio in the SAXS measurement.

DIFFRACTION GEOMETRIES

We will consider firstly geometries, which are present in today's diffractometers. Our next scope will be to consider different geometries of used X-ray beams. Last, but not the least, we shall discuss different geometries of samples.

Instrumental Geometry (θ – 2θ, θ – θ)

The distinction here can be made on various angular motions of the functional parts of the diffractometer. The simplest way to define two types of diffractometers is as follows: in the case of θ –2θ diffractometer, X-ray tube with incident X-ray optics is stationary and there is the angular motion of the specimen (θ) and angular motion of the detector together with diffracted beam X-ray optics (2θ) (Fig. 3). In case of θ – θ diffractometer, both X-ray tube and detector have angular motion, but the specimen is stationary (Fig. 3). Between these two types of diffractometers there is no difference in the

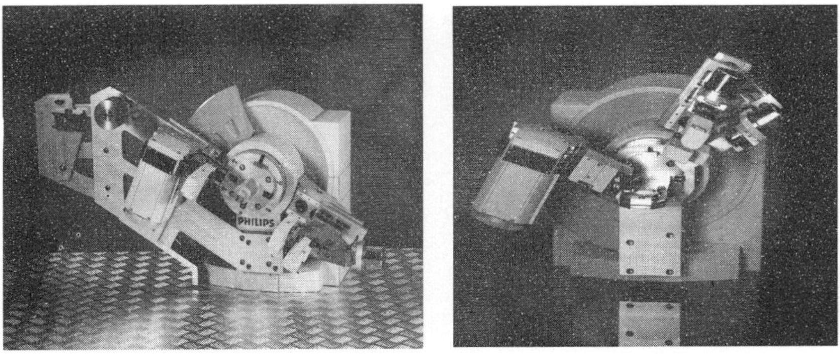

Fig. 3: A photograph of θ – 2θ diffractometer (left) and θ- θ diffractometer (right).

sense of the XRD effects and the choice of use is based on either stationary incident monochromator or the need to accommodate large bulky specimens.

Beam Geometry

In X-ray powder diffraction basic two beam geometries are used: focussing and parallel. Of initially various geometric arrangements developed the two para-focusing geometries have become popular: Bragg-Brentano and Seeman-Bohlin. The majority of commercially available powder diffractometers employ the Bragg-Brentano arrangement since it gives generally a reasonable compromise between mechanical simplicity and performance. The focusing arrangement is not truly focusing because of the finite width of the source and the specimen. That is why this arrangement is called para-focussing one. Although the usual Bragg-Brentano (BB) configuration has many advantages of convenience and relative ease of alignment, for measurements restricted to the back reflection region, where one wishes to vary the incident angle, there are advantages to the Seeman-Bohlin (SB) arrangement.

Classical Para-focussing (BB) Geometry

This beam geometry can be further classified according to two criteria: constant irradiated length with the receiving slit (point detector) or without the receiving slit (line detector), and constant divergence angle: with the receiving slit (point detector) or without the receiving slit (line detector). In case of "infinitely thick" specimen this geometry is preferable, giving good particle statistics and sufficient diffracted intensities. However, this geometry is prone to the sample height errors and in case of rough or non-uniform samples, cannot be used. Diagram of characteristic BB para-focussing geometry is shown in Fig. 4.

Parallel geometry has become available by development of various X-ray optical modules and can be classified in two basic categories: (i) equatorial and (ii) equatorial and axial. In further text both categories are discussed in

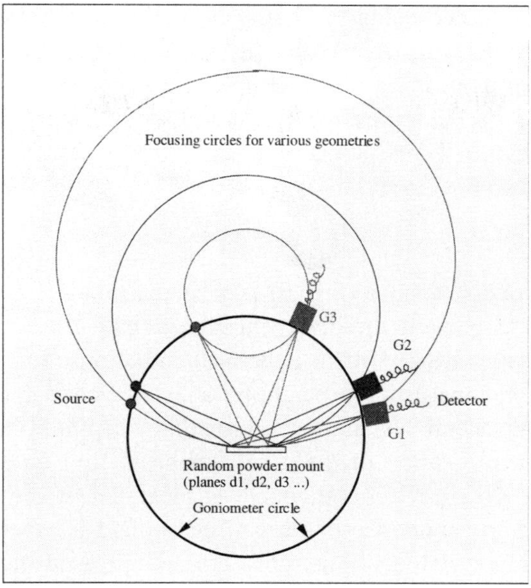

Fig. 4: Diagram of characteristic BB para-focussing geometry. Classical parallel geometry.

details. Various application areas are given in respect to different optical modules.

Equatorial

These are basically optics namely, mirror, hybrid and parallel plate collimator.

X-ray Mirror

The graded multilayer parabolic X-ray mirror is a special kind of beam conditioner, which is able to convert the divergent X-ray beam from a line focus tube to an intense monochromatic quasi-parallel beam (Fig. 5). This is possible because of the combination of the parabolic shape of the mirror and the changing of multilayer period over the length of the mirror. X-ray mirrors are designed to work with a line focus as they are curved in only

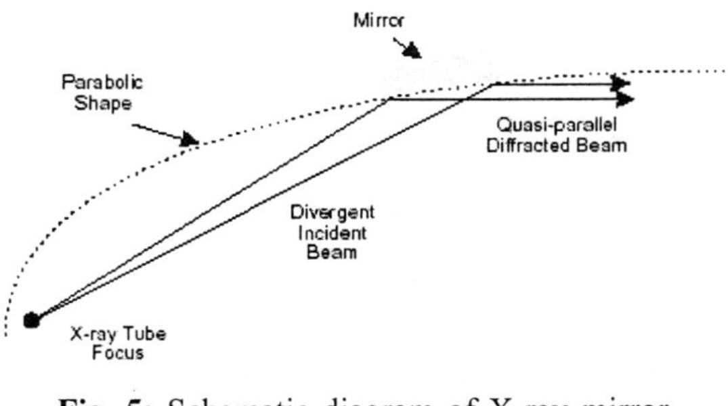

Fig. 5: Schematic diagram of X-ray mirror.

one dimension. When using the X-ray mirror in combination with parallel beam optics on the diffracted beam side, the measurements are not sensitive to the exact height position of the sample. This geometry is also suitable for irregularly shaped samples. The mirror can be applied for measuring polycrystalline thin films, stressed materials, powders, irregularly shaped samples and epitaxial layers.

Hybrid

The hybrid monochromator (Fig. 6) is a special kind of beam conditioner, which is able to convert the divergent X-ray beam from a line tube focus to a quasi-parallel beam with a pure Kα_1 radiation component. A hybrid monochromator consists of a parabolic shaped graded multi layer (X-ray mirror) and a special channel-cut germanium crystal. The X-ray mirror performs a precollimation of the divergent beam to a quasi-parallel beam. Thereafter the Kα_1 spectral component is separated by the monochromator crystal. The hybrid monochromator can be applied for measuring epitaxial layers, thin polycrystalline films, powders and stressed materials.

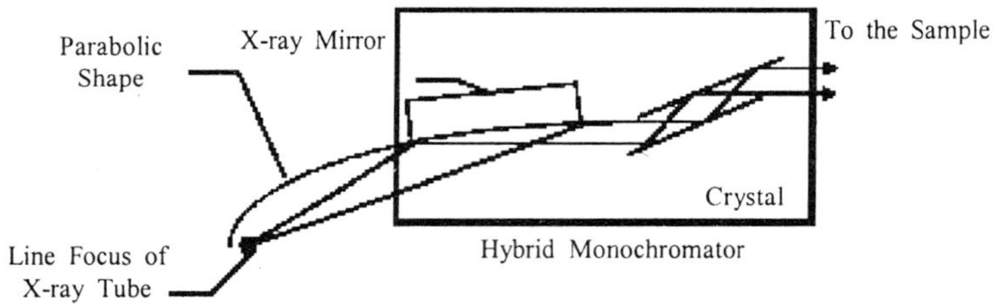

Fig. 6: Schematic diagram of the hybrid monochromator.

Parallel Plate Collimator

Parallel Plate Collimator is a diffracted beam optical module that consists of a set of parallel plates, which define the equatorial acceptance angle as seen by the detector.

Equatorial and Axial

These are basically optics namely, X-ray lens (polycapillary), 4-crystal monochromator and monocapillary.

X-ray Lens (Polycapillary)

The quest for increased resolution and higher beam intensity has led in the recent years to the development of new diffraction optics for laboratory instruments. In particular, multilayer mirrors and polycapillary collimators can now be considered as "standard" components for state-of-art stress/texture diffractometers. Both optical systems allow up to a couple orders of magnitude gain in intensity, further providing either focussed or quasi-parallel beams. The knowledge of the instrumental response of a parallel beam diffractometer based on polycapillary optics increases the accuracy of the collected pole figures and assures a reliable calculation of the orientation distribution function. More reliable data are also expected from residual stress measurements conducted at high sample-tilting where most of the information is generally contained. Wrong information on the texture, in fact, leads to wrong average elastic constants, key data in the residual stress analysis. A schematic diagram of an equipment for texture and stress measurement consisting of X-ray lens is shown in Fig. 7.

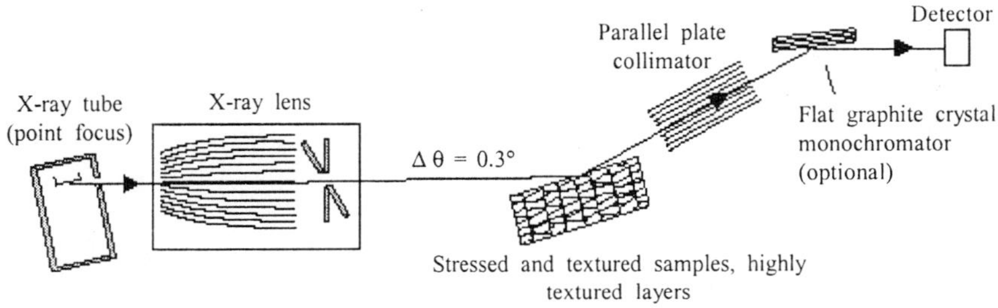

Fig. 7: A schematic diagram of an equipment for texture and stress measurement consisting of X-ray lens and other optics.

Monocapillary

The monocapillary is a hollow glass tube. X-rays are transported through this tube using the effect of total external reflection inside the capillary. The principle is shown in figure below. The diameter of the X-ray beam at the end of a monocapillary ranges from 10 μm to 2 mm. The tube acts in effect as a wave guide which results in the diverging X-ray beam being turned into a narrow quasi-parallel beam. It is designed to be used with the point focus of the X-ray tube. The monocapillary acts as a low pass energy filter (suppresses the white noise background). We only need to use a diffracted beam monochromator if there is strong fluorescence from the sample or to reduce Kβ radiation. The main application of the mono-capillary is XRD analysis of small spots on a sample, micro-stress and micro-texture.

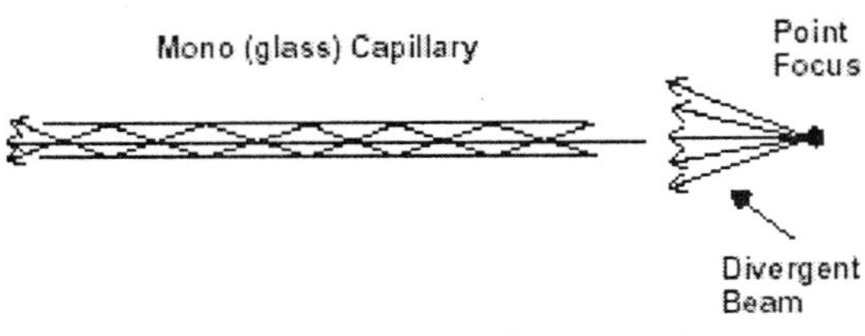

Fig. 8: Principle of monocapillary.

4-crystal Monochromator

High resolution monochromators are based on two U-shaped blocks of single-crystal nearly perfect germanium. The crystals have (110) surfaces and are preset for Ge (220) or Ge (440) $CuK\alpha_1$ reflections. The principle of high resolution monochromators according to Bartels can be described as below.

The source of an X-ray tube with a copper anode emits a full X-ray spectrum, including the CuK doublet. The first U-block in the monochromator (crystal 1 in Fig. 9) is set at the correct Bragg angle for the CuKα_1 line. The X-rays are diffracted twice at the surfaces of the first crystal towards the second germanium channel cut crystal. The second U-block is set at the same Bragg angle but the channel is not parallel to that of the first crystal. The CuKα_1 line is diffracted twice at the surfaces of the second crystal and emerges from the monochromator housing parallel to the original input beam. A typical application for high resolution monochromators are rocking curve measurements and reciprocal space maps.

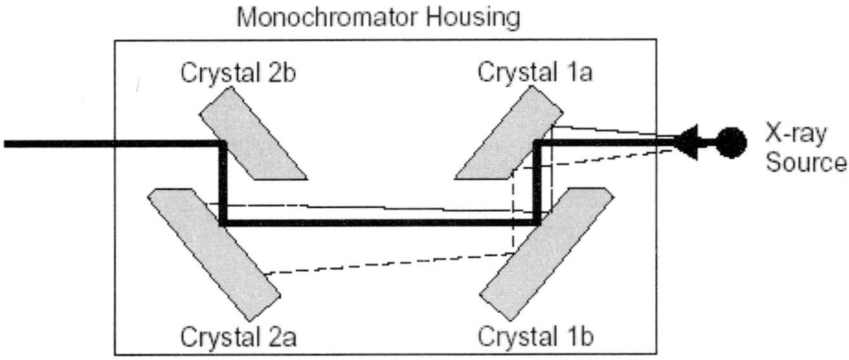

Fig. 9: A schematic diagram of 4-crystal monochromator.

Sample Geometry

Depending on the specimen character, its properties, especially amount, absorption/transparency, sensitivity to air/moisture, shape and size, various geometrical modes are employed. The reflection mode is the most frequently used geometry and it was discussed earlier. In case of small amounts or air/moisture sensitive samples the transmission geometry is applied. This geometry will be discussed further later as Debye–Scherrer geometry. In case of the analysis of thin polycrystalline layers when penetration depth is the main concern, the grazing X-ray incidence geometry is applied.

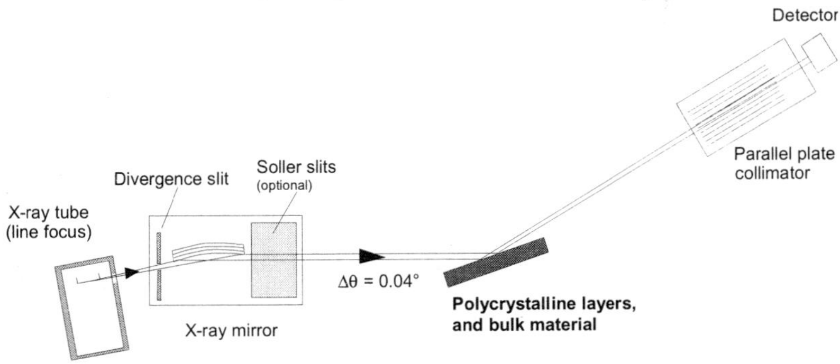

Fig. 10: A typical grazing incidence phase analysis configuration.

Debye–Scherrer Geometry

This can be basically described as transmission method with parallel beam, employing two different geometries of X-ray mirrors. The parabolic mirror geometry has been discussed earlier with the respective areas of application.

Elliptical mirror has all the advantages of transmission geometry for samples with low atomic number (organics, pharmaceuticals, polymers and plastics) or samples on or in foils and glass capillaries. The biggest advantage is that fast linear detectors can be used for data collection and that resolution does not depend on capillary diameter like in the case of parabolic mirror.

In-plane Diffraction

Much of the XRD using conventional diffractometers is carried out using the standard "Bragg" or "reflection" geometry. Scattering occurs such that the incident and reflected beams enter and emerge from the same surface of the sample. "In-plane scattering" refers to incident and scattered beam directions that are both parallel to this surface plane. For diffraction planes that are oriented such that their normals are in the plane parallel to the surface, the usual experimental set-up cannot be used and so the in-plane scattering geometry is required (Fig. 11). In the strictest sense, for the "Bragg" or "reflection" geometry the X-ray beam would enter and exit through one of the edges of the sample. For thin film samples only a very small volume of the material can be probed in this way (we shall call this "edge" scattering, this method is treated elsewhere). To make better use of the volume of material available in thin films, the in-plane scattering geometry can take place through the top surface of the sample. When the incident angle is close to the critical angle for reflectivity, there is a small penetration of the beam into the surface layers of the sample. This is the consequence of the fact that the refractive index of the material is slightly less than 1. Bragg scattering can then occur from planes within this very shallow region. Typical penetration depths for in-plane scattering are typically within the range 1-100 nm.

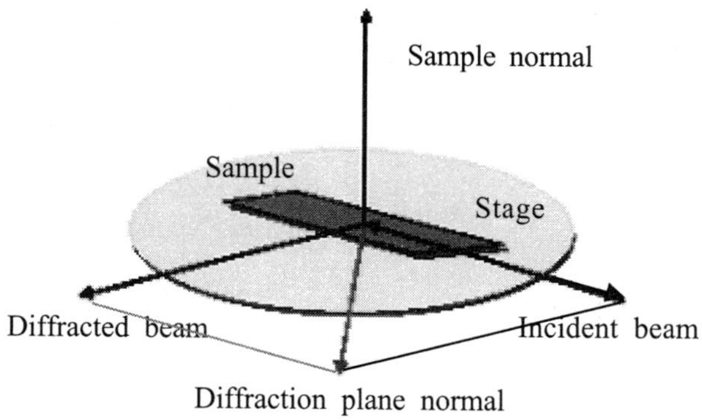

Fig. 11: A schematic of in-plane scattering geometry.

CONCLUSIONS

With the fast development of X-ray technology in recent years all or almost all diffraction geometries are available on one system and can be interchanged. Development of modern X-ray optical modules enable different diffraction geometries to be interchangeable and the best instrument configuration can be chosen for given application or type of the specimen. The final choice of diffraction geometry depends on particular application, required quality of data and results and properties of the specimen.

REFERENCES

Cullity, B.D. (1978). Elements of X-ray Diffraction, 3rd Edition, Addison-Wesley Publishing Co.

Fewster, P.F. (1997). Reciprocal Space Mapping (Critical Reviews in *Solid State and Materials Sciences*), **22(2):** 69.

Guinier, A. (1963). X-ray Diffraction, Dover Publications, New York, USA.

Hammond, C. (1997). The Basics of Crystallography and Diffraction, Oxford University Press, New York, USA.

Pickworth Glusker, J. and Trueblood, K.N. (1985). Crystal Structure Analysis, JP Publishers, New York, USA.

Schultz, J.M. (1989). Diffraction of Materials Scientists, Prentice-Hall Inc., New Jersey, USA.

Schwarzer, R.A. (1989). Proc. 12 Intern. Congr. on X-ray Optics and Microanalysis (12 ICXOM), Cracow, **1:** 205.

Tanner, B.K. (1976). X-ray Diffraction Topography, Pergamon Press, New York, USA.

Taylor, C. (1987). Diffraction: Institute of Physics Publishing, UK.

Young, R.A. (2000). The Rietveld Method ICUR. Oxford Science Pub., New York, USA.

Index